Neuro-Fuzzy and Soft Computing

A Computational Approach to Learning and Machine Intelligence

Jyh-Shing Roger Jang
jang@cs.nthu.edu.tw
Computer Science Department, Tsing Hua University, Taiwan
The MathWorks, Natick, Massachusetts, USA

Chuen-Tsai Sun
ctsun@cis.nctu.edu.tw
Department of Computer and Information Science
National Chiao Tung University, Hsinchu, Taiwan

Eiji Mizutani
eiji@joho.kansai.co.jp
Information Systems Department, Kansai Paint Co., Ltd.
4-3-6 Fushimi Chuo-ku, Osaka, 541, Japan

Prentice Hall Upper Saddle River, NJ 07458

Library of Congress Cataloging-in-Publication Data

Jang, Jyh-Shing Roger.
Neuro-fuzzy and soft computing : a computational approach to learning and machine intelligence / Jyh-Shing Roger Jang, Chuen-Tsai Sun, Eiji Mizutani.
p. cm.
Includes bibliographical references and index.
ISBN 0-13-261066-3
1. Soft Computing. 2. Neural networks (Computer science) 3. Fuzzy systems. I. Sun, Chuen-Tsai. II. Mizutani, Eiji. III. Title.
QA76.9.S63J36 1997 96-29050
006.3--dc20 CIP

Acquisitions Editor: Tom Robbins
Production Editor: Joseph Scordato
Director Prod. & Mfg.: David Riccardi
Production Manager: Bayani Mendoza DeLeon
Cover Designer: Bruce Kenselaar
Copy Editor: Patricia Daly
Buyer: Donna Sullivan
Editorial Assistant : Nancy Garcia

A Pearson Education Company
Upper Saddle River, NJ 07458

Printed in the United States of America

10 9 8 7 6

ISBN 0-13-261066-3

Prentice-Hall International (UK) Limited, *London*
Prentice-Hall of Australia Pty. Limited, *Sydney*
Prentice-Hall Canada Inc., *Toronto*
Prentice-Hall Hispanoamericana, S.A., *Mexico*
Prentice-Hall of India Private Limited, *New Delhi*
Prentice-Hall of Japan, Inc., *Tokyo*
Prentice-Hall Asia Pte. Ltd., *Singapore*
Editora Prentice-Hall do Brasil, Ltda., *Rio de Janerio*

Contents

Foreword

Lotfi Zadeh

Among my many Ph.D. students, some have forged new tools in their work. Jyh-Shing Roger Jang and Chuen-Tsai Sun fall into this category. Neuro-Fuzzy and Soft Computing makes visible their mastery of the subject matter, their insightfulness, and their expository skill. Their coauthor, Eiji Mizutani, has made an important contribution by bringing to the writing of the text his extensive experience in dealing with real-world problems in an industrial setting.

Neuro-Fuzzy and Soft Computing is one of the first texts to focus on soft computing — a concept which has direct bearing on machine intelligence. In this connection, a bit of history is in order.

The concept of soft computing began to crystallize during the past several years and is rooted in some of my earlier work on soft data analysis, fuzzy logic, and intelligent systems. Today, close to four decades after artificial intelligence (AI) was born, it can finally be said with some justification that intelligent systems are becoming a reality. Why did it take so long for the era of intelligent systems to arrive?

In the first place, the AI community had greatly underestimated the difficulty of attaining the ambitious goals which were on its agenda. The needed technologies were not in place and the conceptual tools in AI's armamentarium — mainly predicate logic and symbol manipulation techniques — were not the right tools for building machines which could be called intelligent in a sense that matters in real world applications.

Today we have the requisite hardware, software, and sensor technologies at our disposal for building intelligent systems. But, perhaps more important, we are also in possession of computational tools which are far more effective in the conception and design of intelligent systems than the predicate-logic-based methods, which form the core of traditional AI. The tools in question derive from a collection of methodologies which fall under the rubric of what has come to be known as soft computing (SC). In large measure, the employment of soft computing techniques underlies the rapid growth in the variety and visibility of consumer products and industrial systems which qualify to be assessed as possessing a significantly high

MIQ (machine intelligence quotient).

The essence of soft computing is that unlike the traditional, hard computing, soft computing is aimed at an accommodation with the pervasive imprecision of the real world. Thus, the guiding principle of soft computing is to exploit the tolerance for imprecision, uncertainty, and partial truth to achieve tractability, robustness, low solution cost, and better rapport with reality. In the final analysis, the role model for soft computing is the human mind.

Soft computing is not a single methodology. Rather, it is a partnership. The principal partners at this juncture are fuzzy logic (FL), neurocomputing (NC), and probabilistic reasoning (PR), with the latter subsuming genetic algorithms (GA), chaotic systems, belief networks, and parts of learning theory. The pivotal contribution of FL is a methodology for computing with words; that of NC is system identification, learning, and adaptation; that of PR is propagation of belief; and that of GA is systematized random search and optimization.

In the main, FL, NC, and PR are complementary rather than competitive. For this reason, it is frequently advantageous to use FL, NC, and PR in combination rather than exclusively, leading to so-called hybrid intelligent systems. At this juncture, the most visible systems of this type are neuro-fuzzy systems. We are also beginning to see fuzzy-genetic, neuro-genetic, and neuro-fuzzy-genetic systems. Such systems are likely to become ubiquitous in the not distant future.

In coming years, the ubiquity of intelligent systems is certain to have a profound impact on the ways in which human-made systems are conceived, designed, manufactured, employed, and interacted with. This is the perspective in which the contents of Neuro-Fuzzy and Soft Computing should be viewed.

Taking a closer look at the contents of Neuro-Fuzzy and Soft Computing, what should be noted is that today most of the applications of fuzzy logic involve what might be called the calculus of fuzzy rules, or CFR for short. To a considerable degree, CFR is self-contained. Furthermore, CFR is relatively easy to master because it is close to human intuition. Taking advantage of this, the authors focus their attention on CFR and minimize the time and effort needed to acquire sufficient expertise in fuzzy logic to apply it to real-world problems.

One of the central issues in CFR is the induction of rules from observations. In this context, neural network techniques and genetic algorithms play pivotal roles, which are discussed in Neuro-Fuzzy and Soft Computing, in considerable detail and with a great deal of insight. In the application of neural network techniques, the main tool is that of gradient programming. By contrast, in the application of genetic algorithms, simulated annealing, and random search methods, the existence of a gradient is not assumed. The complementarity of gradient programming and gradient-free methods provides a basis for the conception and design of neuro-genetic systems.

A notable contribution of Neuro-Fuzzy and Soft Computing, is the exposition of ANFIS (Adaptive Neuro Fuzzy Inference System) — a system developed by the authors which is finding numerous applications in a variety of fields. ANFIS and its variants and relatives in the realms of neural, neuro-fuzzy, and reinforcement learning systems represent a direction of basic importance in the conception and design of intelligent systems with high MIQ.

Neuro-Fuzzy and Soft Computing is a thoroughly up-to-date text with a wealth of information which is well-organized, clearly presented, and illustrated by many examples. It is required reading for anyone who is interested in acquiring a solid background in soft computing — a partnership of methodologies which play pivotal roles in the conception, design, and application of intelligent systems.

Lotfi A. Zadeh

Preface

During the past few years, we have witnessed a rapid growth in the number and variety of applications of fuzzy logic and neural networks, ranging from consumer electronics and industrial process control to decision support systems and financial trading. Neuro-fuzzy modeling, together with a new driving force from stochastic, gradient-free optimization techniques such as genetic algorithms and simulated annealing, forms the constituents of so-called soft computing, which is aimed at solving real-world decision-making, modeling, and control problems. These problems are usually imprecisely defined and require human intervention. Thus, neuro-fuzzy and soft computing, with their ability to incorporate human knowledge and to adapt their knowledge base via new optimization techniques, are likely to play increasingly important roles in the conception and design of hybrid intelligent systems.

This book provides the first comprehensive treatment of the constituent methodologies underlying neuro-fuzzy and soft computing, an evolving branch within the scope of computational intelligence that is drawing increasingly more attention as it develops. Its main features include fuzzy set theory, neural networks, data clustering techniques, and several stochastic optimization methods that do not require gradient information. In particular, we put equal emphases on theoretical aspects of covered methodologies, as well as empirical observations and verifications of various applications in practice.

AUDIENCE

This book is intended for use as a text in courses on computational intelligence at either the senior or first-year graduate level. It is also suitable for use as a self-study guide by students and researchers who want to learn basic and advanced neuro-fuzzy and soft computing within the framework of computational intelligence. Prerequisites are minimal; the reader is expected to have basic knowledge of elementary calculus and linear algebra.

ORGANIZATION

Chapter 1 gives an overview of neuro-fuzzy and soft computing. Brief historical traces of relevant techniques are described to direct our first step toward the neuro-fuzzy and soft computing world. The remainder of the book is organized into the eight parts described next.

Part I (Chapters 2 through 4) presents a detailed introduction to the theory and terminology of fuzzy disciplines, including fuzzy sets, fuzzy rules, fuzzy reasoning, and fuzzy inference systems.

Part II (Chapters 5 through 7) provides an overview of system identification and optimization techniques that prove to be effective for neural-fuzzy and soft computing. Chapter 5 introduces least-squares methods in the context of system identification. Chapter 6 describes derivative-based nonlinear optimization techniques, including nonlinear least-squares methods. These two chapters are inevitably mathematically oriented, and for the first reading, many sections labeled with an asterisk ('*') can be omitted. Chapter 7 discusses derivative-free optimization techniques, including genetic algorithms, simulated annealing, the downhill Simplex method, and random search.

Part III (Chapters 8 through 11) introduces a variety of important neural network paradigms found in the literature, including adaptive networks as the most generalized framework for model construction, supervised learning neural networks for data regression and classification, reinforcement learning for infrequent and delayed evaluative signals, unsupervised learning neural networks for data clustering, and some other networks that do not belong to any of the aforementioned categories.

Part IV (Chapters 12 and 13) explains how to build ANFIS (Adaptive Neuro-Fuzzy Inference Systems) and CANFIS (Coactive Neuro-Fuzzy Inference Systems) as core neuro-fuzzy models that can incorporate human expertise as well as adapt themselves through repeated training.

Part V (Chapters 14 through 16) covers structure identification techniques for neural networks and fuzzy modeling, including the CART (Classification and Regression Tree) method, which is quite popular in multivariate analysis of statistics; several data clustering algorithms aimed at batch-mode model building, and efficient rulebase formulation and organization via tree partitioning of input space.

Part VI (Chapter 17 and 18) considers various approaches to the design of neuro-fuzzy controllers, including expert control, inverse learning, specialized learning, backpropagation through time, real-time recurrent learning, reinforcement learning, genetic algorithms, gain scheduling, and feedback linearization (in conjunction with sliding mode control).

The last part, Part VII (Chapters 19 through 22), gives a variety of application examples in different domains, such as printed character recognition, inverse kinematics problems in robotics, adaptive channel equalization, multivariate nonlinear regression, adaptive noise cancellation, nonlinear system identification, plasma spectrum analysis, hand-written numeral recognition, game playing, and color recipe prediction.

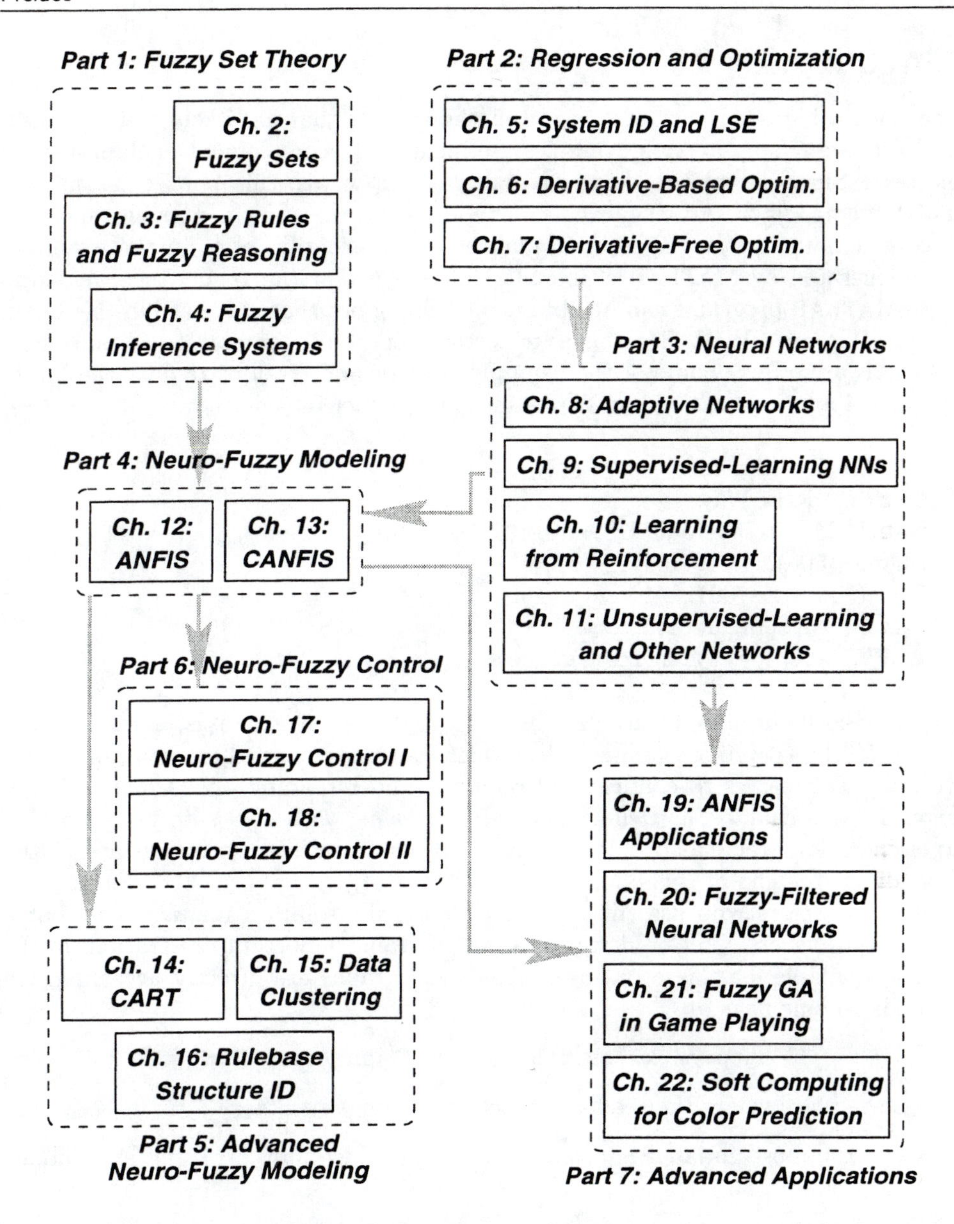

Figure 0.1. *Prerequisite dependencies among chapters of this book.*

The prerequisite dependencies among the individual chapters are shown in Figure 0.1. This diagram arranges chapters with respect to the level of advancement, so the reader has some flexibility in studying the whole book. Sections marked with an asterisk (*) can be skipped for the first reading.

FEATURES

The orientation of the book is toward methodologies that are likely to be of practical use; many step-by-step examples are included to complement explanations in the text. Since one picture is worth thousands of words, this book contains specially designed figures to visualize as many ideas and concepts as possible, and thus help readers understand them at a glance. Most scientific plots in the examples were generated by MATLAB© and SIMULINK©[1]. For the reader's convenience, these MATLAB programs can be obtained by filling out the reply card in this book, or via FTP or WWW. See the next section for details of how to obtain these MATLAB programs. Some of the examples and demonstrations require the Fuzzy Logic Toolbox^{TM} by The MathWorks Inc; the contact information is

The MathWorks, Inc.
24 Prime Park Way
Natick, MA 01760-1500, USA
Phone: (508) 647-7000
Fax: (508) 647-7001
E-mail: `info@mathworks.com`
WWW: `http://www.mathworks.com`

Chapters 2 through 18 are each followed by a set of exercises; some of them involve MATLAB programming tasks, which can be expanded into suitable term projects. This serves to confirm and reinforce understanding of the material presented in each chapter, as well as to equip the reader with hands-on programming experiences for practical problem solving. Hints to selected exercises are in the appendix at the end of this book.

For instructors who use this book as a text, the solution manual is available from the publisher. A set of viewgraphs that contains important illustrations in the book is available for classroom use. These viewgraphs are directly accessible via the book's home page at

`http://www.cs.nthu.edu.tw/~jang/soft.htm`

This is the place where the reader can do other things such as:

- Get the most updated information (such as addendum, erratum, etc.) about the book.
- Get enhancements and bug-fixes of MATLAB programs.
- Give comments and suggestions.
- View the statistics of comments and suggestions of other readers.
- Link to the authors' WWW home pages and other Internet resources.

[1]MATLAB and SIMULINK are registered trademarks of The MathWorks, Inc.

A reference list is given at the end of each chapter that contains references to the research literature. This enables readers to pursue individual topics in greater depth. Moreover, neuro-fuzzy and soft computing is a relatively new field and continues to evolve rapidly within the scope of computational and artificial intelligence. These references provide an entry point from the well-defined core knowledge contained in this book to another dimension of innovative and challenging research and applications.

OBTAINING THE EXAMPLE PROGRAMS

For people without Internet access, the easiest way to get the free MATLAB programs used in this book is to fill the reply card (bound in the book) and send it back to the MathWorks. The MathWorks will send you a floppy disk containing the MATLAB files, free of charge. (Some of the MATLAB programs rely on the Fuzzy Logic Toolbox, which is not freely available.)

For people with Internet access, the example MATLAB programs are available electronically in two ways: by **FTP** (file transfer protocol) or **WWW** (worldwide web). The FTP address is

`ftp.mathworks.com`

The files are at

`/pub/books/jang/*`

For WWW access to the FTP site, the URL (universal resource locator) address is

`ftp://ftp.mathworks.com/pub/books/jang/`

You can also access it via the book's home page at

`http://www.cs.nthu.edu.tw/~jang/soft.htm`

A sample FTP session is shown next, with what you should type in boldface.

unix> **ftp ftp.mathworks.com**
Connected to ftp.mathworks.com.
220 ftp FTP server (Version wu-2.4(2) Tue Aug 1 10:31:36 EDT 1995) ready.
Name (ftp.mathworks.com:slin): **anonymous**
331 Guest login ok, send your complete e-mail address as password.
Password: **slin@lsil.com** (*Use your email address here.*)
230 Guest login ok, access restrictions apply.
ftp> **cd /pub/books/jang**
250 CWD command successful.
ftp> **binary** (*You must specify binary transfer for some data files.*)
200 Type set to I.
ftp> **prompt** (*So you don't need to transfer each file interactively.*)

```
Interactive mode off.
ftp> mget * (Get every file in the directory.)
200 PORT command successful.
150 Opening BINARY mode data connection for addvec.m (691 bytes).
226 Transfer complete.
local: addvec.m remote: addvec.m
691 bytes received in 0.011 seconds (62 Kbytes/s)
200 PORT command successful.
...
ftp> bye
221 Goodbye.
```

You may want to use a mirror site near you:
`ftp://unix.hensa.ac.uk/mirrors/matlab`
`ftp://ftp.ask.uni-karlsruhe.de/pub/matlab`
`ftp://novell.felk.cvut.cz/pub/mirrors/mathwork`
`ftp://ftp.u-aizu.ac.jp:/pub/vendor/mathworks`

ACKNOWLEDGMENTS

This book could not have been finished without the assistance of many individuals. First, we would like to acknowledge Professor Lotfi A. Zadeh at the EECS (electrical engineering and computer science) Department of the University of California at Berkeley; his constant support and encouragement was the major driving force for the birth of this book. We would also like to thank him for offering the title of this book.

We appreciate the constructive and encouraging comments of the manuscript reviewers: Siva Chittajallu (Purdue University), Irena Nagisetty (the Ford Motor Company), Hao Ying, (University of Texas Medical Branch), and Mark J. Wierman (Creighton University).

We would like to thank the Prentice Hall editorial staff—especially Tom Robbins, who offered insightful suggestions and led us through the maze of details associated with publishing a book. We are also greatly indebted to Phyllis Morgan, who meticulously handled correspondence and the reviewing processes. We also wish to thank our proof editor, Patricia Daly, who spent hours polishing our English.

We are deeply grateful to Mr. Issei Nakashio, who designed the lovely cover for the book.

Individual acknowledgments of each author follow.

Acknowledgments of J.-S. Roger Jang

This book was submitted for publication while I was with The MathWorks, Inc., and published after I joined the Department of Computer Science, National Tsing Hua

University, Hsinchu, Taiwan. I'd like to thank The MathWorks, Inc. for making available a stimulating and fun working environment. I am also grateful to the Department of Computer Science at Tsing Hua University for providing a scholarly environment for both teaching and research.

Most of all, the book project was recommended by my Ph.D. advisor, Professor Lotfi Zadeh, and initiated while I was a research associate at the University of California at Berkeley back in 1992. I'd like to express my heart-felt gratitude to Professor Zadeh for his advice and continuous support.

Last but not least, I am indebted to my wife Sue and my children Timmy and Annie. In particular, without Sue's taking care of all the family matters, this book would never have been possible.

Acknowledgments of Chuen-Tsai Sun

First, I greatly appreciate the support of Professor Lotfi Zadeh, who has been encouraging me to do research on soft computing since my Ph.D. years. I am deeply grateful to many scholars in fuzzy systems, neural networks, and evolutionary computing, from whom I have learned over the past 10 years. I want to express my appreciation to Dr. Chi Yung Fu from the Lawrence Livermore National Laboratory, who encouraged me to study fuzzy filtering techniques. I also want to thank my graduate students at National Chiao Tung University, whose eager quest for knowledge has been a continual challenge and inspiration to me. Finally, I want to thank my family for their constant support throughout these years.

Acknowledgments of Eiji Mizutani

First, I would like to express my sincere gratitude to Professor Stuart E. Dreyfus (Department of Industrial Engineering and Operations Research, University of California at Berkeley) and Professor David M. Auslander (Department of Mechanical Engineering, University of California at Berkeley) for their insightful comments on our manuscript and their patience for replying to my repeated inquiries through e-mail. More important, Professor Dreyfus expertly guided me in the labyrinth of so-called artificial neural networks and taught me much of what I know about their computational gizmos. In particular, Chapter 10, "Learning from Reinforcement," would never have been completed without his quintessential mathematical guidance. Professor Auslander supported me in a friendly manner, listening patiently to all my frustrations with school and cultural differences. The research commencement with him gave me a ticket for entering this research world.

Furthermore, reflected in this book are the early influences of Professor Kazuo Inoue and Mr. Atusi Ichimura, who directed my first step toward the challenges of cutting edge research in constructing artificially intelligent systems.

I would like to take this opportunity to thank Mr. Kenichi Nishio (Sony Corporation) for supporting me with warm friendship, and Professor Hideyuki Takagi (Kyushu Institute of Design) for general assistance.

I am also deeply indebted to Professor Masayoshi Tomizuka, Professor Toshio

Fukuda, Professor Sigeru Omatu, Professor Nakaji Honda, Dean Mieczkowski, Jack Hassamontr, David Martin, Leo Li, Arun Biligiri, Agnes Concepcion, Sharon Smith, Zenji Nakano, Tohru Hirayama, Akio Kawamoto, Michael Lee, and Masayuki Murakami. All their assistance greatly smoothed the way.

Finally, I wish to thank Professor J.-S. Roger Jang (first author) for his confidence in me that inspired confidence in myself.

HOW TO CONTACT US

We'd like to hear your comments, corrections of errors, and suggestions for future editions. Please send them to

J.-S. Roger Jang: `jang@cs.nthu.edu.tw`
Chuen-Tsai Sun: `sun@cis.nctu.edu.tw`
Eiji Mizutani: `eiji@joho.kansai.co.jp` or
`eiji@biosys2.me.berkeley.edu`

You can also send your comments via the book's WWW home page at

`http://www.cs.nthu.edu.tw/~jang/soft.htm`

Chapter 1

Introduction to Neuro-Fuzzy and Soft Computing

1.1 INTRODUCTION

Soft computing (**SC**), an innovative approach to constructing computationally intelligent systems, has just come into the limelight. It is now realized that complex real-world problems require intelligent systems that combine knowledge, techniques, and methodologies from various sources. These intelligent systems are supposed to possess humanlike expertise within a specific domain, adapt themselves and learn to do better in changing environments, and explain how they make decisions or take actions. In confronting real-world computing problems, it is frequently advantageous to use several computing techniques synergistically rather than exclusively, resulting in construction of complementary hybrid intelligent systems. The quintessence of designing intelligent systems of this kind is **neuro-fuzzy computing**: neural networks that recognize patterns and adapt themselves to cope with changing environments; fuzzy inference systems that incorporate human knowledge and perform inferencing and decision making. The integration of these two complementary approaches, together with certain derivative-free optimization techniques, results in a novel discipline called **neuro-fuzzy and soft computing**.

As a prelude, we shall provide a bird's-eye view of relevant intelligent system approaches, along with bits of their history, and discuss the features of neuro-fuzzy and soft computing.

1.2 SOFT COMPUTING CONSTITUENTS AND CONVENTIONAL ARTIFICIAL INTELLIGENCE

Soft computing is an emerging approach to computing which parallels the remarkable ability of the human mind to reason and learn in an environment of uncertainty and imprecision. (Lotfi A. Zadeh, 1992 [12])

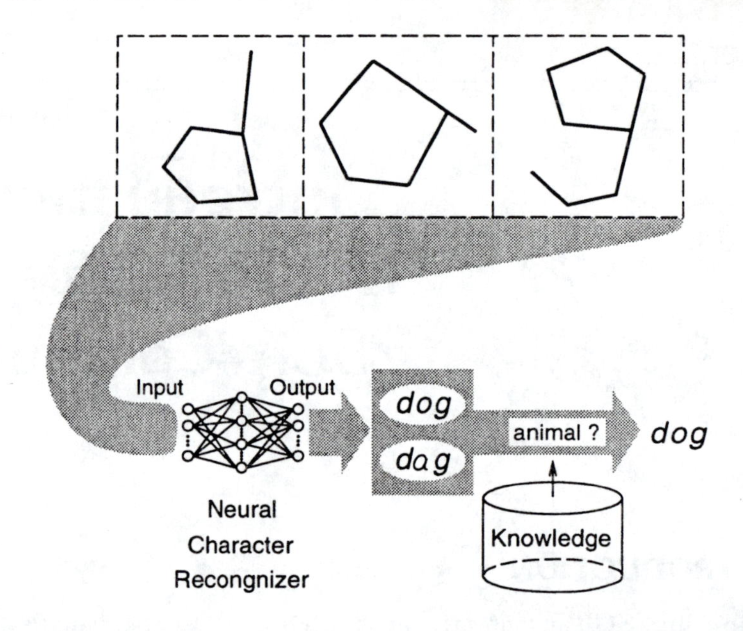

Figure 1.1. *A neural character recognizer and a knowledge base cooperate in responding to three hand-written characters that form a word "dog."*

Table 1.1. *Soft computing constituents (the first three items) and conventional artificial intelligence.*

Methodology	Strength
Neural network	Learning and adaptation
Fuzzy set theory	Knowledge representation via fuzzy if-then rules
Genetic algorithm and simulated annealing	Systematic random search
Conventional AI	Symbolic manipulation

Soft computing consists of several computing paradigms, including neural networks, fuzzy set theory, approximate reasoning, and derivative-free optimization methods such as genetic algorithms and simulated annealing. Each of these constituent methodologies has its own strength, as summarized in Table 1.1. The seamless integration of these methodologies forms the core of soft computing; the synergism allows soft computing to incorporate human knowledge effectively, deal with imprecision and uncertainty, and learn to adapt to unknown or changing environment for better performance. For learning and adaptation, soft computing requires extensive computation. In this sense, soft computing shares the same characteristics as computational intelligence.

In general, soft computing does not perform much symbolic manipulation, so we can view it as a new discipline that complements conventional artificial intelligence (AI) approaches, and vice versa. For instance, Figure 1.1 illustrates a situation in which a neural character recognizer and a knowledge base are used together to determine the meaning of a hand-written word. The neural character recognizer generates two possible answers "dog" and "dag," since the middle character could be either an "o" or "a." If the knowledge base provides an extra piece of information that the given word is related to animals, then the answer "dog" is picked up correctly.

Figure 1.2 is a list of conventional AI approaches and each of the soft computing constituents in chronological order. We discuss the features of conventional AI in Section 1.2.1, and those of soft computing constituents in Sections 1.2.2 through 1.2.4. In Section 1.3, we summarize the neuro-fuzzy and soft computing characteristics.

1.2.1 From Conventional AI to Computational Intelligence

Humans usually employ *natural languages* in reasoning and drawing conclusions. Conventional AI research focuses on an attempt to mimic human intelligent behavior by expressing it in language forms or symbolic rules. Conventional AI basically manipulates symbols on the assumption that such behavior can be stored in symbolically structured knowledge bases. This is the so-called *physical symbol system hypothesis* [3, 5]. Symbolic systems provide a good basis for modeling human experts in some narrow problem areas if explicit knowledge is available. Perhaps the most successful conventional AI product is the knowledge-based system or expert system (ES); it is represented in a schematic form in Figure 1.3.

Conventional AI literature reflects earlier work on intelligent systems. Many AI precursors defined AI in light of their own philosophy; some representative AI definitions are listed next along with a couple of ES definitions.

- "AI is the study of *agents* that exist in an environment and perceive and act." (S. Russell and P. Norvig) [6]
- "AI is the art of making computers do smart things." (Waldrop) [9]
- "AI is a programming style, where programs operate on data according to rules in order to accomplish goals." (W. A. Taylor) [8]
- "AI is the activity of providing such machines as computers with the ability to display behavior that would be regarded as intelligent if it were observed in humans." (R. McLeod) [4]
- "ES is a computer program using expert knowledge to attain high levels of performance in a narrow problem area." (D. A. Waterman) [10]

	Conventional AI	Neural networks	Fuzzy systems	Other methodologies
1940s	1947 Cybernetics	1943 McCulloch-Pitts neuron model		
1950s	1956 Artificial Intelligence	1957 Perceptron		
1960s	1960 Lisp language	1960s Adaline Madaline	1965 Fuzzy sets	
1970s	mid- 1970s Knowledge Engineering (expert systems)	1974 Birth of Back-propagation algorithm 1975 Cognitron Neocognitron	1974 Fuzzy controller	1970s Genetic algorithm
1980s		1980 Self-organizing map 1982 Hopfield Net 1983 Boltzmann machine 1986 Backpropagation algorithm boom	1985 Fuzzy modeling (TSK model)	mid- 1980s Artificial life Immune modeling
1990s			1990s Neuro-fuzzy modeling 1991 ANFIS 1994 CANFIS	1990 Genetic programming

Figure 1.2. *A historical sketch of soft computing constituents and conventional AI approaches.*

- "ES is a caricature of the human expert, in the sense that it knows almost everything about almost nothing." (A. R. Mirzai) [2]

These definitions provide a conspicuous AI framework although they may be somewhat ephemeral because the conceptual framework is metamorphosing rapidly. The reader may well wonder, "Has AI become obsolete already?"

Calling *soft computing constituents* "parts of modern AI" inevitably depends on personal judgment. It is true that today many books on modern AI describe neural networks and perhaps other soft computing components, as seen in [6, 11].

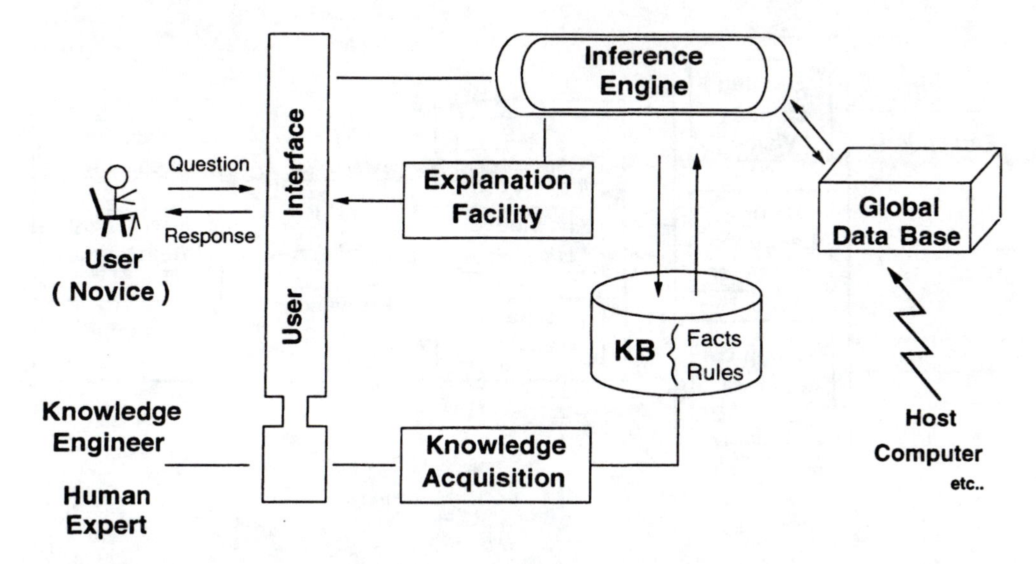

Figure 1.3. *An expert system: one of the most successful (conventional) AI products.*

This means that the AI field is steadily expanding; the boundary between AI and soft computing is becoming indistinct and, obviously, successive generations of AI methodologies will be growing more sophisticated. Further discussion of these philosophical AI territories [7] is beyond the scope of this book.

In practice, the symbolic manipulations limit the situations to which the conventional AI theories can be applied because knowledge acquisition and representation are by no means easy, but are arduous tasks. More attention has been directed toward biologically inspired methodologies such as brain modeling, evolutionary algorithms, and immune modeling; they simulate biological mechanisms responsible for generating natural intelligence. These methodologies are somewhat *orthogonal* to conventional AI approaches and generally compensate for the shortcomings of symbolicism.

The long-term goal of AI research is the creation and understanding of **machine intelligence**. From this perspective, soft computing shares the same ultimate goal with AI. Figure 1.4 is a schematic representation of an intelligent system that can sense its environment (perceive) and act on its perception (react). An easy extension of ES may also result in the same ideal computationally intelligent system sought by soft computing researchers. Soft computing is apparently evolving under AI influences that sprang from **cybernetics** (the study of information and control in humans and machines).

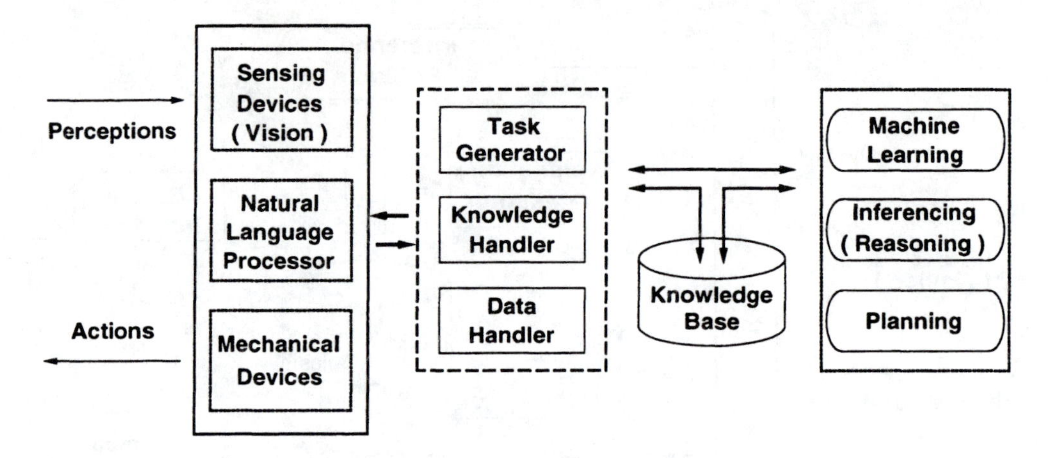

Figure 1.4. *An intelligent system.*

1.2.2 Neural Networks

The human brain is a source of natural intelligence and a truly remarkable parallel computer. The brain processes incomplete information obtained by perception at an incredibly rapid rate. Nerve cells function about 10^6 times slower than electronic circuit gates, but human brains process visual and auditory information much faster than modern computers.

Inspired by biological nervous systems, many researchers, especially brain modelers, have been exploring artificial neural networks, a novel nonalgorithmic approach to information processing. They model the brain as a continuous-time nonlinear dynamic system in connectionist architectures that are expected to mimic brain mechanisms to simulate intelligent behavior. Such connectionism replaces symbolically structured representations with distributed representations in the form of weights between a massive set of interconnected neurons (or processing units). It does not need critical decision flows in its algorithms.

A variety of connectionist approaches have been studied; some representative methodologies and their computational capacities are discussed in subsequent chapters.

1.2.3 Fuzzy Set Theory

The human brain interprets imprecise and incomplete sensory information provided by perceptive organs. Fuzzy set theory provides a systematic calculus to deal with such information linguistically, and it performs numerical computation by using linguistic labels stipulated by membership functions. Moreover, a selection of fuzzy if-then rules forms the key component of a fuzzy inference system (FIS) that can effectively model human expertise in a specific application.

Although the fuzzy inference system has a structured knowledge representation in the form of fuzzy if-then rules, it lacks the adaptability to deal with changing external environments. Thus, we incorporate neural network learning concepts in fuzzy inference systems, resulting in *neuro-fuzzy modeling*, a pivotal technique in soft computing.

We discuss fuzzy sets, fuzzy rules, and fuzzy inference systems in Chapters 2, 3, and 4. Approaches to neuro-fuzzy modeling are described in Chapters 12 and 13.

1.2.4 Evolutionary Computation

Natural intelligence is the product of millions of years of biological evolution. Simulating complex biological evolutionary processes may lead us to discover how evolution propels living systems toward higher-level intelligence. Greater attention is thus being paid to evolutionary computing techniques such genetic algorithms (GAs), which are based on the evolutionary principle of natural selection. *Immune modeling* and *Artificial Life* are similar disciplines and are based on the assumption that chemical and physical laws may be able to explain living intelligence. In particular, Artificial Life, an inclusive paradigm, attempts to realize lifelike behavior by imitating the processes that occur in the development or mechanics of life [1].

Heuristically informed search techniques are employed in many AI applications. When a search space is too large for an exhaustive (blind, brute-force) search and it is difficult to identify knowledge that can be applied to reduce the search space, we have no choice but to use other, more efficient search techniques to find less-than-optimum solutions. The GA is a candidate technique for this purpose; it offers the capacity for population-based systematic random searches. *Simulated annealing* and *random search* are other candidates that explore the search space in a stochastic manner. Those optimization methods are discussed in Chapter 7.

1.3 NEURO-FUZZY AND SOFT COMPUTING CHARACTERISTICS

With neuro-fuzzy modeling as a backbone, the characteristics of soft computing can be summarized as follows:

Human expertise Soft computing utilizes human expertise in the form of fuzzy if-then rules, as well as in conventional knowledge representations, to solve practical problems.

Biologically inspired computing models Inspired by biological neural networks, artificial neural networks are employed extensively in soft computing to deal with perception, pattern recognition, and nonlinear regression and classification problems.

New optimization techniques Soft computing applies innovative optimization methods arising from various sources; they are genetic algorithms (inspired by the evolution and selection process), simulated annealing (motivated by thermodynamics), the random search method, and the downhill Simplex method. These optimization methods do not require the gradient vector of an objective function, so they are more flexible in dealing with complex optimization problems.

Numerical computation Unlike symbolic AI, soft computing relies mainly on numerical computation. Incorporation of symbolic techniques in soft computing is an active research area within this field.

New application domains Because of its numerical computation, soft computing has found a number of new application domains besides that of AI approaches. These application domains are mostly computation intensive and include adaptive signal processing, adaptive control, nonlinear system identification, nonlinear regression, and pattern recognition.

Model-free learning Neural networks and adaptive fuzzy inference systems have the ability to construct models using only target system sample data. Detailed insight into the target system helps set up the initial model structure, but it is not mandatory.

Intensive computation Without assuming too much background knowledge of the problem being solved, neuro-fuzzy and soft computing rely heavily on high-speed number-crunching computation to find rules or regularity in data sets. This is a common feature of all areas of computational intelligence.

Fault tolerance Both neural networks and fuzzy inference systems exhibit fault tolerance. The deletion of a neuron in a neural network, or a rule in a fuzzy inference system, does not necessarily destroy the system. Instead, the system continues performing because of its parallel and redundant architecture, although performance quality gradually deteriorates.

Goal driven characteristics Neuro-fuzzy and soft computing are goal driven; the path leading from the current state to the solution does not really matter as long as we are moving toward the goal in the long run. This is particularly true when used with derivative-free optimization schemes, such as genetic algorithms, simulated annealing, and the random search method. Domain-specific knowledge helps reduces the amount of computation and search time, but it is not a requirement.

Real-world applications Most real-world problems are large scale and inevitably incorporate built-in uncertainties; this precludes using conventional approaches that require detailed description of the problem being solved. Soft computing is an integrated approach that can usually utilize specific techniques within subtasks to construct generally satisfactory solutions to real-world problems.

The field of soft computing is evolving rapidly; new techniques and applications are constantly being proposed. We can see that a firm foundation for soft computing is being built through the collective efforts of researchers in various disciplines all over the world. The underlying driving force is to construct highly automated, intelligent machines for a better life tomorrow, which is already just around the corner.

REFERENCES

[1] C. G.Langton, editor. *Artificial life*, volume 6. Addison-Wesley, Reading, MA, 1989.

[2] A. R. Mirzai. *Artificial intelligence : concepts and applications in engineering.* MIT Press, Cambridge, MA., 1990.

[3] A. Newell and H. A. Simon. Computer science as empirical inquiry: Symbols and search. *Communications of the ACM*, 19(3):113–126, 1976.

[4] Jr Raymond McLeod. *Management information systems.* Science Research Associates, Chicago, 1979.

[5] E. Rich and K. Knight. *Artificial intelligence.* McGraw-Hill, New York, 2nd edition, 1991.

[6] S. Russell and P. Norvig. *Artificial intelligence: a modern approach.* Prentice Hall, Upper Saddle River, NJ, 1995.

[7] P. Smolensky. On the proper treatment of connectionism. *Behavioral and Brain Sciences*, 2(1), 1988.

[8] W. A. Taylor. *What every engineers should know about AI.* MIT Press, Cambridge, MA., 1988.

[9] Mitchell M. Waldrop. *Man-made minds : the promise of artificial intelligence.* Walker, New York, 1987.

[10] Donald A. Waterman. *A guide to expert systems.* Addison-Wesley, Reading, MA, 1986.

[11] Patrick H. Winston. *Artificial intelligence.* Addison-Wesley, Reading, MA, 3rd edition, 1992.

[12] Lotfi A. Zadeh. Fuzzy logic, neural networks and soft computing. One-page course announcement of CS 294-4, Spring 1993, the University of California at Berkeley, November 1992.

Part I

Fuzzy Set Theory

Chapter 2

Fuzzy Sets

J.-S. R. Jang

This chapter introduces the basic definitions, notation, and operations for fuzzy sets that will be needed in the following chapters. Since research on fuzzy sets and their applications has been underway for almost 30 years now, it is impossible to cover all aspects of current developments in this field. Therefore, the aim of this chapter is to provide a concise introduction to and a summary of the basic concepts central to the study of fuzzy sets. Detailed treatments of specific subjects can be found in the reference list at the end of this chapter.

2.1 INTRODUCTION

A classical set is a set with a crisp boundary. For example, a classical set A of real numbers greater than 6 can be expressed as

$$A = \{x \mid x > 6\}, \tag{2.1}$$

where there is a clear, unambiguous boundary 6 such that if x is greater than this number, then x belongs to the set A; otherwise x does not belong to the set. Although classical sets are suitable for various applications and have proven to be an important tool for mathematics and computer science, they do not reflect the nature of human concepts and thoughts, which tend to be abstract and imprecise. As an illustration, mathematically we can express the set of tall persons as a collection of persons whose height is more than 6 ft; this is the set denoted by Equation (2.1), if we let A = "tall person" and x = "height." Yet this is an unnatural and inadequate way of representing our usual concept of "tall person." For one thing, the dichotomous nature of the classical set would classify a person 6.001 ft tall as a tall person, but not a person 5.999 ft tall. This distinction is intuitively unreasonable. The flaw comes from the sharp transition between inclusion and exclusion in a set.

In contrast to a classical set, a **fuzzy set**, as the name implies, is a set without a crisp boundary. That is, the transition from "belong to a set" to "not belong to a set" is gradual, and this smooth transition is characterized by membership functions that give fuzzy sets flexibility in modeling commonly used linguistic expressions, such as "the water is hot" or "the temperature is high." As Zadeh pointed out in 1965 in his seminal paper entitled "Fuzzy Sets" [11], such imprecisely defined sets or classes "play an important role in human thinking, particularly in the domains of pattern recognition, communication of information, and abstraction." Note that the fuzziness does not come from the randomness of the constituent members of the sets, but from the uncertain and imprecise nature of abstract thoughts and concepts.

Let us now set forth several basic definitions concerning fuzzy sets.

2.2 BASIC DEFINITIONS AND TERMINOLOGY

Let X be a space of objects and x be a generic element of X. A classical set A, $A \subseteq X$, is defined as a collection of elements or objects $x \in X$, such that each x can either belong or not belong to the set A. By defining a **characteristic function** for each element x in X, we can represent a classical set A by a set of ordered pairs $(x, 0)$ or $(x, 1)$, which indicates $x \notin A$ or $x \in A$, respectively.

Unlike the aforementioned conventional set, a fuzzy set [11] expresses the degree to which an element belongs to a set. Hence the characteristic function of a fuzzy set is allowed to have values between 0 and 1, which denotes the degree of membership of an element in a given set.

Definition 2.1 *Fuzzy sets and membership functions*

If X is a collection of objects denoted generically by x, then a **fuzzy set** A in X is defined as a set of ordered pairs:

$$A = \{(x, \mu_A(x)) \mid x \in X\}, \tag{2.2}$$

where $\mu_A(x)$ is called the **membership function** (or **MF** for short) for the fuzzy set A. The MF maps each element of X to a membership grade (or membership value) between 0 and 1.

□

Obviously, the definition of a fuzzy set is a simple extension of the definition of a classical set in which the characteristic function is permitted to have any values between 0 and 1. If the value of the membership function $\mu_A(x)$ is restricted to either 0 or 1, then A is reduced to a classical set and $\mu_A(x)$ is the characteristic function of A. For clarity, we shall also refer to classical sets as ordinary sets, crisp sets, nonfuzzy sets, or just sets.

Usually X is referred to as the **universe of discourse**, or simply the **universe**, and it may consist of discrete (ordered or nonordered) objects or continuous space. This can be clarified by the following examples.

Example 2.1 *Fuzzy sets with a discrete nonordered universe*

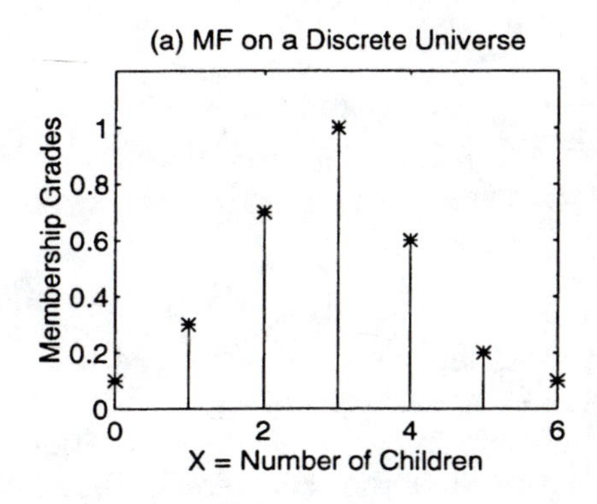

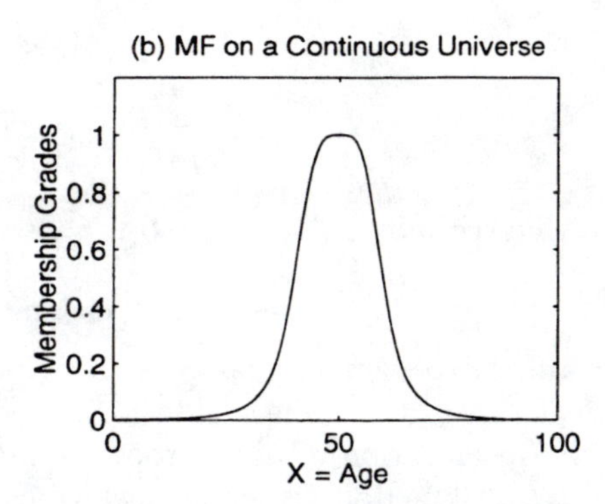

Figure 2.1. *(a) A = "sensible number of children in a family"; (b) B = "about 50 years old."* (MATLAB file: `mf_univ.m`)

Let X = {San Francisco, Boston, Los Angeles} be the set of cities one may choose to live in. The fuzzy set C = "desirable city to live in" may be described as follows:

$$C = \{(\text{San Francisco},\ 0.9),\ (\text{Boston},\ 0.8),\ (\text{Los Angeles},\ 0.6)\}.$$

Apparently the universe of discourse X is discrete and it contains nonordered objects—in this case, three big cities in the United States. As one can see, the foregoing membership grades listed above are quite subjective; anyone can come up with three different but legitimate values to reflect his or her preference.

□

Example 2.2 *Fuzzy sets with a discrete ordered universe*

Let X = {0, 1, 2, 3, 4, 5, 6} be the set of numbers of children a family may choose to have. Then the fuzzy set A = "sensible number of children in a family" may be described as follows:

$$A = \{(0, 0.1), (1, 0.3), (2, 0.7), (3, 1), (4, 0.7), (5, 0.3), (6, 0.1)\}.$$

Here we have a discrete ordered universe X; the MF for the fuzzy set A is shown in Figure 2.1(a). Again, the membership grades of this fuzzy set are obviously subjective measures.

□

Example 2.3 *Fuzzy sets with a continuous universe*

Let $X = R^+$ be the set of possible ages for human beings. Then the fuzzy set B = "about 50 years old" may be expressed as

$$B = \{(x, \mu_B(x)) | x \in X\},$$

where

$$\mu_B(x) = \frac{1}{1 + \left(\frac{x - 50}{10}\right)^4}.$$

This is illustrated in Figure 2.1(b).

□

From the preceding examples, it is obvious that the construction of a fuzzy set depends on two things: the identification of a suitable universe of discourse and the specification of an appropriate membership function. The specification of membership functions is *subjective*, which means that the membership functions specified for the same concept (say, "sensible number of children in a family") by different persons may vary considerably. This subjectivity comes from individual differences in perceiving or expressing abstract concepts and has little to do with randomness. Therefore, the **subjectivity** and **nonrandomness** of fuzzy sets is the primary difference between the study of fuzzy sets and probability theory, which deals with objective treatment of random phenomena.

For simplicity of notation, we now introduce an alternative way of denoting a fuzzy set. A fuzzy set A can be denoted as follows:

$$A = \begin{cases} \sum_{x_i \in X} \mu_A(x_i)/x_i, & \text{if } X \text{ is a collection of discrete objects.} \\ \int_X \mu_A(x)/x, & \text{if } X \text{ is a continuous space (usually the real line } R). \end{cases} \tag{2.3}$$

The summation and integration signs in Equation (2.3) stand for the union of $(x, \mu_A(x))$ pairs; they do not indicate summation or integration. Similarly, "/" is only a marker and does not imply division.

Example 2.4 *Alternative expression*

Using the notation of Equation (2.3), we can rewrite the fuzzy sets in Examples 2.1, 2.2, and 2.3 as

$$C = 0.9/\text{San Francisco} + 0.8/\text{Boston} + 0.6/\text{Los Angeles},$$

$$A = 0.1/0 + 0.3/1 + 0.7/2 + 1.0/3 + 0.7/4 + 0.3/5 + 0.1/6,$$

and

$$B = \int_{R+} \frac{1}{1 + (\frac{x-50}{10})^4} \Big/ x,$$

respectively.

□

In practice, when the universe of discourse X is a continuous space (the real line R or its subset), we usually partition X into several fuzzy sets whose MFs cover X in a more or less uniform manner. These fuzzy sets, which usually carry names

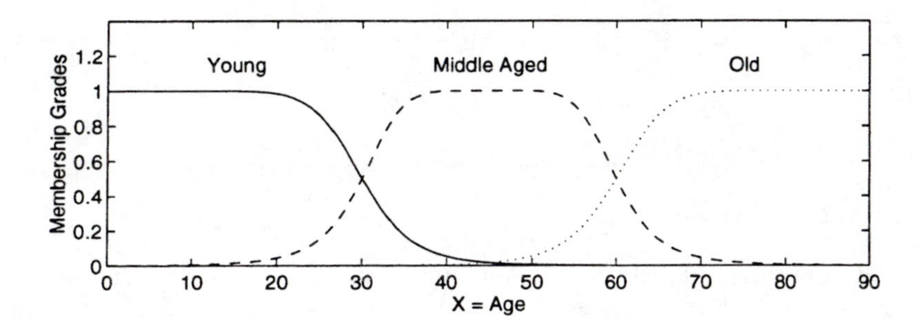

Figure 2.2. *Typical MFs of linguistic values "young," "middle aged," and "old."* (MATLAB file: `lingmf.m`)

that conform to adjectives appearing in our daily linguistic usage, such as "large," "medium," or "small," are called linguistic values or linguistic labels. Thus, the universe of discourse X is often called the linguistic variable. Formal definitions of linguistic variables and linguistic values are given in the next chapter; here we shall give only a simple example.

Example 2.5 *Linguistic variables and linguistic values*

Suppose that X = "age." Then we can define fuzzy sets "young," "middle aged," and "old" that are characterized by MFs $\mu_{old}(x)$, $\mu_{middleaged}(x)$, and $\mu_{old}(x)$, respectively. Just as a variable can assume various values, a linguistic variable "Age" can assume different linguistic values, such as "young," "middle aged," and "old" in this case. If "age" assumes the value of "young," then we have the expression "age is young," and so forth for the other values. Typical MFs for these linguistic values are displayed in Figure 2.2, where the universe of discourse X is totally covered by the MFs and the transition from one MF to another is smooth and gradual.

□

A fuzzy set is uniquely specified by its membership function. To describe membership functions more specifically, we shall define the nomenclature used in the literature. (Unless otherwise specified, we shall assume that the universe of the fuzzy sets under discussion is the real line R or its subset.)

Definition 2.2 *Support*

The **support** of a fuzzy set A is the set of all points x in X such that $\mu_A(x) > 0$:

$$\text{support}(A) = \{x|\mu_A(x) > 0\}. \tag{2.4}$$

□

Definition 2.3 *Core*

The **core** of a fuzzy set A is the set of all points x in X such that $\mu_A(x) = 1$:

$$\text{core}(A) = \{x | \mu_A(x) = 1\}. \tag{2.5}$$

□

Definition 2.4 *Normality*

A fuzzy set A is **normal** if its core is nonempty. In other words, we can always find a point $x \in X$ such that $\mu_A(x) = 1$.

□

Definition 2.5 *Crossover points*

A **crossover point** of a fuzzy set A is a point $x \in X$ at which $\mu_A(x) = 0.5$:

$$\text{crossover}(A) = \{x | \mu_A(x) = 0.5\}. \tag{2.6}$$

□

Definition 2.6 *Fuzzy singleton*

A fuzzy set whose support is a single point in X with $\mu_A(x) = 1$ is called a **fuzzy singleton**.

□

Figures 2.3(a) and 2.3(b) illustrate the cores, supports, and crossover points of the bell-shaped membership function representing "middle aged" and of the fuzzy singleton characterizing "45 years old."

Definition 2.7 *α-cut, strong α-cut*

The **α-cut** or **α-level set** of a fuzzy set A is a crisp set defined by

$$A_\alpha = \{x | \mu_A(x) \geq \alpha\}. \tag{2.7}$$

Strong α-cut or **strong α-level set** are defined similarly:

$$A'_\alpha = \{x | \mu_A(x) > \alpha\}. \tag{2.8}$$

□

Using the notation for a level set, we can express the support and core of a fuzzy set A as

$$\text{support}(A) = A'_0,$$

and

$$\text{core}(A) = A_1,$$

respectively.

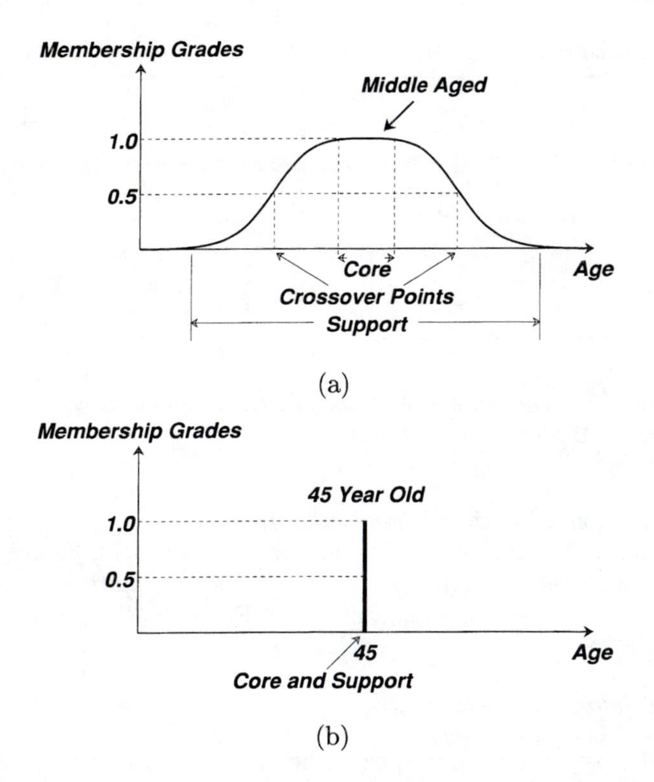

Figure 2.3. *Cores, supports, and crossover points of (a) the fuzzy set "middle aged" and (b) the fuzzy singleton "45 years old."*

Definition 2.8 *Convexity*

A fuzzy set A is **convex** if and only if for any x_1, $x_2 \in X$ and any $\lambda \in [0, 1]$,

$$\mu_A(\lambda x_1 + (1 - \lambda)x_2) \geq \min\{\mu_A(x_1), \mu_A(x_2)\}. \tag{2.9}$$

Alternatively, A is convex if all its α-level sets are convex.

□

A crisp set C in R^n is convex if and only if for any two points $x_1 \in C$ and $x_2 \in C$, their convex combination $\lambda x_1 + (1 - \lambda)x_2$ is still in C, where $0 \leq \lambda \leq 1$. Hence the convexity of a (crisp) level set A_α implies that A_α is composed of a single line segment only.

Note that the definition of convexity of a fuzzy set is not as strict as the common definition of convexity of a function. For comparison, the definition of convexity of a function $f(x)$ is

$$f(\lambda x_1 + (1 - \lambda)x_2) \geq \lambda f(x_1) + (1 - \lambda)f(x_2), \tag{2.10}$$

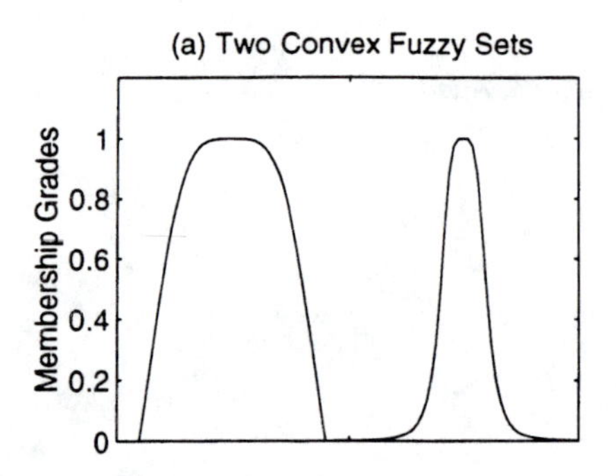

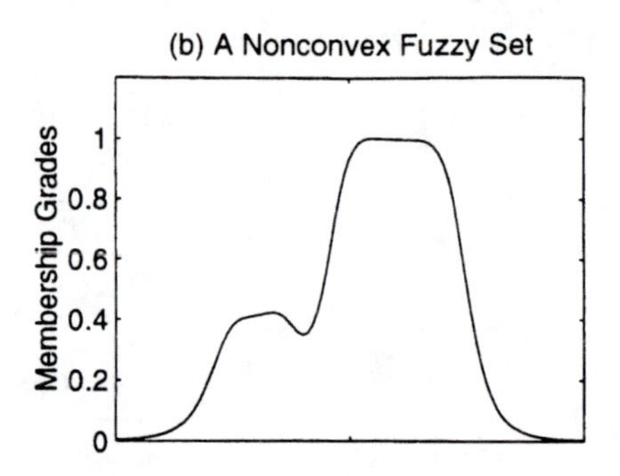

Figure 2.4. *(a) Two convex membership functions; (b) a nonconvex membership function.* (MATLAB file: `convexmf.m`)

which is a tighter condition than Equation (2.9).

Figure 2.4 illustrates the concept of convexity of fuzzy sets; Figure 2.4(a) shows two convex fuzzy sets [the left fuzzy set satisfies both Equations (2.9) and (2.10, while the right one satisfies Equation (2.9) only]; Figure 2.4(b) is a nonconvex fuzzy set.

Definition 2.9 *Fuzzy numbers*

A fuzzy number A is a fuzzy set in the real line (R) that satisfies the conditions for normality and convexity.

□

Most (noncomposite) fuzzy sets used in the literature satisfy the conditions for normality and convexity, so fuzzy numbers are the most basic type of fuzzy sets.

Definition 2.10 *Bandwidths of normal and convex fuzzy sets*

For a normal and convex fuzzy set, the **bandwidth** or **width** is defined as the distance between the two unique crossover points:

$$\text{width}(A) = |x_2 - x_1|, \tag{2.11}$$

where $\mu_A(x_1) = \mu_A(x_2) = 0.5$.

□

Definition 2.11 *Symmetry*

A fuzzy set A is **symmetric** if its MF is symmetric around a certain point $x = c$, namely,

$$\mu_A(c + x) = \mu_A(c - x) \text{ for all } x \in X.$$

□

Definition 2.12 *Open left, open right, closed*

A fuzzy set A is **open left** if $\lim_{x \to -\infty} \mu_A(x) = 1$ and $\lim_{x \to +\infty} \mu_A(x) = 0$; **open right** if $\lim_{x \to -\infty} \mu_A(x) = 0$ and $\lim_{x \to +\infty} \mu_A(x) = 1$; and **closed** if $\lim_{x \to -\infty} \mu_A(x) = \lim_{x \to +\infty} \mu_A(x) = 0$.

□

For instance, the fuzzy set "young" in Figure 2.2 is open left; "old" is open right; and "middle aged" is closed.

2.3 SET-THEORETIC OPERATIONS

Union, intersection, and complement are the most basic operations on classical sets. On the basis of these three operations, a number of identities can be established, as listed in Table 2.1. These identities can be verified using Venn diagrams.

Corresponding to the ordinary set operations of union, intersection, and complement, fuzzy sets have similar operations, which were initially defined in Zadeh's seminal paper [11]. Before introducing these three fuzzy set operations, first we shall define the notion of containment, which plays a central role in both ordinary and fuzzy sets. This definition of containment is, of course, a natural extension of the case for ordinary sets.

Definition 2.13 *Containment or subset*

Fuzzy set A is **contained** in fuzzy set B (or, equivalently, A is a **subset** of B, or A is smaller than or equal to B) if and only if $\mu_A(x) \leq \mu_B(x)$ for all x. In symbols,

$$A \subseteq B \Longleftrightarrow \mu_A(x) \leq \mu_B(x). \tag{2.12}$$

□

Figure 2.5 illustrates the concept of $A \subseteq B$.

Definition 2.14 *Union (disjunction)*

The **union** of two fuzzy sets A and B is a fuzzy set C, written as $C = A \cup B$ or $C = A$ OR B, whose MF is related to those of A and B by

$$\mu_C(x) = \max(\mu_A(x), \mu_B(x)) = \mu_A(x) \vee \mu_B(x). \tag{2.13}$$

□

As pointed out by Zadeh [11], a more intuitive but equivalent definition of union is the "smallest" fuzzy set containing both A and B. Alternatively, if D is any fuzzy set that contains both A and B, then it also contains $A \cup B$. The intersection of fuzzy sets can be defined analogously.

Table 2.1. *Basic identities of classical sets, where A, B, and C are crisp sets; $\overline{A}$, $\overline{B}$, and $\overline{C}$ are their corresponding complements; X is the universe; and $\emptyset$ is the empty set.*

Law	Identity
Law of contradiction	$A \cap \overline{A} = \emptyset$
Law of the excluded middle	$A \cup \overline{A} = X$
Idempotency	$A \cap A = A,\ A \cup A = A$
Involution	$\overline{\overline{A}} = A$
Commutativity	$A \cap B = B \cap A,\ A \cup B = B \cup A$
Associativity	$(A \cup B) \cup C = A \cup (B \cup C)$ $(A \cap B) \cap C = A \cap (B \cap C)$
Distributivity	$A \cup (B \cap C) = (A \cup B) \cap (A \cup C)$ $A \cap (B \cup C) = (A \cap B) \cup (A \cap C)$
Absorption	$A \cup (A \cap B) = A$ $A \cap (A \cup B) = A$
Absorption of complement	$A \cup (\overline{A} \cap B) = A \cup B$ $A \cap (\overline{A} \cup B) = A \cap B$
DeMorgan's laws	$\overline{A \cup B} = \overline{A} \cap \overline{B}$ $\overline{A \cap B} = \overline{A} \cup \overline{B}$

Definition 2.15 *Intersection (conjunction)*

The **intersection** of two fuzzy sets A and B is a fuzzy set C, written as $C = A \cap B$ or $C = A$ AND B, whose MF is related to those of A and B by

$$\mu_C(x) = \min(\mu_A(x), \mu_B(x)) = \mu_A(x) \wedge \mu_B(x). \tag{2.14}$$

□

As in the case of the union, it is obvious that the intersection of A and B is the "largest" fuzzy set which is contained in both A and B. This reduces to the ordinary intersection operation if both A and B are nonfuzzy.

Definition 2.16 *Complement (negation)*

The **complement** of fuzzy set A, denoted by $\overline{A}$ ($\neg A$, NOT A), is defined as

$$\mu_{\overline{A}}(x) = 1 - \mu_A(x). \tag{2.15}$$

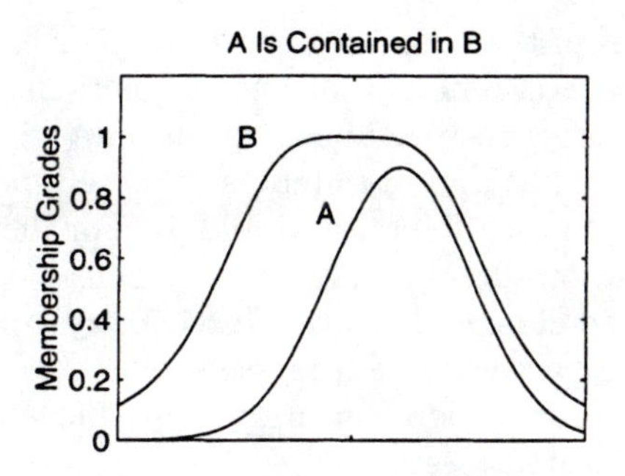

Figure 2.5. *The concept of $A \subseteq B$.* (MATLAB file: `subset.m`)

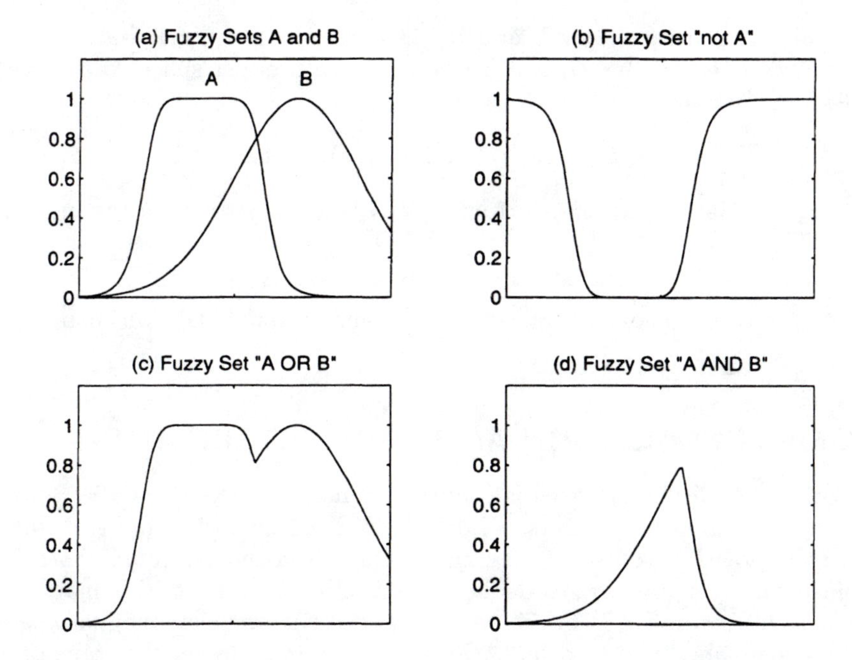

Figure 2.6. *Operations on fuzzy sets: (a) two fuzzy sets A and B; (b) $\overline{A}$; (c) $A \cup B$; (d) $A \cap B$.* (MATLAB file: `fuzsetop.m`)

□

Figure 2.6 demonstrates these three basic operations: Figure 2.6(a) illustrates two fuzzy sets A and B; Figure 2.6(b) is the complement of A; Figure 2.6(c) is the union of A and B; and Figure 2.6(d) is the intersection of A and B.

Equations (2.13), (2.14), and (2.15) perform exactly as the corresponding operations for ordinary sets if the values of the membership functions are restricted to either 0 or 1. However, it is understood that these functions are not the only possible generalizations of the crisp set operations. For each of the aforementioned three set operations, several different classes of functions with desirable properties have been proposed subsequently in the literature; we will introduce some of these functions in Section 2.5. The appropriateness of these functions can be checked via the identities in Table 2.1 (see Exercises 3 and 4). For distinction, the max [Equation (2.13)], min [Equation (2.14)], and the complement operator [Equation (2.15)] will be referred to as the **classical** or **standard fuzzy operators** for intersection, union, and negation, respectively, on fuzzy sets.

Next we define other operations on fuzzy sets which are also direct generalizations of operations on ordinary sets.

Definition 2.17 *Cartesian product and co-product*

Let A and B be fuzzy sets in X and Y, respectively. The **Cartesian product** of A and B, denoted by $A \times B$, is a fuzzy set in the product space $X \times Y$ with the membership function

$$\mu_{A\times B}(x, y) = \min(\mu_A(x), \mu_B(y)). \tag{2.16}$$

Similarly, the **Cartesian co-product** $A + B$ is a fuzzy set with the membership function

$$\mu_{A+B}(x, y) = \max(\mu_A(x), \mu_B(y)). \tag{2.17}$$

Both $A \times B$ and $A + B$ are characterized by two-dimensional MFs, which are explored in greater detail in Section 2.4.2.

2.4 MF FORMULATION AND PARAMETERIZATION

As mentioned earlier, a fuzzy set is completely characterized by its MF. Since most fuzzy sets in use have a universe of discourse X consisting of the real line R, it would be impractical to list all the pairs defining a membership function. A more convenient and concise way to define an MF is to express it as a mathematical formula, as in Example 2.3. In this section we describe the classes of parameterized functions commonly used to define MFs of one and two dimensions. MFs of higher dimensions can be defined similarly. Moreover, we give the derivatives of some of the MFs with respective to their inputs and parameters. These derivatives are important for fine-tuning a fuzzy inference system to achieve a desired input/output mapping; techniques for fine-tuning fuzzy inference systems are discussed in detail in Chapter 4.

2.4.1 MFs of One Dimension

First we define several classes of parameterized MFs of one dimension—that is, MFs with a single input.

Definition 2.18 *Triangular MFs*

A **triangular MF** is specified by three parameters $\{a, b, c\}$ as follows:

$$\text{triangle}(x; a, b, c) = \begin{cases} 0, & x \leq a. \\ \dfrac{x-a}{b-a}, & a \leq x \leq b. \\ \dfrac{c-x}{c-b}, & b \leq x \leq c. \\ 0, & c \leq x. \end{cases} \tag{2.18}$$

By using min and max, we have an alternative expression for the preceding equation:

$$\text{triangle}(x; a, b, c) = \max\left(\min\left(\frac{x-a}{b-a}, \frac{c-x}{c-b}\right), 0\right) \tag{2.19}$$

The parameters $\{a, b, c\}$ (with $a < b < c$) determine the x coordinates of the three corners of the underlying triangular MF.

□

Figure 2.7(a) illustrates a triangular MF defined by triangle$(x; 20, 60, 80)$.

Definition 2.19 *Trapezoidal MFs*

A **trapezoidal MF** is specified by four parameters $\{a, b, c, d\}$ as follows:

$$\text{trapezoid}(x; a, b, c, d) = \begin{cases} 0, & x \leq a. \\ \dfrac{x-a}{b-a}, & a \leq x \leq b. \\ 1, & b \leq x \leq c. \\ \dfrac{d-x}{d-c}, & c \leq x \leq d. \\ 0, & d \leq x. \end{cases} \tag{2.20}$$

An alternative concise expression using min and max is

$$\text{trapezoid}(x; a, b, c, d) = \max\left(\min\left(\frac{x-a}{b-a}, 1, \frac{d-x}{d-c}\right), 0\right). \tag{2.21}$$

The parameters $\{a, b, c, d\}$ (with $a < b \leq c < d$) determine the x coordinates of the four corners of the underlying trapezoidal MF.

□

Figure 2.7(b) illustrates a trapezoidal MF defined by trapezoid$(x; 10, 20, 60, 95)$. Note that a trapezoidal MF with parameter $\{a, b, c, d\}$ reduces to a triangular MF when b is equal to c.

Due to their simple formulas and computational efficiency, both triangular MFs and trapezoidal MFs have been used extensively, especially in real-time implementations. However, since the MFs are composed of straight line segments, they are not smooth at the corner points specified by the parameters. In the following we introduce other types of MFs defined by smooth and nonlinear functions.

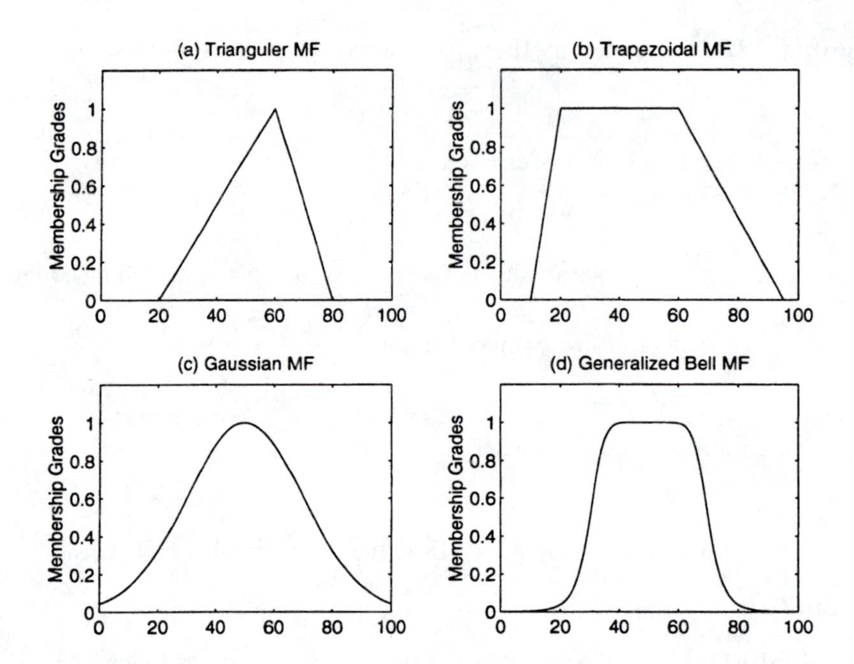

Figure 2.7. *Examples of four classes of parameterized MFs: (a) triangle*$(x; 20, 60, 80)$*; (b) trapezoid*$(x; 10, 20, 60, 95)$*; (c) gaussian*$(x; 50, 20)$*; (d) bell*$(x; 20, 4, 50)$. (MATLAB file: `disp_mf.m`)

Definition 2.20 *Gaussian MFs*

A **Gaussian MF** is specified by two parameters $\{c, \sigma\}$:

$$\text{gaussian}(x; c, \sigma) = e^{-\frac{1}{2}\left(\frac{x-c}{\sigma}\right)^2}. \tag{2.22}$$

□

A Gaussian MF is determined completely by c and σ; c represents the MFs center and σ determines the MFs width. Figure 2.7(c) plots a Gaussian MF defined by gaussian$(x; 50, 20)$.

Definition 2.21 *Generalized bell MFs*

A **generalized bell MF** (or **bell MF**) is specified by three parameters $\{a, b, c\}$:

$$\text{bell}(x; a, b, c) = \frac{1}{1 + \left|\frac{x-c}{a}\right|^{2b}}, \tag{2.23}$$

where the parameter b is usually positive. (If b is negative, the shape of this MF becomes an upside-down bell.) Note that this MF is a direct generalization of

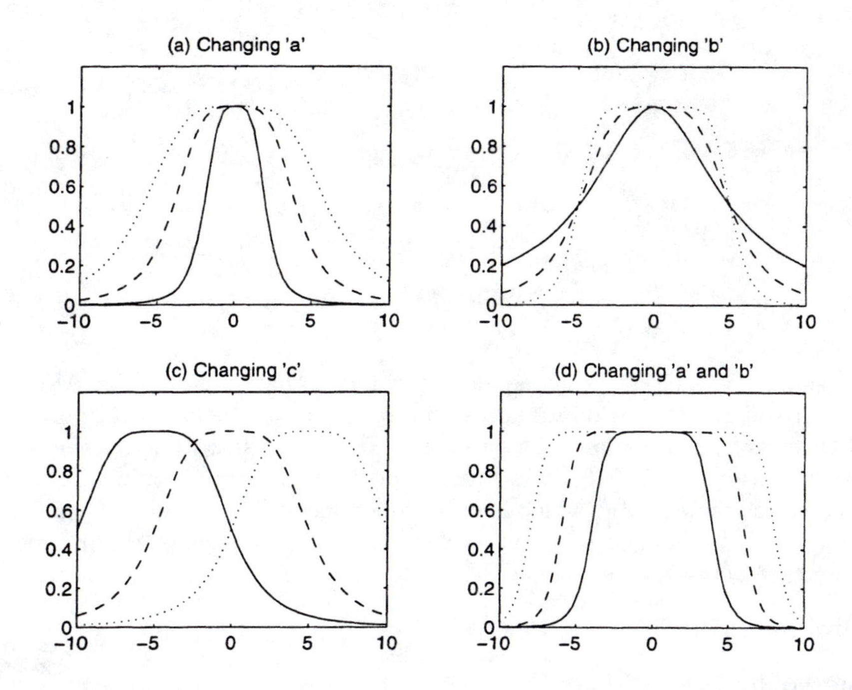

Figure 2.8. *The effects of changing parameters in bell MFs: (a) changing parameter a; (b) changing parameter b; (c) changing parameter c; (d) changing a and b simultaneously but keeping their ratio constant.* (MATLAB file: `allbells.m`)

the Cauchy distribution used in probability theory, so it is also referred to as the **Cauchy MF**.

□

Figure 2.7(d) illustrates a generalized bell MF defined by bell$(x; 20, 4, 50)$. A desired generalized bell MF can be obtained by a proper selection of the parameter set $\{a, b, c\}$. Specifically, we can adjust c and a to vary the center and width of the MF, and then use b to control the slopes at the crossover points. Figure 2.9 shows the physical meanings of each parameter in a bell MF. Figure 2.8 further illustrates the effects of changing each parameter. To obtain hands-on experience of these effects, the reader is encouraged to run the MATLAB file `bellmanu.m`, which is available via FTP and WWW (see page xxiii). Another file `bellanim.m` gives more vivid visual effects by animating two bell MFs when their parameters are changing.

Because of their smoothness and concise notation, Gaussian and bell MFs are becoming increasingly popular for specifying fuzzy sets. Gaussian functions are well known in probability and statistics, and they possess useful properties such as invariance under multiplication (the product of two Gaussians is a Gaussian with a scaling factor) and Fourier transform (the Fourier transform of a Gaussian is still a

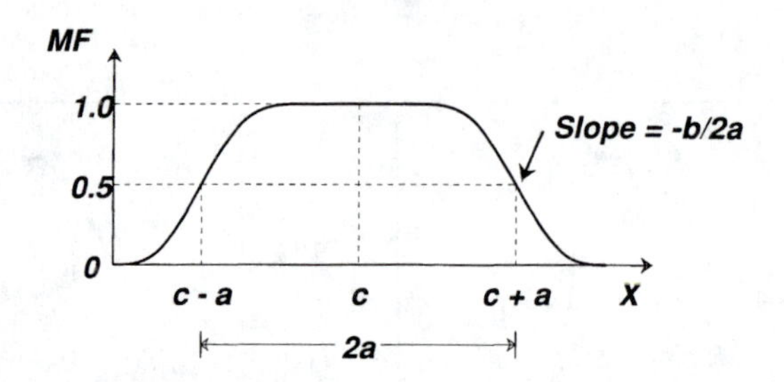

Figure 2.9. *Physical meaning of parameters in a generalized bell MF.*

Gaussian). The bell MF has one more parameter than the Gaussian MF, so it has one more degree of freedom to adjust the steepness at the crossover points.

Although the Gaussian MFs and bell MFs achieve smoothness, they are unable to specify asymmetric MFs, which are important in certain applications. Next we define the sigmoidal MF, which is either open left or right. Asymmetric and close MFs can be synthesized using either the absolute difference or the product of two sigmoidal functions, as explained next.

Definition 2.22 *Sigmoidal MFs*

A **sigmoidal MF** is defined by

$$\mathrm{sig}(x; a, c) = \frac{1}{1 + \exp[-a(x - c)]}, \tag{2.24}$$

where a controls the slope at the crossover point $x = c$.

□

Depending on the sign of the parameter a, a sigmoidal MF is inherently open right or left and thus is appropriate for representing concepts such as "very large" or "very negative." Sigmoidal functions of this kind are employed widely as the activation function of artificial neural networks. Therefore, for a neural network to simulate the behavior of a fuzzy inference system (more on this in later chapters), the first problem we face is how to synthesize a close MF through a sigmoidal function. Two simple ways for achieving this are shown in the following example.

Example 2.6 *Close and asymmetric MFs based on sigmoidal functions*

Figure 2.10(a) shows two sigmoidal functions $y_1 = \mathrm{sig}(x; 1, -5)$ and $y_2 = \mathrm{sig}(x; 2, 5)$; a close and asymmetric MF can be obtained by taking their difference $|y_1 - y_2|$, as shown in Figure 2.10(b). Figure 2.10(c) shows an additional sigmoidal MF defined as $y_3 = \mathrm{sig}(x; -2, 5)$; another way to form a close and asymmetric MF is to take their product $y_1 y_3$, as shown in Figure 2.10(d). The reader is encouraged to try the MATLAB file `siganim.m` (available via FTP and WWW, see page xxiii), which display the animation of two composite MFs based on sigmoidal functions.

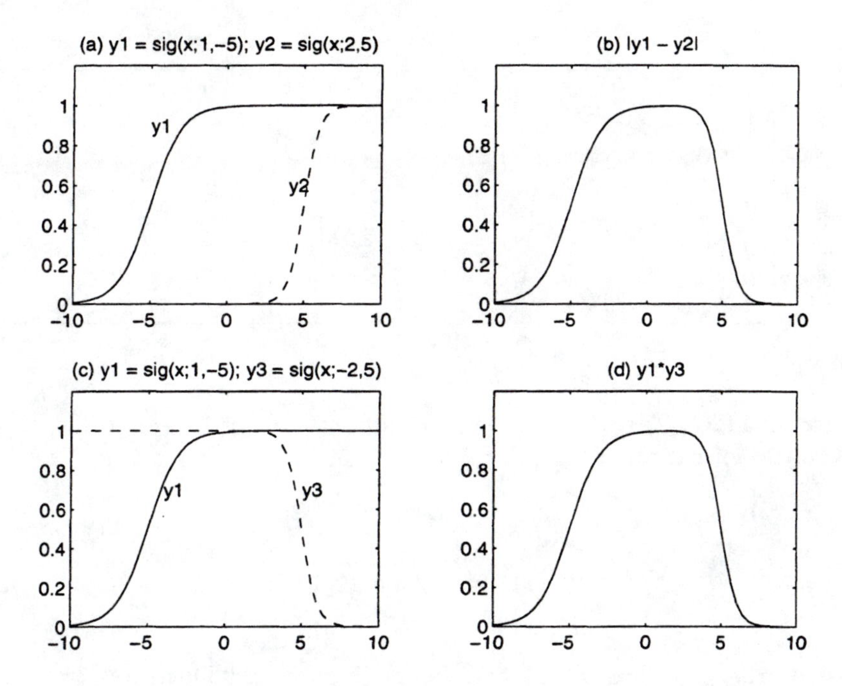

Figure 2.10. *(a) Two sigmoidal functions y_1 and y_2; (b) a close MF obtained from $|y_1 - y_2|$; (c) two sigmoidal functions y_1 and y_3; (d) a close MF obtained from $y_1 y_3$.* (MATLAB file: `disp_sig.m`)

□

In the following we define a much more general type of MF, the left-right MF. This type of MF, although extremely flexible in specifying fuzzy sets, is not used often in practice because of its unnecessary complexity.

Definition 2.23 *Left-right MF (L-R MF)*

A **left-right MF** or **L-R MF** is specified by three parameters $\{\alpha, \beta, c\}$:

$$\mathrm{LR}(x; c, \alpha, \beta) = \begin{cases} F_L\left(\dfrac{c-x}{\alpha}\right), & x \le c. \\ F_R\left(\dfrac{x-c}{\beta}\right), & x \ge c, \end{cases} \tag{2.25}$$

where $F_L(x)$ and $F_R(x)$ are monotonically decreasing functions defined on $[0, \infty)$ with $F_L(0) = F_R(0) = 1$ and $\lim_{x\to\infty} F_L(x) = \lim_{x\to\infty} F_R(x) = 0$.

□

Example 2.7 *L-R MF*

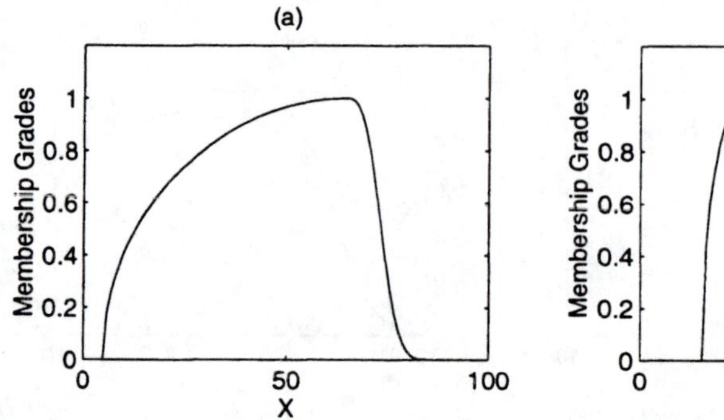

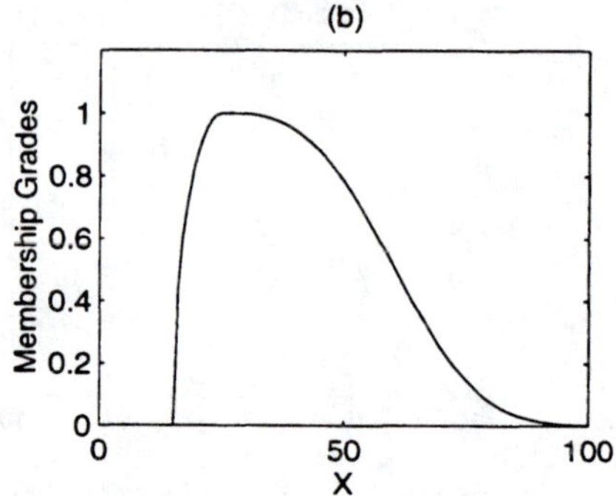

Figure 2.11. *Two L-R MFs: (a) $LR(x; 65, 60, 10)$; (b) $LR(x; 25, 10, 40)$.* (MATLAB file: `difflr.m`)

Let

$$\begin{aligned} F_L(x) &= \sqrt{\max(0, 1 - x^2)}, \\ F_R(x) &= e^{-|x|^3}. \end{aligned}$$

Based on the preceding $F_L(x)$ and $F_R(x)$, Figure 2.11 illustrates two L-R MFs specified by $\text{LR}(x; 65, 60, 10)$ and $\text{LR}(x; 25, 10, 40)$.

□

The list of MFs introduced in this section is by no means exhaustive; other specialized MFs can be created for specific applications if necessary. In particular, any type of continuous probability distribution functions can be used as an MF here, provided that a set of parameters is given to specify the appropriate meanings of the MF.

Several other types of parameterized MFs, such as S, Z, π, and two-sided Gaussian MFs, are examined in Exercises 6, 7, 8, and 10, respectively.

2.4.2 MFs of Two Dimensions

Sometimes it is advantageous or necessary to use MFs with two inputs, each in a different universe of discourse. MFs of this kind are generally referred to as two-dimensional MFs, whereas ordinary MFs (MFs with one input) are referred to as one-dimensional MFs. One natural way to extend one-dimensional MFs to two-dimensional ones is via cylindrical extension, defined next.

Definition 2.24 *Cylindrical extensions of one-dimensional fuzzy sets*

If A is a fuzzy set in X, then its **cylindrical extension** in $X \times Y$ is a fuzzy set $c(A)$ defined by

$$c(A) = \int_{X \times Y} \mu_A(x)/(x, y). \tag{2.26}$$

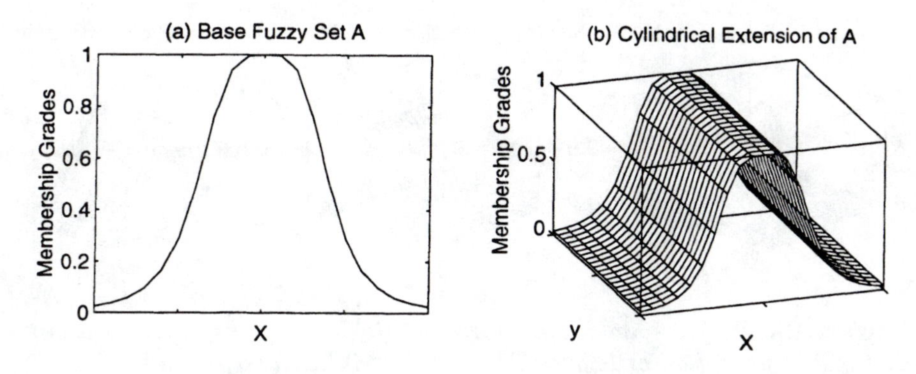

Figure 2.12. *(a) Base set A; (b) its cylindrical extension c(A).* (MATLAB file: cyl_ext.m)

(Usually A is referred to as a **base set**.)

□

The concept of cylindrical extension is quite straightforward; it is illustrated in Figure 2.12. The operation of projection, on the other hand, decreases the dimension of a given (multidimensional) membership function.

Definition 2.25 *Projections of fuzzy sets*

Let R be a two-dimensional fuzzy set on $X \times Y$. Then the **projections** of R onto X and Y are defined as

$$R_X = \int_X [\max_y \mu_R(x, y)]/x$$

and

$$R_Y = \int_Y [\max_x \mu_R(x, y)]/y,$$

respectively.

□

Figure 2.13(a) shows the MF for fuzzy set R; Figure 2.13(b) and Figure 2.13(c) are the projections of R onto X and Y, respectively.

Generally speaking, MFs of two dimensions fall into two categories: composite and noncomposite. If an MF of two dimensions can be expressed as an analytic expression of two MFs of one dimension, then it is composite; otherwise it is noncomposite. An example is given next.

Example 2.8 *Composite and noncomposite MFs*

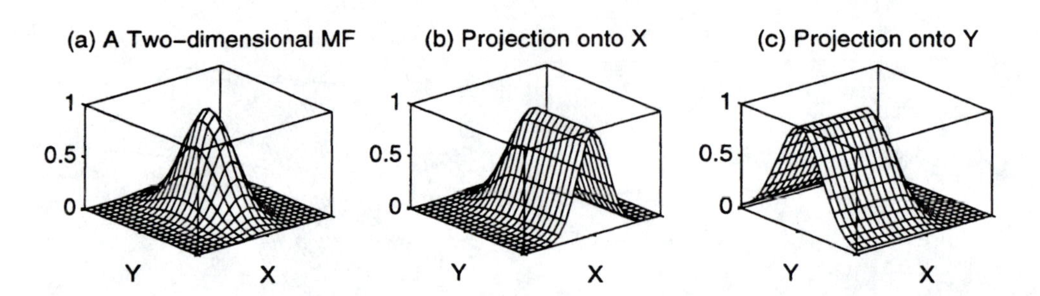

Figure 2.13. *(a) Two-dimensional fuzzy set R; (b) R_X (projection of R onto X); and (c) R_Y (projection of R onto Y).* (MATLAB file: `project.m`)

Suppose that fuzzy set A = "$(x,\ y)$ is near (3, 4)" is defined by

$$\mu_A(x, y) = \exp\left[-\left(\frac{x-3}{2}\right)^2 - (y-4)^2\right].$$

Then this two-dimensional MF is composite, since it can be decomposed into two Gaussian MFs:

$$\begin{aligned}\mu_A(x, y) &= \exp\left[-\left(\tfrac{x-3}{2}\right)^2\right] \exp\left[-\left(\tfrac{y-4}{1}\right)^2\right] \\ &= \text{gaussian}(x; 3, 2)\ \text{gaussian}(y; 4, 1).\end{aligned}$$

Note that we can view the fuzzy set A as two statements joined by the connective AND: "x is near 3 AND y is near 4," where the first statement is defined by

$$\mu_{\text{near } 3}(x) = \text{gaussian}(x; 3, 2),$$

and the second statement is defined by

$$\mu_{\text{near } 4}(y) = \text{gaussian}(y; 4, 1).$$

Thus the multiplication of these two MFs is used to interpret the AND operation of these two statements.

On the other hand, if this fuzzy set is defined by

$$\mu_A(x, y) = \frac{1}{1 + |x-3|\ |y-4|^{2.5}}, \tag{2.27}$$

then it is noncomposite.

□

As demonstrated in the preceding example, a composite two-dimensional MF is usually the result of two statements joined by the AND or OR connectives. Under

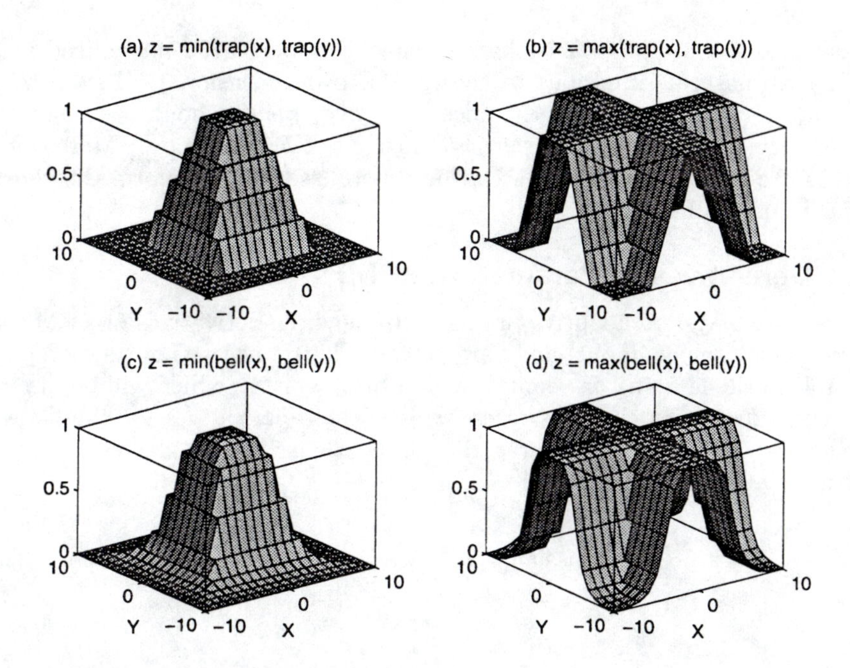

Figure 2.14. *Two-dimensional MFs defined by the min and max operators: (a)* $z = \min(trap(x), trap(y))$*; (b)* $z = \max(trap(x), trap(y))$*; (c)* $z = \min(bell(x), bell(y))$*; (d)* $z = \max(bell(x), bell(y))$. (MATLAB file: `mf2d.m`)

this condition, the two-dimensional MF is defined as the AND or OR aggregation of its two constituent MFs. Classical AND and OR operations on fuzzy sets are min and max (see Section 2.3); their effects on generating two-dimensional MFs are illustrated in the following example.

Example 2.9 *Composite two-dimensional MFs based on* min *and* max *operators*

Let $\text{trap}(x) = \text{trapezoid}(x; -6, -2, 2, 6)$ and $\text{trap}(y) = \text{trapezoid}(y; -6, -2, 2, 6)$ be two trapezoidal MFs on X and Y, respectively. After applying the *min* and *max* operators, we have two-dimensional MFs on $X \times Y$, as shown in Figures 2.14(a) and (b). Figure 2.14(c) and (d) repeat the same plots, except that the trapezoidal MFs are replaced by bell MFs $\text{bell}(x) = \text{bell}(x; 4, 3, 0)$ and $\text{bell}(x) = \text{bell}(y; 4, 3, 0)$.

□

When the min operator is used to aggregate one-dimensional MFs, the resulting two-dimensional MF can be viewed either as the result of applying classical fuzzy intersection (Definition 2.15) to the cylindrical extensions of each one-dimensional MF, or as a Cartesian product of two one-dimensional fuzzy sets (Definition 2.17). Similar interpretations apply to the max operator.

It is obvious that the definitions for one-dimensional MFs introduced in Section 2.2 have natural extensions to the case of two-dimensional MFs, with some appropriate adjustments. (For instance, crossover points should be changed to crossover curves, α-cut is a crisp set in R^2 instead of R, and so on.) Moreover, the concepts introduced in this section can be generalized easily to form the concepts of n-dimensional MFs.

2.4.3 Derivatives of Parameterized MFs

To make a fuzzy system adaptive, we need to know the derivatives of an MF with respect to its argument (input) and parameters. This derivative information plays a central role in the learning or adaptation of a fuzzy system, which will be discussed in depth in subsequent chapters. Here we list these derivatives for the Gaussian and bell MFs; the reader is encouraged to derive them independently.

For the Gaussian MF, let

$$y = \text{gaussian}(x; \sigma, c) = e^{-\frac{1}{2}\left(\frac{x-c}{\sigma}\right)^2}. \tag{2.28}$$

Then

$$\frac{\partial y}{\partial x} = -\frac{x-c}{\sigma^2} y. \tag{2.29}$$

$$\frac{\partial y}{\partial \sigma} = \frac{(x-c)^2}{\sigma^3} y. \tag{2.30}$$

$$\frac{\partial y}{\partial c} = \frac{x-c}{\sigma^2} y. \tag{2.31}$$

In the preceding expressions, the derivatives are arranged to include y and thus save computation. For the bell MF, let

$$y = \text{bell}(x; a, b, c) = \frac{1}{1 + \left|\frac{x-c}{a}\right|^{2b}}. \tag{2.32}$$

Then

$$\frac{\partial y}{\partial x} = \begin{cases} -\frac{2b}{x-c} y(1-y), & \text{if } x \neq c. \\ 0, & \text{if } x = c. \end{cases} \tag{2.33}$$

$$\frac{\partial y}{\partial a} = \frac{2b}{a} y(1-y). \tag{2.34}$$

$$\frac{\partial y}{\partial b} = \begin{cases} -2 \ln\left|\frac{x-c}{a}\right| y(1-y), & \text{if } x \neq c. \\ 0, & \text{if } x = c. \end{cases} \tag{2.35}$$

$$\frac{\partial y}{\partial c} = \begin{cases} \frac{2b}{x-c} y(1-y), & \text{if } x \neq c. \\ 0, & \text{if } x = c. \end{cases} \tag{2.36}$$

Derivation of these formulas are left as Exercises 12 and 13.

2.5 MORE ON FUZZY UNION, INTERSECTION, AND COMPLEMENT*

This section discusses more advanced topics concerning complement, union, and intersection operations on fuzzy sets. For a first-time reading, this section may be omitted without discontinuity. Throughout this book, sections or subsections marked with a star (*) can be skipped at first reading.

Although the classical fuzzy set operators [Equations (2.13], (2.14), and (2.15)) possess more rigorous axiomatic properties (as shown in this section), they are not the only ways to define reasonable and consistent operations on fuzzy sets. This section examines other viable definitions of the fuzzy complement, intersection, and union operators.

2.5.1 Fuzzy Complement*

A fuzzy complement operator is a continuous function $N : [0,1] \rightarrow [0,1]$ which meets the following axiomatic requirements:

$$\begin{array}{ll} N(0) = 1 \text{ and } N(1) = 0 & \text{(boundary)} \\ N(a) \geq N(b) \; if \; a \leq b & \text{(monotonicity).} \end{array} \qquad (2.37)$$

All functions satisfying these requirements form the general class of fuzzy complements. It is evident that violation of any of these requirements would add to this class some functions which are unacceptable as complement operators. Specifically, a violation of the boundary conditions would include functions that do not conform to the ordinary complement for crisp sets. The monotonic decreasing requirement is essential since we intuitively expect that an increase in the membership grade of a fuzzy set must result in a decrease in the membership grade of its complement. These two requirements are the basic requirements that a fuzzy complement operator should meet. Another optional requirement imposes **involution** on a fuzzy complement:

$$N(N(a)) = a \quad \text{(involution)}, \qquad (2.38)$$

which guarantees that the double complement of a fuzzy set is still the set itself. The following examples of fuzzy complements satisfy two of the basic requirements in Equation (2.37) as well as the aforementioned optional one.

Example 2.10 *Sugeno's complement*

One class of fuzzy complements is **Sugeno's complement** [8], defined by

$$N_s(a) = \frac{1-a}{1+sa}, \qquad (2.39)$$

where s is a parameter greater than -1. For each value of the parameter s, we obtain a particular fuzzy complement operator, as shown in Figure 2.15(a).

□

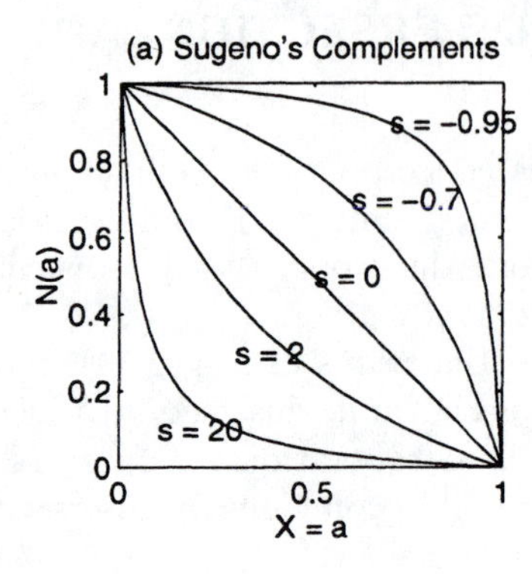

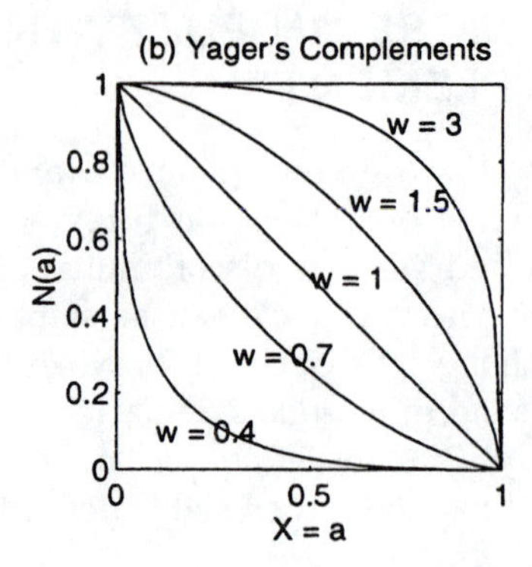

Figure 2.15. *Sugeno's and Yager's complements.* (MATLAB file: `negation.m`)

Example 2.11 *Yager's complement*

Another class of fuzzy complements is **Yager's complement** [10] defined by

$$N_w(a) = (1 - a^w)^{1/w}, \tag{2.40}$$

where w is a positive parameter. Figure 2.15(b) demonstrates this class of functions for various values of w. Note that due to the involution requirement, both Sugeno's and Yager's complements are symmetric about the 45-degree straight line connecting (0, 0) and (1, 1).

□

Obviously, these axiomatic requirements for fuzzy complements do not determine $N(\cdot)$ uniquely. However, $N(a)$ is equal to $1-a$ (the classical fuzzy complement) if the following requirement is introduced [1]:

$$\mu_A(x_1) - \mu_A(x_2) = \mu_{\overline{A}}(x_2) - \mu_{\overline{A}}(x_1). \tag{2.41}$$

This requirement ensures that a change in the membership value in A should have a corresponding effect on the membership in $\overline{A}$. This requirement together with the basic requirements (boundary and monotonicity) for fuzzy complements entails $N(a) = 1 - a$, which automatically satisfies the involution requirement.

2.5.2 Fuzzy Intersection and Union*

The intersection of two fuzzy sets A and B is specified in general by a function $T : [0,1] \times [0,1] \to [0,1]$, which aggregates two membership grades as follows:

$$\mu_{A\cap B}(x) = T(\mu_A(x), \mu_B(x)) = \mu_A(x) \tilde{*} \mu_B(x), \tag{2.42}$$

where $\tilde{*}$ is a binary operator for the function T. This class of fuzzy intersection operators, which are usually referred to as T-norm (triangular norm) operators, meets the following basic requirements.

Definition 2.26 *T-norm*

A **T-norm** operator [3] is a two-place function $T(\cdot,\cdot)$ satisfying

$$\begin{array}{ll} T(0,0)=0,\ T(a,1)=T(1,a)=a & \text{(boundary)} \\ T(a,b)\le T(c,d) \text{ if } a\le c \text{ and } b\le d & \text{(monotonicity)} \\ T(a,b)=T(b,a) & \text{(commutativity)} \\ T(a,T(b,c))=T(T(a,b),c) & \text{(associativity).} \end{array} \tag{2.43}$$

□

The first requirement imposes the correct generalization to crisp sets. The second requirement implies that a decrease in the membership values in A or B cannot produce an increase in the membership value in $A \cap B$. The third requirement indicates that the operator is indifferent to the order of the fuzzy sets to be combined. Finally, the fourth requirement allows us to take the intersection of any number of sets in any order of pairwise groupings. The following example illustrates four of the most frequently encountered T-norm operators.

Example 2.12 *Four T-norm operators*

Four of the most frequently used T-norm operators are

$$\begin{array}{ll} \textbf{Minimum:} & T_{min}(a,b)=\min(a,b)=a\wedge b. \\ \textbf{Algebraic product:} & T_{ap}(a,b)=ab. \\ \textbf{Bounded product:} & T_{bp}(a,b)=0\vee(a+b-1). \\ \textbf{Drastic product:} & T_{dp}(a,b)=\begin{cases} a, & \text{if } b=1. \\ b, & \text{if } a=1. \\ 0, & \text{if } a,b<1. \end{cases} \end{array} \tag{2.44}$$

With the understanding that a and b are between 0 and 1, we can draw surface plots of these four T-norm operators as functions of a and b; see the first row of Figure 2.16. The second row of Figure 2.16 shows the corresponding surfaces when $a = \mu_A(x) = \text{trapezoid}(x;3,8,12,17)$ and $b = \mu_B(y) = \text{trapezoid}(y;3,8,12,17)$; these two-dimensional MFs can be viewed as the Cartesian product of A and B under four different T-norm operators.

From Figure 2.16, it can be observed that

$$T_{dp}(a,b) \le T_{bp}(a,b) \le T_{ap}(a,b) \le T_{min}(a,b). \tag{2.45}$$

This can be verified mathematically.

□

Like fuzzy intersection, the fuzzy union operator is specified in general by a function $S: [0,1]\times[0,1]\to[0,1]$. In symbols,

$$\mu_{A\cup B}(x) = S(\mu_A(x),\mu_B(x)) = \mu_A(x) \tilde{+} \mu_B(x), \tag{2.46}$$

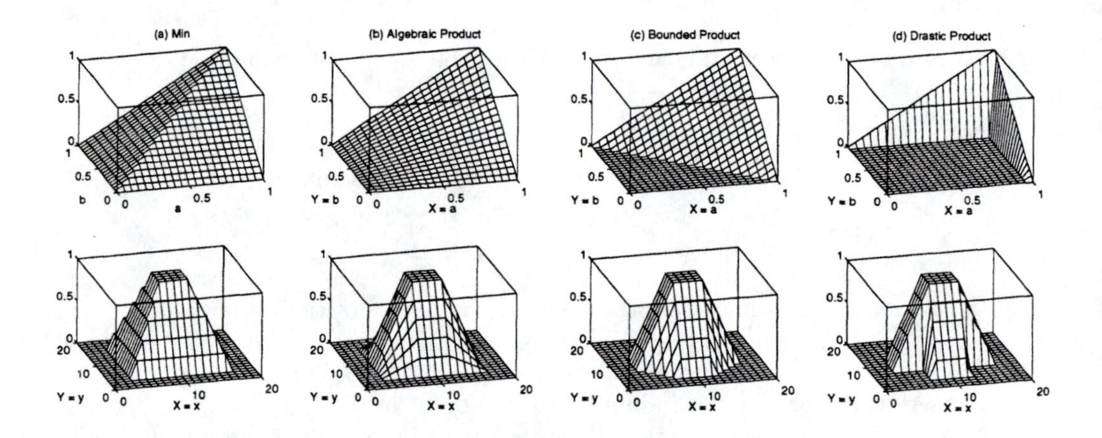

Figure 2.16. *(First row) Four T-norm operators* $T_{min}(a,b)$, $T_{ap}(a,b)$, $T_{bp}(a,b)$, *and* $T_{dp}(a,b)$*; (second row) the corresponding surfaces for* $a = trapezoid(x, 3, 8, 12, 17)$ *and* $b = trapezoid(y, 3, 8, 12, 17)$. (MATLAB file: `tnorm.m`)

where $\tilde{+}$ is a binary operator for the function S. This class of fuzzy union operators, which are often referred to as T-conorm (or S-norm) operators, satisfy the following basic requirements.

Definition 2.27 *T-conorm (S-norm)*

A **T-conorm** (or **S-norm**) operator [3] is a two-place function $S(\cdot,\cdot)$ satisfying

$$\begin{array}{ll} S(1,1)=1,\ S(0,a)=S(a,0)=a & \text{(boundary)} \\ S(a,b)\le S(c,d) \text{ if } a\le c \text{ and } b\le d & \text{(monotonicity)} \\ S(a,b)=S(b,a) & \text{(commutativity)} \\ S(a,S(b,c))=S(S(a,b),c) & \text{(associativity).} \end{array} \tag{2.47}$$

□

The justification of these basic requirements is similar to that of the requirements for T-norm operators.

Example 2.13 *Four T-conorm operators*

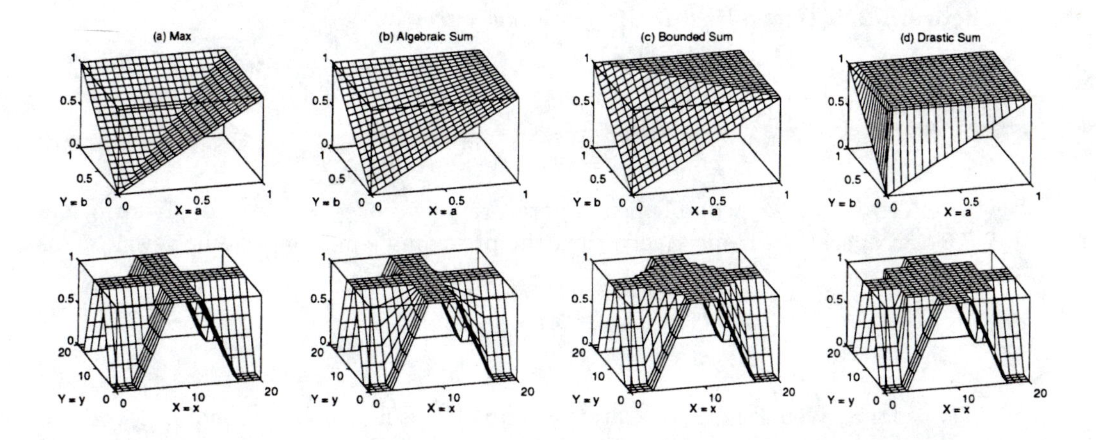

Figure 2.17. *(First row) Four T-conorm operators* $S_{min}(a,b)$, $S_{ap}(a,b)$ $S_{bp}(a,b)$ *and* $S_{dp}(a,b)$; *(second row) the corresponding surfaces for* $a =$ *trapezoid*$(x, 3, 8, 12, 17)$ *and* $b =$ *trapezoid*$(y, 3, 8, 12, 17)$. (MATLAB file: `tconorm.m`)

Corresponding to the four T-norm operators in the previous example, we have the following four T-conorm operators.

$$
\begin{array}{ll}
\textbf{Maximum:} & S(a,b) = \max(a,b) = a \vee b. \\
\textbf{Algebraic sum:} & S(a,b) = a + b - ab. \\
\textbf{Bounded sum:} & S(a,b) = 1 \wedge (a+b). \\
\textbf{Drastic sum:} & S(a,b) = \begin{cases} a, & \text{if } b = 0. \\ b, & \text{if } a = 0. \\ 1, & \text{if } a, b > 0. \end{cases}
\end{array}
\tag{2.48}
$$

The first row of Figure 2.17 shows the surface plot of these T-conorm operators. The second row demonstrates the corresponding two-dimensional MFs when $a = \mu_A(x) = \text{trapezoid}(x; 3, 8, 12, 17)$ and $b = \mu_B(x) = \text{trapezoid}(y; 3, 8, 12, 17)$; these MFs are the Cartesian coproduct of A and B using these four T-conorm operators. It can also be verified that

$$
S_{max}(a,b) \leq S_{ap}(a,b) \leq S_{bp}(a,b) \leq S_{dp}(a,b). \tag{2.49}
$$

□

Note that these essential requirements for T-norm and T-conorm operators cannot uniquely determine the classical fuzzy intersection and union—namely, the min and max operators. Stronger restrictions have to be taken into consideration to pinpoint the min and max operators. For a detailed treatment of this subject, see [1].

Theorem 2.1 *Generalized DeMorgan's law*

T-norms $T(\cdot,\cdot)$ and T-conorms $S(\cdot,\cdot)$ are duals which support the generalization of **DeMorgan's law**:

$$\begin{aligned} T(a,b) &= N(S(N(a),N(b))),\\ S(a,b) &= N(T(N(a),N(b))), \end{aligned} \tag{2.50}$$

where $N(\cdot)$ is a fuzzy complement operator. If we use $\tilde{*}$ and $\tilde{+}$ for T-norm and T-conorm operators, respectively, then the preceding equations can be rewritten as

$$\begin{aligned} a\tilde{*}b &= N(N(a)\tilde{+}N(b)),\\ a\tilde{+}b &= N(N(a)\tilde{*}N(b)). \end{aligned} \tag{2.51}$$

□

Thus for a given T-norm operator, we can always find a corresponding T-conorm operator through the generalized DeMorgan's law, and vice versa. (In fact, the four T-norm and T-conorm operators in Examples 2.12 and 2.13, respectively, are dual in the sense of the generalized DeMorgan's law. The reader is encouraged to verify this.)

2.5.3 Parameterized T-norm and T-conorm*

Several **parameterized T-norms** and dual T-conorms have been proposed in the past, such as those of Yager [9], Dubois and Prade [4], Schweizer and Sklar [7], and Sugeno [8]. For instance, Schweizer and Sklar's T-norm operator can be expressed as

$$\begin{aligned} T_{SS}(a,b,p) &= [\max\{0,(a^{-p}+b^{-p}-1)\}]^{-\frac{1}{p}}\\ S_{SS}(a,b,p) &= 1-[\max\{0,((1-a)^{-p}+(1-b)^{-p}-1)\}]^{-\frac{1}{p}} \end{aligned} \tag{2.52}$$

It is observed that

$$\begin{aligned} &\lim_{p\to 0} T_{SS}(a,b,p) = ab,\\ &\lim_{p\to\infty} T_{SS}(a,b,p) = \min(a,b), \end{aligned} \tag{2.53}$$

which correspond to two of the more commonly used T-norms for the fuzzy AND operation.

To give a general idea of how the parameter p affects the T-norm and T-conorm operators, Figure 2.18(a) shows typical membership functions of fuzzy sets A and B; Figure 2.18(b) and Figure 2.18(c) are $T_{SS}(a,b,p)$ and $S_{SS}(a,b,p)$, respectively, with $p=\infty$ (solid line), 1 (dashed line), 0 (dotted line) and -1 (dash-dotted line). Note that the bell-shaped membership functions of A and B in Figure 2.18(a) are defined as follows:

$$\mu_A(x) = \text{bell}(x;-5,2,7.5) = \frac{1}{1+(\frac{x+5}{7.5})^4}, \tag{2.54}$$

$$\mu_B(x) = \text{bell}(x;5,1,5) = \frac{1}{1+(\frac{x-5}{5})^2}. \tag{2.55}$$

For completeness, other types of parameterized T-norms are given next.

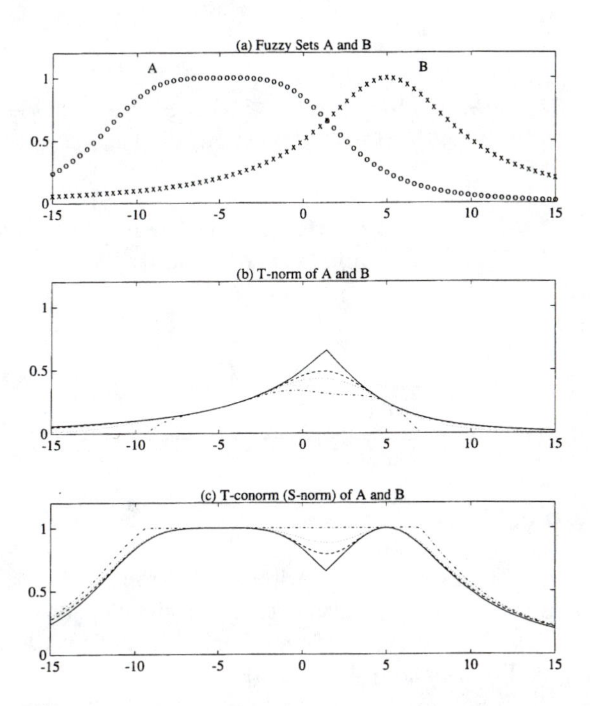

Figure 2.18. *Schweizer and Sklar's parameterized T-norms and T-conorms: (a) membership functions for fuzzy set A and B; (b)* $T_{SS}(a,b,p)$ *and (c)* $S_{SS}(a,b,p)$ *with* $p=\infty$ *(solid line), 1 (dashed line), 0 (dotted line) and* -1 *(dash-dotted line).* (MATLAB file: `sstnorm.m`)

Yager [9]: For $q > 0$,

$$\begin{cases} T_Y(a,b,q) = 1 - \min\{1, [(1-a)^q + (1-b)^q]^{1/q}\}, \\ S_Y(a,b,q) = \min\{1, (a^q + b^q)^{1/q}\}. \end{cases} \tag{2.56}$$

Dubois and Prade [4]: For $\alpha \in [0,1]$,

$$\begin{cases} T_{DP}(a,b,\alpha) = ab/\max\{a,b,\alpha\}, \\ S_{DP}(a,b,\alpha) = [a+b-ab-\min\{a,b,(1-\alpha)\}/\max\{1-a,1-b,\alpha\}. \end{cases} \tag{2.57}$$

Hamacher [6]: For $\gamma > 0$,

$$\begin{cases} T_H(a,b,\gamma) = ab/[\gamma + (1-\gamma)(a+b-ab)], \\ S_H(a,b,\gamma) = [a+b+(\gamma-2)ab]/[1+(\gamma-1)ab]. \end{cases} \tag{2.58}$$

Frank [5]: For $s > 0$,

$$\begin{cases} T_F(a,b,s) = \log_s[1 + (s^a - 1)(s^b - 1)/(s-1)], \\ S_F(a,b,s) = 1 - \log_s[1 + (s^{1-a} - 1)(s^{1-b} - 1)/(s-1)]. \end{cases} \tag{2.59}$$

Sugeno [8]: For $\lambda \geq -1$,

$$\begin{cases} T_S(a,b,\lambda) = \max\{0, (\lambda+1)(a+b-1) - \lambda ab\}, \\ S_S(a,b,\lambda) = \min\{1, a+b-\lambda ab\}. \end{cases} \tag{2.60}$$

Dombi [2]: For $\lambda > 0$,

$$\begin{cases} T_D(a,b,\lambda) = \dfrac{1}{1 + [(a^{-1}-1)^{\lambda} + (b^{-1}-1)^{\lambda}]^{1/\lambda}}, \\ S_D(a,b,\lambda) = \dfrac{1}{1 + [(a^{-1}-1)^{-\lambda} + (b^{-1}-1)^{-\lambda}]^{-1/\lambda}}. \end{cases} \tag{2.61}$$

2.6 SUMMARY

This chapter introduces the basic definitions, notation, and operations for fuzzy sets, including their membership function representations, set-theoretic operations (AND, OR, and NOT), various types of membership functions, and advanced fuzzy set operators such as T-norms and T-conorms.

Most membership functions are determined by domain experts. The human-determined membership functions, however, may not be precise enough for certain applications. Therefore, it is always advisable to apply optimization techniques to fine-tune parameterized membership functions for better performance. In the discussion of neuro-fuzzy modeling in the subsequent chapters, we shall come across parameterized membership functions again and use their derivatives for derivative-based optimization.

Fuzzy sets lay the foundation for the entire fuzzy set theory and related disciplines. In the next chapter, we shall introduce the use of fuzzy sets in fuzzy if-then rules and fuzzy reasoning.

EXERCISES

1. Sometimes it is useful to decompose an MF into a combination of its α-level sets' MFs; this is the **resolution principle**, which states

$$\mu_A(x) = \max_{\alpha} \min[\alpha,\ \mu_{A_\alpha}(x)], \tag{2.62}$$

where A_α is the α-cut (level set) of fuzzy set A and $\mu_{A_\alpha}(x)$ is the MF of A_α. [Remember that A_α is a crisp set and thus $\mu_{A_\alpha}(x)$ can take only values in

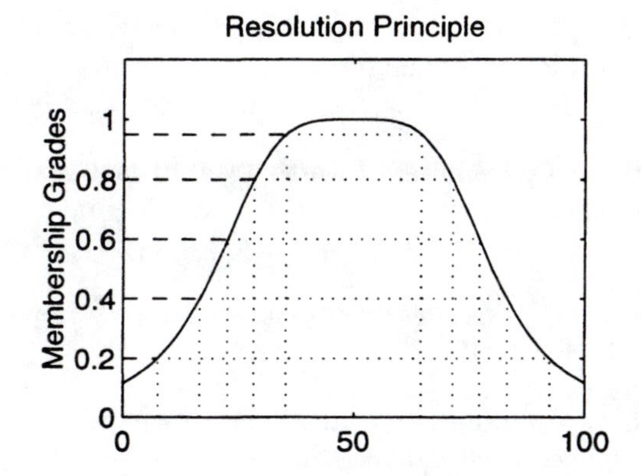

Figure 2.19. *Resolution principle.* (MATLAB file: `resolut.m`)

$\{0, 1\}$.] Figure 2.19 illustrates the concept of the resolution principle, where each rectangle under the MF represents $\min[\alpha; \mu_{A_\alpha}(x)]$ for a specific α between 0 and 1. (a) Adapt Equation (2.62) to a fuzzy set with a discrete universe of discourse. (b) Put the MF of fuzzy set A in Example 2.2 into the resolution format.

2. Verify the identities in Table 2.1 using Venn diagrams.

3. Determine if the classical fuzzy operators [Equations (2.13), (2.14), and (2.15)], hold for each identity in Table 2.1. Explain why by giving simple proofs or counterexamples.

4. Repeat Exercise 3, assuming that the fuzzy union and intersection are defined differently:

$$\mu_{A\cup B}(x) = \mu_A(x) + \mu_B(x) - \mu_A(x)\mu_B(x) \tag{2.63}$$

$$\mu_{A\cap B}(x) = \mu_A(x)\mu_B(x). \tag{2.64}$$

5. Suppose that fuzzy set A is described by $\mu_A(x) = \text{bell}(x; a, b, c)$. Show that the classical fuzzy complement of A is described by $\mu_{\overline{A}}(x) = \text{bell}(x; a, -b, c)$.

6. The **S-MF** with two parameters l and r $(l < r)$ is an S-shaped open-right MF defined by

$$\text{S}(x; l, r) = \begin{cases} 0, & \text{for } x \le l. \\ 2(\frac{x-l}{r-l})^2, & \text{for } l < x \le \frac{l+r}{2}. \\ 1 - 2(\frac{r-x}{r-l})^2, & \text{for } \frac{l+r}{2} < x \le r. \\ 1, & \text{for } r < x. \end{cases} \tag{2.65}$$

(a) Write a MATLAB function to implement this MF. (b) Plot instances of this MF with various values of parameters. (c) Find the crossover point of $\mathrm{S}(x; l, r)$. (d) Prove that the derivative $\mathrm{S}(x; l, r)$ with respect to x is continuous.

7. The **Z-MF** with two parameters l and r ($l < r$) is a Z-shaped open-left MF defined by

$$\mathrm{Z}(x; l, r) = 1 - \mathrm{S}(x; l, r), \tag{2.66}$$

where $\mathrm{S}(x; l, r)$ is the S-MF in the previous exercise. Repeat (a) through (d) of Exercise 6 with the Z-MF.

8. The π-MF with two parameters a and c is a π-shaped MF defined via the S and Z-MFs introduced earlier, as follows:

$$\pi(x; a, c) = \begin{cases} \mathrm{S}(x; c-a, c), & \text{for } x \leq c \\ \mathrm{Z}(x; c, c+a), & \text{for } x > c, \end{cases} \tag{2.67}$$

where c is the center and a (> 0) is the spread on each side of the MF. (a) Write a MATLAB function to implement this MF. (b) Plot instances of this MF with various values of parameters. (c) Find the crossover points and width of $\pi(x; l, r)$.

9. The **two-sided π-MF** is an extension of the π-MF introduced previously; it is defined with four parameters a, b, c, and d:

$$\text{ts\_}\pi(x; a, b, c, d) = \begin{cases} 0, & \text{for } x \leq a. \\ \mathrm{S}(x, a, b), & \text{for } a < x < b. \\ 1, & \text{for } b \leq x \leq c. \\ \mathrm{Z}(x, c, d), & \text{for } c < x < d. \\ 0, & \text{for } d \leq x. \end{cases} \tag{2.68}$$

(a) Write a MATLAB function to implement this MF. (b) Plot instances of this MF with various values of parameters. (c) Find the crossover points and width of $\text{ts\_}\pi(x; a, b, c, d)$.

10. The **two-sided Gaussian MF** is defined by

$$\text{ts\_gaussian}(x; c_1, \sigma_1, c_2, \sigma_2) = \begin{cases} \exp\left[-\frac{1}{2}\left(\frac{x-c_1}{\sigma_1}\right)^2\right], & \text{for } x \leq c_1. \\ 1, & \text{for } c_1 < x < c_1. \\ \exp\left[-\frac{1}{2}\left(\frac{x-c_2}{\sigma_2}\right)^2\right], & \text{for } c_2 \leq x. \end{cases} \tag{2.69}$$

(a) Write a MATLAB function to implement this MF. (b) Plot instances of this MF with various values of parameters. (c) Find the crossover points and width of this MF.

11. Find the level set A_α and its width for a fuzzy set A defined by $\mu_A(x) = \text{trapezoid}(x; a, b, c, d)$.

12. Derive the partial derivatives of a Gaussian MF $y = \text{gaussian}(x; c, \sigma)$ with respect to its argument x and parameters σ and c, and thus verify Equations (2.29) to (2.31).

13. Derive the partial derivatives of a bell MF $y = \text{bell}(x; a, b, c)$ with respect to its argument x and parameters a, b, and c, and thus verify Equations (2.33) to (2.36).

14. Let the fuzzy set A be defined by a Gaussian MF $\text{gaussian}(x, c, \sigma)$. Show that $\text{width}(A_{0.99})/\text{width}(A_{0.01})$ is a constant independent of the parameters c and σ.

15. Let the fuzzy set A be defined by a generalized bell MF $\text{bell}(x, a, b, c)$. Show that $\text{width}(A_{0.99})/\text{width}(A_{0.01})$ is a function of parameter b only.

16. Let the fuzzy set A be defined by a bell MF $\text{bell}(x, a, b, c)$. Demonstrate that for all $\alpha \in [0, 1)$,

$$\lim_{b \to \infty} \text{width}(A_\alpha) = 2a.$$

This demonstrates that a bell MF will approach a characteristic function of a crisp set if $b \to \infty$.

17. Show that the Sugeno's and Yager's complement operators (Examples 2.10 and 2.11) satisfy the involution requirement of Equation (2.38).

18. Verify that the four T-norm and T-conorm operators in Examples 2.12 and 2.13 are dual to each other in the sense of the generalized DeMorgan's law.

19. Prove the inequalities of Equations (2.45) and (2.49).

20. Show that (a) $T_{SS}(a, b, p) = ab$ when $p \to 0$; (b) $T_{SS}(a, b, p) = \min(a, b)$ when $p \to \infty$.

21. Show that the following operators on fuzzy sets satisfy DeMorgan's law: (a) Dombi's T-norm and T-conorm, with $N(a) = 1 - a$; (b) Hamacher's T-norm and T-conorm, with $N(a) = 1 - a$; (c) max and min, with $N(a)$ as Sugeno's complement.

22. Show that the two-dimensional MF defined by

$$\mu_A(x, y) = \frac{1}{1 + \left[\left(\frac{x - c_x}{a_x}\right)^2 + \left(\frac{y - c_y}{a_y}\right)^2\right]^b}$$

is a composite MF based on the one-dimensional generalized bell MF aggregated by Dombi's T-norm operator.

23. Use `bellmanu.m` as a template to write new MATLAB files that allow the manual tuning of the following parameterized MFs: (a) triangular MFs; (b) trapezoidal MFs; (c) Gaussian MFs; (d) sigmoidal MFs; (e) S-MFs (see Exercise 6); and (f) π-MFs (also see Exercise 8).

24. Use `bellanim.m` (or `siganim.m`) as a template to write new MATLAB files that show the animation of the following parameterized MFs: (a) triangular MFs; (b) trapezoidal MFs; (c) Gaussian MFs; (d) S-MFs (see Exercise 6); and (e) π-MFs (see Exercise 8).

REFERENCES

[1] R. E. Bellman and M. Giertz. On the analytic formalism of the theory of fuzzy sets. *Information Sciences*, 5:149–156, 1973.

[2] J. Dombi. A general class of fuzzy operators, the De Morgan class of fuzzy operators and fuzziness measures induced by fuzzy operators. *Fuzzy Sets and Systems*, 8:149–163, 1982.

[3] D. Dubois and H. Prade. *Fuzzy sets and systems: theory and applications.* Academic press, New York, 1980.

[4] D. Dubois and H. Prade. New results about properties and semantics of fuzzy set theoretic operators. In P. P. Wang and S. K. Chang, editors, *fuzzy sets: theory and applications to policy analysis and information systems*, pages 59–75. Plenum, New York, 1980.

[5] M. J. Frank. On the simultaneous associativity of F(x,y) and x+y-F(x,y). *Aequationes Math.*, 19:194–226, 1979.

[6] H. Hamacher. Über logische verknupfungen unscharfer aussagen und deren zugehörige bewertungs funktionen. In R. Trappl, G. J. Klir, and Ricciardi L, editors, *Progress in cybernetics and systems research, III*, pages 276–288. Hemisphere, New York, 1975.

[7] B. Schweizer and A. Sklar. Associative functions and abstract semi-groups. *Publ. Math. Debrecen*, 10:69–81, 1963.

[8] M. Sugeno. Fuzzy measures and fuzzy integrals: a survey. In M. M. Gupta, G. N. Saridis, and B. R. Gaines, editors, *Fuzzy automata and decision processes*, pages 89–102. North-Holland, New York, 1977.

[9] R. Yager. On a general class of fuzzy connectives. *Fuzzy Sets and Systems*, 4:235–242, 1980.

[10] R. R. Yager. On the measure of fuzziness and negation, part I: membership in the unit interval. *International Journal of Man-Machine Studies*, 5:221–229, 1979.

[11] L. A. Zadeh. Fuzzy sets. *Information and Control*, 8:338–353, 1965.

Chapter 3

Fuzzy Rules and Fuzzy Reasoning

J.-S. R. Jang

3.1 INTRODUCTION

In this chapter we introduce the concepts of the extension principle and fuzzy relations, which expand the notions and applicability of fuzzy sets introduced previously. Then we present the definition of linguistic variables and linguistic values and explain how to use them in fuzzy rules, which are an efficient tool for quantitative modeling of words or sentences in a natural or artificial language. By interpreting fuzzy rules as appropriate fuzzy relations, we investigate different schemes of fuzzy reasoning, where inference procedures based on the concept of the compositional rule of inference are used to derive conclusions from a set of fuzzy rules and known facts.

Fuzzy rules and fuzzy reasoning are the backbone of fuzzy inference systems, which are the most important modeling tool based on fuzzy set theory. They have been successfully applied to a wide range of areas, such as automatic control, expert systems, pattern recognition, time series prediction, and data classification. In-depth discussion about fuzzy inference systems is provided in Chapter 4.

3.2 EXTENSION PRINCIPLE AND FUZZY RELATIONS

We shall start by giving definitions and examples of the extension principle and fuzzy relations, which are the rationales behind fuzzy reasoning.

3.2.1 Extension Principle

The **extension principle** [4, 8] is a basic concept of fuzzy set theory that provides a general procedure for extending crisp domains of mathematical expressions to fuzzy domains. This procedure generalizes a common point-to-point mapping of a function $f(\cdot)$ to a mapping between fuzzy sets. More specifically, suppose that f is a function from X to Y and A is a fuzzy set on X defined as

$$A = \mu_A(x_1)/x_1 + \mu_A(x_2)/x_2 + \cdots + \mu_A(x_n)/x_n.$$

Then the extension principle states that the image of fuzzy set A under the mapping $f(\cdot)$ can be expressed as a fuzzy set B,

$$B = f(A) = \mu_A(x_1)/y_1 + \mu_A(x_2)/y_2 + \cdots + \mu_A(x_n)/y_n,$$

where $y_i = f(x_i)$, $i = 1, \ldots, n$. In other words, the fuzzy set B can be defined through the values of $f(\cdot)$ in $x_1, \ldots, x_n$. If $f(\cdot)$ is a many-to-one mapping, then there exist $x_1, x_2 \in X$, $x_1 \neq x_2$, such that $f(x_1) = f(x_2) = y^*$, $y^* \in Y$. In this case, the membership grade of B at $y = y^*$ is the maximum of the membership grades of A at $x = x_1$ and $x = x_2$, since $f(x) = y^*$ may result from either $x = x_1$ or $x = x_2$. More generally, we have

$$\mu_B(y) = \max_{x=f^{-1}(y)} \mu_A(x).$$

A simple example follows.

Example 3.1 *Application of the extension principle to fuzzy sets with discrete universes*

Let

$$A = 0.1/-2 + 0.4/-1 + 0.8/0 + 0.9/1 + 0.3/2$$

and

$$f(x) = x^2 - 3.$$

Upon applying the extension principle, we have

$$\begin{aligned} B &= 0.1/1 + 0.4/-2 + 0.8/-3 + 0.9/-2 + 0.3/1 \\ &= 0.8/-3 + (0.4 \vee 0.9)/-2 + (0.1 \vee 0.3)/1 \\ &= 0.8/-3 + 0.9/-2 + 0.3/1, \end{aligned}$$

where $\vee$ represents max. Figure 3.1 illustrates this example.

For a fuzzy set with a continuous universe of discourse X, an analogous procedure applies.

Example 3.2 *Application of the extension principle to fuzzy sets with continuous universes*

Let

$$\mu_A(x) = \text{bell}(x; 1.5, 2, 0.5)$$

and

$$f(x) = \begin{cases} (x-1)^2 - 1, & \text{if } x \geq 0. \\ x, & \text{if } x \leq 0. \end{cases}$$

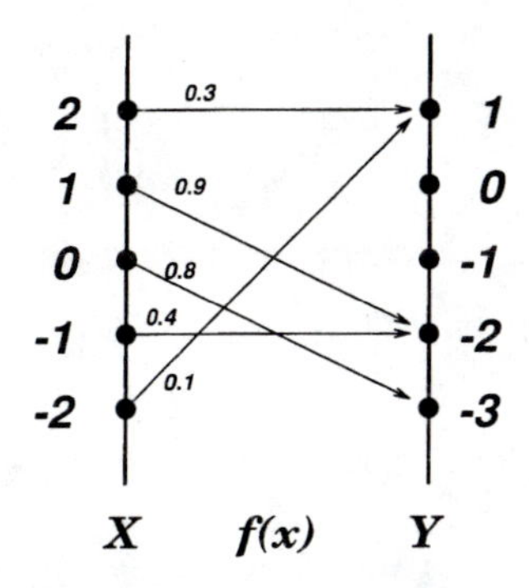

Figure 3.1. *Extension principle on fuzzy sets with discrete universes.*

Figure 3.2(a) is the plot of $y = f(x)$; Figure 3.2(c) is $\mu_A(x)$, the MF of A. After employing the extension principle, we obtain a fuzzy set B; its MF is shown in Figure 3.2(b), where the plot of $\mu_B(y)$ is rotated 90 degrees for easy viewing. Since $f(x)$ is a many-to-one mapping for $x \in [-1,\ 2]$, the max operator is used to obtain the membership grades of B when $y \in [0, 1]$. This causes discontinuities of $\mu_B(y)$ at $y = 0$ and -1. The derivation of $\mu_B(y)$ is left as Exercise 1 at the end of this chapter.

□

Now we consider a more general situation. Suppose that f is a mapping from an n-dimensional product space $X_1 \times \cdots X_n$ to a single universe Y such that $f(x_1, \ldots, x_n) = y$, and there is a fuzzy set A_i in each X_i, $i = 1, \ldots, n$. Since each element in an input vector $(x_1, \ldots, x_n)$ occurs *simultaneously*, this implies an AND operation. Therefore, the membership grade of fuzzy set B induced by the mapping f should be the minimum of the membership grades of the constituent fuzzy set A_i, $i = 1, \ldots, n$. With this understanding, we give a complete formal definition of the extension principle.

Definition 3.1 *Extension principle*

Suppose that function f is a mapping from an n-dimensional Cartesian product space $X_1 \times X_2 \times \cdots X_n$ to a one-dimensional universe Y such that $y = f(x_1, \ldots, x_n)$, and suppose $A_1, \ldots, A_n$ are n fuzzy sets in $X_1, \ldots, X_n$, respectively. Then the extension principle asserts that the fuzzy set B induced by the mapping f is defined by

$$\mu_B(y) = \begin{cases} \max\limits_{(x_1,\ldots,x_n),\ (x_1,\ldots,x_n) = f^{-1}(y)} [\min_i\ \mu_{A_i}(x_i)], & \text{if } f^{-1}(y) \neq \emptyset. \\ 0, & \text{if } f^{-1}(y) = \emptyset. \end{cases} \tag{3.1}$$

□

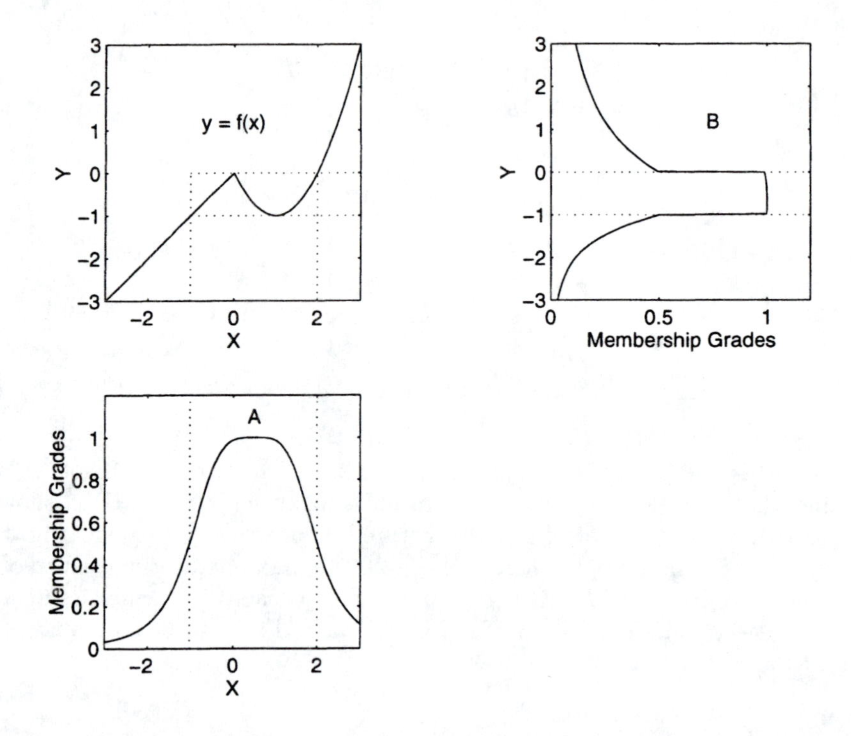

Figure 3.2. *Extension principle on fuzzy sets with continuous universes, as explained in Example 3.2. The lower plot is fuzzy set A; the upper left is the function $y = f(x)$; and the upper right is the fuzzy set B induced via the extension principle.* (MATLAB file: `extensio.m`)

The foregoing extension principle assumes that $y = f(x_1, \ldots, x_n)$ is a crisp function. In cases where f is a fuzzy function [or, more precisely, when $y = f(x_1, \ldots, x_n)$ is a fuzzy set characterized by an $(n+1)$-dimensional MF], then we can employ the compositional rule of inference introduced in Section 3.4.1 (page 63) of the next chapter to find the induced fuzzy set B.

3.2.2 Fuzzy Relations

Binary fuzzy relations [4, 6] are fuzzy sets in $X \times Y$ which map each element in $X \times Y$ to a membership grade between 0 and 1. In particular, unary fuzzy relations are fuzzy sets with one-dimensional MFs; binary fuzzy relations are fuzzy sets with two-dimensional MFs, and so on. Applications of fuzzy relations include areas such as fuzzy control and decision making. Here we restrict our attention to binary fuzzy relations; a generalization to n-ary relations is straightforward.

Definition 3.2 *Binary fuzzy relation*

Let X and Y be two universes of discourse. Then

$$\mathcal{R} = \{(\ (x,y),\ \mu_{\mathcal{R}}(x,y)\ ) \mid (x,y) \in X \times Y\} \tag{3.2}$$

is a **binary fuzzy relation** in $X \times Y$. [Note that $\mu_{\mathcal{R}}(x,y)$ is in fact a two-dimensional MF introduced in Section 2.4.2.]

□

Example 3.3 *Binary fuzzy relations*

Let $X = Y = R^+$ (the positive real line) and $\mathcal{R}$ = "y is much greater than x." The MF of the fuzzy relation $\mathcal{R}$ can be subjectively defined as

$$\mu_{\mathcal{R}}(x,y) = \begin{cases} \dfrac{y-x}{x+y+2}, & \text{if } y > x. \\ 0, & \text{if } y \le x. \end{cases} \tag{3.3}$$

If $X = \{3,4,5\}$ and $Y = \{3,4,5,6,7\}$, then it is convenient to express the fuzzy relation $\mathcal{R}$ as a **relation matrix**:

$$\mathcal{R} = \begin{bmatrix} 0 & 0.111 & 0.200 & 0.273 & 0.333 \\ 0 & 0 & 0.091 & 0.167 & 0.231 \\ 0 & 0 & 0 & 0.077 & 0.143 \end{bmatrix}, \tag{3.4}$$

where the element at row i and column j is equal to the membership grade between the ith element of X and jth element of Y.

□

Other common examples of binary fuzzy relations are as follows:

- x is close to y (x and y are numbers)
- x depends on y (x and y are events)
- x and y look alike (x and y are persons, objects, and so on)
- If x is large, then y is small (x is an observed reading and y is a corresponding action).

The last expression, "If x is A, then y is B," is used repeatedly in a fuzzy inference system. We will explore fuzzy relations of this kind in the following chapter.

Fuzzy relations in different product spaces can be combined through a composition operation. Different composition operations have been suggested for fuzzy relations; the best known is the max-min composition proposed by Zadeh [4].

Definition 3.3 *Max-min composition*

Let $\mathcal{R}_1$ and $\mathcal{R}_2$ be two fuzzy relations defined on $X \times Y$ and $Y \times Z$, respectively. The **max-min composition** of $\mathcal{R}_1$ and $\mathcal{R}_2$ is a fuzzy set defined by

$$\mathcal{R}_1 \circ \mathcal{R}_2 = \{[(x, z), \max_y \min(\mu_{\mathcal{R}_1}(x, y), \mu_{\mathcal{R}_2}(y, z))] | x \in X, y \in Y, z \in Z\}, \tag{3.5}$$

or, equivalently,

$$\begin{aligned} \mu_{\mathcal{R}_1 \circ \mathcal{R}_2}(x, z) &= \max_y \min[\mu_{\mathcal{R}_1}(x, y), \mu_{\mathcal{R}_2}(y, z)] \\ &= \vee_y [\mu_{\mathcal{R}_1}(x, y) \wedge \mu_{\mathcal{R}_2}(y, z)], \end{aligned} \tag{3.6}$$

with the understanding that $\vee$ and $\wedge$ represent max and min, respectively.

□

When $\mathcal{R}_1$ and $\mathcal{R}_2$ are expressed as relation matrices, the calculation of $\mathcal{R}_1 \circ \mathcal{R}_2$ is almost the same as matrix multiplication, except that $\times$ and $+$ are replaced by $\wedge$ and $\vee$, respectively. For this reason, the max-min composition is also called the **max-min product**.

Several properties common to binary relations and max-min composition are given next, where $\mathcal{R}$, $\mathcal{S}$, and $\mathcal{T}$ are binary relations on $X \times Y$, $Y \times Z$, and $Z \times W$, respectively.

$$\begin{array}{ll} \text{Associativity:} & \mathcal{R} \circ (\mathcal{S} \circ \mathcal{T}) = (\mathcal{R} \circ \mathcal{S}) \circ \mathcal{T} \\ \text{Distributivity over union:} & \mathcal{R} \circ (\mathcal{S} \cup \mathcal{T}) = (\mathcal{R} \circ \mathcal{S}) \cup (\mathcal{R} \circ \mathcal{T}) \\ \text{Weak distributivity over intersection:} & \mathcal{R} \circ (\mathcal{S} \cap \mathcal{T}) \subseteq (\mathcal{R} \circ \mathcal{S}) \cap (\mathcal{R} \circ \mathcal{T}) \\ \text{Monotonicity:} & \mathcal{S} \subseteq \mathcal{T} \Longrightarrow \mathcal{R} \circ \mathcal{S} \subseteq \mathcal{R} \circ \mathcal{T} \end{array} \tag{3.7}$$

Although max-min composition is widely used, it is not easily subjected to mathematical analysis. To achieve greater mathematical tractability, max-product composition has been proposed as an alternative to max-min composition.

Definition 3.4 *Max-product composition*

Assuming the same notation as used in the definition of max-min composition, we can define **max-product composition** as follows:

$$\mu_{\mathcal{R}_1 \circ \mathcal{R}_2}(x, z) = \max_y [\mu_{\mathcal{R}_1}(x, y) \mu_{\mathcal{R}_2}(y, z)]. \tag{3.8}$$

□

The following example demonstrates how to apply max-min and max-product composition and how to interpret the resulting fuzzy relations $\mathcal{R}_1 \circ \mathcal{R}_2$.

Example 3.4 *Max-min and max-product composition*

Let

$\mathcal{R}_1$ = "x is relevant to y"

$\mathcal{R}_2$ = "y is relevant to z"

be two fuzzy relations defined on $X \times Y$ and $Y \times Z$, respectively, where $X = \{1, 2, 3\}$, $Y = \{\alpha, \beta, \gamma, \delta\}$, and $Z = \{a, b\}$. Assume that $\mathcal{R}_1$ and $\mathcal{R}_2$ can be expressed as the following relation matrices:

$$\mathcal{R}_1 = \begin{bmatrix} 0.1 & 0.3 & 0.5 & 0.7 \\ 0.4 & 0.2 & 0.8 & 0.9 \\ 0.6 & 0.8 & 0.3 & 0.2 \end{bmatrix}$$

$$\mathcal{R}_2 = \begin{bmatrix} 0.9 & 0.1 \\ 0.2 & 0.3 \\ 0.5 & 0.6 \\ 0.7 & 0.2 \end{bmatrix}.$$

Now we want to find $\mathcal{R}_1 \circ \mathcal{R}_2$, which can be interpreted as a derived fuzzy relation "x is relevant to z" based on $\mathcal{R}_1$ and $\mathcal{R}_2$. For simplicity, suppose that we are only interested in the degree of relevance between 2 ($\in X$) and a ($\in Z$). If we adopt max-min composition, then

$$\begin{aligned} \mu_{\mathcal{R}_1 \circ \mathcal{R}_2}(2, a) &= \max(0.4 \wedge 0.9,\ 0.2 \wedge 0.2,\ 0.8 \wedge 0.5,\ 0.9 \wedge 0.7) \\ &= \max(0.4,\ 0.2,\ 0.5,\ 0.7) \\ &= 0.7 \text{ (by max-min composition).} \end{aligned}$$

On the other hand, if we choose max-product composition instead, we have

$$\begin{aligned} \mu_{\mathcal{R}_1 \circ \mathcal{R}_2}(2, a) &= \max(0.4 \times 0.9,\ 0.2 \times 0.2,\ 0.8 \times 0.5,\ 0.9 \times 0.7) \\ &= \max(0.36,\ 0.04,\ 0.40,\ 0.63) \\ &= 0.63 \text{ (by max-product composition).} \end{aligned}$$

Figure 3.3 illustrates the composition of two fuzzy relations, where the relation between element 2 in X and element a in Z is built up via the four possible paths (solid lines) connecting these two elements. The degree of relevance between 2 and a is the maximum of these four paths' strengths, while each path's strength is the minimum (or product) of the strengths of its constituent links.

□

Both the max-min and max-product composition of two relation matrices can be obtained through the MATLAB file `max_star.m`.

In the previous example, we used max to interpret OR and * to interpret AND, where * can be either min or product. The MATLAB file `max_star.m` (available via FTP or WWW, see page xxiii) can be used to compute the max-* composition of two relation matrices. In general, we can have (T-conorm)-(T-norm) composition that interprets OR and AND using T-conorm and T-norm operators, respectively. These extended meanings for fuzzy OR and AND are discussed in the next section.

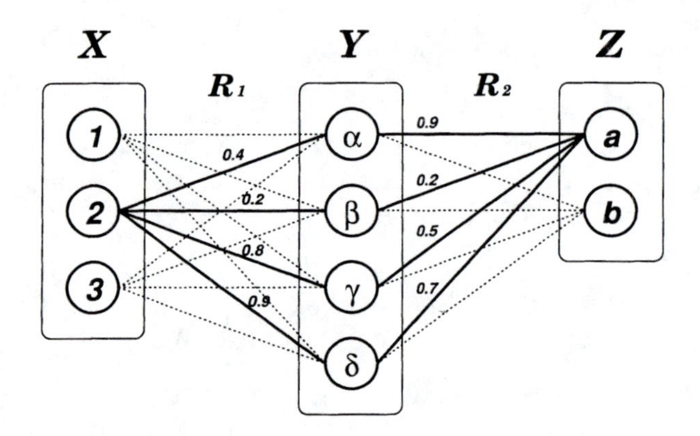

Figure 3.3. *Composition of fuzzy relations.*

3.3 FUZZY IF-THEN RULES

In this section, the definition and examples of linguistic variables are given first. Then we explain two interpretations of fuzzy if-then rules and how to obtain a fuzzy relation that represents the meaning of a given fuzzy rule.

3.3.1 Linguistic Variables

As was pointed out by Zadeh [7], conventional techniques for system analysis are intrinsically unsuited for dealing with humanistic systems, whose behavior is strongly influenced by human judgment, perception, and emotions. This is a manifestation of what might be called the **principle of incompatibility**: "As the complexity of a system increases, our ability to make precise and yet significant statements about its behavior diminishes until a threshold is reached beyond which precision and significance become almost mutually exclusive characteristics" [7]. It was because of this belief that Zadeh proposed the concept of linguistic variables [5] as an alternative approach to modeling human thinking—an approach that, in an approximate manner, serves to summarize information and express it in terms of fuzzy sets instead of crisp numbers. We present the formal definition of linguistic variables next; the example that follows will clarify the definition, which may seem cryptic at the first reading.

Definition 3.5 *Linguistic variables and other related terminology*

A **linguistic variable** is characterized by a quintuple $(x, T(x), X, G, M)$ in which x is the name of the variable; $T(x)$ is the **term set** of x—that is, the set of its **linguistic values** or **linguistic terms**; X is the universe of discourse; G is a **syntactic rule** which generates the terms in $T(x)$; and M is a **semantic rule** which associates with each linguistic value A its meaning $M(A)$, where $M(A)$ denotes a fuzzy set in X.

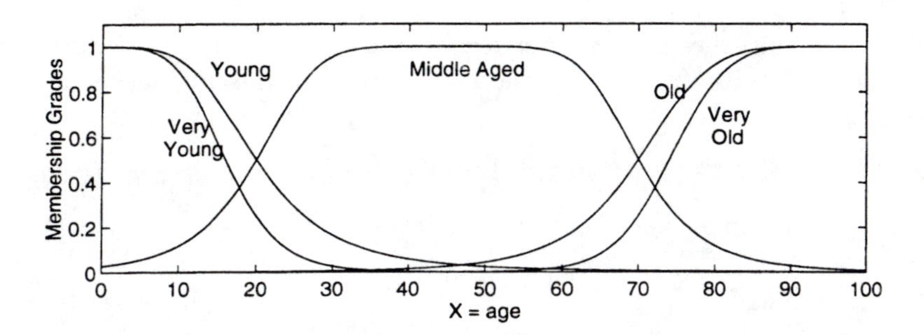

Figure 3.4. *Typical membership functions of the term set $T(age)$.* (MATLAB file: `lv.m`)

□

The following example helps clarify the preceding definition.

Example 3.5 *Linguistic variables and linguistic values*

If *age* is interpreted as a linguistic variable, then its term set $T(age)$ could be

$$\begin{aligned} T(age) \;=\; \{ & \; young,\ not\ young,\ very\ young,\ not\ very\ young,\ ..., \\ & \; middle\ aged,\ not\ middle\ aged,\ ..., \\ & \; old,\ not\ old,\ very\ old,\ more\ or\ less\ old,\ not\ very\ old,\ ..., \\ & \; not\ very\ young\ and\ not\ very\ old,\ ...\ \}, \end{aligned} \tag{3.9}$$

where each term in $T(age)$ is characterized by a fuzzy set of a universe of discourse $X = [0, 100]$, as shown in Figure 3.4. Usually we use "age is young" to denote the assignment of the linguistic value "young" to the linguistic variable *age*. By contrast, when *age* is interpreted as a numerical variable, we use the expression "age = 20" instead to assign the numerical value "20" to the numerical variable *age*. The syntactic rule refers to the way the linguistic values in the term set $T(age)$ are generated. The semantic rule defines the membership function of each linguistic value of the term set; Figure 3.4 displays some of the typical membership functions.

□

From the preceding example, we can see that the term set consists of several **primary terms** (*young, middle aged, old*) modified by the **negation** ("not") and/or the **hedges** (*very, more or less, quite, extremely*, and so forth), and then linked by **connectives** such as *and, or, either*, and *neither*. In the sequel, we shall treat the connectives, the hedges, and the negation as operators that change the meaning of their operands in a specified, context-independent fashion.

Definition 3.6 *Concentration and dilation of linguistic values*

Let A be a linguistic value characterized by a fuzzy set with membership function $\mu_A(\cdot)$. Then A^k is interpreted as a modified version of the original linguistic value expressed as

$$A^k = \int_X [\mu_A(x)]^k / x. \tag{3.10}$$

In particular, the operation of **concentration** is defined as

$$\text{CON}(A) = A^2, \tag{3.11}$$

while that of **dilation** is expressed by

$$\text{DIL}(A) = A^{0.5}. \tag{3.12}$$

□

Conventionally, we take CON(A) and DIL(A) to be the results of applying the hedges *very* and *more or less*, respectively, to the linguistic term A. However, other consistent definitions for these linguistic hedges are possible and well justified for various applications.

Following the definitions in the previous chapter, we can interpret the negation operator NOT and the connectives AND and OR as

$$\begin{aligned} \text{NOT}(A) = \neg A &= \int_X [1 - \mu_A(x)]/x, \\ A \text{ AND } B = A \cap B &= \int_X [\mu_A(x) \wedge \mu_B(x)]/x, \\ A \text{ OR } B = A \cup B &= \int_X [\mu_A(x) \vee \mu_B(x)]/x, \end{aligned} \tag{3.13}$$

respectively, where A and B are two linguistic values whose meanings are defined by $\mu_A(\cdot)$ and $\mu_B(\cdot)$.

Through the use of CON($\cdot$) and DIL($\cdot$) for the linguistic hedges *very* and *more or less*, together with the interpretations of negation and the connectives AND and OR in Equation (3.13), we are now able to construct the meaning of a **composite linguistic term**, such as "not very young and not very old" and "young but not too young."

Example 3.6 *Constructing MFs for composite linguistic terms*

Let the meanings of the linguistic terms *young* and *old* be defined by the following membership functions:

$$\mu_{\text{young}}(x) = \text{bell}(x, 20, 2, 0) = \frac{1}{1 + (\frac{x}{20})^4}, \tag{3.14}$$

$$\mu_{\text{old}}(x) = \text{bell}(x, 30, 3, 100) = \frac{1}{1 + (\frac{x-100}{30})^6}, \tag{3.15}$$

where x is the age of a given person, with the interval $[0, 100]$ as the universe of discourse. Then we can construct MFs for the following composite linguistic terms:

- *more or less old* = DIL(*old*) = $old^{0.5}$

$$= \int_X \sqrt{\frac{1}{1+(\frac{x-100}{30})^6}} \bigg/ x.$$

- *not young and not old* = $\neg young \cap \neg old$

$$= \int_X \left[1 - \frac{1}{1+(\frac{x}{20})^4}\right] \wedge \left[1 - \frac{1}{1+(\frac{x-100}{30})^6}\right] \bigg/ x.$$

- *young but not too young* = $young \cap \neg young^2$

$$= \int_X \left[\frac{1}{1+(\frac{x}{20})^4}\right] \wedge \left[1 - \left(\frac{1}{1+(\frac{x}{20})^4}\right)^2\right] \bigg/ x.$$

- *extremely old*

$$= \text{CON}(\text{CON}(\text{CON}(old))) = ((old^2)^2)^2 = \int_X \left[\frac{1}{1+(\frac{x-100}{30})^6}\right]^8 \bigg/ x.$$

We assume that the meaning of the hedge *too* is the same as that of *very* and the meaning of *extremely* is the same as that of *very very very*. Figure 3.5(a) shows the MFs for the primary linguistic terms *young* and *old*; Figure 3.5(b) shows the MFs for the composite linguistic terms *more or less old*, *not young and not old*, *young but not too young*, and *extremely old*.

□

Another operation that reduces the fuzziness of a fuzzy set A is defined as follows.

Definition 3.7 *Contrast intensification*

The operation of **contrast intensification** on a linguistic value A is defined by

$$\text{INT}(A) = \begin{cases} 2A^2, & \text{for } 0 \le \mu_A(x) \le 0.5, \\ \neg 2(\neg A)^2, & \text{for } 0.5 \le \mu_A(x) \le 1. \end{cases} \tag{3.16}$$

□

The contrast intensifier INT increases the values of $\mu_A(x)$ which are above 0.5 and diminishes those which are below this point. Thus, contrast intensification has the effect of reducing the fuzziness of linguistic value A. The inverse operator of contrast intensifier is contrast diminisher DIM, which is explored in Exercise 9.

Example 3.7 *Contrast intensifier.*

Let A be defined by

$$\mu_A(x) = \text{triangle}(x, 1, 3, 9),$$

which is a triangular MF with the vertex at $x = 3$ and the base located at $x = 1$ to $x = 9$. Figure 3.6 illustrates the results of applying the contrast intensifier INT

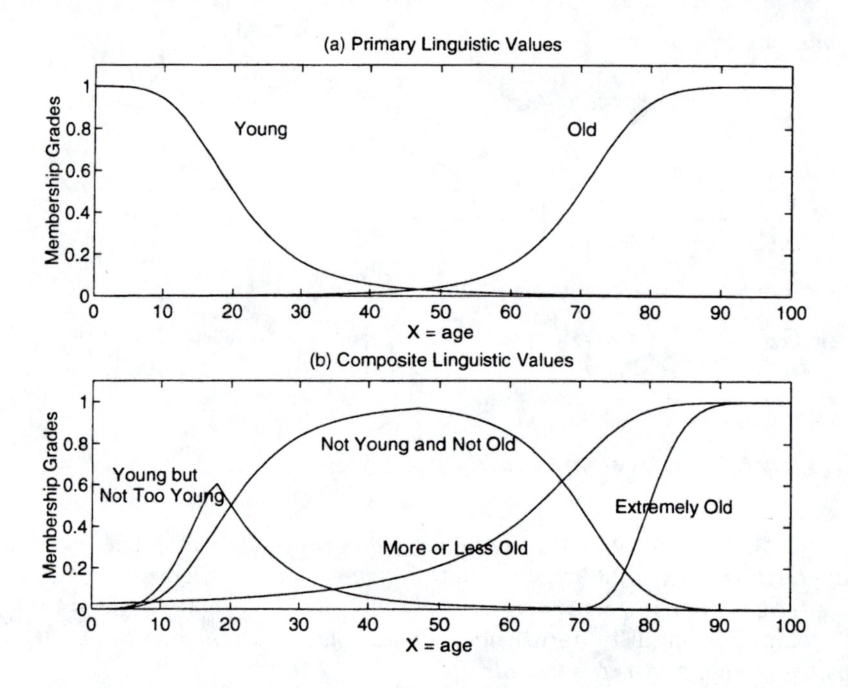

Figure 3.5. *MFs for primary and composite linguistic values in Example 3.6.* (MATLAB file: `complv.m`)

to A several times: the solid line is A; the dotted line is $\mathrm{INT}(A)$; the dashed line is $\mathrm{INT}^2(A) = \mathrm{INT}(\mathrm{INT}(A))$; and the dash-dot line is $\mathrm{INT}^3(A) = \mathrm{INT}(\mathrm{INT}(\mathrm{INT}(A)))$. Thus repeated applications of INT reduces the fuzziness of a fuzzy set; in the extreme case, the fuzzy set becomes a crisp set with boundaries at the crossover points.

□

When we define MFs of linguistic values in a term set, it is intuitively reasonable to have these MFs roughly satisfy the requirement of orthogonality, which is described next.

Definition 3.8 *Orthogonality*

A term set $T = t_1, \ldots, t_n$ of a linguistic variable x on the universe X is **orthogonal** if it fulfills the following property:

$$\sum_{i=1}^{n} \mu_{t_i}(x) = 1, \ \forall x \in X, \tag{3.17}$$

where the t_i's are convex and normal fuzzy sets defined on X and these fuzzy sets make up the term set T.

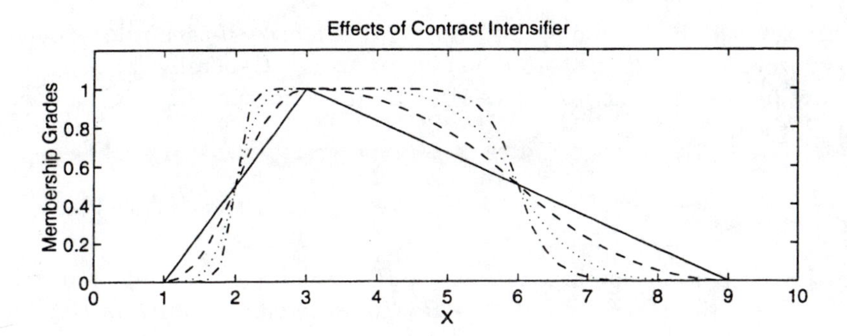

Figure 3.6. *Example 3.7: the effects of the contrast intensifier.* (MATLAB file: `intensif.m`)

□

For the MFs in a term set to be intuitively reasonable, the orthogonality requirement has to be followed to some extent. This is shown in Figure 2.2, where the term set contains three linguistic terms, "young," "middle aged," and "old."

An in-depth exposition of linguistic variables and their applications can be found in [8]. Next, we shall discuss the use of linguistic variables and linguistic values in fuzzy if-then rules.

3.3.2 Fuzzy If-Then Rules

A **fuzzy if-then rule** (also known as **fuzzy rule**, **fuzzy implication**, or **fuzzy conditional statement**) assumes the form

$$\text{if } x \text{ is } A \text{ then } y \text{ is } B, \tag{3.18}$$

where A and B are linguistic values defined by fuzzy sets on universes of discourse X and Y, respectively. Often "x is A" is called the **antecedent** or **premise**, while "y is B" is called the **consequence** or **conclusion**. Examples of fuzzy if-then rules are widespread in our daily linguistic expressions, such as the following:

- If pressure is high, then volume is small.
- If the road is slippery, then driving is dangerous.
- If a tomato is red, then it is ripe.
- If the speed is high, then apply the brake a little.

Before we can employ fuzzy if-then rules to model and analyze a system, first we have to formalize what is meant by the expression "if x is A then y is B", which is sometimes abbreviated as $A \rightarrow B$. In essence, the expression describes a relation between two variables x and y; this suggests that a fuzzy if-then rule be defined as a binary fuzzy relation R on the product space $X \times Y$. Generally speaking, there are two ways to interpret the fuzzy rule $A \to B$. If we interpret $A \to B$ as A **coupled with** B, then

$$R = A \to B = A \times B = \int_{X \times Y} \mu_A(x) \tilde{*} \mu_B(y)/(x, y),$$

where $\tilde{*}$ is a T-norm operator and $A \to B$ is used again to represent the fuzzy relation R. On the other hand, if $A \to B$ is interpreted as A **entails** B, then it can be written as four different formulas:

- Material implication:

$$R = A \to B = \neg A \cup B. \tag{3.19}$$

- Propositional calculus:

$$R = A \to B = \neg A \cup (A \cap B). \tag{3.20}$$

- Extended propositional calculus:

$$R = A \to B = (\neg A \cap \neg B) \cup B. \tag{3.21}$$

- Generalization of modus ponens:

$$\mu_R(x, y) = \sup\{c \mid \mu_A(x) \tilde{*} c \leq \mu_B(y) \text{ and } 0 \leq c \leq 1\}, \tag{3.22}$$

  where $R = A \to B$ and $\tilde{*}$ is a T-norm operator.

Although these four formulas are different in appearance, they all reduce to the familiar identity $A \to B \equiv \neg A \cup B$ when A and B are propositions in the sense of two-valued logic. Figure 3.7 illustrates these two interpretations of a fuzzy rule $A \to B$.

Based on these two interpretations and various T-norm and T-conorm operators, a number of qualified methods can be formulated to calculate the fuzzy relation $R = A \to B$. Note that R can be viewed as a fuzzy set with a two-dimensional MF

$$\mu_R(x, y) = f(\mu_A(x), \mu_B(y)) = f(a, b),$$

with $a = \mu_A(x)$, $b = \mu_B(y)$, where the function f, called the **fuzzy implication function**, performs the task of transforming the membership grades of x in A and y in B into those of (x, y) in $A \to B$.

Suppose that we adopt the first interpretation, "A coupled with B," as the meaning of $A \to B$. Then four different fuzzy relations $A \to B$ result from employing four of the most commonly used T-norm operators.

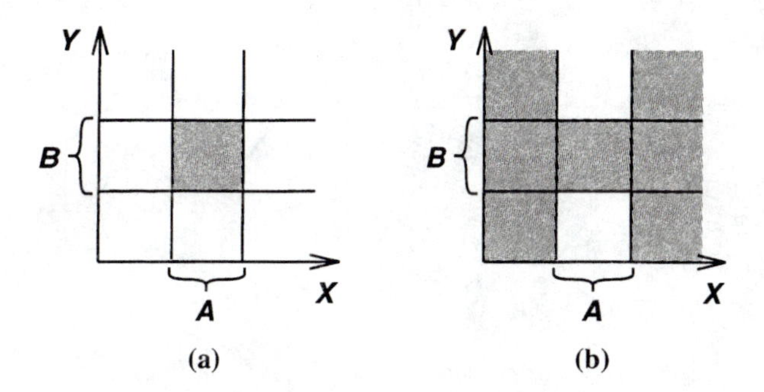

Figure 3.7. *Two interpretations of fuzzy implication: (a) A coupled with B; (b) A entails B.*

- $R_m = A \times B = \int_{X \times Y} \mu_A(x) \wedge \mu_B(y)/(x, y)$, or $f_c(a, b) = a \wedge b$. This relation, which was proposed by Mamdani [3], results from using the min operator for conjunction.

- $R_p = A \times B = \int_{X \times Y} \mu_A(x)\mu_B(y)/(x, y)$, or $f_p(a, b) = ab$. Proposed by Larsen [2], this relation is based on using the algebraic product operator for conjunction.

- $R_{bp} = A \times B = \int_{X \times Y} \mu_A(x) \odot \mu_B(y)/(x, y) = \int_{X \times Y} 0 \vee (\mu_A(x) + \mu_B(y) - 1)/(x, y)$, or $f_{bp}(a, b) = 0 \vee (a + b - 1)$. This formula employs the bounded product operator for conjunction.

- $R_{dp} = A \times B = \int_{X \times Y} \mu_A(x) \hat{\cdot} \mu_B(y)/(x, y)$, or

$$f(a, b) = a \hat{\cdot} b = \begin{cases} a & \text{if } b = 1. \\ b & \text{if } a = 1. \\ 0 & \text{otherwise.} \end{cases}$$

  This formula uses the drastic product operator for conjunction.

The first row of Figure 3.8 shows these four fuzzy implication functions [with $a = \mu_A(x)$ and $b = \mu_B(y)$]; the second row shows the corresponding fuzzy relations R_m, R_p, R_{bp}, and R_{dp} when $\mu_A(x) = \text{bell}(x; 4, 3, 10)$ and $\mu_B(y) = \text{bell}(y; 4, 3, 10)$.

When we adopt the second interpretation, "A entails B," as the meaning of $A \rightarrow B$, again there are a number of fuzzy implication functions that are reasonable candidates. The following four have been proposed in the literature:

- $R_a = \neg A \cup B = \int_{X \times Y} 1 \wedge (1 - \mu_A(x) + \mu_B(y))/(x, y)$, or $f_a(a, b) = 1 \wedge (1 - a + b)$. This is Zadeh's arithmetic rule, which follows Equation (3.19) by using the bounded sum operator for $\cup$.

- $R_{mm} = \neg A \cup (A \cap B) = \int_{X \times Y} (1 - \mu_A(x)) \vee (\mu_A(x) \wedge \mu_B(y))/(x, y)$, or $f_m(a, b) = (1 - a) \vee (a \wedge b)$. This is Zadeh's max-min rule, which follows Equation (3.20) by using min for $\cap$ and max for $\cup$.

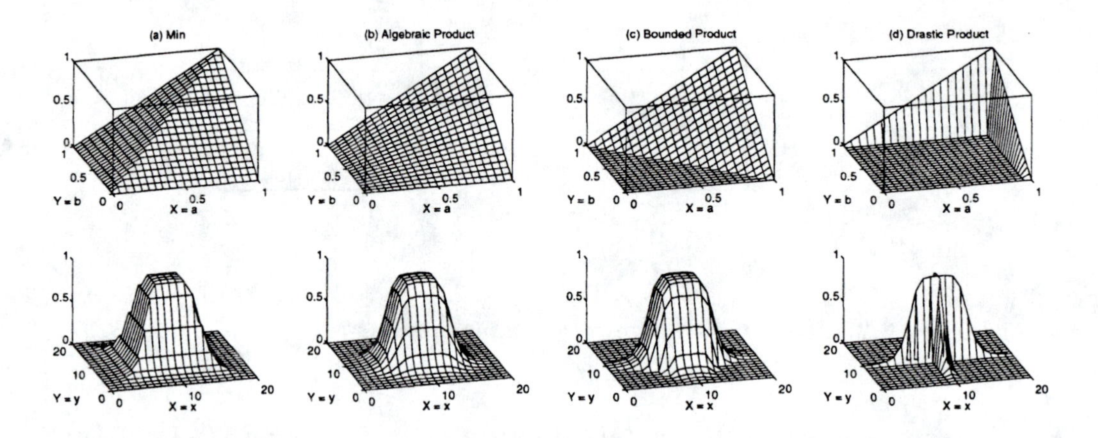

Figure 3.8. *First row: fuzzy implication functions based on the interpretation "A coupled with B"; second row: the corresponding fuzzy relations.* (MATLAB file: `fuzimp.m`)

- $R_s = \neg A \cup B = \int_{X \times Y} (1 - \mu_A(x)) \vee \mu_B(x)$, or $f_s(a, b) = (1 - a) \vee b$. This is Boolean fuzzy implication using max for $\cup$.

- $R_\triangle = \int_{X \times Y} (\mu_A(x) \tilde{<} \mu_B(y))/(x, y)$, where

$$a \tilde{<} b = \begin{cases} 1 & \text{if } a \leq b. \\ b/a & \text{if } a > b. \end{cases}$$

This is Goguen's fuzzy implication, which follows Equation (3.22) by using the algebraic product for the T-norm operator.

Figure 3.9 shows these four fuzzy implication functions [with $a = \mu_A(x)$ and $b = \mu_B(y)$] and the resulting fuzzy relations R_a, R_{mm}, R_s, and $R_\triangle$ when $\mu_A(x) = \text{bell}(x; 4, 3, 10)$ and $\mu_B(y) = \text{bell}(y; 4, 3, 10)$.

It should be kept in mind that the fuzzy implication functions introduced here are by no means exhaustive. Interested readers can find other feasible fuzzy implication functions in [1].

3.4 FUZZY REASONING

Fuzzy reasoning, also known as approximate reasoning, is an inference procedure that derives conclusions from a set of fuzzy if-then rules and known facts. Before introducing fuzzy reasoning, we shall discuss the compositional rule of inference, which plays a key role in fuzzy reasoning.

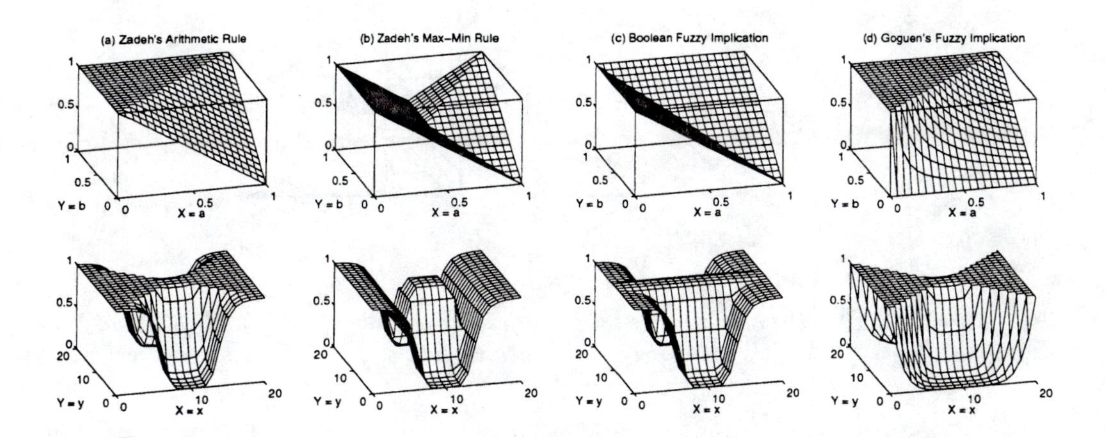

Figure 3.9. *First row: fuzzy implication functions based on the interpretation "A entails B"; second row: the corresponding fuzzy relations.* (MATLAB file: `fuzimp.m`)

3.4.1 Compositional Rule of Inference

The concept behind the compositional rule of inference proposed by Zadeh [7] should not be totally new to the reader; we employed the same idea to explain the max-min composition of relation matrices in Section 3.2.2. Moreover, the extension principle in Section 3.2.1 is actually a special case of the compositional rule of inference.

The compositional rule of inference is a generalization of the following familiar notion. Suppose that we have a curve $y = f(x)$ that regulates the relation between x and y. When we are given $x = a$, then from $y = f(x)$ we can infer that $y = b = f(a)$; see Figure 3.10(a). A generalization of the aforementioned process would allow a to be an interval and $f(x)$ to be an interval-valued function, as shown in Figure 3.10(b). To find the resulting interval $y = b$ corresponding to the interval $x = a$, we first construct a cylindrical extension of a and then find its intersection I with the interval-valued curve. The projection of I onto the y-axis yields the interval $y = b$.

Going one step further in our generalization, we assume that F is a fuzzy relation on $X \times Y$ and A is a fuzzy set of X, as shown in Figures 3.11(a) and 3.11(b). To find the resulting fuzzy set B, again we construct a cylindrical extension $c(A)$ with base A. The intersection of $c(A)$ and F [Figure 3.11(c)] forms the analog of the region of intersection I in Figure 3.10(b). By projecting $c(A) \cap F$ onto the y-axis, we infer y as a fuzzy set B on the y-axis, as shown in Figure 3.11(d).

Specifically, let μ_A, $\mu_{c(A)}$, μ_B, and μ_F be the MFs of A, $c(A)$, B, and F, respectively, where $\mu_{c(A)}$ is related to μ_A through

$$\mu_{c(A)}(x, y) = \mu_A(x).$$

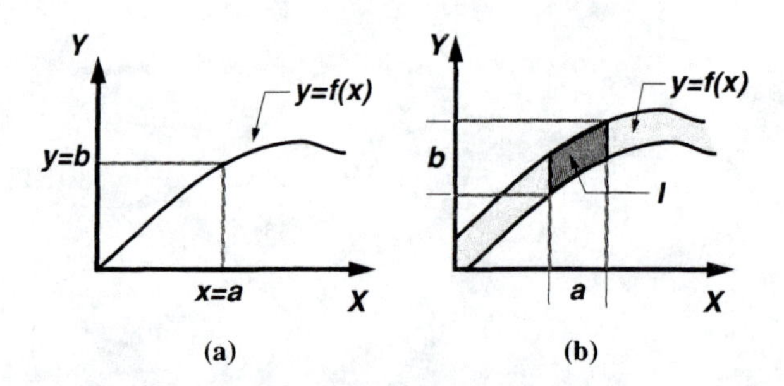

Figure 3.10. *Derivation of $y = b$ from $x = a$ and $y = f(x)$: (a) a and b are points, $y = f(x)$ is a curve; (b) a and b are intervals, $y = f(x)$ is an interval-valued function.*

Then

$$\begin{aligned}\mu_{c(A)\cap F}(x, y) &= \min[\mu_{c(A)}(x, y), \mu_F(x, y)] \\ &= \min[\mu_A(x), \mu_F(x, y)].\end{aligned}$$

By projecting $c(A) \cap F$ onto the y-axis, we have

$$\begin{aligned}\mu_B(y) &= \max_x \min[\mu_A(x), \mu_F(x, y)] \\ &= \vee_x[\mu_A(x) \wedge \mu_F(x, y)].\end{aligned}$$

This formula reduces to the max-min composition (see Definition 3.3 in Section 3.2) of two relation matrices if both A (a unary fuzzy relation) and F (a binary fuzzy relation) have finite universes of discourse. Conventionally, B is represented as

$$B = A \circ F,$$

where $\circ$ denotes the composition operator.

It is interesting to note that the extension principle introduced in Section 3.2.1 of the previous chapter is in fact a special case of the compositional rule of inference. Specifically, if $y = f(x)$ in Figure 3.10 is a common crisp one-to-one or many-to-one function, then the derivation of the induced fuzzy set B on Y is exactly what is accomplished by the extension principle.

Using the compositional rule of inference, we can formalize an inference procedure upon a set of fuzzy if-then rules. This inference procedure, generally called approximate reasoning or fuzzy reasoning, is the topic of the next subsection.

3.4.2 Fuzzy Reasoning

The basic rule of inference in traditional two-valued logic is **modus ponens**, according to which we can infer the truth of a proposition B from the truth of A and the implication $A \rightarrow B$. For instance, if A is identified with "the tomato is red" and B with "the tomato is ripe," then if it is true that "the tomato is red," it is also true that "the tomato is ripe." This concept is illustrated as follows:

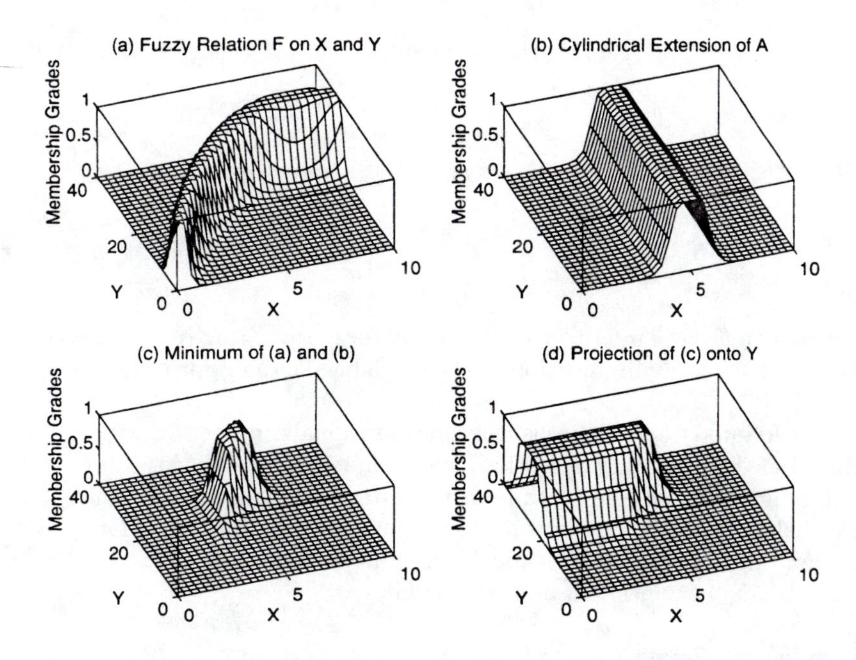

Figure 3.11. *Compositional rule of inference.* (MATLAB file: `cri.m`)

premise 1 (fact):	x is A,
premise 2 (rule):	if x is A then y is B,
consequence (conclusion):	y is B.

However, in much of human reasoning, modus ponens is employed in an approximate manner. For example, if we have the same implication rule "if the tomato is red, then it is ripe" and we know that "the tomato is more or less red," then we may infer that "the tomato is more or less ripe." This is written as

premise 1 (fact):	x is A',
premise 2 (rule):	if x is A then y is B,
consequence (conclusion):	y is B',

where A' is close to A and B' is close to B. When A, B, A', and B' are fuzzy sets of appropriate universes, the foregoing inference procedure is called **approximate reasoning** or **fuzzy reasoning**; it is also called **generalized modus ponens** (**GMP** for short), since it has modus ponens as a special case.

Using the composition rule of inference introduced in the previous subsection, we can formulate the inference procedure of fuzzy reasoning as the following definition.

Definition 3.9 *Approximate reasoning (fuzzy reasoning)*

Let A, A', and B be fuzzy sets of X, X, and Y, respectively. Assume that the fuzzy implication $A \rightarrow B$ is expressed as a fuzzy relation R on $X \times Y$. Then the fuzzy set B induced by "x is A'" and the fuzzy rule "if x is A then y is B" is defined by

$$\begin{aligned} \mu_{B'}(y) &= \max_x \min[\mu_{A'}(x), \mu_R(x,y)] \\ &= \vee_x[\mu_{A'}(x) \wedge \mu_R(x,y)], \end{aligned} \tag{3.23}$$

or, equivalently,

$$B' = A' \circ R = A' \circ (A \rightarrow B). \tag{3.24}$$

□

Now we can use the inference procedure of fuzzy reasoning to derive conclusions, provided that the fuzzy implication $A \rightarrow B$ is defined as an appropriate binary fuzzy relation.

In what follows, we shall discuss the computational aspects of the fuzzy reasoning introduced in the preceding definition, and then extend the discussion to situations in which multiple fuzzy rules with multiple antecedents are involved in describing a system's behavior. However, we will restrict our considerations to Mamdani's fuzzy implication functions and the classical max-min composition, because of their wide applicability and easy graphic interpretation.

Single Rule with Single Antecedent

This is the simplest case, and the formula is available in Equation (3.23). A further simplification of the equation yields

$$\begin{aligned} \mu_{B'}(y) &= [\vee_x(\mu_{A'}(x) \wedge \mu_A(x)] \wedge \mu_B(y) \\ &= w \wedge \mu_B(y). \end{aligned}$$

In other words, first we find the degree of match w as the maximum of $\mu_{A'}(x) \wedge \mu_A(x)$ (the shaded area in the antecedent part of Figure 3.12); then the MF of the resulting B' is equal to the MF of B clipped by w, shown as the shaded area in the consequent part of Figure 3.12. Intuitively, w represents a measure of degree of belief for the antecedent part of a rule; this measure gets propagated by the if-then rules and the resulting degree of belief or MF for the consequent part (B' in Figure 3.12) should be no greater than w.

Single Rule with Multiple Antecedents

A fuzzy if-then rule with two antecedents is usually written as "if x is A and y is B then z is C." The corresponding problem for GMP is expressed as

premise 1 (fact):	x is A' and y is B',
premise 2 (rule):	if x is A and y is B then z is C,
consequence (conclusion):	z is C'.

The fuzzy rule in premise 2 can be put into the simpler form "$A \times B \rightarrow C$." Intuitively, this fuzzy rule can be transformed into a ternary fuzzy relation R_m

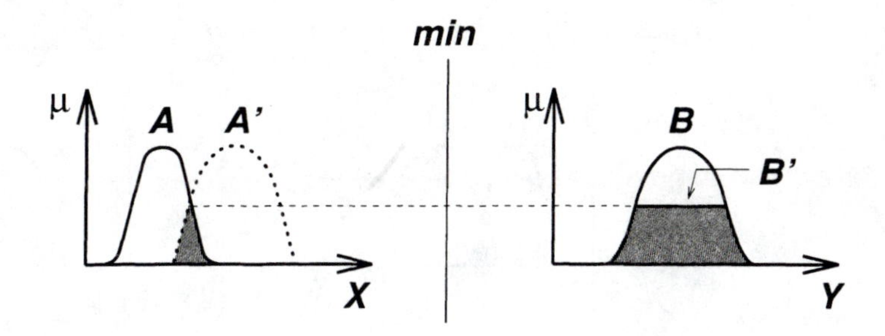

Figure 3.12. *Graphic interpretation of GMP using Mamdani's fuzzy implication and the max-min composition.*

based on Mamdani's fuzzy implication function, as follows:

$$R_m(A, B, C) = (A \times B) \times C = \int_{X \times Y \times Z} \mu_A(x) \wedge \mu_B(y) \wedge \mu_C(z)/(x, y, z).$$

The resulting C' is expressed as

$$C' = (A' \times B') \circ (A \times B \to C).$$

Thus

$$\begin{aligned} \mu_{C'}(z) &= \vee_{x,y}[\mu_{A'}(x) \wedge \mu_{B'}(y)] \wedge [\mu_A(x) \wedge \mu_B(y) \wedge \mu_C(z)] \\ &= \vee_{x,y}\{[\mu_{A'}(x) \wedge \mu_{B'}(y) \wedge \mu_A(x) \wedge \mu_B(y)]\} \wedge \mu_C(z) \\ &= \underbrace{\{\vee_x[\mu_{A'}(x) \wedge \mu_A(x)]\}}_{w_1} \wedge \underbrace{\{\vee_y[\mu_{B'}(y) \wedge \mu_B(y)]\}}_{w_2} \wedge \mu_C(z) \\ &= \underbrace{(w_1 \wedge w_2)}_{\text{firing strength}} \wedge \mu_C(z), \end{aligned} \tag{3.25}$$

where w_1 and w_2 are the maxima of the MFs of $A \cap A'$ and $B \cap B'$, respectively. In general, w_1 denotes the **degrees of compatibility** between A and A'; similarly for w_2. Since the antecedent part of the fuzzy rule is constructed by the connective "and," $w_1 \wedge w_2$ is called the **firing strength** or **degree of fulfillment** of the fuzzy rule, which represents the degree to which the antecedent part of the rule is satisfied. A graphic interpretation is shown in Figure 3.13, where the MF of the resulting C' is equal to the MF of C clipped by the firing strength w, $w = w_1 \wedge w_2$. The generalization to more than two antecedents is straightforward.

An alternative way of calculating C' is explained in the following theorem.

Theorem 3.1 *Decomposition method for calculating B'*

$$\begin{aligned} C' &= (A' \times B') \circ (A \times B \to C) \\ &= [A' \circ (A \to C)] \cap [B' \circ (B \to C)] \end{aligned} \tag{3.26}$$

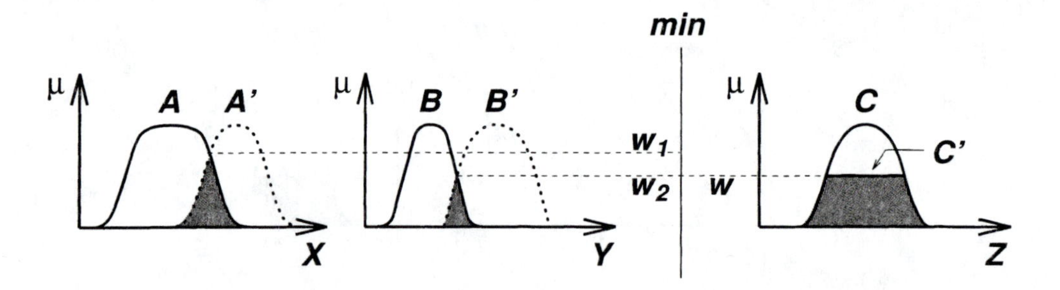

Figure 3.13. *Approximate reasoning for multiple antecedents.*

Proof:

$$
\begin{aligned}
\mu_{C'}(z) &= \vee_{x,y}[\mu_{A'}(x) \wedge \mu_{B'}(y)] \wedge [\mu_A(x) \wedge \mu_B(y) \wedge \mu_C(z)] \\
&= \vee_x\{\mu_{A'}(x) \wedge \mu_B(y) \wedge \mu_C(z) \wedge \vee_y[\mu_{B'}(y) \wedge \mu_B(y) \wedge \mu_C(z)]\} \\
&= \mu_{A'}(x) \wedge \mu_B(y) \wedge \mu_C(z) \wedge \mu_{B'\circ(B\to C)}(z)\} \\
&= \mu_{A'\circ(A\to C)}(y) \wedge \mu_{B'\circ(B\to C)}(y).
\end{aligned}
\tag{3.27}
$$

□

The preceding theorem states that the resulting consequence C' can be expressed as the intersection of $C'_1 = A' \circ (A \to C)$ and $C'_2 = B' \circ (B \to C)$, each of which corresponds to the inferred fuzzy set of a GMP problem for a single fuzzy rule with a single antecedent.

Multiple Rules with Multiple Antecedents

The interpretation of multiple rules is usually taken as the union of the fuzzy relations corresponding to the fuzzy rules. Therefore, for a GMP problem written as

premise 1 (fact):	x is A' and y is B',
premise 2 (rule 1):	if x is A_1 and y is B_1 then z is C_1,
premise 3 (rule 2):	if x is A_2 and y is B_2 then z is C_2,
consequence (conclusion):	z is C',

we can employ the fuzzy reasoning shown in Figure 3.14 as an inference procedure to derive the resulting output fuzzy set C'.

To verify this inference procedure, let $R_1 = A_1 \times B_1 \to C_1$ and $R_2 = A_2 \times B_2 \to C_2$. Since the max-min composition operator $\circ$ is distributive over the $\cup$ operator, it follows that

$$
\begin{aligned}
C' &= (A' \times B') \circ (R_1 \cup R_2) \\
&= [(A' \times B') \circ R_1] \cup [(A' \times B') \circ R_2] \\
&= C'_1 \cup C'_2,
\end{aligned}
\tag{3.28}
$$

where C'_1 and C'_2 are the inferred fuzzy sets for rules 1 and 2, respectively. Figure 3.14 shows graphically the operation of fuzzy reasoning for multiple rules with multiple antecedents.

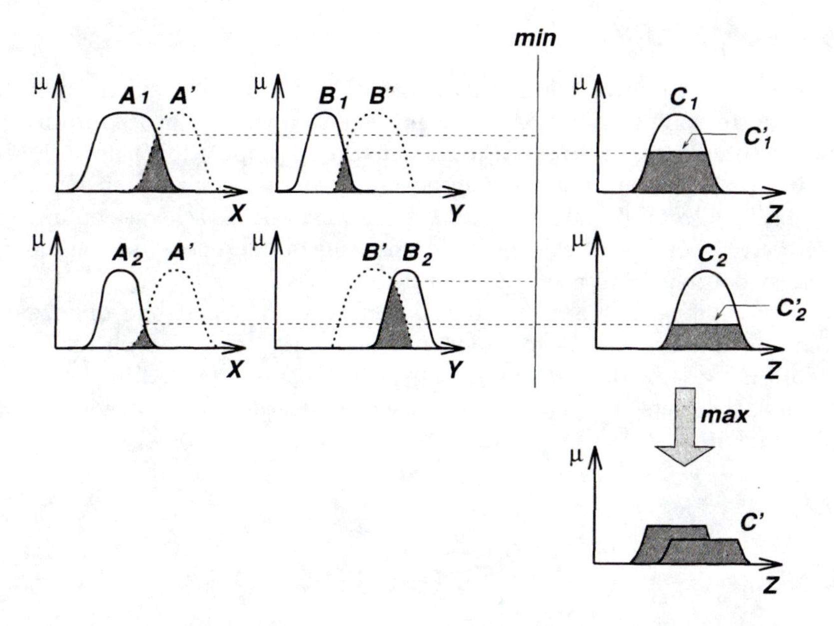

Figure 3.14. *Fuzzy reasoning for multiple rules with multiple antecedents.*

When a given fuzzy rule assumes the form "if x is A or y is B then z is C," then firing strength is given as the maximum of degree of match on the antecedent part for a given condition. This fuzzy rule is equivalent to the union of the two fuzzy rules "if x is A then z is C" and "if y is B then z is C."

In summary, the process of fuzzy reasoning or approximate reasoning can be divided into four steps:

Degrees of compatibility Compare the known facts with the antecedents of fuzzy rules to find the degrees of compatibility with respect to each antecedent MF.

Firing strength Combine degrees of compatibility with respect to antecedent MFs in a rule using fuzzy AND or OR operators to form a firing strength that indicates the degree to which the antecedent part of the rule is satisfied.

Qualified (induced) consequent MFs Apply the firing strength to the consequent MF of a rule to generate a qualified consequent MF. (The qualified consequent MFs represent how the firing strength gets propagated and used in a fuzzy implication statement.)

Overall output MF Aggregate all the qualified consequent MFs to obtain an overall output MF.

These four steps are also employed in a fuzzy inference system, which is introduced in Chapter 4.

3.5 SUMMARY

This chapter introduces the extension principle, fuzzy relations, fuzzy if-then rules, and fuzzy reasoning. The extension principle provides a procedure for mappings between fuzzy sets. Fuzzy relations and their composition rules stipulate fuzzy sets and their combinations and interpretations in a multidimensional space. By interpreting fuzzy if-then rules as fuzzy relations, various schemes of fuzzy reasoning (based on the concept of the compositional rule of inference) are commonly used to derive conclusions from a set of fuzzy if-then rules.

Fuzzy if-then rules and fuzzy reasoning are the backbone of fuzzy inference systems, which are the most important modeling tool based on fuzzy set theory. In-depth discussion about fuzzy inference systems is provided in the next chapter; their adaptive version and corresponding applications are investigated in Chapters 12 and 19.

EXERCISES

1. Apply the extension principle to derive the MF of the fuzzy set B in Example 3.2 (Figure 3.2).

2. Prove the identities in Equation (3.7) for max-min composition.

3. Do the identities in Equation (3.7) hold for other types of composition?

4. Carry out the calculation of $\mathcal{R}_1 \circ \mathcal{R}_2$ in Example 3.4, using both max-min and max-product composition. Double check your results using the MATLAB file `max_star.m`.

5. Repeat Example 3.6 with the new meanings of *old* and *new* defined by the following triangular MFs:

$$\mu_{\text{young}}(x) = \text{gaussian}(x, 0, 20) = e^{-(\frac{x}{20})^2},$$

$$\mu_{\text{old}}(x) = \text{gaussian}(x, 100, 30) = e^{-(\frac{x-100}{30})^2}.$$

   Plot the MFs for the linguistic values in Example 3.6 using MATLAB.

6. Use the MFs of *old* and *small* in Exercise 5 to generate the MFs for the following nonprimary terms:
   (a) *not very young and not very old*
   (b) *very young or very old*
   Plot the MFs for these two linguistic values using MATLAB.

7. Increasing the magnitude (absolute value) of the b parameter of a bell MF has an effect similar to that of the contract intensifier in Equation (3.16)—that is,

it increases the membership grades above 0.5 and diminishes those below 0.5. Explain why.

8. Suppose that the MFs for fuzzy set A and B are a trapezoidal MF trapezoid$(x; a, b, c, d)$ and a two-sided π-MF ts-$\pi(x; a, b, c, d)$ [Equation (2.68)], respectively. Show that INT$(A) = B$.

9. Find an operator **contrast diminisher** DIM that is the inverse of contrast intensifier INT. Namely, for any fuzzy set A, DIM should satisfy the following:

$$\text{DIM}(\text{INT}(A)) = A.$$

Repeat the plot in Figure 3.6 but use the contrast diminisher instead.

10. Verify that Equations (3.19) to (3.22) reduce to the familiar identity $A \to B \equiv \neg A \cup B$ when A and B are propositions in the sense of two-value logic.

11. Repeat the plot of the fuzzy relations in the second rows of Figures 3.8 and 3.9, assuming A and B are defined by
(a) $\mu_A(x) = \text{triangle}(x, 5, 10, 15)$ and $\mu_B(y) = \text{triangle}(y, 5, 10, 15)$.
(b) $\mu_A(x) = \text{trapezoid}(x, 3, 8, 12, 17)$ and $\mu_B(y) = \text{trapezoid}(y, 3, 8, 12, 17)$.

12. Another fuzzy implication function based on the interpretation "A entails B" can be expressed as

$$\begin{aligned} R_s &= A \to B \\ &= \int_{X \times Y} sgn[\mu_B(y) - \mu_A(x)]/(x, y), \end{aligned}$$

or, alternatively,

$$f_s(a, b) = sgn(b - a) = \begin{cases} 1, & \text{if } b \geq a, \\ 0, & \text{otherwise,} \end{cases}$$

where $a = \mu_A(x)$ and $b = \mu_B(y)$. Plot $z = f_s(a, b)$ and $\mu_{R_s}(x, y) = sgn[\mu_B(y) - \mu_A(x)]$, assuming the MFs for A and B are $\mu_A(x) = \text{bell}(x, 4, 3, 10)$ and $\mu_B(y) = \text{bell}(y, 4, 3, 10)$, respectively.

13. Use the identities in the previous exercise to show that the fuzzy rule "if x is A or y is B then z is C" is equivalent to the union of the two fuzzy rules "if x is A then z is C" and "if y is B then z is C" under max-min composition.

REFERENCES

[1] S. Fukami, M. Mizumoto, and K. Tanaka. Some considerations on fuzzy conditional inference. *Fuzzy Sets and Systems*, 4:243–273, 1980.

[2] P. M. Larsen. Industrial applications of fuzzy logic control. *International Journal of Man-Machine Studies*, 12(1):3–10, 1980.

[3] E. H. Mamdani and S. Assilian. An experiment in linguistic synthesis with a fuzzy logic controller. *International Journal of Man-Machine Studies*, 7(1):1–13, 1975.

[4] L. A. Zadeh. Fuzzy sets. *Information and Control*, 8:338–353, 1965.

[5] L. A. Zadeh. Quantitative fuzzy semantics. *Information Sciences*, 3:159–176, 1971.

[6] L. A. Zadeh. Similarity relations and fuzzy ordering. *Information Sciences*, 3:177–206, 1971.

[7] L. A. Zadeh. Outline of a new approach to the analysis of complex systems and decision processes. *IEEE Transactions on Systems, Man, and Cybernetics*, 3(1):28–44, January 1973.

[8] L. A. Zadeh. The concept of a linguistic variable and its application to approximate reasoning, Parts 1, 2 and 3. *Information Sciences*, 8:199-249, 8:301-357, 9:43-80, 1975.

Chapter 4

Fuzzy Inference Systems

J.-S. R. Jang

4.1 INTRODUCTION

The **fuzzy inference system** is a popular computing framework based on the concepts of fuzzy set theory, fuzzy if-then rules, and fuzzy reasoning. It has found successful applications in a wide variety of fields, such as automatic control, data classification, decision analysis, expert systems, time series prediction, robotics, and pattern recognition. Because of its multidisciplinary nature, the fuzzy inference system is known by numerous other names, such as **fuzzy-rule-based system**, **fuzzy expert system** [2], **fuzzy model** [10, 9], **fuzzy associative memory** [3], **fuzzy logic controller** [6, 4, 5], and simply (and ambiguously) **fuzzy system**.

The basic structure of a fuzzy inference system consists of three conceptual components: a **rule base**, which contains a selection of fuzzy rules; a **database** (or **dictionary**), which defines the membership functions used in the fuzzy rules; and a **reasoning mechanism**, which performs the inference procedure (usually the fuzzy reasoning introduced in Section 3.4.2) upon the rules and given facts to derive a reasonable output or conclusion.

Note that the basic fuzzy inference system can take either fuzzy inputs or crisp inputs (which are viewed as fuzzy singletons), but the outputs it produces are almost always fuzzy sets. Sometimes it is necessary to have a crisp output, especially in a situation where a fuzzy inference system is used as a controller. Therefore, we need a method of **defuzzification** to extract a crisp value that best represents a fuzzy set. A fuzzy inference system with a crisp output is shown in Figure 4.1, where the dashed line indicates a basic fuzzy inference system with fuzzy output and the defuzzification block serves the purpose of transforming an output fuzzy set into a crisp single value. An example of a fuzzy inference system without defuzzification block is the two-rule two-input system of Figure 3.14. The function of the defuzzification block is explained in Section 4.2.

With crisp inputs and outputs, a fuzzy inference system implements a nonlinear mapping from its input space to output space. This mapping is accomplished by a number of fuzzy if-then rules, each of which describes the local behavior of the

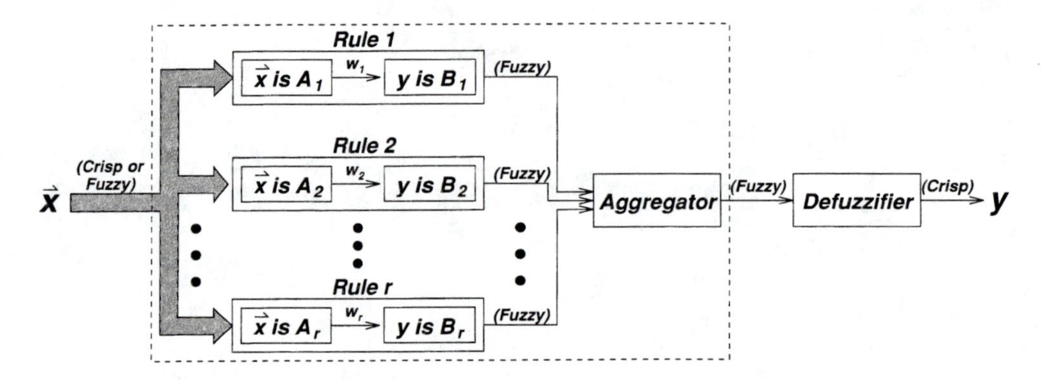

Figure 4.1. *Block diagram for a fuzzy inference system.*

mapping. In particular, the antecedent of a rule defines a fuzzy region in the input space, while the consequent specifies the output in the fuzzy region.

In what follows, we shall first introduce three types of fuzzy inference systems that have been widely employed in various applications. The differences between these three fuzzy inference systems lie in the consequents of their fuzzy rules, and thus their aggregation and defuzzification procedures differ accordingly. Then we will introduce and compare three different ways of partitioning the input space; these partitioning methods can be adopted by any fuzzy inference system, regardless of the structure of the consequents of its rules. Finally, we will address briefly the features and the problems of fuzzy modeling, which is concerned with the construction of fuzzy inference systems for modeling a given target system.

4.2 MAMDANI FUZZY MODELS

The **Mamdani fuzzy inference system** [6] was proposed as the first attempt to control a steam engine and boiler combination by a set of linguistic control rules obtained from experienced human operators. Figure 4.2 is an illustration of how a two-rule Mamdani fuzzy inference system derives the overall output z when subjected to two crisp inputs x and y.

If we adopt max and algebraic product as our choice for the T-norm and T-conorm operators, respectively, and use max-product composition instead of the original max-min composition, then the resulting fuzzy reasoning is shown in Figure 4.3, where the inferred output of each rule is a fuzzy set scaled down by its firing strength via algebraic product. Although this type of fuzzy reasoning was not employed in Mamdani's original paper, it has often been used in the literature. Other variations are possible if we use different T-norm and T-conorm operators.

In Mamdani's application [6], two fuzzy inference systems were used as two controllers to generate the heat input to the boiler and throttle opening of the engine cylinder, respectively, to regulate the steam pressure in the boiler and the

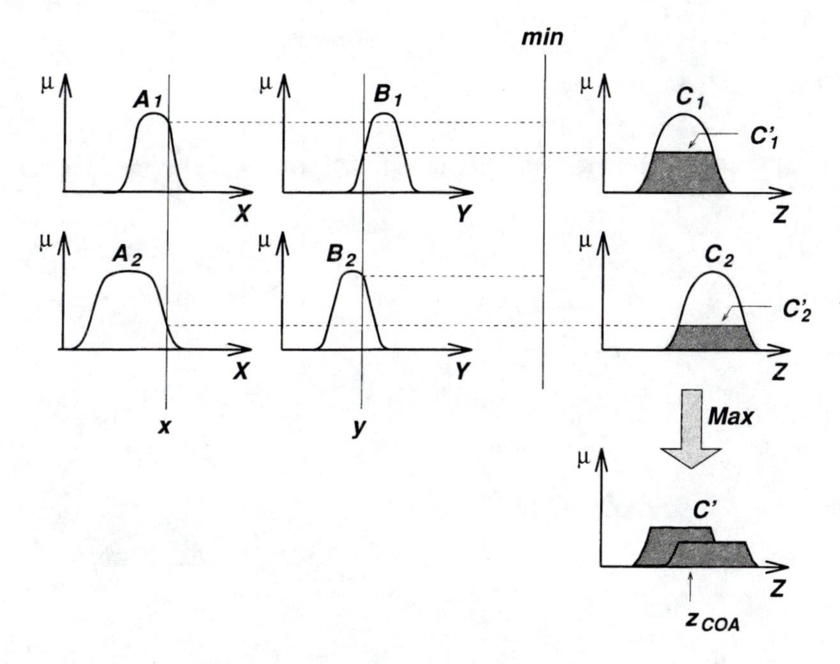

Figure 4.2. *The Mamdani fuzzy inference system using* min *and* max *for T-norm and T-conorm operators, respectively.*

speed of the engine. Since the plant takes only crisp values as inputs, we have to use a defuzzifier to convert a fuzzy set to a crisp value.

Defuzzification

Defuzzification refers to the way a crisp value is extracted from a fuzzy set as a representative value. In general, there are five methods for defuzzifying a fuzzy set A of a universe of discourse Z, as shown in Figure 4.4. (Here the fuzzy set A is usually represented by an aggregated output MF, such as C' in Figures 4.2 and 4.3.) A brief explanation of each defuzzification strategy follows.

- Centroid of area z_{COA}:

$$z_{\mathrm{COA}} = \frac{\int_Z \mu_A(z) z \, dz}{\int_Z \mu_A(z) \, dz}, \tag{4.1}$$

  where $\mu_A(z)$ is the aggregated output MF. This is the most widely adopted defuzzification strategy, which is reminiscent of the calculation of expected values of probability distributions.

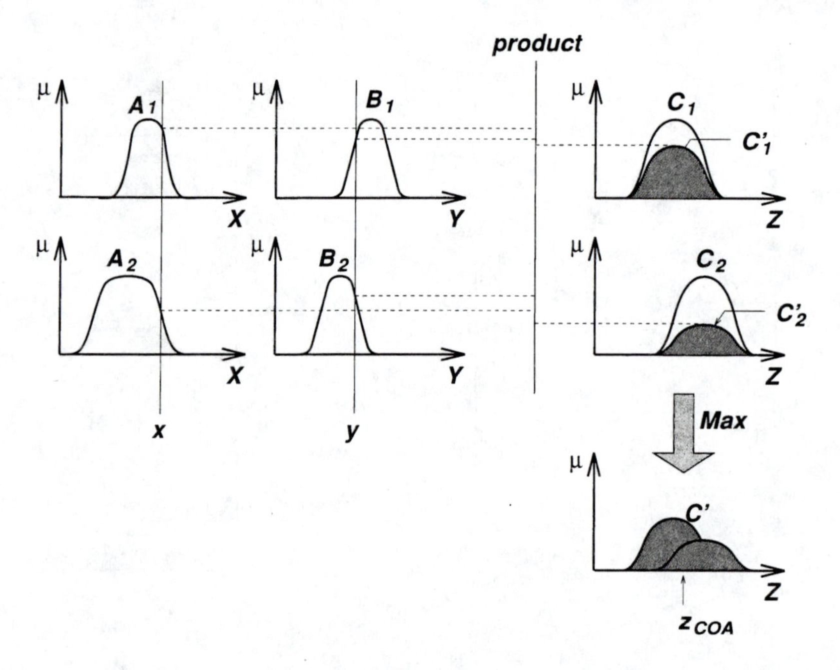

Figure 4.3. *The Mamdani fuzzy inference system using product and* max *for T-norm and T-conorm operators, respectively.*

- Bisector of area z_{BOA}: z_{BOA} satisfies

$$\int_{\alpha}^{z_{\mathrm{BOA}}} \mu_A(z)\,dz = \int_{z_{\mathrm{BOA}}}^{\beta} \mu_A(z)\,dz, \tag{4.2}$$

where $\alpha = \min\{z|z \in Z\}$ and $\beta = \max\{z|z \in Z\}$. That is, the vertical line $z = z_{\mathrm{BOA}}$ partitions the region between $z = \alpha$, $z = \beta$, $y = 0$ and $y = \mu_A(z)$ into two regions with the same area.

- Mean of maximum z_{MOM}: z_{MOM} is the average of the maximizing z at which the MF reach a maximum μ^*. In symbols,

$$z_{\mathrm{MOM}} = \frac{\int_{Z'} z\,dz}{\int_{Z'} dz}, \tag{4.3}$$

where $Z' = \{z \mid \mu_A(z) = \mu^*\}$. In particular, if $\mu_A(z)$ has a single maximum at $z = z^*$, then $z_{\mathrm{MOM}} = z^*$. Moreover, if $\mu_A(z)$ reaches its maximum whenever $z \in [z_{\mathrm{left}}, z_{\mathrm{right}}]$ (this is the case in Figure 4.4), then $z_{\mathrm{MOM}} = (z_{\mathrm{left}} + z_{\mathrm{right}})/2$. The mean of maximum is the defuzzification strategy employed in Mamdani's fuzzy logic controllers [6].

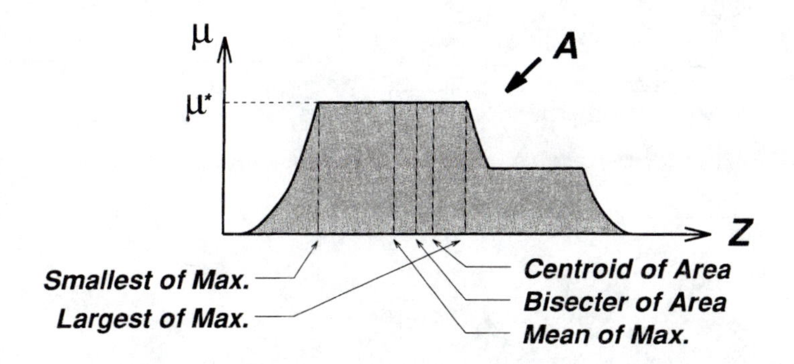

Figure 4.4. *Various defuzzification schemes for obtaining a crisp output.*

- Smallest of maximum z_{SOM}: z_{SOM} is the minimum (in terms of magnitude) of the maximizing z.
- Largest of maximum z_{LOM}: z_{LOM} is the maximum (in terms of magnitude) of the maximizing z. Because of their obvious bias, z_{SOM} and $_{\text{LOM}}$ are not used as often as the other three defuzzification methods.

The calculation needed to carry out any of these five defuzzification operations is time-consuming unless special hardware support is available. Furthermore, these defuzzification operations are not easily subject to rigorous mathematical analysis, so most of the studies are based on experimental results. This leads to the propositions of other types of fuzzy inference systems that do not need defuzzification at all; two of them are introduced in the next section. Other more flexible defuzzification methods can be found in [7, 8, 12].

The following two examples are single-input and two-input Mamdani fuzzy models.

Example 4.1 *Single-input single-output Mamdani fuzzy model*

An example of a single-input single-output Mamdani fuzzy model with three rules can be expressed as

$$\begin{cases} \text{If } X \text{ is small then } Y \text{ is small.} \\ \text{If } X \text{ is medium then } Y \text{ is medium.} \\ \text{If } X \text{ is large then } Y \text{ is large.} \end{cases}$$

Figure 4.5(a) plots the membership functions of input X and output Y, where the input and output universe are $[-10, 10]$ and $[0, 10]$, respectively. With max-min composition and centroid defuzzification, we can find the overall input-output curve, as shown in Figure 4.5(b). Note that the output variable never reaches the maximum (10) and minimum (0) of the output universe. Instead, the reachable minimum and maximum of the output variable are determined by the centroids of the leftmost and rightmost consequent MFs, respectively.

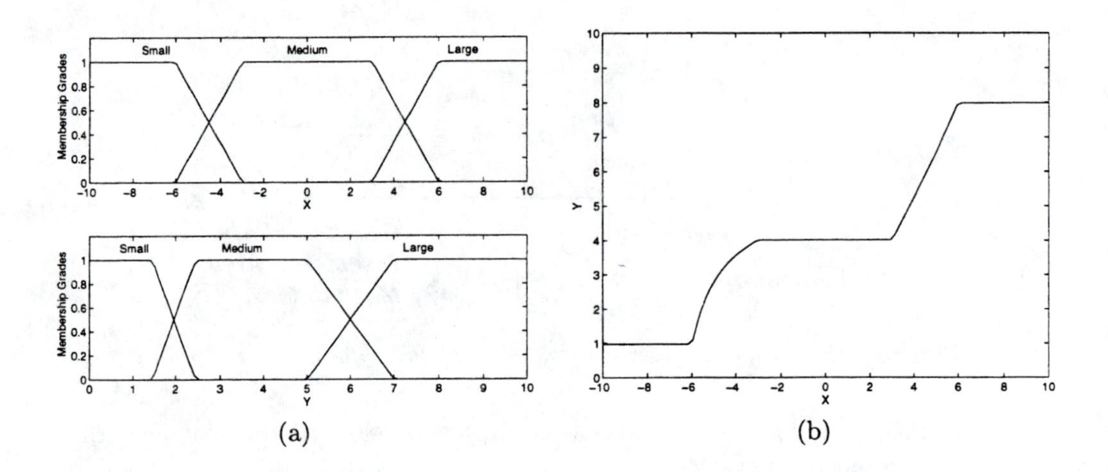

Figure 4.5. *Single-input single-output Mamdani fuzzy model in Example 4.1: (a) antecedent and consequent MFs; (b) overall input-output curve.* (MATLAB file: `mam1.m`)

□

Example 4.2 *Two-input single-output Mamdani fuzzy model*

An example of a two-input single-output Mamdani fuzzy model with four rules can be expressed as

$$\begin{cases} \text{If } X \text{ is small and } Y \text{ is small then } Z \text{ is negative large.} \\ \text{If } X \text{ is small and } Y \text{ is large then } Z \text{ is negative small.} \\ \text{If } X \text{ is large and } Y \text{ is small then } Z \text{ is positive small.} \\ \text{If } X \text{ is large and } Y \text{ is large then } Z \text{ is positive large.} \end{cases}$$

Figure 4.6(a) plots the membership functions of input X and Y and output Z, all with the same universe $[-5, 5]$. With max-min composition and centroid defuzzification, we can find the overall input-output surface, as shown in Figure 4.6(b). For a multiple-input fuzzy model, sometimes it is helpful to have a tool for viewing the process of fuzzy inference; Figure 4.7 is the fuzzy inference viewer available in the Fuzzy Logic Toolbox, where you can change the input values by click and drag the input vertical lines and then see the interactive changes of qualified consequent MFs and overall output MF.

□

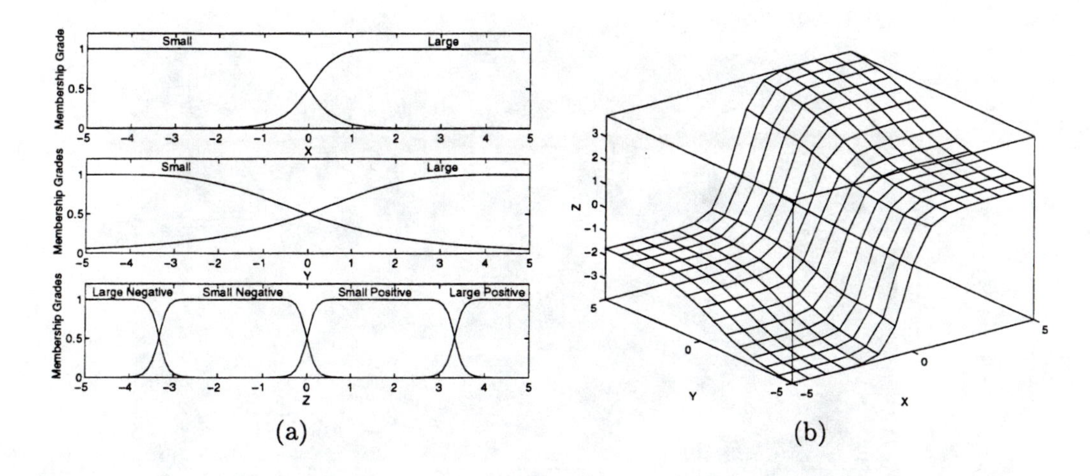

Figure 4.6. *Two-input single-output Mamdani fuzzy model in Example 4.2: (a) antecedent and consequent MFs; (b) overall input-output surface.* (MATLAB file: `mam2.m`)

4.2.1 Other Variants

Figures 4.2 and 4.3 conform to the fuzzy reasoning defined previously. However, in consideration of computation efficiency or mathematical tractability, a fuzzy inference system in practice may have a certain reasoning mechanism that does not follow the strict definition of the compositional rule of inference. For instance, one might use product for computing firing strengths (for rules with AND'ed antecedent), min for computing qualified consequent MFs, and max for aggregating them into an overall output MF. Therefore, to completely specify the operation of a Mamdani fuzzy inference system, we need to assign a function for each of the following operators:

- **AND operator** (usually T-norm) for calculating the firing strength of a rule with AND'ed antecedents.
- **OR operator** (usually T-conorm) for calculating the firing strength of a rule with OR'ed antecedents.
- **Implication operator** (usually T-norm) for calculating qualified consequent MFs based on given firing strength.
- **Aggregate operator** (usually T-conorm) for aggregating qualified consequent MFs to generate an overall output MF.
- **Defuzzification operator** for transforming an output MF to a crisp single output value.

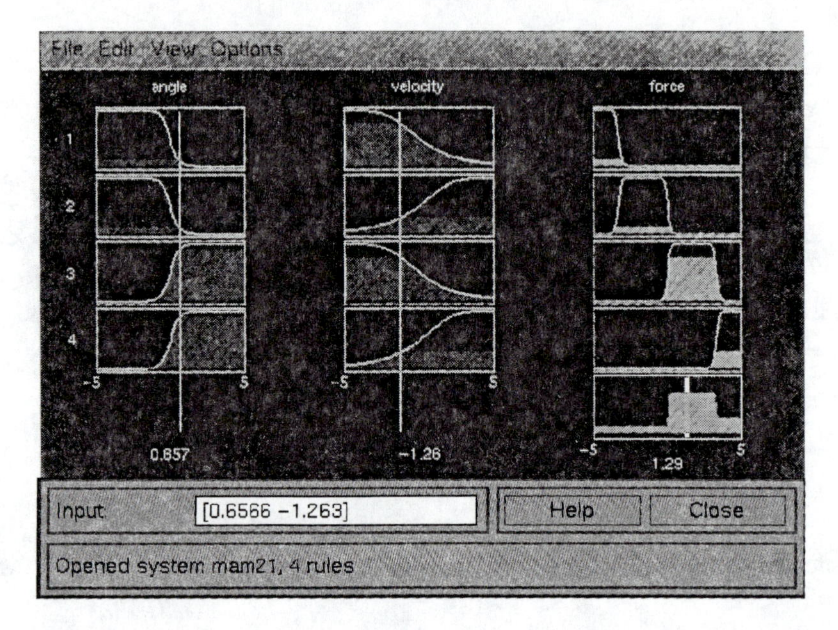

Figure 4.7. *Fuzzy inference viewer in the Fuzzy Logic Toolbox. This is obtained by typing* `ruleview mam21` *within* MATLAB.

One such example is to use product for the implication operator and pointwise summation (sum) for the aggregate operator. (Note that sum is not even a T-conorm operator.) An advantage of this **sum-product composition** [3] is that the final crisp output via centroid defuzzification is equal to the weighted average of the centroids of consequent MFs, where the weighting factor for each rule is equal to its firing strength multiplied by the area of the consequent MF. This is expressed as the following theorem.

Theorem 4.1 *Computation shortcut for Mamdani fuzzy inference systems*

Under sum-product composition, the output of a Mamdani fuzzy inference system with centroid defuzzification is equal to the weighted average of the centroids of consequent MFs, where each of the weighting factors is equal to the product of a firing strength and the consequent MF's area.

Proof: We shall prove this theorem for a fuzzy inference system with two rules (see Figure 4.3). By using product and sum for implication and aggregate operators, respectively, we have

$$\mu_{C'}(z) = w_1 \mu_{C_1}(z) + w_2 \mu_{C_2}(z).$$

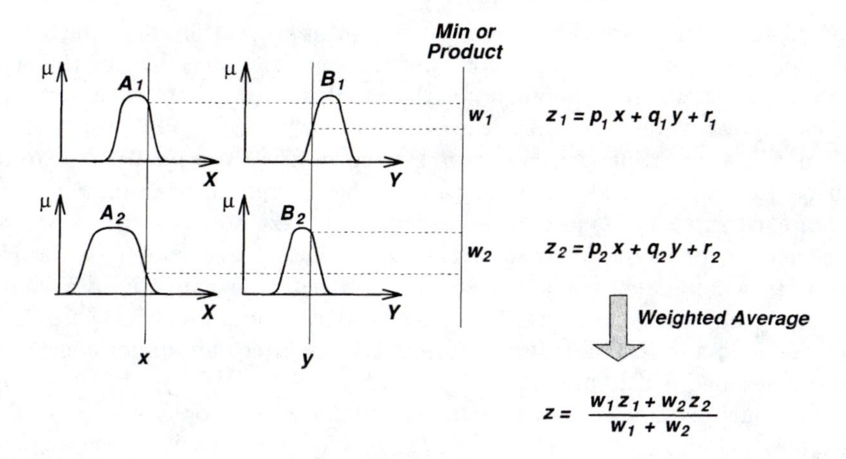

Figure 4.8. *The Sugeno fuzzy model.*

(Note that the preceding MF could have values greater than 1 at certain points.) The crisp output under centroid defuzzification is

$$
\begin{aligned}
z_{\text{COA}} &= \frac{\int_Z \mu_{C'}(z) z dz}{\int_Z \mu_{C'}(z) dz} \\
&= \frac{w_1 \int \mu_{C_1}(z) z dz + w_2 \int \mu_{C_2}(z) z dz}{w_1 \int \mu_{C_1}(z) dz + w_2 \int \mu_{C_2}(z) dz} \\
&= \frac{w_1 a_1 z_1 + w_2 a_2 z_2}{w_1 a_1 + w_2 a_2},
\end{aligned}
$$

where a_i $(= \int_Z \mu_{C_i}(z) dz)$ and z_i $(= \frac{\int_Z \mu_{C_i}(z) z dz}{\int_Z \mu_{C_i}(z) dz})$ are the area and centroid of the consequent MF $\mu_{C_i}(z)$, respectively.

□

By using this theorem, computation is more efficient if we can obtain the area and centroid of each consequent MF in advance.

4.3 SUGENO FUZZY MODELS

The **Sugeno fuzzy model** (also known as the **TSK fuzzy model**) was proposed by Takagi, Sugeno, and Kang [10, 9] in an effort to develop a systematic approach to generating fuzzy rules from a given input-output data set. A typical fuzzy rule in a Sugeno fuzzy model has the form

$$\text{if } x \text{ is } A \text{ and } y \text{ is } B \text{ then } z = f(x, y),$$

where A and B are fuzzy sets in the antecedent, while $z = f(x, y)$ is a crisp function in the consequent. Usually $f(x, y)$ is a polynomial in the input variables x and y, but it can be any function as long as it can appropriately describe the output of the model within the fuzzy region specified by the antecedent of the rule. When $f(x, y)$ is a first-order polynomial, the resulting fuzzy inference system is called a **first-order Sugeno fuzzy model**, which was originally proposed in [10, 9]. When f is a constant, we then have a **zero-order Sugeno fuzzy model**, which can be viewed either as a special case of the Mamdani fuzzy inference system, in which each rule's consequent is specified by a fuzzy singleton (or a pre-defuzzified consequent), or a special case of the Tsukamoto fuzzy model (to be introduced next), in which each rule's consequent is specified by an MF of a step function center at the constant. Moreover, a zero-order Sugeno fuzzy model is functionally equivalent to a radial basis function network under certain minor constraints [1], as will be detailed in Chapter 12.

The output of a zero-order Sugeno model is a smooth function of its input variables as long as the neighboring MFs in the antecedent have enough overlap. In other words, the overlap of MFs in the consequent of a Mamdani model does not have a decisive effect on the smoothness; it is the overlap of the antecedent MFs that determines the smoothness of the resulting input-output behavior.

Figure 4.8 shows the fuzzy reasoning procedure for a first-order Sugeno fuzzy model. Since each rule has a crisp output, the overall output is obtained via **weighted average**, thus avoiding the time-consuming process of defuzzification required in a Mamdani model. In practice, the weighted average operator is sometimes replaced with the **weighted sum** operator (that is, $z = w_1 z_1 + w_2 z_2$ in Figure 4.8) to reduce computation further, especially in the training of a fuzzy inference system. However, this simplification could lead to the loss of MF linguistic meanings unless the sum of firing strengths (that is, $\sum_i w_i$) is close to unity.

Since the only fuzzy part of a Sugeno model is in its antecedent, it is easy to demonstrate the distinction between a set of fuzzy rules and nonfuzzy ones.

Example 4.3 *Fuzzy and nonfuzzy rule set—a comparison*

An example of a single-input Sugeno fuzzy model can be expressed as

$$\begin{cases} \text{If } X \text{ is small then } Y = 0.1X + 6.4. \\ \text{If } X \text{ is medium then } Y = -0.5X + 4. \\ \text{If } X \text{ is large then } Y = X - 2. \end{cases}$$

If "small," "medium," and "large" are nonfuzzy sets with membership functions shown in Figure 4.9(a), then the overall input-output curve is piecewise linear, as shown in Figure 4.9(b). On the other hand, if we have smooth membership functions [Figure 4.9(c)] instead, the overall input-output curve [Figure 4.9(d)] becomes a smoother one.

□

Sometimes a simple Sugeno fuzzy model can generate complex behavior. The following is an example of a two-input system.

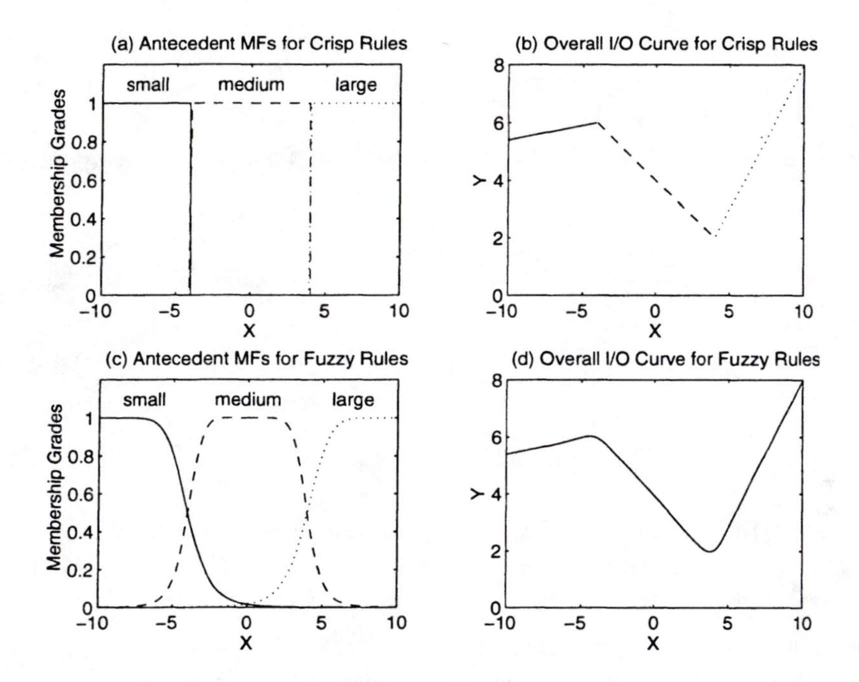

Figure 4.9. *Comparison between fuzzy and nonfuzzy rules in Example 4.3: (a) Antecedent MFs and (b) input-output curve for nonfuzzy rules; (c) Antecedent MFs and (d) input-output curve for fuzzy rules.* (MATLAB file: `sug1.m`)

Example 4.4 *Two-input single-output Sugeno fuzzy model*

An example of a two-input single-output Sugeno fuzzy model with four rules can be expressed as

$$\begin{cases} \text{If } X \text{ is small and } Y \text{ is small then } z = -x + y + 1. \\ \text{If } X \text{ is small and } Y \text{ is large then } z = -y + 3. \\ \text{If } X \text{ is large and } Y \text{ is small then } z = -x + 3. \\ \text{If } X \text{ is large and } Y \text{ is large then } z = x + y + 2. \end{cases}$$

Figure 4.10(a) plots the membership functions of input X and Y, and Figure 4.10(b) is the resulting input-output surface. The surface is complex, but it is still obvious that the surface is composed of four planes, each of which is specified by the output equation of a fuzzy rule.

□

Unlike the Mamdani fuzzy model, the Sugeno fuzzy model cannot follow the compositional rule of inference (Section 3.4.1) strictly in its fuzzy reasoning mechanism. This poses some difficulties when the inputs to a Sugeno fuzzy model are

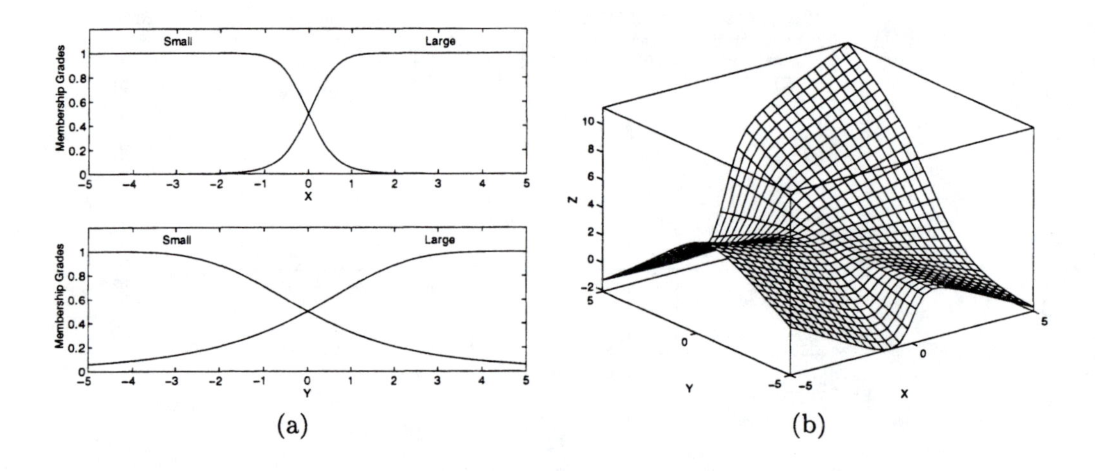

Figure 4.10. *Two-input single-output Sugeno fuzzy model in Example 4.4: (a) antecedent and consequent MFs; (b) overall input-output surface.* (MATLAB file: sug2.m)

fuzzy. Specifically, we can still employ the matching of fuzzy sets, as shown in the antecedent part of Figure 3.14, to find the firing strength of each rule. However, the resulting overall output via either weighted average or weighted sum is always crisp; this is counterintuitive since a fuzzy model should be able to propagate the fuzziness from inputs to outputs in an appropriate manner.

Without the time-consuming and mathematically intractable defuzzification operation, the Sugeno fuzzy model is by far the most popular candidate for sample-data-based fuzzy modeling, which is introduced in Chapter 12.

4.4 TSUKAMOTO FUZZY MODELS

In the **Tsukamoto fuzzy models** [11], the consequent of each fuzzy if-then rule is represented by a fuzzy set with a monotonical MF, as shown in Figure 4.11. As a result, the inferred output of each rule is defined as a crisp value induced by the rule's firing strength. The overall output is taken as the weighted average of each rule's output. Figure 4.11 illustrates the reasoning procedure for a two-input two-rule system.

Since each rule infers a crisp output, the Tsukamoto fuzzy model aggregates each rule's output by the method of weighted average and thus avoids the time-consuming process of defuzzification. However, the Tsukamoto fuzzy model is not used often since it is not as transparent as either the Mamdani or Sugeno fuzzy models. The following is a single-input example.

Example 4.5 *Single-input Tsukamoto fuzzy model*

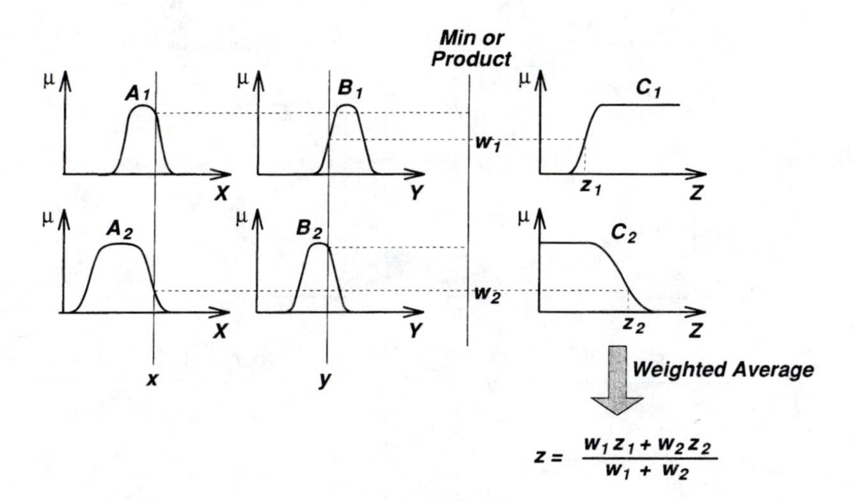

Figure 4.11. *The Tsukamoto fuzzy model.*

An example of a single-input Tsukamoto fuzzy model can be expressed as

$$\begin{cases} \text{If } X \text{ is small then } Y \text{ is } C_1 \\ \text{If } X \text{ is medium then } Y \text{ is } C_2 \\ \text{If } X \text{ is large then } Y \text{ is } C_3, \end{cases}$$

where the antecedent MFs for "small," "medium," and "large" are shown in Figure 4.12(a), and the consequent MFs for "C_1," "C_2," and "C_3" are shown in Figure 4.12(b). The overall input-output curve, as shown in Figure 4.12(d), is equal to $(\sum_{i=1}^{3} w_i f_i)/(\sum_{i=1}^{3} w_i)$, where f_i is the output of each rule induced by the firing strength w_i and MF for C_i. If we plot each rule's output f_i as a function of x, we obtain Figure 4.12(c), which is not quite obvious from the original rule base and MF plots.

□

Since the reasoning mechanism of the Tsukamoto fuzzy model does not follow strictly the compositional rule of inference, the output is always crisp even when the inputs are fuzzy.

4.5 OTHER CONSIDERATIONS

There are certain common issues concerning all the three fuzzy inference systems introduced previously, such as how to partition an input space and how to construct a fuzzy inference system for a particular application. We shall examine these issues in this section.

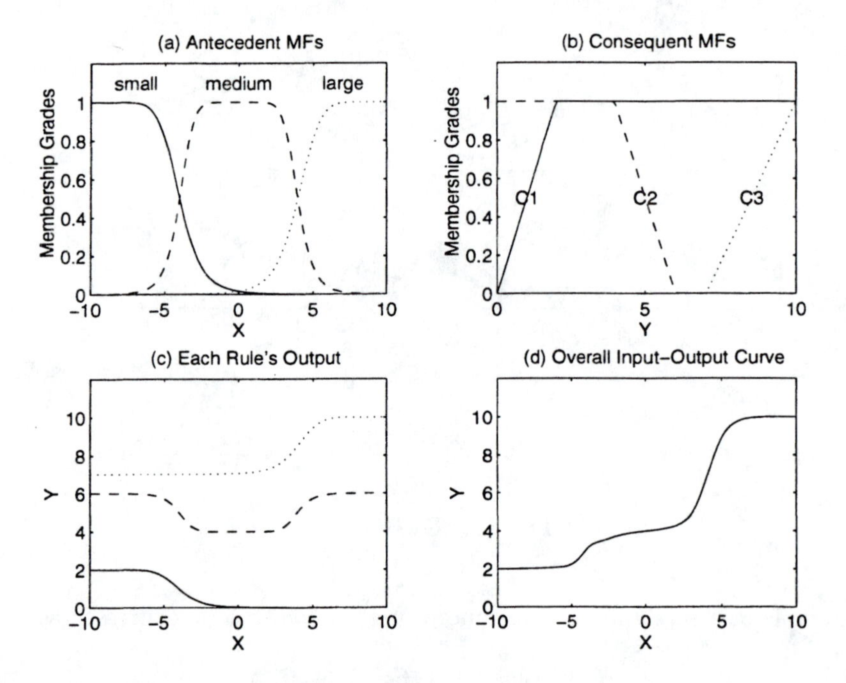

Figure 4.12. *Single-input single output Tsukamoto fuzzy model in Example 4.4: (a) antecedent MFs; (b) consequent MFs; (c) each rule's output curve; (d) overall input-output curve.* (MATLAB file: `tsu1.m`)

4.5.1 Input Space Partitioning

Now it should be clear that the spirit of fuzzy inference systems resembles that of "divide and conquer"—the antecedent of a fuzzy rule defines a local fuzzy region, while the consequent describes the behavior within the region via various constituents. The consequent constituent can be a consequent MF (Mamdani and Tsukamoto fuzzy models), a constant value (zero-order Sugeno model), or a linear equation (first-order Sugeno model). Different consequent constituents result in different fuzzy inference systems, but their antecedents are always the same. Therefore, the following discussion of methods of partitioning input spaces to form the antecedents of fuzzy rules is applicable to all three types of fuzzy inference systems.

- **Grid partition**: Figure 4.13(a) illustrates a typical grid partition in a two-dimensional input space. This partition method is often chosen in designing a fuzzy controller, which usually involves only several state variables as the inputs to the controller. This partition strategy needs only a small number of MFs for each input. However, it encounters problems when we have a moderately large number of inputs. For instance, a fuzzy model with 10 inputs and 2 MFs on each input would result in $2^{10} = 1024$ fuzzy if-then

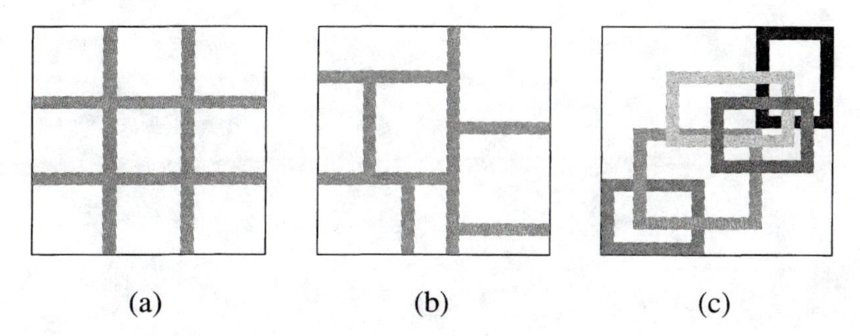

Figure 4.13. *Various methods for partitioning the input space: (a) grid partition; (b) tree partition; (c) scatter partition.*

rules, which is prohibitively large. This problem, usually referred to as the **curse of dimensionality**, can be alleviated by the other partition strategies.

- **Tree partition**: Figure 4.13(b) shows a typical tree partition, in which each region can be uniquely specified along a corresponding decision tree. The tree partition relieves the problem of an exponential increase in the number of rules. However, more MFs for each input are needed to define these fuzzy regions, and these MFs do not usually bear clear linguistic meanings such as "small," "big," and so on. In other words, orthogonality holds roughly in $X \times Y$, but not in either X or Y alone. Tree partition is used by the CART (classification and regression tree) algorithm, as discussed in Chapter 14.

- **Scatter partition**: As shown in Figure 4.13(c), by covering a subset of the whole input space that characterizes a region of possible occurrence of the input vectors, the scatter partition can also limit the number of rules to a reasonable amount. However, the scatter partition is usually dictated by desired input-output data pairs and thus, in general, orthogonality does not hold in X, Y or $X \times Y$. This makes it hard to estimate the overall mapping directly from the consequent of each rule's output.

Note that Figure 4.13 is based on the assumption that MFs are defined on the input variables directly. If MFs are defined on certain transformations of the input variables, we could end up in a more flexible partition style. Figure 4.14 is an example of the input partition when MFs are defined on linear transformations of the input variables.

4.5.2 Fuzzy Modeling

By now the reader should have already developed a clear picture of both the structures and operations of several types of fuzzy inference systems. In general, we design a fuzzy inference system based on the past known behavior of a target system. The fuzzy system is then expected to be able to reproduce the behavior of the

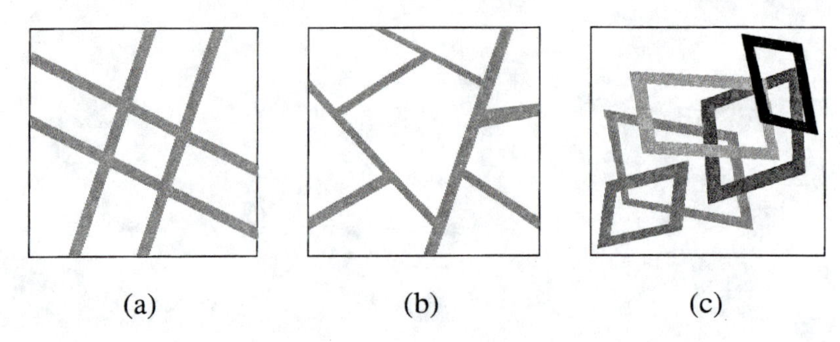

Figure 4.14. *Input space partition when MFs are defined on linear transformations of input variables: (a) grid partition; (b) tree partition; (b) scatter partition.*

target system. For example, if the target system is a human operator in charge of a chemical reaction process, then the fuzzy inference system becomes a fuzzy logic controller that can regulate and control the process. Similarly, if the target system is a medical doctor, then the fuzzy inference becomes a fuzzy expert system for medical diagnosis.

Let us now consider how we might construct a fuzzy inference system for a specific application. Generally speaking, the standard method for constructing a fuzzy inference system, a process usually called **fuzzy modeling**, has the following features:

- The rule structure of a fuzzy inference system makes it easy to incorporate human expertise about the target system directly into the modeling process. Namely, fuzzy modeling takes advantage of **domain knowledge** that might not be easily or directly employed in other modeling approaches.
- When the input-output data of a target system is available, conventional system identification techniques can be used for fuzzy modeling. In other words, the use of **numerical data** also plays an important role in fuzzy modeling, just as in other mathematical modeling methods.

In what follows, we shall summarize some general guidelines concerning fuzzy modeling. Specific examples of fuzzy modeling for various applications can be found in subsequent chapters.

Conceptually, fuzzy modeling can be pursued in two stages, which are not totally disjoint. The first stage is the identification of the **surface structure**, which includes the following tasks:

1. Select relevant input and output variables.
2. Choose a specific type of fuzzy inference system.
3. Determine the number of linguistic terms associated with each input and output variables. (For a Sugeno model, determine the order of consequent equations.)

4. Design a collection of fuzzy if-then rules.

Note that to accomplish the preceding tasks, we rely on our own knowledge (common sense, simple physical laws, and so on) of the target system, information provided by human experts who are familiar with the target system (which could be the human experts themselves), or simply trial and error.

After the first stage of fuzzy modeling, we obtain a rule base that can more or less describe the behavior of the target system by means of linguistic terms. The meaning of these linguistic terms is determined in the second stage, the identification of **deep structure**, which determines the MFs of each linguistic term (and the coefficients of each rule's output polynomial if a Sugeno fuzzy model is used). Specifically, the identification of deep structure includes the following tasks:

1. Choose an appropriate family of parameterized MFs (see Section 2.4).

2. Interview human experts familiar with the target systems to determine the parameters of the MFs used in the rule base.

3. Refine the parameters of the MFs using regression and optimization techniques.

Task 1 and 2 assume the availability of human experts, while task 3 assumes the availability of a desired input-output data set. Various system identification and optimization techniques for parameter identification in task 3 are detailed in Chapters 5, 6, and 7. A specific network structure that facilitates task 3 is covered in Chapter 12. When a fuzzy inference system is used as a controller for a given plant, then the objective in task 3 should be changed to that of searching for parameters that will generate the best performance of the plant; this aspect of fuzzy logic controller design is explored in Chapters 17 and 18.

4.6 SUMMARY

This chapter presents three of the frequently used fuzzy inference systems: the Mamdani, Sugeno, and Tsukamoto fuzzy models. We discuss their strengths and weaknesses and other related issues, such as input space partitioning and fuzzy modeling.

Fuzzy inference systems are the most important modeling tool based on fuzzy set theory. Conventional fuzzy inference systems are typically built by domain experts and have been used in automatic control, decision analysis, and expert systems. Optimization and adaptive techniques expand the applications of fuzzy inference systems to fields such as adaptive control, adaptive signal processing, nonlinear regression, and pattern recognition. Chapter 12 discusses adaptive fuzzy inference systems; their applications are covered in Chapter 19.

EXERCISES

1. Derive the fuzzy reasoning mechanism shown in Figure 4.3 by choosing the max and the algebraic product for the T-norm and T-conorm operators, respectively.

2. Give formulas for the three defuzzification strategies [Equations (4.1) to (4.3)] when we have a finite universe of discourse $X = \{x_1, \ldots, x_n\}$, where $x_1 < \cdots < x_n$.

3. Use the three defuzzification strategies in Equations (4.1), (4.2), and (4.3) to find the representative values of a fuzzy set A defined by

$$\mu_A(x) = \text{trapezoid}(x, 10, 30, 50, 90).$$

4. Repeat the previous exercise, but assume that the universe of discourse X contains integers from 0 to 100.

5. Modify the program `mam1.m` in Example 4.1 such that you can click and drag a corner of a trapezoidal MF to change its shape and see the interactive changes of the overall input-output curves.

6. Change the MFs in Example 4.2 to trapezoidal ones and plot the overall input-output surface.

7. Modify Example 4.3 such that only constant terms are retained in the consequent. Repeat the plots in Figure 4.9.

8. Modify Example 4.3 by adding a second-order term to the consequent equation of each rule. Repeat the plots in Figure 4.9.

9. In Example 4.4, use the following different definitions of MFs:

$$\begin{aligned}
\mu_{\text{small}_x} &= \text{sig}(x, [-5, 0]), \\
\mu_{\text{large}_x} &= \text{sig}(x, [5, 0]), \\
\mu_{\text{small}_y} &= \text{sig}(y, [-2, 0]), \\
\mu_{\text{large}_y} &= \text{sig}(y, [2, 0]),
\end{aligned} \tag{4.4}$$

and repeat the plots in Figure 4.10.

10. Repeat the previous exercise but use weighted sum instead weighted average to derive the final output. Do you get exactly the same input-output surface as that in the previous exercise? Why?

REFERENCES

[1] J.-S. Roger Jang and C.-T. Sun. Functional equivalence between radial basis function networks and fuzzy inference systems. *IEEE Transactions on Neural Networks*, 4(1):156–159, January 1993.

[2] A. Kandel, editor. *Fuzzy expert systems.* CRC Press, Inc., Boca Raton, FL, 1992.

[3] B. Kosko. *Neural networks and fuzzy systems: a dynamical systems approach.* Prentice Hall, Upper Saddle River, NJ, 1991.

[4] C.-C. Lee. Fuzzy logic in control systems: fuzzy logic controller-part 1. *IEEE Transactions on Systems, Man, and Cybernetics*, 20(2):404–418, 1990.

[5] C.-C. Lee. Fuzzy logic in control systems: fuzzy logic controller-part 2. *IEEE Transactions on Systems, Man, and Cybernetics*, 20(2):419–435, 1990.

[6] E. H. Mamdani and S. Assilian. An experiment in linguistic synthesis with a fuzzy logic controller. *International Journal of Man-Machine Studies*, 7(1):1–13, 1975.

[7] N. Pfluger, J. Yen, and R. Langari. A defuzzification strategy for a fuzzy logic controller employing prohibitive information in command formulation. In *Proceedings of IEEE International Conference on Fuzzy Systems*, pages 717–723, San Diego, March 1992.

[8] T. A. Runkler and M. Glesner. Defuzzification and ranking in the context of membership value semantics, rule modality, and measurement theory. In *European Congress on Fuzzy and Intelligent Technologies*, Aachen, September 1994.

[9] M. Sugeno and G. T. Kang. Structure identification of fuzzy model. *Fuzzy Sets and Systems*, 28:15–33, 1988.

[10] T. Takagi and M. Sugeno. Fuzzy identification of systems and its applications to modeling and control. *IEEE Transactions on Systems, Man, and Cybernetics*, 15:116–132, 1985.

[11] Y. Tsukamoto. An approach to fuzzy reasoning method. In Madan M. Gupta, Rammohan K. Ragade, and Ronald R. Yager, editors, *Advances in fuzzy set theory and applications*, pages 137–149. North-Holland, Amsterdam, 1979.

[12] R. R. Yager and D. P. Filev. SLIDE: A simple adaptive defuzzification method. *IEEE Transactions on Fuzzy Systems*, 1(1):69–78, February 1993.

Part II

Regression and Optimization

Chapter 5

Least-Squares Methods for System Identification

J.-S. R. Jang

5.1 SYSTEM IDENTIFICATION: AN INTRODUCTION

The problem of determining a mathematical model for an unknown system (also referred to as the **target system**) by observing its input-output data pairs is generally referred to as **system identification**. The purposes of system identification are multiple:

- To predict a system's behavior, as in time series prediction and weather forecasting.
- To explain the interactions and relationships between inputs and outputs of a system. For example, a mathematical model can be used to examine whether the demand indeed varies proportionally to the supply in an economic system.
- To design a controller based on the model of a system, as in aircraft and ship control. Also to do computer simulation of the system under control, you need a model of the system.

System identification generally involves two top-down steps:

Structure identification In this step, we need to apply *a priori* knowledge about the target system to determine a class of models within which the search for the most suitable model is to be conducted. Usually this class of models is denoted by a parameterized function $y = f(\mathbf{u}; \boldsymbol{\theta})$, where y is the model's output, $\mathbf{u}$ is the input vector, and $\boldsymbol{\theta}$ is the parameter vector. The determination of the function f is problem dependent, and the function is based on the designer's experience and intuition and the laws of nature governing the target system.

Parameter identification In the second step, the structure of the model is known and all we need to do is apply optimization techniques to determine the pa-

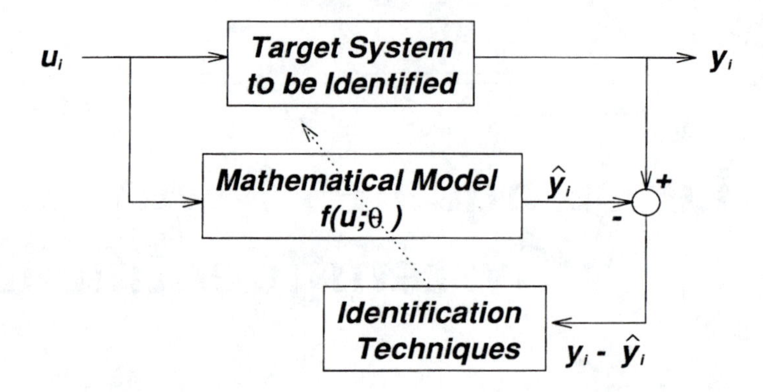

Figure 5.1. *Block diagram for parameter identification.*

rameter vector $\boldsymbol{\theta} = \hat{\boldsymbol{\theta}}$ such that the resulting model $\hat{y} = f(\mathbf{u}; \hat{\boldsymbol{\theta}})$ can describe the system appropriately.

If we do not have any *a priori* knowledge about the target system, then structure identification becomes a difficult problem and we have to select the structure by trial and error. Fortunately, we know a great deal about the structures of most engineering systems and industrial processes; usually it is possible to derive a specific class of models—namely, a parameterized function—that can best describe the target system. Consequently, the system identification problem is usually reduced to that of parameter identification. The problem of parameter identification is thus of great importance, and accordingly this chapter is devoted mostly to this class of problem.

Figure 5.1 illustrates a schematic diagram of parameter identification, where an input u_i is applied to both the system and the model, and the difference between the target system's output y_i and the model's output $\hat{y}_i$ is used in an appropriate manner to update a parameter vector $\boldsymbol{\theta}$ to reduce this difference. Note that the data set composed of m desired input-output pairs $(u_i; y_i)$, $i = 1, \cdots, m$, is often called the **training data set** or **sampled data set**. In the most general case, u_i and y_i represent the desired input and output vectors, respectively.

In general, system identification is not a one-pass process; it needs to do both structure and parameter identification repeatedly until a satisfactory model is found, as follows:

1. Specify and parameterize a class of mathematical models representing the system to be identified.

2. Perform parameter identification to choose the parameters that best fit the training data set.

3. Conduct validation tests to see if the model identified responds correctly to an unseen data set. (This data set is disjoint from the training data set and

is referred to as the **test**, **validating**, or **checking data set**.)

4. Terminate the procedure once the results of the validation test are satisfactory. Otherwise, another class of models is selected and steps 2 through 4 are repeated.

Before delving into the core parts of neuro-fuzzy and soft computing, we shall introduce a class of standard least-squares methods for linear models in system identification. Least-squares methods are powerful and well-developed mathematical tools that have been proposed and used in a variety of areas for decades, including adaptive control, signal processing, and statistics. Nowadays they still prove to be essential and indispensable tools for constructing linear mathematical models. The same fundamental concepts can be extended to nonlinear models as well. Thus it suffices to say that these linear least-squares methods provide the most basic and important mathematical foundation for solving neuro-fuzzy modeling problems in subsequent chapters.

Throughout this chapter, we shall restrict ourselves to the identification of *linear* models and *static* (or memoryless) systems. By linear models, we mean models that are linear in their parameters. Thus a linear model may be nonlinear in its inputs. The least-squares methods provide us with mathematical procedures by which a linear model can achieve a best fit to experimental data in the sense of least-squared error. Nonlinear models that are intrinsically linear can also take advantage of the least-squares methods, as explained in Section 5.8. For intrinsically nonlinear models, a thorough discussion can be found in Section 6.8 of Chapter 6.

By static systems, we mean that the output of the target system depends on its current inputs only; it does not depend on the history of inputs. This assumption does not impair the generality of our discussion since in the discrete time domain, the output of a dynamic system can be treated as a static mapping of its current inputs and (several) previous states, assuming they are available.

In what follows, we begin by briefly reviewing some techniques of matrix manipulation and calculus that will be used throughout this chapter. In Section 5.3 we explain how to find the least-squares estimator of a linear model; the intuitive geometric interpretation of the estimator is discussed in Section 5.4. Sections 5.5 and 5.6 give on-line formulas for the least-squares methods to save computation time and deal with time-varying target systems. Statistical properties of the least-squares estimator and its close relationship with the maximum likelihood estimator are discussed in Section 5.7, which is optional for the first reading.

5.2 BASICS OF MATRIX MANIPULATION AND CALCULUS

Since formulas for the least-squares estimator and derivative-based optimization methods are much more concise in matrix notation, it would be helpful for us to review briefly a few matrix manipulation techniques. (However, this chapter is not intended as an introduction to matrix theory; the reader is expected to have basic knowledge about linear algebra.) Most of the lemmas introduced here are straightforward and thus no proofs are given. The reader is encouraged to contrive simple examples to try out the lemmas introduced.

To minimize ambiguity and enhance readability, our matrix notation follows these guidelines:

- Matrices are represented by bold capital letters, such as **A**, **B**, **X**, and **Y**.
- Vectors are always assumed to be column vectors unless otherwise specified, and they are represented either by lowercase boldface letters (such as **a**, **b**, **x**, and **y**) or by lowercase letters with explicit vector symbols (such as $\vec{a}$, $\vec{b}$, $\vec{x}$, and $\vec{y}$).
- Scalars and scalar-valued functions are represented in nonbold letters, such as a, B, x, and Y.

In what follows, we introduce several definitions and lemmas concerning matrix manipulation and calculus.

Lemma 5.1 *A property of matrix transpose*

Let **A** and **B** be compatible matrices. Then

$$(\mathbf{AB})^T = \mathbf{B}^T\mathbf{A}^T,$$

where the superscript t denotes the transpose of a matrix.

□

Lemma 5.2 *A property of matrix inverse*

Let **A** and **B** be compatible and nonsingular matrices. Then

$$(\mathbf{AB})^{-1} = \mathbf{B}^{-1}\mathbf{A}^{-1},$$

where the superscript -1 denotes the inverse of a matrix.

□

Definition 5.1 *Block form*

A matrix in **block form** is regarded as a matrix containing blocks of smaller matrices; these blocks are dictated by the applications in which the matrix arises.

□

For example, the matrix

$$\mathbf{A} = \begin{bmatrix} 1 & 2 & 3 \\ 4 & 5 & 6 \\ 7 & 8 & 9 \end{bmatrix}$$

can be written in block form as

$$\mathbf{A} = \begin{bmatrix} \mathbf{A}_1 & \mathbf{A}_2 \\ \mathbf{A}_3 & \mathbf{A}_4 \end{bmatrix},$$

where

$$\mathbf{A}_1 = \begin{bmatrix} 1 & 2 \\ 4 & 5 \end{bmatrix},$$

$$\mathbf{A}_2 = \begin{bmatrix} 3 \\ 6 \end{bmatrix},$$

$$\mathbf{A}_3 = \begin{bmatrix} 7 & 8 \end{bmatrix},$$

and

$$\mathbf{A}_4 = \begin{bmatrix} 9 \end{bmatrix}.$$

Note that any $m \times n$ matrix can be viewed either as a row of n column vectors or as a column of m row vectors. In symbols,

$$\mathbf{A} = \begin{bmatrix} \mathbf{a}_1 & \cdots & \mathbf{a}_n \end{bmatrix},$$

where $\mathbf{a}_i$ is the ith column of A, or

$$\mathbf{A} = \begin{bmatrix} \mathbf{a}_i^T \\ \vdots \\ \mathbf{a}_m^T \end{bmatrix},$$

where $\mathbf{a}_i^T$ is the ith row of $\mathbf{A}$. (Thus $\mathbf{a}_i$ may denote either the ith column of $\mathbf{A}$ or the transpose of the ith row of $\mathbf{A}$; its meaning should be obvious from the context.)

Lemma 5.3 *Transpose and product of matrices in block form*

Let $\mathbf{A}$ and $\mathbf{B}$ be two matrices in block form:

$$\mathbf{A} = \begin{bmatrix} \mathbf{A}_1 & \mathbf{A}_2 \\ \mathbf{A}_3 & \mathbf{A}_4 \end{bmatrix}, \quad \mathbf{B} = \begin{bmatrix} \mathbf{B}_1 & \mathbf{B}_2 \\ \mathbf{B}_3 & \mathbf{B}_4 \end{bmatrix}.$$

Then

$$\mathbf{A}^T = \begin{bmatrix} \mathbf{A}_1^T & \mathbf{A}_3^T \\ \mathbf{A}_2^T & \mathbf{A}_4^T \end{bmatrix},$$

and

$$\mathbf{AB} = \begin{bmatrix} \mathbf{A}_1\mathbf{B}_1 + \mathbf{A}_2\mathbf{B}_3 & \mathbf{A}_1\mathbf{B}_2 + \mathbf{A}_2\mathbf{B}_4 \\ \mathbf{A}_3\mathbf{B}_1 + \mathbf{A}_4\mathbf{B}_3 & \mathbf{A}_3\mathbf{B}_2 + \mathbf{A}_4\mathbf{B}_4 \end{bmatrix}.$$

Namely, the matrices can be transposed and multiplied in the same way as if the blocks were scalars, provided that the individual products are defined.

Definition 5.2 *Gradient of a scalar function.*

Let $\mathbf{x} = [x_1, \cdots, x_n]^T$ and let $f(\mathbf{x})$ be a scalar function of $\mathbf{x}$. Then the derivative of $f(\mathbf{x})$ with respect to $\mathbf{x}$, called the **gradient vector** or **gradient** of $f(\mathbf{x})$, is a column vector denoted by

$$\nabla f(\mathbf{x}) = \begin{bmatrix} \partial f(\mathbf{x})/\partial x_1 \\ \vdots \\ \partial f(\mathbf{x})/\partial x_n \end{bmatrix}.$$

□

Definition 5.3 *Jacobian of a vector function*

Let $\mathbf{x} = [x_1, \cdots, x_n]^T$ and let $\mathbf{f}(\mathbf{x})$ be a vector function of $\mathbf{x}$, denoted by $\mathbf{f}(\mathbf{x}) = [f_1(\mathbf{x}), \cdots, f_m(\mathbf{x})]^T$. Then the derivative of $\mathbf{f}(\mathbf{x})$ with respect to $\mathbf{x}$, called the **Jacobian matrix** or **Jacobian** of $\mathbf{f}(\mathbf{x})$, is an $m \times n$ matrix denoted by

$$\mathbf{J_f} = \begin{bmatrix} \frac{\partial f_1}{\partial x_1} & \cdots & \frac{\partial f_1}{\partial x_n} \\ \vdots & \vdots & \vdots \\ \frac{\partial f_m}{\partial x_1} & \cdots & \frac{\partial f_m}{\partial x_n} \end{bmatrix} = \begin{bmatrix} \nabla^T f_1(\mathbf{x}) \\ \vdots \\ \nabla^T f_m(\mathbf{x}) \end{bmatrix} = \begin{bmatrix} \mathbf{g}_{f_1}^T \\ \vdots \\ \mathbf{g}_{f_m}^T \end{bmatrix}.$$

□

The Jacobian of a vector function $\mathbf{f}(\mathbf{x})$ is sometimes denoted by $\partial \mathbf{f}(\mathbf{x})/\partial \mathbf{x}^T$.

Definition 5.4 *Hessian of a scalar function*

Let $\mathbf{x} = [x_1, \cdots, x_n]^T$ and let $f(\mathbf{x})$ be a scalar function of $\mathbf{x}$. Then the second derivative of $\mathbf{f}(\mathbf{x})$ with respect to $\mathbf{x}$, called the **Hessian matrix** or **Hessian** of $f(\mathbf{x})$, is an $n \times n$ matrix denoted by

$$\mathbf{H}_f = \begin{bmatrix} \frac{\partial^2 f}{\partial x_1^2} & \frac{\partial^2 f}{\partial x_1 \partial x_2} & \cdots & \frac{\partial^2 f}{\partial x_1 \partial x_n} \\ \frac{\partial^2 f}{\partial x_2 \partial x_1} & \frac{\partial^2 f}{\partial x_2^2} & \cdots & \frac{\partial^2 f}{\partial x_2 \partial x_n} \\ \vdots & \vdots & \ddots & \vdots \\ \frac{\partial^2 f}{\partial x_n \partial x_1} & \frac{\partial^2 f}{\partial x_n \partial x_2} & \cdots & \frac{\partial^2 f}{\partial x_n^2} \end{bmatrix}.$$

In other words, the Hessian can be expressed as

$$\mathbf{H}_f = \begin{bmatrix} \frac{\partial}{\partial x_1}\left(\frac{\partial f}{\partial x_1}\right) & \cdots & \frac{\partial}{\partial x_n}\left(\frac{\partial f}{\partial x_1}\right) \\ \vdots & \vdots & \vdots \\ \frac{\partial}{\partial x_1}\left(\frac{\partial f}{\partial x_n}\right) & \cdots & \frac{\partial}{\partial x_n}\left(\frac{\partial f}{\partial x_n}\right) \end{bmatrix} = \begin{bmatrix} \nabla^T \frac{\partial f}{\partial x_1} \\ \vdots \\ \nabla^T \frac{\partial f}{\partial x_n} \end{bmatrix} = \mathbf{J}_{\mathbf{g}_f}.$$

□

The preceding equation states that the Hessian of a scalar function $f(\mathbf{x})$ is the Jacobian of the gradient of the function. Other commonly used notation for a Hessian matrix is $\nabla^2 f(\mathbf{x})$ or $\frac{\partial^2 f(\mathbf{x})}{\partial \mathbf{x} \partial \mathbf{x}^T}$. Note that $\mathbf{H}_f$ is symmetric if $f(\mathbf{x})$ is continuous.

Example 5.1 *Gradient of a linear function*

Let $\mathbf{c} = [c_1, \cdots, c_n]^T$ and $\mathbf{x} = [x_1, \cdots, x_n]^T$. Then the gradient of a linear scalar function $f(\mathbf{x}) = \mathbf{c}^T\mathbf{x} = \mathbf{x}^T\mathbf{c}$ is

$$\mathbf{g}_f = \nabla f(\mathbf{x}) = \mathbf{c}.$$

□

Definition 5.5 *Quadratic form*

Let $\mathbf{A} = [a_{ij}]_{n \times n}$ be square matrix and $\mathbf{x} = [x_1, \cdots, x_n]^T$ be a column vector. Then the **quadratic form** in $\mathbf{x}$ with matrix $\mathbf{A}$ is

$$\mathbf{x}^T\mathbf{A}\mathbf{x} = \sum_{i=1}^{n} a_{ii}x_i^2 + \sum_{i=1}^{n} \sum_{j=1, j\neq i}^{n} a_{ij}x_i x_j.$$

□

Note that $\mathbf{A}$ can be assumed symmetric without loss of generality; this is shown by the following identity:

$$\mathbf{x}^T\mathbf{A}\mathbf{x} = \mathbf{x}^T \underbrace{\frac{\mathbf{A} + \mathbf{A}^T}{2}}_{\text{symmetric}} \mathbf{x}.$$

□

Lemma 5.4 *Gradient and Hessian of a quadratic form*

Suppose $f(\mathbf{x}) = \mathbf{x}^T\mathbf{A}\mathbf{x}$ is the quadratic form in Definition 5.5. Then the gradient vector is

$$\mathbf{g}_f = \begin{cases} (\mathbf{A} + \mathbf{A}^T)\mathbf{x}, & \text{if } \mathbf{A} \text{ is not symmetric,} \\ 2\mathbf{A}\mathbf{x}, & \text{if } \mathbf{A} \text{ is symmetric.} \end{cases}$$

The Hessian matrix is

$$\mathbf{H}_f = \begin{cases} \mathbf{A} + \mathbf{A}^T, & \text{if } \mathbf{A} \text{ is not symmetric,} \\ 2\mathbf{A}, & \text{if } \mathbf{A} \text{ is symmetric.} \end{cases}$$

□

Definition 5.6 *Positive definite matrices*

A matrix $\mathbf{A}$ is **positive definite**, denoted by $\mathbf{A} > 0$, if $\mathbf{x}^T\mathbf{A}\mathbf{x} > 0$ for all $\mathbf{x} \neq 0$. Alternatively, $\mathbf{A} > 0$ if all its eigenvalues are positive.

□

Conversely, a matrix $\mathbf{A}$ is negative definite if $\mathbf{x}^T\mathbf{A}\mathbf{x} < 0$ for all $\mathbf{x} \neq 0$, or if all its eigenvalues are negative.

Lemma 5.5 *Optimum of a quadratic function*

Any quadratic functions in $f(\mathbf{x})$ can be expressed in matrix notation as follows:

$$f(\mathbf{x}) = \mathbf{x}^T\mathbf{A}\mathbf{x} + 2\mathbf{b}^T\mathbf{x} + c, \tag{5.1}$$

where $\mathbf{A}$ is a symmetric matrix. If $\mathbf{A}$ is positive definite, then $f(\mathbf{x})$ achieves its minimum at $\mathbf{x} = -\mathbf{A}^{-1}\mathbf{b}$. On the other hand, $f(\mathbf{x})$ achieves its maximum at $\mathbf{x} = -\mathbf{A}^{-1}\mathbf{b}$ if $\mathbf{A}$ is negative definite.

□

The preceding lemma can be verified directly by setting the derivative of $f(\mathbf{x})$ to zero. Another method for finding the minimum of $f(\mathbf{x})$ is explored in Exercise 2.

The definitions of the gradient, the Jacobian and the Hessian can lead to a number of identities. The following are formulas for finding gradient vectors (with respect to $\mathbf{x}$) of scalar functions. These useful formulas are to be referred in the subsequent chapters.

$$\nabla(\mathbf{x}^T\mathbf{y}) = \nabla(\mathbf{y}^T\mathbf{x}) = \mathbf{y} \tag{5.2}$$

$$\nabla(\mathbf{x}^T\mathbf{x}) = 2\mathbf{x} \tag{5.3}$$

$$\nabla(\mathbf{x}^T\mathbf{A}\mathbf{y}) = \mathbf{A}\mathbf{y} \tag{5.4}$$

$$\nabla(\mathbf{y}^T\mathbf{A}\mathbf{x}) = \mathbf{A}^T\mathbf{y} \tag{5.5}$$

$$\nabla(\mathbf{x}^T\mathbf{A}\mathbf{x}) = (\mathbf{A} + \mathbf{A}^T)\mathbf{x} \tag{5.6}$$

$$\nabla[\mathbf{f}^T(\mathbf{x})\mathbf{g}(\mathbf{x})] = \nabla[\mathbf{g}^T(\mathbf{x})\mathbf{f}(\mathbf{x})] = \mathbf{J}_\mathbf{f}^T\mathbf{g} + \mathbf{J}_\mathbf{g}^T\mathbf{f} \tag{5.7}$$

$$\nabla[\mathbf{g}^T(\mathbf{x})\mathbf{Q}\mathbf{g}(\mathbf{x})] = 2\mathbf{J}_\mathbf{g}^T\mathbf{Q}\mathbf{g}(\mathbf{x}) \text{ if } \mathbf{Q} \text{ is symmetric} \tag{5.8}$$

The following matrix inversion formula is useful when we want to find the inverse of a matrix of a specific form.

Lemma 5.6 *Matrix inversion formula*

Let $\mathbf{A}$ and $\mathbf{I} + \mathbf{C}\mathbf{A}^{-1}\mathbf{B}$ be nonsingular square matrices. Then

$$(\mathbf{A} + \mathbf{B}\mathbf{C})^{-1} = \mathbf{A}^{-1} - \mathbf{A}^{-1}\mathbf{B}(\mathbf{I} + \mathbf{C}\mathbf{A}^{-1}\mathbf{B})^{-1}\mathbf{C}\mathbf{A}^{-1}. \tag{5.9}$$

Proof: By direct substitution, we have

$$\begin{aligned}
& (\mathbf{A} + \mathbf{B}\mathbf{C})[\mathbf{A}^{-1} - \mathbf{A}^{-1}\mathbf{B}(\mathbf{I} + \mathbf{C}\mathbf{A}^{-1}\mathbf{B})^{-1}\mathbf{C}\mathbf{A}^{-1}] \\
= \; & (\mathbf{A} + \mathbf{B}\mathbf{C})\mathbf{A}^{-1} - (\mathbf{A} + \mathbf{B}\mathbf{C})\mathbf{A}^{-1}\mathbf{B}(\mathbf{I} + \mathbf{C}\mathbf{A}^{-1}\mathbf{B})^{-1}\mathbf{C}\mathbf{A}^{-1} \\
= \; & \mathbf{I} + \mathbf{B}\mathbf{C}\mathbf{A}^{-1} - (\mathbf{B} + \mathbf{B}\mathbf{C}\mathbf{A}^{-1}\mathbf{B})(\mathbf{I} + \mathbf{C}\mathbf{A}^{-1}\mathbf{B})^{-1}\mathbf{C}\mathbf{A}^{-1} \\
= \; & \mathbf{I} + \mathbf{B}\mathbf{C}\mathbf{A}^{-1} - \mathbf{B}(\mathbf{I} + \mathbf{C}\mathbf{A}^{-1}\mathbf{B})(\mathbf{I} + \mathbf{C}\mathbf{A}^{-1}\mathbf{B})^{-1}\mathbf{C}\mathbf{A}^{-1} \\
= \; & \mathbf{I} + \mathbf{B}\mathbf{C}\mathbf{A}^{-1} - \mathbf{B}\mathbf{C}\mathbf{A}^{-1} \\
= \; & \mathbf{I}.
\end{aligned}$$

□

Almost all derivative-based optimization techniques (see Chapter 6) employ the concept of the Taylor series expansion, which is defined next.

Definition 5.7 *Taylor series expansion*

Let $f(\mathbf{x})$ be a real-valued differentiable scalar function of a vector $\mathbf{x} = [\mathbf{x_1}, \cdots, \mathbf{x_n}]^\mathbf{T}$. Then the **Taylor series expansion** of $f(\cdot)$ at $\mathbf{x}$, with respect to a small deviation $\mathbf{d} = [d_1, \cdots, d_n]^T$, can be expressed as

$$f(\mathbf{x} + \mathbf{d}) = f(\mathbf{x}) + \sum_{i=1}^{n} \frac{\partial f(\mathbf{x})}{\partial x_i} d_i + \frac{1}{2} \sum_{i=1}^{n} \sum_{j=1}^{n} \frac{\partial^2 f(\mathbf{x})}{\partial x_i \partial x_j} d_i d_j + \text{H.O.T.} \tag{5.10}$$

H.O.T. means "higher-order terms"; they are seldom used in practical situations since they can be neglected if the deviation $||\mathbf{d}||$ is sufficiently small. If we use the gradient $\mathbf{g}$ and the Hessian $\mathbf{H}$ of the function $f(\cdot)$ at $\mathbf{x}$, and omit the higher order terms, then the preceding equation can be rewritten as

$$f(\mathbf{x} + \mathbf{d}) \approx f(\mathbf{x}) + \mathbf{g}^T\mathbf{d} + \frac{1}{2}\mathbf{d}^T\mathbf{H}\mathbf{d}. \tag{5.11}$$

This states that when the deviation $\mathbf{d}$ from $\mathbf{x}$ is small, then the behavior of function $f(\cdot)$ near $\mathbf{x}$ is close to a quadratic function in terms of the deviation $\mathbf{d}$.

□

An interesting example for demonstrating Taylor series expansion of a single variable function is the MATLAB file `taylor.m`, which is available via FTP or WWW (see page xxiii). After executing this program within MATLAB you can click any point in the figure window to display the first-order (a straight line), second-order (a hyperbola), and third-order Taylor series approximations to the original curve. Moreover, you can click and drag on any of the ten control points of the curve to change its shape interactively. (In fact, the curve is the ninth-order least-squares polynomial of these ten control points. This is explained in the next section.)

5.3 LEAST-SQUARES ESTIMATOR

In the general least-squares problem, the output of a linear model y is given by the linearly parameterized expression

$$y = \theta_1 f_1(\mathbf{u}) + \theta_2 f_2(\mathbf{u}) + \cdots + \theta_n f_n(\mathbf{u}), \tag{5.12}$$

where $\mathbf{u} = [u_1, \cdots, u_p]^T$ is the model's input vector, $f_1, \cdots, f_n$ are known functions of $\mathbf{u}$, and $\theta_1, \cdots, \theta_n$ are unknown parameters to be estimated. In statistics, the task of fitting data using a linear model is referred to as **linear regression.** Thus Equation (5.12) is also called the **regression function**, and the θ_i's are called the **regression coefficients.**

To identify the unknown parameters θ_i, usually we have to perform experiments to obtain a **training data set** composed of data pairs $\{(\mathbf{u}_i; y_i), i = 1, \cdots, m\}$; they represent desired input-output pairs of the target system to be modeled. Substituting each data pair into Equation (5.12) yields a set of m linear equations:

$$\left\{ \begin{array}{lcl} f_1(\mathbf{u}_1)\theta_1 + f_2(\mathbf{u}_1)\theta_2 + \cdots + f_n(\mathbf{u}_1)\theta_n & = & y_1, \\ f_1(\mathbf{u}_2)\theta_1 + f_2(\mathbf{u}_2)\theta_2 + \cdots + f_n(\mathbf{u}_2)\theta_n & = & y_2, \\ \vdots & \vdots & \vdots \\ f_1(\mathbf{u}_m)\theta_1 + f_2(\mathbf{u}_m)\theta_2 + \cdots + f_n(\mathbf{u}_m)\theta_n & = & y_m. \end{array} \right. \tag{5.13}$$

Using matrix notation, we can rewrite the preceding equations in a concise form:

$$\mathbf{A}\boldsymbol{\theta} = \mathbf{y}, \tag{5.14}$$

where $\mathbf{A}$ is an $m \times n$ matrix (sometimes called the **design matrix**):

$$\mathbf{A} = \begin{bmatrix} f_1(\mathbf{u}_1) & \cdots & f_n(\mathbf{u}_1) \\ \vdots & \vdots & \vdots \\ f_1(\mathbf{u}_m) & \cdots & f_n(\mathbf{u}_m) \end{bmatrix},$$

$\boldsymbol{\theta}$ is an $n \times 1$ unknown parameter vector:

$$\boldsymbol{\theta} = \begin{bmatrix} \theta_1 \\ \vdots \\ \theta_n \end{bmatrix},$$

and $\mathbf{y}$ is an $m \times 1$ output vector:

$$\mathbf{y} = \begin{bmatrix} y_1 \\ \vdots \\ y_m \end{bmatrix}.$$

The ith row of the joint data matrix $[\mathbf{A} \vdots \mathbf{y}]$, denoted by $[\mathbf{a}_i^T \vdots y_i]$, is related to the ith input-output data pair $(\mathbf{u}_i;\ y_i)$ through

$$\mathbf{a}_i^T = [f_1(\mathbf{u}_i),\ \cdots,\ f_n(\mathbf{u}_i)].$$

Since most of our calculation is based on matrices $\mathbf{A}$ and $\mathbf{y}$, sometimes we loosely refer to $(\mathbf{a}_i^T; y_i)$ as the ith data pair of the training data set.

To identify uniquely the unknown vector $\boldsymbol{\theta}$, it is necessary that $m \geq n$. If $\mathbf{A}$ is square $(m = n)$ and nonsingular, then we can solve $\mathbf{x}$ from Equation (5.14) by

$$\boldsymbol{\theta} = \mathbf{A}^{-1}\mathbf{y}. \tag{5.15}$$

However, usually m is greater than n, indicating that we have more data pairs than fitting parameters. In this case, an exact solution satisfying all the m equations is not always possible, since the data might be contaminated by noise, or the model might not be appropriate for describing the target system. Thus Equation (5.14) should be modified by incorporating an **error vector e** to account for random noise or modeling error, as follows:

$$\mathbf{A}\boldsymbol{\theta} + \mathbf{e} = \mathbf{y}. \tag{5.16}$$

Now, instead of finding the exact solution to Equation (5.14), we want to search for a $\boldsymbol{\theta} = \hat{\boldsymbol{\theta}}$ which minimizes the **sum of squared error** defined by

$$E(\boldsymbol{\theta}) = \sum_{i=1}^{m} (y_i - \mathbf{a}_i^T\boldsymbol{\theta})^2 = \mathbf{e}^T\mathbf{e} = (\mathbf{y} - \mathbf{A}\boldsymbol{\theta})^T(\mathbf{y} - \mathbf{A}\boldsymbol{\theta}), \tag{5.17}$$

where $\mathbf{e} = \mathbf{y} - \mathbf{A}\boldsymbol{\theta}$ is the error vector produced by a specific choice of $\boldsymbol{\theta}$. Note that $E(\boldsymbol{\theta})$ is in quadratic form and has a unique minimum at $\boldsymbol{\theta} = \hat{\boldsymbol{\theta}}$. The following theorem states a necessary condition satisfied by the least-squares estimator $\hat{\boldsymbol{\theta}}$.

Theorem 5.1 *Least-squares estimator*

The squared error in Equation (5.17) is minimized when $\boldsymbol{\theta} = \hat{\boldsymbol{\theta}}$, called the **least-squares estimator** (**LSE** for short), which satisfies the **normal equation**

$$\mathbf{A}^T\mathbf{A}\hat{\boldsymbol{\theta}} = \mathbf{A}^T\mathbf{y}. \tag{5.18}$$

If $\mathbf{A}^T\mathbf{A}$ is nonsingular, $\hat{\boldsymbol{\theta}}$ is unique and is given by

$$\hat{\boldsymbol{\theta}} = (\mathbf{A}^T\mathbf{A})^{-1}\mathbf{A}^T\mathbf{y}. \tag{5.19}$$

Proof: There are a number of methods available in the literature for finding the least-squares estimator for Equation (5.14). One straightforward approach is to set the derivative of $E(\boldsymbol{\theta})$ with respect to $\boldsymbol{\theta}$ equal to zero. Noting that $\boldsymbol{\theta}^T\mathbf{A}^T\mathbf{y} = \mathbf{y}^T\mathbf{A}\boldsymbol{\theta}$ is a scalar, we can expand $E(\boldsymbol{\theta})$:

$$E(\boldsymbol{\theta}) = (\mathbf{y}^T - \boldsymbol{\theta}^T\mathbf{A}^T)(\mathbf{y} - \mathbf{A}\boldsymbol{\theta}) = \boldsymbol{\theta}^T\mathbf{A}^T\mathbf{A}\boldsymbol{\theta} - 2\mathbf{y}^T\mathbf{A}\boldsymbol{\theta} + \mathbf{y}^T\mathbf{y}. \tag{5.20}$$

Then the derivative of $E(\boldsymbol{\theta})$ is

$$\frac{\partial E(\boldsymbol{\theta})}{\partial \boldsymbol{\theta}} = 2\mathbf{A}^T\mathbf{A}\boldsymbol{\theta} - 2\mathbf{A}^T\mathbf{y}. \tag{5.21}$$

By setting $\frac{\partial E(\boldsymbol{\theta})}{\partial \boldsymbol{\theta}} = 0$ at $\boldsymbol{\theta} = \hat{\boldsymbol{\theta}}$, we obtain the normal equation

$$\mathbf{A}^T\mathbf{A}\hat{\boldsymbol{\theta}} = \mathbf{A}^T\mathbf{y}. \tag{5.22}$$

If $\mathbf{A}^T\mathbf{A}$ is nonsingular, then $\hat{\boldsymbol{\theta}}$ can be solved uniquely:

$$\hat{\boldsymbol{\theta}} = (\mathbf{A}^T\mathbf{A})^{-1}\mathbf{A}^T\mathbf{y}. \tag{5.23}$$

□

LSE can also be obtained directly from Lemma 5.5 since $E(\boldsymbol{\theta})$ in Equation (5.20) is a quadratic function of $\boldsymbol{\theta}$. The least-squared error achieved by $\boldsymbol{\theta} = \hat{\boldsymbol{\theta}}$ can be found to be

$$E(\hat{\boldsymbol{\theta}}) = (\mathbf{y} - \mathbf{A}\hat{\boldsymbol{\theta}})^T(\mathbf{y} - \mathbf{A}\hat{\boldsymbol{\theta}}) = \mathbf{y}^T\mathbf{y} - \mathbf{y}^T\mathbf{A}(\mathbf{A}^T\mathbf{A})^{-1}\mathbf{A}^T\mathbf{y}. \tag{5.24}$$

However, if $\mathbf{A}^T\mathbf{A}$ is singular, then the LSE is not unique and we have to employ the concept of **generalized inverse** to find $\hat{\boldsymbol{\theta}}$. Readers interested in this aspect of the problem will find a detailed treatment of it in the literature, such as in Chapter 5 of [10]. Without loss of generality, we shall assume that $\mathbf{A}^T\mathbf{A}$ is nonsingular throughout this chapter.

The foregoing derivation is based on the assumption that every element of the error vector $\mathbf{e}$ has the same weight toward the overall squared error. A further generalization is to let each error term be weighted differently. Specifically, let $\mathbf{W}$ be the desired weighting matrix, which is symmetric and positive definite. Then the weighted squared error is

$$E_{\mathbf{W}}(\boldsymbol{\theta}) = (\mathbf{y} - \mathbf{A}\boldsymbol{\theta})^T\mathbf{W}(\mathbf{y} - \mathbf{A}\boldsymbol{\theta}). \tag{5.25}$$

Table 5.1. *Training data for the spring example.*

Experiment	Force (newtons)	Length of Spring (inches)
1	1.1	1.5
2	1.9	2.1
3	3.2	2.5
4	4.4	3.3
5	5.9	4.1
6	7.4	4.6
7	9.2	5.0

Minimizing $E_{\mathbf{W}}(\boldsymbol{\theta})$ with respect to $\boldsymbol{\theta}$ yields the **weighted least-squares estimator $\hat{\boldsymbol{\theta}}_{\mathbf{W}}$**:

$$\hat{\boldsymbol{\theta}}_{\mathbf{W}} = (\mathbf{A}^T\mathbf{W}\mathbf{A})^{-1}\mathbf{A}^T\mathbf{W}\mathbf{y}. \tag{5.26}$$

Obviously, $\hat{\boldsymbol{\theta}}_{\mathbf{W}}$ reduces to $\hat{\boldsymbol{\theta}}$ when $\mathbf{W}$ is chosen as an identity matrix.

Example 5.2 *Least-squares estimator*

From Hooke's law, we know that when a force is applied to a spring constructed of uniform material, the change in the length of the spring is proportional to the force applied. Therefore, we have the following expression governing the relationship between a spring's length l and an applied force f:

$$l = k_0 + k_1 f, \tag{5.27}$$

where k_0 represents the length of the spring with no force applied and k_1 (the spring constant) represents the change in length when a unit of force is applied. To identify k_0 and k_1 for a particular spring, ideally we can apply two different forces and observe the corresponding lengths of the spring. Then the values of k_0 and k_1 can be determined uniquely by solving two simultaneous linear equations in two unknowns. However, this approach is sensitive to measurement error or noise, and therefore is not preferred. To identify k_0 and k_1 accurately, usually we apply several different forces and record the corresponding lengths of the spring. Here we suppose that the data pairs obtained are as listed in Table 5.1.

Substituting each row of Table 5.1 into Equation (5.27) and incorporating an error vector $\mathbf{e}$, we have

$$\underbrace{\begin{bmatrix} 1 & 1.1 \\ 1 & 1.9 \\ 1 & 3.2 \\ 1 & 4.4 \\ 1 & 5.9 \\ 1 & 7.4 \\ 1 & 9.2 \end{bmatrix}}_{\mathbf{A}} \begin{bmatrix} k_0 \\ k_1 \end{bmatrix} + \underbrace{\begin{bmatrix} e_1 \\ e_2 \\ e_3 \\ e_4 \\ e_5 \\ e_6 \\ e_7 \end{bmatrix}}_{\mathbf{e}} . = \underbrace{\begin{bmatrix} 1.5 \\ 2.1 \\ 2.5 \\ 3.3 \\ 4.1 \\ 4.6 \\ 5.0 \end{bmatrix}}_{\mathbf{y}}$$

Therefore, the least-squares estimator of $[k_0,\ k_1]^T$ which minimizes $\mathbf{e}^T\mathbf{e} = \sum_{i=1}^{7} e_i^2$ is equal to

$$\begin{bmatrix} \hat{k}_0 \\ \hat{k}_1 \end{bmatrix} = (\mathbf{A}^T\mathbf{A})^{-1}\mathbf{A}^T\mathbf{y} = \begin{bmatrix} 1.20 \\ 0.44 \end{bmatrix}.$$

Figure 5.2(a) shows the least-squares line that minimizes the squared error. It is obvious that as we have more data points, the resulting least-squares estimator is less susceptible to measurement error or noise.

□

In the preceding example, we determine the structure of the model [Equation (5.27)] according to Hooke's law. If the current model is not suitable for describing the spring's behavior, then we can increase the model's degrees of freedom by introducing terms of higher orders:

$$l = k_0 + k_1 f + k_2 f^2 + \cdots + k_n f^n. \tag{5.28}$$

The same identification procedure can be performed to find the LSE for $\boldsymbol{\theta} = [k_0, k_1, \cdots, k_n]^T$, which results in a **least-squares polynomial** that minimizes the squared error. Figures 5.2(b) through 5.2 (d) are least-squares polynomials with order 2, 3, and 4, respectively.

Note that the squared error always decreases as the order of the least-squares polynomial increases. However, although it fits the training data better, a polynomial with a higher order does not always reflect the true characteristics of the system in question. This caveat is demonstrated in Figures 5.2(c) and 5.2(d), where the spring's length is getting shorter when subject to a force 10 N or more. This is an obvious contradiction to our empirical knowledge of a spring's behavior.

Another demonstration of least-squares polynomials is the MATLAB program `taylor.m`, where you can click on any of the ten control points to change the shape of the ninth-order least-squares polynomial. The least-squares polynomial always fit the data perfectly, but it is not robust—a small amount of noise in the data set could change the whole curve dramatically and make it untrustworthy.

An easy way to select a polynomial of suitable order is to apply another input-output data set, called the **validating** or **test data set**, that was not used in

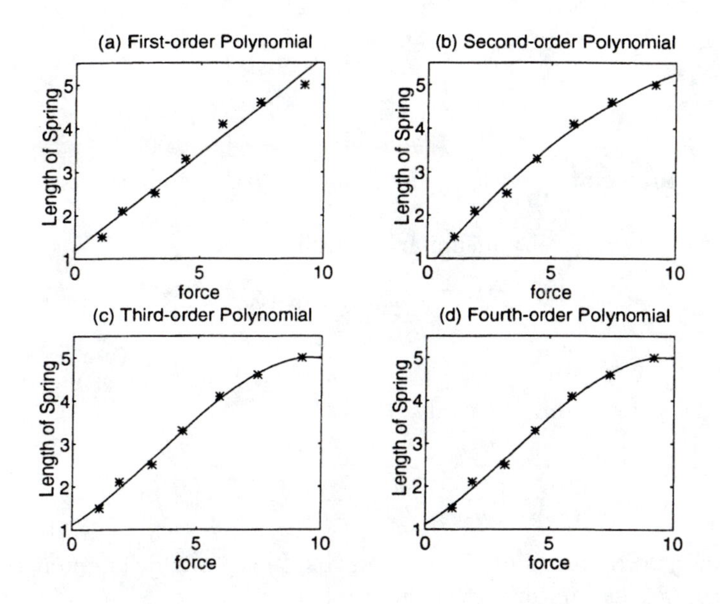

Figure 5.2. *Fitting data through least-squares polynomials.* (MATLAB file: `spring.m`)

constructing the least-squares polynomial. This test data set can verify the **generalization capability** of the resulting models and thus provide an unbiased index for selecting the best model. Many other approaches to determining a model's order have been proposed in the literature. For a thorough treatment, see, for example, Chapter 6 of [2].

If the target system has q outputs, expressed as $\mathbf{y} = [y_1, \cdots y_q]^T$ with $q > 1$, then we have a set of linear equations in matrix form:

$$\mathbf{A\Theta} + \mathbf{E} = \mathbf{Y},$$

where $\mathbf{A}$ is an $m \times n$ matrix, as introduced previously:

$$\mathbf{A} = \begin{bmatrix} f_1(\mathbf{u}_1) & \cdots & f_n(\mathbf{u}_1) \\ \vdots & \vdots & \vdots \\ f_1(\mathbf{u}_m) & \cdots & f_n(\mathbf{u}_m) \end{bmatrix},$$

$\mathbf{\Theta}$ is an $n \times q$ unknown parameter matrix:

$$\mathbf{\Theta} = \begin{bmatrix} \theta_{11} & \cdots & \theta_{1q} \\ \vdots & \vdots & \vdots \\ \theta_{n1} & \cdots & \theta_{nq} \end{bmatrix},$$

and

$$\mathbf{Y} = \begin{bmatrix} y_{11} & \cdots & y_{1q} \\ \vdots & \vdots & \vdots \\ y_{m1} & \cdots & y_{mq} \end{bmatrix}$$

is an $m \times q$ output matrix with y_{ij} denoting the jth output value in the ith data pairs.

Now we want to minimize a similar squared error

$$\begin{aligned} E(\mathbf{\Theta}) &= \sum_{i=1}^{m} \sum_{j=1}^{q} e_{ij}^2 \\ &= \sum_{j=1}^{q} \left(\sum_{i=1}^{m} e_{ij}^2 \right). \end{aligned}$$

Note that $\sum_{i=1}^{m} e_{ij}^2$ is the squared length of the jth column of matrix $\mathbf{E}$, which depends on the jth column of $\mathbf{\Theta}$ only. Hence

$$\min_{\mathbf{\Theta}} \sum_{j=1}^{q} \left(\sum_{i=1}^{m} e_{ij} \right) = \sum_{j=1}^{q} \left(\min_{\boldsymbol{\theta}_j} \sum_{i=1}^{m} e_{ij} \right),$$

where $\boldsymbol{\theta}_j$ is the jth column of $\mathbf{\Theta}$. In other words, $\boldsymbol{\theta}_j = \hat{\boldsymbol{\theta}}_j$, which minimizes $\sum_{i=1}^{m} e_{ij}^2$, is the least-squares estimator to the subproblem

$$\mathbf{A}\boldsymbol{\theta}_j + \mathbf{e}_j = \mathbf{y}_j,$$

where $\mathbf{e}_j$ and $\mathbf{y}_j$ are the jth columns of $\mathbf{E}$ and $\mathbf{Y}$, respectively. As a result,

$$\hat{\boldsymbol{\theta}}_j = (\mathbf{A}^T\mathbf{A})^{-1}\mathbf{A}^T\mathbf{y}_j,$$

and

$$\hat{\mathbf{\Theta}} = (\mathbf{A}^T\mathbf{A})^{-1}\mathbf{A}^T\mathbf{Y}.$$

This implies that the optimality occurs when each column of $\mathbf{A\Theta}$ is equal to the projection of the corresponding column of $\mathbf{Y}$ onto the space spanned by the columns of $\mathbf{A}$ (see the following section on the geometric interpretation of the LSE), so the calculation of each $\hat{\boldsymbol{\theta}}_j$ can be performed independently.

5.4 GEOMETRIC INTERPRETATION OF LSE

Now we shall discuss the geometric interpretation of the least-squares estimator. Let $\mathbf{A}$ be expressed as a row of n column vectors of size $m \times 1$, as follows:

$$\mathbf{A} = \begin{bmatrix} \mathbf{a}_1 & \cdots & \mathbf{a}_n \end{bmatrix}.$$

Then we have

$$\mathbf{A}\boldsymbol{\theta} = \begin{bmatrix} \mathbf{a}_1 & \cdots & \mathbf{a}_n \end{bmatrix} \begin{bmatrix} \theta_1 \\ \vdots \\ \theta_n \end{bmatrix} = \theta_1 \begin{bmatrix} \mathbf{a}_1 \end{bmatrix} + \cdots + \theta_n \begin{bmatrix} \mathbf{a}_n \end{bmatrix}. \tag{5.29}$$

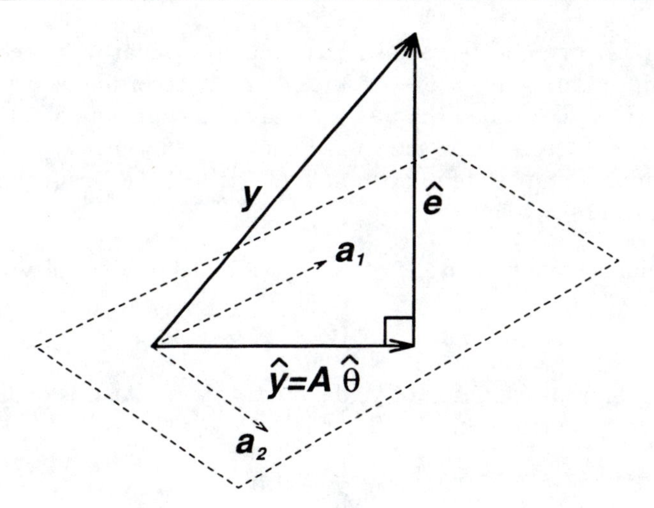

Figure 5.3. *Geometrical interpretation of the least-squares estimator.*

In other words, $\mathbf{A}\boldsymbol{\theta}$ is a linear combination of the basis vector $\{\mathbf{a}_1, \ldots, \mathbf{a}_n\}$ in an m-dimensional space. For $\mathbf{A}\boldsymbol{\theta}$ to approximate $\mathbf{y}$ in a least-squares sense, clearly $\mathbf{A}\boldsymbol{\theta}$ should be equal to the projection of $\mathbf{y}$ onto the space spanned by $\{\mathbf{a}_1, \ldots, \mathbf{a}_n\}$.

To illustrate this concept, Figure 5.3 shows the situation where $n = 2$ and $m = 3$. Here we have

$$\mathbf{A} = \begin{bmatrix} \mathbf{a}_1 & \mathbf{a}_2 \end{bmatrix}$$

and $\mathbf{A}\boldsymbol{\theta} = \theta_1\mathbf{a}_1 + \theta_2\mathbf{a}_2$. Note that $\mathbf{A}\boldsymbol{\theta}$ always stays on the plane spanned by $\mathbf{a}_1$ and $\mathbf{a}_2$. Thus for $\mathbf{e} = \mathbf{y} - \mathbf{A}\boldsymbol{\theta}$ to achieve a minimum length, $\mathbf{A}\boldsymbol{\theta}$ must be equal to the projection of $\mathbf{y}$ onto the plane spanned by $\mathbf{a}_1$ and $\mathbf{a}_2$. This optimality occurs when $\mathbf{e} = \mathbf{y} - \mathbf{A}\boldsymbol{\theta}$ is orthogonal to both $\mathbf{a}_1$ and $\mathbf{a}_2$, and it is called **principle of orthogonality**. In symbols,

$$\mathbf{a}_1^T(\mathbf{y} - \mathbf{A}\hat{\boldsymbol{\theta}}) = 0, \tag{5.30}$$

$$\mathbf{a}_2^T(\mathbf{y} - \mathbf{A}\hat{\boldsymbol{\theta}}) = 0. \tag{5.31}$$

These two conditions reduce to

$$\mathbf{A}^T(\mathbf{y} - \mathbf{A}\hat{\boldsymbol{\theta}}) = 0, \tag{5.32}$$

which is exactly the normal equation [see Equation (5.22), Theorem 5.1] derived by direct differentiation.

If we use $\text{proj}_{\mathbf{A}}(\mathbf{y})$ to denote the projection of vector $\mathbf{y}$ onto the space spanned by the columns of $\mathbf{A}$, then

$$\text{proj}_{\mathbf{A}}(\mathbf{y}) = \mathbf{A}\hat{\boldsymbol{\theta}} = \mathbf{A}(\mathbf{A}^T\mathbf{A})^{-1}\mathbf{A}^T\mathbf{y}. \tag{5.33}$$

$\mathbf{H} = \mathbf{A}(\mathbf{A}^T\mathbf{A})^{-1}\mathbf{A}^T$ can be viewed as a **projection operator**; the complementary **orthogonal operator** is $\mathbf{M} = \mathbf{I} - \mathbf{A}(\mathbf{A}^T\mathbf{A})^{-1}\mathbf{A}^T$. Accordingly, we have $\hat{\mathbf{y}} = \mathbf{H}\mathbf{y}$ and $\hat{\mathbf{e}} = \mathbf{M}\mathbf{y}$. ($\mathbf{H}$ is also called the **hat operator** since it puts a hat on y.)

The geometric interpretation shown in Figure 5.3 is intuitive, yet it is indispensable in confirming a number of important properties of the least-squares estimator $\hat{\boldsymbol{\theta}}$, some of which are as follows:

- The minimal error measure is equal to the inner product of $\mathbf{y}$ and $\hat{\mathbf{e}}$ at optimality. That is,

$$E(\hat{\boldsymbol{\theta}}) = \mathbf{y}^T\hat{\mathbf{e}}. \tag{5.34}$$

- The optimal approximation of $\mathbf{y}$, denoted by $\hat{\mathbf{y}}$ ($= \mathbf{A}\hat{\boldsymbol{\theta}}$), is orthogonal to the minimal-length error vector $\hat{\mathbf{e}}$. That is,

$$\hat{\mathbf{y}}^T\hat{\mathbf{e}} = 0. \tag{5.35}$$

- The minimal-length error vector $\hat{\mathbf{e}}$ is equal to the result of applying the orthogonal operator $\mathbf{M}$ on $\mathbf{e}$. That is,

$$\hat{\mathbf{e}} = \mathbf{M}\mathbf{e} = [\mathbf{I} - \mathbf{A}(\mathbf{A}^T\mathbf{A})^{-1}\mathbf{A}^T](\mathbf{y} - \mathbf{A}\boldsymbol{\theta}). \tag{5.36}$$

- Each column of $\mathbf{A}$ is invariant under the projection operator $\mathbf{H}$. That is,

$$\mathbf{H}\mathbf{A} = \mathbf{A}, \tag{5.37}$$

  or equivalently,

$$\mathbf{M}\mathbf{A} = 0. \tag{5.38}$$

- The projection operator $\mathbf{H} = \mathbf{A}(\mathbf{A}^T\mathbf{A})^{-1}\mathbf{A}^T$ is idempotent. That is, for any integer k greater than one, $\mathbf{H}$ satisfies

$$\mathbf{H}^k = \mathbf{H}. \tag{5.39}$$

  In other words, for any vector $\mathbf{y}$, we have

$$\underbrace{\mathbf{H}\cdot\mathbf{H}\cdots\mathbf{H}}_{k \text{ times}}\mathbf{y} = \mathbf{H}\mathbf{y},$$

  which means that once a vector has been obtained as the projection onto the subspace spanned by the columns of $\mathbf{A}$, it cannot be changed by any further application of $\mathbf{H}$.

- The orthogonal operator $\mathbf{M} = \mathbf{I} - \mathbf{H} = \mathbf{I} - \mathbf{A}(\mathbf{A}^T\mathbf{A})^{-1}\mathbf{A}^T$ is also idempotent, for a reason similar to that given previously.

Mathematical derivation of these properties is left as Exercise 7.

5.5 RECURSIVE LEAST-SQUARES ESTIMATOR

The least-squares estimator derived in the previous section can be expressed as

$$\boldsymbol{\theta}_k = (\mathbf{A}^T\mathbf{A})^{-1}\mathbf{A}^T\mathbf{y}, \tag{5.40}$$

where we have left out the hat (ˆ) for simplicity. Here we assume the row dimensions for $\mathbf{A}$ and $\mathbf{y}$ are k; thus a subscript k is added in the preceding equation to denote the number of data pairs used for the estimator $\boldsymbol{\theta}$. The k also can be looked as a measure of time if the data pairs become available in sequential order. Suppose that a new data pair $(\mathbf{a}^T;\ y)$ becomes available as the $(k+1)$th entry in the data set. Then instead of using all the $k+1$ available data pairs to recalculate the least-squares estimator $\boldsymbol{\theta}_{k+1}$, we want to find a way of taking advantage of the $\boldsymbol{\theta}_k$ already available to obtain $\boldsymbol{\theta}_{k+1}$ with a minimum of effort. In other words, our task is to find a way of using the new data pair $(\mathbf{a}^T;\ y)$ to update $\boldsymbol{\theta}_k$ appropriately to find $\boldsymbol{\theta}_{k+1}$. This problem is called **recursive least-squares identification** and has been fully addressed in the literature [3, 7, 9].

Obviously, $\boldsymbol{\theta}_{k+1}$ can be expressed as

$$\boldsymbol{\theta}_{k+1} = \left(\begin{bmatrix}\mathbf{A}\\ \mathbf{a}^T\end{bmatrix}^T\begin{bmatrix}\mathbf{A}\\ \mathbf{a}^T\end{bmatrix}\right)^{-1}\begin{bmatrix}\mathbf{A}\\ \mathbf{a}^T\end{bmatrix}^T\begin{bmatrix}\mathbf{y}\\ y\end{bmatrix}. \tag{5.41}$$

To simplify the notation, we introduce two $n \times n$ matrices $\mathbf{P}_k$ and $\mathbf{P}_{k+1}$ defined by

$$\mathbf{P}_k = (\mathbf{A}^T\mathbf{A})^{-1} \tag{5.42}$$

$$\begin{aligned}\mathbf{P}_{k+1} &= \left(\begin{bmatrix}\mathbf{A}\\ \mathbf{a}^T\end{bmatrix}^T\begin{bmatrix}\mathbf{A}\\ \mathbf{a}^T\end{bmatrix}\right)^{-1}\\ &= \left(\begin{bmatrix}\mathbf{A}^T & \mathbf{a}\end{bmatrix}\begin{bmatrix}\mathbf{A}\\ \mathbf{a}^T\end{bmatrix}\right)^{-1}\\ &= (\mathbf{A}^T\mathbf{A} + \mathbf{a}\mathbf{a}^T)^{-1}.\end{aligned} \tag{5.43}$$

These two matrices are related by

$$\mathbf{P}_k^{-1} = \mathbf{P}_{k+1}^{-1} - \mathbf{a}\mathbf{a}^T. \tag{5.44}$$

Using $\mathbf{P}_k$ and $\mathbf{P}_{k+1}$, we have

$$\begin{cases}\boldsymbol{\theta}_k &= \mathbf{P}_k\mathbf{A}^T\mathbf{y},\\ \boldsymbol{\theta}_{k+1} &= \mathbf{P}_{k+1}(\mathbf{A}^T\mathbf{y} + \mathbf{a}y).\end{cases} \tag{5.45}$$

To express $\boldsymbol{\theta}_{k+1}$ in terms of $\boldsymbol{\theta}_k$, we have to eliminate $\mathbf{A}^T\mathbf{y}$ in Equation (5.45). From the first equation in (5.45), we have

$$\mathbf{A}^T\mathbf{y} = \mathbf{P}_k^{-1}\boldsymbol{\theta}_k. \tag{5.46}$$

By plugging this expression into the second equation in (5.45) and applying Equation (5.44), we have

$$\begin{aligned} \boldsymbol{\theta}_{k+1} &= \mathbf{P}_{k+1}(\mathbf{P}_k^{-1}\boldsymbol{\theta}_k + \mathbf{a}y) \\ &= \mathbf{P}_{k+1}[(\mathbf{P}_{k+1}^{-1} - \mathbf{a}\mathbf{a}^T)\boldsymbol{\theta}_k + \mathbf{a}y] \\ &= \boldsymbol{\theta}_k + \mathbf{P}_{k+1}\mathbf{a}(y - \mathbf{a}^T\boldsymbol{\theta}_k). \end{aligned} \tag{5.47}$$

Thus $\boldsymbol{\theta}_{k+1}$ can be expressed as a function of the old estimate $\boldsymbol{\theta}_k$ and the new data pairs $(\mathbf{a}^T;\ y)$. It is interesting to note that Equation (5.47) has an intuitive interpretation: The new estimator $\boldsymbol{\theta}_{k+1}$ is equal to the old estimator $\boldsymbol{\theta}_k$ plus a correcting term based on the new data $(\mathbf{a}^T;\ y)$; this correcting term is equal to an **adaptation gain vector** $\mathbf{P}_{k+1}\mathbf{a}$ multiplied by the prediction error produced by the old estimator—that is, $y - \mathbf{a}^T\boldsymbol{\theta}_k$.

We are not done yet, however. Calculating $\mathbf{P}_{k+1}$ by Equation (5.43) involves the inversion of an $n \times n$ matrix. This is computationally expensive and requires us to find an incremental formula for $\mathbf{P}_{k+1}$. From Equation (5.44), we have

$$\mathbf{P}_{k+1} = (\mathbf{P}_k^{-1} + \mathbf{a}\mathbf{a}^T)^{-1}. \tag{5.48}$$

Applying the matrix inversion formula in Lemma 5.6 with $\mathbf{A} = \mathbf{P}_k^{-1}$, $\mathbf{B} = \mathbf{a}$, and $C = \mathbf{a}^T$, we obtain the following incremental formula for $\mathbf{P}_{k+1}$:

$$\begin{aligned} \mathbf{P}_{k+1} &= \mathbf{P}_k - \mathbf{P}_k\mathbf{a}(\mathbf{I} + \mathbf{a}^T\mathbf{P}_k\mathbf{a})^{-1}\mathbf{a}^T\mathbf{P}_k \\ &= \mathbf{P}_k - \frac{\mathbf{P}_k\mathbf{a}\mathbf{a}^T\mathbf{P}_k}{1 + \mathbf{a}^T\mathbf{P}_k\mathbf{a}}. \end{aligned} \tag{5.49}$$

In summary, the **recursive least-squares estimator** for the problem of $\mathbf{A}\boldsymbol{\theta} = \mathbf{y}$, where the kth $(1 \le k \le m)$ row of $[\mathbf{A} \vdots \mathbf{y}]$, denoted by $[\mathbf{a}_k^T \vdots y_k]$, is sequentially obtained, can be calculated as follows:

$$\begin{cases} \mathbf{P}_{k+1} = \mathbf{P}_k - \dfrac{\mathbf{P}_k\mathbf{a}_{k+1}\mathbf{a}_{k+1}^T\mathbf{P}_k}{1 + \mathbf{a}_{k+1}^T\mathbf{P}_k\mathbf{a}_{k+1}}, \\ \boldsymbol{\theta}_{k+1} = \boldsymbol{\theta}_k + \mathbf{P}_{k+1}\mathbf{a}_{k+1}(y_{k+1} - \mathbf{a}_{k+1}^T\boldsymbol{\theta}_k), \end{cases} \tag{5.50}$$

where k ranges from 0 to $m-1$ and the overall LSE $\hat{\boldsymbol{\theta}}$ is equal to $\boldsymbol{\theta}_m$, the estimator using all m data pairs.

To start the algorithm in Equation (5.50), we need to select the initial values of $\boldsymbol{\theta}_0$ and $\mathbf{P}_0$. One way to avoid determining these initial values is to collect the first n data points and solve $\boldsymbol{\theta}_n$ and $\mathbf{P}_n$ directly from

$$\begin{cases} \mathbf{P}_n = (\mathbf{A}_n^T\mathbf{A}_n)^{-1}, \\ \boldsymbol{\theta}_n = \mathbf{P}_n\mathbf{A}_n^T\mathbf{y}_n, \end{cases}$$

where $[\mathbf{A}_n \vdots \mathbf{y}_n]$ is the data matrix composed of the first n data pairs. We can then start iterating the algorithm from the $(n+1)$th data point. However, sometimes it is more convenient to use the recursive formulas in Equation (5.50) throughout the identification process. To do so, notice that

$$\mathbf{P}_k = (\mathbf{P}_0 + \mathbf{A}_k^T \mathbf{A}_k)^{-1}, \tag{5.51}$$

and the corresponding $\boldsymbol{\theta}_k$ is (see Exercise 10)

$$\boldsymbol{\theta}_k = \mathbf{P}_k(\mathbf{A}_k \mathbf{y}_k + \mathbf{P}_0^{-1}\boldsymbol{\theta}_0), \tag{5.52}$$

where $[\mathbf{A}_k \vdots \mathbf{y}_k]$ is the data matrix composed of k data pairs. By choosing

$$\mathbf{P}_0 = \alpha \mathbf{I},$$

we have

$$\lim_{\alpha \to \infty} \mathbf{P}_0^{-1} = \lim_{\alpha \to \infty} \frac{1}{\alpha}\mathbf{I} = 0.$$

Therefore, by setting α equal to a large number, we can force Equations (5.51) and (5.52) arbitrarily close to Equation (5.50), regardless of $\boldsymbol{\theta}_0$. In practice, $\boldsymbol{\theta}_0$ is usually a zero matrix for convenience.

Remarks

- The matrix $\mathbf{P}_k$ is proportional to the covariance of the estimators. (See the Gauss-Markov theorem in Section 5.7 for more details.)

- The least-squares estimator can be interpreted as a Kalman filter [4, 6] for the process

$$\begin{cases} \boldsymbol{\theta}(k+1) & = & \boldsymbol{\theta}(k), \\ y(k) & = & \mathbf{a}^T(k)\boldsymbol{\theta}(k) + e(k), \end{cases}$$

  where k is a time index, $e(k)$ is random noise, $\boldsymbol{\theta}(k)$ is the state to be estimated, and $y(k)$ is the observed output. [Note that $\mathbf{a}^T(k)$ and $y(k)$ are equal to $\mathbf{a}_k^T$ and y_k, respectively, in Equation (5.50).]

- When extra parameters are introduced for better performance, $\hat{\boldsymbol{\theta}}$ will have more components and there will be additional columns in matrix $\mathbf{A}$. It is possible to reduce the complexity of the calculation by employing recursion in the number of parameters. Interested readers are referred to Section 3.6 of [5].

- The recursive LSE for systems with multiple outputs can be derived almost identically:

$$\begin{cases} \mathbf{P}_{k+1} & = & \mathbf{P}_k - \dfrac{\mathbf{P}_k \mathbf{a}_{k+1} \mathbf{a}_{k+1}^T \mathbf{P}_k}{1 + \mathbf{a}_{k+1}^T \mathbf{P}_k \mathbf{a}_{k+1}}, \\ \boldsymbol{\Theta}_{k+1} & = & \boldsymbol{\Theta}_k + \mathbf{P}_{k+1} \mathbf{a}_{k+1} (\mathbf{y}_{k+1}^T - \mathbf{a}_{k+1}^T \boldsymbol{\Theta}_k), \end{cases} \tag{5.53}$$

  where $(\mathbf{a}_k^T ; \mathbf{y}_k^T)$ is the kth data pair.

5.6 RECURSIVE LSE FOR TIME-VARYING SYSTEMS*

From Equation (5.42), we have $\mathbf{P}_k = (\mathbf{A}^T\mathbf{A})^{-1}$, where k is the number of data pairs encountered so far, and it is also the row dimension of $\mathbf{A}$. If k is greater than the number of fitting parameters and the data pairs contain rich enough information, then $\mathbf{A}^T\mathbf{A}$ is usually positive definite, and, as k goes to infinity, $\frac{1}{k}\mathbf{A}^T\mathbf{A}$ approaches a nonsingular constant matrix. Therefore, we have

$$\lim_{k\to\infty} \mathbf{P}_k = \lim_{k\to\infty} \frac{1}{k}\left(\frac{1}{k}\mathbf{A}^T\mathbf{A}\right)^{-1} = 0,$$

which indicates that the adaptation gain $\mathbf{P}_{k+1}\mathbf{a}_{k+1}$ in Equation (5.50) decreases at each iteration. This is a direct consequence of the squared error defined in Equation (5.17), which treats each error component equally. For time-invariant systems, this works well, since the decreasing adaptation means we are getting closer to the optimal point in the parameter space. However, for time-varying systems, this is not appropriate, since the decreasing adaptation gain cannot track the changing optimal parameters. One simple way to resolve this problem is to reset the matrix $\mathbf{P}_k$ to $\mathbf{P}_0$ occasionally, since the LSE converges rapidly to the current optimal parameters. Of course, the obvious time to reset $\mathbf{P}_k$ is when we suspect that a significant parameter change has occurred.

Another way to deal with time-varying systems is to introduce a **forgetting factor** λ that places heavier emphasis on more recent data:

$$E(\boldsymbol{\theta}) = \sum_{i=1}^{m} \lambda^{m-i}(y_i - \mathbf{a}_i^T\boldsymbol{\theta})^2 = (\mathbf{y} - \mathbf{A}\boldsymbol{\theta})^T\mathbf{W}(\mathbf{y} - \mathbf{A}\boldsymbol{\theta}), \tag{5.54}$$

where $\mathbf{W}$ is a diagonal matrix:

$$\mathbf{W} = \begin{bmatrix} \lambda^{m-1} & 0 & \cdots & 0 \\ 0 & \lambda^{m-2} & \ddots & \vdots \\ \vdots & \ddots & \ddots & 0 \\ 0 & \cdots & 0 & 1 \end{bmatrix},$$

and $0 < \lambda \leq 1$. From Equation (5.26), the corresponding LSE that minimizes the preceding weighted error measure is defined by

$$\hat{\boldsymbol{\theta}} = (\mathbf{A}^T\mathbf{W}\mathbf{A})^{-1}\mathbf{A}^T\mathbf{W}\mathbf{y}. \tag{5.55}$$

To derive the formulas for a recursive LSE with a forgetting factor, again we define

$$\boldsymbol{\theta}_k = (\mathbf{A}^T\mathbf{W}\mathbf{A})^{-1}\mathbf{A}^T\mathbf{W}\mathbf{y},$$

where $[\mathbf{A} \vdots \mathbf{y}]$ contains k data pairs and $\boldsymbol{\theta}_k$ is the LSE using these k data pairs.

Then we have

$$\begin{aligned}\boldsymbol{\theta}_{k+1} &= \left(\begin{bmatrix} \mathbf{A} \\ \mathbf{a}^T \end{bmatrix}^T \begin{bmatrix} \lambda \mathbf{W} & 0 \\ 0 & 1 \end{bmatrix} \begin{bmatrix} \mathbf{A} \\ \mathbf{a}^T \end{bmatrix} \right)^{-1} \begin{bmatrix} \mathbf{A} \\ \mathbf{a}^T \end{bmatrix}^T \begin{bmatrix} \lambda \mathbf{W} & 0 \\ 0 & 1 \end{bmatrix} \begin{bmatrix} \mathbf{y} \\ y \end{bmatrix} \\ &= (\lambda \mathbf{A}^T \mathbf{W} \mathbf{A} + \mathbf{a}\mathbf{a}^T)^{-1} (\lambda \mathbf{A}^T \mathbf{W} \mathbf{y} + \mathbf{a} y),\end{aligned}$$

where $(\mathbf{a}^T;\ y)$ is the $(k+1)$th data pair, which has just become available. To simplify the notation, we again introduce $\mathbf{P}_k$ and $\mathbf{P}_{k+1}$, which are defined slightly differently from before:

$$\mathbf{P}_k = (\mathbf{A}^T \mathbf{W} \mathbf{A})^{-1},$$

$$\mathbf{P}_{k+1} = \left(\begin{bmatrix} \mathbf{A} \\ \mathbf{a}^T \end{bmatrix}^T \begin{bmatrix} \lambda \mathbf{W} & 0 \\ 0 & 1 \end{bmatrix} \begin{bmatrix} \mathbf{A} \\ \mathbf{a}^T \end{bmatrix} \right)^{-1} = (\lambda \mathbf{A}^T \mathbf{W} \mathbf{A} + \mathbf{a}\mathbf{a}^T)^{-1},$$

where $\mathbf{P}_k$ and $\mathbf{P}_{k+1}$ are related through

$$\lambda p_k^{-1} = \mathbf{P}_{k+1}^{-1} - \mathbf{a}\mathbf{a}^T. \tag{5.56}$$

Using $\mathbf{P}_k$ and $\mathbf{P}_{k+1}$, we can rewrite $\boldsymbol{\theta}_k$ and $\boldsymbol{\theta}_{k+1}$ as follows:

$$\boldsymbol{\theta}_k = \mathbf{P}_k \mathbf{A}^T \mathbf{W} \mathbf{y}, \tag{5.57}$$

and

$$\boldsymbol{\theta}_{k+1} = \mathbf{P}_{k+1} (\lambda \mathbf{A}^T \mathbf{W} \mathbf{y} + \mathbf{a} y). \tag{5.58}$$

If we eliminate $\mathbf{A}^T \mathbf{W} \mathbf{y}$ and $\mathbf{P}_k$ from the preceding three equations, we have the familiar recursive formula for $\boldsymbol{\theta}_{k+1}$:

$$\boldsymbol{\theta}_{k+1} = \boldsymbol{\theta}_k + \mathbf{P}_{k+1} \mathbf{a} (y - \mathbf{a}^T \boldsymbol{\theta}_k), \tag{5.59}$$

where $\mathbf{P}_{k+1}$ can be expanded from Equation (5.56) using the matrix inversion formula in Lemma 5.6:

$$\mathbf{P}_{k+1} = \frac{1}{\lambda} \left(\mathbf{P}_k - \frac{\mathbf{P}_k \mathbf{a}\mathbf{a}^T \mathbf{P}_k}{\lambda + \mathbf{a}^T \mathbf{P}_k \mathbf{a}} \right). \tag{5.60}$$

It is obvious that Equation (5.60) reduces to Equation (5.49) if $\lambda = 1$. If λ is small, then recent data are weighted more and the algorithm is more capable of tracking time-varying parameters. However, at the same time the estimators may also fluctuate to reflect noise and disturbance. Therefore, the value of λ is often task dependent and has to be determined experimentally.

The bootstrapping techniques introduced in the previous section for initializing $\mathbf{P}_0$ and $\boldsymbol{\theta}_0$ also apply here for recursive LSE with forgetting factors.

5.7 STATISTICAL PROPERTIES AND THE MAXIMUM LIKELIHOOD ESTIMATOR*

To examine the statistical qualities of the LSE derived in the preceding sections, we have to investigate the equation $\mathbf{y} = \mathbf{A}\boldsymbol{\theta} + \mathbf{e}$ in a statistical framework. In other words, we assume $\mathbf{e}$ is a random vector and $\boldsymbol{\theta}$ is the true parameter vector; thus $\mathbf{y}$ is also a random vector depending on $\mathbf{e}$. In particular, we shall introduce the Gauss-Markov conditions and demonstrate that the least-squares estimator $\hat{\boldsymbol{\theta}}$ is unbiased, consistent, and has minimum variance. Moreover, we shall explain the concept of maximum likelihood estimation and establish an equivalence between the maximum likelihood estimator and the least-squares estimator under certain assumptions.

First, let us begin by defining unbiased, consistent, and minimum variance estimators.

Definition 5.8 *Unbiased estimator*

An estimator $\hat{\boldsymbol{\theta}}$ of the parameter $\boldsymbol{\theta}$ is **unbiased** if $E[\hat{\boldsymbol{\theta}}] = \boldsymbol{\theta}$, where $E[\cdot]$ indicates the statistical expectation.

□

Definition 5.9 *Consistent estimator*

An estimator $\hat{\boldsymbol{\theta}}_k$ is a consistent estimator of $\boldsymbol{\theta}$ if

$$\lim_{k\to\infty} P(\|\hat{\boldsymbol{\theta}}_k - \boldsymbol{\theta}\| \geq \varepsilon) = 0 \text{ for any } \varepsilon > 0.$$

Here $P(\cdot)$ is the probability function and $\hat{\boldsymbol{\theta}}_k$ is the estimator using k input-output data pairs.

□

Definition 5.10 *Minimum variance estimator*

An estimator $\hat{\boldsymbol{\theta}}$ is a minimum variance estimator of $\boldsymbol{\theta}$ if for any other estimator $\boldsymbol{\theta}^*$:

$$\text{cov}(\hat{\boldsymbol{\theta}}) \leq \text{cov}(\boldsymbol{\theta}^*),$$

where $\text{cov}(\boldsymbol{\theta})$ represents the covariance matrix of the random vector $\boldsymbol{\theta}$.

□

The **Gauss-Markov conditions** state that the error vector $\mathbf{e}$ that accounts for the measurement noise and/or modeling error is a random vector satisfying

1. $E[\mathbf{e}] = 0$
2. $E[\mathbf{e}\mathbf{e}^T] = \sigma^2\mathbf{I}$.

This is equivalent to the statement that the error vector $\mathbf{e}$ is a vector of m uncorrelated random variables, each with zero mean and the same variance, σ^2. Under these conditions, we have the Gauss-Markov theorem, as follows.

Theorem 5.2 *Gauss-Markov theorem: LSE is unbiased and minimum variance*

Under the Gauss-Markov conditions, we have the **Gauss-Markov theorem**, which states that the LSE $\hat{\boldsymbol{\theta}}$ is unbiased and has minimum variance when compared with all other unbiased estimators that are linear combinations of the observation y_i.
Proof: Taking expectation of $\mathbf{y} = \mathbf{A}\boldsymbol{\theta} + \mathbf{e}$ leads to $E[\mathbf{y}] = E[\mathbf{A}\boldsymbol{\theta}] + E[\mathbf{e}] = \mathbf{A}\boldsymbol{\theta}$. Hence

$$\begin{aligned} E[\hat{\boldsymbol{\theta}}] &= E[(\mathbf{A}^T\mathbf{A})^{-1}\mathbf{A}^T\mathbf{y}] \\ &= (\mathbf{A}^T\mathbf{A})^{-1}\mathbf{A}^T E[\mathbf{y}] \\ &= (\mathbf{A}^T\mathbf{A})^{-1}\mathbf{A}^T\mathbf{A}\boldsymbol{\theta} \\ &= \boldsymbol{\theta}, \end{aligned}$$

which shows that $\hat{\boldsymbol{\theta}}$ is unbiased. (Note that the proof of unbiasedness does not require the second assumption of the Gauss-Markov conditions.) The proof of minimum variance is a bit lengthy; it can be found, for example, in Section 2.9 of [8].

□

Due to the Gauss-Markov theorem, the LSE $\hat{\boldsymbol{\theta}}$ is often referred to as the **best linear unbiased estimator** (**BLUE** for short), where *best* implies minimum variance. The following theorem establishes the consistency of the LSE $\hat{\boldsymbol{\theta}}$.

Theorem 5.3 *Gauss-Markov theorem: LSE is consistent*

Under the Gauss-Markov conditions, the LSE $\hat{\boldsymbol{\theta}}$ is a consistent estimator of $\boldsymbol{\theta}$ if $(\mathbf{A}^T\mathbf{A})^{-1} \to 0$ as $m \to \infty$, where m is the row dimension of $\mathbf{A}$.
Proof: Note that

$$\begin{aligned} \hat{\boldsymbol{\theta}} &= (\mathbf{A}^T\mathbf{A})^{-1}\mathbf{A}^T\mathbf{y} \\ &= (\mathbf{A}^T\mathbf{A})^{-1}\mathbf{A}^T(\mathbf{A}\boldsymbol{\theta} + \mathbf{e}) \\ &= \boldsymbol{\theta} + (\mathbf{A}^T\mathbf{A})^{-1}\mathbf{A}^T\mathbf{e}. \end{aligned}$$

Thus

$$\begin{aligned} \text{cov}(\hat{\boldsymbol{\theta}}) &= E[(\hat{\boldsymbol{\theta}} - \boldsymbol{\theta})(\hat{\boldsymbol{\theta}} - \boldsymbol{\theta})^T] \\ &= E[(\mathbf{A}^T\mathbf{A})^{-1}\mathbf{A}^T\mathbf{e}\mathbf{e}^T\mathbf{A}(\mathbf{A}^T\mathbf{A})^{-1}] \\ &= (\mathbf{A}^T\mathbf{A})^{-1}\mathbf{A}^T E[\mathbf{e}\mathbf{e}^T]\mathbf{A}(\mathbf{A}^T\mathbf{A})^{-1} \\ &= (\mathbf{A}^T\mathbf{A})^{-1}\mathbf{A}^T(\sigma^2\mathbf{I})\mathbf{A}(\mathbf{A}^T\mathbf{A})^{-1} \\ &= \sigma^2(\mathbf{A}^T\mathbf{A})^{-1}. \end{aligned}$$

This implies $\text{cov}(\hat{\boldsymbol{\theta}}) \to 0$ as $m \to \infty$ if $(\mathbf{A}^T\mathbf{A})^{-1} \to 0$, which completes the proof.

□

The rest of this section will be devoted to exploring the relationship between the maximum likelihood estimator and the least-squares estimator under certain assumptions. Maximum likelihood estimation is one of the most widely used techniques for estimating the parameters of a statistical distribution. To introduce this method, let us assume that we have a random variable x whose probability density function (or probability function, if x assumes discrete values) is $f(x;\theta)$, where θ is the parameter to be estimated. For a sample of n observations of this variable $x_1, \cdots x_n$, the **likelihood function**, L, is defined by

$$L = f(x_1;\theta)f(x_2;\theta) \cdots f(x_n;\theta). \tag{5.61}$$

Without any prior information about the true value of θ, we would naturally pick a value of θ that provides a high probability of obtaining the actual observed data $x_1, \cdots x_n$. Thus the **maximum likelihood estimator** (abbreviated as **MLE**), $\hat{\theta}$, is defined as the value of θ which maximizes L:

$$\left.\frac{\partial L}{\partial \theta}\right|_{\theta=\hat{\theta}} = 0. \tag{5.62}$$

Or, equivalently,

$$\left.\frac{\partial \ln L}{\partial \theta}\right|_{\theta=\hat{\theta}} = 0, \tag{5.63}$$

since ln is a monotonic function and the value of θ which maximizes L will be the same as the value of θ which maximizes $\ln L$. The following two examples illustrate the method of estimating the parameters for an exponential and a normal distribution.

Example 5.3 *MLE for exponential distribution*

Suppose a random variable x has an exponential distribution

$$f(x;\theta) = \theta^{-1}e^{-x/\theta}.$$

Then the likelihood function for m observations $x_1, \cdots, x_m$ takes the form

$$\begin{aligned} L &= (\theta^{-1}e^{-x_1/\theta})(\theta^{-1}e^{-x_2/\theta})\cdots(\theta^{-1}e^{-x_m/\theta}) \\ &= \theta^{-m}exp(-\theta^{-1}\textstyle\sum x_i), \end{aligned}$$

and

$$\ln L = -m\ln\theta - \frac{1}{\theta}\sum x_i.$$

Differentiating the preceding equation with respect to θ gives

$$\frac{\partial \ln L}{\partial \theta} = -\frac{m}{\theta} + \frac{\sum x_i}{\theta^2} = 0.$$

Therefore,

$$\hat{\theta} = \frac{\sum x_i}{m}.$$

□

Example 5.4 *MLE for normal distributions*

If a random variable x has a normal distribution

$$f(x;\mu,\sigma) = \frac{1}{\sqrt{2\pi}\sigma} \exp\left[-\frac{1}{2}\left(\frac{x-\mu}{\sigma}\right)^2\right],$$

where μ (mean) and σ^2 (variance) are undetermined parameters. For m observations $x_1, \cdots, x_n$, we have

$$L = \left(\frac{1}{\sqrt{2\pi}\sigma}\right)^m \exp\left[-\frac{1}{2\sigma^2}(x_i-\mu)^2\right],$$

and

$$\ln L = -m\ln(\sqrt{2\pi}\sigma) - \frac{1}{2\sigma^2}\sum(x_i-\mu)^2.$$

Differentiating with respect to μ and σ yields

$$\frac{\partial \ln L}{\partial \mu} = 0 \Longrightarrow \frac{1}{\sigma^2}\sum(x_i-\mu) = 0 \Longrightarrow \hat{\mu} = \frac{\sum x_i}{m}$$

$$\frac{\partial \ln L}{\partial \sigma} = 0 \Longrightarrow -\frac{m}{\sigma} - \frac{1}{2}(-2\sigma^{-3})\sum(x_i-\mu)^2 = 0 \Longrightarrow \hat{\sigma}^2 = \frac{\sum(x_i-\hat{\mu})^2}{m}.$$

□

Now we consider the relationship between the maximum likelihood estimator and the least-squares estimator, which is regulated by the following theorem:

Theorem 5.4 *Equivalence between the LSE and the MLE*

In addition to the Gauss-Markov conditions, if we assume that each element of the error vector $\mathbf{e}$ is a random variable with normal distribution, then the LSE of $\boldsymbol{\theta}$ is precisely equal to the MLE of $\boldsymbol{\theta}$.

Proof: Since each e_i $[= y_i - \sum_j f_j(\vec{u}_i)\theta_j]$ of the error vector $\mathbf{e}$ is uncorrelated with every other e_i and normally distributed with zero mean and the same variance, σ^2, we have the following likelihood function:

$$\begin{aligned} L &= \frac{1}{(\sqrt{2\pi}\sigma)^m} \exp\left\{-\frac{1}{2\sigma^2}\sum_{i=1}^{m}[y_i - \sum_{j=1}^{n} f_j(\vec{u}_i)\theta_j]\right\} \\ &= \frac{1}{(\sqrt{2\pi}\sigma)^m} \exp\left[-\frac{1}{2\sigma^2}(\mathbf{y}-\mathbf{A}\boldsymbol{\theta})^T(\mathbf{y}-\mathbf{A}\boldsymbol{\theta})\right]. \end{aligned}$$

Taking the natural logarithm of the preceding likelihood function gives us

$$\ln L = -\frac{1}{2\sigma^2}(\mathbf{y} - \mathbf{A}\boldsymbol{\theta})^T(\mathbf{y} - \mathbf{A}\boldsymbol{\theta}) - m\ln(\sqrt{2\pi}\sigma).$$

Minimizing $\ln L$ with respect to $\boldsymbol{\theta} = [\theta_1 \cdots \theta_n]^T$ yields

$$\frac{\partial}{\partial\boldsymbol{\theta}}(\mathbf{y} - \mathbf{A}\boldsymbol{\theta})^T(\mathbf{y} - \mathbf{A}\boldsymbol{\theta})\big|_{\boldsymbol{\theta}=\hat{\boldsymbol{\theta}}} = 0.$$

The solution of $\hat{\boldsymbol{\theta}}$ from the preceding equation is identical to the least-squares estimator given by Equation (5.19). Thus the MLEs of the regression coefficients are precisely the same as the least-squares estimators of these coefficients. This establishes the well-known principle that the LSE is equivalent to the MLE under uncorrelated Gaussian noise with zero mean and the same variance.

□

5.8 LSE FOR NONLINEAR MODELS

Although the least-squares methods for linear models are the most widely used techniques for fitting a set of observation data, occasionally it is appropriate to assume that the data are related through a model with nonlinear parameters. Nonlinear models (that is, nonlinear in the parameters to be estimated) can be divided into two types, which we will refer to as **intrinsically linear** and **intrinsically nonlinear** models, respectively. Through appropriate transformations of its input-output variables and fitting parameters, an intrinsically linear model can be expressed in the standard form of a linear model represented by Equation (5.12). Thus we can apply standard least-squares methods to *approximate* the optimal parameters effectively. (Due to the use of transformations, the solution is not exactly optimal in minimizing the squared error measure.)

If a nonlinear model cannot be expressed in a linear form after transformation, then it is intrinsically nonlinear. For nonlinear models of this type, we can apply *nonlinear* least-squares methods, described in Section 6.8 of Chapter 6.

This section gives several examples of transformation methods that can be applied to nonlinear models that are intrinsically linear. One such nonlinear model describes how the radioactivity y of certain chemicals decays with respect to time t:

$$y = ae^{bt}, \tag{5.64}$$

where a and b (< 0) are model parameters to be determined. According to the given training data set $\{(t_i; y_i),\ i = 1, \ldots, m\}$, the squared error measure takes the form

$$E(a, b) = \sum_{i=1}^{m}(y_i - ae^{bt_i})^2. \tag{5.65}$$

Table 5.2. *Nonlinear models that are intrinsically linear.*

Nonlinear models	Transformation	Linear forms
$y = ae^{bx}$	Natural logarithm	$\ln y = \ln a + bx$
$y = ax^b$	Natural logarithm	$\ln y = \ln a + b \ln x$
$y = \frac{ax}{b+x}$	Reciprocal	$\frac{1}{y} = \frac{1}{a} + \frac{b}{a}\frac{1}{x}$
$y = \frac{a}{b+x}$	Reciprocal	$\frac{1}{y} = \frac{1}{a} + \frac{b}{a}x$

Attempting to minimize this error measure yields

$$\begin{cases} \frac{\partial E}{\partial a} = 2\sum_{i=1}^{m}(y_i - ae^{-bt_i})(-e^{bt_i}) = 0, \\ \frac{\partial E}{\partial b} = 2\sum_{i=1}^{m}(y_i - ae^{-bt_i})(-at_i e^{bt_i}) = 0. \end{cases}$$

Unfortunately, these are simultaneous nonlinear equations and it is hard, if not impossible, to find an analytic closed-form solution.

If we take the natural logarithm of Equation (5.64), we obtain a linear model that relates t and $\ln y$ through linear parameters $\ln b$ and a:

$$\ln y = \ln a + bt,$$

which indicates that the original nonlinear model is intrinsically linear. Consequently, through appropriate transformations, a nonlinear model that is intrinsically linear can be converted into a linear one and thus the LSE techniques developed earlier can be applied. Table 5.2 lists some nonlinear models that are intrinsically linear.

On the other hand, for an intrinsically nonlinear model, such as

$$y = a_0 + a_1 x^{b_1} + a_2 x^{b_2},$$

there exists no transformation techniques that can put all the fitting parameters $\boldsymbol{\theta} = [a_0, a_1, a_2, b_1, b_2]^T$ or their transformed quantities into a linear form. In fact, it is obvious that among the five fitting parameters, $[a_0, a_1, a_2]$ are linear parameters while $[b_1, b_2]$ are nonlinear ones. Thus we can apply LSE for the linear parameters and the iterative optimization methods (in Chapter 6) for the nonlinear ones. This leads to so-called **hybrid learning**, which is detailed in Section 8.5.

Sometimes it is necessary to take iterated transformations to reduce a complicated nonlinear model to a linear one. The following example illustrates such a case.

Example 5.5 *Iterated transformations*

Table 5.3. *Training data for Example 5.6.*

t_i	0	0.80	1.84	2.90	4.06	4.81	6.07	7.06	8.15	8.87	9.98
y_i	0.98	0.69	0.47	0.46	0.29	0.16	0.23	0.10	0.03	0.12	0.01

Suppose that we have a nonlinear model given by

$$y = \frac{1}{1 + a x_1^b e^{c x_2}}, \tag{5.66}$$

where x_1 and x_2 are inputs and a, b, and c are parameters. Taking reciprocals, subtracting 1, and taking the natural logarithm of both sides, we can convert the original model into a linear one:

$$\ln(y^{-1} - 1) = \ln a + b \ln x_1 + c x_2, \tag{5.67}$$

which states that the transformed output $\ln(y^{-1} - 1)$ is explicitly expressed as a linear function of the (transformed) parameters $\ln a$, b, and c. Other examples of iterated transformations are explored in the exercises at the end of this chapter.

□

It should be kept in mind, however, that by applying this transformation method to Equation (5.64), we are not searching for the parameters that minimize the error measure v in Equation (5.65). Instead, the parameters we obtained minimize a new error measure defined by

$$\begin{aligned} E'(a, b) &= \sum_{i=1}^{m} (\ln y_i - \ln a - b t_i)^2 \\ &= \sum_{i=1}^{m} \left(\ln \frac{y_i}{a e^{b t_i}} \right). \end{aligned} \tag{5.68}$$

From empirical studies, it is known that the parameters $a = a'$ and $b = b'$, which minimize $E'(a, b)$, will not be too different from the optimal parameters $a = \hat{a}$ and $b = \hat{b}$ that minimize $E(a, b)$ in Equation (5.65), as long as the transformation is monotonic and the observation data consistently conform to the underlying model. Thus most of the time we can take a' and b' as good approximations of the optimal parameters $\hat{a}$ and $\hat{b}$, respectively. In cases where a' and b' cannot be used to describe the observed data satisfactorily, we can use (a', b') as an initial guess and apply other iterative optimization methods discussed in Chapter 7 to improve the fitting. The following example shows the difference between minimizing v in Equation (5.65) and E' in Equation (5.68).

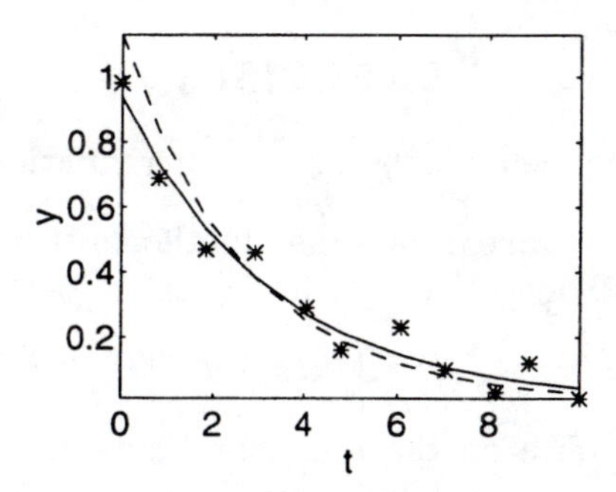

Figure 5.4. *Transformation method (dashed curve) versus iterative optimization (solid curve).* (MATLAB file: `transform.m`)

Example 5.6 *Transformation method versus nonlinear optimization*

Suppose that the underlying model describing the data set in Table 5.3 is known to be

$$y = ae^{bt}, \tag{5.69}$$

which is an intrinsically linear model. Figure 5.4 illustrates the training data set (marked with *), the solid curve that minimizes $E(a, b)$, and the dashed curve that minimizes $E'(a, b)$. The solid curve is obtained through a simple iterative gradient method to be introduced in Chapter 6; the dashed curve is derived by the transformation plus the LSE method. In this example, these two curves do not differ too much. However, if the data set does not conform to the underlying model consistently, these two curves may be quite different in shape. Two such examples can be found in Exercise 13 of Section 8.1 and Exercise 7 of Section 10.3 in [1].

□

5.9 SUMMARY

This chapter presents standard least-squares methods for linear models in system identification. Although our scope is confined to linear models, the underlying concepts can be extended to nonlinear models, as explained in Section 6.8.

Least-squares techniques play a pivotal role in the literatures of adaptive control, adaptive signal processing, regression, and statistics. In the subsequent chapters, we shall see how these techniques can be applied effectively to adaptive networks (Chapter 8), supervised learning neural networks (Chapter 9), and adaptive neuro-fuzzy inference systems (Chapters 12 and 13).

EXERCISES

1. Verify Lemma 5.5 on page 102 by direct differentiation.

2. **Completing the square** is another method of finding the optimum of a quadratic function. Show that $f(\mathbf{x})$ in Equation (5.1) can be reorganized as

$$f(\mathbf{x}) = (\mathbf{x} + \mathbf{A}^{-1}\mathbf{b})^T \mathbf{A}(\mathbf{x} + \mathbf{A}^{-1}\mathbf{b}) + c - \mathbf{b}^T\mathbf{A}^{-1}\mathbf{b}. \qquad (5.70)$$

Note that the first term on the right-hand side of the preceding equation is non-negative if $\mathbf{A}$ is positive definite, and non-positive if $\mathbf{A}$ is negative definite. Therefore, $f(\mathbf{x})$ achieves its optimum $c - \mathbf{b}^T\mathbf{A}^{-1}\mathbf{b}$ at $\mathbf{x} = -\mathbf{A}^{-1}\mathbf{b}$.

3. Explain how the derivation of the previous exercise must be modified if $\mathbf{A}$ is positive definite but not symmetric.

4. Solve the least-squares problem by completing the square. Specifically, arrange $E(\boldsymbol{\theta})$ in Equation (5.17) into the format of Equation (5.70) and thus verify the LSE formula in Equation (5.19) and the minimum error in Equation (5.24).

5. Derive the weighted LSE in Equation (5.26) directly by setting the derivative of the weighted error measure in Equation (5.25) to zero.

6. Find the least-squares polynomials in Figures 5.2(b) through 5.2(d).

7. Prove Equations (5.34) through (5.39) to confirm their intuitive geometrical interpretations.

8. Show that the recursive LSE formulas in Equation (5.50) are equivalent to

$$\begin{cases} \boldsymbol{\theta}_{k+1} = \boldsymbol{\theta}_k + \dfrac{\mathbf{P}_k\mathbf{a}_{k+1}}{1 + \mathbf{a}_{k+1}^T\mathbf{P}_k\mathbf{a}_{k+1}}(y_{k+1} - \mathbf{a}_{k+1}^T\boldsymbol{\theta}_k), \\ \mathbf{P}_{k+1} = \mathbf{P}_k - \dfrac{\mathbf{P}_k\mathbf{a}_{k+1}\mathbf{a}_{k+1}^T\mathbf{P}_k}{1 + \mathbf{a}_{k+1}^T\mathbf{P}_k\mathbf{a}_{k+1}}. \end{cases} \qquad (5.71)$$

9. The recursive LSE in Equation (5.47) can be derived in another way. From Exercise 4, it is clear that the error measure after k data pairs have been observed can be expressed as

$$E_k(\boldsymbol{\theta}) = E_k(\hat{\boldsymbol{\theta}}_k) + (\boldsymbol{\theta} - \hat{\boldsymbol{\theta}}_k)^T\mathbf{A}^T\mathbf{A}(\boldsymbol{\theta} - \hat{\boldsymbol{\theta}}_k).$$

Based on $E_k(\boldsymbol{\theta})$, the error measure after observing the kth data pair $(\mathbf{a}; y)$ can be formulated as

$$\begin{aligned} E_{k+1}(\boldsymbol{\theta}) &= E_k(\boldsymbol{\theta}) + (y - \mathbf{a}^T\boldsymbol{\theta})^T(y - \mathbf{a}^T\mathbf{x}) \\ &= E_k(\hat{\boldsymbol{\theta}}_k) + (\boldsymbol{\theta} - \hat{\boldsymbol{\theta}}_k)^T\mathbf{A}^T\mathbf{A}(\mathbf{x} - \hat{\boldsymbol{\theta}}_k) + (y - \mathbf{a}^T\boldsymbol{\theta})^T(y - \mathbf{a}^T\boldsymbol{\theta}). \end{aligned}$$

Show that Equation (5.47) can be obtained alternatively by

$$\left.\frac{\partial E_{k+1}(\boldsymbol{\theta})}{\partial \boldsymbol{\theta}}\right|_{\boldsymbol{\theta}=\hat{\boldsymbol{\theta}}_{k+1}} = 0.$$

10. Derive Equations (5.51) and (5.52), which show the dependency of $\hat{\boldsymbol{\theta}}_k$ and $\mathbf{P}_k$ on their initial values.

11. Show that $\mathbf{P}_k$ in Equation (5.51) is positive definite.

12. Let $(\mathbf{a}; y)$ be the $(k+1)$th data pair and define a priori and a posteriori prediction errors as

$$\begin{aligned} e_{\text{prior}} &= y - \mathbf{a}^T \hat{\boldsymbol{\theta}}_k, \\ e_{\text{post}} &= y - \mathbf{a}^T \hat{\boldsymbol{\theta}}_{k+1}. \end{aligned}$$

Show that

$$\|e_{\text{post}}\| = \frac{\|e_{\text{prior}}\|}{\mathbf{a}^T \mathbf{P}_k \mathbf{a}} < \|e_{\text{prior}}\|.$$

13. Derive the recursive LSE for multiple-output systems, as shown in Equation (5.53).

14. Derive in detail the formulas for the recursive LSE with forgetting factor λ in Equations (5.59) and (5.60).

15. Show that the following two nonlinear models are intrinsically linear: (a) $y = \frac{1}{1+\exp(\frac{ax}{b+x})}$; (b) $y = \ln a + x - \ln(b + e^x)$. (In both cases, a and b are fitting parameters.)

REFERENCES

[1] R. L. Burden and J. D. Faires. *Numerical analysis.* PWS-Kent Pub. Co., Boston, 5th edition, 1993.

[2] N. R. Draper and H. Smith. *Applied regression analysis.* John Wiley & Sons, New York, 2nd edition, 1981.

[3] G. C. Goodwin and K. S. Sin. *Adaptive filtering prediction and control.* Prentice Hall, Upper Saddle River, NJ, 1984.

[4] S. S. Haykin. *Adaptive filter theory.* Prentice Hall, Upper Saddle River, NJ, 2nd edition, 1991.

[5] T. C. Hsia. *System identification: least-squares methods.* D. C. Heath and Company, 1977.

[6] R. E. Kalman. A new approach to linear filtering and prediction problems. *Journal of Basic Engineering*, pages 35–45, March 1960.

[7] L. Ljung. *System identification: theory for the user.* Prentice Hall, Upper Saddle River, NJ, 1987.

[8] A. Sen and M. Srivastava. *Regression analysis: theory, methods, and applications.* Springer-Verlag, London, 1990.

[9] P. Strobach. *Linear prediction theory: a mathematical basis for adaptive systems.* Springer-Verlag, London, 1990.

[10] G. B. Wetherill, P. Duncombe, M. Kenward, J. Köllerstörm, S. R. Paul, and B. J. Vowden. *Regression analysis with applications.* Monographs on Statistics and Applied Probability. Chapman and Hall Ltd., New York, 1986.

Chapter 6

Derivative-based Optimization

E. Mizutani and J.-S. R. Jang

6.1 INTRODUCTION

This chapter reviews a fundamental class of **gradient-based optimization techniques**, capable of determining *search directions* according to an objective function's **derivative** information. We discuss the preliminary concepts that sustain the *descent algorithms* used for solving minimization problems, as well as their relevant procedures. We begin with the **steepest descent method** and **Newton's method**, which form the foundation of many gradient-based algorithms. Actually, many instrumental algorithms can be regarded as a form of compromise between steepest descent and Newton's methods. We also describe their relevant techniques (e.g., **conjugate gradient methods** for practical advances in large problems, and the **Gauss-Newton method** and its **Levenberg-Marquardt** variant as a nonlinear extension of the *least-squares methods* described in Chapter 5).

A class of the gradient-based methods can be applied to optimizing nonlinear neuro-fuzzy models, thereby allowing such models to play a prominent role in the framework of soft computing. In fact, steepest descent and conjugate gradient methods are major algorithms used for *neural network learning* in conjunction with *back-error propagating* process. The least-squares estimation is another widely employed algorithm because the sum of squared errors is chosen as the object function to be minimized in many cases. Hence, we discuss *nonlinear least-squares problems* with particular emphasis placed on *Gauss-Newton methods with Levenberg-Marquardt notions*. Those methods are commonly used in data fitting and regression involving nonlinear models. Therefore, the gradient-based methods are closely related to neuro-fuzzy and soft computing techniques covered in the subsequent chapters.

6.2 DESCENT METHODS

In this chapter, we focus on minimizing a real-valued objective function E defined on an n-dimensional input space $\boldsymbol{\theta} = [\theta_1, \theta_2, \ldots, \theta_n]^T$. Finding a (possibly local) minimum point $\boldsymbol{\theta} = \boldsymbol{\theta}^*$ that minimizes $E(\boldsymbol{\theta})$ is of primary concern.

In general, a given objective function E may have a nonlinear form with respect to an adjustable parameter $\boldsymbol{\theta}$. Due to the complexity of E, we often resort to an **iterative** algorithm to explore the input space efficiently. In iterative descent methods, the next point $\boldsymbol{\theta}_{next}$ is determined by a step down from the current point $\boldsymbol{\theta}_{now}$ in a **direction vector d**:

$$\boldsymbol{\theta}_{next} = \boldsymbol{\theta}_{now} + \eta \mathbf{d}, \tag{6.1}$$

where η is some positive **step size** regulating to what extent to proceed in that direction. In neuro-fuzzy literature, the term **learning rate** is used for the *step size* η. For our convenience, we alternatively use the following formula:

$$\boldsymbol{\theta}_{k+1} = \boldsymbol{\theta}_k + \eta_k \mathbf{d}_k \quad (k = 1, 2, 3, \ldots), \tag{6.2}$$

where k denotes the current iteration number, and $\boldsymbol{\theta}_{now}$ and $\boldsymbol{\theta}_{next}$ represent two consecutive elements in a generated sequence of solution candidates $\{\boldsymbol{\theta}_k\}$. The $\boldsymbol{\theta}_k$ is intended to converge to a (local) minimum $\boldsymbol{\theta}^*$.

The iterative descent methods compute the kth step $\eta_k \mathbf{d}_k$ through two procedures: first determining *direction* $\mathbf{d}$, and then calculating *step size* η. The next point $\boldsymbol{\theta}_{next}$ should satisfy the following inequality:

$$E(\boldsymbol{\theta}_{next}) = E(\boldsymbol{\theta}_{now} + \eta \mathbf{d}) < E(\boldsymbol{\theta}_{now}). \tag{6.3}$$

The principal differences between various descent algorithms lie in the first procedure for determining successive directions. Once the decision is reached, all algorithms call for movement to a (local) minimum point on the line determined by the current point $\boldsymbol{\theta}_{now}$ and the direction $\mathbf{d}$. That is, for the second procedure, the optimum step size can be determined by **line minimization**:

$$\eta^* = \arg \min_{\eta > 0} \phi(\eta), \tag{6.4}$$

where

$$\phi(\eta) = E(\boldsymbol{\theta}_{now} + \eta \mathbf{d}). \tag{6.5}$$

The search of η^* is accomplished by **line search** (or **one-dimensional search**) methods, as described in Section 6.5.

6.2.1 Gradient-based Methods

When the straight downhill direction $\mathbf{d}$ is determined on the basis of the **gradient** ($\mathbf{g}$) of an objective function E, such descent methods are called *gradient-based* descent methods.

The **gradient** of a differentiable function $E : R^n \rightarrow R$ at $\boldsymbol{\theta}$ is the vector of first derivatives of E, denoted as $\mathbf{g}$. That is,

$$\mathbf{g}(\boldsymbol{\theta}) \; (= \nabla E(\boldsymbol{\theta})) \stackrel{\text{def}}{=} \left[\frac{\partial E(\boldsymbol{\theta})}{\partial \theta_1}, \frac{\partial E(\boldsymbol{\theta})}{\partial \theta_2}, \ldots, \frac{\partial E(\boldsymbol{\theta})}{\partial \theta_n} \right]^T. \tag{6.6}$$

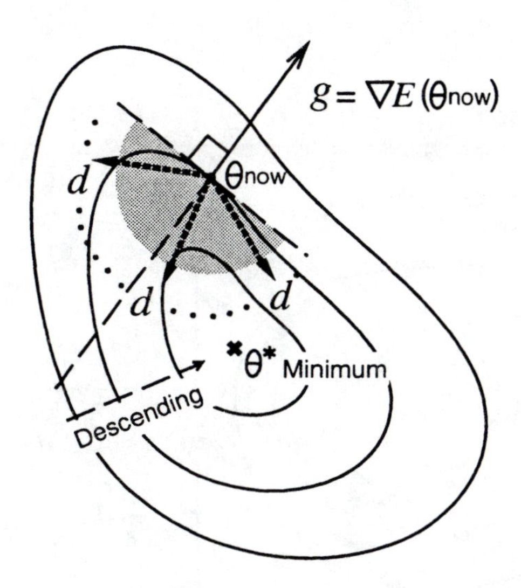

Figure 6.1. *Feasible descent directions. Directions from the starting point* $\boldsymbol{\theta}_{now}$ *in the shaded area are possible descent vector candidates. When* $\mathbf{d} = -\mathbf{g}$, $\mathbf{d}$ *is the steepest descent direction at a local point* $\boldsymbol{\theta}_{now}$.

For simplicity, we frequently use $\mathbf{g}$ by suppressing the argument $\boldsymbol{\theta}$ in $\mathbf{g}(\boldsymbol{\theta})$.

In general, based on a given gradient, downhill directions adhere to the following *condition for feasible descent directions*[1]:

$$\phi'(0) = \left.\frac{dE(\boldsymbol{\theta}_{\text{now}} + \eta\mathbf{d})}{d\eta}\right|_{\eta=0} = \mathbf{g}^T\mathbf{d} = \|\mathbf{g}^T\|\;\|\mathbf{d}\|\cos(\xi(\boldsymbol{\theta}_{\text{now}})) < 0, \qquad (6.7)$$

where ξ signifies the angle between $\mathbf{g}$ and $\mathbf{d}$, and $\xi(\boldsymbol{\theta}_{\text{now}})$ denotes the angle between $\mathbf{g}_{\text{now}}$ and $\mathbf{d}$ at the current point $\boldsymbol{\theta}_{\text{now}}$, as illustrated in Figures 6.1 and 6.2. This can be verified by the Taylor series expansion of E:

$$E(\boldsymbol{\theta}_{\text{now}} + \eta\mathbf{d}) = E(\boldsymbol{\theta}_{\text{now}}) + \eta\mathbf{g}^T\mathbf{d} + O(\eta^2). \qquad (6.8)$$

The second term on the right-hand side will dominate the third and other higher-order terms of η when $\eta \to 0$. With such a small positive η, the inequality (6.3) clearly holds when $\mathbf{g}^T\mathbf{d} < 0$. The shaded area in Figure 6.1 denotes all feasible descent directions that satisfy the condition (6.7). Notably, the gradient directions are always perpendicular to the contour curves (see Exercise 2).

A class of *gradient-based descent methods* has the following fundamental form, in which feasible descent directions can be determined by deflecting the gradients

[1]This *descent direction condition* (6.7) does not guarantee **convergence** of the algorithms.

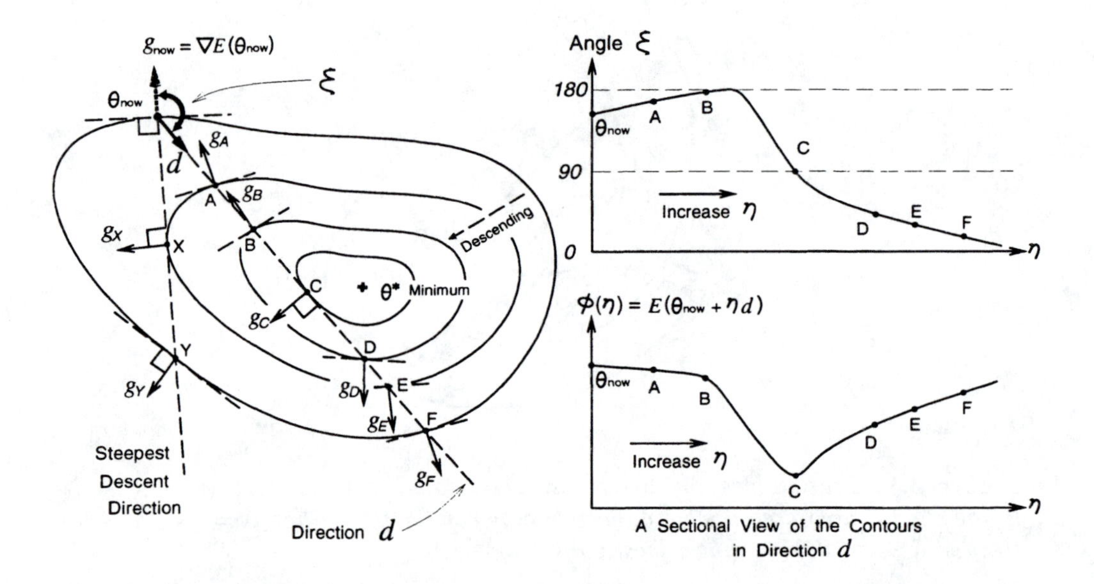

Figure 6.2. *Angle ξ between gradient directions $\mathbf{g}$ and a descent direction $\mathbf{d}$, which is determined by a certain algorithm at the current point $\boldsymbol{\theta}_{now}$. Let N be the set of all possible next points; $N \supset \{A, B, C, D, E, F, X, Y\}$. In the one-way downhill direction $\mathbf{d}$, the next point $\boldsymbol{\theta}_{next}$ may be one of six points—A, B, C, D, E, or F—or be in the vicinity of them, depending on step sizes. By comparison, in the steepest descent direction, $\boldsymbol{\theta}_{next}$, may be either X or Y, or close to them.*

through multiplication by $\mathbf{G}$ (i.e., *deflected gradients*):

$$\boldsymbol{\theta}_{\text{next}} = \boldsymbol{\theta}_{\text{now}} - \eta \mathbf{G}\mathbf{g}, \tag{6.9}$$

with some positive step size η and some positive definite matrix $\mathbf{G}$. Clearly, when $\mathbf{d} = -\mathbf{G}\mathbf{g}$, the descent direction condition (6.7) holds since $\mathbf{g}^T\mathbf{d} = -\mathbf{g}^T\mathbf{G}\mathbf{g} < 0$. Many other variants of gradient-based methods (e.g., Newton's method and the Levenberg-Marquardt method) possess the aforementioned form to bias the negative gradient direction ($-\mathbf{g}$) for a better choice. Those variants are discussed in the subsequent sections.

Ideally, we wish to find a value of $\boldsymbol{\theta}_{\text{next}}$ that satisfies the following[2]:

$$\mathbf{g}(\boldsymbol{\theta}_{\text{next}}) = \left.\frac{\partial E(\boldsymbol{\theta})}{\partial \boldsymbol{\theta}}\right|_{\boldsymbol{\theta}=\boldsymbol{\theta}_{\text{next}}} = 0. \tag{6.10}$$

In practice, however, it is difficult to solve Equation (6.10) analytically. For minimizing the objective function, the descent procedures are typically repeated until one of the following stopping criteria is satisfied:

1. The objective function value is sufficiently small;

2. The length of the gradient vector **g** is smaller than a specified value; or

3. The specified computing time is exceeded.

6.3 THE METHOD OF STEEPEST DESCENT

The **method of steepest descent**, also known as **gradient method**, is one of the oldest techniques for minimizing a given function defined on a multidimensional input space. This method forms the basis for many direct methods used in optimizing both constrained and unconstrained problems. Moreover, despite its slow convergence, the method is the most frequently used nonlinear optimization technique due to its simplicity.

When $\mathbf{G} = \mathbf{I}$ (the identity matrix), Equation (6.9) becomes the well-known steepest descent formula:

$$\boldsymbol{\theta}_{\text{next}} = \boldsymbol{\theta}_{\text{now}} - \eta \mathbf{g}. \tag{6.11}$$

In light of Equations (6.7) and (6.8), if $\cos\xi = -1$—that is, **d** points the same direction as the negative gradient direction ($-\mathbf{g}$)—the objective function E can be decreased *locally* by the most amount at the current point $\boldsymbol{\theta}_{\text{now}}$. This finding implies that the negative gradient direction ($-\mathbf{g}$) points to the *locally steepest downhill* direction. From a *global* perspective, going in the negative gradient direction may not be a shortcut to reach the minimum point $\boldsymbol{\theta}^*$ [Figures 6.1 and 6.2].

If the steepest descent method employs *line minimization* in Equation (6.4)—that is, if the minimum point η^* in a direction **d** is obtained at each iteration—we have

$$\begin{aligned} \phi'(\eta) &= \frac{dE(\boldsymbol{\theta}_{\text{now}} - \eta \mathbf{g}_{\text{now}})}{d\eta} \\ &= \nabla^T E(\boldsymbol{\theta}_{\text{now}} - \eta \mathbf{g}_{\text{now}})\, \mathbf{g}_{\text{now}} \\ &= \mathbf{g}_{\text{next}}^T\, \mathbf{g}_{\text{now}} \\ &= 0, \end{aligned} \tag{6.12}$$

where $\mathbf{g}_{\text{next}}$ is the gradient vector at the next point. The preceding equation indicates that the next gradient vector $\mathbf{g}_{\text{next}}$ is always **orthogonal** to the current gradient vector $\mathbf{g}_{\text{now}}$. Figure 6.2 depicts this situation at point X, where $\mathbf{g}_{\text{next}} = \mathbf{g}_X$.

[2] Note that Equation (6.10) is just the necessary condition because the gradient **g** is zero at any **stationary point**; namely, a maximum, a minimum, or a saddle point.

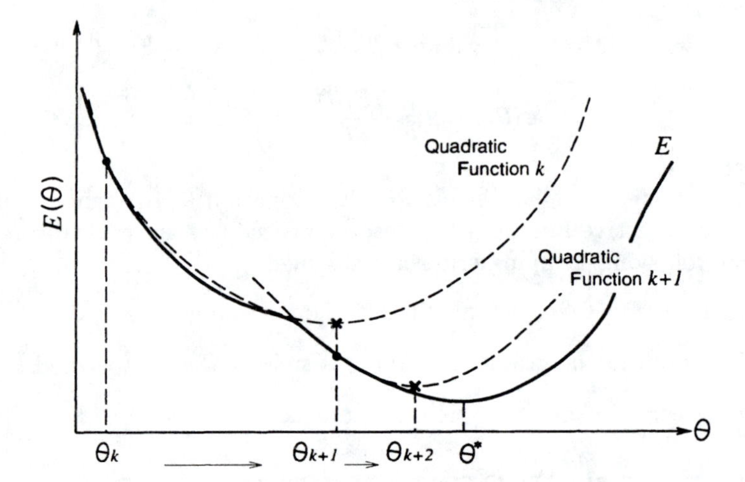

Figure 6.3. *Newton's (or Newton-Raphson) method for minimizing a general objective function E, which is approximated locally as a quadratic form; this approximate function is minimized exactly.*

where $\mathbf{g}_{\text{next}}$ is the gradient vector at the next point. The preceding equation indicates that the next gradient vector $\mathbf{g}_{\text{next}}$ is always **orthogonal** to the current gradient vector $\mathbf{g}_{\text{now}}$. Figure 6.2 depicts this situation at point X, where $\mathbf{g}_{\text{next}} = \mathbf{g}_X$. Section 6.6 also describes related situation. For a quadratic objective function, as discussed in Section 6.7, the method of the steepest descent with line minimization generates only two mutually orthogonal directions that are determined by the starting point [Figure 6.14(a)]. (Previous investigators note that even for a general n-input objective function, the steepest descent method tends to search asymptotically merely in some lower-dimensional subspace [1, 12, 30].) Only if the contours of the objective function E form hyperspheres (or circles in a two-dimensional space), the steepest descent method leads to the minimum [Figure 6.4(a)] in a single step (cf. Theorem 6.1). Otherwise, the method does not necessarily direct toward the minimum point [Figure 6.4(b)].

6.4 NEWTON'S METHODS

6.4.1 Classical Newton's Method

The descent direction $\mathbf{d}$ can be determined by using the *second derivatives* of the objective function E, if available. For a general continuous objective function, the contours may be nearly elliptical in the immediate vicinity of the minimum. If the starting position $\boldsymbol{\theta}_{\text{now}}$ is sufficiently close to a local minimum, the objective function E is expected to be approximated by a quadratic form:

$$E(\boldsymbol{\theta}) \approx E(\boldsymbol{\theta}_{\text{now}}) + \mathbf{g}^T(\boldsymbol{\theta} - \boldsymbol{\theta}_{\text{now}}) + \frac{1}{2}(\boldsymbol{\theta} - \boldsymbol{\theta}_{\text{now}})^T\mathbf{H}(\boldsymbol{\theta} - \boldsymbol{\theta}_{\text{now}}), \tag{6.13}$$

where $\mathbf{H}$ is the **Hessian** matrix, consisting of the *second partial derivatives* of $E(\boldsymbol{\theta})$. The preceding equation is the Taylor series expansion of $E(\boldsymbol{\theta})$ up to the second-order terms. Higher-order terms are omitted due to the assumption that $\|\boldsymbol{\theta} - \boldsymbol{\theta}_{\text{now}}\|$ is sufficiently small.

Since Equation (6.13) is a quadratic function of $\boldsymbol{\theta}$, we can simply find its minimum point $\hat{\boldsymbol{\theta}}$ by differenting Equation (6.13) and setting it to zero. This subsequently leads to a set of linear equations:

$$0 = \mathbf{g} + \mathbf{H}(\hat{\boldsymbol{\theta}} - \boldsymbol{\theta}_{\text{now}}). \tag{6.14}$$

If the inverse of $\mathbf{H}$ exists, we have a unique solution. When the minimum point $\hat{\boldsymbol{\theta}}$ of the approximated quadratic function is chosen as the next point $\boldsymbol{\theta}_{\text{now}}$, we have the so-called **Newton's method** or the **Newton-Raphson method**:

$$\hat{\boldsymbol{\theta}} = \boldsymbol{\theta}_{\text{now}} - \mathbf{H}^{-1}\mathbf{g}. \tag{6.15}$$

The step $-\mathbf{H}^{-1}\mathbf{g}$ is called the **Newton step**, and its direction is called the **Newton direction**. The general gradient-based formula in Equation (6.9) reduces to Newton's method when $\mathbf{G} = \mathbf{H}^{-1}$ and $\eta = 1$. If $\mathbf{H}$ is positive definite and $E(\boldsymbol{\theta})$ is quadratic, then Newton's method directly gets to a local minimum *in the single Newton step*. If $E(\boldsymbol{\theta})$ is not quadratic, then the minimum may not be reached in a single stride, and Newton's method should be repeatedly employed; Figure 6.3 illustrates the progress of repeated application of Newton's method to a single-variable objective function. In this manner, Newton's method proceeds to the minimum point, based on a second-order truncated Taylor series approximation defined in Equation (6.13).

Figure 6.4 compares the steepest descent direction and the Newton direction when the objective functions are quadratic. When the contours are circles in Figure 6.4(a), both directions are indicated by the solid arrow. In contrast, when the contours are ellipsoids in Figure 6.4(b), the Newton direction (shaded arrow) points directly toward the unique minimum point whereas the steepest descent direction (dotted arrow) does not. In any case, the steepest descent direction is always perpendicular to the contour line AB at any point S.

Assume that a linear transformation $\mathbf{T}$ can be introduced:

$$\boldsymbol{\theta}' = \mathbf{T}\boldsymbol{\theta}, \tag{6.16}$$

such that the elliptical contours can be transformed to circular ones. Consequently, the steepest descent direction points toward the unique minimum, and a line search can be employed to find the minimum. For instance, when $E(x, y) = x^2 + 5y^2$, it

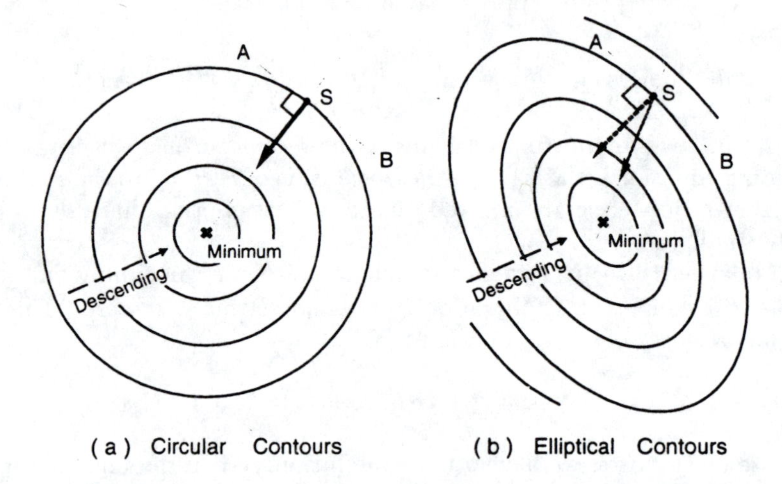

Figure 6.4. *Comparisons between steepest descent direction and the Newton direction in two-dimensional space: (a) Both directions (solid arrow) are the same when the contours are circles, and (b) when the contours are elliptical, the Newton direction (shaded arrow) points toward the minimum point whereas the steepest descent direction (dotted arrow) does not. In any case, the steepest descent direction is always perpendicular to the contour line AB at a point S. The Newton direction is said to be conjugate to line AB in (b); see Section 6.6.1.*

can be transformed to $E'(x, z) = x^2 + z^2$, with $\boldsymbol{\theta}' = (x, z)^T$, $\boldsymbol{\theta} = (x, y)^T$, and

$$\mathbf{T} = \begin{bmatrix} 1 & 0 \\ 0 & \sqrt{5} \end{bmatrix}.$$

In the transformed space, the steepest descent method can be used (cf. Section 6.3). When $\mathbf{T}$ is a diagonal matrix as in the foregoing, such a transformation is called **scaling** [cf. Equation (6.78)]. In this sense, the Newton direction is theoretically *scale invariant* [45]. However, finding a successful transformation or scaling is difficult in many cases [20].

A major disadvantage of Newton's method is that calculating the inverse of the Hessian matrix is computationally intensive and may introduce numerical problems due to round-off errors. Section 6.4.3 introduces some alternative routines that estimate the Hessian (or its inverse).

6.4.2 Modified Newton's Methods*

The classical Newton's method defined in Equation (6.15) frequently requires some refinements before implemented. If the current point $\boldsymbol{\theta}_{\text{now}}$ is remote from a local minimum $\boldsymbol{\theta}^*$, the method may not yield a descent direction due to the truncated higher-order terms in the Taylor series expansion of function $E(\cdot)$.

The following modifications make Newton's method more robust and reliable.

Adaptive Step Length

Even if the Hessian is positive definite, the quadratic approximation may not be satisfactory. That is, the *direct Newton step* with $\eta = 1$ in Equation (6.15) may be too long to decrease $E(\boldsymbol{\theta})$.

A simple modification entails introducing a positive search parameter (or step length) adaptation. Consequently, Equation (6.15) can be modified to

$$\boldsymbol{\theta}_{\text{next}} = \boldsymbol{\theta}_{\text{now}} - \eta \mathbf{H}^{-1}\mathbf{g}, \tag{6.17}$$

where η is selected to minimize E [32]. Near the solution, η is expected to be close to 1 as in Equation (6.15). We can determine η to satisfy $E(\boldsymbol{\theta}_{\text{next}}) < E(\boldsymbol{\theta}_{\text{now}})$ in a heuristic manner. For instance,

$$\eta_{k+1} = \frac{1}{2}\eta_k, \quad k = 0, 1, 2, 3, \ldots$$

where the starting η_0 can be chosen to be 1.0 or a smaller value. This is the so-called **step-halving** procedure. For the same reason as in Equation (6.8), the procedure of this type can be guaranteed to guide the descent direction. Section 6.5 provides more complex step length determination schemes.

Levenberg-Marquardt Modifications

Furthermore, if the Hessian matrix is not positive definite, the Newton direction may point toward a local maximum, or a saddle point.

The Hessian can be altered by adding a positive definite matrix $\mathbf{P}$ to $\mathbf{H}$ to make $\mathbf{H}$ *positive definite*. Levenberg [31] and Marquardt [33] introduced this notion in least-squares problems, as are described in Section 6.8.2. Later, Goldfeld et al. [22] first applied this concept to the Newton's method. When $\mathbf{P} = \lambda\mathbf{I}$, Equation (6.15) will be

$$\boldsymbol{\theta}_{\text{next}} = \boldsymbol{\theta}_{\text{now}} - (\mathbf{H} + \lambda\mathbf{I})^{-1}\mathbf{g}, \tag{6.18}$$

where $\mathbf{I}$ is the identity matrix and λ is some nonnegative value. Figure 6.5 depicts this notion. Depending on the magnitude of λ, the method transits smoothly between the two extremes: Newton's method ($\lambda \to 0$) and well-known steepest descent method ($\lambda \to \infty$). A variety of Levenberg-Marquardt algorithms differ in the selection of λ. Goldfeld et al. computed eigenvalues of $\mathbf{H}$ and set λ to a little larger than the magnitude of the most negative eigenvalue. See also Section 6.8.2.

Moreover, when λ increases, $\|\boldsymbol{\theta}_{\text{next}} - \boldsymbol{\theta}_{\text{now}}\|$ decreases (see Theorem 6.2). In other words, λ plays the same role as an adjustable step length η in Equation (6.17). That is, with some appropriately large λ, the inequality (6.3) holds. Of course, the step size η can be further introduced and can be determined in conjunction with line search methods:

$$\boldsymbol{\theta}_{\text{next}} = \boldsymbol{\theta}_{\text{now}} - \eta(\mathbf{H} + \lambda\mathbf{I})^{-1}\mathbf{g}. \tag{6.19}$$

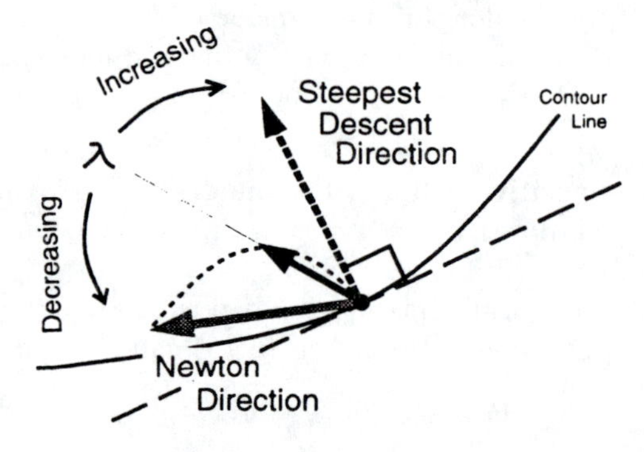

Figure 6.5. *Levenberg-Marquardt step shown in the highlighted arrow; The Levenberg-Marquardt method regulates a descent direction based on the Newton direction (in shaded arrow) and the steepest descent direction (in dotted arrow). The λ also controls the magnitude of the step $||\boldsymbol{\theta}_{next} - \boldsymbol{\theta}_{now}||$ along the dotted curve. When $\lambda = 0$, the step is the Newton step.*

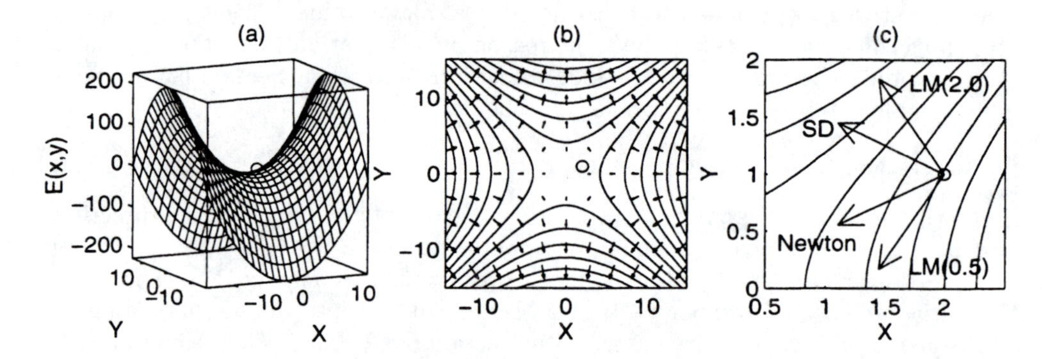

Figure 6.6. *(a) A quadratic surface $E(\boldsymbol{\theta}) = E(x, y) = x^2 - y^2$; (b) its gradient vector field and contour lines; (c) two Levenberg-Marquardt directions: LM(1) and LM(4), the Newton direction, and the steepest descent (SD) direction. These directions are normalized for displaying purpose.* (MATLAB file: `descent.m`)

Now we show a trivial example of Levenberg-Marquardt directions. Figure 6.6 illustrates the **hyperbolic paraboloid** defined in $E(\boldsymbol{\theta}) = E(x, y) = x^2 - y^2$. Its Hessian matrix is **indefinite**; the Newton direction $(-x, -y)^T$ always points toward the saddle point $(0, 0)$. Figure 6.6(c) describes two representative Levenberg-Marquardt directions: $LM(1)$ **when** $\lambda = 1$ **and** $LM(4)$ **when** $\lambda = 4$. **(Notice** that those directions are normalized for displaying purpose.) This shows how the Levenberg-Marquardt modification guides a downhill direction in the vicinity of the saddle point. Gill et al.[20] discussed how to deal with such an indefinite Hessian based on Cholesky factorization to decide a better descent direction.

Trust-region Methods

Recall the second-order truncated Taylor series approximation to the objective function E in Equation (6.13):

$$\begin{aligned} E(\boldsymbol{\theta}_{\text{now}} + \boldsymbol{\delta}) &\approx E(\boldsymbol{\theta}_{\text{now}}) + \mathbf{g}^T\boldsymbol{\delta} + \tfrac{1}{2}\boldsymbol{\delta}^T\mathbf{H}\boldsymbol{\delta} \\ &\equiv q(\boldsymbol{\delta}), \end{aligned} \tag{6.20}$$

where $\boldsymbol{\delta} = \boldsymbol{\theta}_{\text{next}} - \boldsymbol{\theta}_{\text{now}}$. The effectiveness of this quadratic approximation can be measured by ν_{now}, which is defined as

$$\nu_{\text{now}} = \frac{E(\boldsymbol{\theta}_{\text{now}}) - E(\boldsymbol{\theta}_{\text{now}} + \boldsymbol{\delta})}{E(\boldsymbol{\theta}_{\text{now}}) - q(\boldsymbol{\delta})}. \tag{6.21}$$

In conjunction with the same formula as Equation (6.18), **trust-region** (or **restricted step**) methods, attempt to minimize $q(\boldsymbol{\delta})$ only in a designated area, called a *trust region*, which may be defined as

$$\{\boldsymbol{\delta} : ||\boldsymbol{\delta}|| \leq R_{\text{now}}\} \quad (R_{\text{now}} > 0).$$

The value of R_{now} determines the size of the trust region, according to rules of thumb [15]:

If ν_{now} is small (e.g., $\nu_{\text{now}} \leq 0.2$), decrease R_{now}.
If ν_{now} is large (e.g., $\nu_{\text{now}} > 0.8$), reduce R_{now}.
Otherwise, keep the same size of R_{now}.

This method can be implemented even when the Hessian matrix is indefinite or singular. Moreover, this regional search in conjunction with Equation (6.18) can be considered as an extension of the Levenberg-Marquardt idea [20].

6.4.3 Quasi-Newton Methods*

Differentiating Equation (6.13) yields

$$\mathbf{H}_k(\boldsymbol{\theta}_{k+1} - \boldsymbol{\theta}_k) = \mathbf{g}_{k+1} - \mathbf{g}_k, \tag{6.22}$$

where k denotes the current iteration number. This formula intuitively indicates that Hessian $\mathbf{H}$ can be interpreted as the changing rate of the gradients between $\mathbf{g}(\boldsymbol{\theta}_{\text{now}})$ and $\mathbf{g}(\boldsymbol{\theta}_{\text{next}})$. Thus, $\mathbf{H}$ can be approximated on the basis of information of $\Delta\mathbf{g}_k \equiv (\mathbf{g}_{k+1} - \mathbf{g}_k)$ and $\Delta\boldsymbol{\theta}_k \equiv (\boldsymbol{\theta}_{k+1} - \boldsymbol{\theta}_k)$. This is the concept of **quasi-Newton** methods, also known as **variable metric** methods. Based on such derivative information, the quasi-Newton methods attempt to construct gradually an approximation $\mathbf{M}$ to

- *the Hessian matrix* $\mathbf{H}$ (e.g., the **Gill-Murray** method [19, 20]), or
- *the inverse of the Hessian matrix* $\mathbf{H}^{-1}$,

as iterations progress. We discuss the second scheme as below.

The approximations $\mathbf{M}$ ideally converge to $\mathbf{H}^{-1}$ near the solution point:

$$\boldsymbol{\theta}_{k+1} - \boldsymbol{\theta}_k \approx \mathbf{M}_k(\mathbf{g}_{k+1} - \mathbf{g}_k).$$

However, $\mathbf{M}_k$ determines $\Delta\boldsymbol{\theta}_k$; hence $\mathbf{M}_{k+1}$ is used to satisfy the **quasi-Newton condition** (or **secant condition**) given by

$$\boldsymbol{\theta}_{k+1} - \boldsymbol{\theta}_k = \mathbf{M}_{k+1}(\mathbf{g}_{k+1} - \mathbf{g}_k) \tag{6.23}$$

or

$$\Delta\boldsymbol{\theta}_k = \mathbf{M}_{k+1}\Delta\mathbf{g}_k.$$

The initial $\mathbf{M}_0$ is often chosen as $\mathbf{I}$. How to update $\mathbf{M}$ at each iteration is of priority. Two widely used updating schemes are the **Davidon-Fletcher-Powell (DFP)** [16] and the **Broyden-Fletcher-Goldfard-Shanno (BFGS)** [7, 8, 13, 21, 46] updating formulas.

DFP formula

$$\mathbf{M}_{k+1} = \mathbf{M}_k + \frac{\Delta\boldsymbol{\theta}_k \Delta\boldsymbol{\theta}_k^T}{\Delta\boldsymbol{\theta}_k^T \Delta\mathbf{g}_k} - \frac{\mathbf{M}_k \Delta\mathbf{g}_k \Delta\mathbf{g}_k^T \mathbf{M}_k}{\Delta\mathbf{g}_k^T \mathbf{M}_k \Delta\mathbf{g}_k} \tag{6.24}$$

BFGS formula

$$\begin{aligned} \mathbf{M}_{k+1} &= \mathbf{M}_k + \left(1 + \frac{\Delta\mathbf{g}_k^T \mathbf{M}_k \Delta\mathbf{g}_k}{\Delta\boldsymbol{\theta}_k^T \Delta\mathbf{g}_k}\right)\frac{\Delta\boldsymbol{\theta}_k \Delta\boldsymbol{\theta}_k^T}{\Delta\boldsymbol{\theta}_k^T \Delta\mathbf{g}_k} - \frac{\Delta\boldsymbol{\theta}_k \Delta\mathbf{g}_k^T \mathbf{M}_k + \mathbf{M}_k \Delta\mathbf{g}_k \Delta\boldsymbol{\theta}_k^T}{\Delta\boldsymbol{\theta}_k^T \Delta\mathbf{g}_k} \\ &= \left(\mathbf{I} - \frac{\Delta\boldsymbol{\theta}_k \Delta\mathbf{g}_k^T}{\Delta\boldsymbol{\theta}_k^T \Delta\mathbf{g}_k}\right)\mathbf{M}_k\left(\mathbf{I} - \frac{\Delta\mathbf{g}_k \Delta\boldsymbol{\theta}_k^T}{\Delta\boldsymbol{\theta}_k^T \Delta\mathbf{g}_k}\right) + \frac{\Delta\boldsymbol{\theta}_k \Delta\boldsymbol{\theta}_k^T}{\Delta\boldsymbol{\theta}_k^T \Delta\mathbf{g}_k} \end{aligned} \tag{6.25}$$

With the foregoing updating procedures, if

$$\Delta\boldsymbol{\theta}_k \Delta\mathbf{g}_k > 0 \tag{6.26}$$

and $\mathbf{M}_k$ is positive definite, then the subsequent $\mathbf{M}_{k+1}$ is positive definite, which is called **hereditary positive definiteness**. Due to Equation (6.2), $\Delta\boldsymbol{\theta}_k = \eta_k \mathbf{d}_k$. Hence, the preceding inequality is equivalent to

$$\mathbf{g}_{k+1}\mathbf{d}_k > \mathbf{g}_k \mathbf{d}_k. \tag{6.27}$$

This condition will hold if $\mathbf{d}_k$ is chosen to a descent direction with an appropriate step size η_k, as illustrated in Figure 6.2. Therefore, if such a suitable step size is determined, both conditions (6.3) and (6.26) will be satisfied.

BFGS is generally known to be more tolerant of inaccuracy of line minimization than DFP [40]. Moreover, DFP and BFGS are mutually *complementary* (or *dual*) formulas; that is, when the methods attempt to develop an approximation $\mathbf{B}$ to *the Hessian matrix* $\mathbf{H}$, DFP has the same form as Equation (6.25) except that $\mathbf{M}$ is replaced with $\mathbf{B}$ and $\Delta\boldsymbol{\theta}_k$ and $\Delta\mathbf{g}_k$ are interchanged. Similarly, BFGS has the same form as Equation (6.24).

6.5 STEP SIZE DETERMINATION

Recall the formula of a class of gradient-based descent methods given by Equation (6.9):

$$\boldsymbol{\theta}_{\text{next}} = \boldsymbol{\theta}_{\text{now}} + \eta \mathbf{d} = \boldsymbol{\theta}_{\text{now}} - \eta \mathbf{G}\mathbf{g}.$$

This formula entails effectively determining the step size η. The efficiency of the step size determination affects the entire minimization process. For a general function E, analytically solving Equation (6.4) as in

$$\phi'(\eta) = 0, \quad \text{where} \quad \phi(\eta) = E(\boldsymbol{\theta}_{\text{now}} + \eta \mathbf{d}).$$

is often impossible. That is, the univariate function $\phi(\eta)$ should be minimized on the line determined by the current point $\boldsymbol{\theta}_{\text{now}}$ and the direction $\mathbf{d}$. This is accomplished by **line search** (or **one-dimensional search**) methods.

In the rest of this section, we discuss the *line minimization* methods and their *stopping criteria* to prevent greedy search schemes from slowing down the entire minimization algorithm.

6.5.1 Initial Bracketing

The line search methods discussed in subsequent sections basically assume that the search area, or the specified interval, contains a single relative minimum; that is, the function E is **unimodal** over the closed interval. Determining the initial interval in which a relative minimum must lie is of critical importance. To begin with line searches, some routine must be employed for initially **bracketing** an assumed minimum into the starting interval. This kind of procedure can be roughly categorized into two schemes [41]:

1. A scheme, by function evaluations, for finding three points to satisfy

$$E(\boldsymbol{\theta}_{k-1}) > E(\boldsymbol{\theta}_k) < E(\boldsymbol{\theta}_{k+1}), \quad \boldsymbol{\theta}_{k-1} < \boldsymbol{\theta}_k < \boldsymbol{\theta}_{k+1}.$$

2. A scheme, by taking the first derivatives, for finding two points to satisfy

$$E'(\boldsymbol{\theta}_k) < 0, \quad E'(\boldsymbol{\theta}_{k+1}) > 0, \quad \boldsymbol{\theta}_k < \boldsymbol{\theta}_{k+1}.$$

For scheme 1, the common algorithm can be outlined as follows:

Algorithm 6.1 *A initial bracketing procedure for searching three points $\boldsymbol{\theta}_1$, $\boldsymbol{\theta}_2$, and $\boldsymbol{\theta}_3$*

(1) Given a starting point $\boldsymbol{\theta}_0$ and $h \in R$, let $\boldsymbol{\theta}_1$ be $\boldsymbol{\theta}_0 + h$. Evaluate $E(\boldsymbol{\theta}_1)$.

If $E(\boldsymbol{\theta}_0) \geq E(\boldsymbol{\theta}_1)$, $i \leftarrow 1$,
(i.e., go downhill) go to (2).

Otherwise, $h \leftarrow -h$, (i.e., set backward direction)
(i.e., go uphill) $E(\boldsymbol{\theta}_{-1}) \leftarrow E(\boldsymbol{\theta}_1)$,
$\boldsymbol{\theta}_1 \leftarrow \boldsymbol{\theta}_0 + h$,
$i \leftarrow 0$,
go to (3).

(2) Set the next point by: $h \leftarrow 2h, \quad \boldsymbol{\theta}_{i+1} \leftarrow \boldsymbol{\theta}_i + h.$

(3) Evaluate $E(\boldsymbol{\theta}_{i+1})$;

If $E(\boldsymbol{\theta}_i) \geq E(\boldsymbol{\theta}_{i+1})$, $i \leftarrow i + 1$,
(i.e., still go downhill) go to 2.

Otherwise, Arrange $\boldsymbol{\theta}_{i-1}, \boldsymbol{\theta}_i$, and $\boldsymbol{\theta}_{i+1}$ in the decreasing order.
Then, we obtain the three points: $(\boldsymbol{\theta}_1, \boldsymbol{\theta}_2, \boldsymbol{\theta}_3)$.
Stop.

□

The algorithm based on scheme 2 is left to the reader (Exercise 4). The following sections of line search methods assume that *initial bracketing* procedures of these types can adequately find several starting points.

6.5.2 Line Searches

The process of determining η^* that minimizes a one-dimensional function $\phi(\eta)$ is achieved by searching on the line for the minimum. The method of **line searches** (or **one-dimensional searches**) is important [5, 32, 44, 49] because *higher dimensional problems are ultimately solved by repeating line searches.* Also, line search algorithms usually include two components: sectioning (or bracketing), and polynomial interpolation.

Newton's Method

When $\phi(\eta_k)$, $\phi'(\eta_k)$, and $\phi''(\eta_k)$ are available, the classical Newton method in Equation (6.15) can be applied to solving the equation $\phi'(\eta_k) = 0$:

$$\eta_{k+1} = \eta_k - \frac{\phi'(\eta_k)}{\phi''(\eta_k)}. \tag{6.28}$$

Figure 6.7(a) shows that the preceding formula determines the next step size η_{k+1}.

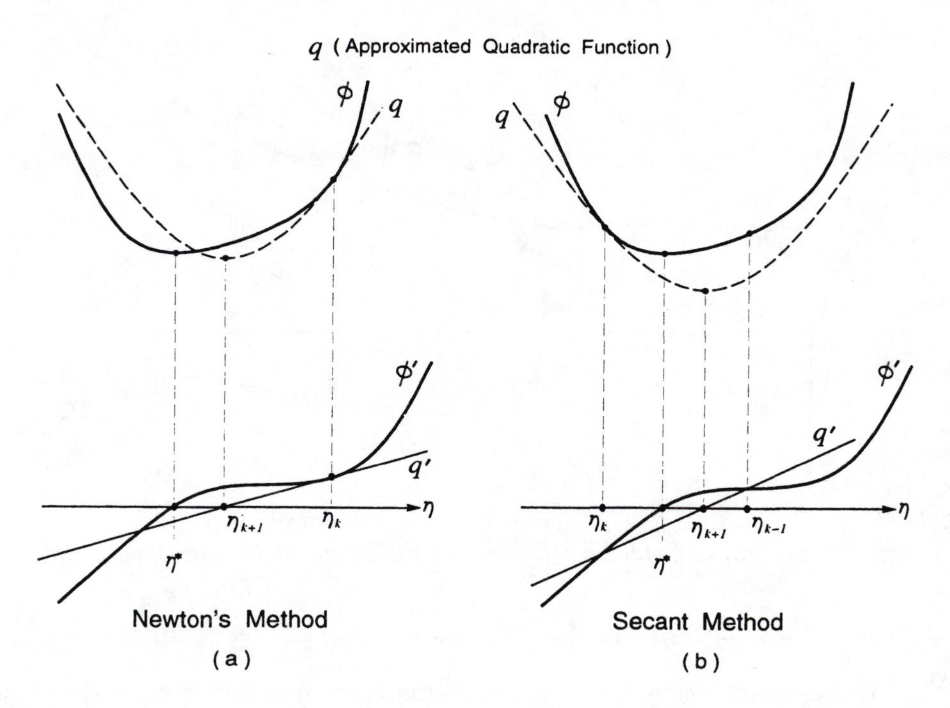

Figure 6.7. *Newton's method (left) and the secant method (right) to determine the step size.*

Secant Method

If we use both η_k and η_{k-1} to approximate the second derivative in Equation (6.28), and if the first derivatives alone are available then we have an estimated η_{k+1}:

$$\eta_{k+1} = \eta_k - \frac{\phi'(\eta_k)}{\frac{\phi'(\eta_k)-\phi'(\eta_{k-1})}{\eta_k-\eta_{k-1}}}. \tag{6.29}$$

This is the so-called **method of false position** or the **secant method**, as illustrated in Figure 6.7(b). (Also refer to Section 6.4.3.)

Sectioning methods

A sectioning algorithm begins with an interval $[a_1, b_1]$ in which the minimum η^* must lie, and then reduces the length of the interval at each iteration by evaluating the value of ϕ at a certain number of points. The two endpoints a_1 and b_1 can be found by the initial bracketing described in Section 6.5.1.

The **bisection method** is one of the simplest sectioning methods for solving $\phi'(\eta^*) = 0$, if first derivatives are available. Let $\phi'(\eta)$ be $\varphi(\eta)$ for simplicity. The algorithm is shown next.

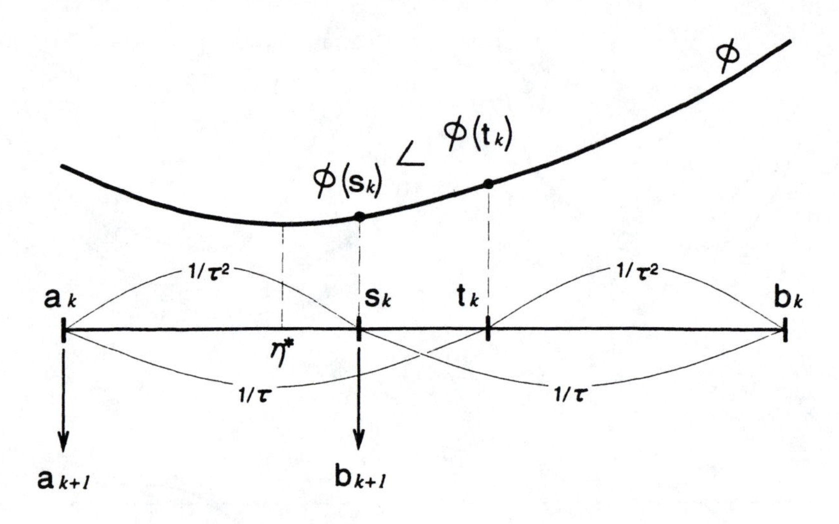

Figure 6.8. *Golden section search to determine the step length.* .

Algorithm 6.2 *Bisection method*

(1) Given a small value $\varepsilon \in R$ and an initial interval with two endpoints a_1 and a_2 such that $a_1 < a_2$ and $\varphi(a_1)\varphi(a_2) < 0$:

$$\begin{aligned} \eta_{\text{left}} &\leftarrow a_1 \\ \eta_{\text{right}} &\leftarrow a_2 \end{aligned}$$

(2) Calculate the midpoint η_{mid}; That is, $\eta_{\text{mid}} \leftarrow \frac{(\eta_{\text{right}} + \eta_{\text{left}})}{2}$.

$$\text{If } \varphi(\eta_{\text{right}})\varphi(\eta_{\text{mid}}) < 0, \quad \eta_{\text{left}} \leftarrow \eta_{\text{mid}}.$$

$$\text{Otherwise,} \qquad \eta_{\text{right}} \leftarrow \eta_{\text{mid}}.$$

(3) Check if $|\eta_{\text{left}} - \eta_{\text{right}}| < \varepsilon$. If it holds, terminate the algorithm. Otherwise, go to (2).

□

The *bisection method* replaces the right or left endpoint by the interval's midpoint based on the function evaluation at the midpoint. The length of the bracketing interval is halved at each iteration.

Next, we describe the **golden section search** method, which efficiently reduces the interval length based on function evaluations alone. The golden section requires ϕ to be neither a continuous function nor a differentiable one.

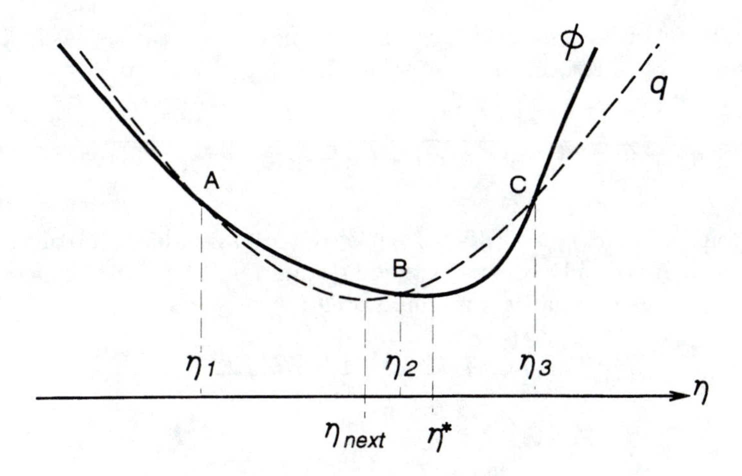

Figure 6.9. *Quadratic interpolation* .

Given an initial interval $[a_1, b_1]$ $(\ni \eta^*)$, the next trial points (s_k, t_k) within the interval are determined by using the **golden section ratio** τ:

$$s_k = b_k - \frac{1}{\tau}(b_k - a_k) = b_k + \frac{\tau - 1}{\tau}(b_k - a_k),$$
$$t_k = a_k + \frac{1}{\tau}(b_k - a_k),$$

where $\tau = \frac{1+\sqrt{5}}{2.} \approx 1.618$. This procedure guarantees that $a_k < s_k < t_k < b_k$, as shown in Figure 6.8.

The algorithm generates a sequence of the two ends, a_k and b_k, according to

If $\phi(s_k) > \phi(t_k)$, $a_{k+1} = s_k$, $b_{k+1} = b_k$.
Otherwise, $a_{k+1} = a_k$, $b_{k+1} = t_k$.

The minimum point η^* is bracketed to an interval just $\frac{1}{\tau}$ (≈ 0.618, i.e., approximately two-third) times the length of the preceding interval. After the kth iteration, the length of the bracketing interval shrinks to $(b_1 - a_1)\left(\frac{1}{\tau}\right)^{k-1}$.

Polynomial Interpolation

Polynomial interpolation methods are based on curve-fitting procedures, which work well when the objective function possesses a certain degree of smoothness.

A **quadratic interpolation** method constructs a smooth quadratic curve q that passes through three evaluated points, (η_1, ϕ_1), (η_2, ϕ_2), and (η_3, ϕ_3):

$$q(\eta) = \sum_{i=1}^{3} \phi_i \frac{\Pi_{j \neq i}(\eta - \eta_j)}{\Pi_{j \neq i}(\eta_i - \eta_j)}, \tag{6.30}$$

where $\phi_i \equiv \phi(\eta_i)$, $i = 1, 2, 3$.

The quadratic function has a unique minimum point, which can be easily determined by solving $q'(\eta) = 0$. Hence, the next trial point η_{next} is given by

$$\eta_{next} = \frac{1}{2}\frac{(\eta_2^2 - \eta_3^2)\phi_1 + (\eta_3^2 - \eta_1^2)\phi_2 + (\eta_1^2 - \eta_2^2)\phi_3}{(\eta_2 - \eta_3)\phi_1 + (\eta_3 - \eta_1)\phi_2 + (\eta_1 - \eta_2)\phi_3}. \quad (6.31)$$

(See Figure 6.9.)

When four values $\phi(\eta_1)$, $\phi'(\eta_1)$, $\phi(\eta_2)$, $\phi'(\eta_2)$ are available, a **cubic interpolation** method can not only construct a cubic equation, but can also determine the next point η_{next} as its relative minimum point:

$$\eta_{next} = \eta_2 - (\eta_2 - \eta_1)\{\frac{\phi'(\eta_2) - \beta + \gamma}{\phi'(\eta_2) - \phi'(\eta_1) + 2\gamma}\}, \quad (6.32)$$

where

$$\begin{aligned} \beta &= \phi'(\eta_1) + \phi'(\eta_2) - 3\,\tfrac{\phi(\eta_1) - \phi(\eta_2)}{\eta_1 - \eta_2}, \text{ and} \\ \gamma &= \sqrt{\beta^2 - \phi'(\eta_1)\phi'(\eta_2)}. \end{aligned}$$

6.5.3 Termination Rules*

In practice, it is nearly impossible to obtain the exact minimum point of the function ϕ by the aforementioned methods of line searches. Therefore, giving up exhaustive line searches at the expense of accuracy is normally desired to accelerate the entire minimization process. That is, a reasonable stopping criterion must be established to terminate the search procedures before they have converged. In the following, we describe some stopping rules widely applicable to line search methods.

Goldstein Test

Because $\phi(\eta) = E(\boldsymbol{\theta}_{now} + \eta\mathbf{d})$ $(= E(\boldsymbol{\theta}_{next}))$, Equation (6.3) can be rewritten as

$$E(\boldsymbol{\theta}_{next}) = \phi(\eta) < \phi(0) = E(\boldsymbol{\theta}_{now}). \quad (6.33)$$

A value of η is considered to be *not too large* if, with a given μ $(0 < \mu < \frac{1}{2})$,

$$\phi(\eta) \leq \phi(0) + \mu\phi'(0)\eta. \quad (6.34)$$

Due to the *feasible descent direction condition* (6.7),

$$\phi'(0) = \mathbf{g}^T\mathbf{d} < 0,$$

we can obtain, from the inequality (6.33),

$$\phi(\eta) \leq \phi(0) + \mu\phi'(0)\eta < \phi(0),$$

where μ and η are positive. That is, Equation (6.34) automatically guarantees that the direction is downhill. In addition, a value of η is considered to be *not too small* if

$$\phi(\eta) > \phi(0) + (1 - \mu)\phi'(0)\eta. \quad (6.35)$$

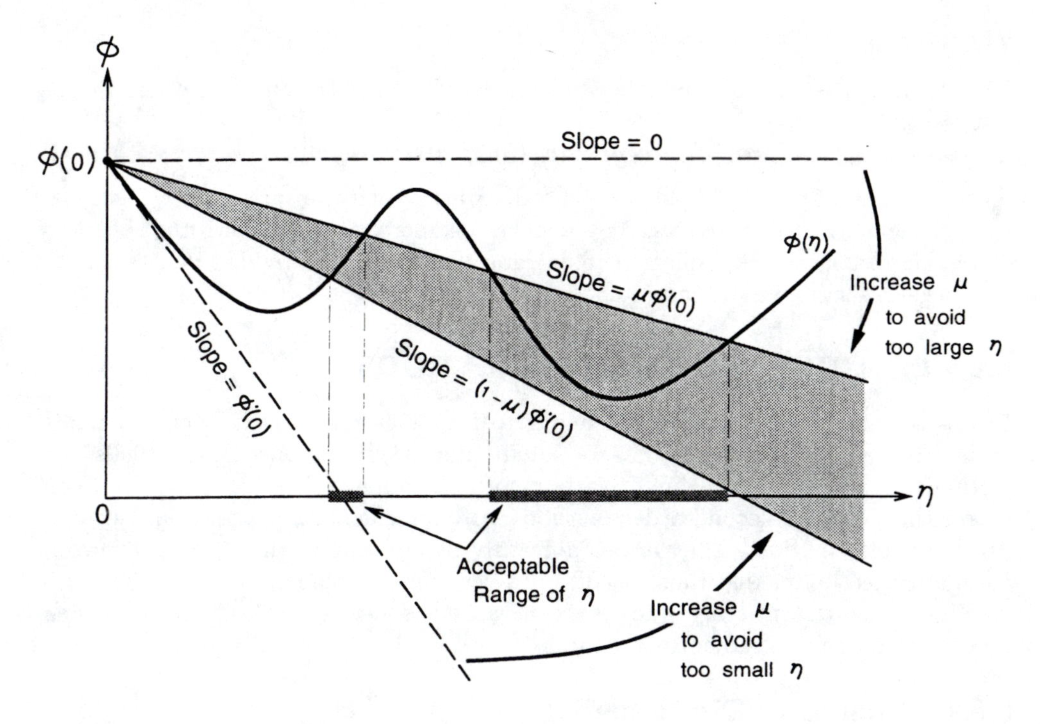

Figure 6.10. *Goldstein test.*

From the preceding two inequalities, we have

$$(1-\mu)\phi'(0)\eta \leq \phi(\eta) - \phi(0)(= E(\boldsymbol{\theta}_{\text{next}}) - E(\boldsymbol{\theta}_{\text{now}})) \leq \mu\phi'(0)\eta.$$

This can be rewritten as

$$0 < \mu \leq \frac{E(\boldsymbol{\theta}_{\text{next}}) - E(\boldsymbol{\theta}_{\text{now}})}{\eta \mathbf{gd}} \leq 1 - \mu < 1.$$

The **Goldstein test** requires that η satisfy the preceding condition. Geometrically speaking, $\phi(\eta)$ must lie in the shaded region between the two lines in Figure 6.10. With the smaller μ, the acceptable range of η becomes wider [23].

Wolfe Test

When the starting η is chosen to satisfy Equation (6.34) with a given μ ($0 < \mu < \frac{1}{2}$), the **Wolfe test** [50] calls for the following condition to ensure that η is not too small:

$$\phi'(\eta) \geq (1-\mu)\phi'(0). \tag{6.36}$$

(See Exercise 5.)

Armijo Test

The **Armijo test** [2] uses the following rule to guarantee that η is considered to be not too small:

$$\phi(\zeta\eta) > \phi(0) + \mu\phi'(0)\zeta\eta \quad (0 < \mu < 1), \tag{6.37}$$

where a value of $\zeta > 1$ should be selected. Starting with an arbitrary η, if Equation (6.34) holds with a given μ $(0 < \mu < 1)$, η is increased by ζ, until the rule does not hold. In contrast, if Equation (6.34) does not hold, η is divided by ζ until the decreased η satisfies Equation (6.34).

6.6 CONJUGATE GRADIENT METHODS*

In this section, we describe **conjugate gradient** methods, which originally made their debut as iterative algorithms for solving linear systems [27]. In the 1960s, the methods became widely used multivariate optimization algorithms because they can retain the power of second-order methods without calculating or storing second-derivative information. They are significantly less expensive than (quasi-)Newton methods, thereby making them useful for a very large problem.

First, we show that **conjugacy** is a generalized concept of **orthogonality**. Then we derive conjugate gradient algorithms.

6.6.1 Conjugate Directions*

Conjugate direction methods are also based on the second-order approximation of the objective function E defined in Equation (6.13). As revealed in Figure 6.4, the Newton direction is trustworthy if the objective function E has a approximately quadratic form. Although the steepest descent direction [the dotted arrow in Figure 6.4(b)] does not necessarily point toward a minimum point, the Newton direction (the shaded arrow) does. Such a Newton direction can be considered to be **conjugate** to the contour line AB at a point S, whereas the steepest direction is **orthogonal** to the contour line AB in Figure 6.4.

Generally, given a symmetric $(n \times n)$ matrix $\mathbf{Q}$, two n-dimensional (direction) vectors $\mathbf{d}_j$ and $\mathbf{d}_k$ are mutually **conjugate with respect to Q**, or **Q-orthogonal**, if the following equation holds:

$$\mathbf{d}_j^T \mathbf{Q} \mathbf{d}_k = 0. \tag{6.38}$$

Especially if $\mathbf{Q} = \mathbf{I}$, then $\mathbf{d}_j$ and $\mathbf{d}_k$ are mutually *orthogonal* [see Equation (6.12)].

Lemma 6.1 *If* $\mathbf{Q}$ *is positive definite,* n *mutually conjugate (nonzero) vectors* $\mathbf{d}_k$ *are* **linearly independent**.

Proof: Assume that the set of nonzero vectors $\mathbf{d}_k$ $(k = 1, 2, \ldots, n)$ are *linearly dependent*. By using scalars c_k that are not all zero, $\sum_{k=1}^{n} c_k \mathbf{d}_k = \mathbf{0}$. Due to Equation (6.38), we can have $c_j \mathbf{d}_j^T \mathbf{Q} \mathbf{d}_j = 0$. This finding implies that all c_j are zero because $\mathbf{Q}$ is positive definite. This is a contradiction.

□

By using this principle, a vector space spanned by a set of n mutually conjugate vectors $\mathbf{d}_k$ is R^n because n vectors $\mathbf{d}_k$ are linearly independent. (For a detailed discussion of sets of vectors in Euclidean n-space R^n, refer to refs. [18] and [42].) The quadratic objective function can be minimized in n-dimensional space in at most n search iterations in conjugate directions $\mathbf{d}_k$ ($k = 1, 2, \ldots, n$). The key is how to generate (linearly independent) conjugate direction vectors.

6.6.2 From Orthogonality to Conjugacy*

We consider an iterative step in a conjugate direction $\mathbf{d}_k$ to minimize such a quadratic function as Equation (6.13). From Equation (6.22), we have

$$\mathbf{g}_{k+1} - \mathbf{g}_k = \eta_k \mathbf{H} \mathbf{d}_k. \tag{6.39}$$

In a quadratic case, we can analytically specify the kth step size due to Equation (6.55) in Section 6.7:

$$\eta_k = -\frac{\mathbf{g}_k^T \mathbf{d}_k}{\mathbf{d}_k^T \mathbf{H} \mathbf{d}_k}. \tag{6.40}$$

By using the preceding formula (6.40), we consider *conjugate descent searches*, starting with *orthogonal descent searches*.

Coordinate Descent Searches

Coordinate directions can be used to realize *orthogonal descent searches*. That is, we consider a special case in which orthonormal coordinate bases $\mathbf{e}_k$ ($k = 1, 2, \ldots$) are used for search direction vectors $\mathbf{d}_k$. Note that $\mathbf{e}_k$ is a unit vector of zeros except for the kth element (e.g., $\mathbf{e}_1 = [1, 0, \ldots, 0]^T$, $\mathbf{e}_2 = [0, 1, \ldots, 0]^T$, ... etc.) Simply, at the kth step, the kth downhill direction can be regarded as either $\mathbf{e}_k$ or $-\mathbf{e}_k$. This is known as **coordinate descent methods**, whereby E may be ultimately minimized by sequentially changing the only kth component: i.e., $\min_{\boldsymbol{\theta}_k} E(\boldsymbol{\theta}_1, \boldsymbol{\theta}_2, \ldots, \boldsymbol{\theta}_n)$. Figure 6.11 depicts this concept. Furthermore, if the largest component (in absolute value) of the gradient vector $\mathbf{g}_k$ is chosen as the kth coordinate to search, this is called the **Gauss-Southwell method**. These coordinate descent searches are known to be less efficient than the steepest descent methods.

In a quadratic case, Equation (6.40) can be rewritten as

$$\eta_k = -\frac{\mathbf{g}_k^T \mathbf{e}_k}{\mathbf{e}_k^T \mathbf{H} \mathbf{e}_k}, \tag{6.41}$$

and

$$\boldsymbol{\theta}_{k+1} = \boldsymbol{\theta}_k + \eta_k \mathbf{e}_k.$$

This equation corresponds to the well-known **Gauss-Seidel iteration** for solving a system of linear equations.

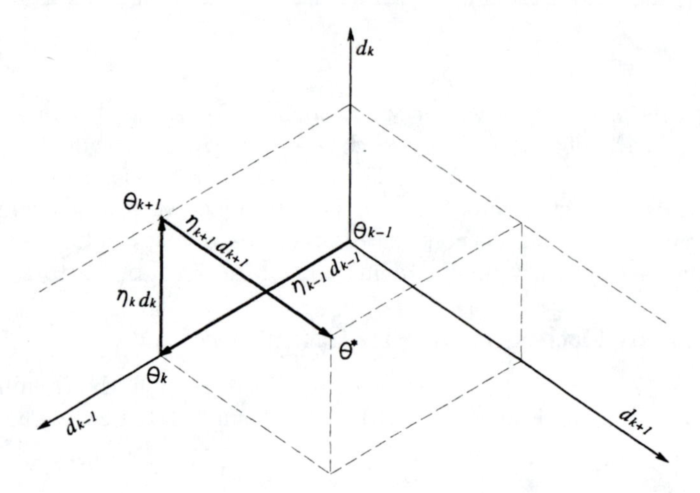

Figure 6.11. *Concept of orthogonal searches toward the minimum point* $\hat{\boldsymbol{\theta}}$. *The point* $\boldsymbol{\theta}_k$ *minimizes the objective function over the subspace spanned by* $\{\mathbf{d}_0, \mathbf{d}_0, \ldots, \mathbf{d}_{k-1}\}$.

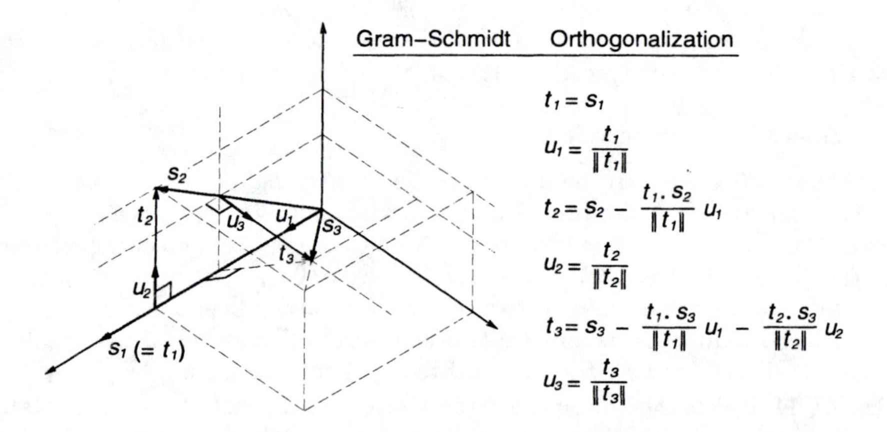

Figure 6.12. *The procedures of the Gram-Schmidt orthogonalization. As an example, given three vectors* $\mathbf{s}_1$, $\mathbf{s}_2$, *and* $\mathbf{s}_3$, *three mutually orthogonal vectors* $\mathbf{u}_1$, $\mathbf{u}_2$, *and* $\mathbf{u}_3$ *are produced. In other words, the step-by-step Gram-Schmidt process in Equation (6.42) converts an arbitrary basis* $\{\mathbf{s}_1, \mathbf{s}_2, \mathbf{s}_3\}$ *into an orthonormal basis* $\{\mathbf{u}_1, \mathbf{u}_2, \mathbf{u}_3\}$.

Gram-Schmidt Orthogonalization

Alternatively, given a set of direction vectors ($\mathbf{s}_k$), **Gram-Schmidt orthogonalization** can be performed to make $\mathbf{u}_k$ *orthogonal* to all previous directions

$\mathbf{u}_j$ $(j < k)$, as follows:

$$\mathbf{u}_k = \mathbf{s}_k - \sum_{j=0}^{k-1} \frac{\mathbf{u}_j \mathbf{s}_k}{\mathbf{u}_j \mathbf{u}_j} \mathbf{u}_j. \tag{6.42}$$

Figure 6.12 presents a simple example, in which the step-by-step *Gram-Schmidt process* in Equation (6.42) converts an arbitrary basis $\{\mathbf{s}_1, \mathbf{s}_2, \mathbf{s}_3\}$ into an orthonormal basis $\{\mathbf{u}_1, \mathbf{u}_2, \mathbf{u}_3\}$.

Such a $\mathbf{u}_k$ can also be set to the kth search direction $\mathbf{d}_k$ Since conjugacy is a generalized concept of orthogonality, we assume that the following equation inspired by Equation (6.42) can generate a direction $\mathbf{d}_k$ *conjugate* to all previous descent directions $\mathbf{d}_j$ $(j < k)$, with respect to $\mathbf{H}$:

$$\mathbf{d}_k = \mathbf{s}_k - \sum_{j=0}^{k-1} \frac{\mathbf{d}_j^T \mathbf{H} \mathbf{s}_k}{\mathbf{d}_j^T \mathbf{H} \mathbf{d}_j} \mathbf{d}_j. \tag{6.43}$$

If $\mathbf{s}_k$ is set to the unit vector $\mathbf{e}_k$ used in the *Gauss-Seidel iteration*, the preceding formula will be equivalent to a **Gaussian elimination procedure** with $\mathbf{d}_0 = \mathbf{e}_1$.

Lemma 6.2 *In a conjugate direction algorithm, the gradients* $\mathbf{g}_k$ $(k = 1, 2, \ldots, n)$ *satisfy*

$$\mathbf{g}_k \mathbf{d}_j = 0 \;\; for \;\; j < k, \tag{6.44}$$

where directions denoted by $\mathbf{d}_j$ $(j = 0, 1, \ldots, k-1)$ *are mutually conjugate with respect to* $\mathbf{H}$.

Proof by Mathematical Induction:

Verify that this is true for $k = 1$. Multiplied by $\mathbf{d}_0$, Equation (6.39) with $k = 0$ becomes

$$\mathbf{d}_0 \mathbf{g}_1 - \mathbf{d}_0 \mathbf{g}_0 = \eta_0 \mathbf{d}_0 \mathbf{H} \mathbf{d}_0.$$

With Equation (6.40),

$$\eta_0 = -\frac{\mathbf{g}_0^T \mathbf{d}_0}{\mathbf{d}_0^T \mathbf{H} \mathbf{d}_0},$$

we have $\mathbf{g}_1 \mathbf{d}_0 = 0$.

Now assume Equation (6.44) is true for k. Multiplied by $\mathbf{d}_k$, Equation (6.39) becomes

$$\mathbf{d}_k^T \mathbf{g}_{k+1} = \mathbf{d}_k^T \mathbf{g}_k + \eta_k \mathbf{d}_k^T \mathbf{H} \mathbf{d}_k. \tag{6.45}$$

Using Equation (6.40) yields the following equation:

$$\begin{aligned} \mathbf{d}_k^T \mathbf{g}_{k+1} &= \mathbf{d}_k^T \mathbf{g}_k - \tfrac{\mathbf{g}_k^T \mathbf{d}_k}{\mathbf{d}_k \mathbf{H} \mathbf{d}_k} \mathbf{d}_k^T \mathbf{H} \mathbf{d}_k \\ &= 0. \end{aligned}$$

Similarly, multiplied by $\mathbf{d}_j$ $j < k$, Equation (6.39) produces

$$\mathbf{d}_j^T \mathbf{g}_{k+1} = \mathbf{d}_j^T \mathbf{g}_k + \eta_k \mathbf{d}_j^T \mathbf{H} \mathbf{d}_k = 0$$

because of the induction assumption and conjugacy of the $\mathbf{d}_j$ and $\mathbf{d}_k$.

□

This lemma means that point C is chosen as the next point $\boldsymbol{\theta}_{\text{next}}$ in Figure 6.2.

6.6.3 Conjugate Gradient Algorithms*

When the gradient vector is used to determine conjugate directions, the algorithm is particularly called a **conjugate gradient** method. More specifically, by setting $\mathbf{s}_k$ to the negative gradient $-\mathbf{g}_k$, Equation (6.43) becomes

$$\mathbf{d}_k = -\mathbf{g}_k + \sum_{j=0}^{k-1} \frac{\mathbf{d}_j^T \mathbf{H} \mathbf{g}_k}{\mathbf{d}_j^T \mathbf{H} \mathbf{d}_j} \mathbf{d}_j. \tag{6.46}$$

Although Equation (6.43) requires the memory of all directions $\mathbf{d}_k$, this disadvantage vanishes from this formula (6.46) because $\mathbf{s}_k = -\mathbf{g}_k$. This occurrence can be shown in the following manner.

Let α_{kj} be the second term on the right-hand side of Equation (6.46):

$$\alpha_{kj} = \sum_{j=0}^{k-1} \frac{\mathbf{d}_j^T \mathbf{H} \mathbf{g}_k}{\mathbf{d}_j^T \mathbf{H} \mathbf{d}_j} \mathbf{d}_j.$$

Using Equation (6.39) yields

$$\alpha_{kj} = \sum_{j=0}^{k-1} \frac{\mathbf{g}_k^T (\mathbf{g}_{j+1} - \mathbf{g}_j)}{\mathbf{d}_j^T (\mathbf{g}_{j+1} - \mathbf{g}_j)} \mathbf{d}_j.$$

Due to Lemma (6.2), we obtain

$$\alpha_{kj} = \begin{cases} 0 & \text{if } \mathrm{j} < \mathrm{k} - 1 \\ \alpha_{k,k-1} & \text{otherwise,} \end{cases} \tag{6.47}$$

where

$$\alpha_{k,k-1} = \frac{\mathbf{g}_k^T (\mathbf{g}_k - \mathbf{g}_{k-1})}{\mathbf{d}_{k-1}^T (\mathbf{g}_k - \mathbf{g}_{k-1})} \mathbf{d}_k.$$

Hence, Equation (6.46) reduces to

$$\mathbf{d}_k = -\mathbf{g}_k + \beta_k \mathbf{d}_{k-1}, \tag{6.48}$$

where

$$\beta_k = \frac{\mathbf{g}_k^T (\mathbf{g}_k - \mathbf{g}_{k-1})}{\mathbf{d}_{k-1}^T (\mathbf{g}_k - \mathbf{g}_{k-1})}. \tag{6.49}$$

This is the so-called **Beale-Sorenson's formula** [4, 47] with such a specific β_k. The term β_k can be defined in various ways. A family of conjugate gradient methods generates the direction vectors according to the basic formula (6.48).

Furthermore, by switching k to $k-1$ in Equation (6.46), we have

$$\mathbf{d}_{k-1} = -\mathbf{g}_{k-1} + \sum_{j=0}^{k-2} \frac{\mathbf{d}_j^T \mathbf{H} \mathbf{g}_{k-1}}{\mathbf{d}_j^T \mathbf{H} \mathbf{d}_j} \mathbf{d}_j.$$

By plugging this into Equation (6.49), we obtain **Polak-Ribiere's formula** [38]:

$$\beta_k = \frac{\mathbf{g}_k^T(\mathbf{g}_k - \mathbf{g}_{k-1})}{\mathbf{g}_{k-1}^T \mathbf{g}_{k-1}}. \tag{6.50}$$

Now Equation (6.48) shows that $\mathbf{g}_k$ is in the subspace spanned by $\{\mathbf{d}_0, \mathbf{d}_0, \ldots, \mathbf{d}_k\}$. Therefore, due to Equation (6.44), we have the following property:

$$\mathbf{g}_k^T \mathbf{g}_j = 0 \quad \text{for } j < k. \tag{6.51}$$

This shows $\mathbf{g}_{k+1}^T \mathbf{g}_k = 0$ and, thus, Equation (6.50) leads to **Fletcher-Reeves's formula** [17]:

$$\beta_k = \frac{\mathbf{g}_k^T \mathbf{g}_k}{\mathbf{g}_{k-1}^T \mathbf{g}_{k-1}}. \tag{6.52}$$

Note that the first direction $\mathbf{d}_0$ is set to the steepest descent direction $-\mathbf{g}$, when $\beta_0 = 0$ in those formulas.

These three formulas attributed to Fletcher-Reeves, Polak-Ribiere, and Beale-Sorenson, respectively, are listed as follows:

$$\begin{aligned}
\text{Fletcher-Reeves:} \quad \beta_k &= \frac{\mathbf{g}_k^T \mathbf{g}_k}{\mathbf{g}_{k-1}^T \mathbf{g}_{k-1}} \\
\text{Polak-Ribiere:} \quad \beta_k &= \frac{\mathbf{g}_k^T (\mathbf{g}_k - \mathbf{g}_{k-1})}{\mathbf{g}_{k-1}^T \mathbf{g}_{k-1}} \\
\text{Beale-Sorenson:} \quad \beta_k &= \frac{\mathbf{g}_k^T (\mathbf{g}_k - \mathbf{g}_{k-1})}{\mathbf{d}_{k-1}^T (\mathbf{g}_k - \mathbf{g}_{k-1})}.
\end{aligned}$$

From Equation (6.40), we have derived those conjugate gradient formulas by assuming that η_k can be determined analytically. Thus, for applications to general objective functions, the methods require line minimization. Since Fletcher-Reeves's formula (6.52) assumes that the property $\mathbf{g}_{k+1}\mathbf{g}_k = 0$ holds, Polak-Ribiere's formula (6.50) often works better in practical applications [39]. Obviously, for a quadratic objective function, the two formulas (6.50) and (6.52) are identical, but not for a general function.

To enhance the performance of steepest descent, Fletcher and Reeves [17] introduced a systematic conjugate gradient algorithm with line minimization to determine step sizes. They also introduced a **restart algorithm**. Because of errors in computing directions and step sizes, the generated set of n direction vectors $\mathbf{d}_k$ ($k = 1, 2, \ldots, n$) may not be mutually conjugate. Therefore, the conjugate gradient methods are frequently implemented with some sort of **restart algorithms**,

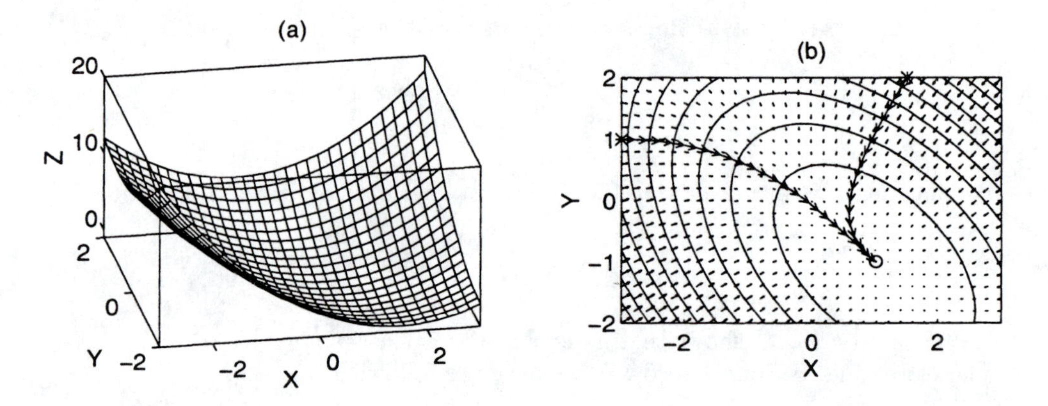

Figure 6.13. *(a) A quadratic surface $z = E(\boldsymbol{\theta}) = E(x, y) = x^2 + xy + y^2 - x + y$; (b) its gradient vector field and two downhill gradient paths.* (MATLAB file: `gdss1.m`)

wherein the direction vector is reset to the steepest descent direction ($\mathbf{d} \leftarrow -\mathbf{g}$) after n or $n + 1$ iterations. Yet it is known that the frequency of restarts depends on the objective function E. Based on Beale's technique [4], Powell [39] proposed a more effective restart algorithm; however, it requires more storage space of vectors.

6.7 ANALYSIS OF QUADRATIC CASE

Analyzing quadratic objective functions is of particular importance, because a general function can be well approximated by a quadratic function in the neighborhood of a (local) minimum, due to a consequence of Taylor's theorem.

Allow the objective function E to have a quadratic form:

$$E(\boldsymbol{\theta}) = \frac{1}{2}\boldsymbol{\theta}^T \mathbf{A}\boldsymbol{\theta} + \mathbf{b}^T\boldsymbol{\theta} + c. \tag{6.53}$$

The gradient of $E(\boldsymbol{\theta})$ can be expressed as

$$\mathbf{g}(\boldsymbol{\theta}) = \mathbf{A}\boldsymbol{\theta} + \mathbf{b}. \tag{6.54}$$

The step size plays a critical role in a class of descent methods. For a quadratic objective function, the line minimization problem in Equation (6.4) can be solved analytically. To minimize $\phi(\eta) = E(\boldsymbol{\theta}_{\text{now}} + \eta\mathbf{d})$ in a general descent direction $\mathbf{d}$, we set the derivative of $\phi(\eta)$ to zero:

$$\phi'(\eta^*) = \frac{dE(\boldsymbol{\theta}_{\text{now}} + \eta\mathbf{d})}{d\eta} = [\mathbf{A}(\boldsymbol{\theta} + \eta^*\mathbf{d}) + \mathbf{b}]^T(\mathbf{d}) = 0.$$

Due to Equation (6.54), this leads to

$$\eta^* = -\frac{\mathbf{d}^T\mathbf{g}}{\mathbf{d}^T\mathbf{A}\mathbf{d}}. \tag{6.55}$$

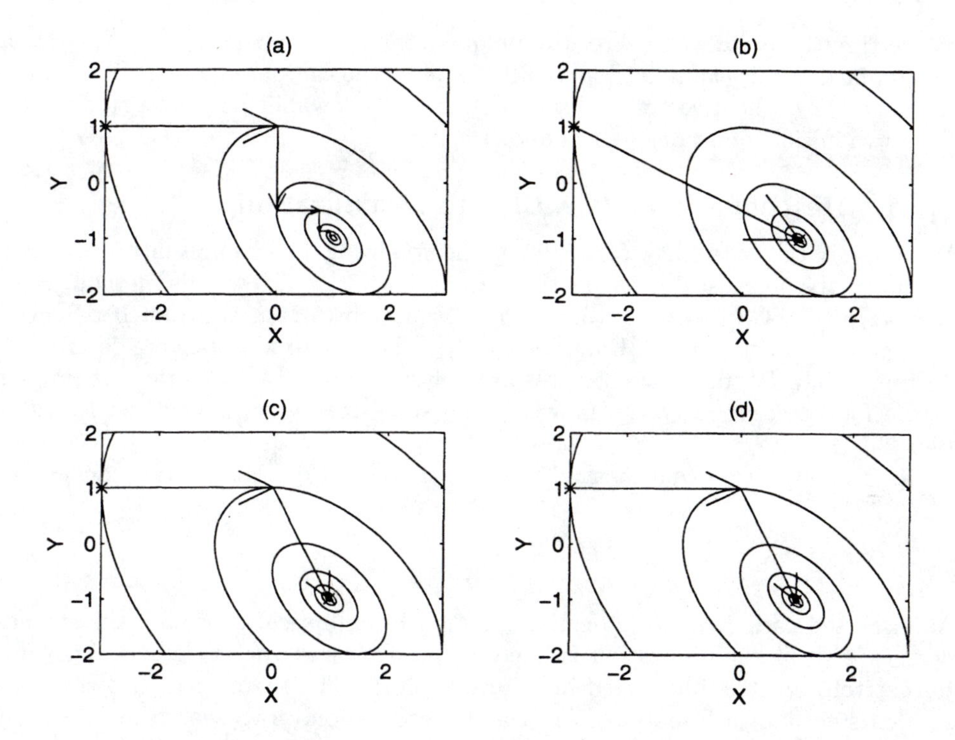

Figure 6.14. *Progress of four descent algorithms with line searches for minimizing a quadratic function* $E(x, y) = x^2 + xy + y^2 - x + y$. *(a) the steepest descent method.* (MATLAB file: `hem.m`) *(b) Newton's method; (c) BFGS quasi-Newton method* (MATLAB file: `bfgs.m`); *(d) Fletcher-Reeves's conjugate gradient method.* (MATLAB file: `cg.m`)

In the quadratic case, the general gradient-based formula in Equation (6.9) takes the following form:

$$\boldsymbol{\theta}_{\text{next}} = \boldsymbol{\theta}_{\text{now}} - \frac{\mathbf{g}^T \mathbf{G} \mathbf{g}}{\mathbf{g}^T \mathbf{G} \mathbf{A} \mathbf{G} \mathbf{g}} \mathbf{g}. \tag{6.56}$$

If we successfully make $\mathbf{G}$ close to $\mathbf{A}^{-1}$, the rapid convergence property of Newton's method can be drawn [see Figure 6.14(b)].

In the following, we consider a minimization problem in which the objection function, as shown in Figure 6.13(a), is a two-dimensional quadratic function defined by

$$E(\boldsymbol{\theta}) = E(x, y) = x^2 + xy + y^2 - x + y, \tag{6.57}$$

where $\boldsymbol{\theta} = [x, y]$. The gradient of this function is

$$\nabla E(x, y) = [2x + y - 1, \;\; x + 2y + 1]^T. \tag{6.58}$$

By setting the gradient to zero, the unique minimum point $(x^*, y^*) = (1, -1)$ can be obtained in this example. Figure 6.13(b) plots the contour curves (solid ellipses), the negative gradient vector field (arrows), and two downhill gradient paths starting from two initial conditions $[-3, 1]$ and $[1.5, 2]$.

6.7.1 Descent Methods with Line Minimization

We present the feasibility of applying the following four representative descent methods: the steepest descent method [Equation (6.11)], Newton's method [Equation (6.15)], BFGS quasi-Newton method [Equation (6.25)], and Fletcher-Reeves's conjugate gradient method [Equation (6.52)]. (These four methods are described in Sections 6.3, 6.4.1, 6.4.3, and 6.6, respectively.) Figure 6.14 illustrates the progress of the four descent methods, in which the step sizes were determined by Equation (6.55).

In the steepest descent method whereby $\mathbf{d} = -\mathbf{g}$, Equation (6.55) reduces to

$$\eta^* = \frac{\mathbf{g}^T\mathbf{g}}{\mathbf{g}^T\mathbf{A}\mathbf{g}}. \tag{6.59}$$

As discussed earlier, if the objective function has elliptical contours, the steepest descent with line minimization is likely to produce a zigzag trajectory, known as **hemstitching**; it is illustrated in Figure 6.14(a). The search path is orthogonal at each step due to Equation (6.12), and there are only two search directions. If the contours are circular, the search would be a one-step process instead of the inefficient zigzag one shown here (see Figure 6.4).

On the other hand, the minimum was reached in a single stride by Newton's method [Figure 6.14(b)]. Quasi-Newton and conjugate gradient methods produced the same two-step trajectory to reach the minimum, as shown in Figures 6.14(c) and 6.14(d).

Here we determined the step sizes analytically by using Equation (6.55). In general, determining the optimal step size in Equation (6.5) for a general function $E(\boldsymbol{\theta})$ is not a trivial task, as discussed in Section 6.5. In the next subsection, based on the steepest descent method, we discuss *fixed* or *heuristically updating* step size determination schemes that are less computational and are applicable to any complicated objective functions.

6.7.2 Steepest Descent Method without Line Minimization

In recalling Equation (6.11), we first use a small fixed step size η:

$$\boldsymbol{\theta}_{\text{next}} = \boldsymbol{\theta}_{\text{now}} - \eta\mathbf{g}. \tag{6.60}$$

A slightly different version of Equation (6.60) can be obtained by normalizing the gradient:

$$\boldsymbol{\theta}_{\text{next}} = \boldsymbol{\theta}_{\text{now}} - \kappa\frac{\mathbf{g}}{\|\mathbf{g}\|}, \tag{6.61}$$

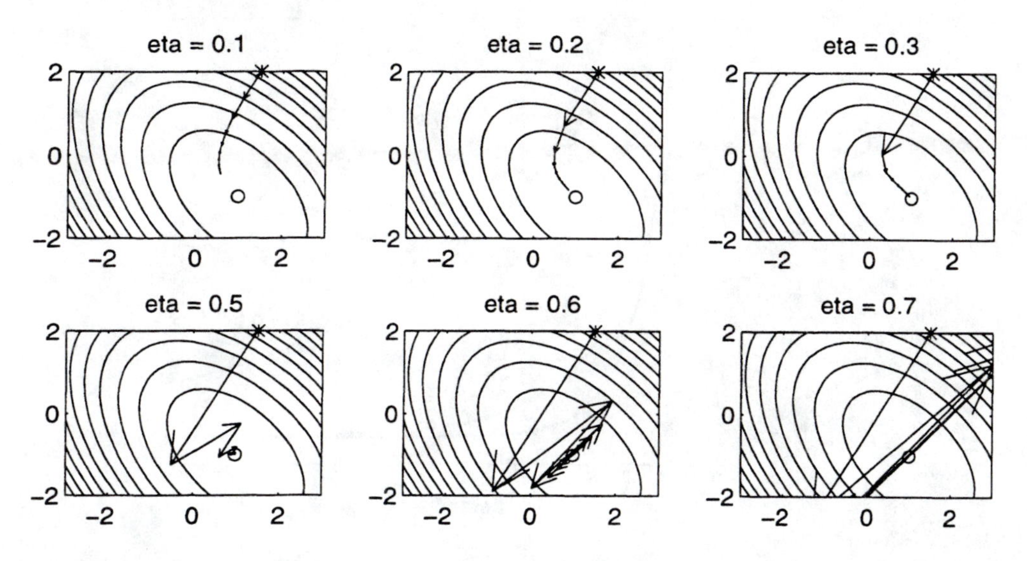

Figure 6.15. *The effects of step sizes η by the steepest descent method defined in Equation(6.60). The search is inefficient when $\eta \leq 0.1$ and unstable when $\eta \geq 0.7$.* (MATLAB file: `gdss1.m`)

where κ is the *real* step size, indicating the Euclidean distance of the transition from $\boldsymbol{\theta}_{\text{now}}$ to $\boldsymbol{\theta}_{\text{next}}$:

$$\kappa = \|\boldsymbol{\theta}_{\text{next}} - \boldsymbol{\theta}_{\text{now}}\|.$$

To differentiate Equations (6.60) and (6.61), we refer to Equation (6.60) as *simple steepest descent*, and Equation (6.61) as a *normalized version of simple steepest descent* in this section.

The magnitude of the step $\eta\mathbf{g}$ in Equation (6.60) with a fixed η automatically changes at each iteration due to different gradients of $\mathbf{g}$. If the minimum point lies in the plateau landscape, then $\mathbf{g}$ tends to be infinitesimally small and the simple steepest descent in Equation (6.60) has slow convergence. On the other hand, the normalized version of simple steepest descent in Equation (6.61) with a fixed κ always makes the same strides, neglecting how steep the slope is. For comparison, we apply Equations (6.60) and (6.61) to the same quadratic problem in the next subsections.

Using a Small Fixed η

Figure 6.15 summarizes the results of applying Equation (6.60) with six different values of η. The search is inefficient when η is less than 0.2. When the step size η is 0.6, the search path exhibits oscillatory behavior. If the step size exceeds 0.6, the search path diverges and the method fails.

Figure 6.16 shows comparisons in Euclidean distance from the minimum point

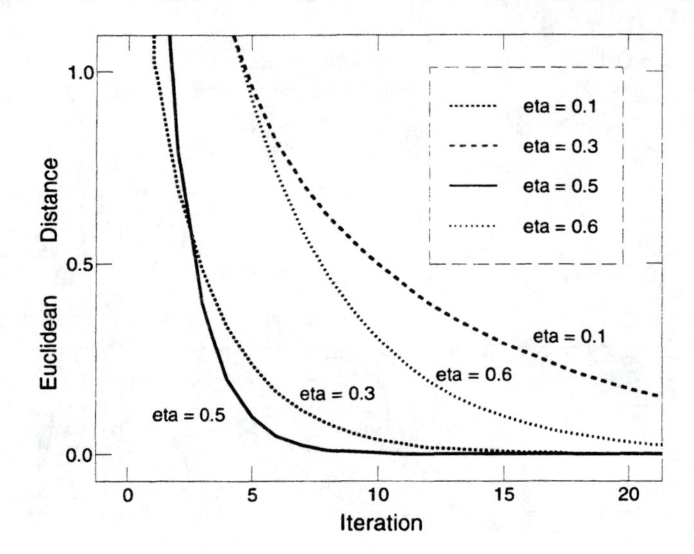

Figure 6.16. *Comparisons in Euclidean distance from the minimum point at each iteration of the simple steepest descent methods.*

at each iteration of the four simple steepest descent methods when $\eta = 0.1, 0.3, 0.5$, and 0.6. As indicated in Figures 6.15 and 6.16, an ideal value of η may be close to 0.5.

How to Update η

Instead of using a fixed small step size, Chan and Fallside [9] introduced a heuristic method capable of updating η in Equation (6.11) during the steepest descent process. They used the well-known **backpropagation learning rule with a momentum term**, as contrived by Rumelhart et al [43], to enhance the simple steepest descent method. The formula with an embedded **momentum** term ω generates the direction vectors of

$$\mathbf{d}_k = -\mathbf{g}_k + \omega \mathbf{d}_{k-1}. \tag{6.62}$$

(The momentum term ω regulates the influence of the previous descent direction. See also Section 9.4.2.) Chan and Fallside proposed a strategy to adjust the step size η:

$$\eta_k = \eta_{k-1}(1 + \frac{1}{2} \cos \alpha_k) \tag{6.63}$$

where

$$\cos \alpha_k = \frac{\mathbf{g}_k \mathbf{d}_{k-1}}{\|\mathbf{g}_k\| \, \|\mathbf{d}_{k-1}\|}.$$

This formula aims to increase the step size when the direction looks good (according to the angle α). The formula correlates with the situation in Figure 6.15 when

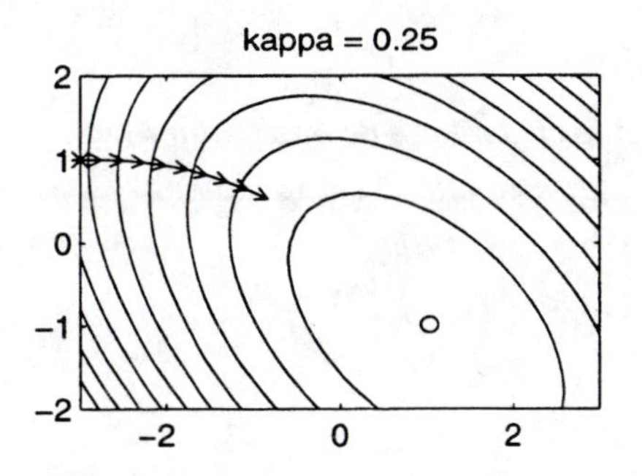

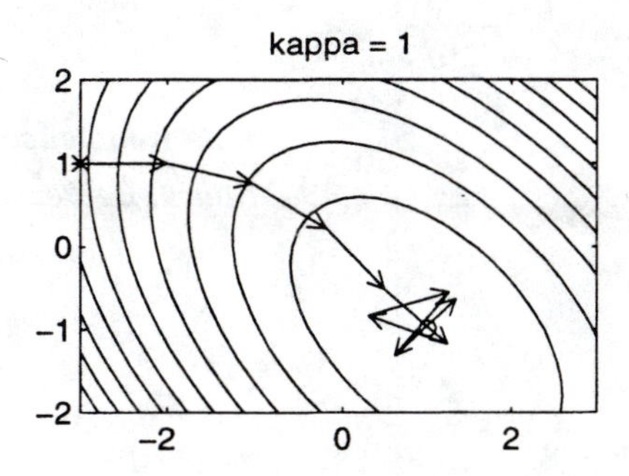

Figure 6.17. *The effects of step sizes κ in the normalized version of simple steepest descent method. (a) Inefficient search when κ is 0.25; (b) oscillatory behavior when κ is 1.* (MATLAB file: `gdss2.m`)

$\eta = 0.1$ or 0.2. Many steps were taken in the same direction because $\|\mathbf{g}\|$ decreased when closer to the minimum in this example.

Equation (6.62) is similar to a class of conjugate gradient methods defined in Equation (6.48). Chan and Shatin [10] presented an advantage of this method over Fletcher-Reeves's conjugate gradient method (Section 6.6).

Using a Small Fixed κ

When a normalized version of the simple steepest descent method [Equation (6.61)] is used, the actual step size κ also affects the search results. To demonstrate the effects, we tested two small fixed values of *kappa* to minimize the quadratic function in Figure 6.13. Figure 6.17 summarizes those results; Figure 6.17(a) is the search path when κ is 0.25, and Figure 6.17(b) is the search path when κ is 1. A small value of κ obviously leads to an inefficient search unless it is already in the vicinity of the minimum. On the other hand, a large value of κ allows the search process to approach the minimum efficiently, but it then oscillates around the local minimum and cannot pinpoint it precisely.

How to Update κ

The preceding example calls for an adaptive strategy to adjust the step size κ dynamically. Based on empirical observations, the step size κ can be updated according to the following two heuristic rules [29]:

1. If the objective function undergoes m consecutive reductions, increase κ by $p\%$.
2. If the objective function undergoes n consecutive combinations of one increase and one reduction, decrease κ by $q\%$.

Figure 6.18 shows that typical values for m, n, p, and q are 4, 2, 10, and 10, respectively. These typical values are more or less chosen arbitrarily. This updating

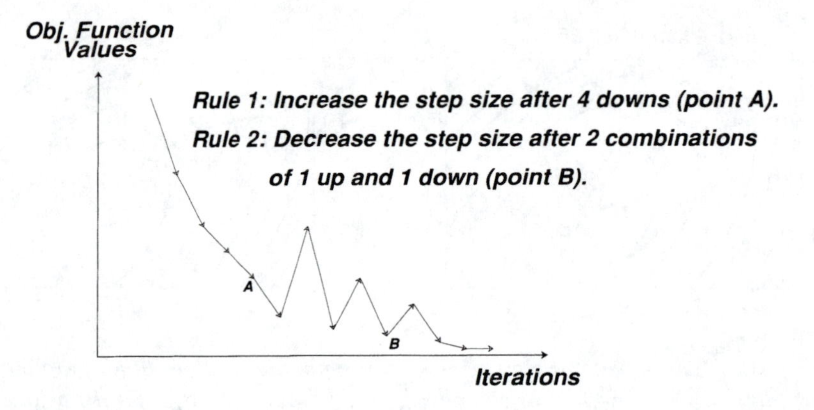

Figure 6.18. *Two heuristic rules for updating step size κ.*

strategy is incorporated in the hybrid learning described in Chapters 12 and 19.

6.8 NONLINEAR LEAST-SQUARES PROBLEMS

In the following discussion of least-squares problems, we wish to optimize a model by minimizing a squared error measure between desired outputs and the model's outputs. We assume a t-input single-output nonlinear model with n modifiable parameters:

$$y = f(\mathbf{x}, \boldsymbol{\theta}), \tag{6.64}$$

where $\mathbf{x}$ is the input vector of size t, y is the model's scalar output, and $\boldsymbol{\theta}$ is the parameter vector of size m.

Given a set of m training data pairs $(\mathbf{x}_p; \mathbf{t}_p)$, $p = 1, \ldots, m$, the most common objective in data fitting and nonlinear regression problems is to find the optimal $\boldsymbol{\theta}$ that minimizes the sum of squared errors:

$$\begin{aligned} E(\boldsymbol{\theta}) &= \sum_{p=1}^{m} (t_p - y_p)^2 \\ &= \sum_{p=1}^{m} (t_p - f(\mathbf{x}_p, \boldsymbol{\theta}))^2. \\ &= \sum_{p=1}^{m} r_p(\boldsymbol{\theta})^2 \\ &= \mathbf{r}^T(\boldsymbol{\theta})\mathbf{r}(\boldsymbol{\theta}), \end{aligned} \tag{6.65}$$

where t_p is the desired output when input is $\mathbf{x}_p$; $y_p = f(\mathbf{x}_p, \boldsymbol{\theta})$ is the model's output when input is $\mathbf{x}_p$; $r_p(\boldsymbol{\theta})$ is the difference between t_p and y_p; and $\mathbf{r}(\boldsymbol{\theta})$ is a vector composed of $r_i(\boldsymbol{\theta})$, $i = 1, \ldots, m$. We shall derive the gradient and Hessian of the preceding objective function. Such information will be used in the developments of Gauss-Newton and Levenberg-Marquardt methods. In the remainder, we focus on such methods for solving the nonlinear least-squares problem [33, 28]; that is, $\boldsymbol{\theta}$ is chosen to be the least-squares estimator.

The gradient vector of E at $\boldsymbol{\theta}$ is the vector of the first derivatives of E:

$$\begin{aligned}\mathbf{g} = \mathbf{g}(\boldsymbol{\theta}) &\equiv \frac{\partial E(\boldsymbol{\theta})}{\partial \boldsymbol{\theta}} \\ &= 2\textstyle\sum_{p=1}^{m} r_p(\boldsymbol{\theta})\frac{\partial r_p(\boldsymbol{\theta})}{\partial \boldsymbol{\theta}} \\ &= 2\mathbf{J}^T\mathbf{r},\end{aligned} \tag{6.66}$$

where $\mathbf{J}$ is the **Jacobian matrix** of $\mathbf{r}$. [See Equation (5.7) on page 103.] Since $r_p(\boldsymbol{\theta}) = t_p - f(\mathbf{x}_p, \boldsymbol{\theta})$, the pth row of $\mathbf{J}$ is equal to $-\nabla_{\boldsymbol{\theta}}^T f(\mathbf{x}_p, \boldsymbol{\theta})$.

Taking additional partial derivatives yields the Hessian matrix $\mathbf{H}$ at $\boldsymbol{\theta}$:

$$\begin{aligned}\mathbf{H} = \mathbf{H}(\boldsymbol{\theta}) &\equiv \frac{\partial^2 E(\boldsymbol{\theta})}{\partial \boldsymbol{\theta}\partial \boldsymbol{\theta}^T} \\ &= 2\textstyle\sum_{p=1}^{m}\left[\frac{\partial r_p(\boldsymbol{\theta})}{\partial \boldsymbol{\theta}}\frac{\partial r_p(\boldsymbol{\theta})}{\partial \boldsymbol{\theta}^T} + r_p(\boldsymbol{\theta})\frac{\partial^2 r_p(\boldsymbol{\theta})}{\partial \boldsymbol{\theta}\partial \boldsymbol{\theta}^T}\right] \\ &= 2(\mathbf{J}^T\mathbf{J} + \mathbf{S}),\end{aligned} \tag{6.67}$$

where $\mathbf{S}$ is defined by

$$\mathbf{S} = \mathbf{S}(\boldsymbol{\theta}) = \sum_{p=1}^{m} r_p(\boldsymbol{\theta})\frac{\partial^2 r_p(\boldsymbol{\theta})}{\partial \boldsymbol{\theta}\partial \boldsymbol{\theta}^T}.$$

6.8.1 Gauss-Newton Method

The **Gauss-Newton method**, also known as the **linearization method**, first uses a Taylor series expansion to obtain a linear model that approximates the original nonlinear model. Then the ordinary least-squares methods discussed in Chapter 5 are employed to estimate the model's parameters. More specifically, if we allow the current parameters to be denoted by $\boldsymbol{\theta}_{\text{now}}$, the nonlinear model f in Equation (6.64) can be expanded in a Taylor series around $\boldsymbol{\theta} = \boldsymbol{\theta}_{\text{now}}$ and only the linear terms are retained:

$$y = f(\mathbf{x}, \boldsymbol{\theta}_{\text{now}}) + \sum_{i=1}^{n}\left(\left.\frac{\partial f(\mathbf{x}, \boldsymbol{\theta})}{\partial \theta_i}\right|_{\boldsymbol{\theta}=\boldsymbol{\theta}_{\text{now}}}\right)(\theta_i - \theta_{i,\text{now}}). \tag{6.68}$$

Inspection of the preceding equation reveals that the translated output $y - f(\mathbf{x}, \boldsymbol{\theta}_{\text{now}})$ is a linear function of the translated parameters $\theta_i - \theta_{i,\text{now}}$.

Plugging this approximated linear model into E in Equation (6.65) yields

$$\begin{aligned}E(\boldsymbol{\theta}) &= \left\|\mathbf{t} - f(\mathbf{x}, \boldsymbol{\theta}_{\text{now}}) - \frac{\partial f(\mathbf{x}, \boldsymbol{\theta}_{\text{now}})}{\partial \boldsymbol{\theta}}(\boldsymbol{\theta} - \boldsymbol{\theta}_{\text{now}})\right\|^2 \\ &= \|\mathbf{r} + \mathbf{J}^T(\boldsymbol{\theta} - \boldsymbol{\theta}_{\text{now}})\|^2 \\ &= \|\mathbf{r} + \mathbf{J}^T\boldsymbol{\delta}\|^2,\end{aligned} \tag{6.69}$$

where $\boldsymbol{\delta} = \boldsymbol{\theta} - \boldsymbol{\theta}_{\text{now}}$. We thus obtain $\boldsymbol{\theta}_{\text{next}}$ to minimize the preceding equation by solving Equation (6.10). That is,

$$\left.\frac{\partial E(\boldsymbol{\theta})}{\partial \boldsymbol{\theta}}\right|_{\boldsymbol{\theta}=\boldsymbol{\theta}_{\text{next}}} = \mathbf{J}^T\{\mathbf{r} + \mathbf{J}(\boldsymbol{\theta}_{\text{next}} - \boldsymbol{\theta}_{\text{now}})\} = 0.$$

Therefore, we have the following *Gauss-Newton formula*, expressed as

$$\begin{aligned}\boldsymbol{\theta}_{\text{next}} &= \boldsymbol{\theta}_{\text{now}} - (\mathbf{J}^T\mathbf{J})^{-1}\mathbf{J}^T\mathbf{r} \\ &= \boldsymbol{\theta}_{\text{now}} - \tfrac{1}{2}(\mathbf{J}^T\mathbf{J})^{-1}\mathbf{g},\end{aligned} \tag{6.70}$$

where $\mathbf{g}$ is the gradient defined in Equation (6.66). The preceding equation conforms to the general form of gradient-based descent in Equation (6.9), with $\mathbf{G} = \mathbf{J}^T\mathbf{J}$ being positive definite unless $\mathbf{J}$ is not of full rank.

Other Derivations

The Gauss-Newton method is named after the fact, introduced by Gauss in 1809, that Equation (6.70) can be obtained by modifying the Newton method. In light of Equations (6.66) and (6.67), Equation (6.15) (i.e., Newton's method) can be rewritten as

$$\begin{aligned}\boldsymbol{\theta}_{\text{next}} &= \boldsymbol{\theta}_{\text{now}} - \tfrac{1}{2}\mathbf{H}^{-1}\mathbf{g} \\ &= \boldsymbol{\theta}_{\text{now}} - (\mathbf{J}^T\mathbf{J} + \mathbf{S})^{-1}\mathbf{J}^T\mathbf{r}.\end{aligned} \tag{6.71}$$

The *Gauss-Newton* method is based on the assumption that $\mathbf{S}$ is smaller than $\mathbf{J}^T\mathbf{J}$. This assumption amounts to neglecting $\mathbf{S}$, which involves the second derivatives. We then obtain Equation (6.70). The strength of the Gauss-Newton algorithms from the Newton methods resides in that they require only the first derivatives.

Notably, with the same notations used in Chapter 5, Equation (6.70) can be represented as

$$\begin{aligned}\boldsymbol{\theta}_{\text{next}} &= \boldsymbol{\theta}_{\text{now}} + (\mathbf{A}^T\mathbf{A})^{-1}\mathbf{A}^T\Delta\mathbf{y} \\ &= \boldsymbol{\theta}_{\text{now}} + \Delta\boldsymbol{\theta},\end{aligned} \tag{6.72}$$

where $\Delta y_i = y_i - f(\mathbf{x}_i, \boldsymbol{\theta}_{\text{now}})$, $\Delta\boldsymbol{\theta} = (\mathbf{A}^T\mathbf{A})^{-1}\mathbf{A}^T\Delta\mathbf{y}$, and $\mathbf{J}$ corresponds to $-\mathbf{A}$. This occurrence can be shown in the following manner. After plugging all training data into Equation (6.68), we have the following matrix equation:

$$\mathbf{A}\Delta\boldsymbol{\theta} = \mathbf{r},$$

where the ith element of $\mathbf{r}$ is $r_i = t_i - f(\mathbf{x}_i, \boldsymbol{\theta}_{\text{now}})$, $\Delta\boldsymbol{\theta} = \boldsymbol{\theta}_{\text{next}} - \boldsymbol{\theta}_{\text{now}}$, and the element at row i and column j of matrix $\mathbf{A}$ is $\left.\dfrac{\partial f(\mathbf{x}_i, \boldsymbol{\theta})}{\partial \theta_j}\right|_{\boldsymbol{\theta}=\boldsymbol{\theta}_{\text{now}}}$. By using the standard least-squares method described in Chapter 5, we have

$$\Delta\boldsymbol{\theta} = (\mathbf{A}^T\mathbf{A})^{-1}\mathbf{A}^T\mathbf{r}.$$

Implementations

Calculating $(\mathbf{J}^T\mathbf{J})^{-1}$ in the Gauss-Newton method in Equation (6.70) may not be numerically stable for a certain $\mathbf{J}$. To overcome this limitation, $\boldsymbol{\delta}$ can be obtained by solving the *linear* least-squares problem

$$\text{minimize}_{\boldsymbol{\delta}} \ \|\mathbf{r} + \mathbf{J}^T\boldsymbol{\delta}\|^2, \tag{6.73}$$

based on decompositions of $\mathbf{J}$, such as QR decompositions [24] and singular-value decompositions.

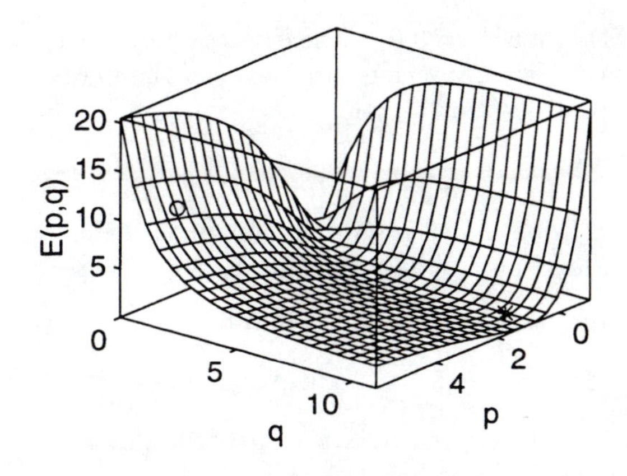

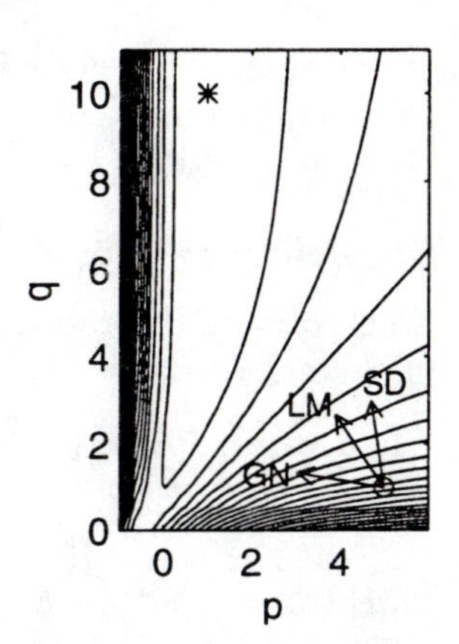

Figure 6.19. *The left figure illustrates the surface of the objective function defined in Equation (6.75). The right figure shows a Levenberg-Marquardt (LM) direction, the Gauss-Newton (GN) direction, and the steepest descent (SD) direction. The circle signifies the starting point, and the * denotes the optimal minimum.*

Hartley's Method

The Gauss-Newton method can be modified by introducing the step size, resulting in the form similar to Equation (6.17). The step length can be determined in conjunction with line searches to satisfy Equation (6.7). This modified method is called *Hartley's method* [25, 26] or the *dumped Gauss-Newton method.*

It is known that Hartley's modification does not make the Gauss-Newton method as robust as the Levenberg-Marquardt modification, which follows.

6.8.2 Levenberg-Marquardt Concepts

The Levenberg-Marquardt concept discussed in Section 6.4.2 can be applied to the Gauss-Newton method. This method can handle well ill-conditioned matrices $\mathbf{J}^T\mathbf{J}$ by altering Equation (6.70) to

$$\boldsymbol{\theta}_{\text{next}} = \boldsymbol{\theta}_{\text{now}} - (\mathbf{J}^T\mathbf{J} + \lambda\mathbf{I})^{-1}\mathbf{g}_h, \tag{6.74}$$

where λ is some nonnegative value and $\mathbf{g}_h \equiv \frac{1}{2}\mathbf{g}$ for simplicity.

For instance, we consider a trivial nonlinear model $f(\mathbf{x}_k, \boldsymbol{\theta})$ with two adjustable parameters p and q; i.e., $\boldsymbol{\theta} = (p, q)^T$. The objective is to find the optimal $\boldsymbol{\theta}^* = (1, 10)$ that minimizes the following sum of squared errors.

$$E(\boldsymbol{\theta}) = \sum_{k=1}^{10}(t_k - f(\mathbf{x}_k, \boldsymbol{\theta}))^2, \tag{6.75}$$

where t_k is the desired output when the input is $0.1k$ [6]. Figure 6.19 illustrates the surface of the objective function E, and Levenberg-Marquardt direction that is determined by using Equation (6.74) when $\lambda = 0.07$. Clearly, the Levenberg-Marquardt direction (in highlighted arrow) is an intermediate between the Gauss-Newton direction ($\lambda \to 0$) and the steepest descent direction ($\lambda \to \infty$).

Theoretical aspects

To clarify the theoretical aspects behind Equation (6.74) we quote a series of three important theorems, which are due to Morrison [35] and Marquardt [33].

Theorem 6.1 *Let $\lambda \geq 0$ be arbitrary and let $\boldsymbol{\delta}_0$ satisfy the equation*

$$(\mathbf{J}^T\mathbf{J} + \lambda\mathbf{I})\boldsymbol{\delta}_0 = -\mathbf{g}_h, \tag{6.76}$$

which corresponds to Equation (6.74). Then $\boldsymbol{\delta}_0$ minimizes the approximated $\mathbf{E}(\boldsymbol{\theta})$ defined in Equation (6.69), on the **sphere** *whose radius $\|\boldsymbol{\delta}\|$ satisfies*

$$\|\boldsymbol{\delta}\|^2 = \|\boldsymbol{\delta}_0\|^2.$$

(For proof, see refs. [33] and [35].)

Theorem 6.2 *Let $\boldsymbol{\delta}(\lambda)$ be the solution of Equation (6.76) for a given value of λ. Then $\|\boldsymbol{\delta}(\lambda)\|^2$ is a continuously decreasing function of λ such that as $\lambda \to \infty$, $\|\boldsymbol{\delta}(\lambda)\|^2 \to 0$.*

Proof: A rough proof can be given as follows. Notably, taking derivatives of Equation (6.76) with respect to λ yields

$$\boldsymbol{\delta}(\lambda) + (\mathbf{J}^T\mathbf{J} + \lambda\mathbf{I})\frac{d\boldsymbol{\delta}(\lambda)}{d\lambda} = 0.$$

This equation can be rewritten as

$$\frac{d\boldsymbol{\delta}(\lambda)}{d\lambda} = -(\mathbf{J}^T\mathbf{J} + \lambda\mathbf{I})^{-1}\boldsymbol{\delta}(\lambda).$$

Thus, we obtain

$$\begin{aligned}\frac{d\|\boldsymbol{\delta}(\lambda)\|^2}{d\lambda} &= 2\boldsymbol{\delta}^T(\lambda)\frac{d\|\boldsymbol{\delta}(\lambda)\|^2}{d\lambda} \\ &= -2\boldsymbol{\delta}^T(\lambda)(\mathbf{J}^T\mathbf{J} + \lambda\mathbf{I})^{-1}\boldsymbol{\delta}(\lambda) \\ &< 0.\end{aligned}$$

This inequality shows that $\|\boldsymbol{\delta}(\lambda)\|^2$ is a continuously decreasing function of λ. (For detailed proof, see refs. [33] and [35].)

□

Theorem 6.3 *Let γ be the angle between $\boldsymbol{\delta}_0$ and the negative gradient of E $(-\mathbf{g})$. Then γ is a continuous monotone decreasing function λ such that as $\lambda \to \infty$, $\gamma \to 0$. Since $\mathbf{g}$ is independent of λ, it follows that $\boldsymbol{\delta}_0$ rotates toward the negative gradient $(-\mathbf{g})$ as $\lambda \to \infty$. (For proof, see ref. [33].)*

The situation of Theorem 6.3 is illustrated in Figure 6.5.

In proving Theorem 6.2, Marquardt [33] used the concept of **scaling** [as in Equation (6.16)]. Now we show that some scaling rewrites Equation (6.76) as

$$(\mathbf{J}^T\mathbf{J} + \lambda\mathbf{D})\boldsymbol{\delta}_0 = -\mathbf{g}_h. \tag{6.77}$$

The diagonal matrix $\mathbf{D}$ can be determined by

$$\mathbf{D}_{ii} = (\mathbf{J}^T\mathbf{J})_{ii} + p,$$

where p is some nonnegative value [31, 36, 37]. For instance,

$$\mathbf{D} = \text{diag}(\mathbf{J}^T\mathbf{J}) + \mathbf{I}.$$

Consider the following transformation:

$$\boldsymbol{\theta}' = \sqrt{\mathbf{D}}\boldsymbol{\theta}, \quad \mathbf{J} = \mathbf{J}'\sqrt{\mathbf{D}}, \quad \mathbf{g}_h = \sqrt{\mathbf{D}}\mathbf{g}'_h. \tag{6.78}$$

Equation (6.77) can be rewritten in the transformed space $\boldsymbol{\theta}'$:

$$(\mathbf{J}'^T\mathbf{J}' + \lambda\mathbf{I})\boldsymbol{\delta}'_0 = -\mathbf{g}'_h, \tag{6.79}$$

which is the same form as Equation (6.76). This implies that solving Equation (6.77) automatically accomplishes some scaling. When the columns of $\mathbf{J}^T\mathbf{J}$ have significantly different norms, Equation (6.77) may work better than Equation (6.76).

Implementaions

The Gauss-Newton-based Levenberg-Marquardt method works well in practice and has become the standard of nonlinear least-squares routines. There are many variations in the Levenberg-Marquardt procedures. The following algorithm is one of them.

Algorithm 6.3 *Main loop of the Gauss-Newton based Levenberg-Marquardt algorithm*

(1) $d \leftarrow 0.001$, and $factor \leftarrow 10.0$

(2) Evaluate $E(\boldsymbol{\theta}_{\text{now}})$.

(3) $h_{\text{max}} \leftarrow \max\{\ \text{diag}(\ [\mathbf{J}^T\mathbf{J}_{kk}])\ \}, k = 1, 2, \ldots.$

(4) $\lambda \leftarrow dh_{\text{max}}$.

(5) Solve Equation (6.76) or (6.77) for $\boldsymbol{\delta}(\equiv \boldsymbol{\theta}_{\text{next}} - \boldsymbol{\theta}_{\text{now}})$.

(6) Evaluate $E(\boldsymbol{\theta}_{\text{next}})$(i.e.,$E(\boldsymbol{\theta}_{\text{now}} + \boldsymbol{\delta})$).

(7)

$$\begin{array}{ll} \text{If } E(\boldsymbol{\theta}_{\text{next}}) < E(\boldsymbol{\theta}_{\text{now}}), & d \leftarrow d/factor, \\ & \boldsymbol{\theta}_{\text{now}} \leftarrow \boldsymbol{\theta}_{\text{next}}, \\ & \text{go to (4).} \\ \text{Otherwise,} & d \leftarrow d \cdot factor, \\ & \text{go to (4).} \end{array}$$

□

Initial parameter values, which are problem dependent, must be found by a process of trial and error; the initial λ may be 0.01, and *factor* can be set to a smaller value, such as 2.0. In place of procedures (3) and (4), we can simply use an alternative procedure: $\lambda \leftarrow d$. Also, to adapt λ, a ***trust-region*** approach [14, 34] or a line search [36] can be incorporated.

As discussed in solving Equation (6.70), the linear least-squares problem in (6.73) can be considered. Likewise, instead of calculating $\mathbf{J}^T\mathbf{J}$, numerically stable methods can be applied to solving Equation (6.76) or (6.77) in procedure (5). That is, $\boldsymbol{\delta}$ can be obtained as the solution of the following linear least-squares problem:

$$\text{minimize}_{\delta} \quad \left\| \begin{pmatrix} \mathbf{r} \\ 0 \end{pmatrix} + \begin{pmatrix} \mathbf{J} \\ \sqrt{\lambda \mathbf{D}} \end{pmatrix} \boldsymbol{\delta} \right\|^2. \tag{6.80}$$

This new Jacobian matrix $\begin{bmatrix} \mathbf{J} \\ \sqrt{\lambda \mathbf{D}} \end{bmatrix}$ can be regarded as an expanded matrix of $\mathbf{J}$ by assuming that n training data are added.

6.9 INCORPORATION OF STOCHASTIC MECHANISMS

Virtually no gradient-based descent algorithm is guaranteed to find the global optimum of a complex objective function within a finite period of time. All descent methods discussed so far are **deterministic** in the sense that they inevitably lead to convergence to the nearest local minimum. Figure 6.20 displays a typical behavior common to deterministic gradient-based descent methods, where Figure 6.20(a) illustrates a bimodal surface containing two minima, and Figure 6.20(b) shows that two negative or downhill gradient paths, starting from two close but different initial points $(2, 3)$ and $(2.5, 3)$, converge to different minima. Without further explanation, selecting initial positions for the deterministic methods clearly has a decisive effect on the final results.

In practice, however, knowing good starting points is nearly impossible. If the starting point is to be randomly selected, then it would be advisable to employ a random method to perturb the final positions where the method converges and begin the optimization process all over again. That is, the approach must somehow include **stochastic** nature.

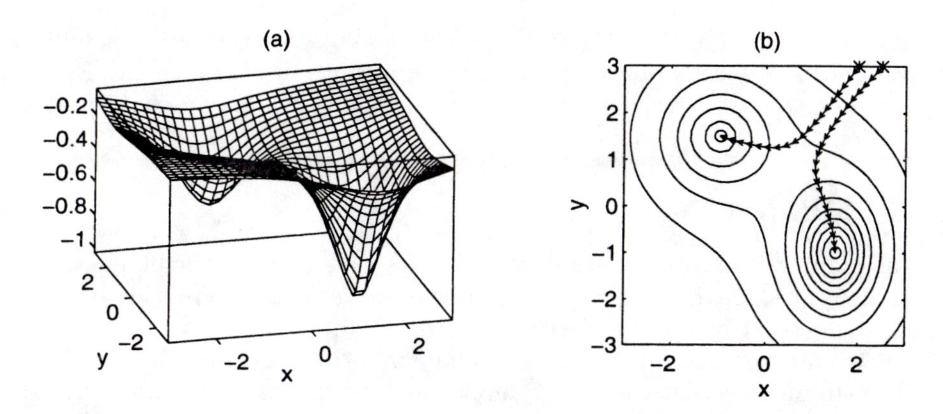

Figure 6.20. *The deterministic descent methods are sensitive to initial conditions: (a) a bimodal surface; (b) contour curves, and two negative gradient paths starting from two different points.* (MATLAB file: `gdconv.m`)

Also, if calculating the gradient is time consuming or difficult due to the complexity of the objective function, we should resort to *stochastic* or derivative-free optimization methods, which are introduced in Chapter 7 (e.g., genetic algorithms, simulated annealing, the random search method, and the downhill Simplex method). The stochastic derivative-free methods require many function evaluations for the attempt to descend toward minima without derivative information. Thus, they usually require more computation than for deterministic derivative-based methods to reach a satisfactory level. Hence, it may be better to approximate the gradient by evaluating a finite difference in a single step if the gradient is not available. For those reasons, constructing an algorithm that combines derivative information and stochastic nature is quite natural.

There must be many algorithmic variations for fusing deterministic methods and stochastic ones, depending on their goals. One possible goal is to guarantee that the current point is a local minimum (i.e., neither a saddle point nor a local maximum [45]). Another possible goal is to escape a local minimum in an attempt to reach a global minimum. For those goals, many things must be considered. For instance, some criterion should be set up to determine when to switch a deterministic method to a stochastic one, and vice versa. The timing for switching must be important because a stochastic optimization scheme usually imposes a significant computational expense.

Baba et al. proposed a hybrid algorithm that combines a conjugate gradient method and a random search method [3]. Since the neural network (NN) may possibly construct many small local minima, such unifying algorithms have attracted much attention for NN learning [11, 48].

Furthermore, as long as squashing functions are employed in an NN architecture, they act as *implicit constraints* because the net inputs to hidden nodes may get driven to their limits (saturation). In this sense, NN learning is not fully categorized into the *unconstrained* optimization, which we discussed so far. Even with a sophisticated unconstrained optimization technique, NN learning may fail due to saturation (implicit constraint).

6.10 SUMMARY

In this chapter, we have addressed important aspects of widely employed gradient-based descent algorithms. The primary differences among them reside in selecting successive descent directions. Once the downhill direction is determined, all algorithms require a step down toward the minimum in the corresponding line.

The steepest descent method is important for practical application. It can be employed as a touchstone for discovering the intrinsic difficulty of a given task and for establishing a certain reference for performance comparison. Particularly when the task is a large and complex problem, it must be worthwhile due to its simplicity. Also, we have presented a class of Newton methods. The usual practice is to modify the pure Newton's (or Newton-Rafson) method since evaluating the Hessian may not be worthwhile due to its heavy computational requirements. Many promising algorithms can be regarded as a sort of intermediate between steepest descent and Newton methods (e.g., conjugate gradient methods and Levenberg-Marquardt methods, which are widely used and considered as good general-purpose algorithms for solving nonlinear least-squares problems).

For comparison, we tabulate a compendium of several gradient-based descent methods already discussed in this chapter:

Descent Method	$\mathbf{G}$
Steepest Descent	$\mathbf{I}$
Newton-Raphson	$(\mathbf{J}^T\mathbf{J} + \mathbf{S})^{-1}$
(+ Levenberg-Marquardt)	$(\mathbf{J}^T\mathbf{J} + \mathbf{S} + \lambda\mathbf{I})^{-1}$
Gauss-Newton	$(\mathbf{J}^T\mathbf{J})^{-1}$
(+ Levenberg-Marquardt)	$(\mathbf{J}^T\mathbf{J} + \lambda\mathbf{I})^{-1}$

They are compared in terms of matrix $\mathbf{G}$ in their general formula (6.9). The matrices $\mathbf{J}$ and $\mathbf{S}$ are components of the Hessian matrix, as defined in Equations (6.66) and (6.67).

The described methods can be applied to *neuro-fuzzy learning* in subsequent chapters. Thus, we have explored several variations of useful algorithms in an attempt to find a better algorithmic variation for soft computing methodologies.

EXERCISES

1. Prove the "feasible descent directions" condition 6.7 by the mean value theorem.

2. Prove that the gradient paths and contour curves of a continuous function $E(\boldsymbol{\theta})$ are always perpendicular to each other.

3. In the Polak-Ribiere conjugate gradient algorithm [see Equation (6.50)], the direction vectors are generated by

$$\text{Polak-Ribiere: } \mathbf{d}_k = -\mathbf{g}_k + \frac{\mathbf{g}_k^T(\mathbf{g}_k - \mathbf{g}_{k-1})}{\mathbf{g}_{k-1}^T \mathbf{g}_{k-1}} \mathbf{d}_{k-1}.$$

Verify that the kth direction vector ($\mathbf{d}_k$) and the rest $(k-1)$ of the direction vectors are mutually conjugate, by using Equation (6.38).

4. In light of Algorithm 6.1 for the initial bracketing scheme 1 in Section 6.5.1, sketch another algorithm for scheme 2.

5. To clarify the concept of the Wolfe condition in Section 6.5.3, draw a figure of a Wolfe test, similar to Figure 6.10.

6. Apply the classical Newton method to the simple quadratic problem discussed in Section 6.7, and verify that the Newton method can locate the minimum point in a single step. (Again, the Newton direction is theoretically scale invariant.)

REFERENCES

[1] Hirotugu Akaike. On a successive transformation of probability distribution and its application to the analysis of the optimun gradient method. *Ann. Inst. Statist. Math.*, 11:1–17, 1959.

[2] L. Armijo. Minimization of functions having Lipschitz-continuous first partial derivatives. *Pacific J. Math.*, 16:1–3, 1966.

[3] N. Baba. A new approach for finding the global minimum of error function of neural networks. *Neural Networks*, 2:367–373, 1989.

[4] E. M. L. Beale. A derivation of conjugate gradients. In F. A. Lootsma, editor, *Nemerical methods for nonlinear optimization*, pages 39–43. Academic Press, London, 1972.

[5] C. S. Beightler, D. T. Phillips, and D. J. Wilde. *Foundations of optimization.* Prentice Hall, Upper Saddle River, NJ, 2nd edition, 1979.

[6] M. J. Box. A comparison of several current optimization methods and the use of transformations in constrained problems. *Computer Journal*, 9:67–77, 1966.

[7] C. G. Broyden. Quasi-newton methods and their applications to function minimization. *Math. Comp.*, 21:368–381, 1967.

[8] C. G. Broyden. The convergence of a class of double rank minimization algorithms: Part 1 and 2. *J. Inst. Math. Appl.*, 6:76–90 and 222–231, 1970.

[9] L.-W. Chan and F. Fallside. An adaptive training algorithm for back propagation networks. *Computer Speech and Language*, 2:205–218, 1987.

[10] L.-W. Chan and N. T. Shatin. Efficacy of different learning algorithms of the back-propagation network. In *Proceedings of IEEE Region 10 Conference on Computer and Comunication systems*, pages 23–27, September 1990.

[11] S. L. Chiu. Fuzzy model identification based on cluster estimation. *Journal of Intelligent and Fuzzy Systems*, 2(3), 1994.

[12] H. B. Curry. The method of steepest descent for nonlinear minimization problems. *Quart. Journal of Applied Mathematics*, 2:258–261, 1944.

[13] R Fletcher. A new approach to variable metric algorithms. *Computer Journal*, 13(3):317–322, 1970.

[14] R Fletcher. A modified Marquardt subroutine for nonlinear least squares. Technical Report AERE-R6799, Harwell Report, 1971.

[15] R Fletcher. *Practical methods of optimization: unconstrained optimization.* John Wiley & Sons, New York, 1980.

[16] R Fletcher and M. J. D. Powell. A rapidly convergent descent method for minimization. *Computer Journal*, 6:163–168, 1963.

[17] R Fletcher and C. M. Reeves. Function minimization by conjugate gradients. *Computer Journal*, 7:149–154, 1964.

[18] S. H. Friedberg, A. J. Insel, and L. E. Spence. *Linear algebra.* Prentice Hall, Upper Saddle River, NJ, 1989. 2nd Ed.

[19] P. E. Gill and W. Murray. Quasi-newton methods for unconstrained minimization. *J. Inst. Math. Appl.*, 9:91–108, 1972.

[20] P. E. Gill, W. Murray, and M. H. Wright. *Practical optimization.* Academic Press, New York, 1981.

[21] D. Goldfarb. A family of variable metric methods derived by variational means. *Math. Comp.*, 24:23–26, 1970.

[22] S. M. Goldfeld, R. E. Quandt, and H. F. Trotter. Maximization by quadratic hill climbing. *Econometrica*, 34:541–551, 1966.

[23] A. A. Goldstein. *Constructive real analysis.* Harper & Row, 1968.

[24] G. H. Golub and C. F. Van Loan. *Matrix computation.* Johns Hopkins University Press, Baltimore, 2nd edition, 1989.

[25] H. O. Hartley. The modified gauss-newton method for the fitting of non-linear regression function by least squares. *Technometrics*, 3:269–280, 1961.

[26] H. O. Hartley and A. Booker. Nonlinear least squares estimation. *Ann. Math. Stat.*, 36:638–650, 1965.

[27] M. R. Hestenes and E. Stiefel. Method of conjugate gradients for solving linear systems. *J. Res. NBS*, 49:409–436, 1952.

[28] Jr J. E. Dennis. Nonlinear least squares. In D. Jacobs, editor, *State of the art in numerical analysis*, pages 269–312. Academic Press, London, 1977.

[29] J.-S. Roger Jang. ANFIS: Adaptive-Network-based Fuzzy Inference Systems. *IEEE Transactions on Systems, Man, and Cybernetics*, 23(03):665–685, May 1993.

[30] J. Kowalik and M. R. Osborne. *Methods for unconstrained optimization problems.* Elsevier Science, New York, 1968.

[31] K. Levenberg. A method for the solution of certain problems in least squares. *Quart. Apl. Math.*, 2:164–168, 1944.

[32] David G. Luenberger. *Linear and nonlinear programming.* Addison-Wesley, Reading, MA, 1989. 2nd Ed.

[33] Donald W. Marquardt. An algorithm for least squares estimation of nonlinear parameters. *Journal of the Society of Industrial and Applied Mathematics*, 11:431–441, 1963.

[34] J. J. More. The levenberg-marquardt algorithm: implementation and theory. In G. A. Watson, editor, *Nemerical analysis, lecture notes in mathematics 630*, pages 105–116. Springer-Verlag, London, 1977.

[35] D. D. Morrison. Methods for nonlinear least squares problems and convergence proofs, tracking programs and orbit determination. In *Proceedings Jet Propulsion Laboratory Seminar*, pages 1–9, 1960.

[36] J. C. Nash. Minimizing a nonlinear sum of squares function on a small computer. *J. Inst. Math. Appl.*, 19:231–237, 1977.

[37] J. C. Nash. *Compact numerical methods for computers: linear algebra and function minimization.* Adam Hilger, 1990. 2nd Ed.

[38] E. Polak. *Computational methods in optimization.* Academic Press, 1971.

[39] M. J. D. Powell. Restart procedures for the conjugate gradient method. *Mathematical Programming*, 12:241–254, 1977.

[40] W. H. Press, B. P. Flannery, S. A. Teukolsky, and W. T. Vetterling. *Nemerical recipes in C.* Cambridge University Press, 1991.

[41] H. Rosenbrock. An automatic method for finding the greatest or least value of a function. *The Computer Journal*, 3:175–184, 1960.

[42] Walter Rudin. *Principles of mathematical analysis.* McGraw-Hill, New York, 1976. 3rd Ed.

[43] D. E. Rumlhart, G. E. Hinton, and R. J. Williams. Learning internal representations by error propagation. In D. E. Rumlhart and James L. McClelland, editors, *Parallel distributed processing: volume 1*, chapter 8, pages 318–362. MIT Press, Cambridge, MA., 1986.

[44] L. E. Scales. *Introduction to nonlinear optimization.* Macmillan, London, 1985.

[45] G. A. F. Seber and C. J. Wild. *Nonlinear regression.* John Wiley & Sons, 1989.

[46] D. F. Shanno. Conditioning of Quasi-Newton methods for function minimization. *Math. Comp.*, 24:647–656, 1970.

[47] H. W. Sorenson. Comparison of some conjugate direction precedures for function minimization. *Journal of Franklin Inst.*, 288:421–441, 1969.

[48] M. A. Styblinski and T.-S. Tang. Experiments in nonconvex optimization: Stochastic approximation with function smoothing and simulated annealing. *IEEE Transactions on Neural Networks*, 4:599–613, 1991.

[49] D. A. Wismer and R. Chattergy. *Introduction to nonlinear optimization: a problem solving approach*, chapter 6, pages 139–162. North-Holland, Amsterdam, 1978.

[50] P. Wolfe. A method of conjugate subgradients for minimizing nondifferentialbe functions. *Math. Prog. Study*, 3:145–173, 1975.

Chapter 7

Derivative-Free Optimization

J.-S. R. Jang

7.1 INTRODUCTION

This chapter introduces four of the most popular derivative-free optimization methods: genetic algorithms, simulated annealing, random search method, and downhill simplex search. They have been used extensively for both continuous and discrete optimization problems. Common characteristics shared by these methods are described next.

Derivative freeness These methods do not need functional derivative information to search for a set of parameters that minimize (or maximize) a given objective function. Instead, they rely exclusively on repeated evaluations of the objective function, and the subsequent search direction after each evaluation follows certain heuristic guidelines.

Intuitive guidelines The guidelines followed by these search procedures are usually based on simple intuitive concepts. Some of these concepts are motivated by so-called nature's wisdom, such as evolution and thermodynamics.

Slowness Without using derivatives, these methods are bound to be generally slower than derivative-based optimization methods for continuous optimization problems.

Flexibility Derivative freeness also relieves the requirement for differentiable objective functions, so we can use as complex an objective function as a specific application might need, without sacrificing too much in extra coding and computation time. In some cases, an objective function can even include the structure of a data-fitting model itself, which may be a neural network or fuzzy model. This means that by minimizing (or maximizing) a single objective function of this type, we can do structure and parameter identification at the same time.

Randomness All of these methods (with the probable exception of the standard downhill simplex search) are stochastic, which means that they all use random number generators in determining subsequent search directions. This element of randomness usually gives rise to the overly optimistic view that these methods are "global optimizers" that will find a global optimum given enough computation time. In theory, their random nature does make the probability of finding an optimal solution nonzero over a fixed amount of computation time. In practice, however, it might take a considerable amount of computation time, if not forever, to find the optimal solution of a given problem.

Analytic opacity It is difficult to do analytic studies of these methods, in part because of their randomness and problem-specific nature. Therefore, most of our knowledge about them is based on empirical studies.

Iterative nature Unlike the linear least-squares estimator (Section 5.3), these techniques are iterative in nature and we need certain stopping criteria to determine when to terminate the optimization process. Let k denote an iteration count and f_k denote the best objective function obtained at count k; common stopping criteria for a maximization problem include the following:

- Computation time: a designated amount of computation time, or number of function evaluations and/or iteration counts is reached.
- Optimization goal: f_k is less than a certain preset goal value.
- Minimal improvement: $f_k - f_{k-1}$ is less than a preset value.
- Minimal relative improvement: $(f_k - f_{k-1})/f_{k-1}$ is less than a preset value.

Both genetic algorithms (GAs) and simulated annealing (SA) have been receiving increasing amounts of attention due to their versatile optimization capabilities for both continuous and discrete optimization problems. Moreover, both of them are motivated by so-called *nature's wisdom*: GAs are loosely based on the concepts of natural selection and evolution; while SA originated in the annealing processes found in thermodynamics and metallurgy.

Random search and downhill simplex search are primarily for continuous optimization problems. Random search is the simplest and most intuitive optimization scheme of all; its implementation takes only a few lines in the MATLAB program `randsrch.m`, which is available available via FTP or WWW (see page xxiii). Downhill simplex search is based on heuristic adaptation of a geometric object (a simplex) to explore a performance landscape, efficiently. The file `fmins.m` shipped with MATLAB is an implementation of this optimization method.

All these techniques are often described as "weak" methods by the artificial intelligence community because of their relatively few assumptions about the problems being solved. Fortunately, due to their high degree of flexibility, it is easy to increase their efficiency by incorporating problem-specific heuristics into these methods.

Although the concepts and implementations of random search and downhill simplex search are simpler than those of GAs and SA, it should not be inferred that GAs and SA are better for all problems all the time. In general, one should not expect any single technique to outperform all the others in a given application.

In what follows, we shall describe the basics of these derivative-free optimization methods. The characteristics of each method will be explained in sufficient detail that the reader can decide which method is better suited for his or her needs.

7.2 GENETIC ALGORITHMS

Genetic algorithms (GAs) [4, 7] are derivative-free stochastic optimization methods based loosely on the concepts of natural selection and evolutionary processes. They were first proposed and investigated by John Holland at the University of Michigan in 1975 [7]. As a general-purpose optimization tool, GAs are moving out of academia and finding significant applications in many other venues. Their popularity can be attributed to their freedom from dependence on functional derivatives and to their incorporation of these characteristics:

- GAs are parallel-search procedures that can be implemented on parallel-processing machines for massively speeding up their operations.
- GAs are applicable to both continuous and discrete (combinatorial) optimization problems.
- GAs are stochastic and less likely to get trapped in local minima, which inevitably are present in any practical optimization application.
- GAs' flexibility facilitates both structure and parameter identification in complex models such as neural networks and fuzzy inference systems.

GAs encode each point in a parameter (or solution) space into a binary bit string called a **chromosome**, and each point is associated with a "fitness" value that, for maximization, is usually equal to the objective function evaluated at the point. Instead of a single point, GAs usually keep a set of points as a **population** (or **gene pool**), which is then evolved repeatedly toward a better overall fitness value. In each **generation**, the GA constructs a new population using **genetic operators** such as crossover and mutation; members with higher fitness values are more likely to survive and to participate in mating (crossover) operations. After a number of generations, the population contains members with better fitness values; this is analogous to Darwinian models of evolution by random mutation and natural selection. GAs and their variants are sometimes referred to as methods of **population-based optimization** that improve performance by upgrading entire populations rather than individual members.

Major components of GAs include encoding schemes, fitness evaluations, parent selection, crossover operators, and mutation operators; these are explained next.

Encoding schemes These transform points in parameter space into bit string representations. For instance, a point $(11, 6, 9)$ in a three-dimensional parameter space can be represented as a concatenated binary string:

$$\underbrace{1011}_{11}\underbrace{0110}_{6}\underbrace{1001}_{9}$$

in which each coordinate value is encoded as a **gene** composed of four binary bits using binary coding. Other encoding schemes, such as Gray coding, can also be used and, when necessary, arrangements can be made for encoding negative, floating-point, or discrete-valued numbers. Encoding schemes provide a way of translating problem-specific knowledge directly into the GA framework, and thus play a key role in determining GAs' performance. Moreover, genetic operators, such as crossover and mutation, can and should be designed along with the encoding scheme used for a specific application.

Fitness evaluation The first step after creating a generation is to calculate the fitness value of each member in the population. For a maximization problem, the fitness value f_i of the ith member is usually the objective function evaluated at this member (or point). We usually need fitness values that are positive, so some kind of monotonical scaling and/or translation may be necessary if the objective function is not strictly positive. Another approach is to use the rankings of members in a population as their fitness values. The advantage of this is that the objective function does not need to be accurate, as long as it can provide the correct ranking information.

Selection After evaluation, we have to create a new population from the current generation. The selection operation determines which parents participate in producing offspring for the next generation, and it is analogous to *survival of the fittest* in natural selection. Usually members are selected for mating with a selection probability proportional to their fitness values. The most common way to implement this is to set the selection probability equal to $f_i / \sum_{k=1}^{k=n} f_k$, where n is the population size. The effect of this selection method is to allow members with above-average fitness values to reproduce and replace members with below-average fitness values.

Crossover To exploit the potential of the current gene pool, we use **crossover** operators to generate new chromosomes that we hope will retain good features from the previous generation. Crossover is usually applied to selected pairs of parents with a probability equal to a given **crossover rate. One-point crossover** is the most basic crossover operator, where a crossover point on the genetic code is selected at random and two parent chromosomes are interchanged at this point. In **two-point crossover**, two crossover points are selected and the part of the chromosome string between these two points is then

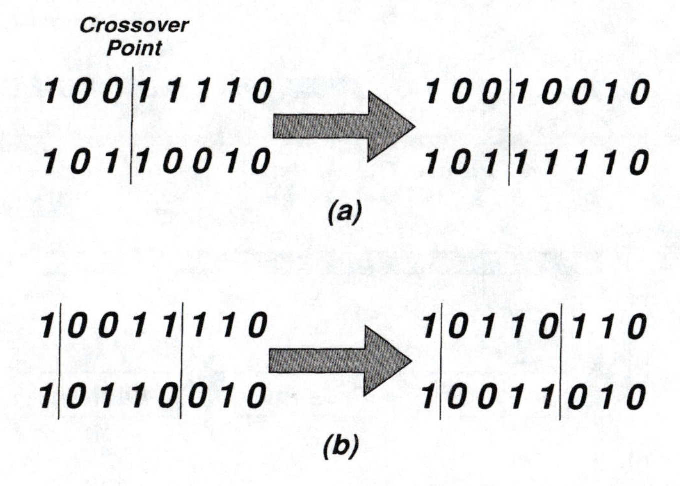

Figure 7.1. *Crossover operators: (a) one-point crossover; (b) two-point crossover.*

swapped to generate two children. We can define n-point crossover similarly. In general, $(n-1)$-point crossover is a special case of n-point crossover. Examples of one- and two-point crossover are shown in Figures 7.1(a) and 7.1(b), respectively.

The effect of crossover is similar to that of mating in the natural evolutionary process, in which parents pass segments of their own chromosomes on to their children. Therefore, some children are able to outperform their parents if they get "good" genes or genetic traits from both parents.

Mutation Crossover exploits current gene potentials, but if the population does not contain all the encoded information needed to solve a particular problem, no amount of gene mixing can produce a satisfactory solution. For this reason, a **mutation** operator capable of spontaneously generating new chromosomes is included. The most common way of implementing mutation is to flip a bit with a probability equal to a very low given **mutation rate**. A mutation operator can prevent any single bit from converging to a value throughout the entire population and, more important, it can prevent the population from converging and stagnating at any local optima. The mutation rate is usually kept low so good chromosomes obtained from crossover are not lost. If the mutation rate is high (above 0.1), GA performance will approach that of a primitive random search. Figure 7.2 provides an example of mutation.

In the natural evolutionary process, selection, crossover, and mutation all occur in the single act of generating offspring. Here we distinguish among them clearly to facilitate implementation of and experimentation with GAs.

Note that this section only gives a general description of the basics of GAs; detailed implementations vary considerably. For instance, we may choose a policy

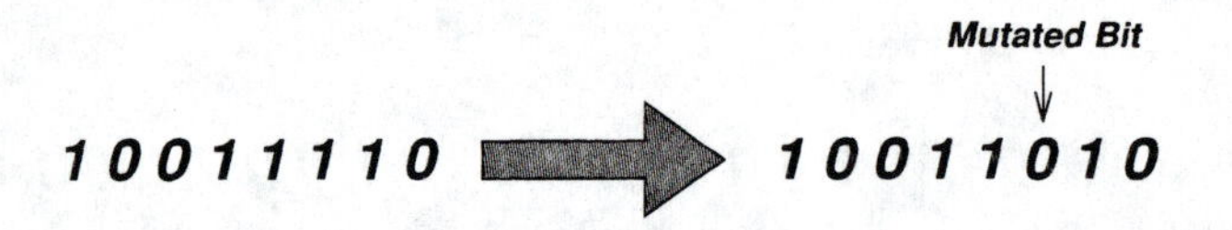

Figure 7.2. *Mutation operator.*

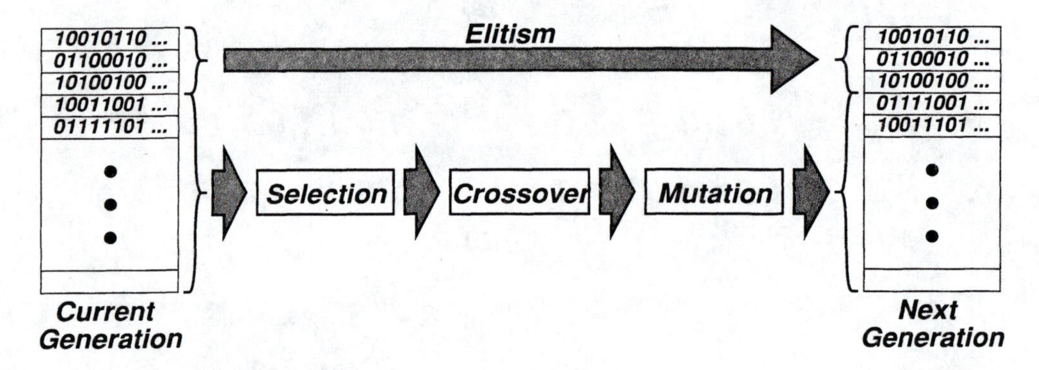

Figure 7.3. *Producing the next generation in GAs.*

of always keeping a certain number of best members when each new population is generated; this principle is usually called **elitism**.

Based on the aforementioned concepts, a simple genetic algorithm for maximization problems is described next.

Step 1: Initialize a population with randomly generated individuals and evaluate the fitness value of each individual.

Step 2:

- **(a)** Select two members from the population with probabilities proportional to their fitness values.
- **(b)** Apply crossover with a probability equal to the crossover rate.
- **(c)** Apply mutation with a probability equal to the mutation rate.
- **(d)** Repeat (a) to (d) until enough members are generated to form the next generation.

Step 3: Repeat steps 2 and 3 until a stopping criterion is met.

Figure 7.3 is a schematic diagram illustrating how to produce the next generation from the current one.

Example 7.1 *Maximization of the "peaks" function using GAs*

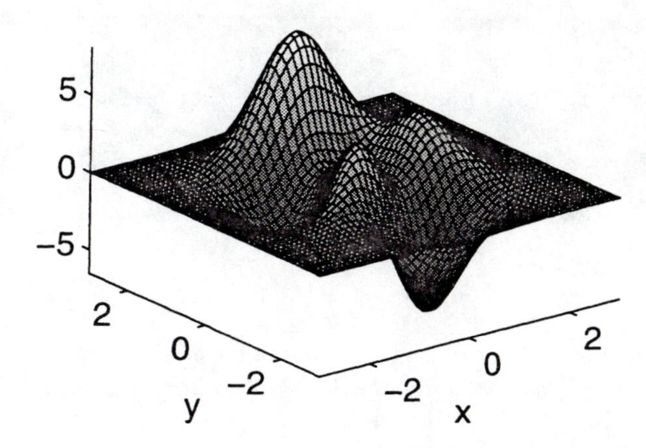

Figure 7.4. *The "peaks" function.* (MATLAB file: `peaks.m`)

The "peaks" function is a two-input function defined as

$$z = f(x,y) = 3(1-x)^2 e^{-x^2-(y+1)^2} - 10(\frac{x}{5} - x^3 - y^5)e^{-x^2-y^2} - \frac{1}{3}e^{-(x+1)^2-y^2}. \quad (7.1)$$

The surface plot of this function, shown in Figure 7.4, can be obtained directly by typing `peaks` within MATLAB. (Here we have changed the color map to "gray" such that a patch's brightness is proportional to its height.)

To use GAs to find the maximum of this function, we first confine the search domain to be a square area of $[-3,3] \times [-3,3]$. We use 8-bit binary coding for each variable, which results in a search space size of $2^8 \times 2^8 = 65,536$. Each generation in our GA implementation contains 20 points or individuals. Each point's fitness is defined as the value of the "peaks" function minus the minimum function value across the population. This guarantees that all fitness values are nonnegative. We use a simple one-point crossover scheme with the crossover rate equal to 1.0, which means that we always do crossover on selected parents. We choose uniform mutation (that is, each bit has the same probability of mutation) with the mutation rate equal to 0.01. We also apply elitism to keep the best two individuals across generations. Figure 7.5(a) is the contour plot of the "peaks" function, with the initial population locations denoted by circles. After the fifth generation, the population starts to converge to the peak containing the maximum, as shown in Figure 7.5(b). Figure 7.5(c) is the population distribution after the tenth generation; the individual that is far away from the main cluster is the result of the mutation operator.

Figure 7.6 is a plot of the best, average, and poorest values of the objective function across 30 generations. Since we are using elitism to keep the best two

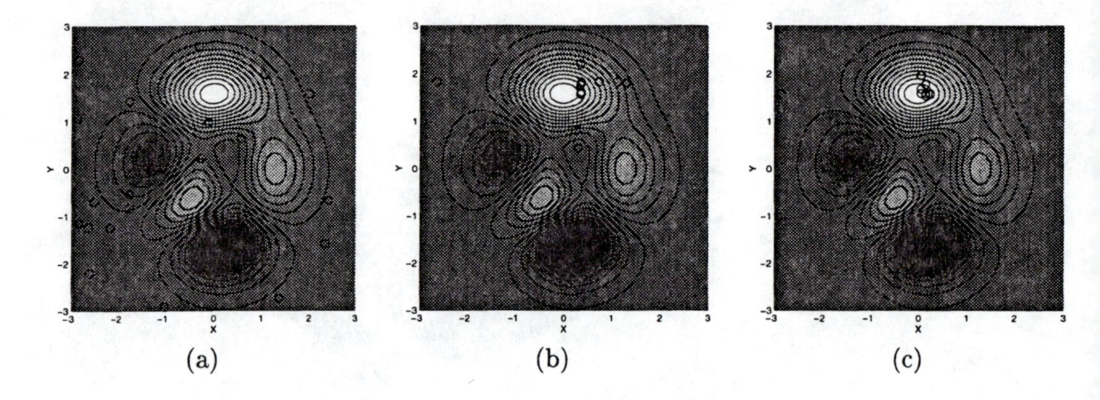

Figure 7.5. *Using GAs to find the maximum of the "peaks" function: (a) initial population; (b) population after the fifth generation; (c) population after the tenth generation.* (MATLAB file: go_ga.m)

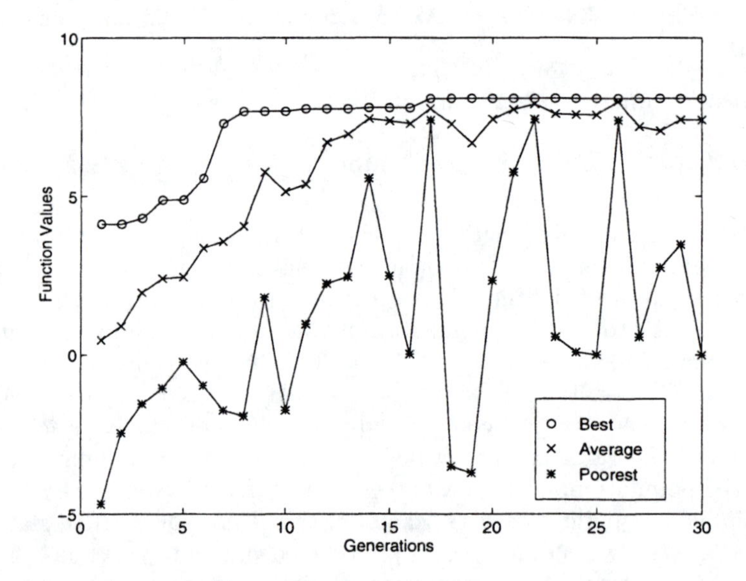

Figure 7.6. *Performance of GAs across generations.* (MATLAB file: go_ga.m)

individuals at each generation, the "best" curve is monotonically increasing with respect to generation numbers. The erratic behavior of the "poorest" curve is due to the mutation operator, which explores the landscape in a somewhat random manner.

□

7.3 SIMULATED ANNEALING

Simulated annealing (SA) [10, 15] is another derivative-free optimization method that has recently drawn much attention for being as suitable for continuous as for discrete (combinatorial) optimization problems. When SA was first proposed [10], it was mostly known for its effectiveness in finding near optimal solutions for large-scale combinatorial optimization problems, such as traveling salesperson problems (finding the shortest cyclical itinerary for a salesperson who must visit each of N cities in turn) and placement problems [18] (finding the layout of a computer chip that minimizes the total area). Recent applications of SA and its variants [9] also demonstrate that this class of optimization approaches can be considered competitive with other approaches when there are continuous optimization problems to be solved.

Simulated annealing was derived from physical characteristics of spin glasses [10, 13]. The principle behind simulated annealing is analogous to what happens when metals are cooled at a controlled rate. The slowly falling temperature allows the atoms in the molten metal to line themselves up and form a regular crystalline structure that has high density and low energy. But if the temperature goes down too quickly, the atoms do not have time to orient themselves into a regular structure and the result is a more amorphous material with higher energy.

In simulated annealing, the value of an objective function that we want to minimize is analogous to the energy in a thermodynamic system. At high temperatures, SA allows function evaluations at faraway points and it is likely to accept a new point with higher energy. This corresponds to the situation in which high-mobility atoms are trying to orient themselves with other nonlocal atoms and the energy state can occasionally go up. At low temperatures, SA evaluates the objective function only at local points and the likelihood of it accepting a new point with higher energy is much lower. This is analogous to the situation in which the low-mobility atoms can only orient themselves with local atoms and the energy state is not likely to go up again.

Obviously, the most important part of SA is the so-called **annealing schedule** or **cooling schedule**, which specifies how rapidly the temperature is lowered from high to low values. This is usually application specific and requires some experimentation by trial-and-error.

Before giving a detailed description of SA, first we shall explain the fundamental terminology of SA.

Objective function An objective function $f(\cdot)$ maps an input vector $\mathbf{x}$ into a scalar E:

$$E = f(\mathbf{x}),$$

where each $\mathbf{x}$ is viewed as a point in an input space. The task of SA is to sample the input space effectively to find an $\mathbf{x}$ that minimizes E.

Generating function A generating function $g(\cdot, \cdot)$ specifies the probability density function of the difference between the current point and the next point to be visited. Specifically, $\Delta\mathbf{x}$ $(= \mathbf{x}_{\text{new}} - \mathbf{x})$ is a random variable with probability density function $g(\Delta\mathbf{x}, T)$, where T is the temperature. For common SA (especially when used in combinatorial optimization applications), $g(\cdot, \cdot)$ is usually a function independent of the temperature T.

Acceptance function After a new point $\mathbf{x}_{\text{new}}$ has been evaluated, SA decides whether to accept or reject it based on the value of an acceptance function $h(\cdot, \cdot)$. The most frequently used acceptance function is the **Boltzmann probability distribution**:

$$h(\Delta E, T) = \frac{1}{1 + \exp(\Delta E/(cT))},$$

where c is a system-dependent constant, T is the temperature, and ΔE is the energy difference between $\mathbf{x}_{\text{new}}$ and $\mathbf{x}$:

$$\Delta E = f(\mathbf{x}_{\text{new}}) - f(\mathbf{x}).$$

The common practice is to accept $\mathbf{x}_{\text{new}}$ with probability $h(\Delta E, T)$. Note that when ΔE is negative, SA tends to accept the new point because it reduces the energy. When ΔE is positive, SA may accept the new point and end up in a higher energy state. In other words, SA can go either uphill or downhill; but the lower the temperature, the less likely SA is to accept any significant uphill actions.

Annealing schedule An annealing schedule regulates how rapidly the temperature T goes from high to low values, as a function of time or iteration counts. The exact interpretation of *high* and *low* and the specification of a good annealing schedule require certain problem-specific physical insights and/or trial-and-error. The easiest way of setting an annealing schedule is to decrease the temperature T by a certain percentage at each iteration.

Having presented this brief guide to clearer understanding of the SA terminology, we now describe the basic steps involved in a general SA method.

Step 1: Choose a start point $\mathbf{x}$ and set a high starting temperature T. Set the iteration count k to 1.

Step 2: Evaluate the objective function:

$$E = f(\mathbf{x}).$$

Step 3: Select $\Delta\mathbf{x}$ with probability determined by the generating function $g(\Delta\mathbf{x}, T)$. Set the new point $\mathbf{x}_{\text{new}}$ equal to $\mathbf{x} + \Delta\mathbf{x}$.

Step 4: Calculate the new value of the objective function:

$$E_{\text{new}} = f(\mathbf{x}_{\text{new}}).$$

Step 5: Set $\mathbf{x}$ to $\mathbf{x}_{\text{new}}$ and E to E_{new} with probability determined by the acceptance function $h(\Delta E, T)$, where $\Delta E = E_{\text{new}} - E$.

Step 6: Reduce the temperature T according to the annealing schedule (usually by simply setting T equal to ηT, where η is a constant between 0 and 1).

Step 7: Increment iteration count k. If k reaches the maximum iteration count, stop the iterating. Otherwise, go back to step 3.

In conventional SA, also known as **Boltzmann machines** [5, 6], the generating function is a Gaussian probability density function:

$$g(\Delta\mathbf{x}, T) = (2\pi T)^{-n/2} \exp[-\|\Delta\mathbf{x}\|^2/(2T)],$$

where $\Delta\mathbf{x}$ $(= \mathbf{x}_{\text{new}} - \mathbf{x})$ is the deviation of the new point from the current one, T is the temperature, and n is the dimension of the space under exploration. It has been proven in refs. [3] that a Boltzmann machine using the aforementioned generating function $g(\cdot,\ \cdot)$ can find a global optimum of $f(\mathbf{x})$ if the temperature T is reduced not faster than $T_0/\ln k$.

For discrete or combinatorial optimization problems, each $\mathbf{x}$ is not necessarily an n-element vector with unconstrained values. Instead, each $\mathbf{x}$ is confined to be one of N points that constitute the solution space or input space. Usually N is finite, but it is very large such that the use of an exhaustive search method is impossible. Thus, adding randomly generated $\Delta\mathbf{x}$ to a current point $\mathbf{x}$ may not generate another legal point in the solution space, so generating functions are rarely used. Instead, to find the next legal point to explore, we usually define a **move set**, denoted by $M(\mathbf{x})$, as the set of legal points available for exploration after $\mathbf{x}$. Usually the move set $M(\mathbf{x})$ represents a set of *neighboring* points of the current point $\mathbf{x}$ in the sense that the objective function at any point of the move set will not differ too much from the objective function at $\mathbf{x}$. The definition of a move set is problem dependent and reflects our knowledge about the problem under consideration. Once the move set is defined, $\mathbf{x}_{\text{new}}$ is usually selected at random from the move set, where every member stands an equal probability of being chosen. A well-known instance of combinatorial optimization is the traveling salesperson problem, which is tackled using the SA technique in the following example.

Example 7.2 *Traveling salesperson problem*

In a typical **traveling salesperson problem (TSP)**, we are given n cities, and the distance (or cost) between all pairs of these cities is an $n \times n$ distance (or cost) matrix $\mathbf{D}$, where the element d_{ij} represents the distance (or cost) of traveling from city i to city j. The problem is to find a a closed tour in which each city, except for the starting one, is visited exactly once, such that the total length (cost) is minimized. The traveling salesperson problem is a well-known problem in combinatorial optimization; it belongs to a class of problems known as **NP-complete**[1] [16], in which the computation time required to find an optimal solution increases exponentially with n. For a TSP with n cities, the number of possible tours is $(n-1)!/2$, which becomes prohibitively large even for a moderate n. For instance, finding the best tour of the state capitals of the United States ($n = 50$) would require many billions of years even with the fastest modern computers.

For a common traveling salesperson problem, we can define at least three move sets for SA:

Inversion Remove two edges from the tour and replace them to make it another legal tour. This is equivalent to removing a section (6-7-8-9) of the tour and then replacing with the same cities running in the opposite order. See Figures 7.7(a) and 7.7(b).

Translation Remove a section (8-7) of the tour and then replace it in between two randomly selected consecutive cities (4 and 5). See Figures 7.7(b) and 7.7(c).

Switching Randomly select two cities (3 and 11) and switch them in a tour. See Figures 7.7(c) and Figures 7.7(d).

Generally speaking, the switching move set tends to rupture the original tour and results in a tour that has a total length (or cost) significantly different from that of the original tour. Comparisons between the inversion and switching move set can be found in refs. [15].

We apply the SA technique with the inversion move set to a TSP with 100 cities; Figure 7.8 is a typical result after 10 minutes of simulation on a SUN SPARC II workstation. The MATLAB file is `tsp.m`. (Note that the function `tsp.m` is not fully optimized for speed yet; it only serves to demonstrate the important aspects of SA techniques.)

□

Variants of Boltzmann machines include the **Cauchy machine** or **fast simulated annealing** [21, 22], where the generating function is the Cauchy distribution:

$$g(\Delta \mathbf{x}) = \frac{T}{(\|\Delta \mathbf{x}\|^2 + T^2)^{(n+1)/2}}.$$

The fatter tail of the Cauchy distribution allows it to explore farther from the current point during the search process.

Another variant of the original SA, the so-called **very fast simulated reannealing (VFSR)** [8], was designed for optimization problems in a constrained

[1]NP stands for non-polynomial.

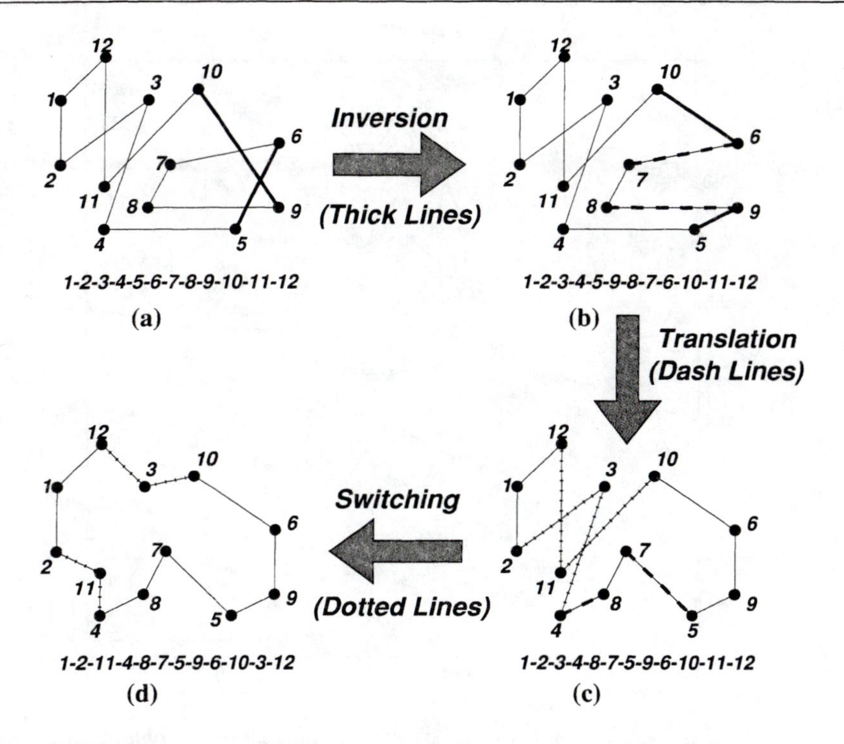

Figure 7.7. *Three operations for generating move sets in the traveling salesperson problem.*

search space. For a parameter $x_i(k)$ in dimension i at annealing time k, the new value is generated by

$$x_i(k+1) = x_i(k) + \lambda_i(x_i^{\max} - x_i^{\min}),$$

where $\lambda_i \in [-1, 1]$, $x_i^{\max}$ and $x_i^{\min}$ are the maximum and minimum of the ith dimension. This is repeated until a legal x_i between $x_i^{\min}$ and $x_i^{\max}$ is generated. The generating function for λ_i is

$$g(\lambda_i, T_i) = \frac{1}{2(|y_i| + T_i)\ln(1 + 1/T_i)}.$$

To generate λ_i according to the preceding distribution, we can apply the following formula:

$$\lambda_i = \text{sgn}(u_i - 0.5)T_i[(1 + 1/T_i)^{|2u_i - 1|} - 1],$$

where u_i is a uniformly distributed random variable between 0 and 1. A global optimum can be obtained statistically if the annealing schedule is

$$T_i(k) = T_i(0)\exp(-c_i k^{1/n}),$$

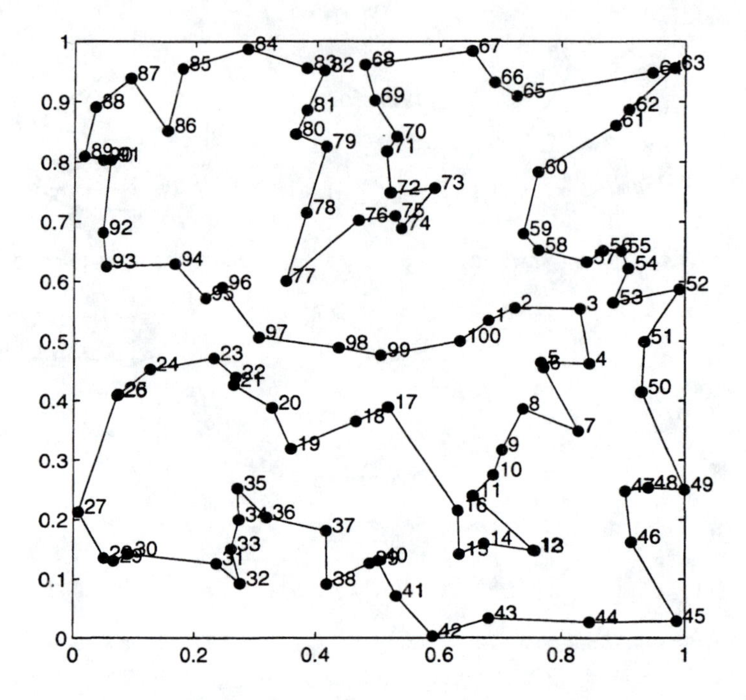

Figure 7.8. *The result of a 100-city traveling salesperson problem using the inversion move set in simulated annealing.* (MATLAB file: `tsp.m`)

where c_i is a user-defined parameter whose value should be selected according to the guidelines in refs. [9]. The same type of annealing schedule should be used for both the generating function $g(\cdot, \cdot)$ and the acceptance function $h(\cdot, \cdot)$.

Reannealing or **temperature rescaling** in the VFSR algorithm periodically rescales the generating temperature in terms of the sensitivities s_i calculated at the most current minimum value of the cost function E^*:

$$s_i = |\partial E^* / \partial x_i|.$$

The annealing time k_i is adjusted according to s_i, based on the heuristic concept that the generating distribution used in a relatively insensitive dimension should be wider than that of the distribution produced in a dimension more sensitive to change. A detailed discussion of the reannealing process can be found in refs. [9].

VFSR was reported to be faster than genetic algorithms on several test problems; see refs. [9] for more detail.

7.4 RANDOM SEARCH

Random search [2, 11, 12, 20] explores the parameter space of an objective function sequentially in a seemingly random fashion to find the optimal point that minimizes (or maximizes) the objective function. Besides being derivative free, the most distinguishing strength of the random search method lies in its simplicity, which makes the method easily understood and conveniently customized for specific applications. Moreover, it has been proved that this method converges to the global optimum with probability 1 on a compact set [1, 20]. However, the theoretical result of convergence to the global optimum is not really important here since the optimization process itself could take a prohibitively long time.

Here we shall start with the most primitive version proposed by Matyas [11]. Following some heuristic guidelines, we shall also present a modified version that is more efficient.

Let $f(\mathbf{x})$ be the objective function to be minimized and $\mathbf{x}$ be the point currently under consideration. The original random search method [11] tries to find the optimal $\mathbf{x}$ by iterating the following four steps:

Step 1: Choose a start point $\mathbf{x}$ as the current point.

Step 2: Add a random vector $\mathbf{dx}$ to the current point $\mathbf{x}$ in the parameter space and evaluate the objective function at the new point at $\mathbf{x} + \mathbf{dx}$.

Step 3: If $f(\mathbf{x} + \mathbf{dx}) < f(\mathbf{x})$, set the current point $\mathbf{x}$ equal to $\mathbf{x} + \mathbf{dx}$.

Step 4: Stop if the maximum number of function evaluations is reached. Otherwise, go back to step 2 to find a new point.

This is a truly random method in the sense that search directions are purely guided by a random number generator. There are several ways to improve this primitive version; these are based on the following observations:

Observation 1: If search in a direction results in a higher objective function, the opposite direction can often lead to a lower objective function.

Observation 2: Successive successful searches in a certain direction should bias subsequent searching toward this direction. On the other hand, successive failures in a certain direction should discourage subsequent searching along this direction.

The first observation leads to a **reverse step** in the original method. The second observation motivates the use of a **bias term** as the center for the random vector. After including these two guidelines, the modified random search method [20] involves the following six steps:

Step 1: Choose a start point $\mathbf{x}$ as the current point. Set initial bias $\mathbf{b}$ equal to a zero vector.

Step 2: Add a bias term $\mathbf{b}$ and a random vector $\mathbf{dx}$ to the current point $\mathbf{x}$ in the input space and evaluate the objective function at the new point at $\mathbf{x}+\mathbf{b}+\mathbf{dx}$.

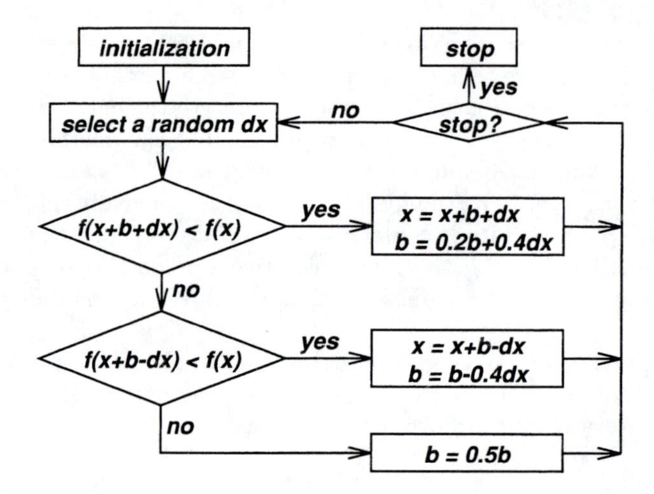

Figure 7.9. *Flow chart for the random search method.*

Step 3: If $f(\mathbf{x}+\mathbf{b}+\mathbf{dx}) < f(\mathbf{x})$, set the current point $\mathbf{x}$ equal to $\mathbf{x}+\mathbf{b}+\mathbf{dx}$ and the bias $\mathbf{b}$ equal to $0.2\mathbf{b}+0.4\mathbf{dx}$; go to step 6. Otherwise, go to the next step.

Step 4: If $f(\mathbf{x}+\mathbf{b}-\mathbf{dx}) < f(\mathbf{x})$, set the current point $\mathbf{x}$ equal to $\mathbf{x}+\mathbf{b}-\mathbf{dx}$ and the bias $\mathbf{b}$ equal to $\mathbf{b}-0.4\mathbf{dx}$; go to step 6. Otherwise, go to the next step.

Step 5: Set the bias equal to $0.5\mathbf{b}$ and go to step 6.

Step 6: Stop if the maximum number of function evaluations is reached. Otherwise go back to step 2 to find a new point.

A detailed flow chart of the preceding steps is shown in Figure 7.9. Usually the initial bias is set to a zero vector. Each component of the random vector **dx** should be a random variable that has a zero mean and a variance proportional to the range of the corresponding parameter; this allows the method to apply the same degree of exploration for each dimension of the parameter space. A further enhancement is to make the variance of each element in **dx** decrease with time; this serves the same purpose as the cooling schedule in simulated annealing. The file `randsrch.m`, available via FTP or WWW (see page xxiii), is an simple implementation of the improved random search method with time-independent variances.

Unlike genetic algorithms and simulated annealing, the random search method is primarily for continuous optimization problems. It is possible to come up with a random search method for discrete or combinatorial optimization problems, but then the preceding observations may no longer be true and we might lose the advantages of the modified method.

Figure 7.10 demonstrates how the random search method tried to find the minimum of the "peaks" function from three different start points; these start points are surrounded by circles while the corresponding endpoints are denoted by crosses.

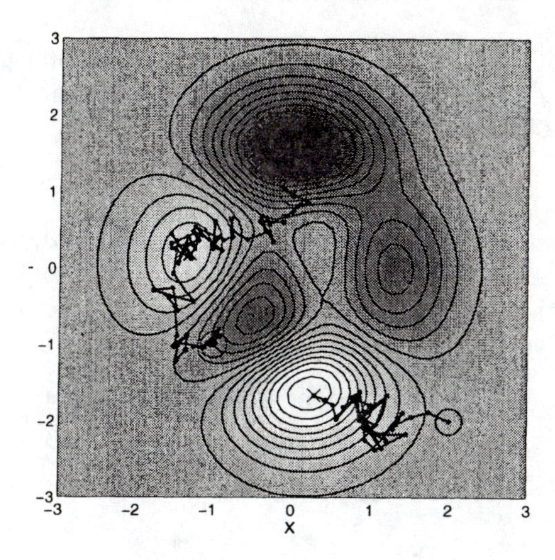

Figure 7.10. *Random search applied to the "peaks" function.* (MATLAB file: go_rand.m)

(For a more distinct illustration, we flipped the color map such that brighter regions represent valleys instead of peaks.) Despite its simplicity, the method performed well on this optimization problem.

7.5 DOWNHILL SIMPLEX SEARCH

Downhill simplex search [14] is a derivative-free method for multidimensional function optimization. As with other derivative-free approaches, this search method is not very efficient compared to derivative-based methods. However, the concept behind downhill simplex search is simple and it has an interesting geometrical interpretation.

We consider the minimization of a function of n variables with no constraints. We start with an initial **simplex**, which is a collection of $n + 1$ points in n-dimensional space. The downhill simplex search repeatedly replaces the point having the highest function value in a simplex with another point. (Note that this method has little to do with the simplex method for linear programming, except that both of them make use of the geometrical concept of a simplex.) When combined with other operations, the simplex under consideration adapts itself to the local landscape, elongating down long inclined planes, changing direction on encountering a valley at an angle, and contracting in the neighborhood of a minimum. These operations are described next.

To start the downhill simplex search, we must initialize a simplex of $n + 1$ points. For example, a simplex is a triangle in two-dimensional space and a tetrahedron in three-dimensional space. Moreover, we would like the simplex to be nondegenerate—that is, it encloses a finite inner n-dimensional volume. An easy way to set up a simplex is to start with an initial starting point $\mathbf{P}_0$ and the other n points can be taken as

$$\mathbf{P}_i = \mathbf{P}_0 + \lambda_i \mathbf{e}_i, \ i = 1, \ldots, n, \tag{7.2}$$

where $\mathbf{e}_i$'s are unit vectors consisting of a basis of the n-dimensional space and λ_i is a constant reflecting the guess of the characteristic length scale of the optimization problem in question.

We write y_i for the function value at $\mathbf{P}_i$ and let

$$\begin{aligned} l &= \arg\min_i(y_i) \ (l \text{ for "low"}), \\ h &= \arg\max_i(y_i) \ (h \text{ for "high"}). \end{aligned} \tag{7.3}$$

In other words, l and h are respectively the indices for the minimum and maximum of y_i. In symbols,

$$\begin{aligned} y_l &= \min_i(y_i) \\ y_h &= \max_i(y_i) \end{aligned} \tag{7.4}$$

Let $\bar{\mathbf{P}}$ be the average (centroid) of these $n+1$ points. Each cycle of this method starts with a reflection point $\mathbf{P}^*$ of $\mathbf{P}_h$. Depending on the function value at $\mathbf{P}^*$, we have four possible operations to change the current simplex to explore the landscape of the function efficiently in multidimensional space. These four operations are (1) reflection away from $\mathbf{P}_h$; (2) reflection and expansion away from $\mathbf{P}_h$; (3) contraction along one dimension connecting $\mathbf{P}_h$ and $\bar{\mathbf{P}}$; and (4) shrinkage toward $\mathbf{P}_l$ along all dimensions. These four operations are shown in Figure 7.11 when f is a two-input function.

Before describing the full cycle of the simplex search, we need to define four intervals to be used in the search process:

- Interval 1: $\{y | y \le y_l\}$
- Interval 2: $\{y | y_l < y \le \max_{i, i \ne h}\{y_i\}\}$
- Interval 3: $\{y | \max_{i, i \ne h}\{y_i\} < y \le y_h\}$
- Interval 4: $\{y | y_h < y\}$.

These intervals are shown in Figure 7.12.

A full cycle of the downhill simplex search involves the following four steps:

Reflection: Define the reflection point $\mathbf{P}^*$ and its value y^* as

$$\begin{aligned} \mathbf{P}^* &= \bar{\mathbf{P}} + \alpha(\bar{\mathbf{P}} - \mathbf{P}_h), \\ y^* &= f(\mathbf{P}^*), \end{aligned} \tag{7.5}$$

where the *reflection coefficient* α is a positive constant. Thus $\mathbf{P}^*$ is on the line joining $\mathbf{P}_h$ and $\bar{\mathbf{P}}$, on the far side of $\bar{\mathbf{P}}$ from $\mathbf{P}_h$. Depending on the value of y^*, we have the following actions:

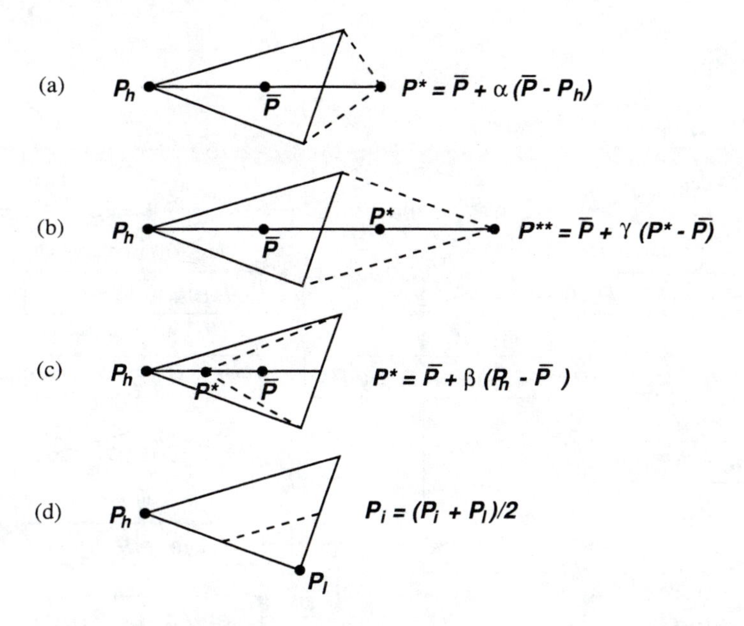

Figure 7.11. *Outcomes for a cycle in the downhill simplex search after (a) reflection away from* $\mathbf{P}_h$*; (b) reflection and expansion away from* $\mathbf{P}_h$*; (c) contraction along one dimension connecting* $\mathbf{P}_h$ *and* $\bar{\mathbf{P}}$*; (4) shrinkage toward* $\mathbf{P}_l$ *alone all dimensions.*

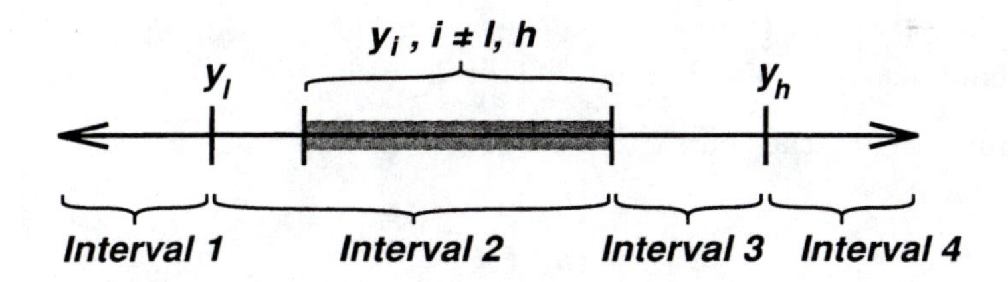

Figure 7.12. *Four intervals used in the downhill simplex search.*

1. If y^* is in interval 1, go to **expansion**.
2. If y^* is in interval 2, replace $\mathbf{P}_h$ with $\mathbf{P}^*$ and finish this cycle.
3. If y^* is in interval 3, replace $\mathbf{P}_h$ with $\mathbf{P}^*$ and go to **contraction**.
4. If y^* is in interval 4, go to **contraction**.

Expansion: Define the expansion point $\mathbf{P}^{**}$ and its value y^{**} as

$$\begin{aligned} \mathbf{P}^{**} &= \bar{\mathbf{P}} + \gamma(\mathbf{P}^* - \bar{\mathbf{P}}), \\ y^{**} &= f(\mathbf{P}^{**}), \end{aligned} \tag{7.6}$$

where the *expansion coefficient* γ is greater than unity. If y^{**} is in interval

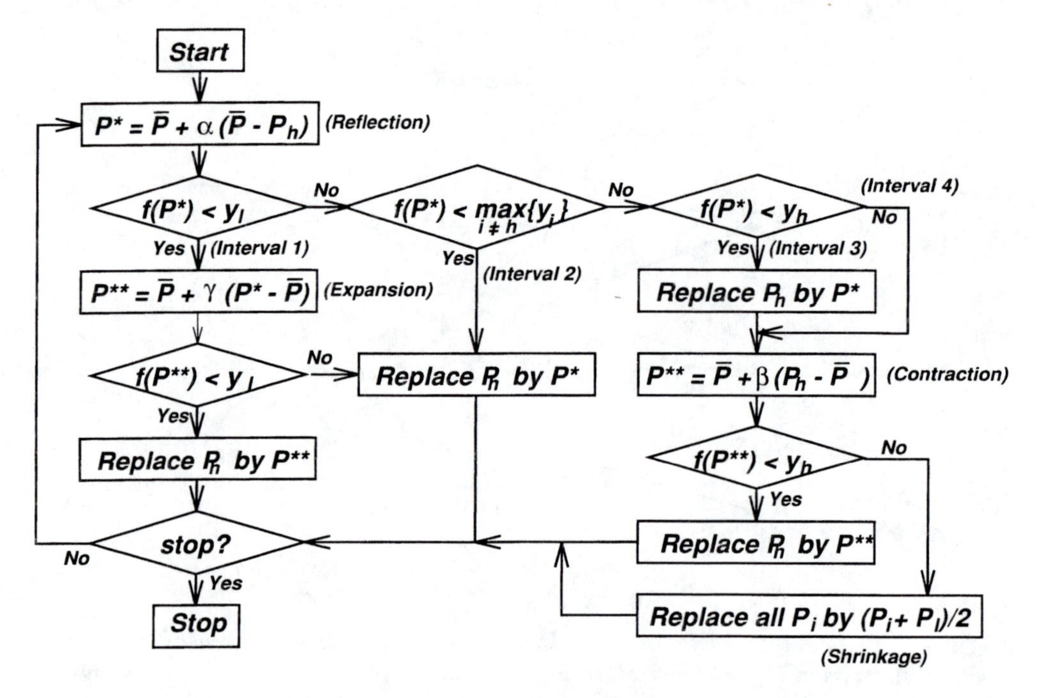

Figure 7.13. *Flow chart for the downhill simplex search.*

1, replace $\mathbf{P}_h$ with $\mathbf{P}^{**}$ and finish this cycle. Otherwise, replace $\mathbf{P}_h$ with the original reflection point $\mathbf{P}^*$ and finish this cycle.

Contraction: Define the contraction point $\mathbf{P}^{**}$ and its value y^{**} as

$$\begin{aligned} \mathbf{P}^{**} &= \bar{\mathbf{P}} + \beta(\mathbf{P}_h - \bar{\mathbf{P}}), \\ y^{**} &= f(\mathbf{P}^{**}), \end{aligned} \tag{7.7}$$

where the *contraction coefficient* β lies between 0 and 1. If y^{**} is in interval 1, 2, or 3, replace $\mathbf{P}_h$ with $\mathbf{P}^{**}$ and finish this cycle. Otherwise, go to **shrinkage**.

Shrinkage: Replace each $\mathbf{P}_i$ with $(\mathbf{P}_i + \mathbf{P}_l)/2$. Finish this cycle.

The foregoing cycle is repeated until a given stopping criterion is met. Common stopping criteria are described in Section 7.1. A complete flow chart of the foregoing steps is given in Figure 7.13.

Before starting using this method, we still need to determine three constants α, β and γ, which are the coefficients for reflection, contraction, and expansion. Generally speaking, the optimal values for these coefficients are application dependent, and the best way to select their values is by doing some trial-and-error experiments. A good starting point is $(\alpha, \beta, \gamma) = (1, 0.5, 2)$; these values are suggested in Nelder and Mead's original paper [14].

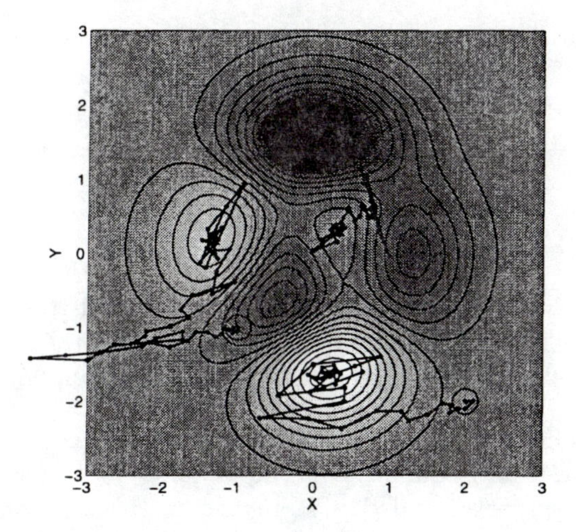

Figure 7.14. *Applying downhill simplex search to the "peaks" function.* (MATLAB file: `go_simp.m`)

Figure 7.14 demonstrates the transition of the current point of the downhill simplex search when minimizing the "peaks" function from three different start points; these start points are surrounded by circles while the corresponding endpoints are denoted by crosses. (Again, we flipped the color map such that brighter regions represent valleys instead of peaks.)

One more thing to note is that this method is deterministic—starting from the same initial point, this method will always lead to the same final point for a given objective function. This implies that the method will potentially lead to a local minimum and stay there forever. To enable this method to get out of local minima, we have to make it stochastic. One way to do this is to make the coefficients α, β, and γ random variables within appropriate ranges, such that the method can explore a wider area of the input domain. Another simple way is to run the search procedure repeatedly from various initial points selected randomly.

The downhill simplex search is implemented as the command `fmins` in MATLAB. To use this command, type `help fmins` at the MATLAB prompt to get more information.

7.6 SUMMARY

This chapter presents four of the most popular derivative-free optimization methods: genetic algorithms, simulated annealing, random search, and downhill simplex search. These techniques rely on modern high-speed computers; they all require a significant amount of computation when compared with derivative-based approaches (Chapters 5 and 6). However, these derivative-free approaches are more flexible in terms of incorporating intuitive guidelines and forming sophisticated objective functions.

EXERCISES

1. For a traveling salesperson problem in which the cost to travel from one city to another is equal to the Euclidean distance between them, a closed tour is not optimal whenever there is a cross in the tour. Explain why.

2. At the beginning of `tsp.m`, we have to fill a distance matrix such that the element at row i and column j in the distance matrix is equal to the travel distance between city i and city j. In `tsp.m`, this is done by two nested `for` loops:

```
for i = 1:NumCity,
        for j = 1:NumCity,
                distance(i, j) = norm(loc(i, :) - loc(j, :));
        end
end
```

   where `NumCity` is the number of cities. Can you vectorize this code to get rid of the `for` loops?

3. Can you speed up `tsp.m`? Explain what can be done to make it run faster. Use the ideas you come up with to modify `tsp.m` and see how much speedup you actually gain.

4. Modify `tsp.m` such that whenever the path comes across a circle centered at $[0.5, 0.5]$ with radius 0.3, an extra cost of 0.5 is incurred. (Think of the circle as a river and you have to pay a toll to cross it.)

5. Repeat Exercise 4, but -0.3 is incurred instead. (Think of the circle as a national boundary that you are smuggling something across whenever you pass it.)

6. Modify the program `tsp.m` to use the translation move set and compare the performance with the original program using the default 30-city problem. Note that SA is a stochastic optimization procedure, so the comparison should be based on a number of simulation runs. Try to do 10 runs on both programs and compare their average running time and objective function values.

7. Repeat Exercise 6, but use the switching move set.

8. Use the command `fmins` to minimize Rosenbrock's parabolic valley [19] (also known as the **banana function**):

$$f(x_1, x_2) = 100(x_2 - x_1^2)^2 + (1 - x_1)^2.$$

The starting point is $(-1.2, 1)$.

9. Repeat Exercise 8 using the random search method. Compare the average running time and objective function value over 10 runs with those of `fmins`.

10. Use the command `fmins` to minimize Powell's quartic function [17]:

$$f(x_1, x_2, x_3, x_4) = (x_1 + 10x_2)^2 + 5(x_3 - x_4)^2 + (x_2 - 2x_3)^4 + 10(x_1 - x4)^4.$$

The starting point is $(3, -1, 0, 1)$.

11. Repeat Exercise 10 using the random search method. Compare the average running time and objective function value over 10 runs with those of `fmins`.

12. Rewrite the random search method `randsrch.m` to include the feature of decreasing variances of **dx** with respect to time. Run `go_rand.m` again and compare the result to Figure 7.10.

13. Repeat Exercise 9 using the modified `randsrch.m` from Exercise 12.

14. Repeat Exercise 11 using the modified `randsrch.m` from Exercise 12.

REFERENCES

[1] N. Baba, T. Shoman, and Y. Sawaragi. A modified convergence theorem for a random optimization method. *Information Science*, 13:159–166, 1977.

[2] Norio Baba, Yoshio Mogami, Motokazu Kohzake, Yasuhiro Shiraishi, and Yutaka Yoshida. A hybrid algorithm for finding the global minimum of error function of neural networks and its applications. *Neural Networks*, 7(8):1253–1265, 1994.

[3] S. Geman and D. Geman. Stochastic relaxation, Gibbs distribution and the Bayesian restoration in images. *IEEE Transactions of Pattern Analysis and Machine Intelligence*, 6(6):721–741, 1984.

[4] D. E. Goldberg. *Genetic algorithms in search, optimization, and machine learning.* Addison-Wesley, Reading, MA, 1989.

[5] G. E. Hinton and T. J. Sejnowski. Learning and relearning in Boltzmann machines. In D. E. Rumerhart, J. L. McClelland, and the PDP research group, editors, *Parallel distributed processing*, chapter 7, pages 282–317. MIT Press, Cambridge, MA., 1986.

[6] G. E. Hinton, T. J. Sejnowski, and D. H. Ackley. Boltzmann machines: constraint satisfaction networks that learn. Technical Report CMU-CS-84-119, Department of Computer Science, Carnegie Mellon University, 1984.

[7] J. H. Holland. *Adaptation in natural and artificial systems.* University of Michigan Press, Michigan, 1975.

[8] L. Ingber. Very fast simulated re-annealing. *Mathematical and Computer Modelling*, 12(8):967–973, 1989.

[9] L. Ingber and B. E. Rosen. Genetic algorithms and very fast simulated reannealing. *Mathematical and Computer Modelling*, 16(11):87–100, 1992.

[10] S. Kirkpatrick, C. D. Gelatt, and M. P. Vecchi. Optimization by simulated annealing. *Science*, 220(4598):671–680, May 1983.

[11] J. Matyas. Random optimization. *Automation and Remote Control*, 26:246–253, 1965.

[12] W. S. Meisel. *Computer-oriented approaches to pattern recognition*, volume 83 of *Mathematics in science and engineering.* Academic Press, New York, 1972.

[13] N. Metropolis, A. W. Rosenbluth, M. N. Rosenbluth, A. H. Teller, and E. Teller. Equation of state calculations by fast computing machines. *Journal of Chemical Physics*, 21(6):1087–1092, 1953.

[14] J. A. Nelder and R. Mead. A simplex method for function minimization. *The Computer Journal*, 7:308–313, 1965.

[15] R. H. J. M. Otten and L. P. P. P. van Ginneken. *The annealing algorithm.* Kluwer Academic, 1989.

[16] C. Papadimitriou and K. Steiglitz. *Combinatorial optimization: algorithms and complexity.* Prentice Hall, Upper Saddle River, NJ, 1982.

[17] M. J. D. Powell. An efficient method for finding the minimum of a function of several variables without calculating derivatives. *The Computer Journal*, 5:147, 1962.

[18] F. I. Romeo. *Simulated annealing: theory and applications to layout problems.* PhD thesis, EECS Department, University of California at Berkeley, May 1989.

[19] H. Rosenbrock. An automatic method for finding the greatest or least value of a function. *The Computer Journal*, 3:175–184, 1960.

[20] F. J. Solis and J. B. Wets. Minimization by random search techniques. *Mathematics of Operations Research*, 6(1):19–30, 1981.

[21] H. Szu. Fast simulated annealing. In J. S. Denker, editor, *Neural networks for computing.* American Institute of Physis, New York, 1986.

[22] H. Szu and R. Hartley. Fast simulated annealing. *Physics Letters*, 122:157–162, 1987.

Part III

Neural Networks

Chapter 8

Adaptive Networks

J.-S. R. Jang

8.1 INTRODUCTION

This chapter describes the architectures and learning procedures of adaptive networks, a unifying framework that subsumes almost all kinds of neural network paradigms with supervised learning capabilities. The fundamentals of adaptive networks will be a key element in understanding other various neural network paradigms (such as multilayer perceptrons and radial basis function networks) introduced in the subsequent chapters.

An adaptive network, as the name indicates, is a network structure consisting of a number of nodes connected through directional links. Each node represents a process unit, and the links between nodes specify the causal relationship between the connected nodes. All or part of the nodes are adaptive, which means the outputs of these nodes depend on modifiable parameters pertaining to these nodes. The learning rule specifies how these parameters should be updated to minimize a prescribed error measure, which is a mathematical expression that measures the discrepancy between the network's actual output and a desired output. In other words, an adaptive network is used for system identification (see Chapter 5), and our task is to find an appropriate network architecture and a set of parameters which can best model an unknown target system that is described by a set of input-output data pairs.

The basic learning rule of the adaptive network is the well-known steepest descent method, in which the gradient vector is derived by successive invocations of the chain rule. This method for systematic calculation of the gradient vector was proposed independently several times, by Bryson and Ho [1], Werbos [16], and Parker [9]. However, because research on artificial neural networks was still in its infancy at those times, these researchers' early work failed to receive the attention it deserved. In 1986, Rumelhart et al. [11] used the same procedure to find the gradient in a multilayer neural network. Their procedure was called the backpropagation learning rule, a name which is now widely known because the work of Rumelhart et al. inspired enormous interest in research on neural networks. In this chapter, we

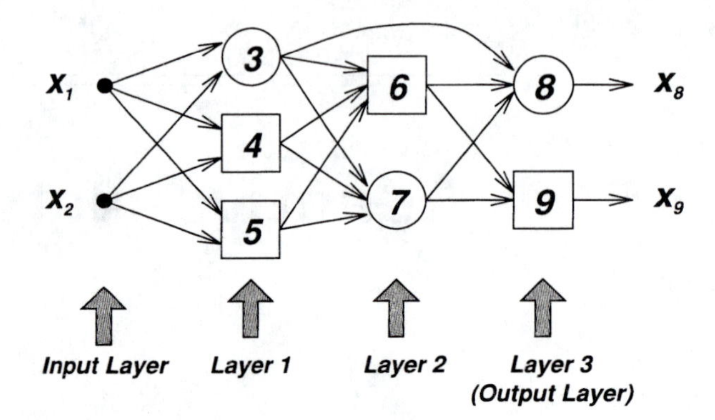

Figure 8.1. *A feedforward adaptive network in layered representation.*

introduce Werbos's original backpropagation method for finding gradient vectors and also present an improved version [3, 4] which speeds up the time-consuming learning process by incorporating the least-squares method.

8.2 ARCHITECTURE

As the name implies, an **adaptive network** (Figure 8.1) is a network structure whose overall input-output behavior is determined by a collection of modifiable parameters. Specifically, the configuration of an adaptive network is composed of a set of nodes connected by directed links, where each node performs a static **node function** on its incoming signals to generate a single **node output** and each link specifies the direction of signal flow from one node to another. Usually a node function is a parameterized function with modifiable parameters; by changing these parameters, we change the node function as well as the overall behavior of the adaptive network.

In the following discussion, we shall assume that each node in an adaptive network performs a static mapping from its input(s) to output. Namely, a node's output depends on its current inputs only; there are no dynamics or internal states in each node. Moreover, to facilitate the development of learning algorithms, we assume that all node functions are differentiable except at a finite number of points. In the most general case, an adaptive network is heterogeneous and each node may have a specific node function different from the others. Links in an adaptive network are merely used to specify the propagation direction of node outputs; generally there are no weights or parameters associated with links. Figure 8.1 is a typical adaptive network with two inputs and two outputs.

The parameters of an adaptive network are distributed into its nodes, so each node has a local parameter set. The union of these local parameter sets is the network's overall parameter set. If a node's parameter set is not empty, then its node function depends on the parameter values; we use a square to represent this

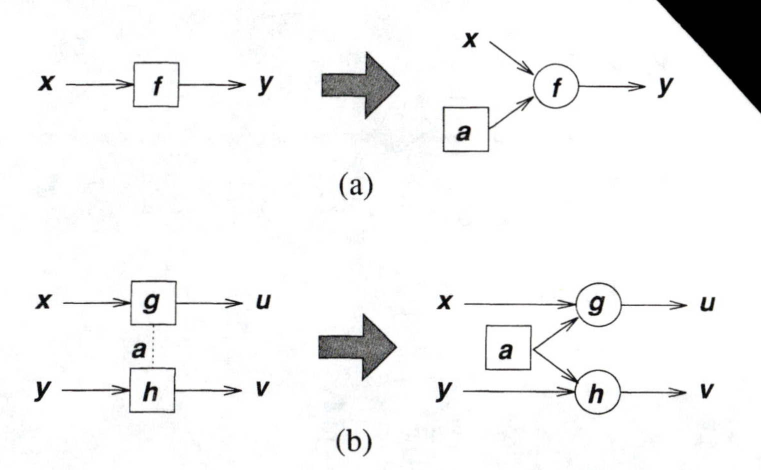

Figure 8.2. *Decomposition of adaptive nodes: (a) a single node; (b) parameter sharing problem.*

kind of **adaptive node**. On the other hand, if a node has an empty parameter set, then its function is fixed; we use a circle to denote this type of **fixed node**. Each adaptive node can be decomposed into a fixed node plus one or several parameter nodes, as illustrated in the following example.

Example 8.1 *Parameter sharing in adaptive networks*

Figure 8.2(a) shows an adaptive network with only one node, which can be represented as $y = f(x, a)$, where x and y are the input and output, respectively, and a is the parameter of the node. An equivalent representation is to move the parameter out of the node and put it into a **parameter node**, as shown in Figure 8.2(a). It is obvious that a parameter node is a special case of an adaptive node in which there are no inputs and the output is the parameter itself. The parameter node is useful in solving certain representation problems, such as the **parameter sharing problem** in Figure 8.2(b), where two adaptive nodes $u = g(x, a)$ and $v = h(y, a)$ share the same parameter a, as denoted by the dotted line linking these two nodes. By taking out the parameter and putting it into a parameter node, we can embed the parameter sharing requirement into the architecture. This simplifies network representation as well as software implementation.

□

Adaptive networks are generally classified into two categories on the basis of the type of connections they have: **feedforward** and **recurrent**. The adaptive network shown in Figure 8.1 is feedforward, since the output of each node propagates from the input side (left) to the output side (right) unanimously. If there is a feedback link that forms a circular path in a network, then the network is recurrent; Figure 8.3

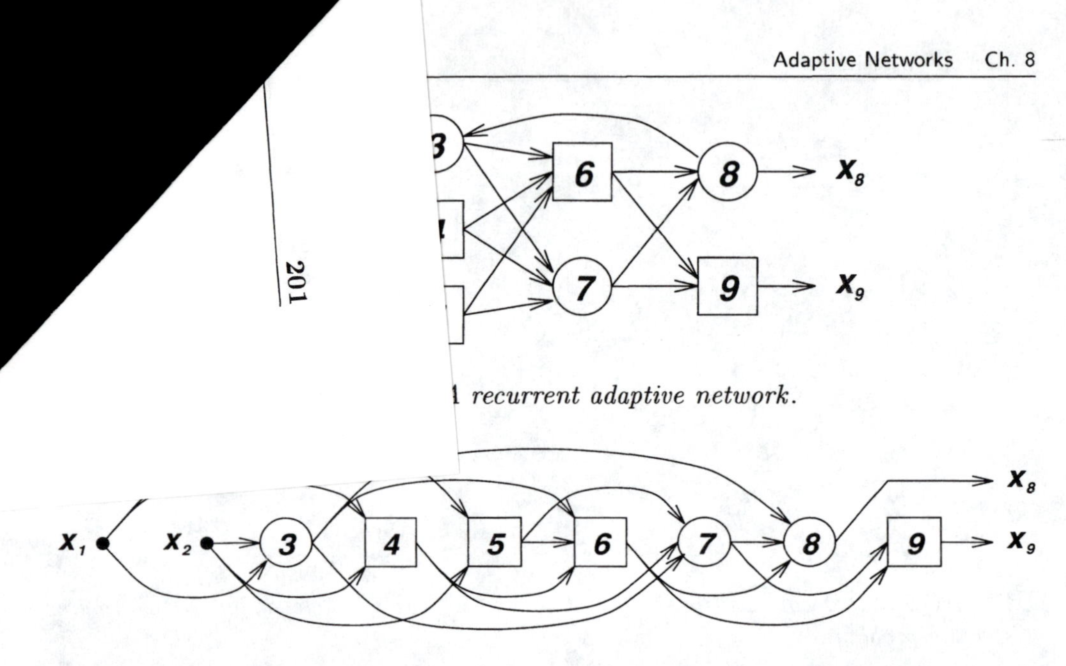

A recurrent adaptive network.

Figure 8.4. *A feedforward adaptive network in topological ordering representation.*

is an example. (From the viewpoint of graph theory, a feedforward network is represented by an *acyclic* directed graph which contains no directed cycles, while a recurrent network always contains at least one directed cycle.)

In the **layered representation** of the feedforward adaptive network in Figure 8.1, there are no links between nodes in the same layer, and outputs of nodes in a specific layer are always connected to nodes in succeeding layers. This representation is usually preferred because of its modularity, in that nodes in the same layer have the same functionality or generate the same level of abstraction about input vectors.

Another representation of feedforward networks is the **topological ordering representation**, which labels the nodes in an ordered sequence $1, 2, 3, \ldots$, such that there are no links from node i to node j whenever $i \geq j$. Figure 8.4 is the topological ordering representation of the network in Figure 8.1. This representation is less modular than the layer representation, but it facilitates the formulation of learning rules, as will be detailed in the next section. (Note that the topological ordering representation is in fact a special case of the layered representation, with one node per layer.)

Conceptually, a feedforward adaptive network is actually a static mapping between its input and output spaces; this mapping may be either a simple linear relationship or a highly nonlinear one, depending on the network structure (node arrangement and connections, and so on) and the functionality for each node. Here our aim is to construct a network for achieving a desired nonlinear mapping that is regulated by a data set consisting of desired input-output pairs of a target system to be modeled. This data set is usually called the **training data set**, and

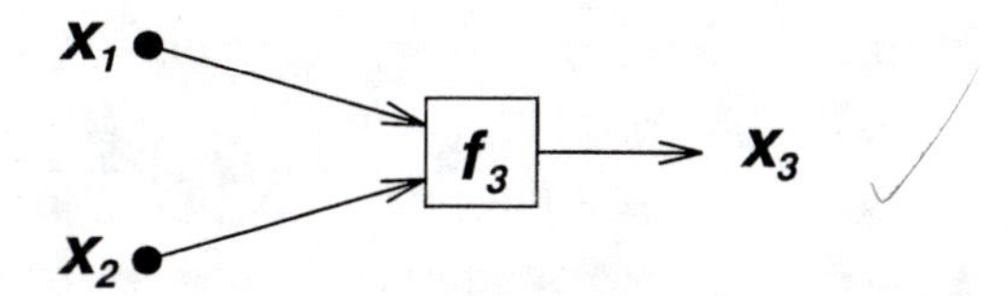

Figure 8.5. *A linear single-node adaptive network.*

the procedures we follow in adjusting the parameters to improve the network's performance are often referred to as the **learning rules** or **adaptation algorithms**. Usually a network's performance is measured as the discrepancy between the desired output and the network's output under the same input conditions. This discrepancy is called the **error measure** and it can assume different forms for different applications. Generally speaking, a learning rule is derived by applying a specific optimization technique to a given error measure.

Before introducing a basic learning algorithm for adaptive networks, we shall present several examples of adaptive networks.

Example 8.2 *An adaptive network with a single linear node*

Figure 8.5 is an adaptive network with a single node specified by

$$x_3 = f_3(x_1, x_2; a_1, a_2, a_3) = a_1x_1 + a_2x_2 + a_3,$$

where x_1 and x_2 are inputs and a_1, a_2, and a_3 are modifiable parameters. The function defines a plane in $x_1 - x_2 - x_3$ space, and by setting appropriate values for the parameters, we can place this plane arbitrarily. By adopting the squared error as the error measure for this network, we can identify the optimal parameters via the linear least-squares estimation method introduced in Chapter 5.

□

Example 8.3 *Perceptron network*

If we add another node to let the output of the adaptive network in Figure 8.5 have only two values 0 and 1; then the nonlinear network shown in Figure 8.6 is obtained. Specifically, the node outputs are expressed as

$$x_3 = f_3(x_1, x_2; a_1, a_2, a_3) = a_1x_1 + a_2x_2 + a_3,$$

and

$$x_4 = f_4(x_3) = \begin{cases} 1 & \text{if } x_3 \geq 0 \\ 0 & \text{if } x_3 < 0 \end{cases},$$

where f_3 is a linearly parameterized function and f_4 is a step function which maps x_3 to either 0 or 1. The overall function of this network can be viewed as a **linear**

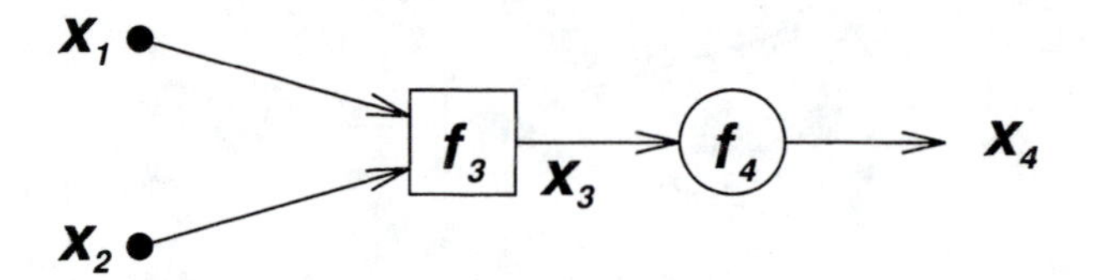

Figure 8.6. *A nonlinear single-node adaptive network.*

classifier: The first node forms a decision boundary as a straight line in $x_1 - x_2$ space, and the second node indicates which half plane the input vector $(x_1,\ x_2)$ resides in. Obviously, we can form an equivalent network with a single node whose function is the composition of f_3 and f_4; the resulting node is the building block of the classical **perceptron** [8, 10].

Since the step function is discontinuous at one point and flat at all the other points, it is not suitable for derivative-based learning procedures. One way to get around this difficulty is to use the **sigmoidal function** as a squashing function that has values between 0 and 1:

$$x_4 = f_4(x_3) = \frac{1}{1 + e^{-x_3}}.$$

This is a continuous and differentiable approximation to the step function. The composition of f_3 and this differentiable f_4 is the building block for the multilayer perceptron in the following example.

□

Example 8.4 *A multilayer perceptron*

Figure 8.7 is a typical architecture for a multilayer perceptron with three inputs, two outputs, and three hidden nodes that do not connect directly to either inputs or outputs. Each node in a network of this kind has the same node function, which is the composition of a linear f_3 and a sigmoidal f_4 in Example 8.3. For instance, the node function of node 7 in Figure 8.7 is

$$x_7 = \frac{1}{1 + \exp[-(w_{4,7}x_4 + w_{5,7}x_5 + w_{6,7}x_6 + t_7)]},$$

where x_4, x_5, and x_6 are outputs from nodes 4, 5, and 6, respectively, and the parameter set of node 7 is denoted by $\{w_{4,7}, w_{5,7}, w_{6,7}, t_7\}$. Usually we view $w_{i,j}$ as the **weight** associated with the link connecting node i and j and t_j as the **threshold** associated with node j. However, this weight-link association is only valid in this type of network. In general, a link only indicates the signal flow direction between connected nodes, as will be shown in other types of adaptive networks in the subsequent discussion.

A more detailed discussion about the structure and learning rules of the multilayer perceptron is in presented Section 9.4.

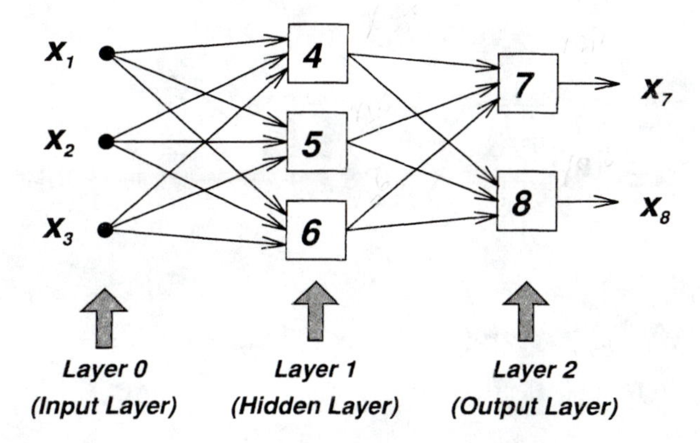

Figure 8.7. *A 3-3-2 neural network.*

□

8.3 BACKPROPAGATION FOR FEEDFORWARD NETWORKS

This section introduces a basic learning rule for adaptive networks, which is in essence the simple steepest descent method discussed in Section 6.3 of Chapter 6. The central part of this learning rule concerns how to recursively obtain a gradient vector in which each element is defined as the derivative of an error measure with respect to a parameter. This is done by means of the chain rule, a basic formula for differentiating composite functions that is covered in every textbook on elementary calculus. The procedure of finding a gradient vector in a network structure is generally referred to as **backpropagation** because the gradient vector is calculated in the direction opposite to the flow of the output of each node. Once the gradient is obtained, a number of derivative-based optimization and regression techniques (Chapters 5 and 6) are available for updating the parameters. In particular, if we use the gradient vector in a simple steepest descent method, the resulting learning paradigm is often referred to as the **backpropagation learning rule**. We shall introduce this learning rule in the rest of this section.

Suppose that a given feedforward adaptive network in the layered representation has L layers and layer l ($l = 0, 1, \ldots, L$; $l = 0$ represents the input layer) has $N(l)$ nodes. Then the output and function of node i [$i = 1, \ldots, N(l)$] in layer l can be represented as $x_{l,i}$ and $f_{l,i}$, respectively, as shown in Figure 8.8(a). Without loss of generality, we assume that there are no jumping links (that is, links connecting nonconsecutive layers). Since the output of a node depends on the incoming signals and the parameter set of the node, we have the following general expression for the node function $f_{l,i}$:

$$x_{l,i} = f_{l,i}(x_{l-1,1}, \ldots x_{l-1,N(l-1)}, \alpha, \beta, \gamma, \ldots), \tag{8.1}$$

where α, β, γ, etc. are the parameters of this node.

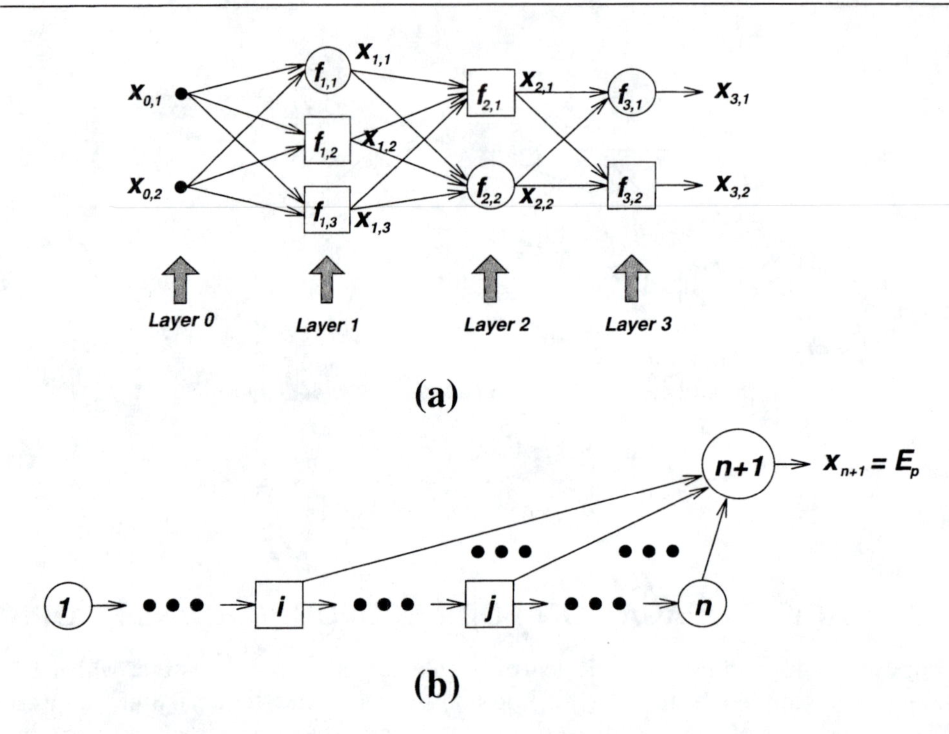

Figure 8.8. *Our notational conventions: (a) layered representation; (b) topological ordering representation.*

Assuming that the given training data set has P entries, we can define an error measure for the pth ($1 \leq p \leq P$) entry of the training data as the sum of squared errors:

$$E_p = \sum_{k=1}^{N(L)} (d_k - x_{L,k})^2, \tag{8.2}$$

where d_k is the kth component of the pth desired output vector and $x_{L,k}$ is the kth component of the actual output vector produced by presenting the pth input vector to the network. (For notational simplicity, we omit the subscript p for both d_k and $x_{L,k}$.) Obviously, when E_p is equal to zero, the network is able to reproduce exactly the desired output vector in the pth training data pair. Thus our task here is to minimize an overall error measure, which is defined as $E = \sum_{p=1}^{P} E_p$.

Remember that the definition of E_p in Equation (8.2) is not universal; other definitions of E_p are possible for specific situations or applications. Therefore, we shall avoid using an explicit expression for the error measure E_p to emphasize the generality. In addition, we assume that E_p depends on the output nodes only; more general situations will be discussed later.

To use steepest descent to minimize the error measure, first we have to obtain the gradient vector. Before calculating the gradient vector, we should observe the following causal relationships:

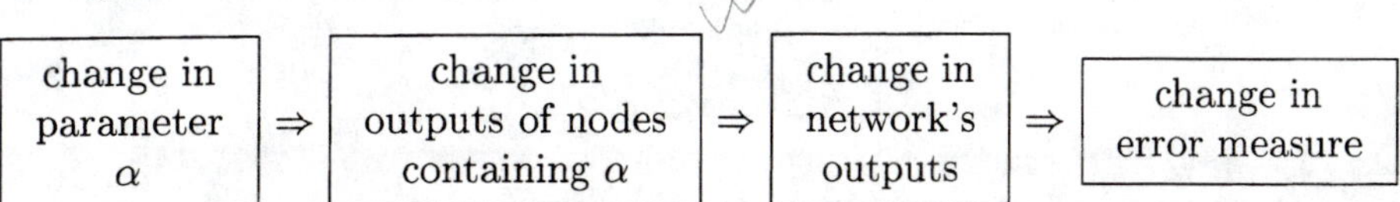

where the arrows ⇒ indicate causal relationships. In other words, a small change in a parameter α will affect the output of the node containing α; this in turn will affect the output of the final layer and thus the error measure. Therefore, the basic concept in calculating the gradient vector is to pass a form of derivative information starting from the output layer and going backward layer by layer until the input layer is reached.

To facilitate the discussion, we define the **error signal** $\epsilon_{l,i}$ as the derivative of the error measure E_p with respect to the output of node i in layer l, taking *both* direct and indirect paths into consideration. In symbols,

$$\epsilon_{l,i} = \frac{\partial^+ E_p}{\partial x_{l,i}}. \tag{8.3}$$

This expression was called the **ordered derivative** by Werbos [16]. The difference between the ordered derivative and the ordinary partial derivative lies in the way we view the function to be differentiated. For an internal node output $x_{l,i}$ (where $l \neq L$), the partial derivative $\frac{\partial E_p}{\partial x_{l,i}}$ is equal to zero, since E_p does not depend on $x_{l,i}$ directly. However, it is obvious that E_p does depend on $x_{l,i}$ indirectly, since a change in $x_{l,i}$ will propagate through indirect paths to the output layer and thus produce a corresponding change in the value of E_p. Therefore, $\epsilon_{l,i}$ can be viewed as the ratio of these two changes when they are made infinitesimal. The following example demonstrates the difference between the ordered derivative and the ordinary partial derivative.

Example 8.5 *Ordered derivatives and ordinary partial derivatives*

Consider a simple adaptive network shown in Figure 8.9, where z is a function of x and y, and y is in turn a function of x:

$$\begin{cases} z = g(x, y), \\ y = f(x). \end{cases}$$

For the ordinary partial derivative $\frac{\partial z}{\partial x}$, we assume that all the other input variables (in this case, y) are constant:

$$\frac{\partial z}{\partial x} = \frac{\partial g(x, y)}{\partial x}.$$

In other words, we assume that the inputs x and y to the function g are independent, without paying attention to the fact that y is actually a function of x. For the

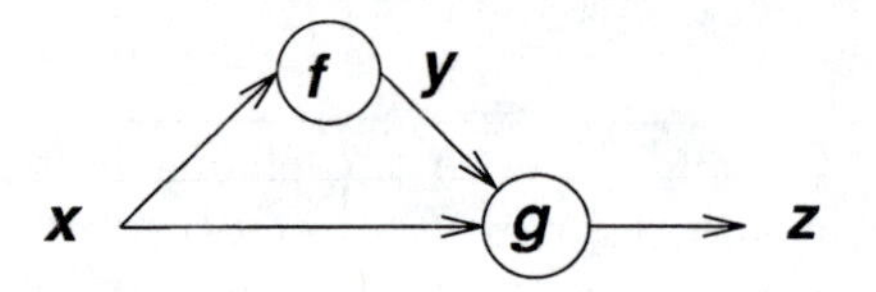

Figure 8.9. *Ordered derivatives and ordinary partial derivatives (Example 8.5).*

ordered derivative, we take this indirect causal relationship into consideration:

$$\begin{aligned}\frac{\partial^+ z}{\partial x} &= \frac{\partial g(x, f(x))}{\partial x} \\ &= \left.\frac{\partial g(x,y)}{\partial x}\right|_{y=f(x)} + \left.\frac{\partial g(x,y)}{\partial y}\right|_{y=f(x)} \frac{\partial f(x)}{\partial x}.\end{aligned}$$

Therefore, the ordered derivative takes into consideration both the direct and indirect paths that lead to the causal relationship.

□

The error signal for the ith output node (at layer L) can be calculated directly:

$$\epsilon_{L,i} = \frac{\partial^+ E_p}{\partial x_{L,i}} = \frac{\partial E_p}{\partial x_{L,i}}. \tag{8.4}$$

This is equal to $\epsilon_{L,i} = -2(d_i - x_{L,i})$ if E_p is defined as in Equation (8.2). For the internal node at the ith position of layer l, the error signal can be derived by the chain rule:

$$\epsilon_{l,i} = \underbrace{\frac{\partial^+ E_p}{\partial x_{l,i}}}_{\substack{\text{error signal}\\ \text{at layer } l}} = \sum_{m=1}^{N(l+1)} \underbrace{\frac{\partial^+ E_p}{\partial x_{l+1,m}}}_{\substack{\text{error signal}\\ \text{at layer } l+1}} \frac{\partial f_{l+1,m}}{\partial x_{l,i}} = \sum_{m=1}^{N(l+1)} \epsilon_{l+1,m} \frac{\partial f_{l+1,m}}{\partial x_{l,i}}, \tag{8.5}$$

where $0 \le l \le L-1$. That is, the error signal of an internal node at layer l can be expressed as a linear combination of the error signal of the nodes at layer $l+1$. Therefore, for any l and i $[0 \le l \le L$ and $1 \le i \le N(l)]$, we can find $\epsilon_{l,i} = \frac{\partial^+ E_p}{\partial x_{l,i}}$ by first applying Equation (8.4) once to get error signals at the output layer, and then applying Equation (8.5) iteratively until we reach the desired layer l. The underlying procedure is called backpropagation since the error signals are obtained sequentially from the output layer back to the input layer.

The gradient vector is defined as the derivative of the error measure with respect to each parameter, so we have to apply the chain rule again to find the gradient vector. If α is a parameter of the ith node at layer l, we have

$$\frac{\partial^+ E_p}{\partial \alpha} = \frac{\partial^+ E_p}{\partial x_{l,i}} \frac{\partial f_{l,i}}{\partial \alpha} = \epsilon_{l,i} \frac{\partial f_{l,i}}{\partial \alpha}. \tag{8.6}$$

Note that if we allow the parameter α to be shared between different nodes, then Equation (8.6) should be changed to a more general form:

$$\frac{\partial^+ E_p}{\partial \alpha} = \sum_{x^* \in S} \frac{\partial^+ E_p}{\partial x^*} \frac{\partial f^*}{\partial \alpha}, \tag{8.7}$$

where S is the set of nodes containing α as a parameter; and x^* and f^* are the output and function, respectively, of a generic node in S.

The derivative of the overall error measure E with respect to α is

$$\frac{\partial^+ E}{\partial \alpha} = \sum_{p=1}^{P} \frac{\partial^+ E_p}{\partial \alpha}. \tag{8.8}$$

Accordingly, for simple steepest descent without line minimization, the update formula for the generic parameter α is

$$\Delta \alpha = -\eta \frac{\partial^+ E}{\partial \alpha}, \tag{8.9}$$

in which η is the **learning rate**, which can be further expressed as

$$\eta = \frac{\kappa}{\sqrt{\sum_\alpha (\frac{\partial E}{\partial \alpha})^2}}, \tag{8.10}$$

where κ is the **step size**, the length of each transition along the gradient direction in the parameter space. Usually we can change the step size to vary the speed of convergence; see Section 6.7.2 of Chapter 6.

When an n-node feedforward network is represented in its topological order, we can envision the error measure E_p as the output of an additional node with index $n+1$, whose node function f_{n+1} can be defined on the outputs of any nodes with smaller index; see Figure 8.8(b). (Therefore, E_p may depend directly on any nodes.) Applying the chain rule again, we have the following concise formula for calculating the error signal $\epsilon_i = \partial E_p / \partial x_i$:

$$\frac{\partial^+ E_p}{\partial x_i} = \frac{\partial f_{n+1}}{\partial x_i} + \sum_{i<j\le n} \frac{\partial^+ E_p}{\partial x_j} \frac{\partial f_j}{\partial x_i}, \tag{8.11}$$

or

$$\epsilon_i = \frac{\partial f_{n+1}}{\partial x_i} + \sum_{i<j\le n} \epsilon_j \frac{\partial f_j}{\partial x_i}, \tag{8.12}$$

where the first term shows the direct effect of x_i on E_p via the direct path from node i to node $n+1$ and each product term in the summation indicates the indirect effect of x_i on E_p. Once we find the error signal for each node, then the gradient vector for the parameters is derived as before.

Another systematic way to calculate the error signals is through the representation of the **error-propagation network** (or **sensitivity model**), which is obtained from the original adaptive network by reversing the links and supplying the error signals at the output layer as inputs to the new network. The following example illustrates the idea.

Example 8.6 *Adaptive network and its error-propagation model*

Figure 8.10(a) is an adaptive network, where each node is indexed by a unique number. Again, we use f_i and x_i to denote the function and output of node i. To calculate the error signals at internal nodes, an error-propagation network is constructed in Figure 8.10(b), where the output of node i is the error signal of this node in the original adaptive network. In symbols, if we choose the squared error measure for E_p, then we have the following:

$$\begin{aligned}
\epsilon_9 &= \frac{\partial^+ E_p}{\partial x_9} = \frac{\partial E_p}{\partial x_9} = -2\,(d_9 - x_9), \\
\epsilon_8 &= \frac{\partial^+ E_p}{\partial x_8} = \frac{\partial E_p}{\partial x_8} = -2\,(d_8 - x_8), \\
\epsilon_7 &= \frac{\partial^+ E_p}{\partial x_7} = \frac{\partial^+ E_p}{\partial x_8}\frac{\partial f_8}{\partial x_7} + \frac{\partial^+ E_p}{\partial x_9}\frac{\partial f_9}{\partial x_7} = \epsilon_8 \frac{\partial f_8}{\partial x_7} + \epsilon_9 \frac{\partial f_9}{\partial x_7}, \\
\epsilon_6 &= \frac{\partial^+ E_p}{\partial x_6} = \frac{\partial^+ E_p}{\partial x_8}\frac{\partial f_8}{\partial x_6} + \frac{\partial^+ E_p}{\partial x_9}\frac{\partial f_9}{\partial x_6} = \epsilon_8 \frac{\partial f_8}{\partial x_6} + \epsilon_9 \frac{\partial f_9}{\partial x_6}.
\end{aligned}$$

Thus nodes 9 and 8 in the error-propagation network are only buffer nodes. Similar expressions can be written for the error signals of nodes 1, 2, 3, 4, and 5. It is interesting to observe that in the error-propagation net, if we associate each link connecting nodes i and j $(i < j)$ with a weight $w_{ij} = \frac{\partial f_i}{\partial x_j}$, then each node performs a linear function and the error-propagation net is actually a linear network.

□

There are two types of learning paradigms that are available to suit the needs for various applications. In **off-line learning** (or **batch learning**), the update formula for parameter α is based on Equation (8.8) and the update action takes place only after the whole training data set has been presented—that is, only after each **epoch** or **sweep**. On the other hand, in **on-line learning** (or **pattern-by-pattern learning**), the parameters are updated immediately after each input-output pair has been presented, and the update formula is based on Equation (8.6). In practice, it is possible to combine these two learning modes and update the parameter after k training data entries have been presented, where k is between 1 and P and it is sometimes referred to as the **epoch size**. These two types of learning paradigms are described in greater detail in Section 8.5, where a hybrid learning rule is introduced.

8.4 EXTENDED BACKPROPAGATION FOR RECURRENT NETWORKS

For recurrent adaptive networks, it is possible to derive an extended version of the backpropagation procedure that finds gradient vectors. To simplify our notation,

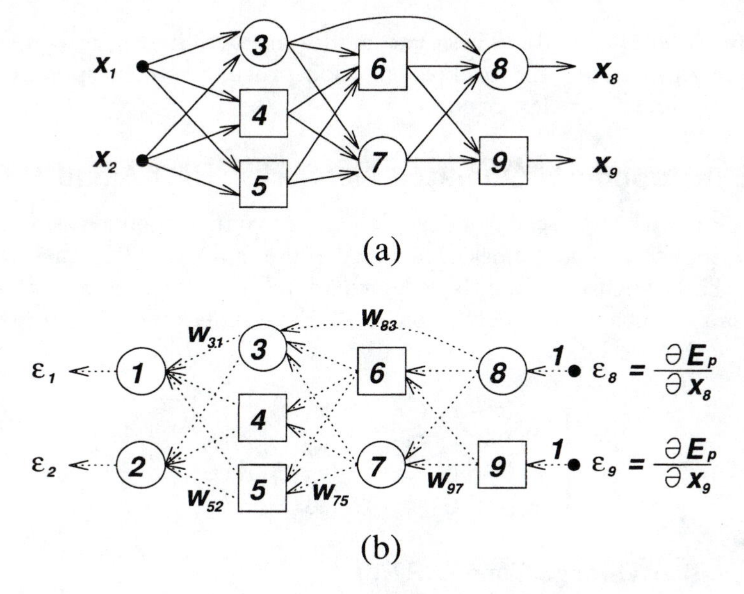

Figure 8.10. *(a) An adaptive network and (b) its error-propagation model.*

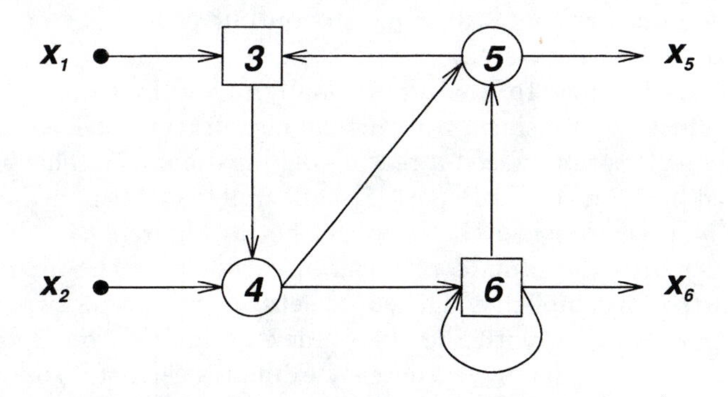

Figure 8.11. *A simple recurrent network.*

we shall use the network in Figure 8.11 for most of our discussion, where x_1 and x_2 are inputs and x_5 and x_6 are output nodes. Because it has directional loops 3-4-5, 3-4-6-5, and 6 (a self-loop), this is a typical recurrent network with node functions denoted as follows:

$$\left\{\begin{array}{lcl} x_3 & = & f_3(x_1, x_5), \\ x_4 & = & f_4(x_2, x_3), \\ x_5 & = & f_5(x_4, x_6), \\ x_6 & = & f_6(x_4, x_6). \end{array}\right. \tag{8.13}$$

To derive correctly the backpropagation procedure for the recurrent net in Figure 8.11, we have to distinguish two operating modes through which the network may satisfy Equation (8.13). These two modes are **synchronous operation** and **continuous operation**; the backpropagation procedures corresponding to these two operating modes are described next.

8.4.1 Synchronously Operated Networks: BPTT and RTRL

If a network is operated synchronously, all nodes change their outputs simultaneously according to a global clock signal and there is a time delay associated with each link. This synchronization is reflected by adding the time t as an argument to the output of each node in Equation (8.13) (assuming there is a unit time delay associated with each link):

$$\begin{cases} x_3(t+1) &= f_3(x_1(t), x_5(t)), \\ x_4(t+1) &= f_4(x_2(t), x_3(t)), \\ x_5(t+1) &= f_5(x_4(t), x_6(t)), \\ x_6(t+1) &= f_6(x_4(t), x_6(t)). \end{cases} \tag{8.14}$$

Backpropagation Through Time (BPTT)

When using synchronously operated networks, we usually are interested in identifying a set of parameters that will make the output of a node (or several nodes) follow a given trajectory (or trajectories) in the discrete time domain. This problem of **tracking** or **trajectory following** is usually solved by using a method called **unfolding of time** to transform a recurrent network into a feedforward one, as long as the time t does not exceed a reasonable maximum T. This idea was originally introduced by Minsky and Papert [7] and combined with backpropagation by Rumelhart, et al. [11]. Consider the recurrent net in Figure 8.11, which is redrawn in Figure 8.12(a) with the same configuration except that the input variables x_1 and x_2 are omitted for simplicity. The same network in a feedforward architecture is shown in Figure 8.12(b), with the time index t running from 1 to 4. In other words, for a recurrent net that synchronously evaluates each of its node functions at $t = 1, 2, \ldots, T$, we can simply duplicate all units T times and arrange the resulting network in a layered feedforward manner.

It is obvious that the two networks in Figures 8.12(a) and 8.12(b) behave identically for $t = 1$ to T, provided that all copies of parameters across different time steps remain identical. For instance, the parameter in node 3 of Figure 8.12(a) must be the same at all time instants. This is the parameter sharing problem addressed in Example 8.1. A quick and dirty solution to this problem is to move the parameters from nodes 3 and 6 into the so-called parameter nodes, which are independent of the time step, as shown in Figure 8.13. (Without loss of generality, we assume that nodes 3 and 6 both have only one parameter, denoted by a and b, respectively.) After setting up the parameter nodes in this way, we can apply the backpropagation procedure as usual to the network (which is still feedforward in nature) in Figure 8.13 without the slightest concern about the parameter sharing constraint. Note that the error signals of a parameter node come from nodes located

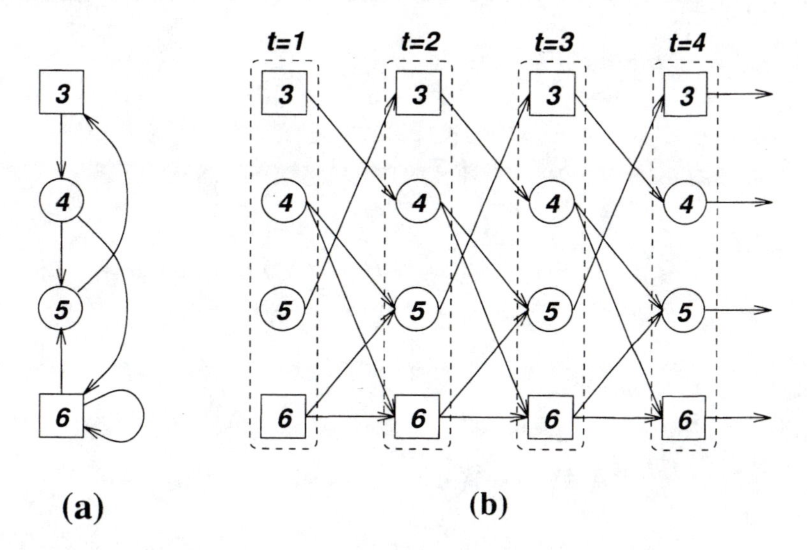

Figure 8.12. *(a) A synchronously operated recurrent network and (b) its feedforward equivalent obtained via unfolding of time.*

at layers across different time instants; thus the backpropagation procedure (and the corresponding steepest descent) for this kind of unfolded network is often called **backpropagation through time (BPTT)**.

Real-Time Recurrent Learning (RTRL)

BPTT generally works well for most problems; the only complication is that it requires extensive computing resources when the sequence length T is large, because the duplication of nodes makes both memory requirements and simulation time proportional to T. Therefore, for long sequences or sequences of unknown length, **real-time recurrent learning** (RTRL) [18] is employed instead to perform on-line learning—that is, to update parameters while the network is running rather than at the end of the presented sequences.

To explain the rationale behind RTRL, we take as an example the simple recurrent network in Figure 8.14(a), where there is only one node with one parameter a. After moving the parameter out of the unfolded architecture, we obtain the feedforward network shown in Figure 8.14(b). Figure 8.14(c) is the corresponding error-propagation network. Here we assume $E = \sum_i^T E_i = \sum_i^T (d_i - x_i)^2$, where i is the index for time and d_i and x_i are the desired and the actual node output, respectively, at time instant i.

To save computation and memory requirements, a sensible choice is to minimize E_i at each time step instead of trying to minimize E at the end of a sequences. To achieve this, we need to calculate $\partial^+ E/\partial a$ recursively at each time step i. For

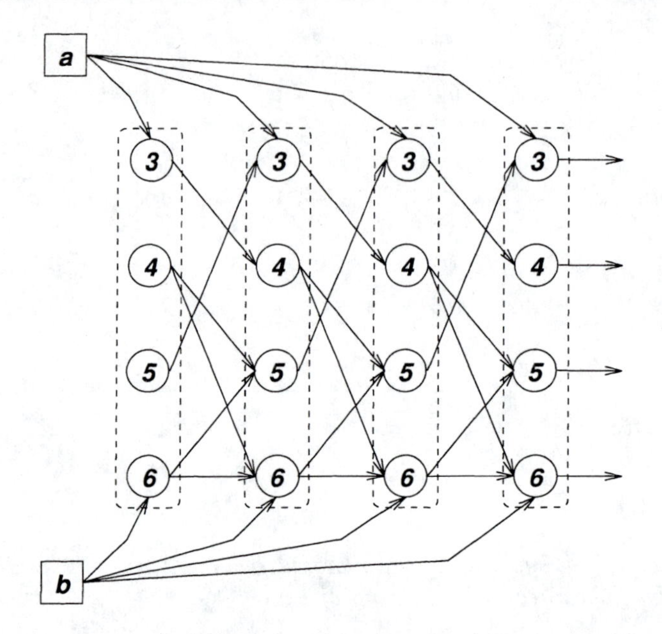

Figure 8.13. *An alternative representation of Figure 8.12(b) that satisfies the parameter sharing requirement.*

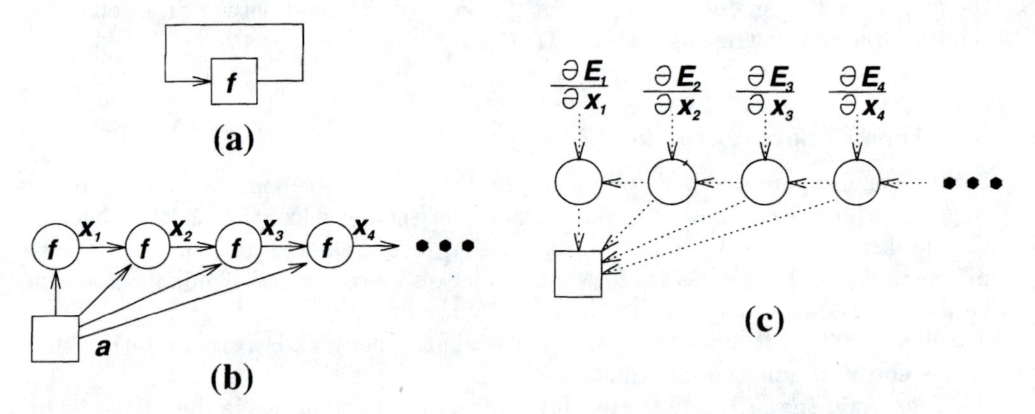

Figure 8.14. *A simple recurrent adaptive network to illustrate RTRL: (a) a recurrent net with single node and single parameter; (b) unfolding-of-time architecture; (c) error-propagation network.*

$i = 1$, the error-propagation network is as shown in Figure 8.15(a) and we have

$$\frac{\partial^+ x_1}{\partial a} = \frac{\partial x_1}{\partial a} \text{ and } \frac{\partial^+ E_1}{\partial a} = \frac{\partial E_1}{\partial x_1}\frac{\partial^+ x_1}{\partial a}. \tag{8.15}$$

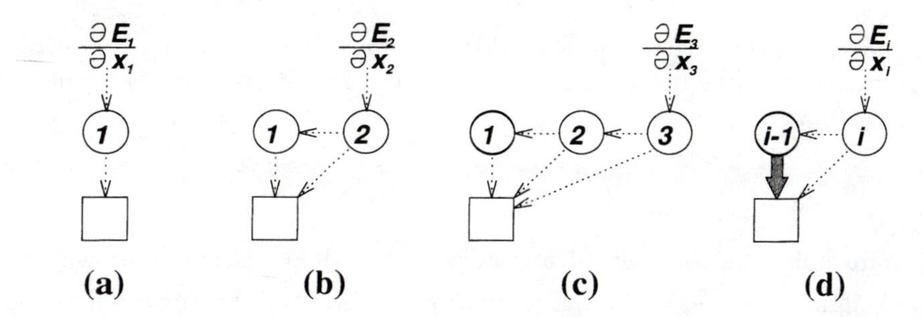

Figure 8.15. *Error-propagation networks at different time steps: (a) $i = 1$; (b) $i = 2$; (c) $i = 3$; (d) a general situation where the thick arrow represents* $\frac{\partial^+ x_{i-1}}{\partial a}$.

For $i = 2$, the error-propagation network is as shown in Figure 8.15(b) and we have

$$\frac{\partial^+ x_2}{\partial a} = \frac{\partial x_2}{\partial a} + \frac{\partial x_2}{\partial x_1}\frac{\partial^+ x_1}{\partial a} \text{ and } \frac{\partial^+ E_2}{\partial a} = \frac{\partial E_2}{\partial x_2}\frac{\partial^+ x_2}{\partial a}. \tag{8.16}$$

For $i = 3$, the error-propagation network is as shown in Figure 8.15(c) and we have

$$\frac{\partial^+ x_3}{\partial a} = \frac{\partial x_3}{\partial a} + \frac{\partial x_3}{\partial x_2}\frac{\partial^+ x_2}{\partial a} \text{ and } \frac{\partial^+ E_3}{\partial a} = \frac{\partial E_3}{\partial x_3}\frac{\partial^+ x_3}{\partial a}. \tag{8.17}$$

In general, for the error-propagation at time instant i, we have

$$\frac{\partial^+ x_i}{\partial a} = \frac{\partial x_i}{\partial a} + \frac{\partial x_i}{\partial x_{i-1}}\frac{\partial^+ x_{i-1}}{\partial a} \text{ and } \frac{\partial^+ E_i}{\partial a} = \frac{\partial E_i}{\partial x_i}\frac{\partial^+ x_3}{\partial a}, \tag{8.18}$$

where $\frac{\partial^+ x_{i-1}}{\partial a}$ is already available from the calculation at the previous time instant. Figure 8.15 shows this general situation, where the thick arrow represents $\frac{\partial^+ x_{i-1}}{\partial a}$, which is already available at the time instant $i - 1$.

Therefore, by trying to minimize each individual E_i, we can recursively find the gradient $\frac{\partial^+ E_i}{\partial a}$ at each time instant; there is no need to wait until the end of the presented sequence. Since this is an approximation of the original BPTT, the learning rate η in the steepest descent update formula

$$\Delta a = -\eta \frac{\partial^+ E_i}{\partial a}$$

should be kept small and, as a result, the learning process usually takes longer.

8.4.2 Continuously Operated Networks: Mason's Gain Formula*

In a network that is operated continuously, all nodes continuously change their outputs until Equation (8.13) is satisfied. This operating mode is of particular interest for analog circuit implementations, where a certain kind of dynamical evolution rule is imposed on the network. For instance, the dynamical formula for node 3 in Figure 8.11 can be written as

$$\tau_3 \frac{dx_3}{dt} + x_3 = f_3(x_1, x_5). \tag{8.19}$$

Similar formulas can be devised for other nodes. It is obvious that when $x_3(t)$ stops changing (i.e., $\frac{dx_3}{dt} = 0$), Equation (8.19) leads to the correct fixed points satisfying Equation (8.13). We assume that at least one such stable fixed point exists for every possible node output in Equation (8.13). Other situations, such as limit cycles, can be either generated or eliminated via the approximation of a synchronously operated network using the techniques introduced in Section 8.4.1.

By assuming that the error measure E is a function of the output nodes—that is, $E = E(x_5, x_6)$—we obtain the following equations via repeated applications of the chain rule:

$$\left\{ \begin{array}{lcl} \frac{\partial^+ E}{\partial x_3} & = & \frac{\partial^+ E}{\partial x_4} \frac{\partial f_4}{\partial x_3}, \\ \frac{\partial^+ E}{\partial x_4} & = & \frac{\partial^+ E}{\partial x_5} \frac{\partial f_5}{\partial x_4} + \frac{\partial^+ E}{\partial x_6} \frac{\partial f_6}{\partial x_4}, \\ \frac{\partial^+ E}{\partial x_5} & = & \frac{\partial^+ E}{\partial x_3} \frac{\partial f_3}{\partial x_5} + \frac{\partial E}{\partial x_5}, \\ \frac{\partial^+ E}{\partial x_6} & = & \frac{\partial^+ E}{\partial x_5} \frac{\partial f_5}{\partial x_6} + \frac{\partial^+ E}{\partial x_6} \frac{\partial f_6}{\partial x_6} + \frac{\partial E}{\partial x_6}. \end{array} \right. \tag{8.20}$$

As before, if the error signal $\partial^+ E / \partial x_i$ is denoted as ϵ_i, Equation (8.20) can be simplified to the following:

$$\left\{ \begin{array}{lcl} \epsilon_3 & = & \epsilon_4 w_{43}, \\ \epsilon_4 & = & \epsilon_5 w_{54} + \epsilon_6 w_{64}, \\ \epsilon_5 & = & \epsilon_3 w_{35} + \frac{\partial E}{\partial x_5}, \\ \epsilon_6 & = & \epsilon_5 w_{56} + \epsilon_6 w_{66} + \frac{\partial E}{\partial x_6}, \end{array} \right. \tag{8.21}$$

or, equivalently,

$$\begin{bmatrix} 1 & -w_{43} & 0 & 0 \\ 0 & 1 & -w_{54} & -w_{64} \\ -w_{35} & 0 & 1 & 0 \\ 0 & 0 & -w_{56} & 1 - w_{66} \end{bmatrix} \begin{bmatrix} \epsilon_3 \\ \epsilon_4 \\ \epsilon_5 \\ \epsilon_6 \end{bmatrix} = \begin{bmatrix} 0 \\ 0 \\ \partial E / \partial x_5 \\ \partial E / \partial x_6 \end{bmatrix}, \tag{8.22}$$

where $w_{ij} = \frac{\partial f_i}{\partial x_j}$. Then ϵ_i can be obtained through the standard method for linear algebraic equations. Once we have ϵ_i, the gradient for a generic parameter α in node i can be found directly:

$$\frac{\partial^+ E}{\partial \alpha} = \frac{\partial^+ E}{\partial x_i} \frac{\partial f_i}{\partial \alpha} = \epsilon_i \frac{\partial f_i}{\partial \alpha}. \tag{8.23}$$

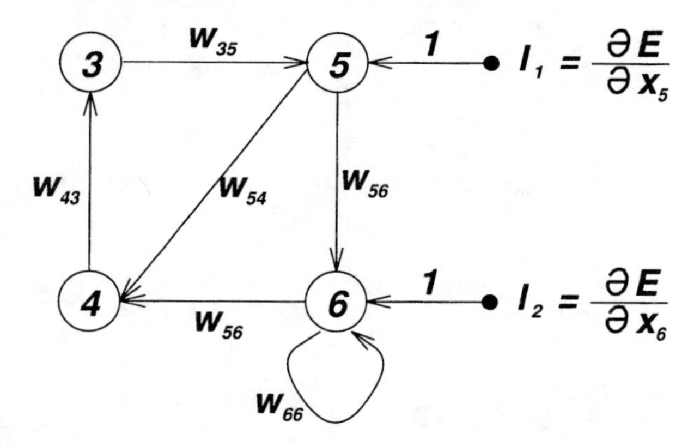

Figure 8.16. *A recurrent error-propagation network corresponding to Figure 8.11.*

It is of interest to note that Equation (8.21) can be represented as the recurrent network shown in Figure 8.16, where the output of node i is the error signal ϵ_i. The topology of this **recurrent error-propagation network** is the same as that of the original network (Figure 8.11), except that the direction of each link is reversed and the link from node i to j is associated with a weight w_{ij}, defined by $\partial f_i / \partial x_j$. Moreover, two quantities $\frac{\partial E}{\partial x_5}$ and $\frac{\partial E}{\partial x_6}$ are provided as the inputs to this network. As a result, if we have hardware-implemented networks, the calculation of ϵ_i would be similar to finding the fixed points of the original network, which is topologically equivalent to its error-propagation network.

Without hardware-implemented networks, solving Equation (8.22) would seem to require lengthy calculation using the Gaussian elimination method or other similar techniques. An alternative approach to obtaining ϵ_i is **Mason's gain formula** [6], which is commonly employed to find transfer functions of linear systems represented in signal flow graphs or block diagrams. The signal flow graph [5] may be regarded as a cause-and-effect representation of a linear system; our recurrent error-propagation network undoubtedly is such a system. Therefore, by applying the following gain formula, we can obtain ϵ_i by mere inspection.

Theorem 8.1 *Mason's gain formula [6] for the recurrent error-propagation network*

The general gain formula between ϵ_i and an input quantity I is

$$M = \frac{\partial \epsilon_i}{\partial I} = \sum_{k=1}^{N} \frac{M_k \Delta_k}{\Delta}, \tag{8.24}$$

where

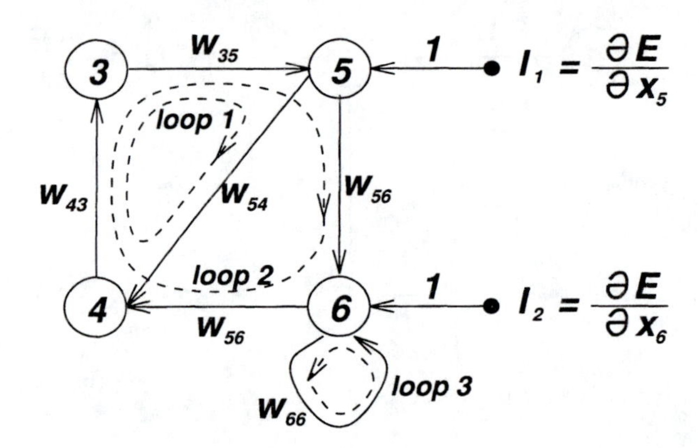

Figure 8.17. *Loops in the recurrent error-propagation network of Figure 8.16. (Note that loop 1 and 3 are nontouching.)*

M	=	gain between ϵ_i and I
ϵ_i	=	output of node i of the recurrent error-propagation network
I	=	input quantity
N	=	total number of forward paths from I to ϵ_i
M_k	=	gain of the kth forward path
Δ	=	$1 - \sum_m P_{m1} + \sum_m P_{m2} - \sum_m P_{m3} + \cdots$
P_{mr}	=	gain product of the mth possible combination of r nontouching loops, that is, loops sharing no common nodes
Δ_k	=	the Δ for that part of the recurrent error-propagation network which is nontouching with the kth forward path

□

At first glance, calculating Mason's gain formula may seem to be a formidable task because of the complex expressions for Δ and Δ_k. In practice, however, this gain formula is straightforward, since recurrent error-propagation networks with a large number of nontouching loops are rare. The following example illustrates how to apply Mason's gain formula for the error-propagation network in Figure 8.16.

Example 8.7 *Mason's gain formula*

To express the error signal ϵ_3 in terms of the inputs I_1 and I_2 in Figure 8.16, we first observe that there are three loops in the recurrent error-propagation network, as shown in Figure 8.17. The gains for these loops are

$$\begin{aligned} \text{loop 1 (5-4-3):} \quad l_1 &= w_{54}w_{43}w_{35}, \\ \text{loop 2 (5-6-4-3):} \quad l_2 &= w_{56}w_{64}w_{43}w_{35}, \\ \text{loop 3 (6):} \quad l_3 &= w_{66}, \end{aligned} \tag{8.25}$$

where loop 1 and loop 3 are nontouching loops, since they do not share a common node. Thus the Δ in Equation (8.24) [which is equal to the determinant of the square matrix in Equation (8.22)] is expressed as

$$\Delta = 1 - (l_1 + l_2 + l_3) + (l_1 l_3). \tag{8.26}$$

To find the gain between ϵ_3 and I_1, note that there are two direct paths between them: I_1-5-4-3 and I_1-5-6-4-3. For the first path, we have $M_1 = w_{54}w_{43}$ and $\Delta_1 = 1 - w_{66}$, since only loop 3 is nontouching with respect to this path. For the second path, we have $M_2 = w_{56}w_{64}w_{43}$ and $\Delta_2 = 1$, since no loop is nontouching with respect to this path. Therefore, the gain g_1 between ϵ_3 and I_1 is

$$\begin{aligned} g_1 &= \frac{M_1\Delta_1 + M_2\Delta_2}{\Delta} \\ &= \frac{w_{54}w_{43}(1 - w_{66}) + w_{56}w_{64}w_{43}}{\Delta}. \end{aligned} \tag{8.27}$$

In contrast, there is only one direct path (I_2-6-4-3) connecting ϵ_3 and I_2. Consequently, we have $M_1 = w_{64}w_{43}$ and $\Delta_1 = 1$, and the gain g_2 between ϵ_3 and I_2 is

$$\begin{aligned} g_2 &= \frac{M_1\Delta_1}{\Delta} \\ &= \frac{w_{64}w_{43}}{\Delta}. \end{aligned} \tag{8.28}$$

Since the recurrent error-propagation network is a linear system, the principle of superposition applies. Thus, by combining Equations (8.27) and (8.28), we obtain the error signal ϵ_3 as follows:

$$\begin{aligned} \epsilon_3 &= g_1 I_1 + g_2 I_2 \\ &= \frac{w_{54}w_{43}(1 - w_{66}) + w_{56}w_{64}w_{43}}{\Delta}\frac{\partial E}{\partial x_5} + \frac{w_{64}w_{43}}{\Delta}\frac{\partial E}{\partial x_6}. \end{aligned} \tag{8.29}$$

□

In essence, the function of the continuously operated recurrent network is still a static mapping when the property of dynamical evolution [see Equation (8.19)] is ignored. There is no theoretical proof or derivation comparing the approximation power of a continuously operated recurrent network with that of a typical ordinary feedforward network. Moreover, calculating the stable attractor defined by Equation (8.13) could be time-consuming in software simulations of the network. As a result, continuously operated recurrent networks are not used as often as the synchronously operated recurrent networks described in Section 8.4.1.

8.5 HYBRID LEARNING RULE: COMBINING STEEPEST DESCENT AND LSE

Although we can apply backpropagation or steepest descent to identify the parameters in an adaptive network, this simple optimization method usually takes a long time before it converges. We may observe, however, that an adaptive network's output (assuming there is only one) is linear in some of the network's parameters; thus we can identify these linear parameters by the linear least-squares method described in Chapter 5. This approach leads to a hybrid learning rule [3, 4] which combines steepest descent (SD) and the least-squares estimator (LSE) for fast identification of parameters.

8.5.1 Off-Line Learning (Batch Learning)

For simplicity, assume that the adaptive network under consideration has only one output represented by

$$o = F(\mathbf{i},\ S), \tag{8.30}$$

where $\mathbf{i}$ is the vector of input variables, S is the set of parameters, and F is the overall function implemented by the adaptive network. If there exists a function H such that the composite function $H \circ F$ is linear in some of the elements of S, then these elements can be identified by the least-squares method. More formally, if the parameter set S can be divided into two sets

$$S = S_1 \oplus S_2, \tag{8.31}$$

(where $\oplus$ represents direct sum) such that $H \circ F$ is linear in the elements of S_2, then upon applying H to Equation (8.30), we have

$$H(o) = H \circ F(Bi,\ S), \tag{8.32}$$

which is linear in the elements of S_2. Now given values of elements of S_1, we can plug P training data into Equation (8.32) and obtain a matrix equation:

$$\mathbf{A}\boldsymbol{\theta} = \mathbf{y} \tag{8.33}$$

where $\boldsymbol{\theta}$ is an unknown vector whose elements are parameters in S_2. Obviously Equation (8.33) is exactly the same as Equation (5.14) in Section 5.3; thus this is a standard linear least-squares problem, and the best solution for $\boldsymbol{\theta}$, which minimizes $||\mathbf{A}\boldsymbol{\theta} - \mathbf{y}||^2$, is the least-squares estimator (LSE) $\boldsymbol{\theta}^*$:

$$\boldsymbol{\theta}^* = (\mathbf{A}^T\mathbf{A})^{-1}\mathbf{A}^T\mathbf{y}, \tag{8.34}$$

where $\mathbf{A}^T$ is the transpose of $\mathbf{A}$ and $(\mathbf{A}^T\mathbf{A})^{-1}\mathbf{A}^T$ is the pseudoinverse of $\mathbf{A}$ if $\mathbf{A}^T\mathbf{A}$ is nonsingular. Of course, we can also employ the recursive LSE formula introduced in Section 5.5 to calculate $\boldsymbol{\theta}^*$. Specifically, let the ith row vector of matrix $\mathbf{A}$ defined in Equation (8.33) be $\mathbf{a}_i^T$ and the ith element of $\mathbf{y}$ be y_i^T; then $\boldsymbol{\theta}$ can be calculated iteratively as follows:

$$\left.\begin{aligned} \boldsymbol{\theta}_{i+1} &= \boldsymbol{\theta}_i + \mathbf{P}_{i+1}\mathbf{a}_{i+1}(y_{i+1}^T - \mathbf{a}_{i+1}^T\boldsymbol{\theta}_i) \\ \mathbf{P}_{i+1} &= \mathbf{P}_i - \frac{\mathbf{P}_i\mathbf{a}_{i+1}\mathbf{a}_{i+1}^T\mathbf{P}_i}{1 + \mathbf{a}_{i+1}^T\mathbf{P}_i\mathbf{a}_{i+1}}, \quad i = 0, 1, \cdots, P-1 \end{aligned}\right\}, \tag{8.35}$$

where the least-squares estimator $\boldsymbol{\theta}^*$ is equal to $\boldsymbol{\theta}_P$. The initial conditions needed to bootstrap Equation (8.35) are $\boldsymbol{\theta}_0 = 0$ and $\mathbf{P}_0 = \gamma \mathbf{I}$, where γ is a positive large number and $\mathbf{I}$ is the identity matrix of dimension $M \times M$. The effects of these initial conditions on the identification of $\boldsymbol{\theta}^*$ are described in Section 5.5. When we are dealing with adaptive networks with multiple outputs [o in Equation (8.30) is a column vector], Equation (8.35) still applies except that y_i^T is the ith row of matrix $\mathbf{y}$.

Now we can combine steepest descent and the least-squares estimator to update the parameters in an adaptive network. For hybrid learning to be applied in a batch mode, each epoch is composed of a **forward pass** and a **backward pass**. In the forward pass, after an input vector is presented, we calculate the node outputs in the network layer by layer until a corresponding row in the matrices $\mathbf{A}$ and $\mathbf{y}$ in Equation (8.33) is obtained. This process is repreated for all the training data pairs to form the complete $\mathbf{A}$ and $\mathbf{y}$; then parameters in S_2 are identified by either the pseudoinverse formula in Equation (8.34) or the recursive least-squares formulas in Equation (8.35). After the parameters in S_2 are identified, we can compute the error measure for each training data pair. In the backward pass, the error signals [the derivative of the error measure with respect to each node output, see Equations (8.4) and (8.5)] propagate from the output end toward the input end; the gradient vector is accumulated for each training data entry. At the end of the backward pass for all training data, the parameters in S_1 are updated by steepest descent in Equation (8.9).

For given fixed values of the parameters in S_1, the parameters in S_2 thus found are guaranteed to be the global optimum point in the S_2 parameter space because of the choice of the squared error measure. Not only can this hybrid learning rule decrease the dimension of the search space explored by the original steepest descent method, but, in general, it will also substantially reduce the time needed to reach convergence. However, sometimes the parameters in S_2 are concealed and we need a certain transformation method to recover them. In the following example, we use a multilayer perceptron with one hidden layer to explain how the linear parameters may be recovered.

Example 8.8 *Recovery of linear parameters in a multilayer perceptron*

For a single-hidden-layer perceptron with p output units, the o in Equation (8.30) is a column vector. If these output units are linear and there are no squashing functions to limit the output range, then the outputs are linear in the weights and thresholds of the output layer and these parameters can be identified by the least-squares method. On the other hand, if these output units have a sigmoidal activation function, then we can apply the inverse sigmoid function

$$H(x) = \ln\left(\frac{x}{1-x}\right), \tag{8.36}$$

such that the $H(o)$ in Equation (8.32) becomes a linear (vector) function in the parameters (weights and thresholds) of the output layer. In other words,

S_1 = weights and thresholds of hidden layer,
S_2 = weights and thresholds of output layer.
Therefore, we can apply backpropagation (or equivalently, steepest descent) to tune the parameters in the hidden layer, and the parameters in the output layer can be identified by the least-squares method.

□

As mentioned in Section 5.8, however, it should be kept in mind that by using the least-squares method on the data transformed by $H(\cdot)$, the obtained parameters are optimal in terms of the transformed squared error measure instead of the original one. In practice, usually this does not cause a problem as long as $H(\cdot)$ is monotonically increasing and the training data are not too noisy. See Section 5.8 for a more detailed treatment of other transformation methods.

8.5.2 On-Line Learning (Pattern-By-Pattern Learning)

If the parameters are updated after each data presentation, we have the scheme of on-line or pattern-by-pattern learning. This learning strategy is vital to on-line parameter identification for systems with changing characteristics. To modify the batch learning rule to obtain an on-line version, it is obvious that steepest descent should be based on E_p [see Equation (8.6)] instead of E. Strictly speaking, this is not a truly gradient search procedure for minimizing E, yet it will approximate one if the learning rate is small.

For the recursive least-squares formula to account for the time-varying characteristics of the incoming data, the effects of old data pairs must decay as new data pairs become available. Again, this problem is well studied in the adaptive control and system identification literature, and a number of solutions are available [2]. One simple method is to formulate the squared error measure as a weighted version that gives higher weighting factors to more recent data pairs. This amounts to the addition of a forgetting factor λ to the original recursive formula:

$$\left.\begin{aligned} \boldsymbol{\theta}_{i+1} &= \boldsymbol{\theta}_i + \mathbf{P}_{i+1}\mathbf{a}_{i+1}(y_{i+1}^T - \mathbf{a}_{i+1}^T\boldsymbol{\theta}_i) \\ \mathbf{P}_{i+1} &= \frac{1}{\lambda}[\mathbf{P}_i - \frac{\mathbf{P}_i\mathbf{a}_{i+1}\mathbf{a}_{i+1}^T\mathbf{P}_i}{\lambda + \mathbf{a}_{i+1}^T\mathbf{P}_i\mathbf{a}_{i+1}}] \end{aligned}\right\}, \tag{8.37}$$

where the typical value of λ in practice is between 0.9 and 1. The smaller λ is, the faster the effects of old data decay. A small λ sometimes causes numerical instability, however, and thus should be avoided. [For a complete discussion and derivation of Equation (8.37), the reader is referred to Section 5.6 of Chapter 5.]

8.5.3 Different Ways of Combining Steepest Descent and LSE

The computational complexity of the least-squares estimator (LSE) is usually higher than that of steepest descent (SD) for one-step adaptation. However, for achieving a prescribed performance level, the LSE is usually much faster. Consequently, depending on the available computing resources and required level of performance, we can choose from among at least five types of hybrid learning rules combining SD and LSE in different degrees, as follows:

1. One pass of LSE only: Nonlinear parameters are fixed while linear parameters are identified by one-time application of LSE.

2. SD only: All parameters are treated as nonlinear and updated by SD iteratively.

3. One pass of LSE followed by SD: LSE is employed only once at the beginning to obtain the initial values of linear parameters, and then SD takes over to update all parameters iteratively.

4. SD and LSE: Linear and nonlinear parameters are distinguished first. Each iteration (epoch) of SD used to update the nonlinear parameters is followed by LSE to identify the linear parameters.

5. LSE only: The outputs of an adaptive network are linearized with respect to its parameters, and then the extended Kalman filter algorithm, the Gauss-Newton method, or the Levenberg-Marquardt method is employed to update all parameters. These methods have also been used in the neural network literature [12, 13, 14]. (See also Section 6.8 of Chapter 6.)

The choice of one of the foregoing methods should be based on a trade-off between computational complexity and performance. Note that the linear parameters can also be updated by the Widrow-Hoff LMS algorithm [17], as reported in refs. [15]. The Widrow-Hoff algorithm requires less computation and is suitable for parallel hardware implementation, but it converges relatively slowly compared with the least-squares estimator. See also Section 9.3 of Chapter 9.

8.6 SUMMARY

This chapter describes the architectures and learning procedures of adaptive networks, a unifying framework that subsumes all supervised learning neural networks (Chapter 9), such as perceptrons, Adalines, multilayer perceptrons, radial basis function networks and modular networks. Understanding adaptive networks also paves the avenue to neuro-fuzzy modeling paradigms, such as ANFIS and CANFIS, which are presented in Chapters 12 and 13, respectively.

EXERCISES

1. Finish Example 8.6 by giving the expressions of the error signals for nodes 1, 2, 3, 4, and 5.

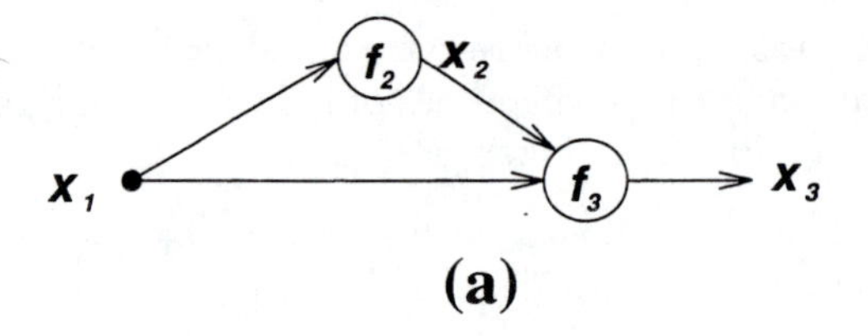

(a)

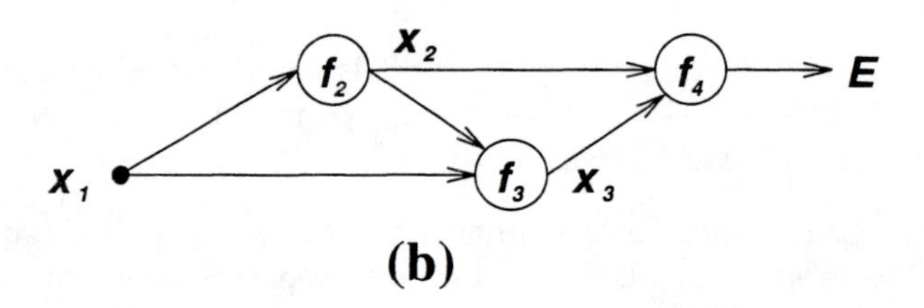

(b)

Figure 8.18. *Adaptive networks used in the exercises.*

2. Figure 8.18(a) is a simple adaptive network in which the error measure E is defined as a function of x_2 and x_3. (a) Give the formulas for each ϵ_i. (b) Draw the error-propagation network.

3. In the previous exercise, suppose that E is expressed as $E = f_4(x_2, x_3)$ and it is added as a new node to form the new network shown in Figure 8.18(b). Draw the corresponding error-propagation network. Is it the same as what you obtained in the previous exercise?

4. Verify that the Δ in Equation (8.26) is equal to the determinant of the square matrix in Equation (8.22).

5. Apply Mason's gain formula to express ϵ_4, ϵ_5, and ϵ_6 in terms of I_1 and I_2 in Example 8.7.

6. Solve Equation (8.22) algebraically and compare the answer with the formulas obtained via Mason's gain formula in the previous exercise.

REFERENCES

[1] A. E. Bryson and Y.-C. Ho. *Applied optimal control.* Blaisdell, New York, 1969.

[2] G. C. Goodwin and K. S. Sin. *Adaptive filtering prediction and control.* Prentice Hall, Upper Saddle River, NJ, 1984.

[3] J.-S. Roger Jang. Fuzzy modeling using generalized neural networks and Kalman filter algorithm. In *Proceedings of the Ninth National Conference on Artificial Intelligence (AAAI-91)*, pages 762–767, July 1991.

[4] J.-S. Roger Jang. ANFIS: Adaptive-Network-based Fuzzy Inference Systems. *IEEE Transactions on Systems, Man, and Cybernetics*, 23(03):665–685, May 1993.

[5] S. J. Mason. Feedback theory — some properties of signal flow graphs. *Proceedings IRE*, 41(9):1144–1156, September 1953.

[6] S. J. Mason. Feedback theory — further properties of signal flow graphs. *Proceedings IRE*, 44(7):920–926, July 1956.

[7] M. Minsky and S. Papert. *Perceptrons.* MIT Press, Cambridge, MA., 1969.

[8] N. J. Nilsson. *Learning machines: foundations of trainable pattern classifying systems.* McGraw-Hill, New York, 1965.

[9] D. B. Parker. Learning logic. Invention Report S81-64, File 1, Office of Technology Licensing, Standford University, October 1982.

[10] F. Rosenblatt. *Principles of neurodynamics: perceptrons and the theory of brain mechanisms.* Spartan, New York, 1962.

[11] D. E. Rumelhart, G. E. Hinton, and R. J. Williams. Learning internal representations by error propagation. In D. E. Rumelhart and James L. McClelland, editors, *Parallel distributed processing: explorations in the microstructure of cognition, volume 1*, chapter 8, pages 318–362. MIT Press, Cambridge, MA., 1986.

[12] S. Shah, F. Palmieri, and M. Datum. Optimal filtering algorithms for fast learning in feedforward neural networks. *Neural Networks*, 5(5):779–787, 1992.

[13] S. Shar and F. Palmieri. MEKA — a fast, local algorithm for training feedforward neural networks. In *Proceedings of the International Joint Conference on Neural Networks*, pages 41–46 (Volume III), 1990.

[14] S. Singhal and L. Wu. Training multilayer perceptrons with the extended kalman algorithm. In David S. Touretzky, editor, *Advances in neural information processing systems I*, pages 133–140. Morgan Kaufmann, San Mateo, CA, 1989.

[15] S. M. Smith and D. J. Comer. Automated calibration of a fuzzy logic controller using a cell state space algorithm. *IEEE Control Systems Magazine*, 11(5):18–28, August 1991.

[16] P. Werbos. *Beyond regression: New tools for prediction and analysis in the behavioral sciences.* PhD thesis, Harvard University, 1974.

[17] B. Widrow and D. Stearns. *Adaptive signal processing.* Prentice Hall, Upper Saddle River, NJ, 1985.

[18] R. J. William and D. Zipser. A learning algorithm for continually running fully recurrent neural networks. *Neural Computation*, 1:270–280, 1989.

Chapter 9

Supervised Learning Neural Networks

J.-S. R. Jang and E. Mizutani

9.1 INTRODUCTION

Artificial neural networks, or simply **neural networks** (NNs), have been studied for more than three decades since Rosenblatt [47] first applied single-layer *perceptrons* to pattern classification learning in the late 1950s. However, because Minsky and Papert [33] pointed out that single-layer systems were limited and expressed pessimism over multilayer systems, interest in NNs dwindled in the 1970s. The recent resurgence of interest in the field of NNs has been inspired by new developments in NN learning algorithms [10, 40, 48, 59], analog VLSI (very large scale integrated) circuits, and parallel processing techniques [25].

Quite a few NN models have been proposed and investigated in recent years. These NN models can be classified according to various criteria, such as their learning methods (supervised versus unsupervised), architectures (feedforward versus recurrent), output types (binary versus continuous), node types (uniform versus hybrid), implementations (software versus hardware), connection weights (adjustable versus hardwired), operations (biologically motivated versus psychologically motivated), and so on. In this chapter, we confine our scope to modeling problems with desired input-output data sets, so the resulting networks must have adjustable parameters that are updated by a supervised learning rule. Such networks are often referred to as **supervised learning** or **mapping networks**, since we are interested in shaping the input-output mappings of the networks according to a given training data set. (For details on unsupervised learning networks that try to cluster a given data set, see the next chapter.)

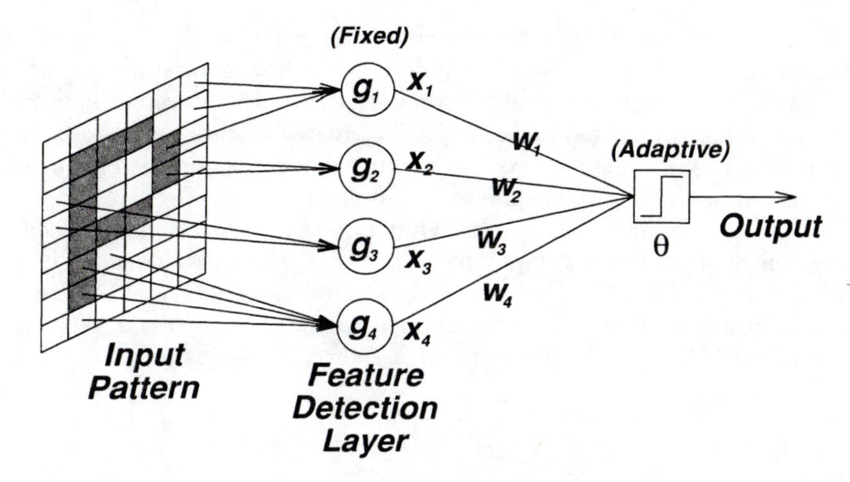

Figure 9.1. *The perceptron.*

Figure 9.2. *Introduction of the bias connection weight; the bias term* w_0 $(= \ -\theta)$ *can be viewed as the connection weight between the output unit and a "dummy" incoming signal* x_0 *that is always equal to 1.*

9.2 PERCEPTRONS

9.2.1 Architecture and Learning Rule

The **perceptron** represents one of the early attempts to build intelligent and self-learning systems using simple components. It was derived from a biological brain neuron model introduced by McCulloch and Pitts [32] in 1943. Later, Rosenblatt [47] designed the perceptron with a view toward explaining and modeling pattern-recognition abilities of biological visual systems. Although the goal is ambitious, the network paradigm is simple. Figure 9.1 is a typical perceptron setup for pattern-recognition applications, in which visual patterns are represented as matrices of elements between 0 and 1. The first layer of the perceptron acts as a set of "feature detectors" that are hardwired to the input signals to detect specific features. The second (output) layer takes the outputs of the "feature detectors" in the first layer and classifies the given input pattern. Learning is initiated by making adjustments to the relevant connection strengths (i.e., weights w_i) and a threshold value θ. For a two-class problem (for instance, determining whether the given pattern in Figure 9.1 is a "P" or not), the output layer usually has only a single node. For an n-class problem with n greater than or equal to 3, the output layer usually has n nodes, each corresponding to a class, and the output node with the largest value indicates which class the input vector belongs to.

Each function g_i in layer 1 is a fixed function that has to be determined *a priori*; it maps all or a part of the input pattern into a **binary** value $x_i \in \{-1, 1\}$ or a **bipolar** value $x_i \in \{0, 1\}$. The term x_i is referred to as **active** or **excitatory** if its value is 1, **inactive** if its value is 0, and **inhibitory** if its value is -1. The output unit is a **linear threshold element** with a threshold value θ:

$$\begin{aligned} o &= f\left(\textstyle\sum_{i=1}^{n} w_i x_i - \theta\right), \\ &= f\left(\textstyle\sum_{i=1}^{n} w_i x_i + w_0\right),\ w_0 \equiv -\theta, \\ &= f\left(\textstyle\sum_{i=0}^{n} w_i x_i\right),\ x_0 \equiv 1. \end{aligned} \tag{9.1}$$

where w_i is a modifiable weight associated with an incoming signal x_i; and w_0 ($= -\theta$) is the bias term. For computational efficiency, we can introduce the bias connection weight w_0 in place of the threshold value θ; Equation (9.1) shows that the threshold can be viewed as the connection weight between the output unit and a "dummy" incoming signal x_0 that is always equal to 1, as illustrated in Figure 9.2. In Equation (9.1), $f(\cdot)$ is the **activation function** of the perceptron and it is typically either a **signum function** $\text{sgn}(x)$ or **step function** $\text{step}(x)$:

$$\text{sgn}(x) = \begin{cases} 1 & \text{if } x > 0, \\ -1 & \text{otherwise,} \end{cases}$$

$$\text{step}(x) = \begin{cases} 1 & \text{if } x > 0, \\ 0 & \text{otherwise.} \end{cases}$$

Note that the "feature detector" g_i can be any function of the input pattern, but the learning procedure only adjusts the connection weights to the output unit (in the last layer). Since only the weights leading to the last layer are modifiable, the perceptron in Figure 9.1 is usually treated as a **single-layer perceptron**. Starting with a set of random connection weights, the basic learning algorithm for a single-layer perceptron repeats the following steps until the weights converge:

1. Select an input vector $\mathbf{x}$ from the training data set.

2. If the perceptron gives an incorrect response, modify all connection weights w_i according to

$$\Delta w_i = \eta t_i x_i,$$

where t_i is a target output and η is a learning rate.

The foregoing learning rule can be applied as well to updating a threshold θ ($= -w_0$) according to Equation (9.1). The value for the learning rate η can be a constant throughout training; or it can be a varying quantity proportional to the error. An η that is proportional to the error usually leads to faster convergence but can cause unstable learning.

The preceding learning algorithm above is roughly based on gradient descent; Rosenblatt [47] proved that there exists a method for tuning the weights that is guaranteed to converge to provide the required output if and only if such a set of weights exist. This is called the **perceptron convergence theorem**. Moreover, depending on the functions g_i, perceptrons can be grouped into different families; a number of these families and their properties are described in refs. [47].

In the early 1960s, perceptrons created a great deal of interest and optimism directed toward building real self-learning intelligent systems. However, the initial enthusiasm waned after the publication of Minsky and Papert's *Perceptrons* [33] in 1969, in which they analyzed the perceptron extensively and concluded that single-layer perceptrons can only be used for toy problems. One of their most discouraging results shows that a single-layer perceptron cannot represent a simple exclusive-OR function, as explained next.

9.2.2 Exclusive-OR Problem

The simplest and most well-known pattern recognition problem in neural network literature is the **exclusive-OR (XOR)** problem. The task is to classify a binary input vector to class 0 if the vector has an even number of 1's, or assign it to class 1. For a two-input binary XOR problem, the desired behavior is regulated by a truth table:

	X	Y	Class
Desired i/o pair 1	0	0	0
Desired i/o pair 2	0	1	1
Desired i/o pair 3	1	0	1
Desired i/o pair 4	1	1	0

A bipolar XOR problem is similarly defined except that all instances of 1 in the truth table are replaced with -1.

The XOR problem is not **linearly separable**; this can easily be observed from the plot in Figure 9.3. In other words, we cannot use a single-layer perceptron [Figure 9.4(a)] to construct a straight line to partition the two-dimensional input space into two regions, each containing only data points of the same class. Symbolically, using a single-layer perceptron to solve this problem requires satisfying the following four inequalities:

$$
\begin{aligned}
0 \times w_1 + 0 \times w_2 + w_0 \leq 0 &\iff w_0 \leq 0, \\
0 \times w_1 + 1 \times w_2 + w_0 > 0 &\iff w_0 > -w_2, \\
1 \times w_1 + 0 \times w_2 + w_0 > 0 &\iff w_0 > -w_1, \\
1 \times w_1 + 1 \times w_2 + w_0 \leq 0 &\iff w_0 \leq -w_1 - w_2.
\end{aligned}
$$

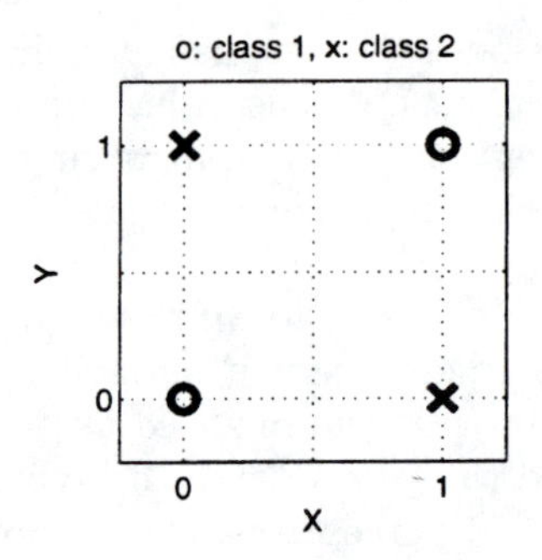

Figure 9.3. *XOR problem.* (MATLAB file: `xordata.m`)

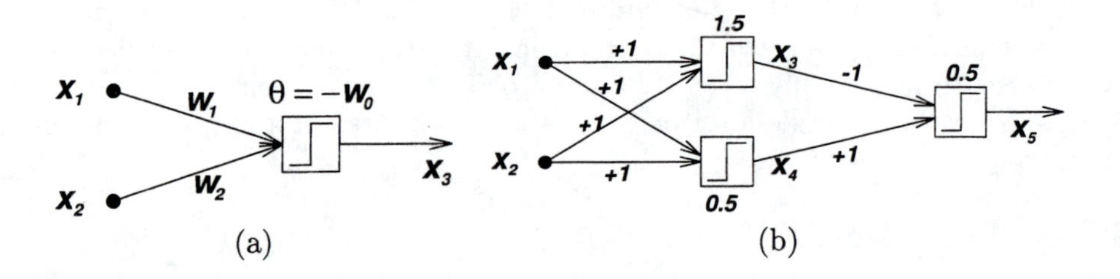

Figure 9.4. *Perceptrons for the two-input exclusive-OR problem: (a) the single-layer perceptron, and (b) the two-layer perceptron. Both use the step function as the activation function for each node.*

However, the set of inequalities is self-contradictory when considered as a whole.

It is possible to solve the problem with the two-layer perceptron illustrated in Figure 9.4(b), in which the connection weights and thresholds are indicated. More specifically, we can plot the output of each neuron as a surface of its two inputs, as shown in Figures 9.5(a) through 9.5(d). Figure 9.5(d) is the overall input-output plot of the two-layer perceptron, indicating that it has solved the XOR problem.

In summary, multilayer perceptrons can solve nonlinearly separable problems and are thus much more powerful than the original single-layer version. In Section 9.4, we discuss a learning method that can be used to find appropriate connection weights and thresholds in multilayer perceptrons.

9.3 ADALINE

The **adaptive linear element** (or **Adaline**), suggested by Widrow and Hoff [62], represents a classical example of the simplest intelligent self-learning system that can adapt itself to achieve a given modeling task. Figure 9.6 is a schematic diagram for such a network. It has a purely linear output unit; hence the network output o

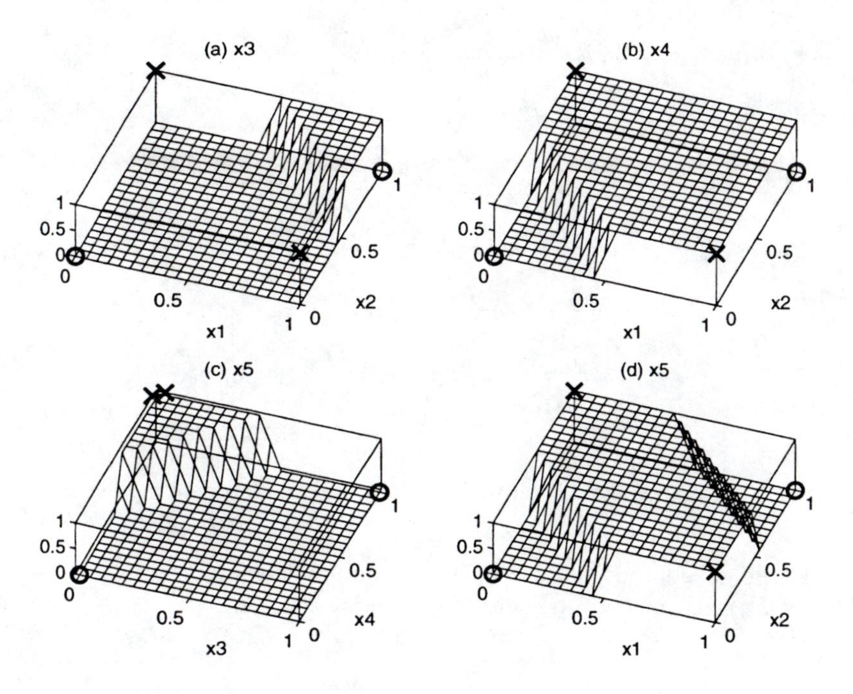

Figure 9.5. *Node outputs as surfaces of their inputs in Figure 9.4: (a) x_3; (b) x_4; (c) x_5 (as a function of x_3 and x_4); (d) x_5 (as a function of x_1 and x_2).* (MATLAB file: `xorsurf1.m`)

is a weighted linear combination of the inputs plus a constant term:

$$o = \sum_{i=1}^{n} w_i x_i + w_0. \tag{9.2}$$

In a simple physical implementation, the input signals x_i are voltages and the w_i are conductances of controllable resistors; the network's output is the summation of the currents caused by the input voltages. The problem is to find a suitable set of conductances (or weights) such that the input-output behavior of the Adaline is close to a set of desired input-output data points.

It is obvious that the preceding Adaline equation is an exactly linear model with $n + 1$ linear parameters, so we can employ the least-squares methods introduced in Chapter 5 to minimize the error in the sense of least squares. However, most of the least-squares methods require extensive calculations, which are not possible in a physical system with simple components. To overcome this, Widrow and Hoff introduced the **delta rule** for adjusting the weights. For the pth input-output pattern, the error measure of a single-output Adaline can be expressed as

$$E_p = (t_p - o_p)^2,$$

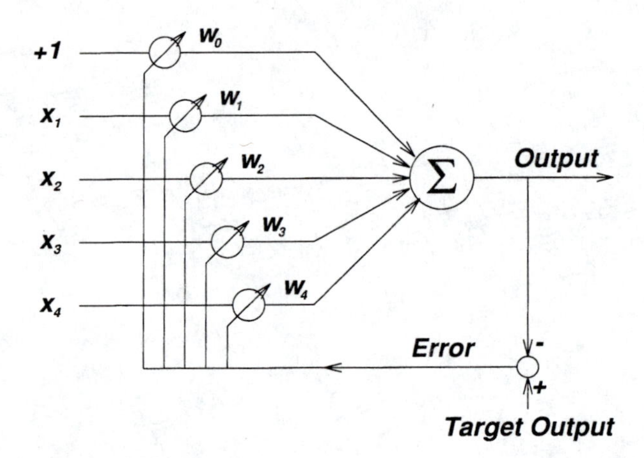

Figure 9.6. *Adaline (adaptive linear element).*

where t_p is the target output and o_p is the actual output of the Adaline. The derivative of E_p with respect to each weight w_i is

$$\frac{\partial E_p}{\partial w_i} = -2(t_p - o_p)x_i.$$

Therefore, to decrease E_p by gradient descent, the update formula for w_i on the pth input-output pattern is

$$\Delta_p w_i = \eta(t_p - o_p)x_i. \tag{9.3}$$

This update formula has strong intuitive appeal. It essentially states that when $t_p > o_p$, we want to boost o_p by increasing $w_i x_i$; therefore, we should increase w_i if x_i is positive and decrease w_i if x_i is negative. Similar reasoning holds when $t_p < o_p$. Since the delta rule tries to minimize squared errors, it is also referred to as the **least mean square (LMS) learning procedure** or **Widrow-Hoff learning rule**. The features of the delta rule are as follows:

- Simplicity: This is obvious from Equation (9.3).
- Distributed learning: Learning is not reliant on central control of the network; it can be performed locally at each node level; see Equation (9.3).
- On-line (or pattern-by-pattern) learning: Weights are updated after presentation of each pattern.

These features make Adaline, with the delta rule, suitable for simple hardware implementation.

In the 1960s, two or more Adaline components were integrated to develop **Madaline** (many Adalines) in an attempt to implement nonlinearly separable logic functions. To solve the previously mentioned XOR problem, for instance, two Adalines were connected to an AND logic device (Madaline unit) to provide an output [65]. However, the Adaline and Madaline systems were limited in that they had only one layer with adjustable weights, just like single-layer perceptrons.

Adaline and Madline have been used for adaptive noise cancellation [61] and adaptive inverse control [64]. In adaptive noise cancellation, the objective is to filter out an interference component by identifying a linear model of a measurable noise source and the corresponding unmeasurable interference; such applications include interference canceling in electrocardiograms (ECGs), echo elimination from long-distance telephone transmission lines, and antenna sidelobe interference canceling [64]; for more details on adaptive inverse control, see refs. [64] or Section 17.4 of this text. For details on Adaline, Madline, and LMS methods, refer to refs. [64] and [63].

9.4 BACKPROPAGATION MULTILAYER PERCEPTRONS

As mentioned earlier, the lack of suitable training methods for **multilayer perceptrons (MLPs)** led to a waning of interest in neural networks in the 1960s and 1970s. This was not changed until the reformulation of the **backpropagation** training method for MLPs in the mid-1980s by Rumelhart et al. [48]. (The derivation of the backpropagation method in a more general framework can be found in Section 8.3 of Chapter 8.)

The single-layer perceptron discussed in Section 9.2 is a principal NN component and provides the grounds for current understanding and most applications of NNs. However, because of the nondifferentiability of the hard-limiter activation function, the learning strategies of early multilayer perceptrons with signum or step activation functions are not obvious unless continuous activation functions are employed.

A backpropagation MLP, as already mentioned in Examples 8.3 and 8.4, is an adaptive network whose nodes (or neurons) perform the same function on incoming signals; this node function is usually a composite of the weighted sum and a differentiable nonlinear activation function, also known as the **transfer function**. Figure 9.7 depicts three of the most commonly used activation functions in backpropagation MLPs:

$$\begin{aligned}
&\text{Logistic function:} && f(x) = \frac{1}{1+e^{-x}}\\
&\text{Hyperbolic tangent function:} && f(x) = \tanh(x/2) = \frac{1-e^{-x}}{1+e^{-x}}\\
&\text{Identity function:} && f(x) = x.
\end{aligned}$$

Both the hyperbolic tangent and logistic functions approximate the signum and step function, respectively, and yet provide smooth, nonzero derivatives with respect to input signals. Sometimes these two activation functions are referred to as **squashing functions** since the inputs to these functions are squashed to the range $[0, 1]$ or $[-1, 1]$. They are also called **sigmoidal functions** because their S-shaped curves exhibit smoothness and asymptotic properties. (Sometimes the hyperbolic tangent function are referred to as *bipolar* sigmoidal and the logistic function are

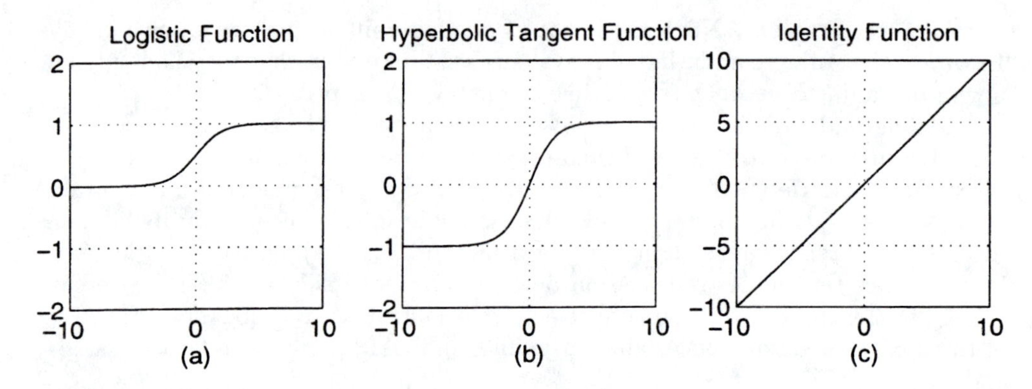

Figure 9.7. *Activation functions for backpropagation MLPs: (a) logistic function; (b) hyperbolic tangent function; (c) identity function.* (MATLAB file: `activati.m`)

referred to as *binary* sigmoidal.) Both of these activations are used often on regression and classification problems. Other activation functions are discussed in Section 13.3.3 of Chapter 13.

For a neural network to approximate a continuous-valued function not limited to the interval $[0,1]$ or $[-1,1]$, we usually let the node function for the output layer be a weighted sum with no squashing functions. This is equivalent to a situation in which the activation function is an identity function, and output nodes of this type are often called **linear nodes.**

Backpropagation MLPs are by far the most commonly used NN structures for applications in a wide range of areas, such as pattern recognition, signal processing, data compression, and automatic control. Some of the well-known instances of applications include NETtalk [51, 52], which trained an MLP to pronounce English text, Carnegie Mellon University's ALVINN (Autonomous Land Vehicle in a Neural Network) [42, 43], which used an MLP for steering an autonomous vehicle; and optical character recognition (OCR) [23, 49].

9.4.1 Backpropagation Learning Rule

For simplicity, we assume that the backpropagation MLP in question uses the logistic function as its activation function; the reader is encouraged to derive the backpropagation procedure when other types of continuous activation functions are used.

The **net input** $\bar{x}$ of a node is defined as the weighted sum of the incoming signals plus a bias term. For instance, the net input and output of node j in Figure 9.8 are

$$\begin{aligned} \bar{x}_j &= \textstyle\sum_i w_{ij} x_i + w_j, \\ x_j &= f(\bar{x}_j) = \frac{1}{1+\exp(-\bar{x}_j)}, \end{aligned} \tag{9.4}$$

where x_i is the output of node i located in any one of the previous layers, w_{ij} is

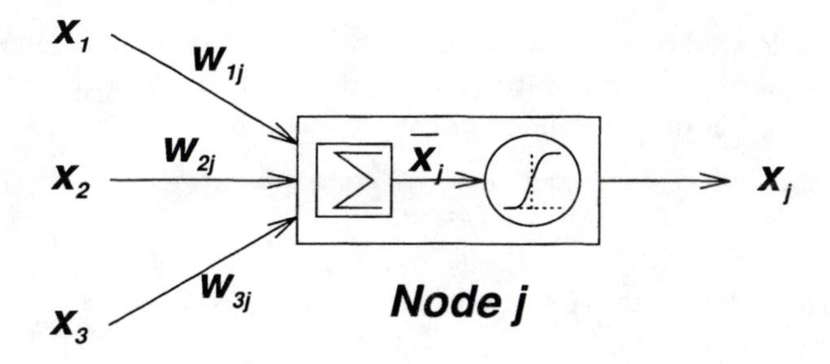

Figure 9.8. *Node j of a backpropagation MLP.*

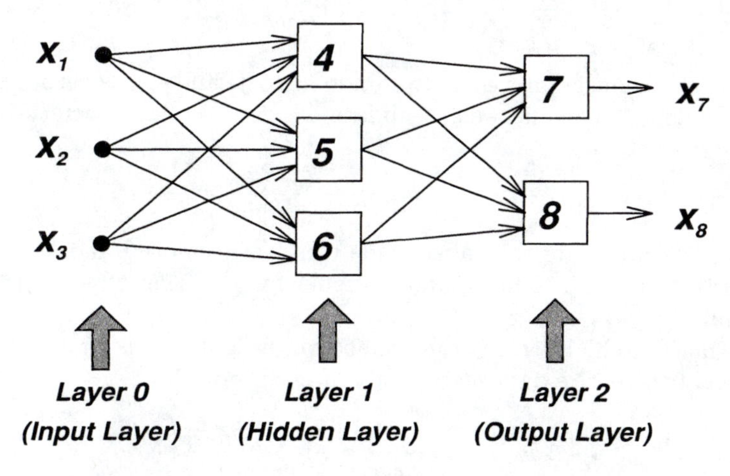

Figure 9.9. *A 3-3-2 backpropagation MLP.*

the weight associated with the link connecting nodes i and j, and w_j is the bias of node j. Since the weights w_{ij} are actually internal parameters associated with each node j, changing the weights of a node will alter the behavior of the node and in turn alter the behavior of the whole backpropagation MLP. Figure 9.9 shows a two-layer backpropagation MLP with three inputs to the input layer, three neurons in the hidden layer, and two output neurons in the output layer. For simplicity, this backpropagation MLP will be referred to as a 3-3-2 network, corresponding to the number of nodes in each layer. (Note that the input layer is composed of three buffer nodes for distributing the input signals; therefore, this layer is conventionally not counted as a physical layer of a backpropagation MLP.)

The **backward error propagation**, also known as the **backpropagation** (**BP**) or the **generalized delta rule (GDR)**, is explained next. First, a squared error measure for the pth input-output pair is defined as

$$E_p = \sum_k (d_k - x_k)^2, \tag{9.5}$$

where d_k is the desired output for node k, and x_k is the actual output for node k when the input part of the pth data pair is presented. To find the gradient vector, an error term $\bar{\epsilon}_i$ for node i is defined as

$$\bar{\epsilon}_i = \frac{\partial^+ E_p}{\partial \bar{x}_i}. \tag{9.6}$$

By the chain rule, the recursive formula for $\bar{\epsilon}_i$ can be written as

$$\bar{\epsilon}_i = \begin{cases} -2(d_i - x_i)\frac{\partial x_i}{\partial \bar{x}_i} = -2(d_i - x_i)x_i(1 - x_i) & \text{if node } i \text{ is a output node,} \\ \frac{\partial x_i}{\partial \bar{x}_i} = \sum_{j,i<j} \frac{\partial^+ E_p}{\partial \bar{x}_j}\frac{\partial \bar{x}_j}{\partial x_i} = x_i(1 - x_i)\sum_{j,i<j} \bar{\epsilon}_j w_{ij} & \text{otherwise,} \end{cases} \tag{9.7}$$

where w_{ij} is the connection weight from node i to j; and w_{ij} is zero if there is no direct connection. Then the weight update w_{ki} for on-line (pattern-by-pattern) learning is

$$\Delta w_{ki} = -\eta \frac{\partial^+ E_p}{\partial w_{ki}} = -\eta \frac{\partial^+ E_p}{\partial \bar{x}_i}\frac{\partial \bar{x}_i}{\partial w_{ki}} = -\eta \bar{\epsilon}_i x_k, \tag{9.8}$$

where η is a learning rate that affects the convergence speed and stability of the weights during learning. The update formula for the bias of each node can be derived similarly.

For off-line (batch) learning, the connection weight w_{ki} is updated only after presentation of the entire data set, or only after an **epoch**:

$$\Delta w_{ki} = -\eta \frac{\partial^+ E}{\partial w_{ki}} = -\eta \sum_p \frac{\partial^+ E_p}{\partial w_{ki}}, \tag{9.9}$$

or, in vector form,

$$\Delta \mathbf{w} = -\eta \frac{\partial^+ E}{\partial \mathbf{w}} = -\eta \nabla_{\mathbf{w}} E, \tag{9.10}$$

where $E = \sum_p E_p$. This corresponds to a way of using the true gradient direction based on the entire data set.

9.4.2 Methods of Speeding Up MLP Training

Quite a few ad hoc methods exist to speed up MLP backpropagation training. Some of them are applicable to general backpropagation gradient descent, while others are only effective for MLP backpropagation.

One way to speed up off-line training is to use the so-called **momentum term** [48]:

$$\Delta \mathbf{w} = -\eta \nabla_{\mathbf{w}} E + \alpha \Delta \mathbf{w}_{\text{prev}}, \tag{9.11}$$

where $\mathbf{w}_{\text{prev}}$ is the previous update amount, and the **momentum constant** α, in practice, is usually set to something between 0.1 and 1. The addition of the momentum term smoothes weight updating and tends to resist erratic weight changes due to gradient noise or high spatial frequencies in the error surface. However, the use of momentum terms does not always seem to speed up training; it is more or less application dependent [63].

Another useful technique is normalized weight updating:

$$\Delta \mathbf{w} = -\kappa \frac{\nabla_{\mathbf{w}} E}{||\nabla_{\mathbf{w}} E||}. \tag{9.12}$$

This causes the network's weight vector to move the same Euclidean distance κ in the weight space with each update, which allows control of the distance κ based on the history of error measures. One of the adaptation strategies for varying the step size κ is explained in Section 6.7.2 of Chapter 6. Other methods for speeding up MLP backpropagation training include the quick-propagation algorithm [9], the delta-bar-delta approach [14], the extended Kalman filter method [54, 55], second-order optimization [58], and the optimal filtering approach [53].

Generally, an MLP with hyperbolic tangent functions can be trained more rapidly than one with logistic functions. This is only an empirical observation and it has exceptions (for instance, the encoding problem in ref. [9]), but it is advisable to try both types of MLP when encountering a new application.

Sometimes it is desirable to do **data scaling** on the raw training data and then use the processed data to train the MLP. In **output scaling**, the range of target values is constrained to remain within the range of the sigmoidal activation function. For instance, for an MLP with hyperbolic tangent functions, the target values must be within, say, $[-0.9, 0.9]$, instead of within the usual activation function range $[-1, 1]$. This prevents backpropagation from driving some of the connection weights to infinity and slowing down training. A similar approach using *modified sigmoid functions* is discussed in Section 13.3.3. In **input scaling**, the range of each input is scaled (usually linearly) to the range of the activation function used. This scaling allows the connection weights to have the same order of magnitude during training.

The initial values of the connection weights and biases in an MLP should be uniformly distributed across a small range, usually $[-1, 1]$. If the initial values of these modifiable parameters are too large, then some of the neurons might get saturated and produce small error signals. On the other hand, if the range is too small, then the gradient vector is also small and learning will be very slow initially. Note that when all the free parameters are zeros, the gradient vector is always zero since it happens to be a *saddle point* in the error landscape. Another and better way to initialize network parameters is to choose the weight and bias values such that the "slope" parts of neurons in the hidden layer can cover the input space; see ref. [37] for details.

All neurons in an MLP should get updated at approximately the same rate. However, the error signals at the the output layer tend to be larger than those at the front-end layer of the network. This can be seen directly in Equation (9.7), where $x_i(1 - x_i)$ appears once in the output layer, twice in the layer next to the output, and so on. Note that the term $x_i(1 - x_i)$ is always less than or equal to

0.25 (see dashed lines in Figure 13.8), so the error signals at front-end layers tend to be smaller due to multiple multiplication of this term. Therefore, the learning rate η of the front-end layers should be larger than that of the output layer. This is called the **learning rate rescaling**; see ref. [46] for details.

9.4.3 MLP's Approximation Power

The approximation power of backpropagation MLPs has been explored by some researchers. Yet there is very little theoretical guidance for determining network size in terms of, say, the number of hidden nodes and hidden layers it should contain. Cybenko [8] showed that a backpropagation MLP, with one hidden layers and any fixed continuous sigmoidal nonlinear function, can approximate any continuous function arbitrarily well on a compact set. When used as a binary-valued neural network with the hard-limiter (step) activation function, a backpropagation MLP with two hidden layers can form arbitrary complex decision regions to separate different classes, as Lippmann [25] pointed out. For function approximation as well as data classification, two hidden layers may be required to learn a piecewise-continuous function [29]. In their book, Hertz et al. [13] introduced an intuitive explanation that MLPs with two hidden layers may be able to construct localized receptive fields out of logistic functions. Thus, two-layer MLPs may have abilities comparable to radial basis function networks, which are discussed next.

9.5 RADIAL BASIS FUNCTION NETWORKS

9.5.1 Architectures and Learning Methods

Locally tuned and overlapping receptive fields are well-known structures that have been studied in regions of the cerebral cortex, the visual cortex, and others. Drawing on knowledge of biological receptive fields, Moody and Darken [34, 35] proposed a network structure that employs local receptive fields to perform function mappings. Similar schemes have been proposed by Powell [44], Broomhead and Lowe [4], and many others in the areas of **interpolation** and **approximation theory**; these schemes are collectively called radial basis function approximations. Here we shall call the network structure the **radial basis function network** or **RBFN**.

Figure 9.10(a) shows a schematic diagram of an RBFN with four receptive field units; the activation level of the ith receptive field unit (or hidden unit) is

$$w_i = R_i(\mathbf{x}) = R_i(\|\mathbf{x} - \mathbf{u}_i\|/\sigma_i),\ i = 1,\ 2,\ \ldots,\ H, \tag{9.13}$$

where $\mathbf{x}$ is a multidimensional input vector, $\mathbf{u}_i$ is a vector with the same dimension as $\mathbf{x}$, H is the number of radial basis functions (or, equivalently, receptive field units), and $R_i(\cdot)$ is the ith radial basis function with a single maximum at the origin. There are no connection weights between the input layer and the hidden layer. Typically, $R_i(\cdot)$ is a Gaussian function

$$R_i(\mathbf{x}) = \exp\left(-\frac{\|\mathbf{x}-\mathbf{u}_i\|^2}{2\sigma_i^2}\right) \tag{9.14}$$

or a logistic function

$$R_i(\mathbf{x}) = \frac{1}{1+\exp[\|\mathbf{x}-\mathbf{u}_i\|^2/\sigma_i^2]}. \tag{9.15}$$

Thus, the activation level of radial basis function w_i computed by the ith hidden unit is maximum when the input vector $\mathbf{x}$ is at the center $\mathbf{u}_i$ of that unit.

The output of an RBFN can be computed in two ways. In the simpler method, as shown in Figure 9.10(a), the final output is the **weighted sum** of the output value associated with each receptive field:

$$d(\mathbf{x}) = \sum_{i=1}^{H} c_i w_i = \sum_{i=1}^{H} c_i R_i(\mathbf{x}), \tag{9.16}$$

where c_i is the output value associated with the ith receptive field. We can also view c_i as the connection weight between the receptive field i and the output unit. A more complicated method for calculating the overall output is to take the **weighted average** of the output associated with each receptive field:

$$d(\mathbf{x}) = \frac{\sum_{i=1}^{H} c_i w_i}{\sum_{i=1}^{H} w_i} = \frac{\sum_{i=1}^{H} c_i R_i(\mathbf{x})}{\sum_{i=1}^{H} R_i(\mathbf{x})}. \tag{9.17}$$

Weighted average has a higher degree of computational complexity, but it is advantageous in that points in the areas of overlap between two or more receptive fields will have a well-interpolated overall output between the outputs of the overlapping receptive fields. An example is presented in Section 9.5.4.

For representation purposes, if we change the radial basis function $R_i(\mathbf{x})$ in each node of layer 2 in Figure 9.10(a) to its **normalized** counterpart $R_i(\mathbf{x})/\sum_i R_i(\mathbf{x})$, then the overall output is specified by Equation (9.17). A more explicit representation is shown in Figure 9.10(b), where the division of the weighted sum ($\sum_i c_i w_i$) by the activation total ($\sum_i w_i$) is indicated in the division node in the last layer. In Figure 9.10, plots (c) and (d) are the two-output counterparts of the RBFNs in (a) and (b).

Moody-Darken's RBFN may be extended by assigning a linear function to the output function of each receptive field—that is, making c_i a linear combination of the input variables plus a constant:

$$c_i = \mathbf{a}_i^T \mathbf{x} + b_i, \tag{9.18}$$

where $\mathbf{a}_i$ is a parameter vector and b_i is a scalar parameter. Stokbro et al. [57] used this structure to model the Mackey-Glass chaotic time series [27] and found

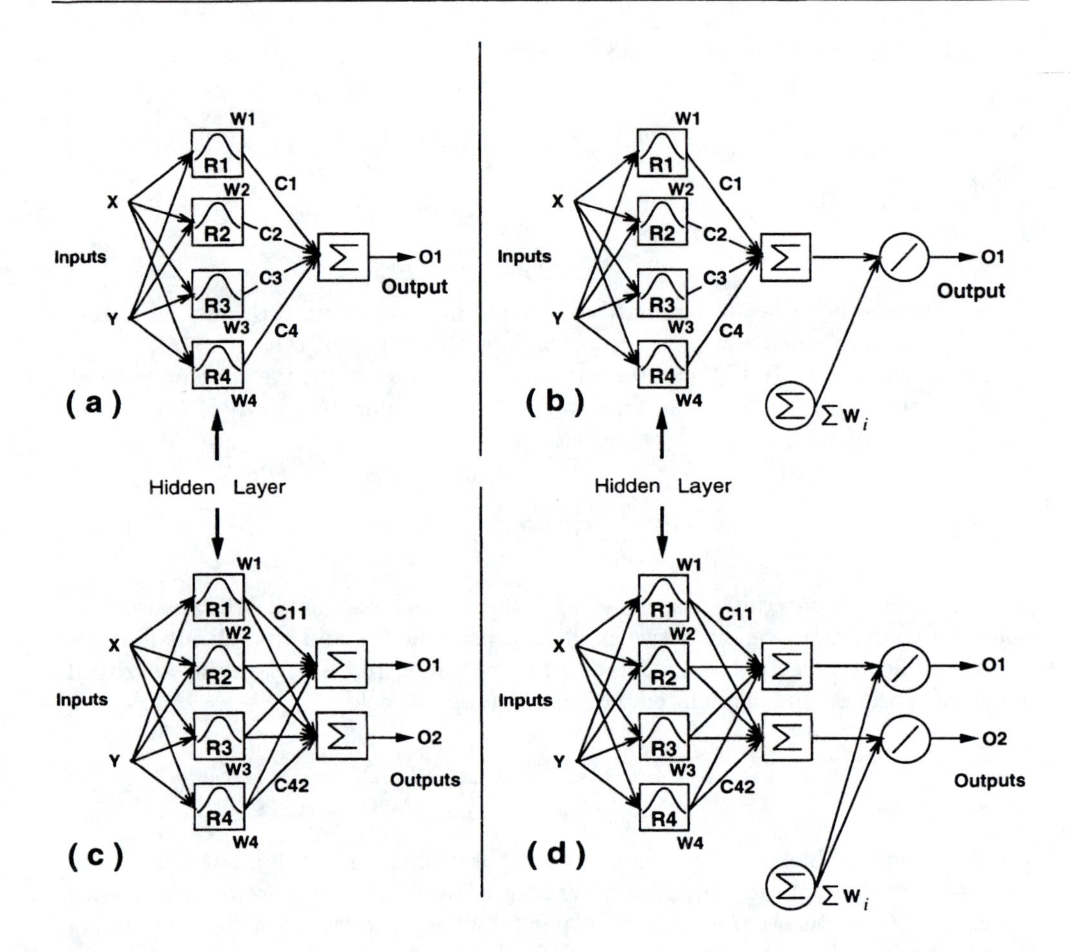

Figure 9.10. *Four RBFNs that possess four basis functions: (a) single-output RBFN that uses weighted sum; (b) single-output RBFN that uses weighted average; (c) two-output RBFN that uses weighted sum; (d) two-output RBFN that uses weighted average. The network in (d) is equivalent to Figure 13.3 (upper right). [Note that in (b) and (d), four connections to the lower summation unit are omitted for simplicity.]*

that this extended version performed better than the original RBFN with the same number of fitting parameters.

An RBFN's approximation capacity may be further improved with supervised adjustments of the center and shape of the receptive field (or radial basis) functions [24, 60]. Several learning algorithms have been proposed to identify the parameters ($\mathbf{u}_i$, σ_i, and c_i) of an RBFN. Besides using a supervised learning scheme alone to update all modifiable parameters, a variety of sequential training algorithms for RBFNs have been reported. The receptive field functions are first fixed, and then the weights of the output layer are adjusted. Several schemes have been proposed to determine the center positions ($\mathbf{u}_i$) of the receptive field functions. Lowe [26] proposed a way to determine the centers based on standard deviations of training data. Moody and Darken [34, 35] selected the centers $\mathbf{u}_i$ by means of data clustering techniques (see Chapter 15) that assume that similar input vectors produce similar outputs; σ_i's are then obtained heuristically by taking the average distance to the several nearest neighbors of u_i's. In another variation, Nowlan [38] employed the so-called *soft competition* among Gaussian hidden units to locate the centers. This soft competition method is based on a "maximum likelihood estimate" for the centers, in contrast to the so-called *hard* competitions such as the k-means winner-take-all algorithm.

Once these nonlinear parameters are fixed and the receptive fields are frozen, the linear parameters (i.e., the weights of the output layer) can be updated using either the least-squares method or the gradient method. Alternatively, we can apply the pseudoinverse method in solving Equation (9.23) to determine these weights [5]. Chen et al. [6] used another method that employs the orthogonal least-squares algorithm to determine the $\mathbf{u}_i$'s and c_i's while keeping the σ_i's at predetermined values. There are many other schemes as well, such as generalization properties [2], and sequential adaptation [22], among others [19, 36].

9.5.2 Functional Equivalence to FIS

An extension of the originally proposed Moody-Darken's RBFN is to assign a linear function as the output function of each receptive field; that is, C_i is a linear function of the input variables instead of a constant:

$$C_i = \vec{a}_i \cdot \vec{x} + b_i, \tag{9.19}$$

where $\vec{a}_i$ is a parameter vector and b_i is a scalar parameter. Stokbro et al. [57] used this structure to model the Mackey-Glass chaotic time series [27] and found that this extended version performed better than the originally proposed RBFN with the same number of fitting parameters. Using Equation (9.19), the extended RBFN response given by Equation (9.16) or Equation (9.17) is identical to the response produced by the first-order Sugeno fuzzy inference system (FIS) discussed in Chapter 4, provided that the membership functions, the radial basis functions, and certain operators are choose correctly.

While the RBFN consists of radial basis functions, the FIS comprises a certain number of membership functions. With those radially shaped functions, both FIS and RBFN have a mechanism whereby they can produce a center-weighted response to small receptive fields, localizing the primary input excitation. Although the FIS and the RBFN were developed on different bases, they are essentially rooted in the same soil. Just as the RBFN enjoys quick convergence, the FIS can evolve to recognize some feature in a training data set quickly compared with simple back-propagation MLPs (see Chapter 12).

The conditions under which an RBFN and a FIS are functionally equivalent are summarized as follows [18].

- Both the RBFN and the FIS under consideration use the same aggregation method (namely, either weighted average or weighted sum) to derive their overall outputs.
- The number of receptive field units in the RBFN is equal to the number of fuzzy if-then rules in the FIS.
- Each radial basis function of the RBFN is equal to a multidimensional composite MF of the premise part of a fuzzy rule in the FIS. One way to achieve this is to use Gaussian MFs with the same variance in a fuzzy rule, and apply product to calculate the firing strength. The multiplication of these Gaussian MFs becomes a multidimensional Gaussian function—a radial basis function in RBFN. (See Figure 13.3 for more details.)
- Corresponding radial basis function and fuzzy rule should have the same response function. That is, they should have the same constant terms (for the original RBFN and zero-order Sugeno FIS) or linear equations (for the extended RBFN and first-order Sugeno FIS)..

The functional equivalence between FIS and RBFN cross-fertilizes both computing paradigms. Further details are presented in Section 12.4.

9.5.3 Interpolation and Approximation RBFNs

Assuming that there is no noise in the training data set, we need to estimate a function $d(\cdot)$ that yields exact desired outputs for all training data. This task is usually called an "interpolation" problem, and the resultant function $d(\cdot)$ should pass through all of the training data points. When we use an RBFN with the same number of basis functions as we have training patterns, we have a so-called **interpolation RBFN**, where each neuron in the hidden layer reponds to one particular training input pattern.

Consider a Gaussian basis function centered at u_i with a width parameter σ:

$$w_i = R_i(\|\mathbf{x} - \mathbf{u}_i\|) = \exp\left[-\frac{(\mathbf{x} - \mathbf{u}_i)^2}{2\sigma_i^2}\right]. \tag{9.20}$$

Each training input x_i serves as a center for for the basis function, R_i. Thus, from Equation (9.16), we have a Gaussian interpolation RBFN:

$$d(\mathbf{x}) = \sum_{i=1}^{n} c_i \exp\left[-\frac{(\mathbf{x} - \mathbf{x}_i)^2}{2\sigma_i^2}\right]. \tag{9.21}$$

For given σ_i, $i = 1, \ldots, n$, we obtain the following n simultaneous linear Equations for the n unknown weight coefficients, c_i:

$$
\begin{aligned}
d_1 &= c_1 \exp\left[-\frac{\|\mathbf{x}_1 - \mathbf{x}_1\|^2}{2\sigma_1^2}\right] + \cdots + c_n \exp\left[-\frac{\|\mathbf{x}_1 - \mathbf{x}_n\|^2}{2\sigma_n^2}\right], \\
d_2 &= c_1 \exp\left[-\frac{\|\mathbf{x}_2 - \mathbf{x}_1\|^2}{2\sigma_1^2}\right] + \cdots + c_n \exp\left[-\frac{\|\mathbf{x}_2 - \mathbf{x}_n\|^2}{2\sigma_n^2}\right], \\
&\vdots \\
d_n &= c_1 \exp\left[-\frac{\|\mathbf{x}_n - \mathbf{x}_1\|^2}{2\sigma_1^2}\right] + \cdots + c_n \exp\left[-\frac{\|\mathbf{x}_n - \mathbf{x}_n\|^2}{2\sigma_n^2}\right].
\end{aligned}
$$

Writing them in matrix form, we obtain

$$
\begin{bmatrix} d_1 \\ d_2 \\ \vdots \\ d_n \end{bmatrix} = \begin{bmatrix} \exp\left[-\frac{\|\mathbf{x}_1-\mathbf{x}_1\|^2}{2\sigma_1^2}\right] & \cdots & \exp\left[-\frac{\|\mathbf{x}_1-\mathbf{x}_n\|^2}{2\sigma_n^2}\right] \\ \exp\left[-\frac{\|\mathbf{x}_2-\mathbf{x}_1\|^2}{2\sigma_1^2}\right] & \cdots & \exp\left[-\frac{\|\mathbf{x}_2-\mathbf{x}_n\|^2}{2\sigma_n^2}\right] \\ \vdots & \vdots & \vdots & \vdots \\ \exp\left[-\frac{\|\mathbf{x}_n-\mathbf{x}_1\|^2}{2\sigma_1^2}\right] & \cdots & \exp\left[-\frac{\|\mathbf{x}_n-\mathbf{x}_n\|^2}{2\sigma_n^2}\right] \end{bmatrix} \begin{bmatrix} c_1 \\ c_2 \\ \vdots \\ c_n \end{bmatrix}. \tag{9.22}
$$

Rewriting the preceding in a compact form, we have

$$
\mathbf{D} = \mathbf{G}\mathbf{C}, \tag{9.23}
$$

where

$$
\mathbf{D} = [d_1,\ d_2,\ \ldots,\ d_n]^T,\ \mathbf{C} = [c_1,\ c_2,\ \ldots,\ c_n]^T.
$$

When the matrix $\mathbf{G}$ is nonsingular, we have a unique solution:

$$
\mathbf{C} = \mathbf{G}^{-1}\mathbf{D}, \tag{9.24}
$$

where $\mathbf{G}^{-1}$ denotes the inverse matrix of $\mathbf{G}$. In practice, however, $\mathbf{G}$ may be ill-conditioned (close to singularity), especially when the training set is large. The regularization approach [12, 41] allows for such cases by modifying $\mathbf{G}$ to $\mathbf{G} + \lambda\mathbf{I}$, where λ is a positive real number (regularization parameter) and $\mathbf{I}$ is the identity matrix. (Similar modifications can be found in the Levenberg-Marquardt method to make the Hessian matrix positive definite; see Chapter 6 or refer to refs. [28, 30, 50].) Poggio and Girosi [41] took this approach in implementing a regularization network. Specht [56] presents a general regression neural network (GRNN) using a (nonlinearly) weighted average of the given training samples; the GRNN can be described by a topology identical to the RBFN with weighted average in Figure 9.10(b).

When there are fewer basis functions than there are available training samples, an initial guess is required to determine their center positions. We then have an **approximation RBFN**. In this case, matrix $\mathbf{G}$ is not square and the least-squares methods described in Chapter 5 are commonly used to find the matrix $\mathbf{C}$ in Equation (9.23). Note that the matrix $\mathbf{G}$ may be ill-conditioned and limited precision may be encountered even when the pseudoinverse can be approximated by singular value decomposition [45].

Thorough discussions of both interpolation and approximation nets based on Poggio's work [41] can be found in ref. [66].

9.5.4 Examples

RBFN Fitting

Figures 9.11 and 9.12 provide the results of a small interpolation problem with five data points. We tested two interpolation RBFNs: the RBFN with Gaussian basis functions in Equation (9.20), and an RBFN with exponential basis functions:

$$w_i = R_i(||\mathbf{x} - \mathbf{x}_i||) = \exp(-\sigma||\mathbf{x} - \mathbf{x}_i||), \tag{9.25}$$

where we set σ to 1.0 for both the exponential and the Gaussian basis functions. The presented results were obtained by solving the matrix Equation (9.24), so the resultant interpolation curves pass through all training data points. In Figures 9.11 and 9.12, the results based on Equations (9.16) (weighted sum) and (9.17) (weighted average) are displayed in tandem with the five basis functions; when the basis functions do not have enough overlap, the weighted sum of the hidden outputs based on Equation (9.16) may generate curves that are not smooth enough. (Notice that the first two basis functions do not have enough overlap.) Output normalization (or weighted average) surely helps the normalized basis functions cover the input space, and therefore may lead to smoother curves. Notice that we cannot determine which curve provides the best values for intermediate points because we have no certain knowledge of any points other than the five given data points. However, in terms of smoothness, weight average does provides better performance than weighted sum.

Polynomial Fitting

For comparison purposes, Figure 9.13 shows the results obtained by the following four methods: linear interpolation, cubic spline interpolation, fourth-order polynomial interpolation, and third-order polynomial approximation. All the methods except for the first one can generate smooth curves. However, it is cumbersome to extend spline and polynomial interpolation to high-dimensional data sets that can be handled by the interpolation RBFN easily.

Backpropagation MLP Fitting

Figure 9.14 presents the results obtained by a backpropagation MLP. In the MLP results, six choices for the number of hidden units are considered to see how the overall responses are affected. (With respect to bias and variance in connection with the number of hidden units, refer to ref. [11]; it has been reported that an NN with few hidden neurons may exhibit high bias whereas an NN with many hidden neurons may have high variance. We did not examine bias and variance in this

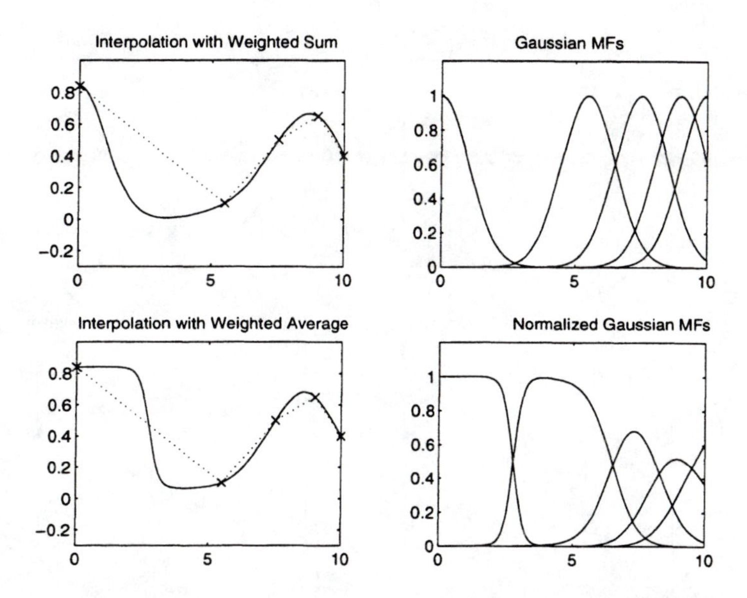

Figure 9.11. *Interpolation results obtained by a Gaussian interpolation network.*

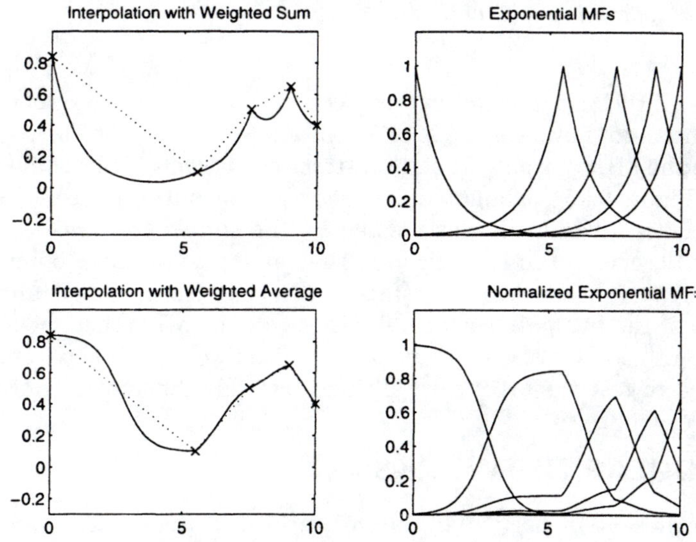

Figure 9.12. *Interpolation results obtained by an interpolation network with the exponential basis functions defined in Equation (9.25).*

simulation.) The curves generated by MLP are quite smooth and they all pass the five given data points. However, the positions of these curves between data points are strongly influenced by the initial weights of the MLP. Moreover, as more hidden

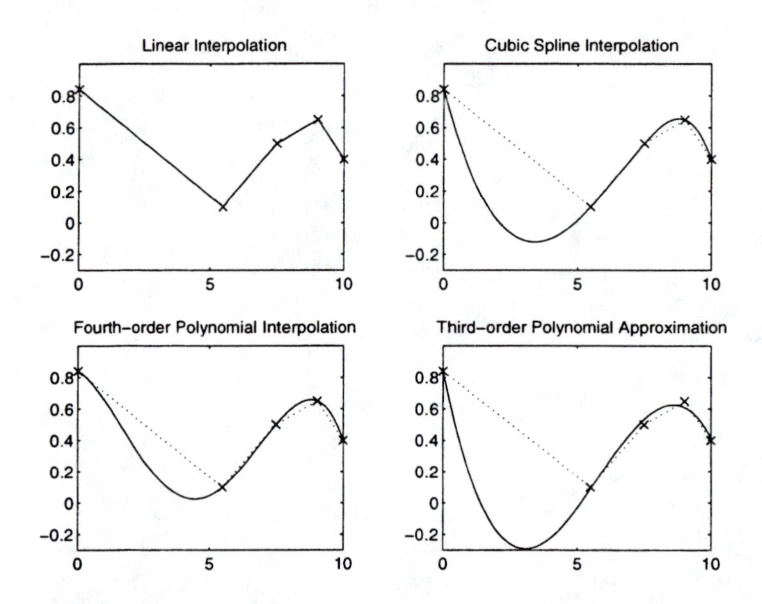

Figure 9.13. *Interpolation and approximation results obtained by four polynomial interpolation/approximation methods.*

nodes are used, the variations of curves between data points are more pronounced.

Furthermore, to show how the MLP evolution proceeds to fit the five data points, we illustrate the MLP outputs at eight distinct error levels (or learning stages) from (a) to (h) in Figure 9.15; highlighted lines show the output responses of an MLP with two hidden units. The MLP was trained by the simple steepest-descent learning scheme. By observing the lines, we see that at least two data points are always passed through after the learning stage (c). That is, the MLP almost finished learning two of the five patterns at error level (c); the MLP then tried to learn the rest. This learning habit also depends on the initial weights. Therefore, it is usually recommended to test the MLP several times by using different sets of weights.

9.6 MODULAR NETWORKS

This section presents a particular class of **modular networks**, which have a hierarchical organization comprising multiple neural networks; the architecture basically consists of two principal components: **local experts** and an **integrating unit** (or **expert networks** and a **gating network**, if they are expressed in the network form), as illustrated in Figure 9.16. A variety of modular connectionist architectures has been discussed, and thus such diverse names as **committees of networks**, **adaptive mixtures**, and **hierarchical mixtures of experts** have all been mentioned. (For simplicity, we shall call a collection of such variations modular networks.)

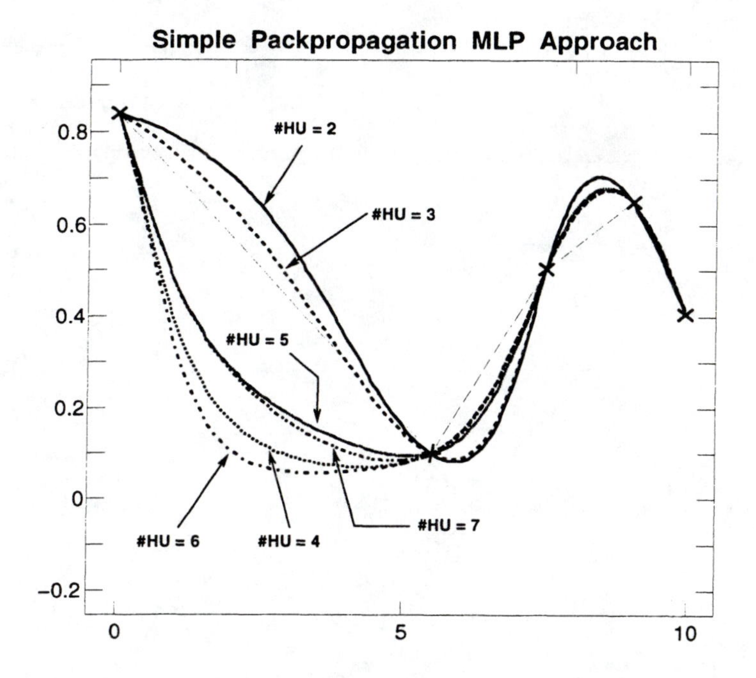

Figure 9.14. *Interpolation results obtained by simple backpropagation MLPs. "# HU" denotes "number of hidden units."*

In general, the basic concept resides in the idea that combined (or averaged) estimators may be able to exceed the limitation of a single estimator. Clemen has shown that the principle of combining a certain number of estimators has a long history and he cited more than 200 papers in his review [7]. The idea also shares conceptual links with the **divide-and-conquer** methodology. Divide-and-conquer algorithms attack a complex problem by dividing it into simpler problems whose solutions can be combined to yield a solution to the complex problem [20]. In other words, the central idea is **task decomposition**. When using a modular network, a given task is split up among several local expert NNs. The average load on each NN is reduced in comparison with a single NN that must learn the entire original task, and thus the combined model may be able to surpass the limitation of a single NN. The outputs of a certain number of local experts (O_i) are mediated by an integrating unit. The integrating unit puts those outputs together using estimated combination weights (g_i). The overall output Y of the modular network is given by

$$Y_i = \sum_{i=1}^{K} g_i O_i.$$

The *task decomposition* idea can be also found in FIS because the outputs are mediated by fuzzy membership functions in a similar way. Thus, the Sugeno-type FIS

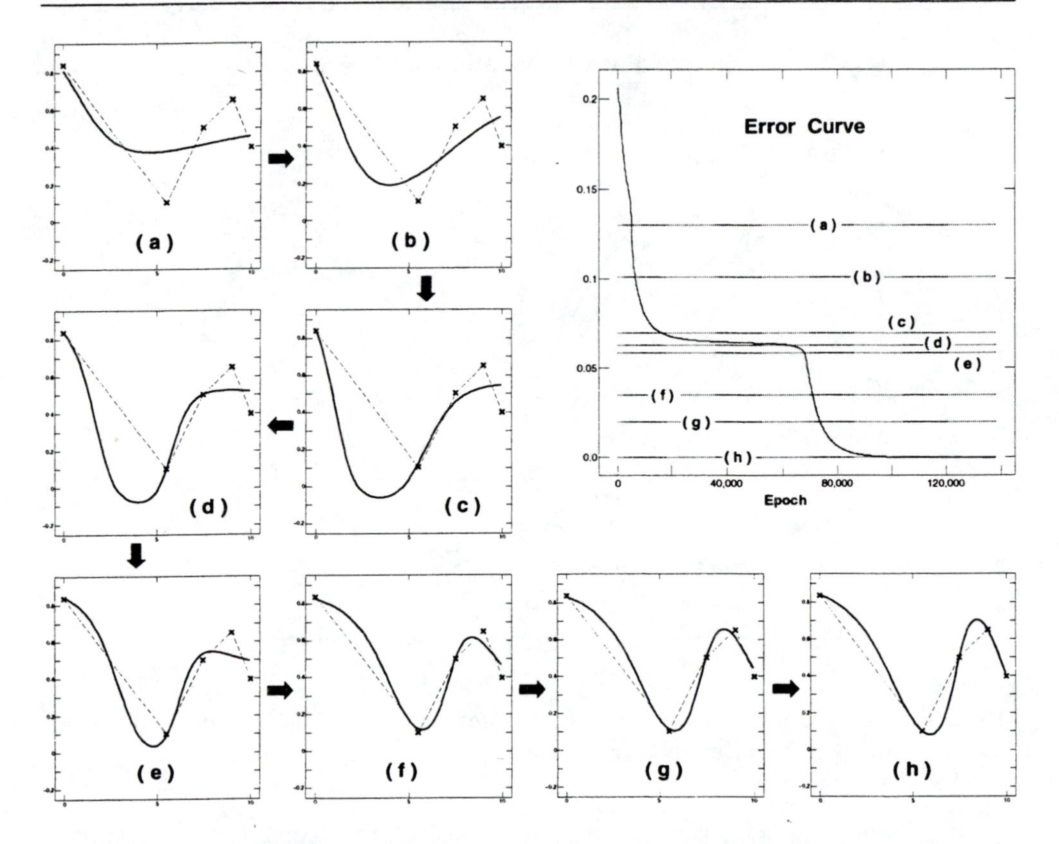

Figure 9.15. *Backpropagation MLP approximations of five sample data at the eight distinct learning stages: (a) to (h). Highlighted lines show the output responses of an MLP based on the simple backpropagation learning scheme.*

discussed in Chapter 4 can be viewed as a variation of the modular network illustrated in Figure 9.16 if each local expert is expressed in the linear function defined in Equation (9.19), and the integrating unit is replaced with a fuzzy membership value generator. (See Figure 13.6 for more details.)

Nowlan, Jacobs, Hinton, and Jordan [15, 17, 38, 39] described modular networks from a *competitive* mixture perspective, which is surveyed in ref. [12]. That is, in the gating network, they used the **softmax** activation function, which was introduced by McCullagh and Nelder [31] and Bridle [3]. More precisely, the gating network uses a softmax activation g_i of the ith output unit given by

$$g_i = \frac{\exp(ku_i)}{\sum_j \exp(ku_j)}, \tag{9.26}$$

where u_i is the weighted sum of the inputs flowing to the ith output neuron of the gating network. It is illustrated in Figure 9.17. Use of the softmax activation

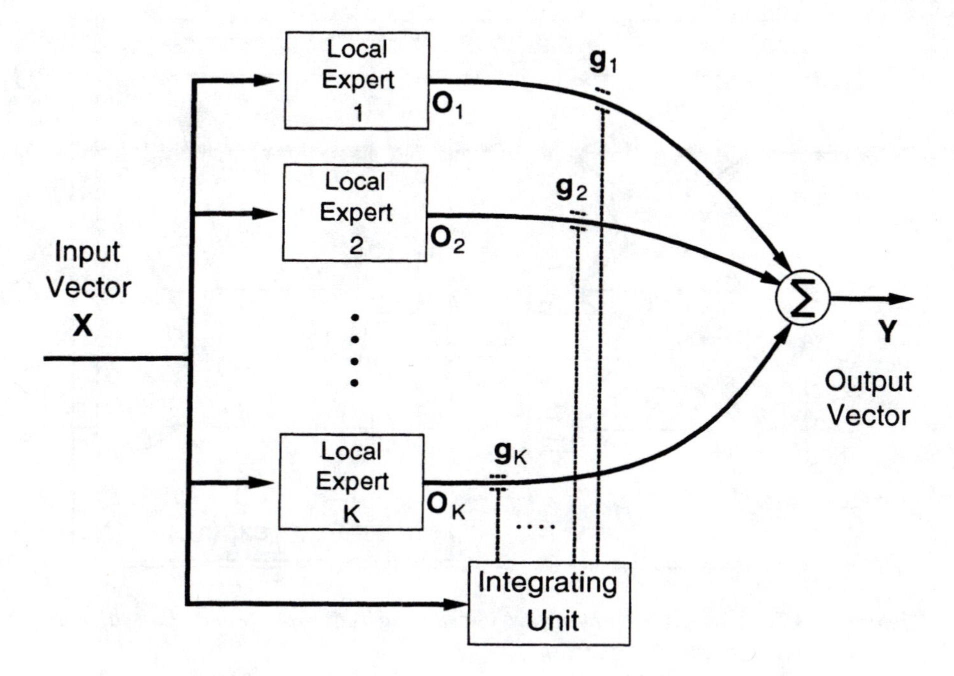

Figure 9.16. *A typical modular network architecture.*

function in modular networks provides a sort of "competitive" mixing perspective because the ith local expert's output O_i with a minor activation u_i does not have a great impact on the overall output Y_i due to Equation (9.26). A feature of a modular network approach, for example, has been discussed from the "competitive" standpoint [12, 16], by using a trivial task example that requires a fit to the pointed corner of a discontinuous (piecewise-linear) function $g(x)$:

$$g(x) = \begin{cases} x, & \text{if } x > 0, \\ -x, & \text{if } x \leq 0. \end{cases}$$

It is claimed that it would be preferable to

- split that function into two separate pieces,
- use two local expert NNs to learn each piece separately in the modular construction, and
- combine the two experts' outputs by using the values given by softmax functions.

This claim conforms with the "competitive" mixing idea. In contrast, fuzzy membership functions in FIS attempt to split the task into pieces more *softly* than the

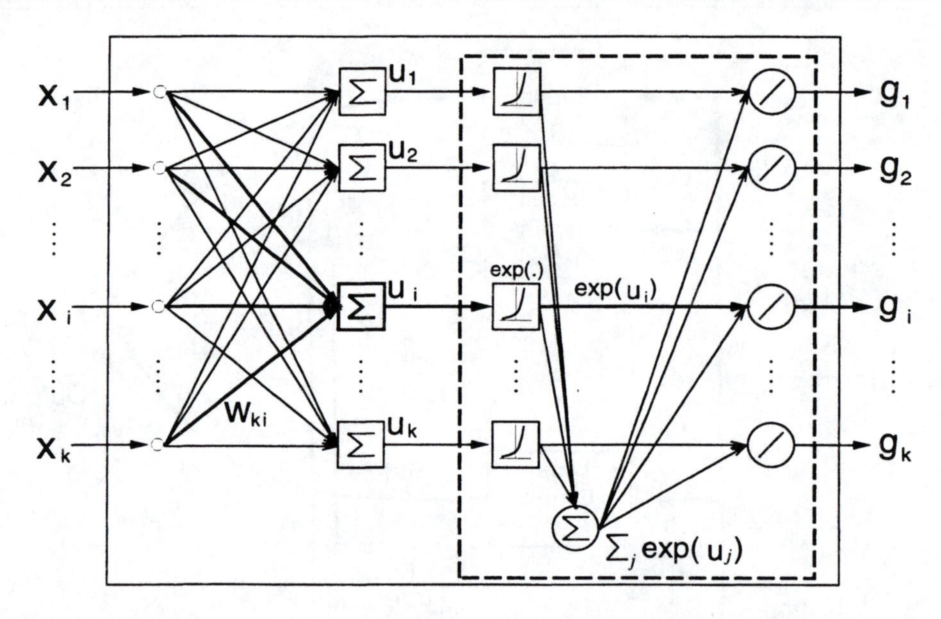

Figure 9.17. *The gating network. The dotted rectangle shows the softmax activation function defined in Equation (9.26).*

softmax functions. That is, FIS stands on a sort of "complementary" mixture viewpoint due to the *weighted average* of fuzzy membership functions' outputs. As we will see in Section 13.5 of Chapter 13, adaptive FIS (that is, ANFIS and CANFIS described in Chapters 12 and 13) learn to fit such a pointed corner in different ways from modular networks with softmax functions. The results shown in Section 13.5 also suggest that RBFN with linear functions of the input variables discussed in Section 9.5.2 may be able to fit well discontinuous functions.

Jordan and Jacobs [20] and Jordan and Xu [21] discussed an application of the expectation-maximization (EM) algorithm for maximum likelihood estimation [1] to train their modular networks.

9.7 SUMMARY

This chapter discusses supervised learning neural networks, including perceptrons, Adalines, backpropagation multilayer perceptrons, radial basis function networks, and modular neural networks. These networks employ optimization techniques (or learning rules) to fine-tune their parameters to match a given data set produced by a target system to be modeled. Since the data set always contains desired outputs to be reproduced by the networks, the underlying learning rule is referred to as *supervised*, as compared to recording, reinforcement, and unsupervised learning, which are discussed in other chapters. Table 9.1 charts our voyage into other types of neural network learning rules.

Table 9.1. *Learning modes of artificial neural networks and relevant chapters.*

Learning mode	Characteristics of available information for learning	Relevant chapters
Supervised	Instructive information on desired responses, explicitly specified by a teacher	9
Recording	*A priori* design information for memory storing	11
Reinforcement	Partial information about desired responses, or only "right" or "wrong," evaluative information	10
Unsupervised	No information about desired responses	11

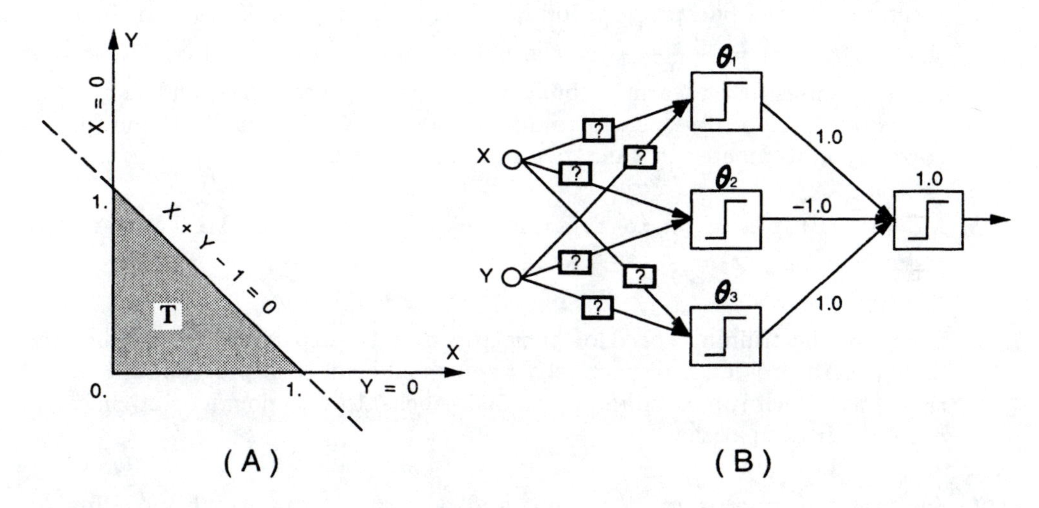

Figure 9.18. *(a) A classification problem, and (b) the perceptron designed to solve the problem.*

EXERCISES

1. Finish designing the perceptron with step-function threshold units in Figure 9.18 by determining threshold values θ_i ($i = 1, 2, 3$) and six connection

weights between the input layer and the hidden layer that enable your perceptron to recognize correctly whether an arbitrary point (x, y) is in the shaded triangular area T or not. (Note that the points on the three dividing boundary lines are considered to be outside area T.) The perceptron is supposed to produce the output o:

$$o = \begin{cases} 1 & \text{if } (x, y) \in T, \\ 0 & \text{otherwise.} \end{cases}$$

2. Demonstrate that the backpropagation learning rule for an MLP in Equation (9.7) can be derived from either Equation (8.5) or (8.12).

3. Modify the backpropagation learning rule in Equation (9.7) to accommodate making the hyperbolic tangent function the activation function.

4. The MATLAB function `tanmlp.m` (available via FTP, see page xxiii) is an implementation of the backpropagation MLP with the hyperbolic tangent function $f(x) = \tanh(x) = \frac{e^x - e^{-x}}{e^x + e^{-x}}$ as the network's activation function. Note that `tanmlp.m` uses batch learning, but the code is fully vectorized and there is no for-loop to cycle through each training data pair. Verify that Equation (9.7) is correctly implemented in `tanmlp.m` in matrix forms.

5. Modify `tanmlp.m` to get `logmlp.m` for the backpropagation MLP with the logistic function $f(x) = \frac{1}{1 + e^{-x}}$.

6. Compare the training speed of `tanmlp.m` and `logmlp.m` for the bipolar and binary XOR problems, respectively. (You should run each program at least 10 times, with each run including up to 500 epochs, to get enough error curves to make a fair comparison.)

7. Set the momentum term α to zero and perform 10 runs (each including up to 500 epochs) of `tanmlp.m` for the bipolar XOR problem to plot the average RMSE (root-mean-squared error) curve. Repeat the same task, but use a couple of nonzero momentum terms. Does the use of momentum terms speed up training? What is the best value of α in your simulation?

8. Explain why an MLP does not learn if the initial weights and biases are all zeros.

9. Use `tanmlp.m` to solve a three-input bipolar XOR problem with a 3-n-1 back-

propagation MLP. The training data matrix is

$$\begin{bmatrix} -1 & -1 & -1 & -1 \\ -1 & -1 & 1 & 1 \\ -1 & 1 & -1 & 1 \\ -1 & 1 & 1 & -1 \\ 1 & -1 & -1 & 1 \\ 1 & -1 & 1 & -1 \\ 1 & 1 & -1 & -1 \\ 1 & 1 & 1 & 1 \end{bmatrix},$$

where the first three columns are inputs, the last column is output, and each row represents a desired input-output data pair. Do some experiments to find the smallest n (number of hidden units) needed to solve the problem.

10. Write a MATLAB script that uses an interpolation RBFN to solve the two-input bipolar XOR problem. Plot the overall input-output surface.

11. Derive the backpropagation learning rule for a single-output RBFN, where the parameters include the center ($\mathbf{u}_i$), the width (σ_i), and the connection weight (c_i) for each receptive field.

12. Implement a MATLAB function `rbfn.m` to do backpropagation in a single-output RBFN. You should follow the input argument convention of `tanmlp.m`. Try not to use for-loops to cycle through the data set; use vectorized operations instead to speed up computation.

REFERENCES

[1] L. E. Baum, T. Petrie, G. Soules, and N. Weiss. A maximization technique occurring in the statistical analysis of probabilistic functions of Markov chains. *The Annals of Mathematical Statistics*, 41:164–171, 1970.

[2] S. M. Botros and C. G. Atkeson. Generalization properties of radial basis functions. In D. S. Touretzky, editor, *Advances in neural information processing systems III*, pages 707–713. Morgan Kaufmann, San Mateo, CA, 1991.

[3] J. Bridle. Probabilistic interpretation of feedforward classification network outputs, with relationships to statistical pattern recognition. In F. Fogelman and J. Herault, editors, *Neuro-computing: algorithms, architectures, and applications*. Springer-Verlag, London, 1989.

[4] D. S. Broomhead and D. Lowe. Multivariable functional interpolation and adaptive networks. *Complex Systems*, 2:321–355, 1988.

[5] D. S. Broomhead and D. Lowe. Multivariable functional interpolation and adaptive networks. *Complex Systems*, 2:321–355, 1988.

[6] S. Chen, C. F. N. Cowan, and P. M. Grant. Orthogonal least squares learning algorithm for radial basis function networks. *IEEE Transactions on Neural Networks*, 2(2):302–309, March 1991.

[7] R. T. Clemen. Combining forecasts: A review and annotated bibliography. *International Journal of Forecasting*, 5:559–583, 1989.

[8] G. Cybenko. Approximation by superpositions of a sigmoidal function. *Mathematics of Control, Signals, and Systems*, 2:303–314, 1989.

[9] S. E. Fahlman. Faster-learning variations on back-propagation: an empirical study. In D. Touretzky, G. Hinton, and T. Sejnowski, editors, *Proceedings of the 1988 Connectionist Models Summer School*, pages 38–51, Carnegic Mellon University, 1988.

[10] S. E. Fahlman and C. Lebiere. The cascade-correlation learning architecture. In D. S. Touretzky, G. Hinton, and T. Sejnowski, editors, *Advances in neural information processing systems II*. Morgan Kaufmann, San Mateo, CA, 1990.

[11] S. Geman, E. Bienenstock, and R. Doursat. Neural networks and the bias/variance dilemma. *Neural Computation*, 4:1–58, 1992.

[12] Simon Haykin. *Neural networks: a comprehensive foundation*. Macmillan College Publishing, 1994.

[13] J. Hertz, A. Krogh, and R. G. Palmer. *Introduction to the theory of neural computation*. Addison-Wesley, Reading, MA, 1991.

[14] R. A. Jacobs. Increased rates of convergence through learning rate adaptation. *Neural Networks*, 1:295–307, 1988.

[15] R. A. Jacobs and M. I. Jordan. A competitive modular connectionist architecture. In R. P. Lippmann, J. E. Moody, and D. J. Touretzky, editors, *Advances in Neural Information Processing Systems 3*, pages 767–773, San Mateo, CA, 1991. Morgan Kaufmann.

[16] R. A. Jacobs, M. I. Jordan, and A. G. Barto. Task decomposition through competition in a modular connectionist architecture: The what and where vision tasks. *Cognitive Science*, 15:219–250, 1991.

[17] R. A. Jacobs, M. I. Jordan, S. J. Nowlan, and G. E. Hinton. Adaptive mixtures of local experts. *Neural Computation*, 3:79–87, 1991.

[18] J.-S. Roger Jang and C.-T. Sun. Functional equivalence between radial basis function networks and fuzzy inference systems. *IEEE Transactions on Neural Networks*, 4(1):156–159, January 1993.

[19] R. D. Jones, Y. C. Lee, C. W. Barnes, G. W. Flake, K. Lee, and P. S. Lewis. Function approximation and time series prediction with neural networks. In *Proceedings of IEEE International Joint Conference on Neural Networks*, pages 649–665 (Volume I), 1990.

[20] M. I. Jordan and R. A. Jacobs. Hierarchical mixtures of experts and the EM algorithm. *Neural Computation*, 6:181–214, 1994.

[21] M. I. Jordan and L. Xu. Convergence results for the EM approach to mixtures of experts architectures. *Neural Networks*. in press.

[22] V. Kadirkamanathan, M. Niranjan, and F. Fallside. Sequential adaptation of radial basis function neural networks. In D. S. Touretzky, editor, *Advances in neural information processing systems III*, pages 721–727. Morgan Kaufmann, San Mateo, CA, 1991.

[23] Y. LeCun, B. Boser, J. S. Denker, D. Henderson, R. E. Howard, W. Hubbbard, and L. D. Jackel. Handwritten digit recognition with a back-propagation network. In D. S. Tourestsky, editor, *Neural information processing system*, pages 143–155. Morgan Kaufmann, San Mateo, CA, 1990.

[24] S. Lee and R. M. Kil. A Gaussian potential function network with hierarchically self-organizing learning. *Neural Networks*, 4(2):207–224, 1991.

[25] R. P. Lippmann. An introduction to computing with neural networks. *IEEE Acoustics, Speech, and Signal Processing Magazine*, 4(2):4–22, 1987.

[26] D. Lowe. Adaptive radial basis function nonlinearities, and the problem of generalization. In *Proceedings of the First IEEE International Conference on Artificial Neural Networks*, pages 171–175, London, UK, 1989.

[27] M. C. Mackey and L. Glass. Oscillation and chaos in physiological control systems. *Science*, 197:287–289, July 1977.

[28] Donald W. Marquardt. An algorithm for least squares estimation of nonlinear parameters. *Journal of the Society of Industrial and Applied Mathematics*, 11:431–441, 1963.

[29] Timothy Masters. *Practical neural network recipe in C++.* Academic Press, Inc., 1993.

[30] Timothy Masters. *Advanced algorithms for neural networks: a C++ sourcebook.* John Wiley & Sons, New York, 1995.

[31] P. McCullagh and J. A. Nelder. *Generalized linear models.* Chapman and Hall, 1983.

[32] W. S. McCulloch and W. Pitts. A logical calculus of the ideas imminent in neural nets. *Bulletin of Mathematical Biophysics*, 5:115–137, 1943.

[33] M. Minsky and S. Papert. *Perceptrons.* MIT Press, Cambridge, MA., 1969.

[34] J. Moody and C. Darken. Learning with localized receptive fields. In D. Touretzky, G. Hinton, and T. Sejnowski, editors, *Proceedings of the 1988 Connectionist Models Summer School*, San Mateo, CA, 1988. Carnegie Mellon University, Morgan Kaufmann.

[35] J. Moody and C. Darken. Fast learning in networks of locally-tuned processing units. *Neural Computation*, 1:281–294, 1989.

[36] M. T. Musavi, W. Ahmed, K. H. Chan, K. B. Faris, and D. M. Hummels. On the training of radial basis function classifiers. *Neural Networks*, 5(4):595–603, 1992.

[37] D. Nguyen and B. Widrow. Improving the learning speed of 2-layer neural networks by choosing initial values of the adaptive weights. In *Proceedings of IEEE International Joint Conference on Neural Networks*, pages 21–26 (Volume III), 1990.

[38] S. J. Nowlan. Maximum likelihood competitive learning. In D. J. Touretzky, editor, *Advances in Neural Information Processing Systems 2*, pages 574–582, San Mateo, CA, 1989. Morgan Kaufmann.

[39] S. J. Nowlan and G. E. Hinton. Evaluation of adaptive mixtures of competing experts. In R. P. Lippmann, J. E. Moody, and D. J. Touretzky, editors, *Advances in Neural Information Processing Systems 3*, pages 774–780, San Mateo, CA, 1991. Morgan Kaufmann.

[40] D. B. Parker. Learning logic. Invention Report S81-64, File 1, Office of Technology Licensing, Standford University, October 1982.

[41] T. Poggio and F. Girosi. Networks for approximation and learning. *The Proceedings of the IEEE*, 78(9):1485–1497, September 1990.

[42] D. A. Pomerleau. Efficient training of artificial neural networks for autonomous navigation. *Neural Computation*, 3:88–97, 1991. 1991.

[43] D. A. Pomerleau. *Neural network perception for mobile robot guidance.* PhD thesis, Department of Computer Science, Carnegie Mellon Univeristy, 1992.

[44] M. J. D. Powell. Radial basis functions for multivariable interpolation: a review. In J. C. Mason and M. G. Cox, editors, *Algorithms for approximation*, pages 143–167. Oxford University Press, 1987.

[45] W. H. Press, B. P. Flannery, S. A. Teukolsky, and W. T. Vetterling. *Nemerical recipes in C.* Cambridge University Press, 1991.

[46] A. K. Rigler, J. M. Irvine, and T. P. Vogl. Rescaling of variables in back propagation learning. *Neural Networks*, 4(2):225–229, 1991.

[47] F. Rosenblatt. *Principles of neurodynamics: perceptrons and the theory of brain mechanisms.* Spartan, New York, 1962.

[48] D. E. Rumelhart, G. E. Hinton, and R. J. Williams. Learning internal representations by error propagation. In D. E. Rumelhart and James L. McClelland, editors, *Parallel distributed processing: explorations in the microstructure of cognition, volume 1*, chapter 8, pages 318–362. MIT Press, Cambridge, MA., 1986.

[49] E. Säckinger, B. E. Boser, J. Bromley, Y. LeCun, and L. D. Jackel. Application of the anna neural network chip to high-speed character recognition. *IEEE Transactions on Neural Networks*, 3:498–505, 1992.

[50] G. A. F. Seber and C. J. Wild. *Nonlinear regression.* John Wiley & Sons, 1989.

[51] Terrence J. Sejnowski and Charles R. Rosenberg. NETtalk: a parallel network that learns to read aloud. JHU/EECS 86/01, Johns Hopkins University, 1986.

[52] Terrence J. Sejnowski and Charles R. Rosenberg. Parallel networks that learn to pronounce english text. *Complex Systems*, 1:145–168, 1987.

[53] S. Shah, F. Palmieri, and M. Datum. Optimal filtering algorithms for fast learning in feedforward neural networks. *Neural Networks*, 5(5):779–787, 1992.

[54] S. Shar and F. Palmieri. MEKA — a fast, local algorithm for training feedforward neural networks. In *Proceedings of the International Joint Conference on Neural Networks*, pages 41–46 (Volume III), 1990.

[55] S. Singhal and L. Wu. Training multilayer perceptrons with the extended kalman algorithm. In David S. Touretzky, editor, *Advances in neural information processing systems I*, pages 133–140. Morgan Kaufmann, San Mateo, CA, 1989.

[56] Donald F. Specht. A general regression neural network. *IEEE Transactions on Neural Networks*, 2(6):568–576, November 1991.

[57] K. Stokbro, D. K. Umberger, and J. A. Hertz. Exploiting neurons with localized receptive fields to learn chaos. *Complex Systems*, 4:603–622, 1990.

[58] R. L. Watrous. Learning algorithms for connectionist network: applied gradient methods of nonlinear optimization. In *Proceedings of IEEE International Conference on Neural Networks*, pages 619–627, 1991.

[59] P. Werbos. *Beyond regression: New tools for prediction and analysis in the behavioral sciences.* PhD thesis, Harvard University, 1974.

[60] D. Wettschereck and T. Dietterich. Improving the performance of radial basis function networks by learning center locations. In J. E. Moody, editor, *Advances in Neural Information Processing Systems 4*, pages 1133–1140, San Mateo, CA, 1992. Morgan Kaufmann.

[61] B. Widrow and J. R. Glover. Adaptive noise cancelling: Principles and applications. *IEEE Proceedings*, 63:1692–1716, 1975.

[62] B. Widrow and M. E. Hoff. Adaptive switching circuits. In *IRE WESCON Convention Record*, pages 96–104, New York, 1960.

[63] B. Widrow and M. A. Lehr. 30 years of adaptive neural networks: Perceptron, madline, and backpropagation. *Proceedings of the IEEE*, 78(9):1415–1442, 1990.

[64] B. Widrow and D. Stearns. *Adaptive signal processing.* Prentice Hall, Upper Saddle River, NJ, 1985.

[65] B. Widrow and R. Winter. Neural nets for adaptive filtering and adaptive pattern recognition. *IEEE Computer*, 21(3):25–39, March 1988.

[66] Patrick H. Winston. *Artificial intelligence.* Addison-Wesley, Reading, MA, 3rd edition, 1992.

Chapter 10

Learning from Reinforcement[1]

E. Mizutani

10.1 INTRODUCTION

Reinforcement learning has been accepted as a fundamental paradigm for **machine learning** with particular emphasis on computational aspects of learning.

Learning from reinforcement is a trial-and-error learning scheme whereby a computational agent learns to perform an appropriate action by receiving evaluative feedback (called a reinforcement signal, performance score, grade, etc.) through interaction with the world (or environment) that includes no explicit teacher for any correct instruction. The learning agent (or learner) reinforces itself from lesson failures; accumulated fiascos will lead it to success but not to disaster. This learning method can be considered a simple way of adjusting behavior, and can be found in animals' learning skills and coping with a physical environment. It also matches our common-sense ideas [7, 76]:

> *If an action is followed by a satisfactory state of affairs, or an improvement in the state of affairs, then the tendency to produce that action is strengthened or reinforced (rewarded). Otherwise, that tendency is weakened or inhibited (penalized).*

There is a vast body of literature on reinforcement learning, also known as *graded learning*. It has been discussed in game playing, autonomous robot control, and so on. Commonly applied reinforcement learning strategies are classified into a few categories from the architectural standpoint; Sutton delineates four basic representative architectures for reinforcement learning [74]:

- Policy-only

[1]The content of this chapter is largely attributable to Professor Stuart E. Dreyfus at the Department of Industrial Engineering and Operations Research (IEOR), University of California at Berkeley. Some of the materials came from the work in his class IEOR 290N (Artificial Neural Networks), spring 1994. Of course, any errors and mistakes are our own.

- Reinforcement comparison
- Adaptive heuristic critic
- Q-learning.

This chapter explores selected topics in reinforcement learning. Besides the aforementioned four learning architectures, we describe the temporal difference notion and dynamic programming principle, and discuss their computational learning powers.

10.2 FAILURE IS THE SUREST PATH TO SUCCESS

An agent usually learns to *map a representation of a state to an appropriate action or a probability distribution over a set of actions*. This mapping is called the **policy**. The simplest reinforcement learning architecture consists solely of an adjustable policy, which is called a **policy-only** architecture [74]. In this section, we first show a simple form of reinforcement learning based on a policy-only architecture that closely resembles ordinary supervised learning. We outline it by means of an example in the next subsection. After that, we discuss credit assignment and evaluation functions in light of the example. They are fundamental and important elements in reinforcement learning.

10.2.1 Jackpot Journey

The objective of reinforcement learning is to find an optimal *policy* for selecting a series of actions by means of a reward-penalty scheme. We consider application of an elementary reward-penalty scheme to a jackpot journey problem that requires finding an optimal path to "gold" in the simple triangular-path network illustrated in Figure 10.1.

We imagine that many travelers start their jackpot journey, wishing to find *gold* from the starting point A. [Hereafter we call a path intersection (point) of our network a vertex.] At each vertex, there is a signpost that has a box with some white and black stones in it. A traveler picks a stone from the signpost box and follows certain instructions; when a "white" stone is picked, "go diagonally upward," denoted by action u. Conversely, when a "black" stone is chosen, "go diagonally downward," signified by action d.

Suppose that the journey is always started at vertex A, and *gold* is placed at the terminal vertex H. Further suppose that travelers strictly follow the signpost instructions.

Each traveler's behavior can be described as follows: At vertex A, pick one stone (selection) and put it on the signpost; according to the stone's instruction, proceed to the next vertex (action). Repeat this selection-action procedure at the second and third vertices. (After the third action, the traveler will reach one of the four terminal vertices: G, H, I, or J.) When the jackpot is hit (success), prepare a *reward*

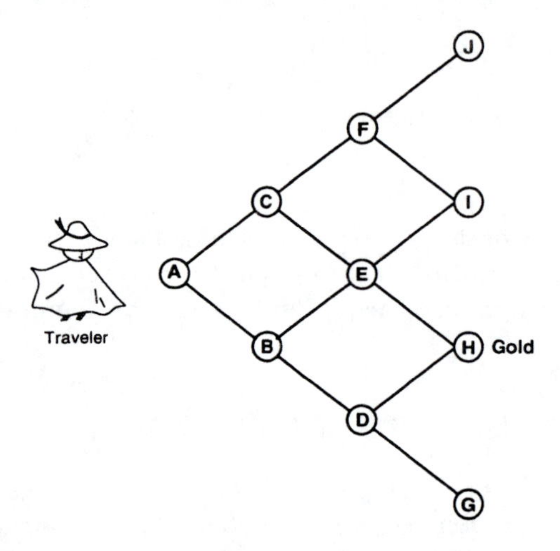

Figure 10.1. *The jackpot journey problem.*

scheme. When the *gold* is not found (failure), prepare a *penalty* scheme. Then trace back to the starting vertex A; at each visited vertex, apply the reward or penalty scheme. That is, put the placed stone back into the signpost box with an additional stone of the same color (reward), or take the placed stone away from the signpost (penalty). (When the traveler returns, the next traveler will hit the road with a bit more hope.) Repeat the same journey many times. Obviously, the probability of finding an optimal policy[2] will increase as more and more journeys are undertaken.

In conformity with the general terminology, we can tabulate the relevant terms as follows:

path network configuration	$\leftrightarrow$	world or environment
traveler	$\leftrightarrow$	agent or learner
vertex (intersection)	$\leftrightarrow$	state
picking a stone	$\leftrightarrow$	selecting an action
sequence of three stones	$\leftrightarrow$	policy[2] (or trajectory).

It is easy to simulate such voyages on computers. We define the probability of action d, P_{down}, for each state as follows:

$$P_{\text{down}} = \frac{Num_{\text{black}}}{Num_{\text{black}} + Num_{\text{white}}}, \tag{10.1}$$

[2]We use the term *policy*, which is usually employed for a **decision-making problem under uncertainty** [26] such as a **stochastic** problem, in which the agent makes a decision first and then is informed by the environment which action to follow. [See Equation (10.13).] Here, the term *path* or *trajectory* may be more appropriate because this jackpot journey is a **deterministic** problem, in which a decision determines the next state with **certainty** even if the decision is chosen randomly. Of course, *policy* is not very important for a *deterministic* problem.

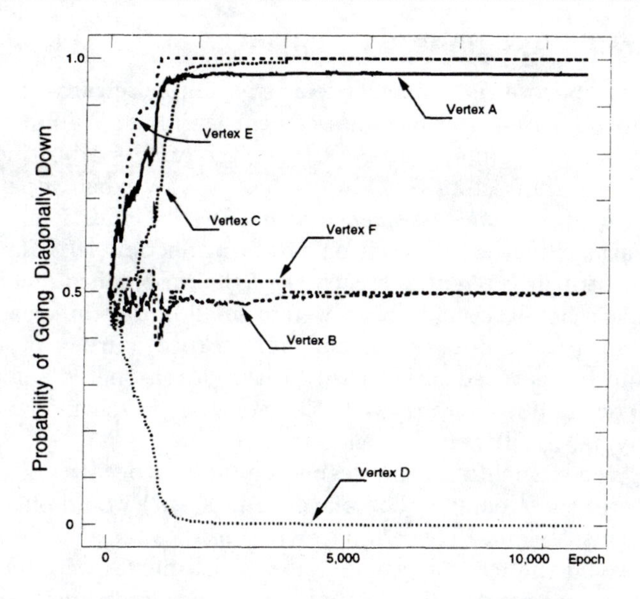

Figure 10.2. *Changing probability of action d, "go diagonally downward," at each non-terminal vertex (A—F) as trials progress in the jackpot journey problem.*

where Num_{black} is the number of black stones and Num_{white} is the number of white stones. Accordingly, the probability of the other action u, P_{up}, can be defined as

$$P_{\text{up}} = \frac{Num_{\text{white}}}{Num_{\text{black}} + Num_{\text{white}}} = 1 - P_{\text{down}}. \tag{10.2}$$

Figure 10.2 shows the changing P_{down} at the six non-terminal vertices (A—–F) as trials progressed. Initially, all signposts had 20 black stones and 20 white ones, and $P_{\text{down}} = P_{\text{up}} = 0.5$. At vertices C and E, the specified action was action d whereas the dictated action was action u at vertex D. High P_{down} at vertex A (the starting point) favors action d. This must be because at the next vertex B, we have no chance to lose a path toward *gold* no matter action we take, unlike the situation at the other alternative vertex C. This simulation result confirms that such a simple reward-penalty scheme is applicable to learning an "optimal policy."

Michie applied this reward-penalty scheme to a tic-tac-toe game, using a few hundred *matchboxes* with colored beads [53]; the matchboxes corresponded to possible game states, and the colored beads specified admissible actions. He called this model MENACE (Matchbox Educable Naughts And Crosses Engine), showing that the scheme drove MENACE to learn an "optimal policy" to win the tic-tac-toe game. Later, Michie proposed the "Boxes" system based on a scheme similar to MENACE to solve the pole-balancing control problem (see Sections 10.5.1 and 18.2).

10.2.2 Credit Assignment Problem

Through the jackpot journey, we have learned a simple reward-and-penalty scheme, which gives us a basic aspect of reinforcement learning. We notice, however, that this strategy is strictly **success or failure driven**; its adjustment scheme (i.e., increasing or decreasing stones) is always applied only when the final outcome becomes known *after the entire sequence of actions.* Thus, it is close to ordinary *supervised* learning methods [Equation (10.7) in Section 10.3.1]. In other words, it ignores the intrinsic *sequential* structure of the problem to make adjustments at each state. The scheme seems to work well in small finite state space problems such as the well-known tic-tac-toe game and our jackpot journey problem; the entire state space can be searched rather easily within a reasonable amount of computation time. Here the question arises: Is this goal-driven scheme really applicable to any game playing? Is all well that ends well?

In playing chess, such a scheme seems impractical because of the huge number of possible states chess entails. The player (i.e., agent) would only receive *feedback* (or a reinforcement signal) concerning *win* or *lose*, after a fairly long sequence of moves. How would the learner (agent) know which moves were inappropriate after an unsuccessful experience? How would the learner know which moves may have been excellent? In chess playing, the learner needs to make *better* moves with no performance indication regarding winning during the game. The problem of rewarding or penalizing each state (or move) individually in such a long sequence toward an eventual victory or loss is called the **temporal** credit assignment problem [55]. In contrast to *temporal* credit assignment, apportioning credit to the internal agent's action structures is called the **structural** credit assignment problem. In structural terms, it is necessary to determine which part and how much should be altered to enhance overall performance. The structural credit assignment problem is concerned with the development of appropriate internal representations. Any distributed learning model, such as a neural network (NN), usually involves both *temporal* and *structural* credit assignment at the same time (Section 10.5.2).

The power of reinforcement learning actually lies in this point that *the agent does not have to wait until it receives feedback at the end to make adjustments.* Sutton clearly defined the theoretical aspects of this key concept in codifying **temporal difference (TD)** methods, which are discussed in Section 10.3. Its precursor, Samuel's checkers-playing program [67], and more recent work, such as Holland's bucket brigade [34] (in Section 10.10.1), also share this fundamental key concept that the states in a sequence should be evaluated and adjusted according to their immediate or near-immediate successors, rather than according to the final outcome [77]. For such evaluation purposes, we can use the computational measure of an **evaluation function. Delayed reinforcement** learning deals with a temporal sequence of input state vectors aimed at optimizing an *evaluation function*, as in many real-world problems that involve *delays* between action and any resultant reinforcement [78].

In comparison, **immediate reinforcement** is determined by the most recent input-output pair alone. Sutton clearly described reinforcement comparison architectures as follows [74]:

> *Reinforcement comparison architectures are effective at optimizing immediate rewards, but not at optimizing total reward in the long run. The problem is that actions have two kinds of consequences—they affect the next reward and they affect the next state, but reinforcement comparison architectures only take the first of these into account. Suppose an action produces high immediate reward but deposits the environment in a state from which only low reward can be obtained? In order to optimize long-term reward, these delayed affects of action must be taken into account.*

Both **adaptive heuristic critic** (**AHC**) and **Q-learning** architectures can take such delayed effects into account, as described later in Sections 10.5 and 10.6. Various algorithms for "reinforcement comparison" are discussed by Dayan [22].

10.2.3 Evaluation Functions

In Section 10.2.1, changing the number of stones affects the probability of an action at a state. As a result, the destination signposts indicate optimal trajectories; MENACE uses similar methods to find winning trajectories in the tic-tac-toe game. Their methods correspond to search strategies for optimal trajectories in artificial intelligence (AI) literature. The principal goal is finding successful trajectories among all admissible world states. The agent searches for states that yield reward by performing a series of actions. *Learning thus can be viewed as a search.*

Evaluation functions, also known as **value functions**, play an important role in learning [44, 60, 63, 97]. The evaluation functions produce scalar values (reinforcement signals) of states to aid in finding optimal trajectories. (Such evaluation functions correspond to the emotions in the biological brain [3].) In the AI field, it is widely accepted that the performance of search algorithms is greatly improved by the use of evaluation functions. *Manhattan Distance,* for instance, is a well-known heuristic evaluation function for the "eight puzzle" problem in which we must move tiles to target positions in a 3×3 square frame containing eight numbered square tiles and a "blank" space [18, 27, 42, 60, 64]. The *Manhattan Distance* heuristic computes the sum of each tile's vertical and horizontal distance from its target position to estimate the number of moves required to reach the goal. It can operate as a heuristic function in the well-known A^* *algorithm* [27, 60, 97], whose evaluation function combines a heuristic (or cost-to-go) function and a cost (or cost-so-far) function [66]. Winston [97] introduced an improved version of A^* with a reflection of the *dynamic programming principle*, which we shall discuss in Section 10.4. Korf [42] proposed a *learning real-time* A^* ($LRTA^*$) algorithm as an extension of the A^* algorithm. Russell and Wefald well treated a family of these algorithms [65], and Barto et al. discussed a close tie between $LRTA^*$ and learning based on real-time dynamic programming [6].

Manhattan Distance is not perfect, but it seems to reflect well the actual chances of reaching a given goal despite its simplicity. Evaluation functions must not require heavy computations; hence there is a dilemma between the accuracy of the evaluation functions and their computation cost. Major concerns regarding the evaluation functions are as follows:

- How to devise effective evaluation functions,
- How to store evaluation functions,
- How to adjust evaluation functions.

Devising effective evaluation functions may require problem-specific knowledge. Evaluation functions can be stored as tables, symbolic rules, decision trees [21], CMACs [3] (Cerebellar Model Arithmetic Computers/Cerebellar Model Articulation Controller) [46, 86] or parametric approximators. Adjusting schemes of evaluation functions may depend on the function forms.

In Section 10.2.1, the agent's behavior was totally **result-driven**; that is, the agent acted based on an **evaluation-free** algorithm. Unlike such a *result-driven agent*, an **evaluation-driven agent** employs the evaluation functions at each visited state to estimate admissible actions and how likely the actions are to lead eventually to choosing the most promising action, without concern for what the value of the final solution is. In other words, evaluation functions are indispensable for *evaluating states in a temporally successive manner*; we elaborate on this in the next section.

10.3 TEMPORAL DIFFERENCE LEARNING

Temporal difference (**TD**) methods are a class of incremental learning procedures specialized for prediction whereby credit is assigned based on the difference between temporally successive predictions [77].

In earlier studies, Samuel [67] employed a TD method in the checkers-playing program to improve parameterized evaluation functions (linear functions of input variables) through experiential tuning; it could learn on the basis of its experience and thus improve its performance by updating parameters. More specifically, Samuel's program used the evaluation difference between a board configuration and a likely future board configuration. It chose the most advantageous position in the absence of complete information at any temporal game state. The TD method was also employed to solve the temporal credit assignment problem in the predictor part of an adaptive critic [8, 5], which is discussed in Section 10.5.

Sutton has formalized these TD methods as general methods for learning to predict arbitrary events (not just goal-related ones).

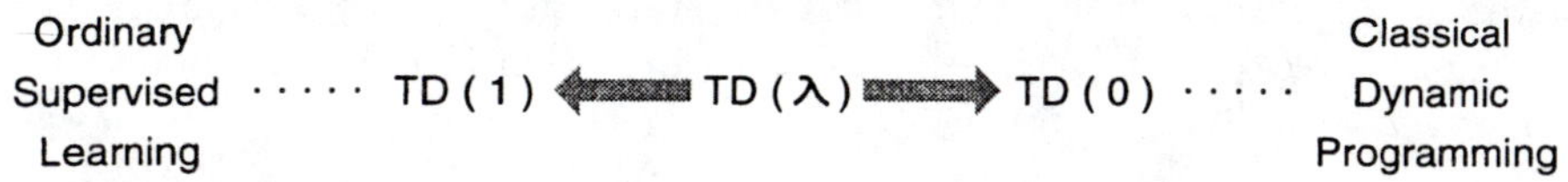

Ordinary Supervised Learning ····· TD (1) ⟸ TD (λ) ⟹ TD (0) ····· Classical Dynamic Programming

Figure 10.3. *TD learning spectrum. TD(λ) migrates various degrees of the TD learning spectrum according to the value of λ.*

10.3.1 TD Formulation

In the general form of TD methods, TD(λ), the modifiable parameters w of the agent's predictor obey the following update rule with a learning rate α:

$$\text{TD}(\lambda): \quad \Delta w_t = \alpha(V_{t+1} - V_t) \sum_{k=1}^{t} \lambda^{t-k} \nabla_w V_k, \tag{10.3}$$

where V_t is the prediction value at time t and λ is a (discounting) recency parameter ranging from 0 to 1. Adjustments to predictions occurring k steps (or stages) in the past are exponentially weighted; more recent predictions make greater weight changes. This may match biological brain strategies for deciding how strongly recently received stimuli should be used in combination with the current stimuli to determine actions. This can be viewed as a usual supervised learning procedure for the pair current prediction V_t ("actual" output) and its subsequent prediction V_{t+1} ("desired" output) in the error term. In this sense, reinforcement (TD) learning can be regarded as a form of supervised error-correction learning, which often minimizes the squared error E_{td} between *final outcome* z and *current prediction* V_t:

$$\begin{aligned} E_{td} &= \tfrac{1}{2}\{z - V_t\}^2 \\ &= \tfrac{1}{2}\{\textstyle\sum_{k=t}^{m}(V_{k+1} - V_k)\}^2, \end{aligned} \tag{10.4}$$

where $V_{m+1} = z$. Figure 10.3 shows that TD(λ) migrates various degrees of the TD learning spectrum according to the value of λ.

In two particularly extreme cases in which $\lambda = 1$ and $\lambda = 0$, we have TD(1) and TD(0), respectively, as follows:

$$\text{TD}(1): \quad \Delta w_t = \alpha(V_{t+1} - V_t) \sum_{k=1}^{t} \nabla_w V_k, \tag{10.5}$$

and

$$\text{TD}(0): \quad \Delta w_t = \alpha(V_{t+1} - V_t)\, \nabla_w V_t. \tag{10.6}$$

Note that here a convention "$0^0 = 1$" is used. In the special case in which TD(1) as defined in Equation (10.5), all past predictions make equal contributions to the weight alterations; this means that all states are equally weighted. Because $\Delta w_t \propto -\nabla_w E_{td} = \nabla_w V_t(z - V_t)$, due to Equation (10.4), Equation (10.5) can be rewritten as follows:

$$\Delta w_t = \alpha(z - V_t)\ \nabla_w V_t. \qquad (10.7)$$

The prediction error is represented as the difference between *final outcome* z and *current prediction* V_t.

Consider the difference between Equation (10.5) and Equation (10.7). Equation (10.7) is actually closer to an **ordinary supervised learning** procedure for the pair current prediction V_t (actual output) and its final consequent z (target output). Using Equation (10.7), Δw_t can be determined only after the whole sequence of actions has been completed and the final outcome z is available; therefore, all state-prediction pairs must be remembered to make this adjustment. In other words, Equation (10.7) cannot be computed incrementally in multiple-step (stage) problems. Recall our previous jackpot journey problem in Section 10.2.1; the signpost model realizes a type of learning similar to that stipulated in Equation (10.7). Past sequences must be remembered to make stone adjustments; that is, the traveler must remember past traces to apply the reward-penalty scheme to each signpost, and all states receive equal reinforcement signals (i.e., one stone). On the other hand, Equation (10.5) offers a way to compute incrementally, which saves memory space required to store past values.

In the other extreme case in which TD(0) is as defined in Equation (10.6), only the most recent prediction affects the alterations. This is closer to a **dynamic programming** (**DP**) procedure discussed in Section 10.4. Application of Equation (10.6) to the jackpot journey problem is discussed in the next subsection.

Sutton [77] introduced a game-playing example, illustrated in Figure 10.4, to clarify the inefficiency of supervised learning methods in comparison with TD methods. Figure 10.4 represents a case in which the trajectory followed from a new state reaches an unusual win via a bad state that has led 90% of the time to a loss and only 10% to a win from past experience. In this case, TD methods reasonably evaluate the new state by adjusting the values of the observed states in a temporally successive manner. Supervised learning methods, on the other hand, associate the new state fully with the observed victory, although the new state is much less promising.

10.3.2 Expected Jackpot

We revisit the jackpot paradise discussed in Section 10.2.1; we show how TD(0) from Equation (10.6) works using the lookup-table perceptron with linearly independent state vectors depicted in Figure 10.5. The TD (perceptron) net has an identity function at the output layer. In this case, we can apply the **linear supervised learning** (**Widrow-Hoff**) rule; we have $V_t = w^T s_t$, and thus $\nabla_w V_t = s_t$. Therefore, Equation (10.6) can be written as follows:

$$\text{Linear TD}(0): \quad \Delta w_t = \alpha(w^T s_{t+1} - w^T s_t)s_t.$$

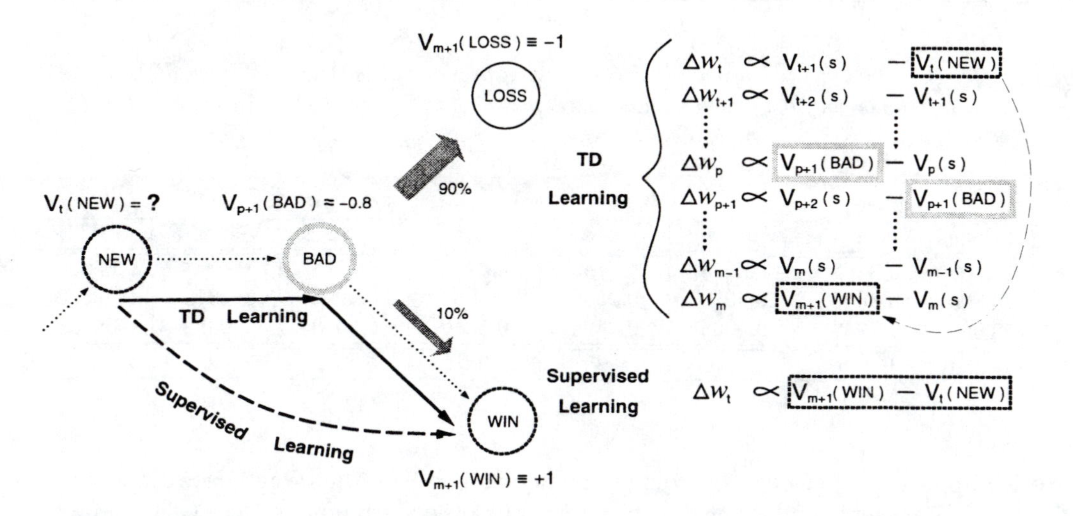

Figure 10.4. *An example (originally introduced by Sutton [77]) where supervised learning methods poorly evaluate a new state, which is reached for the first time, and then follows the trajectory marked by the dotted lines. TD learning methods predict an estimate of a new state,* $V_t(NEW)$*, by considering temporally successive predictions, including* $V_{p+1}(BAD)$*, together with the observed victory (+1), whereas supervised learning methods associate* $V_t(NEW)$ *fully with* $V_{m+1}(WIN)$ $(\equiv +1)$.

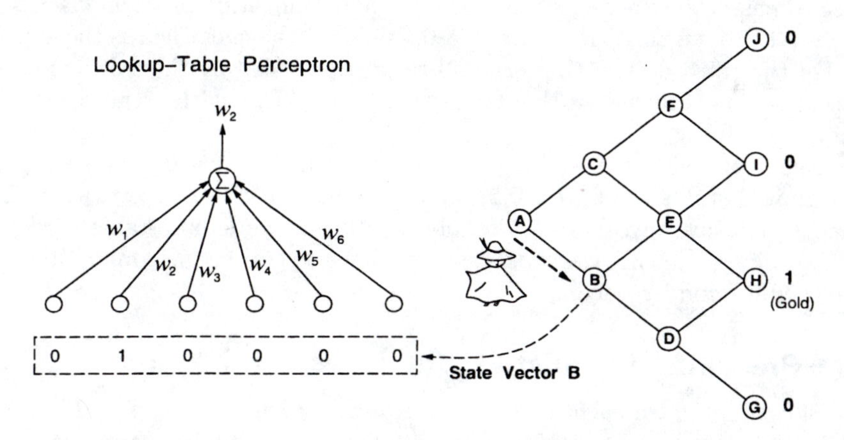

Figure 10.5. *A lookup-table perceptron to approximate expected values in the jackpot journey problem.*

Note that there are six nonterminal vertices in Figure 10.5. The states (i.e., vertices) are expressed as linearly independent unit-basis vectors $[s(i)]$ of length 6; that is, vertex A can be defined as a vector, $(1\ 0\ 0\ 0\ 0\ 0)^T$, and vertex B can be defined as another linearly independent vector, $(0\ 1\ 0\ 0\ 0\ 0)^T$, and so on.

Table 10.1. *Predicted probabilities for six nonterminal vertices at three distinctive learning stages using a lookup-table perceptron with a small learning rate (0.001). RMSE means "root-mean-squared error."*

Epoch	Vertex A	Vertex B	Vertex C	Vertex D	Vertex E	Vertex F	RMSE
10,000	0.3109	0.4248	0.2163	0.4178	0.4718	0.0231	0.05624
20,000	0.3656	0.4922	0.2536	0.5066	0.4995	0.0019	0.00590
40,000	0.3742	0.5008	0.2496	0.5016	0.5033	0.0	0.00155
Target	0.375	0.5	0.25	0.5	0.5	0.0	0

Suppose our journey's outcome is defined as $z = 1$ for the target vertex H where "gold" is present, and defined as $z = 0$ for the other terminal vertices G, I, and J, where no *gold* is present. For this choice of z, the expected value of each vertex is tantamount to the probability of reaching *gold* (vertex H) from that vertex. Sutton discussed a random-walk example [77], in which he introduced a similar probabilistic interpretation.

An agent randomly and equally likely selects either action d, "go diagonally downward," or action u, "go diagonally upward," at each vertex. This probability is not changed based on experience, as it was in the optimization problem discussed in Section 10.2.1. Given this particular "0.5-0.5 policy," the agent learns the expected values. For the third action, the terminal reward provided by the world is used as the outcome value rather than the TD net's output. (This is the realization of the boundary conditions.)

The ideal probabilities of reaching *gold* (i.e., the desired predictions) for each of the nonterminal states are 0.375, 0.5, 0.25, 0.5, 0.5, and 0.0 for vertices A, B, C, D, E, and F, respectively. The results, obtained from the lookup-table perceptron with a small fixed learning rate (0.001), are shown in Table 10.1; many iterations were required for convergence.

10.3.3 Predicting Cumulative Outcomes

Consider a case in which each action in a sequence incurs a cost. For instance, in the simple path network sketched in Figure 10.6, numbers between any two vertices represent the cost of traveling that interval. At each vertex, we need V_t to estimate the remaining cumulative cost rather than the total cost of the sequence. TD methods can be extended to deal with this case [77]. That is, TD methods are not confined to predicting only the final outcome of sequences, but can also be used to estimate quantities that accumulate over sequences.

Let c_{t+1} denote the actual cost incurred between times t and $t+1$. We want V_t to equal the expected value of $z_t = \sum_{k=t}^{m} c_{k+1}$, where m is the number of observation vectors in the sequence.

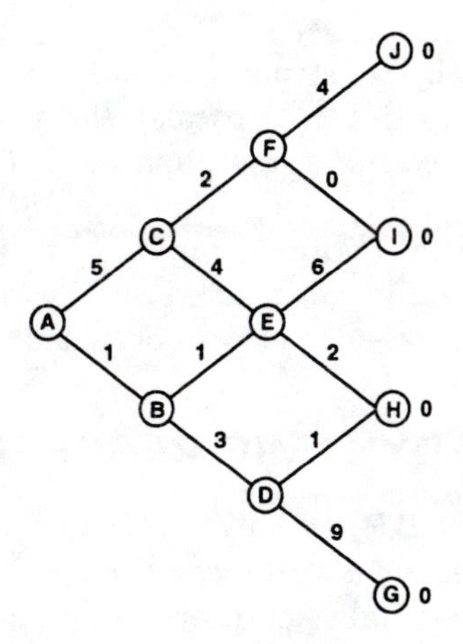

Figure 10.6. *A simple cost path problem.*

As in Equation (10.7), the prediction error can be represented as

$$\begin{aligned}\text{(final outcome)} - \text{(current prediction)} &= z_t - V_t \\ &= \textstyle\sum_{k=t}^{m} (c_{k+1}) - V_t \\ &= \textstyle\sum_{k=t}^{m} (c_{k+1} + V_{k+1} - V_k),\end{aligned}$$

where $V_{m+1} = 0$ (boundary conditions). We can thus derive the update rule for the following cumulative TD(λ):

$$\Delta w_t = \alpha(c_{t+1} + V_{t+1} - V_t) \sum_{k=1}^{t} \lambda^{t-k} \nabla_w V_k. \tag{10.8}$$

We discuss this cost path problem in greater detail in Section 10.7.

In infinite prediction problems, there are no goal states to terminate the sum z_t. To prevent the divergence of this sum, discounting factors can be introduced. Thus, the agent's objective is to minimize the discounted sum of future costs given by

$$\sum_{k=0}^{\infty} \gamma^k c_{t+k+1} = c_{t+1} + \gamma c_{t+2} + \gamma^2 c_{t+3} + \cdots \tag{10.9}$$

In this case, Equation (10.8) will be

$$\Delta w_t = \alpha(c_{t+1} + \gamma V_{t+1} - V_t) \sum_{k=1}^{t} \lambda^{t-k} \nabla_w V_k. \tag{10.10}$$

TD(λ) has been proven to converge only for a linear network with linearly independent state vector inputs [23, 77]. (We described such a linear network in

Section 10.3.2.) However, by constructing a *nonlinear* NN with state vectors not linearly independent that used TD(λ) to play the game of backgammon, Tesauro was able to report the successful game application (TD-gammon) and discussed practical issues concerning TD learning [78, 79, 80, 81].

Bertsekas [17] presented a counterexample to TD learning, in which a representation of the evaluation function constructed by TD(λ) becomes worse as λ changes from 1 to 0; the optimal representation was obtained when $\lambda = 1$.

10.4 THE ART OF DYNAMIC PROGRAMMING

Dynamic programming (DP) *is an optimization procedure that is particularly applicable to problems requiring a sequence of interrelated decisions* [26].

In connection with DP, several researchers have discussed reinforcement learning. Watkins [86, 87] proposed a class of Q-learning (discussed in Section 10.6) that employs "incremental DP," which we shall describe later. We must emphasize the following two aspects of DP with respect to reinforcement learning:

- DP successively approximates optimal evaluation functions by solving *recurrence relations,* instead of conducting searches in state space.
- Backing up state evaluations is a fundamental property of iterative procedures used to solve the recurrence relations.

We clarify these points in the following two subsections.

10.4.1 Formulation of Classical Dynamic Programming

For the purposes of later discussion, we present a brief overview of the basic *art* of DP. For more thorough treatment of classical DP, refer to refs. [11, 26]. Although there are many variations of DP, we focus on major concepts common to all DP procedures, following the definitions of a book entitled *The Art and Theory of Dynamic Programming* by Dreyfus and Law [26].

First, we delineate the **principle of optimality** [11], the backbone of DP, which flows from intuition:

> *An optimal policy has the property that whatever the initial state and initial decision are, the remaining decisions must constitute an optimal policy with regard to the state resulting from the first decision.*

In light of the principle of optimality, the basic procedural insights of (*backward*) DP are as follows [26]:

- Recognize that a given "whole problem" can be solved if the values of the best solutions of certain subproblems can be determined according to the principle of optimality.

- Realize that if one starts at or near the end of the "whole problem," the subproblems are so simple as to have trivial solutions. The principal attraction of (backward) DP resides in such ease.

In the *art* of DP, it is important to define "state space" appropriately and to choose the arguments of the optimal value function (rule of assigning values to various subproblems) so that a real process under consideration has the Markov property [35] and thus the principle of optimality holds. The usual DP formulation consists of determining the four appropriate definitions: (1) an optimal value function, (2) an optimal policy function, (3) a recurrence relation, and (4) boundary conditions.

Recall the cost path problem illustrated in Figure 10.6 in Section 10.3.3. There we sought the expected cost of a given policy using TD methods. (With no action choice to be made, this is not a DP problem.) Now we consider that the objective of the learning agent is to select actions so as to drive the world to a goal state (i.e., terminal vertex) $g \in G \subset S$ at a minimum cumulative cost, where G is a set of goal states and S is the set of all states. We introduce a measure of value [i.e., the optimal value function $V(s)$] as follows:

$$V(s) \equiv \text{the value of the minimum cost of going from a state } s \text{ to a goal } g.$$

The *recurrence relation* given by the principle of optimality uniquely defines these values. That is, the minimal cost of a state s must equal the cost of the best action a, plus the minimal cost of the next state designated by the action a:

$$V(s) = \min_{a \in \text{actions}} \{\text{cost}(s,a) + V(\text{next}(s,a))\}, \quad \forall s \in S - G, \tag{10.11}$$

where $\text{next}(s,a)$ denotes the state subsequent to state s dictated by action a, and $\text{cost}(s,a)$ signifies the incurred cost of the transition following action a. In other words, the evaluation $V(s)$ of state s should be equal to the best of "$\text{cost}(s,a)$ plus the value of the next state $t \equiv \text{next}(s,a)$ that can be reached in one action a." The values of the goal states (or terminal vertices) are defined by

$$V(s) = 0, \quad \forall s \in G. \tag{10.12}$$

Equation (10.12) shows the boundary conditions. (Note that the optimal policy function π is obvious from Equation (10.11); that is, $\pi(s) = a$ such that $V(s) = \min_{a \in \text{actions}} \{\text{cost}(s,a) + V(\text{next}(s,a))\}$.)

For a **stochastic** case of this minimum-cost path problem, we generally assume that when action d is suggested, it is followed with probability P_{d1}; with probability P_{u1} $(= 1 - P_{d1})$, the opposite action u is tried. Likewise, when action u is instructed, it is followed with probability P_{u2}; with probability P_{d2} $(= 1 - P_{u2})$, action d is performed. Now $V(s)$ is the optimal *expected* value function:

$$V(s) \equiv \text{the expected value of the minimum cost of going from a state } s \text{ to a goal } g.$$

Then we can specifically define the recurrence relation in a general form as follows:

$$V(s) = \min \begin{bmatrix} P_{d1}\{\text{cost}(s,d) + V(t_d)\} & + & P_{u1}\{\text{cost}(s,u) + V(t_u)\} \\ P_{u2}\{\text{cost}(s,u) + V(t_u)\} & + & P_{d2}\{\text{cost}(s,d) + V(t_d)\} \end{bmatrix}, \quad (10.13)$$

where $t_d \equiv \text{next}(s,d)$, and $t_u \equiv \text{next}(s,u)$.

TD methods in Section 10.3 approximate the evaluation function for a given policy without knowing the state-transition probabilities and the function determining expected payoff values. Barto et al. [10] mentioned that *the TD learning process is a Monte-Carlo approximation of a successive DP approximation method.* They also discussed that TD methods bear a great resemblance to DP from a learning standpoint [9]. The aforementioned *backward* DP procedure works from the end of a decision task to its beginning. It seems hardly related to animal learning processes, due to this back-to-front processing. Yet Barto et al. showed how TD learning can obtain much the same result as DP by repeated forward passes through a decision task; the computation is incrementally accomplished by moving forward to goal states. This reversed viewpoint can be found in so-called *forward* DP; see also ref. [26].

10.4.2 Incremental Dynamic Programming

Applying classical DP requires knowledge of state-transition probabilities, as demonstrated in Equation (10.13). In the absence of explicit probability information, we can approximate the value function. To improve the value function approximation $\hat{V}(s)$, we can apply incremental DP [6, 16, 73, 75] (also known as approximate DP [91]) given by the following functional equation[3]:

$$\hat{V}(s) \leftarrow \min_{a \in A} \left\{ \text{cost}(s,a) + \hat{V}(\text{next}(s,a)) \right\}, \quad \forall s \in S - G, \quad (10.14)$$

where for $s \in G$, $\hat{V}(s) = V(s) = 0$ (boundary conditions). Sutton [75] claimed that *iterative application of Equation (10.14) to one state after another makes $\hat{V}$ a better and better approximation of V. This operation is, in essence, a one-stage-ahead (or one-ply) search from state s, followed by replacement of the approximate value $\hat{V}(s)$ with its backup value.* This procedure seems convincing, but there is one important caveat that convergence may not be monotonic; that is, the procedure will converge to correct values if done often enough at all possible states. Each value may not necessarily improve at each backup because the approximate value $\hat{V}(s)$ will be worse if a state has a right value and the next state has a wrong one; the value will then temporarily go wrong. In stochastic problems, only the average of many (wrong) values may be right, or may approach correct if the learning rate goes to zero appropriately. In this sense, the performance may improve incrementally while the procedure is being accomplished.

[3] "$a \leftarrow b$" represents "set a equal to b." This notation is also used in Algorithms 10.1 and 10.2.

As the value function approaches the optimal one, the current policy becomes more optimal by degrees. The optimal policy can (it is hoped) be obtained by selecting actions that minimize the right-hand side of Equation (10.11) at each state. That is, the learning process converges when Equation (10.11) holds for all states [73]. Watkins [86, 87] used this key concept to extend TD methods and proposed a class of learning called *Q-learning* (Section 10.6).

When the agent's value predictor is realized by an NN function approximator, the weight parameter update rule simply uses the mismatch between the left-hand side of Equation (10.14) and the right-hand side (i.e., the TD error in recursive equations because such recursive equations can express the desired sequential structural relationships in the given problem). Here we note that *the backing-up property of DP shares basic concepts with TD methods.*

10.5 ADAPTIVE HEURISTIC CRITIC

In this section, we discuss a reinforcement learning strategy based on the so-called **adaptive heuristic critic (AHC)** or **actor-critic** model. It provides a way of *attempting* to find both optimal actions and expected values. The phrase "learning with a critic" was used by Widrow et al. [94] to differentiate it from ordinary supervised learning characterized by the phrase "learning with a teacher." In some situations, where there is less information available on desired outputs, only right or wrong signals will be provided.

The AHC model typically includes two principal components: the critic (evaluation/prediction) module and the action (control) module. The critic generates an estimate of the value (or evaluation) function from state vectors and external reinforcement supplied by the world (or environment) as inputs. That is, the critic plays an important role in predicting the evaluation function. The critic is adaptive because its predictor component is updated using TD methods. The action module *attempts* to learn optimal control or decision-making skills.

10.5.1 Neuron-like Critic

Barto et al. introduced a neuron-like adaptive critic model[4] for a pole-balancing control problem [8]. They discussed their model in conjunction with an earlier study of the Boxes system that Michie and Chambers proposed [54].

The Boxes system was constructed to solve the pole-balancing control problem; its central concept was "task decomposition" whereby state space was partitioned into 162 non-overlapping subspaces called "*boxes*," and different controllers were used in each digitized state space (*box*). No **generalization** was attempted across subspaces (see Section 10.9.1). Like Michie's earlier work with MENACE, discussed in Section 10.2.1, the Boxes model is result-driven—specifically, failure-driven in this case. That is, the model receives no feedback on performance until the pole falls

[4]Concise description of their work can be found in ref. [85]. Barto presented a good overview of the adaptive critic methods in ref. [4].

down (failure). Through many trials, the Boxes system learns to solve the nonlinear control problem; every failure is a stepping stone to success.

Inspired by the Boxes method, Barto et al. constructed an adaptive critic model [8] that consists of ACE (adaptive critic element) and ASE (associative search element) based on an "associative reward-penalty" algorithm [8, 5, 96]; each component is expressed in an ADALINE [95] (ADAptive LINear Element)—what they call a "neuron-like connectionist element." The ASE implements and adjusts the decision policy (or control rules). The ACE learns to provide current evaluations of control decisions by virtue of failure signals. Specifically, the ACE predicts the *internal reinforcement* signal $\hat{r}(t+1)$ (associated with particular input states):

$$\hat{r}(t+1) = r_{t+1} + \gamma V_{t+1} - V_t,$$

where V_t is the current prediction and γ is a discounting factor. This $\hat{r}(t+1)$ corresponds to the TD error discussed in Section 10.3.3 [see Equation (10.10)]. The $\hat{r}(t+1)$ is sent to the ASE to determine a control action.

Performance comparison with the Boxes system pointed out an advantage of Barto et al.'s critic model, "ASE with internal reinforcement supplied by an ACE [8]":

> *The boxes system is restricted in that its design was based on the a priori knowledge that the time until failure was to serve as the evaluation criterion and that the learning process would be divided into distinct trials that would always end with a failure signal. ... The ASE, on the other hand, is capable of working to achieve rewarding events and to avoid punishing events which might occur at any time. It is not exclusively failure-driven, and its operation is specified without reference to the notion of a trial.*

In their adaptive critic model, Barto et al. used 162-component linearly independent state vectors (or standard-unit-basis vectors) as inputs. In other words, this model was equivalent to a big lookup table similar to the lookup-table perceptron shown in Figure 10.5. Thus, this model offered no possibility for generalizing among states [8, 5]. For large state space problems, using lookup tables is impractical, as is visiting all states. Thus, the agent needs to generalize from a limited amount of experience using a compact representation of input state vectors (see Section 10.7.3).

10.5.2 An Adaptive Neural Critic Algorithm

This subsection provides a general description of an AHC algorithm implemented in such parameterized functional forms as NN function approximators (value and action function approximators). The AHC model basically consists of two NNs: the value NN and the action NN. The value NN approximates evaluation functions, mapping states to expected values, whereas the action NN generates a plausible (or legal) action, mapping states to actions [48, 49, 50].

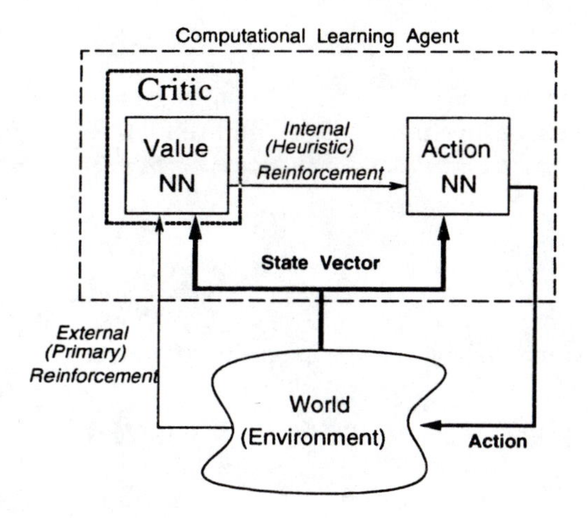

Figure 10.7. *An AHC model that uses neural network function approximators: the value NN and the action NN.*

Figure 10.7 illustrates a block diagram of such an NN-based AHC model. The adaptive critic receives external (primary) reinforcement from the world and transforms it into internal (heuristic) reinforcement.

The weight update rule follows the usual error minimization scheme used in supervised learning by defining the following squared error E_{TD}:

$$E_{\mathrm{TD}} = \frac{1}{2} error^2,$$

with

$$\begin{aligned} error \equiv\ & \text{(value incurred by a selected action)} \\ & + \gamma\text{(expected value of observed successive state produced by value NN)} \\ & - \text{(expected value of current state produced by value NN)}, \end{aligned}$$

where γ is a discount rate [see Equation (10.4)]. Both the action NN and the value NN are trained simultaneously using E_{TD}. Algorithm 10.1 explains an implementation of the AHC concept.

Algorithm 10.1 *Adaptive heuristic critic using neural network function approximators*

1. Observe the current state: $s \leftarrow$ current state s_n.

2. Use the value NN to have $V(s)$: $e \leftarrow V(s)$.

3. Select an action a_n by using the output of the action NN.

4. Execute the action a_n.

5. Observe the successive new state t_n and reinforcement r_n.

6. Use the value NN to compute $V(t_n)$.

7. $E \leftarrow r_n + \gamma\ V(t_n)$.

8. Adjust the value NN by backpropagating the error ($= E - e$).

9. Adjust the action NN according to the error.

□

Recall Equations (10.8) and (10.10); Algorithm 10.1 shows that learning proceeds concurrently with processing; weight coefficients of the action NN and the value NN are adjusted to fit learned values and actions by TD methods.

Algorithm 10.1 demonstrates how to use the output of a value NN to control the output of an action NN by means of backpropagation. To be precise, suppose we want to maximize the expected value (i.e., reward). Each action should be chosen to maximize the sum of all future rewards. Thus given an action a_n, if the value of the next state $V(t_n)$ [or $\gamma\ V(t_n)$] plus the external reinforcement r_n is greater than the value of the current state $V(s_n)$ (i.e., $E - e > 0$), the action looks better than previously expected, and therefore, that action should be reinforced. Conversely, if the reverse inequality holds, the action does not seem better. (This is because we want to maximize the cumulative future rewards.) The action should therefore be inhibited. This procedure demonstrates our intuitive notion presented in the introduction to this chapter. [The intuitive notion can also be seen in Equation (10.14) of the incremental DP discussed in Section 10.4.2.] This is a reward-penalty algorithm. In contrast to it, in the reward-inaction algorithm [5, 59, 96], actions are not updated when the selected action does not look better; this algorithm is based on the concept that before other actions are put into practice, nothing about their effectiveness can be learned from a single experience.

In this way, the action NN and the value NN evolve together in the *attempt* to find an optimal policy. In other words, TD methods are employed to solve *temporal* credit assignment problems, and the backpropagation algorithm is used for *structural* credit assignment problems.

Werbos [36, 88, 89, 90] discussed adaptive critic designs and a relationship between DP and TD methods by using the terms *heuristic dynamic programming* (HDP), *dual heuristic programming* (DHP), *action-dependent heuristic dynamic programming* (AD-HDP), *action-dependent dual heuristic programming* (AD-DHP), and *global dual heuristic programming* (G-DHP). A class of DHPs approximate the *derivatives* of action-value (or value) functions.

10.5.3 Exploration and Action Selection

In the **supervised learning** discussed in Chapter 9, the agent usually acts according to the *gradient vector* provided by a teacher; the teacher implements a reward-penalty scheme to adapt the network's weights. On the other hand, the reinforcement learning agent has no teacher to supply such *directed* information during its learning; it usually receives a *scalar* reinforcement, which is just information about the current evaluation of behavior.

> *The reinforcement learning agent encounters a* **conflict** *between how it has to change behavior in order to obtain directional information, and how the resulting directional information tells it to change its behavior for improvement* [4].

In other words, the following two factors must influence each action selection, and they ordinarily conflict [4, 82, 32];

1. *The desire to acquire more knowledge about actions' consequences to make better selections in the future.*
2. *The desire to use what is already known about the relative merits of the actions.*

The best decision for one is not best necessarily for the other. This dilemma between exploration and exploitation (or conflict between identification and control) is absent in supervised learning.

Usually, a necessary component of any form of reinforcement learning algorithm involves the learner's random search behavior. It would be a cure for the aforementioned dilemma. Such random exploration in the space of possible solutions is important because no direction information toward the right answer is then available. To organize the desired exploratory behavior, stochastic action selection should be considered. The output of the action NN is interpreted as exponents s_i in a **Boltzmann distribution** that yields the probability of an action a_i:

$$\text{Prob}(a_i) = \frac{\exp(\frac{s_i}{T})}{\sum_k \exp(\frac{s_k}{T})}, \tag{10.15}$$

where T is a temperature parameter for an annealing process. Of course, an action favored by the action NN has more chance of being selected. When the cooling-temperature scheme is employed, exploration in the solution space will gradually shrink to favor a deterministic action selection. This sort of stochastic exploration provides a way of estimating directional information to change behavior toward performance enhancement in exploring the environment. Thus, it alleviates the dilemma between exploration and exploitation.

In action NN structural terms, there are several possibilities. For two-action problems such as the jackpot problem in Section 10.2.1 and the cost path problem in Section 10.7.2, there might be three possibilities;

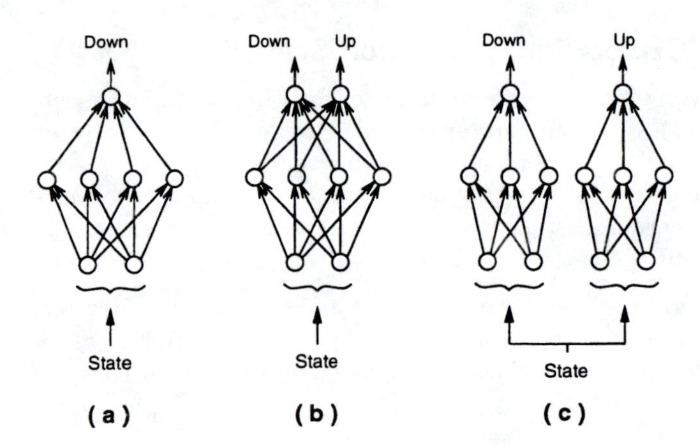

Figure 10.8. *Three possible action NN architectures of the AHC-net for two-action problems such as a cost path problem in Section 10.7.2; "Up" means "go diagonally upward" and "down" denotes "go diagonally downward."*

1. Use one action NN with a single output unit.
2. Use one action NN with two output units, one unit for each action.
3. Use two action NNs with a single output unit; one NN for each action.

These architectures are illustrated in Figure 10.8. For instance, a single-output action NN in Figure 10.8(a) can be applied to a deterministic two-action cost path problem in Section 10.7.2. The action NN produces an output ranging from 0.0 (action u) to 1.0 (action d) [93].

Furthermore, *stochastic units* can be chosen as the output units for action NNs [8, 31, 32, 33]. The stochastic units are usually regarded as the ordinary deterministic threshold units (such as a sigmoidal logistic function) with a random or noisy threshold in a form of $f(x + \text{noise})$.

10.6 Q-LEARNING

Q-learning *is a form of model-free reinforcement learning* [87]. In this section, we investigate "one-step Q-learning," a simple class of Watkins's Q-learning [86, 87]. Throughout the chapter, we use the term *Q-learning.*

10.6.1 Basic Concept

Q-learning is a simple way of solving with incomplete information Markovian action problems based on the **action-value function Q** that *maps state-action pairs to expected returns*. The idea of assigning values to state-action pairs can be seen in DP; Watkins [86] called Q-learning the "incremental version of DP." Q-learning successively improves its evaluations of particular actions at particular states, just as incremental DP, discussed in Section 10.4.2, improves its evaluations of particular states.

Learning proceeds as with TD methods; an agent tries an action at a particular state and evaluates its consequence in terms of the immediate reward or penalty it receives from the world and its estimate of the value of the state resulting from the taken action. The aim of the agent is not merely to maximize its immediate reward in the current state, but to maximize the cumulative reward it receives over some period of future time.

AHC reinforcement learning architecture requires two fundamental memory buffers: one for the evaluation function and one for the policy. On the other hand, Q-learning maintains only one: a pair of state s and action a (i.e., an estimate Q-value of taking a in s). Instead, Q-learning requires additional complexity in determining the policy from the Q-values, as we show in this discussion.

The objective in Q-learning is to estimate the values of an optimal policy. The value of a state can be defined as the value of the state's best state-action pair:

$$V(s) = \max_a Q(s, a). \tag{10.16}$$

The optimal policy is determined according to the policy function π:

$$\pi(s) = a \text{ such that } V(s) = Q(s, a) = \max_{b \in \text{actions}} Q(s, b). \tag{10.17}$$

In one-step Q-learning, only the action-value function Q of the most recent state-action pair is updated after a one-step delay; the update rule for Q-values at the nth stage, where s_n is the current state and a_n is the selected action, is based on TD methods:

$$Q_n(s, a) = \begin{cases} Q_{n-1}(s, a) + \eta_n[r_n + \gamma V_{n-1}(t_n) - Q_{n-1}(s, a)] & \text{if } s = s_n \text{ and } a = a_n, \\ Q_{n-1}(s, a) & \text{otherwise,} \end{cases} \tag{10.18}$$

where

$$V_{n-1}(t) = \max_{b \in \text{actions}} \{Q_{n-1}(t, b)\}.$$

In the early stages of learning, the Q-values may not accurately reflect the policy they implicitly define. By trying all actions in all states repeatedly, the agent learns which are best overall, judged by the long-term discounted reward. In other words, the "current $Q(s, a)$" is the "expected Q-value of taking action a in state s, and then using optimal actions in all future states."

10.6.2 Implementation

Now we consider implementing a Q-learning NN (Q-net). Algorithm 10.2 presents a training procedure for such a Q-net, as illustrated in Figure 10.9.

Algorithm 10.2 *One-step Q-learning using neural network function approximators*

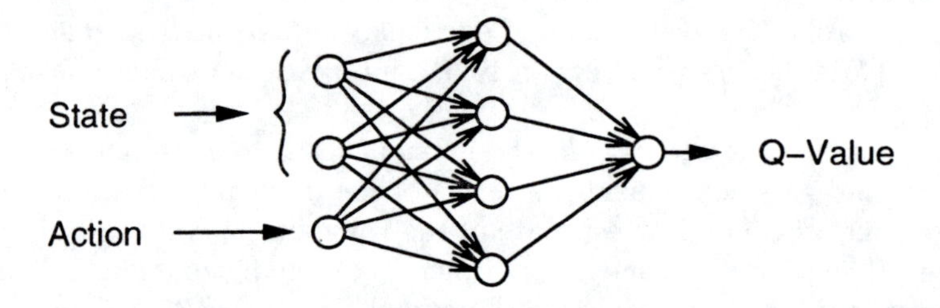

Figure 10.9. *A Q-net architecture.*

1. Observe the current state s_n.
2. Select an action a_n by a stochastic procedure.
3. For the selected action a_n, use the Q-net to compute U_a: $U_a \leftarrow Q_{n-1}(s_n, a_n)$.
4. Execute the action a_n.
5. Observe the resulting new state t_n and reinforcement r_n: $t \leftarrow t_n$.
6. Use the Q-net to compute $Q_{n-1}(t, b)$, $b \in$ actions.
7. $u \leftarrow r_n + \max_{b\in \text{actions}} Q_{n-1}(t, b)$.
8. Adjust the Q-net by backpropagating the one-step error:

$$\Delta U = \begin{cases} u - U_a & \text{if } s = s_n \text{ and } a = a_n, \\ 0 & \text{otherwise.} \end{cases}$$

□

A selected action usually matches $\pi(s)$ as defined in Equation (10.17), but occasionally arrives at an alternative; for instance, the policy is stochastically implemented according to a Boltzmann distribution as in Equation (10.15):

$$\text{Prob}(a_i) = \frac{\exp[\frac{Q(s,a_i)}{T}]}{\sum_k \exp[\frac{Q(s,a_k)}{T}]}, \tag{10.19}$$

where T is the temperature parameter for an annealing process.

Dayan [23] showed that TD(0) corresponds to a special case of Q-learning when there is just one admissible action for each state. Lin discussed the comparison between AHC-nets and Q-nets [48, 50]. Chapman and Kaelbling discussed the "generalization" problem in the Q-learning framework [21]; they described the G algorithm, which is based on recursive splitting of the state space according to statistical measures of differences in reinforcements received. It incrementally builds up a tree-structured table of Q-values.

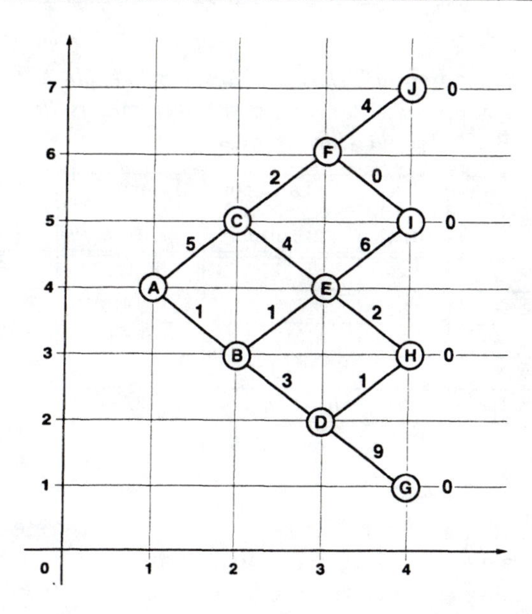

Figure 10.10. *The configuration of a cost path problem in a coordinate system.*

Table 10.2. *Five experimental representations of input states.*

Vertex	Linearly independent 6-D unit-basis vectors	3-D vectors	2-D coordinates (a)	2-D coordinates (b)	1-D scalars
A	(1, 0, 0, 0, 0, 0)	(0, 0, 1)	(1, 4)	(0.01, 0.04)	1
B	(0, 1, 0, 0, 0, 0)	(0, 1, 0)	(2, 3)	(0.02, 0.03)	2
C	(0, 0, 1, 0, 0, 0)	(0, 1, 1)	(2, 5)	(0.02, 0.05)	3
D	(0, 0, 0, 1, 0, 0)	(1, 0, 0)	(3, 2)	(0.03, 0.02)	4
E	(0, 0, 0, 0, 1, 0)	(1, 0, 1)	(3, 4)	(0.03, 0.04)	5
F	(0, 0, 0, 0, 0, 1)	(1, 1, 0)	(3, 6)	(0.03, 0.06)	6

This section concludes with one notice of a convergence theorem: Q-learning based on look-up tables has been proven to converge to optimal values and decisions, whereas the AHC-learning illustrated in Figure 10.7 has *not* been. Watkins and Dayan [87] specifically showed that Q-learning converges to the optimum action-values with probability 1 as long as all actions are repeatedly sampled in all states with discrete action-values, and the learning rate goes to zero appropriately.

10.7 A COST PATH PROBLEM

This section considers a simple cost path problem configured in the triangular three-

Table 10.3. *Five experimental TD-net structures. The fifth TD-net has bias units, whereas the other four do not. Parameter number means the number of adjustable weights in a network. For input state representations, see Table 10.2.*

TD-net	Structure	Bias units	Parameter number	Learning rate α	Input state representations
1	6×1	$\times$	6	0.0007	6-D unit-basis
2	$3 \times 5 \times 1$	$\times$	20	0.00009	3-D vectors
3	$2 \times 4 \times 1$	$\times$	12	0.00007	2-D coordinates (a)
4	$2 \times 4 \times 1$	$\times$	12	0.007	2-D coordinates (b)
5	$1 \times 3 \times 1$	$\circ$	10	0.0000009	1-D scalars

stage path network illustrated in Figure 10.10. It is a discrete-action environment, and each path incurs a cost. We have already discussed the TD formulation of value functions for predicting cumulative outcomes in Section 10.3.3. We first construct NNs based on TD methods (TD-nets) to predict expected costs. We then consider **actions** required to find an optimal minimum-cost path, and investigate both AHC-nets that execute an adaptive heuristic critic concept, and Q-nets that implement Q-learning. Of course, the state-transition probabilities and the incurred-cost data are unknown to the NNs (TD-net, AHC-net, and Q-net).

We examine a lookup-table perceptron with linearly independent 6-D input state vectors, as discussed in Section 10.3.2. For comparison purposes, we also examine NNs with hidden layers, employing input state representations different from those of the linearly independent 6-D vectors in an attempt to draw *generalization capabilities*. Five experimental representations of input states are presented in Table 10.2.

10.7.1 Expected Cost Path Problem by TD methods

In this subsection, we assume that the probability of action d and that of action u are the same, 0.5. The agent's objective is to predict the expected cost of the particular "0.5-0.5 policy" at each vertex using TD-nets. The TD-nets were trained using Equation (10.8) defined in Section 10.3.3. Table 10.3 presents five experimental architectures of the TD-nets in accordance with the five representations of input states in Table 10.2.

For the purposes of testing NN approximation capabilities, this trivial cost path problem has the advantage of enabling us to compute the correct answer readily by

$$V(s) = 0.5\{\text{cost}(s, d) + V(t_d)\} + 0.5\{\text{cost}(s, u) + V(t_u)\},$$

where s, d, u, t_d, t_u, and cost(.) are as defined in Section 10.4.1. (This is not a DP problem because no action choice is to be made.) The expected costs starting at

Table 10.4. *Expected costs for the six nonterminal vertices using five different TD-nets. They were trained by the TD(0.3), in which the recency parameter λ is 0.3. RMSE signifies "root mean squared error."*

TD-net	Vertex A	Vertex B	Vertex C	Vertex D	Vertex E	Vertex F	Required epoch	RMSE
1	9.239	6.508	5.990	5.004	3.965	2.018	29,200	0.018
2	9.301	6.774	5.827	4.578	5.024	1.791	1,865,000	0.479
3	8.939	4.881	7.451	4.224	4.515	5.454	160,000	1.714
4	9.001	6.434	6.523	4.041	4.059	2.975	229,000	0.607
5	9.343	7.987	6.386	4.806	3.535	2.681	9,850,000	0.717
Target	9.25	6.5	6.0	5.0	4.0	2.0	—	0

each vertex can be computed by the preceding equation to be

$$V(F) = 2, \;\; V(E) = 4, \;\; V(D) = 5, \;\; V(C) = 6, \;\; V(B) = 6.5, \;\; V(A) = 9.25.$$

These are the desired predictions.

Table 10.4 shows the results obtained by the five TD-nets. It was observed that when the recency parameter λ was about 0.3, the performance was better than the two extreme cases of TD(0) and TD(1); this finding coincides with results from other problems presented by Sutton [77].

10.7.2 Finding an Optimal Path in a Deterministic Minimum Cost Path Problem

In the following subsections, we consider *actions*, using AHC-nets in Figure 10.7 and Q-nets in Figure 10.9. The agent's goal is to learn the optimal sequence of three actions that minimizes the total cost (i.e., the sum of the costs associated with each of the three steps). A glance at Figure 10.10 shows that the optimal path can readily be found:

$$\text{Optimal path:} \quad \text{vertex A} \;\stackrel{d}{\Longrightarrow}\; \text{vertex B} \;\stackrel{u}{\Longrightarrow}\; \text{vertex E} \;\stackrel{d}{\Longrightarrow}\; \text{vertex H.}$$

We assume that our computational learning agent always starts at vertex A. The agent chooses either action d or action u at each vertex. The environment accordingly informs the agent of the cost associated with the action taken. This process is repeated three times. For the third action, the terminal cost provided by the world (environment) is used as the value rather than the NN's output. This is the realization of the boundary conditions.

AHC-net Simulation

Table 10.5. *Four experimental AHC-nets trained by TD(0). Learning rates denoted by α were chosen by a trial and error process. All action NNs have bias units.*

AHC-net structures		Value NN	Action NN with bias units	State representations (in Table 10.2)
W	size	6×1	$6 \times 4 \times 1$	6-D unit basis
	α	0.007	0.15	
X	size	$3 \times 2 \times 1$	$3 \times 6 \times 1$	3-D
	α	0.00009	0.009	
Y	size	$2 \times 4 \times 1$	$2 \times 4 \times 1$	2-D coordinates (a)
	α	0.0007	0.00001	
Z	size	$2 \times 4 \times 1$	$2 \times 4 \times 1$	2-D coordinates (b)
	α	0.007	0.0007	

Table 10.6. *Results obtained by four AHC-nets whose architectures W, X, Y, and Z, are presented in Table 10.5. P_{down} denotes "probability of action d."*

AHC nets	Expected values	Vertex A	Vertex B	Vertex C	Vertex D	Vertex E	Vertex F	Required epoch
W	P_{down}	0.992	0.034	(0.023)	(0.009)	0.999	(0.998)	300,000
	Costs	4.079	3.019	(2.122)	(1.047)	2.024	(0.014)	
X	P_{down}	0.994	0.0104	(0.006)	(0.050)	0.998	(0.952)	500,000
	Costs	4.030	3.016	(5.716)	(1.013)	2.010	(1.794)	
Y	P_{down}	0.927	0.351	(0.964)	(0.024)	0.759	(0.974)	15,124,500
	Costs	4.153	2.956	(2.614)	(0.710)	1.631	(2.684)	
Z	P_{down}	0.956	0.070	(0.999)	(0.0003)	0.945	(0.999)	8,350,000
	Costs	4.778	3.302	(5.513)	(1.335)	2.117	(3.918)	
Target values	P_{down}	1.0	0.0	(0.0)	(0.0)	1.0	(1.0)	–
	Costs	4.0	3.0	(2.0)	(1.0)	2.0	(0.0)	

We implemented AHC-nets consisting of two feedforward NNs (or multilayer perceptrons): the value NN and the action NN, as illustrated in Figure 10.7. Table 10.5 presents four experimental architectures of the AHC-nets—W, X, Y, and Z—as well as their representations of input states. They were trained by TD(0). In addition, all action NNs had the architecture (a) in Figure 10.8 with *bias units*, and they were trained in conjunction with the well-known *backpropagation learning rule with a momentum term* (0.8). Notice that in structure W of the AHC-net, the value NN (6×1) is equivalent to the first TD-net in Table 10.3 and the lookup-table

Table 10.7. *Two Q-nets' architectures and their resulting convergence performance. These Q-nets have no bias units. Their learning rates were optimized by a process of trial and error.*

Q-net structure (size)	Input state representations	Parameter number	Learning rate	Required epoch
Look-up table method (two "6 × 1" perceptrons)	6-D unit basis	12	0.3	400
A Q-net with one hidden layer (3 × 5 × 1)	2-D coordinates (a) with one "action"	20	0.03	209,500

perceptron we tested in Section 10.3.2.

Algorithm 10.1 in Section 10.5.2 explains implementation of the AHC concept in an NN framework; If the value of the next state $V(t_n)$ plus the external cost r_n was *smaller* than the value of the current state $V(s_n)$ (i.e., if the TD error was *negative*), the action looked *better* because we wanted to *minimize* the cumulative costs. Conversely, if the TD error was positive, the action was inhibited. This was intended as a demonstration of the intuitive notion discussed in Section 10.5.2. Through repeated passes, the AHC-nets learned the optimal actions as well as the expected minimum costs.

Table 10.6 shows the results obtained by the four AHC-nets: W, X, Y, and Z. As seen in the results from (Y), the AHC-net Y did not strongly inform us of the optimal action choices at vertices B and E. Of course, policy (i.e., a *complete* mapping states to actions) is not very important for this *deterministic* problem, but two AHC-nets Y and Z failed to yield the optimal decision at vertex C, whereas AHC-net X optimized the action choice. Additionally, the results from (X) show that the expected costs of the two vertices C and F were not close to the desired expected ones although the action choices were optimized. This is because little learning could take place at those vertices that were rarely visited as decisions became (we hope) more optimal.

Q-learning Simulation

In this subsection, we apply Q-learning to solving the minimum cost path problem. To compare with the look-up table method, we tested the Q-net that has one hidden layer with state-action pairs as inputs illustrated in Figure 10.9. More specifically, the Q-net had three input units: two for the 2-D coordinate state representations (a) in Table 10.2, and one for selected actions; that is, we explicitly represented input vectors by using an extra input unit for an action whereby action d corresponded to -1 and action u to $+1$. For instance, a state of vertex A with action d was expressed in a vector, $(\ 1,\ 4, -1\ )^T$. The experimental Q-nets have no bias units.

Table 10.8. *The desired Q-values at the six non-terminal vertices. Actions u and d denote "go diagonally upward" and "go diagonally downward," respectively.*

Action	Vertex A	Vertex B	Vertex C	Vertex D	Vertex E	Vertex F
Action u	7.0	3.0	2.0	1.0	6.0	4.0
Action d	4.0	4.0	6.0	9.0	2.0	0.0

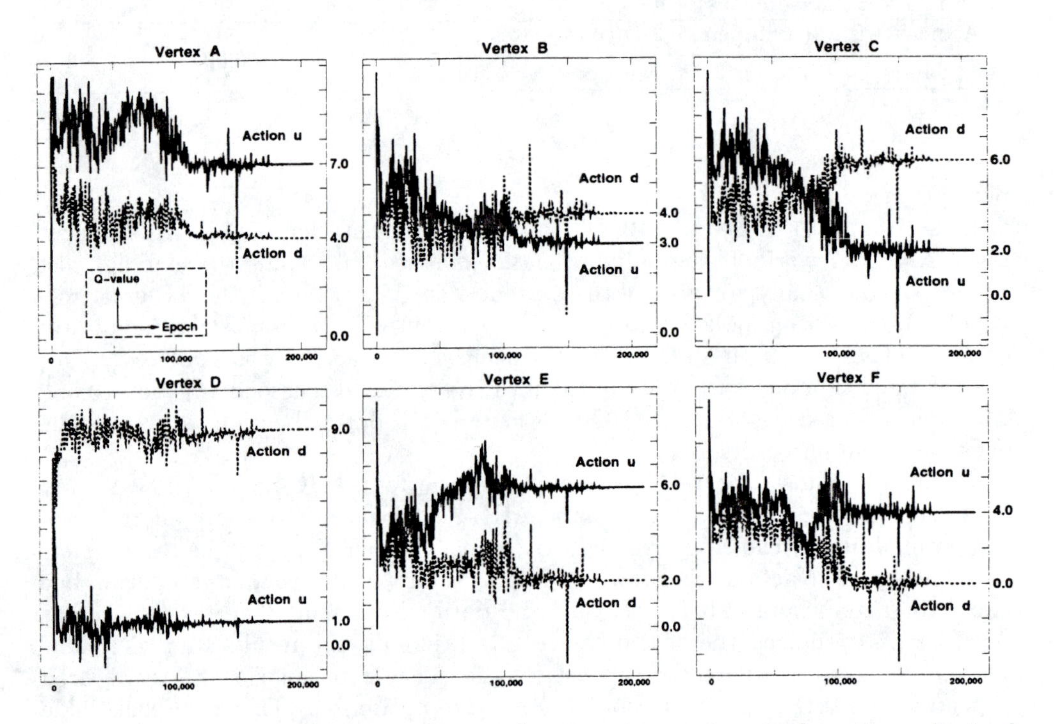

Figure 10.11. *The Q-learning curves for the six non-terminal vertices obtained using a Q-net with one hidden layer (3 × 5 × 1). The horizontal axis represents the epoch number and the vertical axis shows the Q-value.*

Notice that the look-up table method is equivalent to a Q-net composed of two look-up table perceptrons (with linearly independent 6-D input state vectors)—that is, one perceptron for action d and the other for action u.

Algorithm 10.2 in Section 10.6 explains implementation of the Q-net. The path leading to a state with the minimal Q-value had to be chosen for the next action because we wanted to minimize the expected costs.

Table 10.7 presents the convergence performance; the Q-net with one hidden layer required considerably more iterations to converge to the desired Q-values in

Table 10.8, than the look-up table Q-net. Their learning rates were optimized by a process of trial and error. Figure 10.11 shows the Q-learning curves of the six non-terminal vertices obtained from the Q-net with one hidden layer ($3 \times 5 \times 1$).

10.7.3 State Representations for Generalization

We described a trivial minimum cost path problem, implementing TD-learning, an AHC concept, and Q-learning on the basis of NN function approximators. Tables 10.4, 10.6, and 10.7 show that the networks converged quickly and provided outputs close to the desired values when they were set up as lookup-table perceptrons with linearly independent 6-D input state vectors. As we discussed the weakness of the Boxes model in Section 10.5.1, such *tabular representation* for the environment is questionable when it is applied to more realistic environments that entail many possible states. Thus, we have explored more compact representations in pursuit of generalization over input states. However, even for such a trivial problem, when those networks had hidden layers and different input state representations were introduced, they clearly took many more iterations to converge.

Notably, 2-D coordinate state representations (a) in Table 10.2 caused the relatively poor performance of the third TD-net in Table 10.3 and the AHC-net Y in Table 10.5, even though we varied their parameter setups repeatedly. In many of our trial setups, it was observed that the AHC-net Y had converged to *poor* action choices at vertex B, and had attained no precision in matching the expected outputs. That is, the AHC-net Y frequently failed to inform us of the optimal path (i.e., vertex A $\rightarrow$ vertex B $\rightarrow$ vertex E $\rightarrow$ vertex H). The expected values themselves were unrelated to such coordinate state representations. We actually explored other ad hoc 2-D representations of input states. One of the ad hoc 2-D representations was (b) in Table 10.2, wherein coordinates were simply divided by 100. Similarly, coordinates were normalized, or they were mapped onto unit circle inputs, and so on. Yet almost no significant difference in performance was obtained. The difficulties encountered in this small three-stage problem suggest that *state representations* determine the architecture of the networks, and therefore they are crucial to overall performance; the converged values depend strongly on input vector representations. We thus need to choose input state representations carefully as well as parameter setups (e.g., learning rates and number of neurons).

Other issues on state representations can be found in the literature. An autonomous mobile robot should be able to deal with incorrect state descriptions generated by poor sensors [3, 51]. Whitehead and Ballard discussed a similar problem in world state representations [92]; the mapping from world states to the agent's state representations can be many-to-many in a complex system environment. A single state representation may represent multiple world states. Whitehead and Ballard called this overlapping between the world and the agent's state representation "perceptual aliasing."

10.8 WORLD MODELING*

The agent is assumed to be able to observe states, actions, and reinforcement signals. It can therefore model the mapping from *actions* to *reinforcement signals*. More precisely, the world model is intended to model the input-output behavior of the dynamic world (or environment); given a state and an action, it is supposed to predict the received resultant reinforcement and next state [48, 74].

10.8.1 Model-free and Model-based Learning*

From the standpoint of world modeling, reinforcement learning is roughly classified into the following two types:

1. Learning optimal actions and values by sampling the world without attempting to learn a world model

2. Learning a world model by sampling the world, and then basing optimal actions and values on the learned world model.

Method 1 is **model-free** or **direct** learning. Method 2 is **model-based** or **indirect** learning. In conformity with control engineering terminology, this indirect method corresponds to a **system identification** procedure to form a world model [7, 25]. Method 2 describes a sequential training strategy whereby the world model is trained first and then frozen. Alternatively, it can be trained simultaneously with action and value NNs [33, 69].

Classical DP, discussed in Section 10.4.1, is categorized as model-based learning because it uses world models, such as a transition model and a cost (or reward) model. In contrast, model-free learning of optimal actions and values can be viewed as *modern DP* (like incremental DP, discussed in Section 10.4.2), because the scope of DP is basically characterized by seeking the optimal value function without concern for what the final solution value is and by using a backup property of relating a value at a point to values at the next points. This is a modern interpretation of DP in the spirit of model-free learning.

In light of our cost path problem, two agents based on methods 1 and 2 can be described more specifically as follows:

- The *model-free learning agent* attempts to construct policy and evaluation functions directly with no ability to predict transitions or immediate costs resulting from its performed actions, and with no memory to remember more than one past observation explicitly.

- The *model-based learning agent* either knows all relevant probabilistic information about transitions and costs or estimates this information through observation to determine decisions and values by a suitable version of classical dynamic programming or a similar optimization technique.

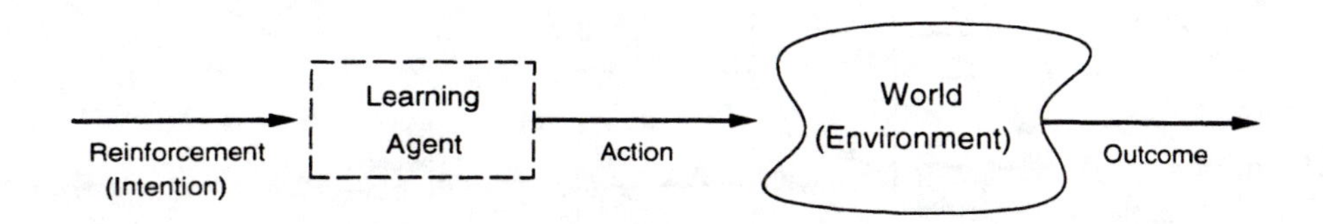

Figure 10.12. *Distal supervised learning; an available target value is the distal outcome but not the proximal action.*

10.8.2 Distal Teacher*

In the framework of *supervised learning*, Werbos discussed a model-based approach that approximated the dynamics of the real world and showed that it augmented the capabilities of *supervised learning* [89]. Similarly, Jordan and Rumelhart demonstrated how **distal supervised learning** algorithms can be applied to an unknown dynamic environment that intervenes between actions and desired outcomes [41]. From the agent, the outcomes can be viewed as "distal" desired values because the agent converts reinforcement signals (called "intentions") into actions, and then the environment transforms the actions into final outcomes; hence this is called "distal supervised learning," as illustrated in Figure 10.12. The learning agent forms a predictive internal model, called a "forward model," by exploring the outcomes associated with particular choices of actions. The forward model outputs a predicted reinforcement signal based on the state and the action; that is, it predicts the consequence of a given action in the context of a given state vector [40, 41]. In biological contrast, Albus [3] mentioned that "any creature with a certain sort of memory can hypothesize an action and receive a mental image of the results of that action before it is performed."

10.8.3 Learning Speed*

In general, reinforcement learning based solely on TD methods is a slow process [48] (the direct/model-free case). To learn a world model and practice with the model can be useful in speeding up the learning process (the indirect/model-based case).

Direct methods can be used as components of model-based methods [9]. Sutton's Dyna architecture [73] implemented one way of blending direct and indirect methods that retains many of the advantages of each approach [7]; Dyna employed the incremental computational DP steps discussed in Section 10.4.2, sometimes taking the actions with the real world and sometimes taking them with the world model based on "relaxation planning" [73, 75]. It has been reported [73, 74] that the use of an internal model can dramatically speed up trial-and-error learning processes, and that planning a sequence of actions to achieve some goal can be done even with incomplete, changing, and often incorrect world models.

Millan and Torras applied a model-based approach to a robot path-finding problem [25]. Mitchell and Thrun [57] tested neural networks with **explanation-based**

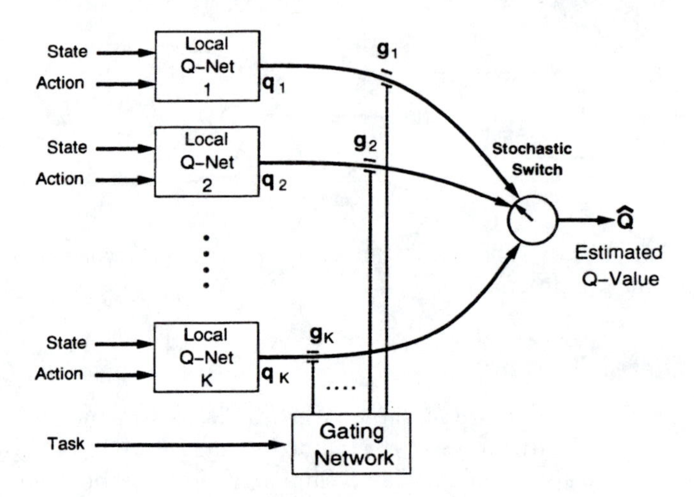

Figure 10.13. *A modular Q-network architecture that can be identical to the CQ-learning model proposed by Singh* [71, 72].

learning [24, 56] for robot control in a discrete action environment, and suggested an advantage over Sutton's Dyna and Jordan/Rumelhart's distal teacher method. Lin compared the performance differences between model-based methods and model-free methods, and implemented modified "relaxation planning" algorithms inspired by Dyna [48, 50]. Lin also noticed that a sufficiently good world model is not easy to obtain [48].

10.9 OTHER NETWORK CONFIGURATIONS*

In Section 10.7, reinforcement learning networks were expressed in the simple form of feedforward (multilayer) perceptrons. In this section, we describe two other important network configurations for realizing reinforcement learning: *modular networks* and *recurrent networks*.

10.9.1 Divide-and-Conquer Methodology*

With respect to continuous state space, the usual practice is to decompose an entire given state space into subsets and apply different evaluation functions under different conditions. The Boxes model in Section 10.5.1 was an example of this divide-and-conquer technique. When such decomposition is used, a form of generalization can be achieved by employing an averaging process over neighboring subsets [8]. Parametric approximators and distributed representations such as neural networks can provide generalization abilities to evaluate states never visited in past experience, as discussed in Section 10.7.3. In this perspective, modular and hierarchical network architectures may be applied to developing more sophisticated

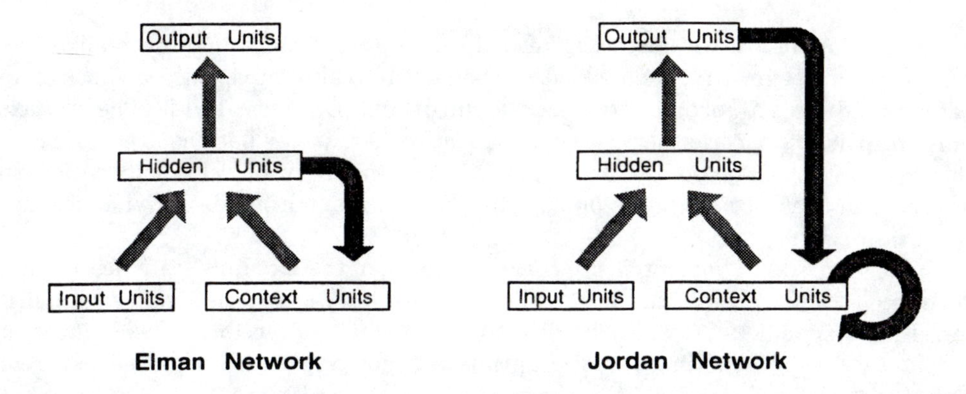

Figure 10.14. *Neural network architectures with context units: an Elman network (left), and a Jordan network (right). Darker-shaded arrows have fixed weights of 1.0.*

reinforcement learning systems, because such network architectures are closely related to the task decomposition concept, as described in Section 9.6. In other words, splitting up a given task may reduce the agent's average learning load to the point of being able to generalize to produce similar actions in similar states.

Mahadevan and Connell [51] employed a switching mechanism in a subsumption architecture so that a Q-learning robot could perform new behaviors in a previously unknown environment. The whole reinforcement learning system had a hierarchical behavior-based structure. Likewise, Singh [71, 72] constructed a CQ-learning (compositional Q-learning) architecture based on the modular network of Jacob et al. [37, 38]. Singh's model learns multiple compositionally structured sequential tasks, whereas Jacobs's model learns multiple non-sequential tasks in accordance with ordinary supervised learning. Figure 10.13 illustrates a schematic diagram that can be identical to Singh's proposed CQ-learning architecture. Several Q-nets (local experts) were mediated by a stochastic switch connected to a gating network.

Radial basis function networks (RBFNs) [45] and fuzzy systems [13, 12, 47] are also discussed in a similar way because of their local generalization abilities. (See also Section 18.2 in Chapter 18.)

10.9.2 Recurrent Networks*

This subsection hints at a conceptually important tie between reinforcement learning and recurrent networks. The recurrence makes it possible for the networks to process sequential inputs. Hence, the recurrent networks may discover an intrinsic temporally successive structure of a given task.

In any environment, the outcome at any given time may be affected arbitrarily by prior actions. Both reinforcements and states may depend arbitrarily on the past history of the agent's outputs. When we use NN function approximators, their outputs must depend on both the current state and on some internal state that, in turn, depends on the historical record. This requirement may be realized by NNs with **context units** embedded in their own architectures, such as an Elman network [28] and a Jordan network [39], illustrated in Figure 10.14. The context units **implicitly** encode the history of the entire past, as far back as it goes. Thus, this type of NN may be able to deal with the sensitivity of value and action to whatever in the prior history is potentially relevant to environmental dynamics and reinforcement.

In Section 10.4.1, we emphasize the importance of the **explicit** state description adequate to allow a DP solution based on the *principle of optimality*. By comparison, Elman-type networks may be able to define a state description *implicitly* and *automatically*, by including the appropriate amount of past history so that a real process under consideration is Markovian.

Schmidhuber discussed recurrent network models for reinforcement learning [69, 70]. Jordan and Rumelhart showed recurrent networks with *forward models* [41].

10.10 REINFORCEMENT LEARNING BY EVOLUTIONARY COMPUTATION*

10.10.1 Bucket Brigade*

Classifier systems are message-passing production systems that learn temporal sequences of string rules (called **classifiers**) through credit assignment based on the **bucket brigade** algorithm and on GA-based rule discovery. The bucket brigade is an algorithm that adjusts the strengths of the classifiers and determines which action should be taken. The alteration equation of classifiers' strengths is not identical but is similar to the Q-value update rule in Equation (10.18) and the TD formula. Rule discovery is done by the GA, and thus it is stochastic in nature. Moreover, the GA forms plausible new classifiers through genetic operations. Detailed mechanisms for classifier systems are discussed in refs [19, 29, 30, 34].

Sutton specified a difference between the TD methods and the bucket brigade [77]; the bucket brigade assigns credit based on rules that activate other rules, whereas the TD methods assign credit based solely on temporal succession. The bucket brigade thus combines both temporal and structural credit assignment in a single rather arbitrary mechanism, although the mechanism may not necessarily correctly solve optimization problems.

10.10.2 Genetic Reinforcers*

Several researchers employed a **genetic algorithm (GA)** in reinforcement learning. Odetayo and McGregor [61] applied GA-based reinforcement learning to the pole-balancing control problem discussed in Section 10.5.1. As with the Boxes system approach, they discretized the state space into 54 regions; each region contained a production rule to specify an action (push left or right). A chromosome consisted of 54 rules each of which represented either "1" (push left) or "0" (push right). The chromosomes were rated for effectiveness in balancing the pole and were subjected to the usual genetic operations: reproduction, crossover, and mutation. It was reported that the GA-based reinforcement learning required fewer iterations than the Boxes, AHC, and CART algorithms.

Whitley et al. [93] investigated the same control task. They used genetic hill climbing to train an NN that produced action probabilities, ranging from 0.0 (push left) to 1.0 (push right). (For general treatment of combinations of GAs and NNs, good surveys are provided in refs. [68] and [98].) Fitness was determined by the amount of time the pole stayed balanced. Whitley et al. reported that the genetic reinforcement learning produced results comparable to AHC.

Ackley and Littman [2] introduced **evolutionary reinforcement learning** (**ERL**), which combines genetic evolution with NN learning and an artificial life "ecosystem." It is based on a *hypothesis* that *evolution and learning progress synergistically.* More specifically, initial weights of the action and value NNs are specified genetically; the weights of the value NN are evolved by a GA, and the value NN outputs are then used to train an action NN by what Ackley and Littman called *complementary reinforcement backpropagation* (CRBP) [1, 2]. CRBP, based on a simple TD algorithm, embodies a heuristic rule that the desired output on *negative* reinforcement is the *complement* of the output generated by the action NN. Their ERL can be viewed as an implementation of a *genetic AHC* concept.

Unemi et al. [83] applied genetic Q-learning to Dyna's navigation task [73]; Q-values as well as other parameters (learning rate and discounting rate) were encoded as binary strings.

10.10.3 Immune Modeling*

Immune models are inspired by biological immune system mechanisms, in which antibodies fight against antigens (intruders). Varela et al. have proposed immune networks (INs) [84] that act mainly on their own rather than in the presence of antigens. (Perelson and Forrest discussed another genetic immune system in ref. [62].) Bersini and Varela formulated an immune recruitment mechanism in conjunction with GAs, calling the result "GIRM: Genetic Immune Recruitment Mechanism" [14, 15]. They also incorporated transitional proximity Q-learning [20] into the recruitment mechanism.

10.11 SUMMARY

This chapter has covered a broad range of reinforcement learning techniques, presenting their fundamental concepts. Departing from the jackpot journey, we have pointed out a diversity of available structures and reviewed their learning operations, including several evolutionary computational models.

As we have discussed, the dynamic programming (DP) notion basically encapsulates the characteristics of reinforcement learning techniques; DP is central to many techniques of reinforcement learning. In particular, emphasis is being placed on features of **totally model-free learning** without recording past trials explicitly, as we have seen in Q-learning. In behavioral ecology, it has been suggested that DP-based computations may be performed by animals for calculating their optimal behavioral policies [43, 52]. Interestingly enough, Montague et al. have reinforced this suggestion; they have constructed an NN model to simulate the behavior of a foraging honeybee in accordance with reinforcement learning [58].

In Section 10.7, we have provided a simulation example for testing TD-learning, an AHC concept, and Q-learning on the basis of NN function approximators. When the networks were designed to possess hidden layers to draw generalization abilities, it was difficult to get them to work even for such a small three-stage minimum-cost path problem. Network configurations and input state representations must be important for better realization. Also, some approximator other than neural networks may be employed for improving overall performance.

Many of the current reinforcement learning techniques are still in the research stage, yielding merely satisfactory results for some problems in narrow areas. But the paradigms are offering great potential. Hence, ongoing explorations may lead them to higher performance techniques and would provide incredibly important mechanisms in machine learning because our learning agents know the axiom, "Failure is the surest path to success!"

EXERCISES

1. Looking at Figure 10.2, specify the reason why the converged P_{down} is 0.5 at vertices B and F.

2. Develop a computer program to simulate the signpost model to solve the jackpot journey problem illustrated in Figure 10.1.

3. Derive Equation (10.5) from Equation (10.7).

4. During the learning process of AHC-nets in Figure 10.7, when the action NN learns extremely slowly compared with the value NN, what do the current values (produced by the value NN) show?

5. Discuss with your classmates how to apply reinforcement learning to the well-known tic-tac-toe game.

REFERENCES

[1] D. Ackley and M. Littman. Generalization and scaling in reinforcement learning. In D. J. Touretzky, editor, *Advances in Neural Information Processing Systems 2*, San Mateo, CA, 1990. Morgan Kaufmann.

[2] D. Ackley and M. Littman. Interactions between learning and evolution. In C. G. Langton, editor, *Artificial Life 2*, pages 487–509, Reading, MA, 1992. Addison-Wesley.

[3] J. S. Albus. *Brains, behavior, and robotics.* BYTE Publications Inc., 1981.

[4] A. G. Barto. Reinforcement learning and adaptive critic methods. In D. A. White and D. A. Sofge, editors, *Handbook of intelligent control: neural, fuzzy, and adaptive approaches*, chapter 12, pages 469–491. Van Nostrand Reinhold, New York, 1992.

[5] A. G. Barto and P. Anandan. Pattern recognizing stochastic learning automata. *IEEE Transactions on Systems, Man, and Cybernetics*, 15:360–375, 1985.

[6] A. G. Barto, S. J. Bradtke, and S. P. Singh. Learning to act using real-time dynamic programming. *Artificial Intelligence*, 72, 1995.

[7] A. G. Barto and S. P. Singh. On the computational economics of reinforcement learning. In D. S. Touretzky, J. L. Elman, T. J. Sejnowski, and G. E. Hinton, editors, *Proceedings of the 1990 Connectionist Models Summer School*, pages 35–44, San Mateo, CA., 1990. Morgan Kaufmann.

[8] A. G. Barto, R. S. Sutton, and C. W. Anderson. Neuronlike adaptive elements that can solve difficult learning control problems. *IEEE Transactions on Systems, Man, and Cybernetics*, 13(5):834–846, 1983.

[9] A. G. Barto, R. S. Sutton, and C. J. C. H. Watkins. Learning and sequential decision making. In *Learning and computational neuroscience: foundations of adaptive Networks*, chapter 13, pages 539–602. MIT Press, Cambridge, MA., 1990.

[10] A. G. Barto, R. S. Sutton, and C. J. C. H. Watkins. Sequential decision problems and neural networks. In D. J. Touretzky, editor, *Advances in neural information processing systems II*, pages 686–693. Morgan Kaufmann, San Mateo, CA., 1990.

[11] R. E. Bellman. *Dynamic programming.* Princeton University Press, Princeton, NJ., 1957.

[12] H. R. Berenji. Fuzzy Q-learning: A new approach for fuzzy dynamic programming. In *Proceedings of IEEE International Conference on Fuzzy Systems*, pages 486–491, July 1994.

[13] H. R. Berenji and P. Khedkar. Learning and tuning fuzzy logic controllers through reinforcements. *IEEE Transactions on Neural Networks*, 3(5):724–740, 1992.

[14] H. Bersini and F. Varela. The immune recruitment mechanism: A selective evolutionary strategy. In *Proceedings of the Fourth International Conference on Genetic Algorithms*, pages 520–526, 1991.

[15] H. Bersini and F. Varela. The immune learning mechanisms: Reinforcement, recruitment and their applications. Technical Report TR/IRIDIA/93-4, IRIDIA-ULB, 1993.

[16] D. P. Bertsekas and J. N. Tsitsiklis. *Parallel distributed processing: numerical methods.* Prentice Hall, Upper Saddle River, NJ, 1989.

[17] Dimitri P. Bertsekas. A Counterexample to Temporal Differences Learning. *Neural Computation*, 7:270–279, 1995.

[18] L. Bolc and J. Cytowski. *Search methods for artificial intelligence.* Academic Press Limited, San Diego, CA., 1992.

[19] L. B. Booker, D. E. Goldberg, and J. H. Holland. Classifier systems and genetic algorithms. *Artificial Intelligence*, 40:235–282, 1989.

[20] R. A. Mc Callum. Using transitional proximity for faster reinforcement learning. In Sleeman and Edwards, editors, *Proceedings of the Ninth International Workshop on Machine Learning*, pages 316–321, San Mateo, CA, 1992. Morgan Kaufmann.

[21] D. Chapman and L. P. Kaelbling. Input generalization in delayed reinforcement learning: An algorithm and performance comparisons. In *The 1991 International Joint Conference on Artificial Intelligence*, pages 726–731, 1991.

[22] P. Dayan. Reinforcement comparison. In D. S. Touretzky, J. L. Elman, T. J. Sejnowski, and G. E. Hinton, editors, *Proceedings of the 1990 Connectionist Models Summer School*, pages 45–51, San Mateo, CA., 1990. Morgan Kaufmann.

[23] P. Dayan. The convergence of TD(λ) for general λ. *Machine Learning*, 8(3):341–362, May 1992.

[24] G. DeJong and R. Mooney. Explanation-based learning: An alternative view. *Machine Learning*, 1(2):145–176, 1986.

[25] J. del R. Millan and C. Torras. A reinforcement connectionist approach to robot path finding in non-maze-like environments. *Machine Learning*, 8(3):363–393, May 1992.

[26] S. E. Dreyfus and A. M. Law. *The art and theory of dynamic programming*, volume 130 of *Mathematics in Science and Engineering*. Academic Press Inc., 1977.

[27] C. L. Dym and R. E. Levitt. *Knowledge-based systems in engineering*. McGraw-Hill, New York, 1991.

[28] J. L. Elman. Finding structure in time. *Cognitive Science*, 14:179–211, 1990.

[29] David E. Goldberg. *Genetic algorithms in search, optimization, and machine learning*. Addison-Wesley, Reading, MA, 1989.

[30] John J. Grefenstette. Credit assignment in rule discovery systems based on genetic algorithms. *Machine Learning*, 3:225–245, 1988.

[31] Mohamad H. Hassoun. *Fundamentals of artificial neural networks*. MIT Press, Cambridge, MA., 1995.

[32] Simon Haykin. *Neural networks: a comprehensive foundation*. Macmillan College Publishing, 1994.

[33] J. Hertz, A. Krogh, and R. G. Palmer. *Introduction to the theory of neural computation*. Addison-Wesley, Reading, MA, 1991.

[34] J. H. Holland. Escaping brittleness: the possibility of general-purpose learning. In R. S. Michalski, J. G. Carbonell, and T. M. Mitchell, editors, *Machine Learning: An Artificial Intelligence Approach, Volume 2*, pages 593–623, San Mateo, CA., 1986. Morgan Kaufmann.

[35] R. A. Howard. *Dynamical programming and Markov processes*. MIT Press, New York, 1960.

[36] W. T. Miller III, R. S. Sutton, and P. J. Werbos, editors. *Neural networks for control*. MIT Press, 1990.

[37] R. A. Jacobs and M. I. Jordan. A competitive modular connectionist architecture. In R. P. Lippmann, J. E. Moody, and D. J. Touretzky, editors, *Advances in Neural Information Processing Systems 3*, pages 767–773, San Mateo, CA, 1991. Morgan Kaufmann.

[38] R. A. Jacobs, M. I. Jordan, S. J. Nowlan, and G. E. Hinton. Adaptive mixtures of local experts. *Neural Computation*, 3:79–87, 1991.

[39] M. I. Jordan. Attractor dynamics and parallelism in a connectionist sequential machine. In *Proceedings of the Eighth Annual Conference of the Cognitive Science Society*, pages 531–546, 1986.

[40] M. I. Jordan and R. A. Jacobs. Learning to control an unstable system with forward modeling. In D. J. Touretzky, editor, *Advances in Neural Information Processing Systems 2*, San Mateo, CA, 1990. Morgan Kaufmann.

[41] M. I. Jordan and D. E. Rumelhart. Forward models: Supervised learning with a distal teacher. *Cognitive Science*, 16:307–354, 1992.

[42] R. E. Korf. Real-time heuristic search. *Artificial Intelligence*, 42(3):189–211, 1990.

[43] J. R. Krebs, A. Kacelnik, and P. Taylor. Test of optimal sampling by foraging great tits. *Nature*, 275:27–31, September 1978.

[44] K-F Lee and S. Mahajan. A pattern classification approach to evaluation function learning. *Artificial Intelligence*, 36:1–25, 1988.

[45] C. S. Lin, Y. H. E. Cheng, and H. Kim. Radial basis function networks for adaptive critic learning. In *Proceedings of IEEE International Conference on Neural Networks*, pages 903–906, July 1994.

[46] C. S. Lin and H. Kim. CMAC-based adaptive critic self-learning control. *IEEE Transactions on Neural Networks*, 2(5):530–533, September 1991.

[47] C.-T. Lin and C.-S. G. Lee. Reinforcement structure/parameter learning for neural-network-based fuzzy logic control systems. *IEEE Transactions on Fuzzy Systems*, 2(1):46–63, 1994.

[48] Long-Ji Lin. Self-improving reactive agents: Case studies of reinforcement learning frameworks. In *Proceedings of the First International Conference on Simulation of Adaptive Behavior: From Animals to Animats*, pages 297–305, 1990.

[49] Long-Ji Lin. Programming robots using reinforcement learning and teaching. In *Proceedings of the Ninth National Conference on Artificial Intelligence (AAAI-91)*, pages 781–786, July 1991.

[50] Long-Ji Lin. Self-improvement based on reinforcement learning, planning and teaching. In L. A. Birnbaum and G. C. Collins, editors, *Machine Learning: Proceedings of the Eighth International Workshop*, pages 323–327, San Mateo, CA., 1991. Morgan Kaufmann.

[51] S. Mahadevan and J. Connell. Scaling reinforcement learning to robotics by exploiting the subsumption architecture. In L. A. Birnbaum and G. C. Collins, editors, *Machine Learning: Proceedings of the Eighth International Workshop*, pages 328–332, San Mateo, CA., 1991. Morgan Kaufmann.

[52] M. Mangel and C. W. Clark. *Dynamic modeling in behavioral ecology.* Princeton University Press, Princeton, NJ, 1988.

[53] D. Michie. Trial and error. *Penguin Science Survey*, 2, 1961.

[54] D. Michie and R. A. Chambers. Boxes: An experiment in adaptive control. In J. T. Tou and R. H. Wilcox, editors, *Machine intelligence*, pages 137–152. Oliver and Boyd, Edinburgh, 1968.

[55] M. Minsky. Steps toward artificial intelligence. In E. A. Feigenbaum and J. Feldman, editors, *Computers and thought.* McGraw-Hill, New York, 1963.

[56] T. M. Mitchell, R. Keller, and S. Kedar-Cabelli. Explanation-based generalization: A unifying view. *Machine Learning*, 1(1):47–80, 1986.

[57] T. M. Mitchell and S. B. Thrun. Explanation-Based Neural Network Learning for Robot Control. In C. L. Giles, S. J. Hanson, and J. D. Cowan, editors, *Advances in Neural Information Processing Systems 5*, San Mateo, CA, 1992. Morgan Kaufmann.

[58] P. R. Montague, P. Dayan, C. person, and T. J. Sejnowski. Bee foraging in uncertain environments using predictive hebbian learning. *Nature*, 377:725–728, October 1995.

[59] K. S. Narendra and M. A. L. Thathatchar. *Learning automata: an introduction.* Prentice Hall, Upper Saddle River, NJ, 1989.

[60] N. J. Nilsson. *Principles of artificial intelligence.* Tioga Publishing Company, Palo Alto, CA., 1980.

[61] M. O. Odetayo and D. R. McGregor. Genetic algorithm for inducing control rules for a dynamic system. In *Proceedings of the Third International Conference on Genetic Algorithms*, pages 177–182, 1989.

[62] A. Perelson and S. Forrest. Genetic algorithms and the immune system. In *Proceedings of the 1990 Workshop on Parallel Problem Solving from Nature*, 1990.

[63] E. Rich and K. Knight. *Artificial intelligence.* McGraw-Hill, New York, 2nd edition, 1991.

[64] S. Russell and P. Norvig. *Artificial intelligence: a modern approach.* Prentice Hall, Upper Saddle River, NJ, 1995.

[65] S. Russell and E. Wefald. *Do the right thing: studies in limited rationality.* MIT Press, Cambridge, MA., 1991.

[66] Stuart Russell. *Heuristic search.* Lecture note of CS 188, Fall 1992, the University of California at Berkeley, Sept. 8, 1992.

[67] Arthur L. Samuel. Some studies in machine learning using the game of checkers. *IBM Journal of Research and Development*, 3(3):210–229, 1959.

[68] J. D. Schaffer, D. Whitely, and L. J. Eshelman. Combinations of genetic algorithms and neural networks: A survey of the state of the art. In *Proceedings of the International Workshop on Combinations of Genetic Algorithms and Neural Networks*, pages 1–37. IEEE Computer Society Press, 1992.

[69] J. Schmidhuber. Learning algorithmes for networks with internal and external feedback. In D. S. Touretzky, J. L. Elman, T. J. Sejnowski, and G. E. Hinton, editors, *Proceedings of the 1990 Connectionist Models Summer School*, pages 52–61, San Mateo, CA., 1990. Morgan Kaufmann.

[70] J. H. Schmidhuber. Temporal-difference-driven learning in recurrent networks. In R. Eckmiller, G. Hartmann, and G. Hauske, editors, *Parallel processing in neural systems and computers*, pages 209–212. North-Holland, Amsterdam, 1990.

[71] S. P. Singh. The efficient learning of mulitple task sequences. In J. E. Moody, S. J. Hanson, and R. P. Lippman, editors, *Advances in neural information processing systems IV*, pages 251–258. Morgan Kaufmann, San Mateo, CA., 1992.

[72] S. P. Singh. Transfer of learning by composing solutions for elemental sequential tasks. *Machine Learning*, 1992.

[73] R. S. Sutton. Integrated architectures for learning, planning, and reacting based on approximating dynamic programming. In *Proceedings of the Seventh International Conference on Machine Learning*, pages 216–224, San Mateo, CA., June 1990. Morgan Kaufmann.

[74] R. S. Sutton. Reinforcement learning architectures for animats. In *Proceedings of the First International Conference on Simulation of Adaptive Behavior: From Animals to Animats*, pages 288–296, 1990.

[75] R. S. Sutton. Planning by incremental dynamic programming. In L. A. Birnbaum and G. C. Collins, editors, *Machine Learning: Proceedings of the Eighth International Workshop*, pages 353–357, San Mateo, CA., 1991. Morgan Kaufmann.

[76] R. S. Sutton, A. G. Barto, and R. J. Williams. Reinforcement learning is direct adaptive optimal control. In *Proceedings of the American Control Conference*, pages 2143–2146, Boston, MA., 1991.

[77] Richard S. Sutton. Learning to predict by the methods of temporal differences. *Machine Learning*, 3:9–44, 1988.

[78] G. Tesauro. Practical issues in temporal difference learning. *Machine Learning*, 8(3/4):257–278, May 1992.

[79] G. Tesauro. Temporal difference learning of backgammon strategy. *Proceedings of Machine Learning*, pages 451–457, 1993.

[80] G. Tesauro. TD-Gammon, a self-teaching backgammon program achieves master-level play. *Neural Computation*, 6, 1994.

[81] G. Tesauro. Temporal difference learning and TD-Gammon. *Communications of the ACM*, 38(3):58–68, March 1995.

[82] S. B. Thrun. The role of exploration in learning control. In D. A. White and D. A. Sofge, editors, *Handbook of intelligent control: neural, fuzzy, and adaptive approaches*, pages 527–559. Van Nostrand Reinhold, New York, 1992.

[83] T. Unemi, M. Nagayoshi, N. Hirayama, T. Nade, K. Yano, and Y. Masujima. Evolutionary differentiation of learning abilities - a case study on optimizing parameter values in Q-learning by a genetic algorithm. In R. A. Brooks and P. Maes, editors, *Artificial Life 4: Proceedings of the Fourth International Workshop on the Synthesis and Simulation of Living Systems*, pages 331–336, Cambridge, MA., 1994. The MIT Press.

[84] F. Varela, A. Coutinho, B. Dupire, and N. Vaz. Cognitive networks: immune, neural and otherwise. In A. Perelson, editor, *Theoretical immunology vol.2*. Addison-Wesley, Reading, MA, 1988. SFI Series on the Science of Complexity.

[85] D. P. Wasserman. *Advanced methods in neural computing*. Van Nostrand Reinhold, New York, 1993.

[86] C. J. C. H. Watkins. *Learning from Delayed Rewards.* PhD thesis, Psychology Department, Cambridge University, 1989.

[87] C. J. C. H. Watkins and P. Dayan. Q-learning. *Machine Learning*, 8:279–292, 1992.

[88] P. J. Werbos. Advanced forecasting methods for global crisis warning and models of intelligence. *General Systems Yearbook*, 22:25–38, 1977.

[89] P. J. Werbos. Building and understanding adaptive systems: A statistical/numerical approach to factory automation and brain research. *IEEE Transactions on Systems, Man, and Cybernetics*, 17(1):7–20, January-February 1987.

[90] P. J. Werbos. Approximate dynamic programming for real-time control and neural modeling. In D. A. White and D. A. Sofge, editors, *Handbook of intelligent control: neural, fuzzy, and adaptive approaches*, pages 493–525. Van Nostrand Reinhold, New York, 1992.

[91] D. A. White and M. I. Jordan. Optimal control: a foundation for intelligent control. In D. A. White and D. A. Sofge, editors, *Handbook of intelligent control: neural, fuzzy, and adaptive approaches*, pages 185–214. Van Nostrand Reinhold, New York, 1992.

[92] S. D. Whitehead and D. H. Ballard. Learning to perceive and act by trial and error. In R. S. Sutton, editor, *Machine learning 7*, pages 45–83. Kluwer Academic, Boston, 1991.

[93] D. Whitley, S. Dominic, and R. Das. Genetic reinforcement learning with multilayer neural networks. In *Proceedings of the Fourth International Conference on Genetic Algorithms*, pages 562–569, 1991.

[94] B. Widrow, N. K. Gupta, and S. Maitra. Punish/reward: learning with a critic in adaptive threshold systems. *IEEE Transactions on Systems, Man, and Cybernetics*, 3:455–465, 1973.

[95] B. Widrow and M. E. Hoff. Adaptive switching circuits. In *IRE WESCON Convention Record*, pages 96–104, New York, 1960.

[96] Ronald J. Williams. Simple statistical gradient-following algorithms for connectionist reinforcement learning. *Machine Learning*, 8:229–256, 1992.

[97] Patrick H. Winston. *Artificial intelligence.* Addison-Wesley, Reading, MA, 3rd edition, 1992.

[98] Xin Yao. A review of evolutionary artificial neural networks. *International Journal of Intelligent Systems*, 8:539–567, 1993.

Chapter 11

Unsupervised Learning and Other Neural Networks

J.-S. R. Jang and E. Mizutani

11.1 INTRODUCTION

In this chapter, we discuss various neural systems that are frequently categorized[1] outside the class of *pure* supervised learning neural networks (NNs) discussed in Chapter 9. In particular, we concentrate on neural networks with two learning modes: **unsupervised learning** and **recording learning**.

When no external teacher or critic's instruction is available, only input vectors can be used for learning. Such an approach is learning without supervision, or what is commonly referred to as *unsupervised learning*. An unsupervised learning system (or agent) evolves to extract features or regularities in presented patterns, without being told what outputs or classes associated with the input patterns are desired. In other words, the learning system detects or categorizes persistent features without any feedback from the environment. Thus, unsupervised learning is frequently employed for data clustering, feature extraction, and similarity detection.

Unsupervised learning NNs attempt to learn to respond to different input patterns with different parts of the network. The network is often trained to strengthen *firing* to respond to frequently occurring patterns, thereby leading to the so-called synonym *probability estimators*. In this manner, the network develops certain internal representations for encoding input patterns. In this chapter, for unsupervised learning paradigms, we describe competitive learning, the Kohonen self-organizing feature map, and principal component analysis.

Another mode of learning, called *recording learning* by Zurada [49], is typically employed for **associative memory networks**. Usually we design an associative memory network by *recording* several ideal patterns into the network's stable states,

[1] Previous literature proposes numerous ways of categorizing neural network paradigms. Our NN classification originates from Barto's taxonomy on page 222 in ref. [36], and Zurada's classification on page 75 in ref. [49], although both are not completely identical. See also Table 9.1 in Chapter 9.

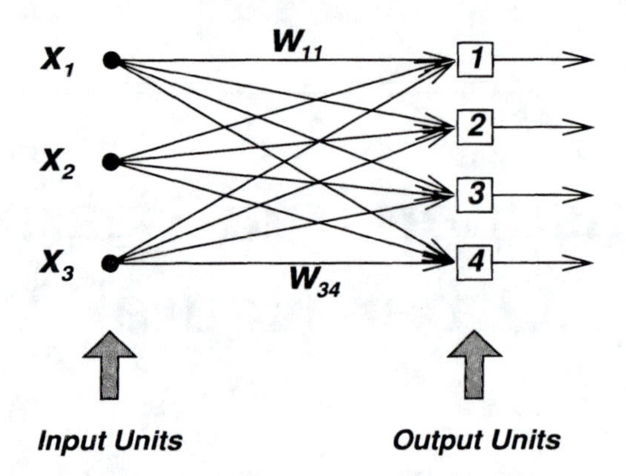

Figure 11.1. *Competitive learning network.*

and we expect the network state to reach one of those patterns when given a pattern (perhaps a contaminated one) as the network's initial state. Stochastic optimization techniques (e.g., simulated annealing) are frequently employed for altering the state transition process of the network. In this chapter, we describe the Hopfield network as an example of *recording* learning systems. The network is also referred to as *content addressable* or *auto-associative*, capable of rectifying and recovering contaminated or incomplete input patterns.

Finally, neural network learning procedures are summarized in light of a general formula proposed by Amari [3].

11.2 COMPETITIVE LEARNING NETWORKS

With no available information regarding the desired outputs, unsupervised learning networks update weights only on the basis of the input patterns. The **competitive learning network** is a popular scheme to achieve this type of unsupervised data clustering or classification; Figure 11.1 presents an example. All input units i are connected to all output units j with weight w_{ij}. The number of inputs is the input dimension, while the number of outputs is equal to the number of clusters that the data are to be divided into. A cluster center's position is specified by the weight vector connected to the corresponding output unit. For the simple network in Figure 11.1, the three-dimensional input data are divided into four clusters, and the cluster centers, denoted as the weights, are updated via the competitive learning rule.

The input vector $\mathbf{x} = [x_1, x_2, x_3]^T$ and the weight vector $\mathbf{w}_j = [w_{1j}, w_{2j}, w_{3j}]^T$ for an output unit j are generally assumed to be normalized to unit length. The activation value a_j of output unit j is then calculated by the inner product of the

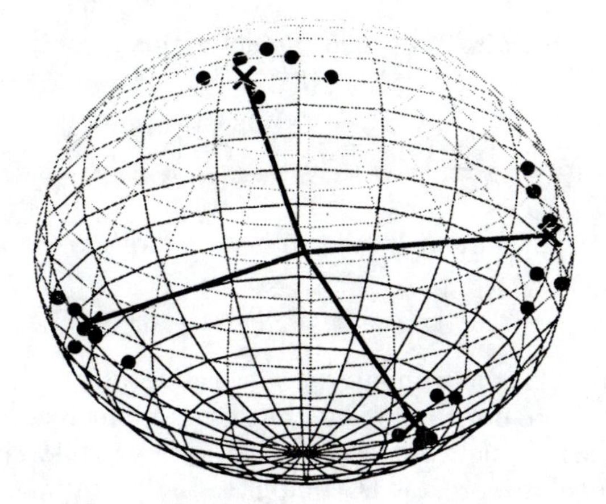

Figure 11.2. *Competitive learning with unit-length vectors. The dots represent the input vectors and the crosses denote the weight vectors for the four output units in Figure 11.1. As the learning continues, the four weight vectors rotate toward the centers of the four input clusters.* (MATLAB command: `compball`)

input and weight vectors:

$$a_j = \sum_{i=1}^{3} x_i w_{ij} = \mathbf{x}^T \mathbf{w}_j = \mathbf{w}_j^T \mathbf{x}. \tag{11.1}$$

Next, the output unit with the highest activation must be selected for further processing, which is what is implied by *competitive*. Assuming that output unit k has the maximal activation, the weights leading to this unit are updated according to the **competitive** or the so-called **winner-take-all** learning rule:

$$\mathbf{w}_k(t+1) = \frac{\mathbf{w}_k(t) + \eta(\mathbf{x}(t) - \mathbf{w}_k(t))}{\|\mathbf{w}_k(t) + \eta(\mathbf{x}(t) - \mathbf{w}_k(t))\|}. \tag{11.2}$$

The preceding weight update formula includes a normalization operation to ensure that the updated weight is always of unit length. Notably, only the weights at the *winner* output unit k are updated; all other weights remain unchanged.

The update formula in Equation (11.2) implements a sequential scheme for finding the cluster centers of a data set of which the entries are of unit length. When an input $\mathbf{x}$ is presented to the network, the weight vector closest to $\mathbf{x}$ *rotates* toward it. Consequently, weight vectors move toward those areas where most inputs appear and, eventually, the weight vectors become the cluster centers for the data set. Figure 11.2 illustrates this dynamic process.

Using the Euclidean distance as a **dissimilarity measure** is a more general scheme of competitive learning, in which the activation of output unit j is

$$a_j = \left(\sum_{i=1}^{3} (x_i - w_{ij})^2 \right)^{0.5} = \|\mathbf{x} - \mathbf{w}_j\|. \tag{11.3}$$

The weights of the output unit with the *smallest* activation are updated according to

$$\mathbf{w}_k(t+1) = \mathbf{w}_k(t) + \eta(\mathbf{x}(t) - \mathbf{w}_k(t)). \tag{11.4}$$

In the preceding equation, the winning unit's weights *shift* toward the input $\mathbf{x}$. In this case, neither the data nor the weights must be of unit length.

A competitive learning network performs an on-line clustering process on the input patterns. When the process is complete, the input data are divided into disjoint clusters such that similarities between individuals in the same cluster are larger than those in different clusters. Here two metrics of similarity are introduced: the similarity measure of inner product in Equation (11.1) and the dissimilarity measure of the Euclidean distance in Equation (11.3). Obviously, other metrics can be used instead, and different selections lead to different clustering results. When the Euclidean distance is adopted, it can be proved that the update formula in Equation (11.4) is actually an on-line version of gradient descent that minimizes the following objection function:

$$E = \sum_{p} \|\mathbf{w}_{f(\mathbf{x}_p)} - \mathbf{x}_p\|^2, \tag{11.5}$$

where $f(\mathbf{x}_p)$ is the winning neuron when input $\mathbf{x}_p$ is presented and $\mathbf{w}_{f(\mathbf{x}_p)}$ is the center of the class where $\mathbf{x}_p$ belongs to. This fact is left as Exercise 1 at the end of this chapter.

A large family of batch-mode (or off-line) clustering algorithms can be used to find cluster centers that minimize Equation (11.5). One of these algorithms is K-means clustering, as explained in detail in Chapter 15.

A limitation of competitive learning is that some of the weight vectors that are initialized to random values may be far from any input vector and, subsequently, it never gets updated. Such a situation can be prevented by initializing the weights to samples from the input data itself, thereby ensuring that all of the weights get updated when all the input patterns are presented. An alternative would be to update the weights of both the winning and losing units, but use a significantly smaller learning rate η for the losers; this is commonly referred to as **leaky learning** [45]. Other methods that prevent weights from not getting updated can be found in ref. [22].

Dynamically changing the learning rate η in the weight update formula of Equation (11.2) or (11.4) is generally desired. An initial large value of η explores the data space widely; later on, a progressively smaller value refines the weights. The operation is similar to the cooling schedule of simulated annealing, as introduced in

Chapter 7. Therefore, one of the following formulas for η is commonly used:

$$\begin{cases} \eta(t) = \eta_0 e^{-\alpha t}, \text{ with } \alpha > 0, \text{ or} \\ \eta(t) = \eta_0 t^{-\alpha}, \text{ with } \alpha \leq 1, \text{ or} \\ \eta(t) = \eta_0 (1 - \alpha t), \text{ with } 0 < \alpha < (\max\{t\})^{-1}. \end{cases}$$

Competitive learning lacks the capability to add new clusters when deemed necessary. Moreover, if the learning rate η is a constant, competitive learning does not guarantee stability in forming clusters; the winning unit that responds to a particular pattern may continue changing during training. On the other hand, η, if decreasing with time, may become too small to update cluster centers when new data of a different probability nature are presented. Carpenter and Grossberg referred to such an occurrence as the **stability-plasticity dilemma**, which is common in designing intelligent learning systems [8]. In general, a learning agent (or system) should be *plastic*, or adaptive in reacting to changing environments; meanwhile, it should be *stable* to preserve knowledge acquired previously. **Adaptive resonance theory** (**ART**), as introduced by Grossberg [17], proposes a solution to this dilemma. Based on ART, Carpenter and Grossberg proposed a series of similar networks, including ART1, ART2 [7], ART3 [9], and ARTMAP [10].

If the output units of a competitive learning network are arranged in a geometric manner (such as in a one-dimensional vector or two-dimensional array), then we can update the weights of the winners as well as the *neighboring* losers. Such a capability corresponds to the notion of Kohonen feature maps, as discussed in the next section.

After competitive learning is finished, the input space is divided into a number of disjoint clusters, each of which is represented by a cluster center. These cluster centers are also known as *template*, *reference vector*, or *codebook vector* [16, 38]. For an input vector, we can use the corresponding template to represent the input vector rather than the vector itself. Such an approach is called **vector quantization** and it has been used for data compression in image processing and communication systems. Section 11.4 introduces a *supervised* version of vector quantization, known as learning vector quantization [33, 34, 35]. Other competitive learning applications include graph bipartitioning [22, 45] and word perception models [45].

11.3 KOHONEN SELF-ORGANIZING NETWORKS

Kohonen self-organizing networks [30, 31], also known as **Kohonen feature maps** or **topology-preserving maps**, are another competition-based network paradigm for data clustering. Networks of this type impose a neighborhood constraint on the output units, such that a certain topological property in the input data is reflected in the output units' weights.

Figure 11.3(a) presents a relatively simple Kohonen self-organizing network with 2 inputs and 49 outputs. The learning procedure of Kohonen feature maps is similar to that of competitive learning networks. That is, a similarity (dissimilarity) measure is selected and the winning unit is considered to be the one with the largest (smallest) activation. For Kohonen feature maps, however, we update not only the

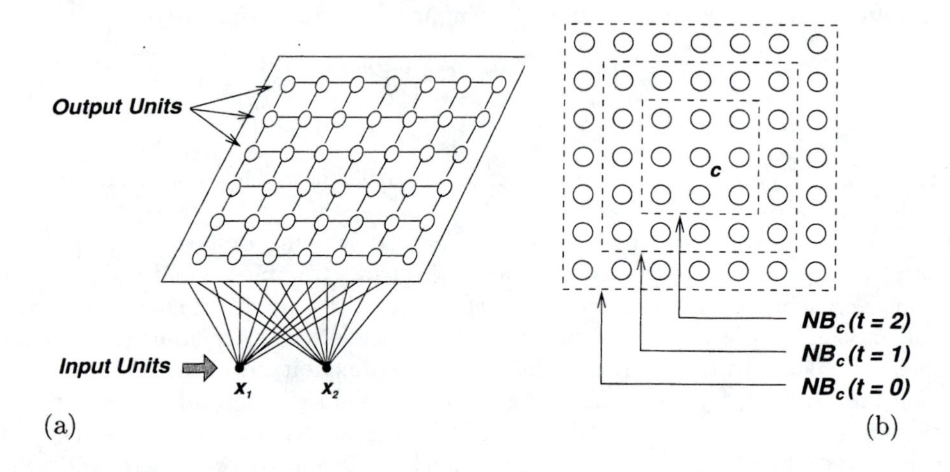

Figure 11.3. *(a) A Kohonen self-organizing network with 2 input and 49 output units; (b) the size of a neighborhood around a winning unit decreases gradually with each iteration.*

winning unit's weights but also all of the weights in a neighborhood around the winning units. The neighborhood's size generally decreases slowly with each iteration, as indicated in Figure 11.3(b). A sequential description of how to train a Kohonen self-organizing network is as follows:

Step1: Select the winning output unit as the one with the largest similarity measure (or smallest dissimilarity measure) between all weight vectors $\mathbf{w}_i$ and the input vector $\mathbf{x}$. If the Euclidean distance is chosen as the dissimilarity measure, then the winning unit c satisfies the following equation:

$$\|\mathbf{x} - \mathbf{w}_c\| = \min_i \|\mathbf{x} - \mathbf{w}_i\|,$$

where the index c refers to the winning unit.

Step2: Let NB_c denote a set of index corresponding to a neighborhood around winner c. The weights of the winner and its neighboring units are then updated by

$$\Delta\mathbf{w}_i = \eta(\mathbf{x} - \mathbf{w}_i), \; i \in NB_c,$$

where η is a small positive learning rate. Instead of defining the neighborhood of a winning unit, we can use a **neighborhood function** $\Omega_c(i)$ around a winning unit c. For instance, the Gaussian function can be used as the neighborhood function:

$$\Omega_c(i) = \exp\left(\frac{-\|p_i - p_c\|^2}{2\sigma^2}\right),$$

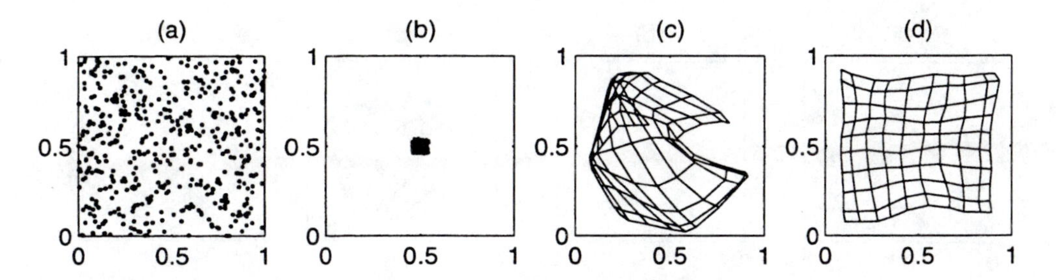

Figure 11.4. *Simulation of the Kohonen self-organizing network: (a) input data uniformly distributed within* $[0, 1] \times [0.1]$*; (b) initial weights; (c) weights after 30 iterations; (d) weights after 1000 iterations.* (MATLAB command: `kfm(1)`)

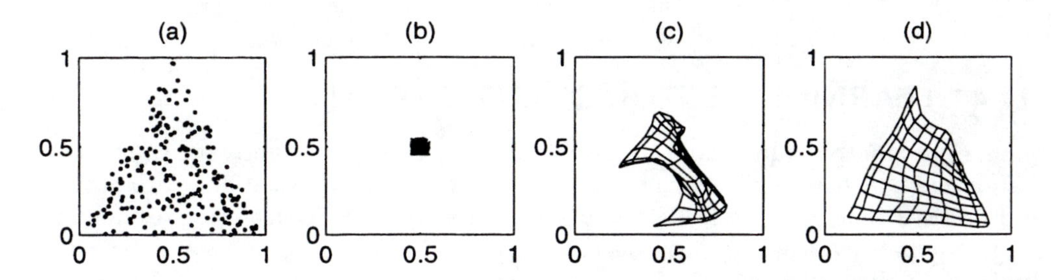

Figure 11.5. *Simulation of the Kohonen self-organizing network: (a) input data uniformly distributed within a triangular region; (b) initial weights; (c) weights after 30 iterations; (d) weights after 1000 iterations.* (MATLAB command: `kfm(2)`)

where p_i and p_c are the positions of the output units i and c, respectively, and σ reflects the scope of the neighborhood. By using the neighborhood function, the update formula can be rewritten as

$$\Delta \mathbf{w}_i = \eta \Omega_c(i)(\mathbf{x} - \mathbf{w}_i), \text{ where } i \text{ is the index for all output units.}$$

To achieve a better convergence, the learning rate η and the size of neighborhood (or σ) should be decreased gradually with each iteration. Figures 11.4, 11.5, and 11.6 present simulation results of Kohonen feature maps with different input data distributions; the output units are arranged in a 10-by-10 two-dimensional mesh. In the simulation, η and σ linearly decreased with the number of iterations.

The most well-known application of Kohonen self-organizing networks is Kohonen's attempt to construct a neural phonetic typewriter [32] that is capable of transcribing speech into written text from an unlimited vocabulary, with an accuracy of 92% to 97%. The network has also been used to learn ballistic arm movements [44].

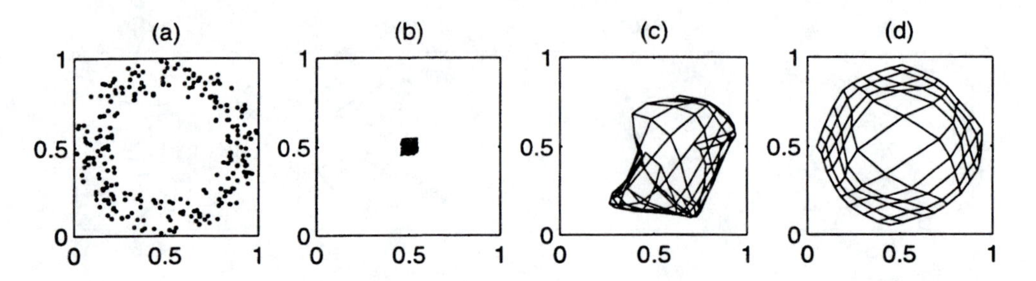

Figure 11.6. *Simulation of the Kohonen self-organizing network: (a) input data uniformly distributed within a doughnut-shaped region; (b) initial weights; (c) weights after 30 iterations; (d) weights after 1000 iterations.* (MATLAB command: `kfm(5)`)

11.4 LEARNING VECTOR QUANTIZATION

Learning vector quantization (**LVQ**) [33, 34, 35] is an adaptive data classification method based on training data with desired class information. Although a *supervised* training method, LVQ employs unsupervised data-clustering techniques (e.g., competitive learning, introduced in Section 11.2) to preprocess the data set and obtain cluster centers.

LVQ's network architecture closely resembles that of a competitive learning network, except that each output unit is associated with a class. Figure 11.7(a) presents an example, where the input dimension is 2 and the input space is divided into six clusters. The first two clusters belong to class 1, while the other four clusters belong to class 2. The LVQ learning algorithm involves two steps. In the first step, an unsupervised learning data clustering method is used to locate several cluster centers without using the class information. In the second step, the class information is used to fine-tune the cluster centers to minimize the number of misclassified cases.

During the first step of unsupervised learning, any of the data clustering techniques introduced in this chapter and Chapter 15 can be used to identify cluster centers (or weight vectors leading to output units) to represent the data set with no class information. The number of clusters can either be specified *a priori* or determined via a cluster technique capable of adaptively adding new clusters when necessary. Once the clusters are obtained, their classes must be labeled before moving to the second step of supervised learning. Such labeling is achieved by the so-called *voting method* (i.e., a cluster is labeled class k if it has data points belonging to class k as a majority within the cluster.) The clustering process for LVQ is based on the general assumption that similar input patterns generally belong to the same class.

During the second step of supervised learning, the cluster centers are fine-tuned to approximate the desired decision hypersurface. The learning method is straightforward. First, the weight vector (or cluster center) $\mathbf{w}$ that is closest to the input

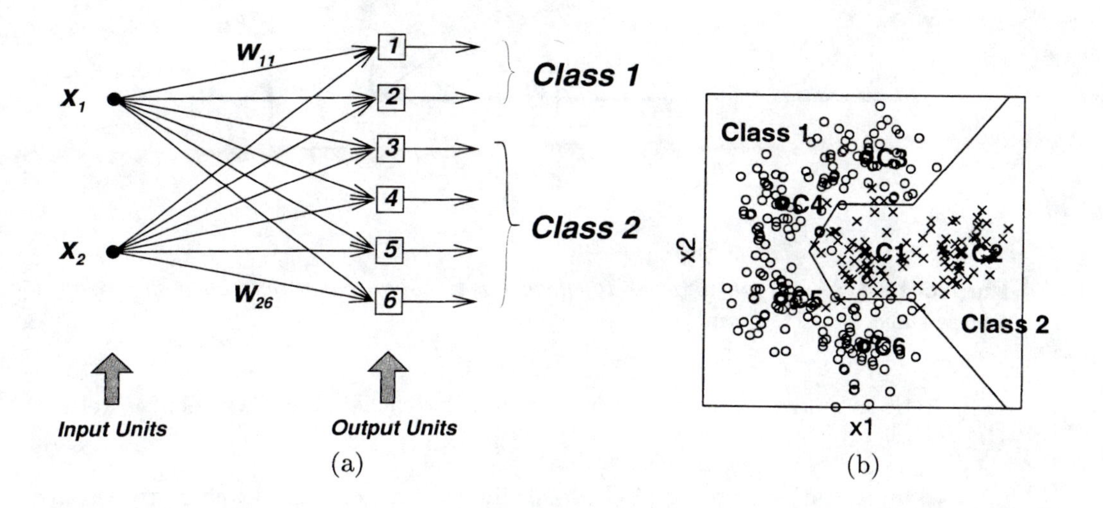

Figure 11.7. *Learning vector quantization (LVQ): (a) network representation; (b) possible data distribution and decision boundary.* [MATLAB command for (b): `lvqdata`]

vector $\mathbf{x}$ must be found. If $\mathbf{x}$ and $\mathbf{w}$ belong to the same class, we move $\mathbf{w}$ toward $\mathbf{x}$; otherwise we move $\mathbf{w}$ away from the input vector $\mathbf{x}$.

After learning, an LVQ network classifies an input vector by assigning it to the same class as the output unit that has the weight vector (cluster center) closest to the input vector. Figure 11.7(b) illustrates a possible distribution of data set and weights after training.

A sequential description of the LVQ method is as follows:

Step 1: Initialize the cluster centers by a clustering method.

Step 2: Label each cluster by the voting method.

Step 3: Randomly select a training input vector $\mathbf{x}$ and find k such that $\|\mathbf{x} - \mathbf{w}_k\|$ is a minimum.

Step 4: If $\mathbf{x}$ and $\mathbf{w}_k$ belong to the same class, update $\mathbf{w}_k$ by

$$\Delta\mathbf{w}_k = \eta(\mathbf{x} - \mathbf{w}_k).$$

Otherwise, update $\mathbf{w}_k$ by

$$\Delta\mathbf{w}_k = -\eta(\mathbf{x} - \mathbf{w}_k).$$

The learning rate η is a positive small constant and should decrease with each iteration.

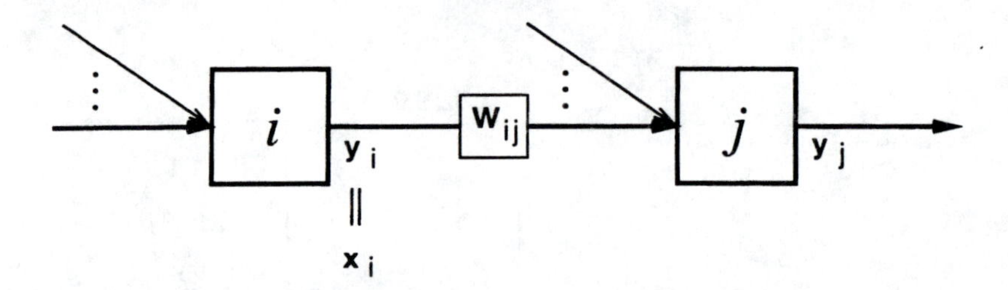

Figure 11.8. *A simple network topology for Hebbian learning; weight w_{ij} resides between two neurons i and j.*

Step 5: If the maximum number of iterations is reached, stop. Otherwise, return to step 3.

Two improved versions of LVQ are available; both of them attempt to use the training data more efficiently by updating the winner and the runner-up (the next closest vector) under a certain condition. The improved versions are called LVQ2 and LVQ3; refs. [34] and [35] provide further details.

11.5 HEBBIAN LEARNING

Hebb [20] described a simple learning method of synaptic weight change. When two cells fire simultaneously (i.e., have strong responses), their connection strength (or weight) increases. Such phenomenon is the so-called **Hebbian learning**, where the weight increase between two neurons is proportional to the frequency at which they fire together. Among various mathematical formulas of this principle, the simplest one is expressed as

$$\Delta w_{ij} = \eta y_i y_j, \tag{11.6}$$

where η is the learning rate. Since weights are adjusted according to the correlation of neuron outputs, the preceding formula is a type of *correlational learning rule*. The ith neuron's output y_i can be regarded as an input (x_i) to another neuron j (Figure 11.8), so Equation (11.6) can be written as

$$\Delta w_{ij} = \eta y_j x_i. \tag{11.7}$$

Restated, a weight is assumed to change proportionately to the *correlation* of the input and output signals. By using a neuron function $f(\cdot)$, y_j is given by

$$y_j = f(\mathbf{w}_j^T\, \mathbf{x}).$$

Thus, Equation (11.7) is equivalent to the following:

$$\Delta w_{ij} = \eta\, f(\mathbf{w}_j^T\, \mathbf{x})\, x_i. \tag{11.8}$$

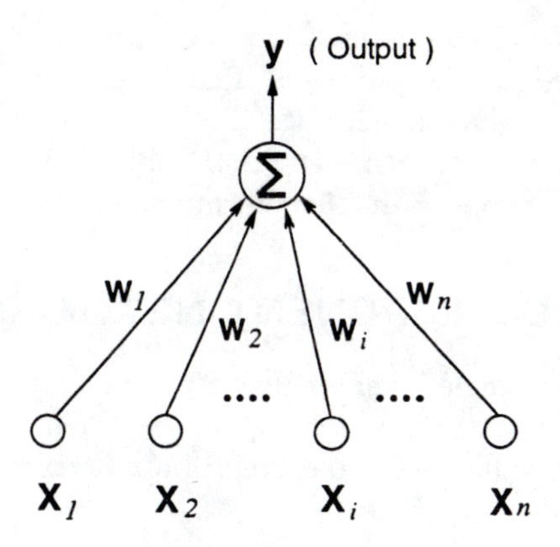

Figure 11.9. *One-layer single-output network with Hebbian learning for principal component analysis.*

A sequence of learning patterns indexed by p is assumed here to be presented to the network; in addition, all initial weights are *zero*. By using Equation (11.6), the update amount of a weight after the entire data set is presented will be

$$w_{ij} = \eta \sum_p y_{ip} y_{jp}. \tag{11.9}$$

Frequent input patterns have the most impact on the weights and, eventually, cause the network to produce the largest outputs. Applying the plain Hebbian learning in Equation (11.6) causes unconstrained growth of the weights. Hence, in some cases, the Hebbian rule is modified to counteract the unlimited growth of weights. Weight normalization, as described in subsection 11.6.2, is one such method.

The rationale behind the Hebbian learning rule is most easily understood via a single-layer n-input one-output neural network with identity activation functions, as shown in Figure 11.9. The output y is equal to $\sum_{i=1}^{n} w_i x_i$, or in matrix form,

$$y = \mathbf{w}^T \mathbf{x} = \mathbf{x}^T \mathbf{w},$$

where $\mathbf{x} = [x_1, \ldots, x_n]^T$ is the input vector and $\mathbf{w} = [w_1, \ldots, w_n]^T$ is the weight vector. The corresponding Hebbian learning rule is

$$\Delta \mathbf{w} = \eta y \mathbf{x}. \tag{11.10}$$

The preceding learning rule is an on-line steepest-ascent scheme that maximizes the objective function

$$J = \frac{1}{2} \sum_p y_p^2 = \frac{1}{2} \sum_p (\mathbf{x}_p^T \mathbf{w})^2, \tag{11.11}$$

where $\mathbf{x}_p$ and y_p are the pth input and corresponding desired output, respectively. If $\mathbf{w}$ is a unit vector, then $\mathbf{x}_p^T \mathbf{w}$ is the projection of the vector $\mathbf{x}_p$ onto the direction specified by a vector **w**. The minimization of J implies finding a unit-length **w** that most accurately represents the *direction* of the entire data set.

Other learning algorithms (such as in the Hopfield network learning and supervised correlation learning) often reflect the Hebbian learning principle [39]; they share the correlational property in their formulas.

11.6 PRINCIPAL COMPONENT NETWORKS

This section describes a single-layer single-output network with Hebbian-type learning that can be used for extracting principal components (eigenvectors corresponding to the largest eigenvalues) of the correlation matrix of the training vectors.

11.6.1 Principal Component Analysis

An important issue in pattern recognition (or data classification) is to select features (or inputs of the training vectors) that have more discriminant power and use the selected features as inputs to a recognition or classification scheme. This **feature selection** process is essential for a real-world data set, in which the number of features generally exceeds 10. Hence we need to determine their priorities and feed the important features to our classification system for effective training.

Similar input patterns likely belong to the same class. Under this observation, the input variables can be normalized to within the unit interval and then selected according to their variances. That is, the larger the variances, the more likely that the inputs variables have better discriminant powers.

For some data sets, combining two features might yield a better discriminant power than either one alone. Therefore, the data set must occasionally be transformed into a more recognizable or trainable form. **Principal component analysis (PCA)** [41] is one approach to combining inputs linearly and identifying their priorities; it is also known as **Karhunen-Loeve transformation** [41] in communication theory and image processing.

Let $\mathbf{x}_i$, $i = 1, \ldots, n$, be the ith entry of the data set under consideration. A unit vector **u** is to be found such that the variance of the data set, after projecting onto **u**, is maximized. Without loss of generality, the data set is assumed to be zero mean:

$$\sum_{i=1}^{n} \mathbf{x}_i = \mathbf{0}.$$

The projection of $\mathbf{x}_i$ onto **u** is defined by the inner product:

$$p_i = \mathbf{x}_i \cdot \mathbf{u} = \mathbf{x}_i^T \mathbf{u} = \mathbf{u}^T \mathbf{x}_i,$$

subject to the constraint that **u** is a unit vector:

$$\|\mathbf{u}\| = \sqrt{\mathbf{u}^T \mathbf{u}} = 1.$$

Since $\mathbf{x}_i$ is zero mean, so is $\mathbf{u}$:

$$\sum_{i=1}^{n} p_i = \sum_{i=1}^{n} \mathbf{u}^T \mathbf{x}_i = \mathbf{u}^T \sum_{i=1}^{n} \mathbf{x}_i = \mathbf{u}^T \cdot \mathbf{0} = 0.$$

The square of p_i can be expressed as

$$p_i^2 = (\mathbf{u}^T \mathbf{x}_i)(\mathbf{x}_i^T \mathbf{u}) = \mathbf{u}^T (\mathbf{x}_i \mathbf{x}_i^T)\mathbf{u}.$$

Therefore, the projection p_i's variance is

$$\begin{aligned} \sigma_p^2(\mathbf{u}) &= \frac{1}{n} \textstyle\sum_{i=1}^{n} p_i^2 \\ &= \mathbf{u}^T \left(\frac{1}{n} \textstyle\sum_{i=1}^{n} \mathbf{x}_i \mathbf{x}_i^T \right) \mathbf{u} \\ &= \mathbf{u}^T \mathbf{R} \mathbf{u}, \end{aligned}$$

where the symmetric matrix $\mathbf{R}$ is the data set's **correlation matrix**. [If the data set is not zero mean, then $\mathbf{R}$ is called the **covariance matrix** and is defined as $\frac{1}{n}\sum_{i=1}^{n}(\mathbf{x}_i - \boldsymbol{\mu})(\mathbf{x}_i - \boldsymbol{\mu})^T$, where $\boldsymbol{\mu}$ is the mean of $\mathbf{x}_i$, $i = 1, \ldots, n$.]

The projection variance $\sigma_p^2(\mathbf{u})$ subject to the unit-vector constraint can be minimized by defining a new objection function using the Lagrange multiplier:

$$J = \mathbf{u}^T \mathbf{R} \mathbf{u} + \lambda(1 - \mathbf{u}^T \mathbf{u}).$$

Differentiating the preceding equation and setting it to zero yield

$$\nabla_{\mathbf{u}} J = 2\mathbf{R}\mathbf{u} - 2\lambda \mathbf{u} = \mathbf{0},$$

or

$$\mathbf{R}\mathbf{u} = \lambda \mathbf{u}.$$

This stationary condition implies that λ is an eigenvalue of the correlation matrix $\mathbf{R}$ and $\mathbf{u}$ is the corresponding vector. At the preceding stationary condition, the projection variance is

$$\sigma_p^2(\mathbf{u}) = \mathbf{u}^T \mathbf{R} \mathbf{u} = \mathbf{u}^T \lambda \mathbf{u} = \lambda \mathbf{u}^T \mathbf{u} = \lambda.$$

Therefore, the projection variance $\sigma_p^2(\mathbf{u})$ has a maximum equal to the largest eigenvalue of the correlation matrix $\mathbf{R}$; this occurs when the projection vector $\mathbf{u}$ is equal to the corresponding eigenvector.

Note that a correlation matrix is symmetric and its eigenvectors are orthogonal to each other. A given vector $\mathbf{x}$ can be expressed using the n eigenvectors of $\mathbf{R}$:

$$\mathbf{x} = \sum_{i=1}^{n} q_i \mathbf{u}_i,$$

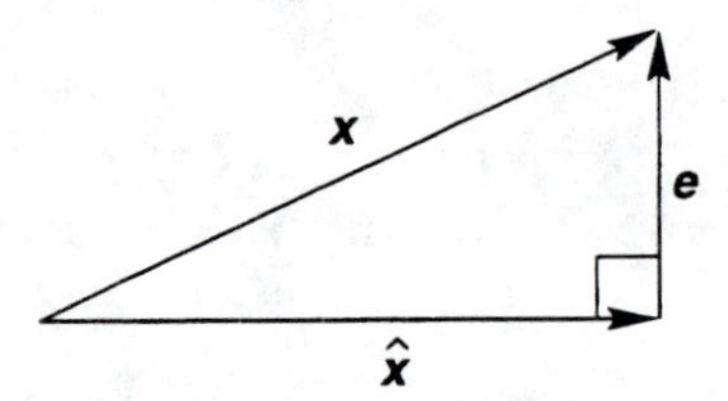

Figure 11.10. *Principle of orthogonality in principal component analysis, in which* $\mathbf{x}$ *is equal to* $\hat{\mathbf{x}} + \mathbf{e}$ *and the truncated version* $\hat{\mathbf{x}}$ *is always orthogonal to the approximation error* $\mathbf{e}$.

where q_i $(= \mathbf{x} \cdot \mathbf{u}_i)$ is the projection of $\mathbf{x}$ onto $\mathbf{u}_i$, and $\mathbf{u}_i$ is the ith unit eigenvector of $\mathbf{R}$. Index i is ordered in such a manner that $\mathbf{u}_i$ belongs to the ith eigenvalue λ_i, $i = 1, \ldots, n$, satisfying the following constraint:

$$\lambda_1 \geq \cdots \geq \lambda_i \geq \cdots \geq \lambda_n.$$

If we want to do **dimensionality reduction** by retaining m inputs with larger variances, $\mathbf{x}$ can be approximated by $\hat{\mathbf{x}}$ by deleting $n-m$ terms containing $\mathbf{u}_{m+1}, \ldots, \mathbf{u}_n$:

$$\hat{\mathbf{x}} = \sum_{i=1}^{m} q_i \mathbf{u}_i.$$

The approximation error is given by

$$\mathbf{e} = \mathbf{x} - \hat{\mathbf{x}} = \sum_{i=m+1}^{n} q_i \mathbf{u}_i.$$

Since $\mathbf{u}_i$ is orthogonal to each other, the error vector $\mathbf{e}$ is orthogonal to the approximating data vector $\hat{\mathbf{x}}$, regardless of the value of m. This is called the **principle of orthogonality**, as graphically represented in Figure 11.10.

If $\mathbf{x}$ is taken to be a random variable and $\mathbf{x}_1, \ldots, \mathbf{x}_n$ are its instances, then the variance of $\mathbf{x}$ is the sum of the variances of $\hat{\mathbf{x}}$ and $\mathbf{e}$. More specifically, the variance of $\mathbf{x}$ is expressed as

$$\sigma_{\mathbf{x}}^2 = \underbrace{\lambda_1 + \cdots + \lambda_m}_{\sigma_{\hat{\mathbf{x}}}^2} + \underbrace{\lambda_{m+1} + \cdots + \lambda n,}_{\sigma_{\mathbf{e}}^2} \tag{11.12}$$

where $\sigma_{\hat{\mathbf{x}}}^2$ and $\sigma_{\mathbf{e}}^2$ are the variances of $\hat{\mathbf{x}}$ and $\mathbf{e}$, respectively. The preceding equation can be proved using the orthogonality among $\mathbf{u}_i$; this is left as an exercise.

Therefore, to achieve dimensionality reduction on a data set, the correlation matrix and its eigenvalues and eigenvectors must be found first. Next, the data set is projected onto the subspace spanned by the eigenvectors belonging to the largest eigenvalues.

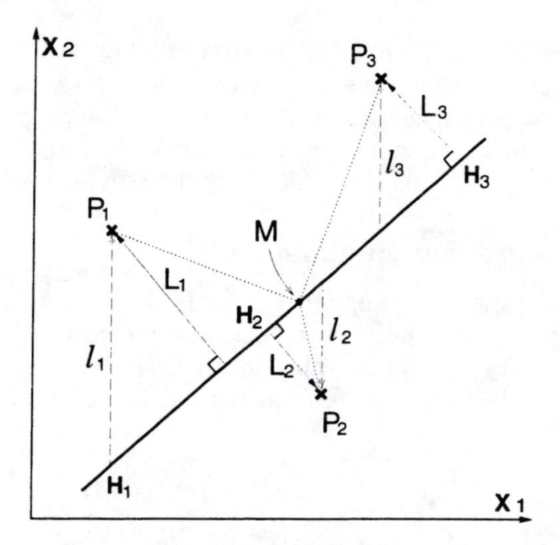

Figure 11.11. *Comparison between principal component analysis and regression analysis. The direction of the first principal component minimizes $\sum_i Li$, the distance in a direction perpendicular to the estimated line. Meanwhile, the regression line minimizes $\sum_i li$, the vertical distance from data points to that line.*

PCA can also be interpreted in terms of data fitting that minimizes the lengths of error vectors perpendicular to the estimated line or surface. As indicated in Figure 11.11, P_i is the location of the ith data point $\mathbf{x}_i$ and M is the location of the sample mean of all the data points. The thick solid line represents the direction of the first principal component passing through the data mean point M, and H_i denotes the projected point of P_i onto the thick line. The principal of orthogonality (see Figure 11.10) implies that the vector from M to H_i is orthogonal to the vector from P_i to H_i. This leads to the following identity:

$$\overline{MP_i}^2 = \overline{MH_i}^2 + \overline{P_iH_i}^2.$$

Summing up the preceding identity over all data points yields

$$\sum_i \overline{MP_i}^2 = \sum_i \overline{MH_i}^2 + \sum_i \overline{P_iH_i}^2 \tag{11.13}$$

In the above equation, the left-hand side is a constant; the first term of the right-hand side is proportional to the variance of principal components; the second term is $\sum_i L_i$, the sum of the distances measured in a direction perpendicular to the estimated line. Therefore, maximizing the variance of principal components (first term) leads to minimizing $\sum_i L_i$ (second term), the sum of the squares of the distances from the points to the line, where the distances are measured in a direction perpendicular to the estimated line. Minimizing $\sum_i L_i$ is the task of the total least-squares (TLS) method [15] for data fitting; the first principal component direction [5, 29, 47] can be used to achieve the same goal. In comparison, the standard least-squares (LS) method, as described in Chapter 5, attempts to find a least-squares regression line that minimizes the vertical distance ($\sum_i li$) from the data points to the line, as indicated in Figure 11.11.

11.6.2 Oja's Modified Hebbian Rule

By using the simple neural network in Figure 11.9, Oja demonstrated that with a Hebbian-type learning rule, the network performs PCA [42].

Employing the plain Hebbian learning rule in Equation (11.10) leads to unlimited growth of the weight vector. One solution is to renormalize the weight vector after each update:

$$w_i(t+1) = \frac{w_i(t) + \eta\, y(t)\, x_i(t)}{\sqrt{\sum_{i=1}^{n} [w_i(t) + \eta\, y(t)\, x_i(t)]^2}}. \tag{11.14}$$

If the learning rate η is small, the preceding equation at $\eta = 0$ can be expanded using Taylor series expansion. Deleting the second- and higher-order terms yields

$$\Delta w_i = \eta y x_i - \eta y^2 w_i = \eta y (x_i - y w_i). \tag{11.15}$$

The preceding equation is the **modified Hebbian learning rule**, which entails adding a *weight decay* proportional to the squared output (y^2) to maintain the weight vector unit length automatically. This Hebbian-type adaptation involves less computation since the normalization operation in Equation (11.14) is not required. Hertz et al. pointed out that Equation (11.15) resembles reverse LMS learning (Table 11.4 in Section 11.8), in which the weight updating is based on the difference between the actual input and the backpropagated output [23].

Oja also indicated that the weight vector $\mathbf{w}$ approaches an eigenvector of the correlation matrix $\mathbf{R}$ with the largest eigenvalue. The larger the eigenvalue, the more precise the direction of the corresponding principal component (or eigenvector). Oja later extended the formulation for multiple-output systems to perform PCA [43].

11.7 THE HOPFIELD NETWORK

In 1982, Hopfield proposed the so-called **Hopfield network**, which possesses *auto-associative* properties. It is a **recurrent** (or *fully interconnected*) network in which all neurons are connected to each other, with the exception that no neuron has any connection to itself. In the network configuration, he embodied the *physical principle*, and set up an **energy function**. The concept derives from a physical system [25]:

> *Any physical system whose dynamics in phase space is dominated by a substantial number of locally stable states to which it is attracted*

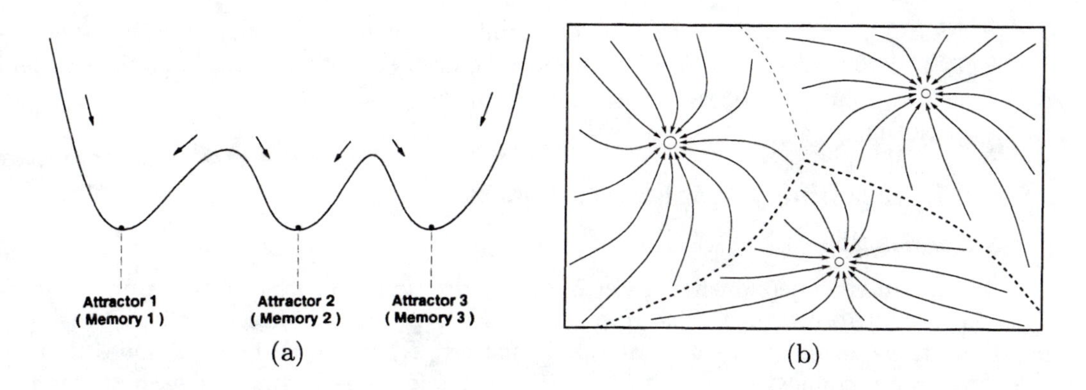

Figure 11.12. *Images of 1-D and 2-D energy terrains configured by three attractors in a Hopfield network. The dots denote stable states where patterns are memorized. The network state moves in the direction of the arrows to one of the attractors, which is determined by the starting point (i.e., the given input pattern).*

> *can therefore be regarded as a general content-addressable memory. The physical system will be a potentially useful memory if, in addition, any prescribed set of states can readily be made the stable states of the system.*

The Hopfield net highlights a content-addressable memory and a tool for solving an optimization problem. These features are discussed next.

11.7.1 Content-Addressable Nature

The Hopfield net normally develops a number of locally stable points in state space. Because the network's dynamics minimize *energy*, other points in state space drain into the stable points (called **attractors** or **wells**), which are (possibly local) energy minima.

The Hopfield net realizes the operation of a *content-addressable (auto-associative) memory* in the sense that newly presented input patterns (or arbitrary initial states) can be connected to the appropriate patterns stored in memories (i.e., attractors or stable states). The presented input pattern vector cannot escape from a region, what we call a **basin of attraction**, configured by each attractor (Figure 11.12). Restated, the network produces a desired memorized pattern in response to the given pattern. The initial network state is assumed to lie within the reach of the fixed attractors' *basins of attraction*. That is, the entire configuration space is divided into different basins of attraction. Notably, an attractor is regarded as a fixed point if it is unique in state space. In general, however, an attractor may have *chaos* or the *limit cycle* of a periodic sequence of states. An advantage of neural networks (NNs) lies in their *fault tolerance*—more specifically, in that NNs are tolerant of a presented pattern's slight distortions. This NN feature is appropriate for performing the primary task of content-addressable memory.

In terms of storage capacity, the number of memories is estimated to be nearly 10% to 20% of the number of neurons in the Hopfield net [12]. Although the Hopfield memory net is not very efficient, its mechanism based on the *energy* concept is worth exploring. The basic binary Hopfield net is first described.

11.7.2 Binary Hopfield Networks

Formulation

Each interconnection has a weight (or connection strength), denoted by T_{ij} from neuron j to neuron i. The Hopfield network considers bidirectionality in the connections, using the symmetric weight matrix, $T_{ij} = T_{ji}$, and also assumes that no neuron is connected to itself ($T_{ii} = 0$). In Hopfield's early analysis, each neuron has a *binary state* of either 0 or 1; those neurons subsequently form the *binary* Hopfield net. At each moment, the entire state of the network can be represented by a binary state vector. The neurons are assumed to be threshold logic units so that a state has one of the two possible values. The following is the *firing rule* of an arbitrary neuron i:

$$V_i = \begin{cases} 1 & \text{if } \sum_{j \neq i} T_{ij} V_j > U_i, \\ 0 & \text{if } \sum_{j \neq i} T_{ij} V_j < U_i. \end{cases}$$

where V_i denotes the output of neuron i and U_i its threshold. This can be rewritten with a neuron function, $f(\cdot)$:

$$\begin{aligned} & V_i = f(\textstyle\sum_{j=1, j \neq i}^{n} T_{ij} V_j - U_i), \\ & \text{where } f(x) = \text{sgn}(x) \equiv \begin{cases} 1 & \text{if } x > 0, \\ 0 & \text{if } x < 0. \end{cases} \end{aligned} \tag{11.16}$$

Each network state has an associated *energy*[2] in a quadratic form:

$$E = -\frac{1}{2} \sum \sum_{j \neq i} T_{ij} V_j V_i + \sum V_i U_i. \tag{11.17}$$

The change in energy (ΔE) with respect to the change of state at neuron i (ΔV_i) is derived from Equation (11.17):

$$\Delta E = -\Delta V_i (\sum_{j \neq i} T_{ij} V_j - U_i). \tag{11.18}$$

When a neuron i alters its state according to the firing rule defined in Equation (11.16), it is thus ensured that ΔE is always negative. In other words, E is a monotonically decreasing function of the network state.

[2]The energy in Equation (11.17) has a quadratic form similar to the kinetic energy E_k of a mass m at velocity v:

$$E_k = \frac{1}{2} m v^2.$$

In many cases, *physical energy* can be represented in such a quadratic form.

Table 11.1. *Transitional behavior per state in the two-neuron Hopfield net. Act_i denotes the activation of neuron i.*

Current states	Energy levels	V_1	V_2	Chosen neuron	Act_1	Act_2		V_1	V_2	Next state
A	0	0	0	1	-0.7	—	$\Rightarrow$	0	0	A
				2	—	0.1	$\Rightarrow$	0	1	B
B	-0.1	0	1	1	-0.4	—	$\Rightarrow$	0	1	B
				2	—	0.1	$\Rightarrow$	0	1	B
C	0.7	1	0	1	-0.7	—	$\Rightarrow$	0	0	A
				2	—	0.4	$\Rightarrow$	1	1	D
D	0.3	1	1	1	-0.4	—	$\Rightarrow$	0	1	B
				2	—	0.4	$\Rightarrow$	1	1	D

Skiing Down the Energy Slope toward a Stable Well

This subsection presents the operational picture of the binary Hopfield net to understand more thoroughly "*state transitions*" with a fixed set of weights. The network can be seen as a dynamical system moving through a sequence of states toward a *stable state* over time. For computational simplicity, we describe a small two-neuron binary Hopfield net illustrated in Figure 11.13(a). We consider a particular weight parameter setup as follows:

$$T_{12} = T_{21} = 0.3, \; U_1 = 0.7, \; U_2 = 0.1.$$

Recall that each neuron can have one of two states. Hence, this two-neuron binary network can have a total of four states: state A, state B, state C, and state D, corresponding to the four pairs of (V_1, V_2): $(0,0)$, $(0,1)$, $(1,0)$, and $(1,1)$, respectively. According to Equation (11.17),

$$E = -T_{12}V_1V_2 + V_1U_1 + V_2U_2 = -0.3V_1V_2 + 0.7V_1 + 0.1V_2.$$

Using the preceding equation, we can calculate the energy levels of the four states: $E = 0$ for state A, $E = -0.1$ for state B, $E = 0.7$ for state C, and $E = 0.3$ for state D.

In light of Equation (11.16), we consider two activations:

$$Act_1 \equiv T_{12}V_2 - U_1 = 0.3V_2 - 0.7,$$

$$Act_2 \equiv T_{21}V_1 - U_2 = 0.3V_1 + 0.1.$$

For state A ($V_1 = V_2 = 0$), Act_1 at neuron 1 is negative (-0.7), and Act_2 at neuron 2 is positive (0.1). Therefore, if neuron 2 starts firing, state A transits to state B ($V_1 = 0$, $V_2 = 1$). If neuron 1 starts firing, however, state A does not transit. Table 11.1 summarizes all state transitions and energy levels.

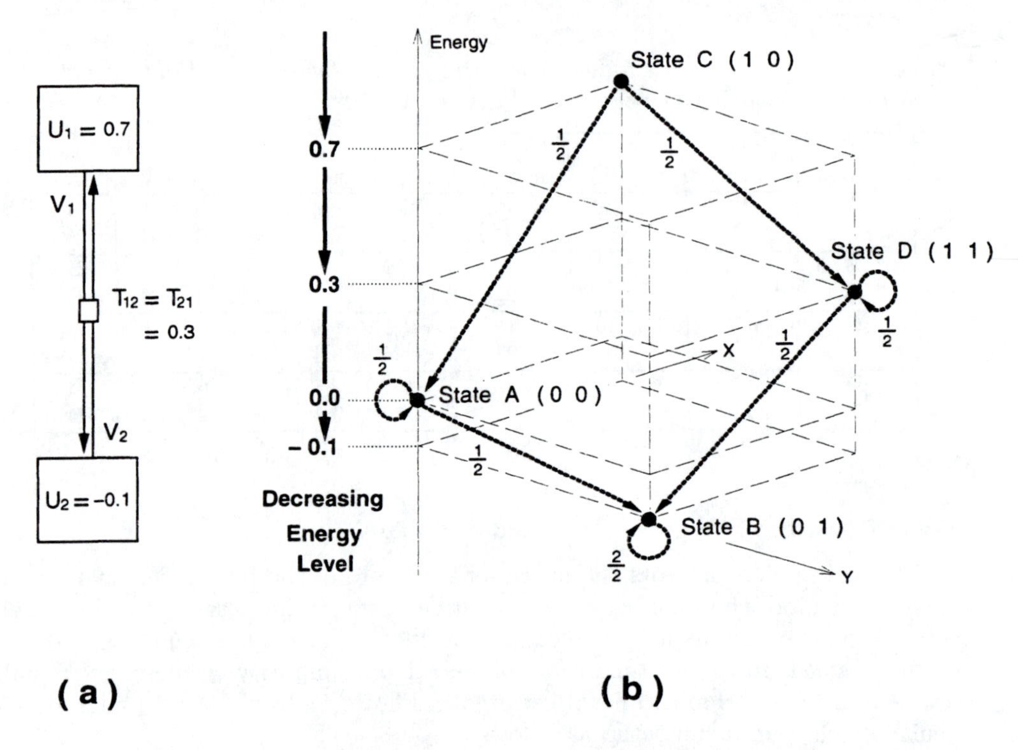

Figure 11.13. *The simplest two-neuron Hopfield net that has one stable state: (a) network topology, and (b) four possible states with their corresponding energy levels and state transitions with their transitional probabilities.*

To delineate the complete state transitions, we need to mention **asynchronous updating**, which Hopfield originally employed in 1982. In this operation, a single neuron is randomly selected to modify its output for a given time, occasionally modifying it and occasionally leaving it the same. Any one of the neurons has a roughly equal probability of firing at each moment. The updating of neurons continues until no more changes can be made. Many updates must be applied to all neurons before the network eventually settles in a stable configuration.

Figure 11.13(b) displays the state transition diagram of the two-neuron binary Hopfield net with the previously specified weight setup. (More detailed discussion of the algorithms and mathematical formulas for calculating weights to store patterns can be found in refs. [2, 23, 49, 19, 18].) Whatever its starting state is, the network eventually settles in state B rather than transits from state to state. Of course, increasingly complicated networks with more than two neurons may develop many stable states (attractors or wells) to memorize several patterns. Therefore, spurious/false states must be dealt with [23, 19]. Due to the concept of physical energy, the successive firing can be interpreted as a sloping down of the energy sur-

Table 11.2. *The comparison between asynchronous updating and continuous updating.*

Network type	Binary	Continuous valued
Updating	Asynchronous	Continuous
Neuron function $f(x)$	Threshold logic unit [e.g., sig(x)]	Sigmoid [e.g., $\frac{1}{2}(1+\tanh(\frac{x}{gain}))$]
Description	Update only one randomly selected neuron's output, according to Equation (11.16)	Update continuously and simultaneously all neurons' outputs toward the values given by Equation (11.19), as well as the net input net_i; see Equations (11.20) and (11.21).

face. The energy decreases until it reaches a (possibly local) minimum. Aleksander and Morton discussed more "memory" examples in ref [2].

11.7.3 Continuous-Valued Hopfield Networks

Hopfield extended the binary net to the continuous-valued net in 1984 [26]. The neuron generates a continuous range of outputs. In the binary net, a neuron function $f(\cdot)$ is typically stipulated by a hard threshold function [Equation (11.16)]. In the continuous-valued Hopfield net, on the other hand, a *sigmoidal function* is typically employed as a neuron function $f(\cdot)$. Updates in time are described continuously (**continuous updating**). For instance, the neurons are governed by the following firing rule:

$$\begin{aligned} V_i &= f(net_i) = \tfrac{1}{2}(1 + tanh(\tfrac{net_i}{gain})) \\ net_i &= \textstyle\sum_{j \neq i} T_{ij} V_j - U_i, \end{aligned} \tag{11.19}$$

where *gain* corresponds to the slope of the sigmoid.

By using a neuron function $f(\cdot)$, the network state is governed by the differential equation:

$$K_i \frac{dV_i}{dt} = -V_i + f(\sum_j T_{ij} V_j - U_i), \tag{11.20}$$

where K_i is a positive constant. The term dV_i/dt represents a directional vector toward an attractor, where $dV_i/dt = 0$. It can be denoted by an arrow, as in Figure 11.12. At an attractor, the net input net_i to neuron i should be equal to $\sum_j T_{ij} V_j - U_i$ to have an ideal output: $V_i (= f(\sum_j T_{ij} V_j - U_i))$. Hence, a dynamical equation similar to Equation (11.20) can be derived with respect to net_i:

$$k_i \frac{dnet_i}{dt} = -net_i + \sum_j T_{ij} V_j - U_i = -net_i + \sum_j T_{ij} f(net_j) - U_i, \qquad (11.21)$$

where k_i is a positive constant. We have discussed so far two modes of updating: asynchronous scheme and continuous one, which are summarized in Table 11.2. Another possible scheme is **synchronous updating**, wherein at each instant of time, all neurons' outputs are simultaneously set to the values given by $f(\cdot)$ [4, 40]. The network may go to the correct memorized pattern after the first updating.

The foregoing dynamics controlled by Equations (11.20) and (11.21) can be found in hardware systems. We show an electrical Hopfield model in the next subsection. Moreover, we describe the application of the continuous net to the traveling salesperson problem.

Electrical Implementation

This subsection discusses an analog electrical circuit to implement the continuous-valued Hopfield net [18, 23, 26, 49]. Figure 11.14 illustrates a circuit diagram that is composed of resistors, capacitors, and nonlinear amplifiers. The ith amplifier produces an output voltage V_i given by $f(u_i)$, where u_i is the input voltage and function f is such a differentiable activation function as defined in Equation (11.19). The conductances $1/R_{ij}$ function as the connection weights T_{ij}. To allow weight values to be *negative*, resistors R_{ij} are connected to $-V_i$. The sign of weights T_{ij} is determined by selecting the positive or negative output of amplifier j. Such selection is realized by an additional inverting amplifier and a negative signal wire, which is omitted for simplicity sake in Figure 11.14, where I_i is any other external input current (or *bias*) to amplifier i.

Figure 11.15 illustrates the ith electrical neuron in the Hopfield circuit. From this figure, the circuit equation can be obtained by employing Ohm's law and Kirchoff's current law. By considering the total current entering capacitor C_i, we have

$$\begin{aligned} C_i \tfrac{du_i}{dt} &= \textstyle\sum_j \tfrac{1}{R_{ij}}(V_j - u_i) + (-\tfrac{u_i}{\rho_i}) + I_i \\ &= \textstyle\sum_j \tfrac{V_j}{R_{ij}} - (\sum_j \tfrac{1}{R_{ij}} + \tfrac{1}{\rho_i}) \cdot u_i + I_i. \end{aligned}$$

This can be rewritten as

$$C_i \frac{du_i}{dt} = -G_i u_i + \sum_j T_{ij} V_j + I_i = -G_i u_i + \sum_j T_{ij} f_j(u_j) + I_i. \qquad (11.22)$$

where

$$G_i = \sum_j \frac{1}{R_{ij}} + \frac{1}{\rho_i} \text{ and } T_{ij} = \frac{1}{R_{ij}}.$$

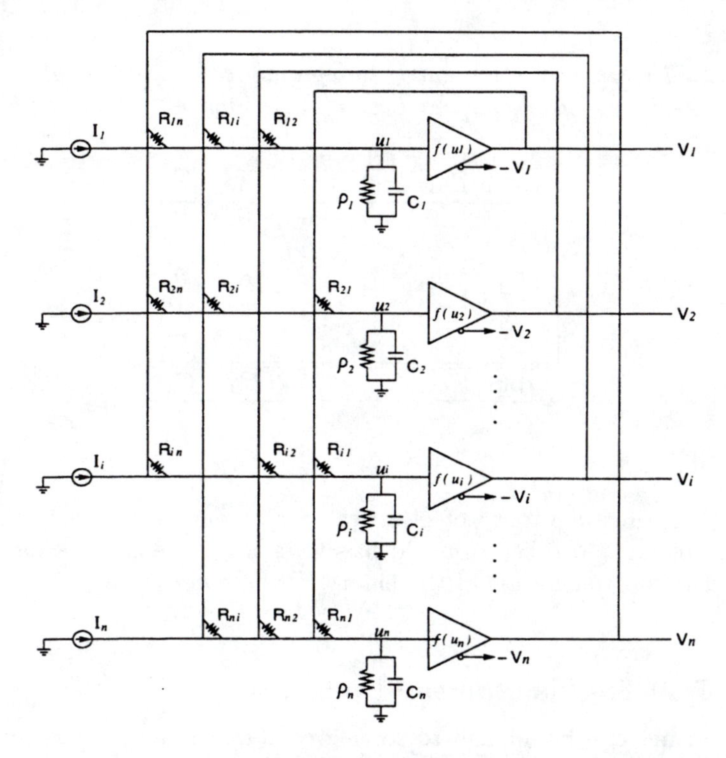

Figure 11.14. *Diagram of an electrical circuit to implement a continuous Hopfield network.*

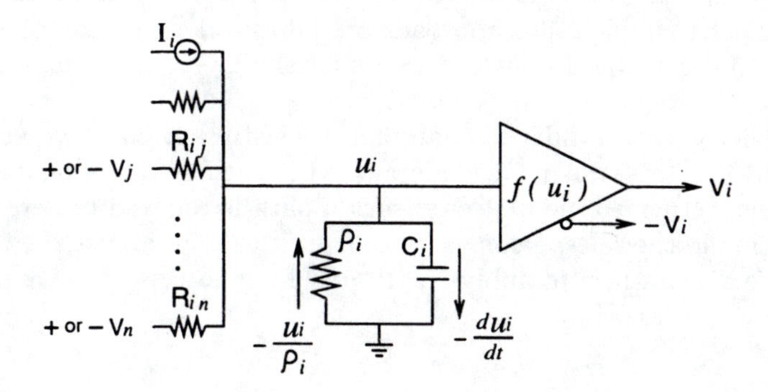

Figure 11.15. *The ith electrical neuron in the Hopfield circuit in Figure 11.14.*

Dynamical Equation (11.22) of the electrical system corresponds equivalently to Equation (11.21); I_i corresponds to U_i in Equation (11.21). (The functional equivalence in terms of *energy* is well treated in refs. [18, 23, 26, 49].) Thus, the circuit depicted in Figure 11.14 functions as a Hopfield net.

Table 11.3. *The network state matrix to represent a specific tour of five cities—*$C_3 \rightarrow C_1 \rightarrow C_5 \rightarrow C_2 \rightarrow C_4$*— for our five-city traveling salesperson problem.*

City name	Position in tour 1	2	3	4	5
C_1	0	1	0	0	0
C_2	0	0	0	1	0
C_3	1	0	0	0	0
C_4	0	0	0	0	1
C_5	0	0	1	0	0
Tour	C_3	C_1	C_5	C_2	C_4

Zurada describes a variety of electrical circuit models in ref. [49]. Some researchers implemented other Hopfield hardware models (e.g., Farhat et al. discussed optical implementation of the binary Hopfield net [13]).

11.7.4 Traveling Salesperson Problem

The Hopfield net can be applied to solving combinatorial optimization problems. This subsection discusses a well-known *traveling salesperson problem* (also discussed in Chapter 7). Hopfield yielded a good solution that is close to the optimal solution [27]. Although practical limitations for this problem are known, designing procedures of the Hopfield net approach are informative and instructive. We first discuss how to design the Hopfield network for the given problem, and then briefly describe how to design the energy function.

For simplicity, we consider a small map consisting of only five cities. A salesperson would like to start at a certain city, visit each of the other four cities only once, and then return to the first city; such a path is referred to here as a *tour*. In what order should the salesperson visit the five cities to minimize the total traveling distance? We discuss how to apply the Hopfield net to this five-city problem.

Representation

We first represent this problem in the network frame; each *neuron* corresponds to a *city with an order to visit*. Since each neuron's output ranges from 0.0 to 1.0, the two extreme values correspond to whether *to visit* or *not to visit*. If a city is to be visited the fourth, for example, the city can be represented by the binary vector (0, 0, 0, 1, 0). The excitatory neuron ij is interpreted here as "*city* C_i *to be visited the* j*th*." For this five-city map, we can construct the Hopfield net that consists of 25 (= 5 × 5) neurons. This means that a tour can be represented by a (5 × 5) state matrix $\mathbf{V}_s$ of neuron's outputs v_{ij}:

$$\mathbf{V}_s \equiv \begin{bmatrix} v_{11} & v_{12} & v_{13} & v_{14} & v_{15} \\ v_{21} & v_{22} & v_{23} & v_{24} & v_{25} \\ v_{31} & v_{32} & v_{33} & v_{34} & v_{35} \\ v_{41} & v_{42} & v_{43} & v_{44} & v_{45} \\ v_{51} & v_{52} & v_{53} & v_{54} & v_{55} \end{bmatrix}. \tag{11.23}$$

The columns of the state matrix $\mathbf{V}_s$ signify different positions in a tour, and the rows identify the cities. By convention, the first index on an element v_{ij} (output of *neuron ij*) denotes its row, the second index its column. Assume that the five cities should be visited in the following order:

$$C_3 \Rightarrow C_1 \Rightarrow C_5 \Rightarrow C_2 \Rightarrow C_4.$$

For this specific tour, we expect to have the following state matrix as a result (as presented in Table 11.3):

$$\mathbf{V}_s = \begin{bmatrix} 0 & 1 & 0 & 0 & 0 \\ 0 & 0 & 0 & 1 & 0 \\ 1 & 0 & 0 & 0 & 0 \\ 0 & 0 & 0 & 0 & 1 \\ 0 & 0 & 1 & 0 & 0 \end{bmatrix}. \tag{11.24}$$

Energy Function

In constructing the energy function, we need to express clearly all constraints and the goal of the given problem.

- *Constraint 1*: The salesperson cannot visit more than one city at the same time.
- *Constraint 2*: The salesperson visits each of the cities only once.
- *Goal*: The salesperson would like to minimize the total traveling distance.

Constraint 1 indicates that each column of matrix $\mathbf{V}_s$ has all zeros except for one element. For a fixed column j, we have

$$\begin{aligned} e_{1j} &= (v_{1j} + v_{2j} + v_{3j} + v_{4j} + v_{5j} - 1)^2, \\ &= (\textstyle\sum_{i=1}^{5} v_{ij} - 1)^2. \end{aligned} \tag{11.25}$$

For all columns, we have

$$E_1 = \sum_{j=1}^{5} e_{1j} = \sum_{j=1}^{5} (\sum_{i=1}^{5} v_{ij} - 1)^2. \tag{11.26}$$

Constraint 2 indicates that each row of matrix $\mathbf{V}_s$ has all zeros except for one element. Considering a fixed row i, we have

$$\begin{aligned} e_{2i} &= (v_{i1} + v_{i2} + v_{i3} + v_{i4} + v_{i5} - 1)^2, \\ &= (\textstyle\sum_{j=1}^{5} v_{ij} - 1)^2. \end{aligned} \tag{11.27}$$

For all rows, we have

$$E_2 = \sum_{i=1}^{5} e_{2i} = \sum_{i=1}^{5} (\sum_{j=1}^{5} v_{ij} - 1)^2. \tag{11.28}$$

To deal with the goal, we let $L(a, b)$ be the distance between *city* a and *city* b. For instance, $L(1, 2)$ denotes the distance between two cities: C_1 and C_2. If C_1 is visited mth, then C_2 should be visited either $(m - 1)$th or $(m + 1)$th. Due to $v_{ij} = 0$ or 1, we can have

$$L(1,2) = L(1,2)\, v_{1,m}\, v_{2,m-1} + L(1,2)\, v_{1,m}\, v_{2,m+1}.$$

In a state presented in Table 11.3, we have $L(1, 2) = 0$, because the two cities C_1 and C_2 are not visited consecutively. By generalizing the preceding equation, the distance sum to be minimized is given by

$$E_3 = \frac{1}{2} \sum_k \sum_i \sum_j L(i,j)\, v_{ik}\, (v_{j,k-1} + v_{j,k+1}) \quad i, j = 1, 2, 3, 4, 5. \tag{11.29}$$

We then obtain the total energy function in the following form:

$$E = k_1\, E_1 + k_2\, E_2 + k_3\, E_3, \tag{11.30}$$

where k_i are constant values. Weights (T_{ij}) and thresholds (U_i) can be determined by equating terms between Equations (11.17) and (11.30). Further details about calculating weights can be found in refs. [11, 27, 49]. Ideally, the states with the shortest tour distances have the lowest energy values.

It is known, however, that this approach has practical limitations for larger problems [6, 21, 28, 48]. Many researchers have discussed how to improve the performance of the Hopfield net. Because the procedure of altering states is performed in an irreversible fashion, the stability can be purely local in the Hopfield net. Stochastic optimization techniques can be employed for overcoming this limitation, as discussed in the next subsection. *Genetic algorithms* are applicable to tuning parameters k_i in Equation (11.30) [37]. For solving a *job shop scheduling problem*, the Hopfield-like net with some stochastic nature has been addressed in ref. [14].

11.7.5 The Boltzmann Machine

To overcome the limitation stemming from the irreversible state-altering manner, **simulated annealing** or (**SA**) can be incorporated; the upgraded network is known as the **Boltzmann machine**, which is considered to be the Hopfield model with a stochastic nature of learning (Chapter 7). More specifically, the firing rule of the binary network may be governed by the following probability function:

$$\mathrm{Prob}(V_i = 1) = \frac{1}{1 + \exp(-net_i/Temp)}$$

$$net_i = \sum_{j \neq i} T_{ij} V_j - U_i$$

$$\mathrm{Prob}(V_i = 0) = 1 - \mathrm{Prob}(V_i = 1),$$

where $Temp$ is a temperature parameter having an effect on the firing probability. Hence, this updating does not necessarily direct the network state toward minimizing energy. In other words, *noise* is introduced to the Hopfield learning mechanism to "shake" the network state out of a local minimum.

In addition, the *Boltzmann machine* possesses the *hidden* units, thereby permitting weight adjustments by *supervised* learning of a stochastic form to minimize the difference between the energies of given states and their desired energies [1, 24, 46].

11.8 SUMMARY

This chapter presents neural networks with two learning modes: unsupervised and recording learning. Unsupervised learning is useful for analyzing data without desired outputs; the networks evolve to capture density characteristics of a data set. Recording learning is employed for designing associative memories; it can be used for combinatorial optimization problems. As a summary of these discussions, we tabulate a compendium of neural network learning formulas in Table 11.4, as explained next.

Amari indicated that most learning rules are generally formulated as follows [3]:

$$\Delta \mathbf{w_i} = \eta r \mathbf{x}, \tag{11.31}$$

where r is a learning signal function. The function r depends on whether the teacher (or target) signal t is available or not:

$$\mathrm{r} = \mathrm{r}(\mathbf{x}, \mathbf{w}, \mathrm{y}) \quad \text{for supervised learning,}$$

or

$$\mathrm{r} = \mathrm{r}(\mathbf{x}, \mathbf{w}) \quad \text{for unsupervised learning.}$$

Equation (11.31) corresponds to the Widrow-Hoff rule when $r = y - \mathbf{w}^T\mathbf{x}$ as discussed previously in Chapter 9. For the Hebbian learning rule, the learning signal r is simply equivalent to the neuron's output.

Recall the winner-take-all learning rule in Equation (11.2). Because this rule does not conform to Equation (11.31), for our convenience, the *learning signal vector* is defined here as

$$\Delta \mathbf{w_i} = \eta(\text{learning signal vector})$$

Table 11.4. *Typical neural network learning formulas, with t the target output, y the neural network's output,* **x** *the input vector, and* **w** *the weight vector.*

Learning algorithm	Learning signal vector	Learning mode
Hebbian	$y\mathbf{x}$	Unsupervised
Correlation	$t\mathbf{x}$	Supervised
Winner-take-all (or competitive)	$\mathbf{x} - \mathbf{w}$	Unsupervised
Outstar	$t - w_i$	Supervised
Perceptron	$\{t - \text{sgn}(\mathbf{w}^T\,\mathbf{x})\}\mathbf{x}$	Supervised
Oja's (reversed LMS)	$(\mathbf{x} - y\,\mathbf{w})y$	Unsupervised
LMS (least mean square) or Widrow-Hoff	$(t - \mathbf{w}^\mathrm{T}\,\mathbf{x})\mathbf{x}$	Supervised
Delta	$\{(t-y)\;f'(net)\}\mathbf{x}$	Supervised

[The learning signal vector corresponds to $r\mathbf{x}$ in Amari's formulation; see Equation (11.31).] By using the learning signal vector, Table 11.4 summarizes typical neural learning rules.

EXERCISES

1. Prove that Equation (11.4) implements an on-line gradient descent scheme for minimizing the objection function in Equation (11.5).

2. Modify the MATLAB program `kfm.m` to generate Figures 11.4, 11.5, and 11.6 such that the output units are organized in a 1-D vector instead of a 2-D array.

3. Prove that the plain Hebbian learning rule in Equation (11.10) implements

an on-line steepest-ascent scheme for maximizing the objection function in Equation (11.11).

4. Derive Equation (11.15) from Equation (11.14) using the Taylor series expansion under the assumption that the learning rate η is small.

5. Use the orthogonality among $\mathbf{u}_i$ to prove Equation (11.12).

6. Consider the application of the Hopfield net to the well-known **knapsack problem**, where the objective is to maximize the total value of n objects (o_i) put in a knapsack subject to a maximum weight constraint, C. This problem may be summarized mathematically as follows:

$$\max \sum_i^n v_i o_i \qquad \text{subject to} \sum_i^n w_i o_i \leq C,$$

where v_i denotes the ith object's value, and w_i the ith object's weight. To solve this knapsack problem, set up the energy function by following the procedure described in Section 11.7.4.

REFERENCES

[1] D. H. Ackley, G. E. Hinton, and T. J. Sejnowski. A learning algorithm for Boltzmann machines. *Cognitive Science*, 9:147–169, 1985.

[2] I. Aleksander and H. Morton. *Neural computing.* Chapman and Hall, 1990.

[3] S. Amari. Mathematical foundations of neurocomputing. *Proceedings of the IEEE*, 78:1443–1463, 1990.

[4] D. H. Amit, H. Gutfreund, and H. Sompolinsky. Spin-glass models of neural networks. *Physical Review*, A32:1007–1018, 1985.

[5] T. W. Anderson. *An introduction to mutivariate statistical analysis.* John Wiley & Sons, New York, 1984.

[6] D. Burr. An improvec elastic net method for the travelling salesman problem. In *Proceedings of the International Conference on Neural Networks*, pages 69–76, 1988.

[7] G. A. Carpenter and S. Grossberg. Art2: Self-organization of stable category recognition codes for analog input patterns. *Applied Optics*, 26(23):4919–4930, 1987.

[8] G. A. Carpenter and S. Grossberg. The art of adaptive pattern recognition by a self-organizing neural network. *IEEE Computer*, 21(3):77–88, March 1988.

[9] G. A. Carpenter and S. Grossberg. Art3: Hierarchical search using chemical transmitters in self-organizing pattern recognition architectures. *Neural Networks*, 3(2):129–152, 1990.

[10] G. A. Carpenter and S. Grossberg. Artmap: Supervised real-time learning and classification of nonstationary data by a self-organizing neural networks. *Neural Networks*, 4(5):565–588, 1991.

[11] J. E. Dayhoff. *Neural network architectures.* Van Nostrand Reinhold, 1990.

[12] J. S. Denker. Neural networks for computation. In J. S. Denker, editor, *AIP Conference Proceedings 151.* American Institute of Physics, 1986.

[13] N. H. Farhat, D. Psaltis, A. Prata, and E. Paek. Optical implementation of the hopfield model. *Applied Optics*, 24:1469–1475, 1985.

[14] Y. S. Foo and Y. Takefuji. Stochastic neural networks for solving job-shop scheduling: Part 1 and 2. In *Proceedings of the International Conference on Neural Networks*, volume 2, pages 275–290, 1988.

[15] G. H. Golub and C. F. Van Loan. *Matrix computation.* Johns Hopkins University Press, Baltimore, 2nd edition, 1989.

[16] R. M. Gray. Vector quantization. *IEEE ASSP Magazine*, 1:4–29, 1984.

[17] S. Grossberg. Adaptive pattern classification and universal recording, i: Parallel development and coding of neural feature detectors. *Biological Cybernetics*, 23:121–134, 1976.

[18] Mohamad H. Hassoun. *Fundamentals of artificial neural networks.* MIT Press, Cambridge, MA., 1995.

[19] Simon Haykin. *Neural networks: a comprehensive foundation.* Macmillan College Publishing, 1994.

[20] D. O. Hebb. *The organization of behavior.* John Wiley & Sons, New York, 1949.

[21] S. U. Hegde, J. L. Sweet, and W. B. Levy. Determination of parameters in a hopfield/tank computational network. In *Proceedings of the International Conference on Neural Networks*, volume 2, pages 291–298, 1988.

[22] J. Hertz, A. Krogh, and R. G. Palmer. *Introduction to the theory of neural computation*, chapter 9, pages 217–250. Addison-Wesley, Reading, MA, 1991.

[23] J. Hertz, A. Krogh, and R. G. Palmer. *Introduction to the theory of neural computation.* Addison-Wesley, Reading, MA, 1991.

[24] G. E. Hinton and T. J. Sejnowski. Optimal perceptual inference. In *Proceedings of the IEEE Computer Society Conference on Computer Vision and Patter Recognition*, pages 448–453, 1983.

[25] J. J. Hopfield. Neural networks and physical systems with emergent collective computational abilities. In *Proceedings of the National Academy of Science, USA*, pages 2554–2558, 1982.

[26] J. J. Hopfield. Neurons with graded response have collective computational properties like those of two-state neurons. In *Proceedings of the National Academy of Science, USA*, pages 3088–3092, 1984.

[27] J. J. Hopfield and D. W. Tank. "neural" computation of decisions in optimization problems. *Biological Cybernetics*, 52:141–152, 1985.

[28] G. Hueter. Solution of the travelling salesman problem with an adaptive ring. In *Proceedings of the International Conference on Neural Networks*, pages 85–92, 1988.

[29] M. G. Kendall. *Mutivariate analysis.* Charles Griffin, 1980.

[30] T. Kohonen. Self-organized formation of topologically correct feature maps. *Biological Cybernetics*, 66:59–69, 1982.

[31] T. Kohonen. *Self-organization and associate memory.* Springer-Verlag, London, 1984.

[32] T. Kohonen. The "neural" phonetic typewriter. *Computer*, 21(3):11–22, 1988.

[33] T. Kohonen. *Self-organization and associate memory.* Springer-Verlag, London, 3rd edition, 1989.

[34] T. Kohonen. Improved versions of learning vector quantization. In *International Joint Conference on Neural Networks*, pages Vol. 1: 545–550, San Diego, 1990.

[35] T. Kohonen. The self-organizing map. *Proceedings of the IEEE*, 78(9):1464–1480, 1990.

[36] Bart Kosko. Fuzzy systems as universal approximators. In *Proceedings of the IEEE International Conference on Fuzzy Systems*, San Diego, March 1992.

[37] W. K. Lai and G. G. Coghill. Genetic breeding of control parameters for the hopfield/tank neural net. In *Proceedings of the International Conference on Neural Networks*, volume 4, pages 618–623, June 1992.

[38] Y. Linde, A. Buzo, and R. M. Gray. An algorithm for vector quantizer design. *IEEE Transactions on Communications*, 28:84–95, 1980.

[39] Ralph Linsker. Self-organization in a perceptual network. *IEEE Computer*, 21(3):105–117, March 1988.

[40] W. A. Little. The existence of persistent states in the brain. *Mathematical Biosciences*, 19:101–120, 1974.

[41] M. Loève. *Probability theory.* Van Nostrand, New York, 3rd edition, 1963.

[42] E. Oja. A simplified neuron model as a principal component analyzer. *Journal of Mathematical Biology*, 15:267–273, 1982.

[43] E. Oja. Neural networks, principal components, and subspaces. *Internatioanl Journal of Neural Systems*, 1:61–68, 1989.

[44] H. Ritter and K. Schulten. Extending kohonen's self-organizing mapping algorithm to learn ballistic movements. In Rolf Eckmiller and Christoph v.d. Malsburg, editors, *Neural computers*, pages 393–406. Springer-Verlag, London, 1987. 1987.

[45] D. E. Rumelhart and D. Zipser. Feature discovery by competitive learning. In D. E. Rumelhart and James L. McClelland, editors, *Parallel distributed processing: explorations in the microstructure of cognition, volume 1*, chapter 5, pages 151–193. MIT Press, Cambridge, MA., 1986.

[46] D. E. Rumlhart, G. E. Hinton, and R. J. Williams. Learning internal representations by error propagation. In D. E. Rumlhart and James L. McClelland, editors, *Parallel distributed processing: volume 1*, chapter 8, pages 318–362. MIT Press, Cambridge, MA., 1986.

[47] G. A. F. Seber. *Multivariate observations.* John Wiley & Sons, New York, 1984.

[48] G. V. Wilson and G. S. Pawley. On the stability of the travelling salesman problem algorithm of hopfield and tank. *Biological Cybernetics*, 58:63–70, 1988.

[49] Jacek M. Zurada. *Introduction to artificial neural systems.* West Publishing, 1992.

Part IV

Neuro-Fuzzy Modeling

Chapter 12

ANFIS: Adaptive Neuro-Fuzzy Inference Systems

J.-S. R. Jang

12.1 INTRODUCTION

The architectures and learning rules of adaptive networks have been described in the previous chapter. Functionally, there are almost no constraints on the node functions of an adaptive network except for the requirement of piecewise differentiability. Structurally, the only limitation on the network configuration is that it should be of the feedforward type (in some cases after unfolding) if we do not want to use the more complex asynchronously operated model. Because of these minimal restrictions, adaptive networks can be employed directly in a wide variety of applications of modeling, decision making, signal processing, and control.

In this chapter, we propose a class of adaptive networks that are functionally equivalent to fuzzy inference systems. The proposed architecture is referred to as **ANFIS**, which stands for **adaptive network–based buzzy inference system** or semantically equivalently, **adaptive neuro fuzzy inference system**. We describe how to decompose the parameter set to facilitate the hybrid learning rule for ANFIS architectures representing both the Sugeno and Tsukamoto fuzzy models. We also demonstrate that under certain minor constraints, the radial basis function network (RBFN) is functionally equivalent to the ANFIS architecture for the Sugeno fuzzy model. The effectiveness of ANFIS with the hybrid learning rule is tested through four simulation examples: Example 1 models a two-dimensional sinc function; Example 2 models a three-input nonlinear function that was used as a benchmark problem for other fuzzy modeling approaches; Example 3 explains how to identify nonlinear components in an on-line control system; and Example 4 predicts the Mackey-Glass chaotic time series. The results from ANFIS are compared extensively with connectionist approaches and conventional statistical methods. More

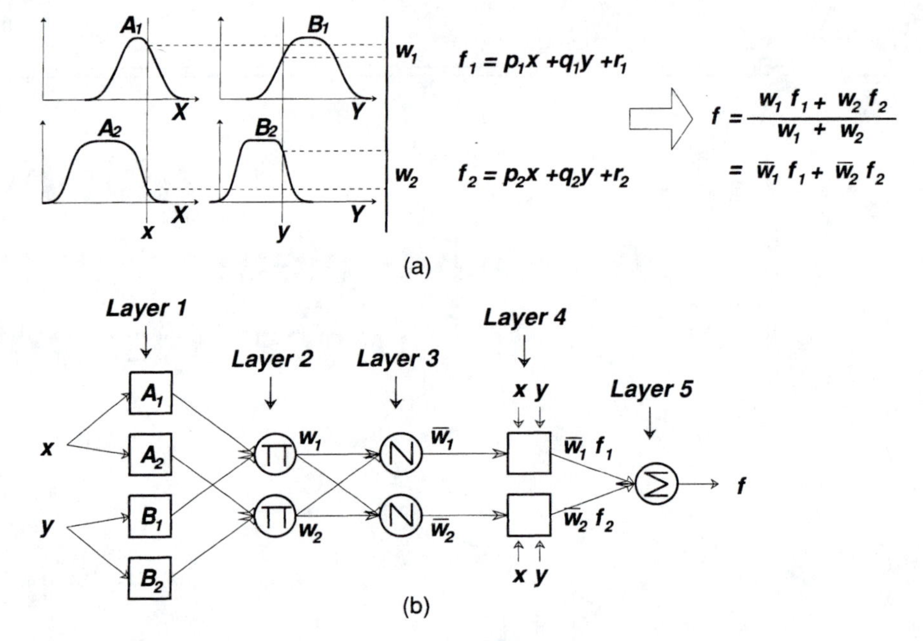

Figure 12.1. *(a) A two-input first-order Sugeno fuzzy model with two rules; (b) equivalent ANFIS architecture.*

ANFIS applications for a number of different domains are described in Chapter 19.

Note that similar network structures were also proposed independently by Lin and Lee [21] and Wang and Mendel [40].

12.2 ANFIS ARCHITECTURE

For simplicity, we assume that the fuzzy inference system under consideration has two inputs x and y and one output z. For a first-order Sugeno fuzzy model [34, 37, 38], a common rule set with two fuzzy if-then rules is the following:

$$\text{Rule 1: If } x \text{ is } A_1 \text{ and } y \text{ is } B_1, \text{ then } f_1 = p_1x + q_1y + r_1,$$
$$\text{Rule 2: If } x \text{ is } A_2 \text{ and } y \text{ is } B_2, \text{ then } f_2 = p_2x + q_2y + r_2.$$

Figure 12.1(a) illustrates the reasoning mechanism for this Sugeno model; the corresponding equivalent ANFIS architecture is as shown in Figure 12.1(b), where nodes of the same layer have similar functions, as described next. (Here we denote the output of the ith node in layer l as $O_{l,i}$.)

Layer 1 Every node i in this layer is an adaptive node with a node function

$$\begin{aligned} O_{1,i} &= \mu_{A_i}(x), && \text{for } i = 1, 2, \text{ or} \\ O_{1,i} &= \mu_{B_{i-2}}(y), && \text{for } i = 3, 4, \end{aligned} \qquad (12.1)$$

where x (or y) is the input to node i and A_i (or B_{i-2}) is a linguistic label (such as "small" or "large") associated with this node. In other words, $O_{1,i}$ is the membership grade of a fuzzy set A ($= A_1$, A_2, B_1 or B_2) and it specifies the degree to which the given input x (or y) satisfies the quantifier A. Here the membership function for A can be any appropriate parameterized membership function introduced in Section 2.4.1, such as the generalized bell function:

$$\mu_A(x) = \frac{1}{1 + \left|\frac{x - c_i}{a_i}\right|^{2b}}, \tag{12.2}$$

where $\{a_i,\ b_i,\ c_i\}$ is the parameter set. As the values of these parameters change, the bell-shaped function varies accordingly, thus exhibiting various forms of membership functions for fuzzy set A. Parameters in this layer are referred to as **premise parameters.**

Layer 2 Every node in this layer is a fixed node labeled Π, whose output is the product of all the incoming signals:

$$O_{2,i} = w_i = \mu_{A_i}(x)\mu_{B_i}(y),\ i = 1, 2. \tag{12.3}$$

Each node output represents the firing strength of a rule. In general, any other T-norm operators (Section 2.5.2) that perform fuzzy AND can be used as the node function in this layer.

Layer 3 Every node in this layer is a fixed node labeled N. The ith node calculates the ratio of the ith rule's firing strength to the sum of all rules' firing strengths:

$$O_{3,i} = \overline{w}_i = \frac{w_i}{w_1 + w_2},\ i = 1, 2. \tag{12.4}$$

For convenience, outputs of this layer are called **normalized firing strengths.**

Layer 4 Every node i in this layer is an adaptive node with a node function

$$O_{4,i} = \overline{w}_i f_i = \overline{w}_i(p_i x + q_i y + r_i), \tag{12.5}$$

where $\overline{w}_i$ is a normalized firing strength from layer 3 and $\{p_i,\ q_i,\ r_i\}$ is the parameter set of this node. Parameters in this layer are referred to as **consequent parameters.**

Layer 5 The single node in this layer is a fixed node labeled Σ, which computes the overall output as the summation of all incoming signals:

$$\text{overall output} = O_{5,1} = \sum_i \overline{w}_i f_i = \frac{\sum_i w_i f_i}{\sum_i w_i} \tag{12.6}$$

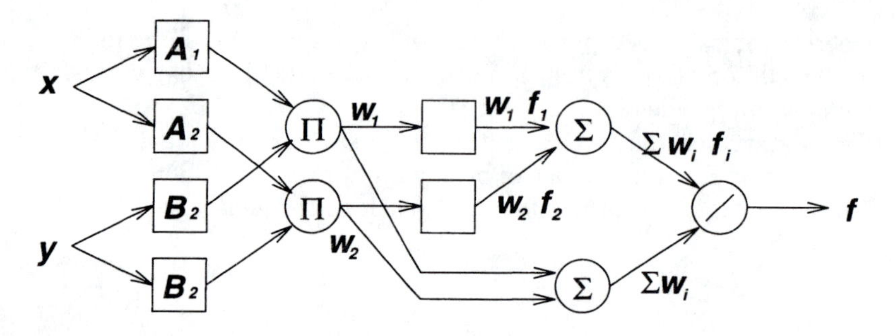

Figure 12.2. *ANFIS architecture for the Sugeno fuzzy model, where weight normalization is performed at the very last layer.*

Thus we have constructed an adaptive network that is functionally equivalent to a Sugeno fuzzy model. Note that the structure of this adaptive network is not unique; we can combine layers 3 and 4 to obtain an equivalent network with only four layers. By the same token, we can perform the weight normalization at the last layer; Figure 12.2 illustrates an ANFIS of this type. In the extreme case, we can even shrink the whole network into a single adaptive node with the same parameter set. Obviously, the assignment of node functions and the network configuration are arbitrary, as long as each node and each layer perform meaningful and modular functionalities.

The extension from Sugeno ANFIS to Tsukamoto ANFIS is straightforward, as shown in Figure 12.3, where the output of each rule (f_i, $i = 1, 2$) is induced jointly by a consequent membership function and a firing strength. For the Mamdani fuzzy inference system with max-min composition, a corresponding ANFIS can be constructed if discrete approximations are used to replace the integrals in the centroid defuzzification scheme introduced in Section 4.2. However, the resulting ANFIS is much more complicated than either Sugeno ANFIS or Tsukamoto ANFIS. The extra complexity in structure and computation of Mamdani ANFIS with max-min composition does not necessarily imply better learning capability or approximation power. If we adopt sum-product composition and centroid defuzzification for a Mamdani fuzzy model, a corresponding ANFIS can be constructed easily based on Theorem 4.1 directly without using any approximations at all. This is left as Exercise 4.

Throughout this chapter, we shall concentrate on the ANFIS architectures for the first-order Sugeno fuzzy model because of its transparency and efficiency.

Figure 12.4(a) is an ANFIS architecture that is equivalent to a two-input first-order Sugeno fuzzy model with nine rules, where each input is assumed to have three associated MFs. Figure 12.4(b) illustrates how the two-dimensional input space is partitioned into nine overlapping fuzzy regions, each of which is governed by a fuzzy if-then rule. In other words, the premise part of a rule defines a fuzzy region, while the consequent part specifies the output within the region.

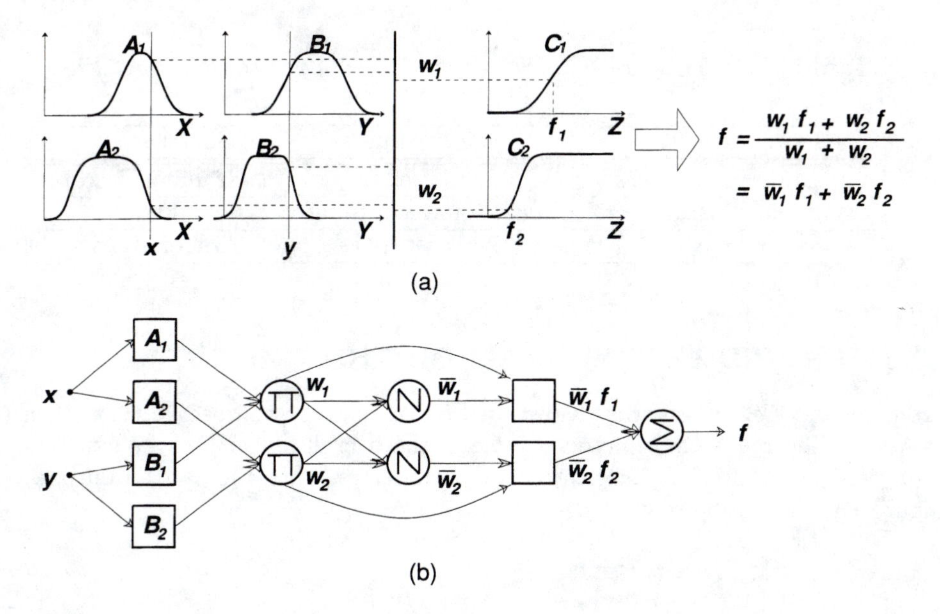

Figure 12.3. *(a) A two-input two-rule Tsukamoto fuzzy model; (b) equivalent ANFIS architecture.*

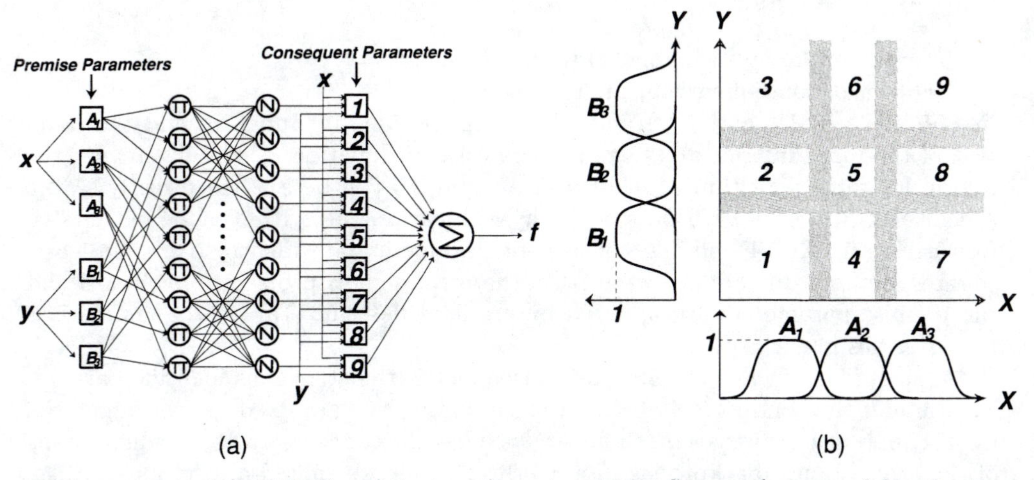

Figure 12.4. *(a) ANFIS architecture for a two-input Sugeno fuzzy model with nine rules; (b) the input space that are partitioned into nine fuzzy regions.*

Next we shall demonstrate how to apply the hybrid learning algorithms developed in Chapter 8 to identify ANFIS parameters.

Table 12.1. *Two passes in the hybrid learning procedure for ANFIS.*

	Forward pass	Backward pass
Premise parameters	Fixed	Gradient descent
Consequent parameters	Least-squares estimator	Fixed
Signals	Node outputs	Error signals

12.3 HYBRID LEARNING ALGORITHM

From the ANFIS architecture shown in Figure 12.1(b), we observe that when the values of the premise parameters are fixed, the overall output can be expressed as a linear combination of the consequent parameters. In symbols, the output f in Figure 12.1(b) can be rewritten as

$$\begin{aligned} f &= \frac{w_1}{w_1+w_2} f_1 + \frac{w_2}{w_1+w_2} f_2 \\ &= \overline{w}_1(p_1 x + q_1 y + r_1) + \overline{w}_2(p_2 x + q_2 y + r_2) \\ &= (\overline{w}_1 x)p_1 + (\overline{w}_1 y)q_1 + (\overline{w}_1)r_1 + (\overline{w}_2 x)p_2 + (\overline{w}_2 y)q_2 + (\overline{w}_2)r_2, \end{aligned} \tag{12.7}$$

which is linear in the consequent parameters p_1, q_1, r_1, p_2, q_2, and r_2. From this observation, we have

S = set of total parameters,
S_1 = set of premise (nonlinear) parameters,
S_2 = set of consequent (linear) parameters

in Equation (8.31); and $H(\cdot)$ and $F(\cdot,\cdot)$ are the identity function and the function of the fuzzy inference system, respectively, in Equation (8.32). Therefore, the hybrid learning algorithm developed in Section 8.5 can be applied directly. More specifically, in the forward pass of the hybrid learning algorithm, node outputs go forward until layer 4 and the consequent parameters are identified by the least-squares method. In the backward pass, the error signals propagate backward and the premise parameters are updated by gradient descent. Table 12.1 summarizes the activities in each pass.

As mentioned in Section 8.5, the consequent parameters thus identified are optimal under the condition that the premise parameters are fixed. Accordingly, the hybrid approach converges much faster since it reduces the search space dimensions of the original pure backpropagation method. Thus we should always look for the possibility of decomposing the parameter set in the first place. For Tsukamoto ANFIS, this can be achieved if the membership function on the consequent part of each rule is replaced by a piecewise linear approximation with two consequent parameters, as shown in Figure 12.5. In this case, again, the consequent parameters constitute the linear parameter set S_2 and the hybrid learning rule can be employed as before.

As discussed in Section 8.5, there are several ways of combining gradient descent

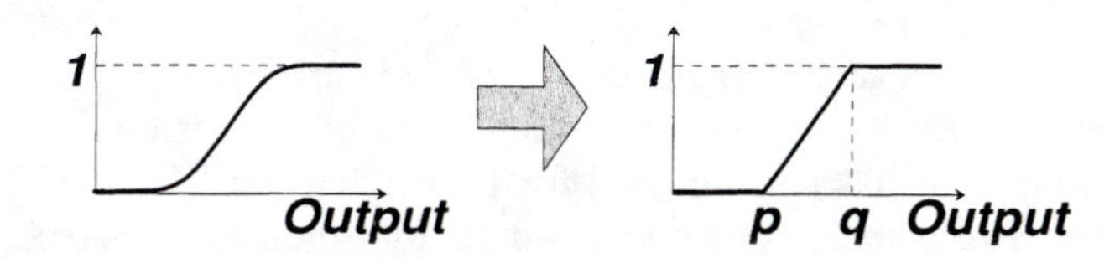

Figure 12.5. *Piecewise linear approximation of consequent MFs in Tsukamoto ANFIS.*

and the least-squares method. We can choose one of these methods according to the available computing resources and required performance level.

As pointed out by a reviewer of the original ANFIS paper [8], the learning mechanisms should not be applied to determine membership functions in Sugeno ANFIS, since they convey a linguistic and subjective descriptions of possibly ill-defined concepts. We think this is a case-by-case situation and the decision should be left to the user. In principle, if the size of the available input-output data set is large, then fine-tuning of the membership functions is recommended (or even necessary), since human-determined membership functions are seldom optimal in terms of reproducing desired outputs. However, if the data set is too small, then it probably does not contain enough information about the target system. In this situation, the human-determined membership functions represent important information that might not be reflected in the data set; therefore, the membership functions should be kept fixed throughout the learning process.

If the membership functions are fixed and only the consequent part is adjusted, Sugeno ANFIS can be viewed as a functional-link network [13, 30], where the "enhanced representations" of the input variables are obtained via the membership functions. These enhanced representations determined by human experts apparently provide more insight into the target system than the functional expansion or the tensor (outer product) models [30]. By updating the membership functions, we are actually tuning this enhanced representation for better performance.

Because the update formulas for the premise and consequent parameters are decoupled in the hybrid learning rule (see Table 12.1), further speedup of learning is possible by using variants of the gradient method or other optimization techniques on the premise parameters, such as conjugate gradient descent, second-order backpropagation [31], quick propagation [3], and many others. See also Chapter 6 for other derivative-based optimization methods.

12.4 LEARNING METHODS THAT CROSS-FERTILIZE ANFIS AND RBFN

As we discussed in Section 9.5.2, under certain minor conditions, an RBFN (radial basis function network) is functionally equivalent to a FIS, and thus adaptive FIS, including ANFIS (introduced in this chapter) and CANFIS (introduced in Chapter 13). This functional equivalence provides a shortcut for better understanding both ANFIS/CANFIS and RBFNs in the sense that developments in either literature cross-fertilize the other. In this section, we briefly describe a variety of adaptive learning mechanisms that can be used for both adaptive FIS and RBFN.

An adaptive FIS usually consists of two distinct modifiable parts: the antecedent part and the consequent part. These two parts can be adapted by different optimization methods, one of which is the hybrid learning procedure combining GD (gradient descent) and LSE (least-squares estimator), as discussed in Section 12.3. Possible combinations of GD and LSE are also discussed in the same section. These learning schemes are equally applicable to RBFNs.

Conversely, the analysis and learning algorithms for RBFNs are also applicable to adaptive FIS (ANFIS/CANFIS). The RBFN approximation capability may be further improved with supervised adjustments of the center and shape of receptive field functions [19, 44]. Besides using a supervised learning scheme alone to update all modifiable parameters, a variety of two-phase training algorithms for RBFNs have been reported. A typical scheme is to fix the receptive field (radial basis) functions first and then adjust the weights of the output layer. There are several schemes proposed to determine the center positions (u_i) of the receptive field functions. Lowe discussed selection of fixed centers based on standard deviations of training data [22]. Moody and Darken discussed unsupervised or self-organized selection of centers u_i by means of vector quantization or clustering techniques [25, 26] (see also Chapters 11 and 15). Then the width parameters σ_i are determined by by taking the average distance to the first several nearest neighbors of u_i's. Nowlan [29] employed the so-called *soft competition* among Gaussian hidden units to locate the centers. (This soft-competitive method is based on the maximum likelihood estimator, in contrast to the so-called *hard* competition such as the k-means winner-take-all algorithm.) Once these nonlinear parameters are fixed and the receptive fields are frozen, the linear parameters (i.e., the weights of the output layer) can be updated by either the least-squares method or the gradient method.

Chen et al. [2] used an alternative method that employs the orthogonal least-squares algorithm to determine the u_i's and C_i's while keeping the σ_i's at a pre-determined constant. Other RBFN analyses, such as generalization properties [1] and sequential adaptation [11], among others [9, 27], are all applicable to adaptive FIS (ANFIS/CANFIS).

12.5 ANFIS AS A UNIVERSAL APPROXIMATOR*

This section explains an intertesting property that when the number of rules is not restricted, a zero-order Sugeno model has unlimited approximation power for matching any nonlinear function arbitrarily well on a compact set. This fact is intuitively reasonable. However, to give a mathematical proof, we need to apply the Stone-Weierstrass theorem [12, 32].

Theorem 12.1 *Stone-Weierstrass theorem*

Let domain D be a compact space of N dimensions, and let $\mathcal{F}$ be a set of continuous real-valued functions on D satisfying the following criteria:

1. **Identity function**: The constant $f(x) = 1$ is in $\mathcal{F}$.

2. **Separability**: For any two points $x_1 \neq x_2$ in D, there is an f in $\mathcal{F}$ such that $f(x_1) \neq f(x_2)$.

3. **Algebraic closure**: If f and g are any two functions in $\mathcal{F}$, then fg and $af + bg$ are in F for any two real numbers a and b.

Then $\mathcal{F}$ is dense in $C(D)$, the set of continuous real-valued functions on D. In other words, for any $\epsilon > 0$ and any function g in $C(D)$, there is a function f in $\mathcal{F}$ such that $|g(x) - f(x)| < \epsilon$ for all $x \in D$.

□

In applications of fuzzy inference systems, the domain in which we operate is almost always compact. It is a standard result in real analysis that every closed and bounded set in R^N is compact. In what follows, we shall describe how to apply the Stone-Weierstrass theorem to show the universal approximation power of the zero-order Sugeno model.

Identity Function

The first hypothesis of the Stone-Weierstrass theorem requires that our fuzzy inference system be able to compute the identity function $f(x) = 1$. An obvious way to compute this function is to set the consequence part of each rule equal to 1. A fuzzy inference system with only one rule suffices to satisfy this requirement.

Separability

The second hypothesis of the Stone-Weierstrass theorem requires that our fuzzy inference system be able to compute functions that have different values for different points. [Without this requirement, the trivial set of functions $f : f(x) = c,\ c \in R$ would satisfy the Stone-Weierstrass theorem.] Again, this is obviously achievable by any fuzzy inference system with appropriate parameters.

Algebraic Closure—Additive

The third hypothesis of the Stone-Weierstrass theorem requires that our fuzzy inference systems be invariant under addition and multiplication. Suppose that we have two fuzzy inference systems S and $\hat{S}$; each of them has two rules, and the final output of each system is specified as

$$S : z = \frac{w_1 f_1 + w_2 f_2}{w_1 + w_2}, \tag{12.8}$$

and

$$\hat{S} : \hat{z} = \frac{\hat{w}_1 \hat{f}_1 + \hat{w}_2 \hat{f}_2}{\hat{w}_1 + \hat{w}_2}. \tag{12.9}$$

Then the sum of z and $\hat{z}$ is equal to

$$\begin{aligned} az + b\hat{z} &= a\frac{w_1 f_1 + w_2 f_2}{w_1 + w_2} + b\frac{\hat{w}_1 \hat{f}_1 + \hat{w}_2 \hat{f}_2}{\hat{w}_1 + \hat{w}_2} \\ &= \frac{w_1\hat{w}_1(af_1 + b\hat{f}_1) + w_1\hat{w}_2(af_1 + b\hat{f}_2) + w_2\hat{w}_1(af_2 + b\hat{f}_1) + w_2\hat{w}_2(af_2 + b\hat{f}_2)}{w_1\hat{w}_1 + w_1\hat{w}_2 + w_2\hat{w}_1 + w_2\hat{w}_2} \end{aligned}$$

Thus we can construct a four-rule fuzzy inference system that computes $az + b\hat{z}$, where the firing strength and the output of each rule are defined by $w_i\hat{w}_j$ and $af_i + b\hat{f}_j$ $(i, j = 1 \text{ or } 2)$, respectively.

Algebraic Closure—Multiplicative

Invariantness under multiplication is the final feature we must demonstrate before we can conclude that the Stone-Weierstrass theorem can be applied to the zero-order Sugeno fuzzy model. The product of the outputs of two fuzzy inference systems z and $\hat{z}$ can be expressed as

$$z\hat{z} = \frac{w_1\hat{w}_1 f_1\hat{f}_1 + w_1\hat{w}_2 f_1\hat{f}_2 + w_2\hat{w}_1 f_2\hat{f}_1 + w_2\hat{w}_2 f_2\hat{f}_2}{w_1\hat{w}_1 + w_1\hat{w}_2 + w_2\hat{w}_1 + w_2\hat{w}_2}. \tag{12.10}$$

Thus we can construct a four-rule fuzzy inference system that computes $z\hat{z}$, where the firing strength and the output of each rule are defined by $w_i\hat{w}_j$ and $f_i\hat{f}_j$ $(i, j = 1 \text{ or } 2)$, respectively.

From the preceding description, we conclude that the ANFIS architectures that compute $az + b\hat{z}$ and $z\hat{z}$ are of the same class as those of S and $\hat{S}$ if and only if the membership functions used are invariant under multiplication. One class of MFs that satisfy this property is the scaled Gaussian membership function [39, 41]:

$$\mu_{A_i}(x) = k_i \exp[-(\frac{x - c_i}{a_i})^2]. \tag{12.11}$$

Another class of MFs that are invariant under the product operator is MFs for crisp sets, which assume values of either 0 or 1. MFs of this kind can be viewed as a special case of either the generalized bell MF with parameter b approaching ∞, or as the trapezoidal MF with $a = b$ and $c = d$ (see Section 2.4.1).

Therefore, with an appropriate class of membership functions, a zero-order Sugeno model can satisfy the four criteria of the Stone-Weierstrass theorem. That is, for any given $\epsilon > 0$ and any real-valued function g, there is a zero-order Sugeno model S such that $|g(\vec{x}) - S(\vec{x})| < \epsilon$ for all $\vec{x}$ in the underlying compact set. The preceding argument of universal approximation power applies to other types of fuzzy models as well, since the zero-order Sugeno model is a special case of the Mamdani fuzzy model, the Tsukamoto fuzzy model, and other higher-order Sugeno models.

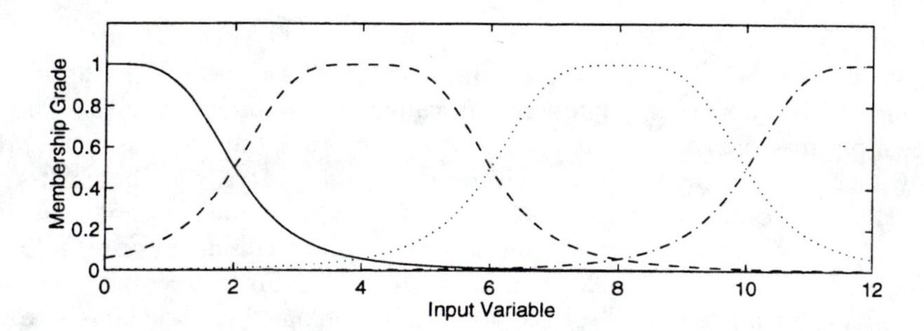

Figure 12.6. *A typical initial MF setting, where input range is assumed to be* [0, 12]. (MATLAB file: `init_mf.m`)

However, caution should be taken in accepting this claim, since there has been no mention of how to construct the Sugeno model according to a given training data set; the Stone-Weierstrass theorem yields only an existence theorem, but not a constructive method.

12.6 SIMULATION EXAMPLES

This section presents simulation results of the ANFIS architecture for the Sugeno fuzzy model (see Figures 12.1 and 12.4). In the first two examples, ANFIS is used to model two nonlinear functions; the results are compared with those achieved by backpropagation MLP approaches (see Section 9.4) and other earlier work on fuzzy modeling. In the third example, we use ANFIS for on-line identification of a nonlinear component in a discrete control system. In the last example, we predict a chaotic time series using ANFIS and demonstrate its superiority to several standard statistical and neural network approaches. The purpose of these examples is to give a detailed description of how to use ANFIS and how it performs; more ANFIS applications for a number of different domains are given in Chapter 19.

12.6.1 Practical Considerations

In a conventional fuzzy inference system, the number of rules is determined by an expert who is familiar with the target system to be modeled. In our simulation, however, no expert is available and the number of MFs assigned to each input variable is chosen empirically—that is, by plotting the data sets and examining them visually, or simply by trial and error. For data sets with more than three inputs, visualization techniques are not very effective and most of the time we have to rely on trial and error. This situation is similar to that of neural networks; there is just no simple way to determine in advance the minimal number of hidden units needed to achieve a desired performance level. (There are several other techniques for determining the numbers of MFs and rules, such as CART and clustering methods. But here we shall not use them since the purpose of this section is to demonstrate the learning capability of ANFIS.)

If we choose the grid partition in Figure 4.13(a) (see page 87), the number of MFs on each input variable uniquely determines the number of rules. The initial values of premise parameters are set in such a way that the centers of the MFs are equally spaced along the range of each input variable. Moreover, these MFs satisfy the condition of **ϵ-completeness** [17, 18] with $\epsilon = 0.5$, which means that given a value x of one of the inputs in the operating range, we can always find a linguistic label A such that $\mu_A(x) \geq \epsilon$. In this manner, the fuzzy inference system can provide smooth transition and sufficient overlap from one linguistic label to another. Although we did not attempt to maintain the ϵ-completeness during the training process, it can be easily achieved by using a constrained gradient method [48]. Figure 12.6 shows a typical initial MF setting when the number of MFs is four and the input range is $[0, 12]$. Throughout the simulation examples presented in this section, all the membership functions used are the generalized bell function defined in Equation (12.2):

$$\mu_A(x) = \text{gbell}(x; a, b, c) = \frac{1}{1 + \left|\frac{x-c}{a}\right|^{2b}}, \tag{12.12}$$

which contains three fitting parameters, a, b, and c. Each of these parameters has a physical meaning: c determines the center of the MF; a is half the width of the MF; and b (together with a) controls the slopes at the crossover points (where the MF value is 0.5). Figure 2.9 in Chapter 2 illustrates these concepts.

We mentioned that the step size κ in Equation (6.61) may influence the speed of convergence. In the simulation reported next, we use two heuristic guidelines to update the step size κ adaptively. These two guidelines (see also Figure 6.18) and the general observations that lead to them are detailed in Section 6.7.2 of Chapter 6.

12.6.2 Example 1: Modeling a Two-Input Sinc Function

In this example, we use ANFIS to model a two-dimensional sinc equation defined by

$$z = \text{sinc}(x, y) = \frac{\sin(x)\sin(y)}{xy}. \tag{12.13}$$

From the evenly distributed grid points of the input range $[-10, 10] \times [-10, 10]$ of the preceding equation, 121 training data pairs were obtained. The ANFIS used here contains 16 rules, with four membership functions assigned to each input variable. The total number of fitting parameters is 72, including 24 premise (nonlinear) parameters and 48 consequent (linear) parameters. (We also tried ANFIS models with four rules and nine rules, but these models are too simple to describe the highly nonlinear sinc function.)

Figure 12.7 shows the **RMSE** (**root mean squared error**) curves for both a 2-18-1 backpropagation MLP and the ANFIS architecture used here. Each curve is the result of averaging 10 error curves from 10 runs. For the MLP, these 10 runs

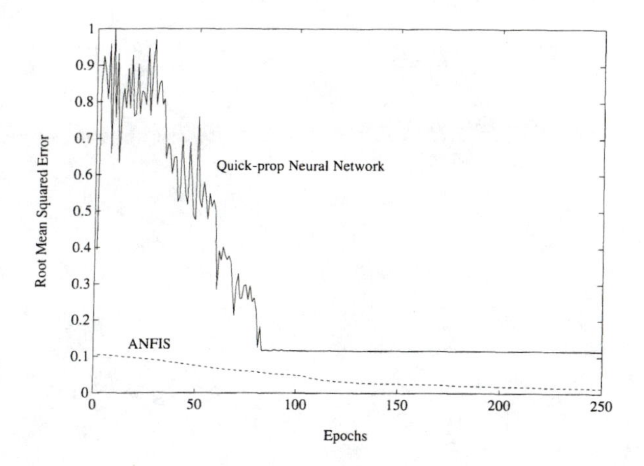

Figure 12.7. *RMSE curves for the MLP and ANFIS.*

were started from different sets of initial random weights. For ANFIS, these 10 runs correspond to 10 κ values ranging from 0.01 to 0.10.

The backpropagation MLP, which contained 73 fitting parameters (connection weights and thresholds), was trained with quick propagation [3], which is considered one of the best learning algorithms for backpropagation MLPs. Figure 12.7 shows how ANFIS approximates a highly nonlinear surface more effectively than an MLP. It should be emphasized that for the same number of epochs (250 in Figure 12.7), the ANFIS model did take longer since the hybrid learning rule involves more computation. However, even we increased the training epochs for the MLP, its performance stayed the same since its error curve levels off after 100 epochs, as shown in Figure 12.7. The poor performance of MLPs seems due to their structure: The learning processes could become trapped in local minima because of the randomly initialized weights, or some neurons could be pushed into saturation during the training. Either of these two situations can significantly decrease the approximation power of MLPs.

The training data and reconstructed surfaces at different epochs during training are depicted in Figure 12.8. Since the error measure is always computed after a forward pass (that is, the first half of a whole epoch) is completed, the epoch numbers shown in the caption of Figure 12.8 always end with .5. (In our later descriptions of ANFIS applications in Chapter 19, we will round these epoch numbers to the next integers for simplicity.) Note that the reconstructed surface after 0.5 epochs is the result after identifying consequent parameters using LSE for the first time; yet it already looks similar to the training data surface.

Figure 12.9 lists the initial and final membership functions. It is interesting to observe that the sharp changes in the training data surface around the origin are accounted for by the membership functions moving toward the origin. Theo-

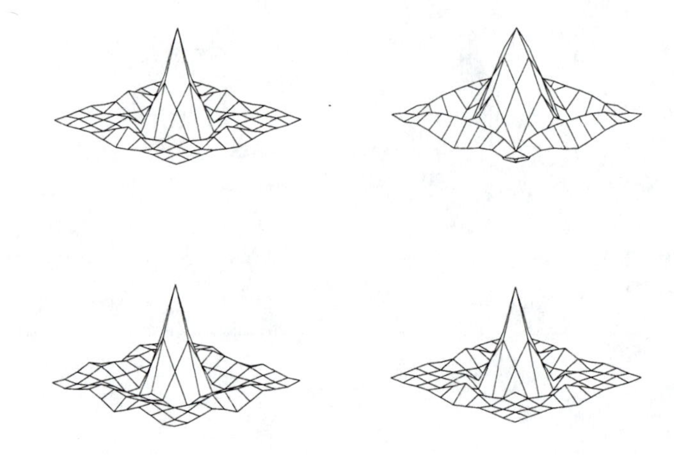

Figure 12.8. *Training data (upper left) and reconstructed surfaces at 0.5 (upper right), 99.5 (lower left), and 249.5 (lower right) epochs in Example 1.* (MATLAB file: `trn_2in.m`)

retically, the final MFs on both x and y should be symmetric with respect to the origin. However, they are not symmetric, due to computer truncation errors and the approximated initial conditions used for bootstrapping the recursive least-squares estimator.

12.6.3 Example 2: Modeling a Three-Input Nonlinear Function

The training data in this example were obtained from a three-input nonlinear equation defined by

$$output = (1 + x^{0.5} + y^{-1} + z^{-1.5})^2. \tag{12.14}$$

This equation was also used by Takagi and Hayashi [36], Sugeno and Kang [34], and Kondo [14] to test their modeling approaches. Here the ANFIS architecture (see Figure 12.10) contains eight rules, with two membership functions assigned to each input variable. A total of 216 training data and 125 checking data were sampled uniformly from the input ranges $[1,6]\times[1,6]\times[1,6]$ and $[1.5,5.5]\times[1.5,5.5]\times[1.5,5.5]$, respectively. The training data were used for training ANFIS, while the checking data were used for verifying the identified ANFIS only. To allow comparison, we use the same performance index adopted in refs. [34, 14]:

$$\text{APE} = \text{average percentage error} = \frac{1}{P}\sum_{i=1}^{P}\frac{|\,T(i) - O(i)\,|}{|\,T(i)\,|}\cdot 100\%, \tag{12.15}$$

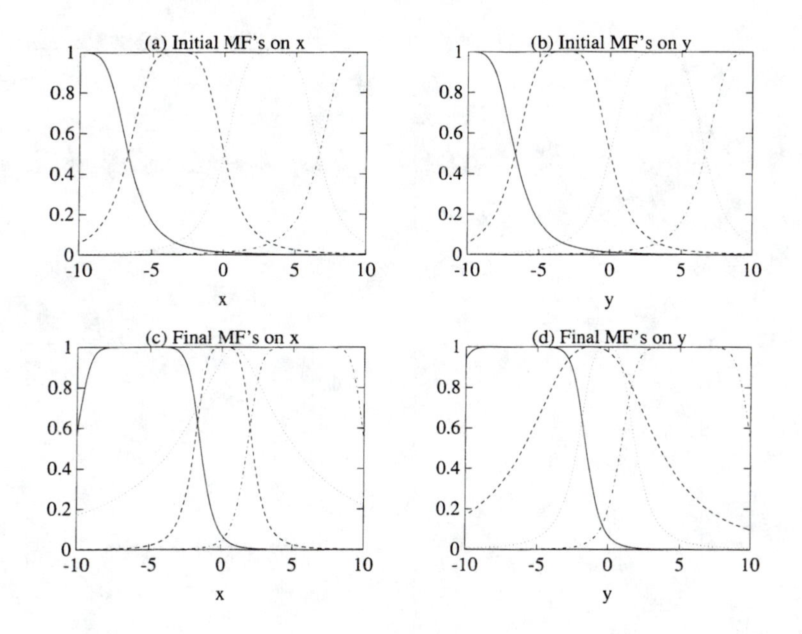

Figure 12.9. *Initial and final MFs in Example 1.* (MATLAB file: `trn_2in.m`)

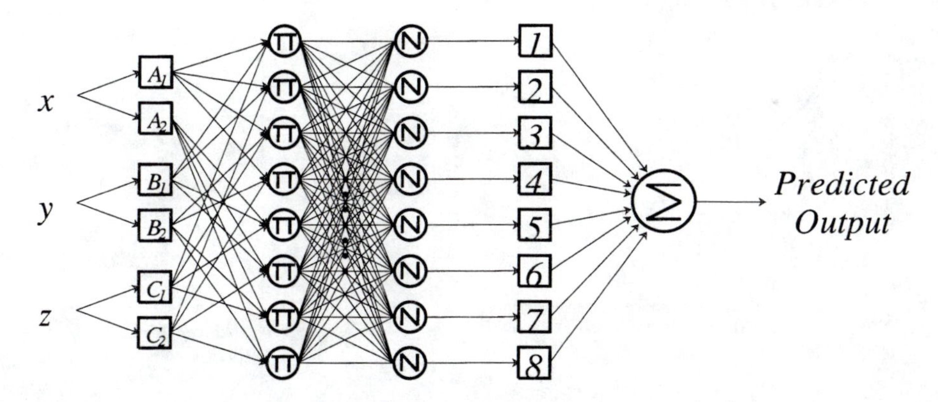

Figure 12.10. *The ANFIS model for Example 2. (The connections from inputs to layer 4 are not shown.)*

where P is the number of data pairs, and $T(i)$ and $O(i)$ are the ith desired output and predicted output, respectively.

Figure 12.11 illustrates the membership functions before and after training. The training error curves with different initial step sizes ($\kappa = 0.01$ to 0.09) are shown in Figure 12.12(a), which indicates that the initial value of κ does not have a critical influence on the final performance as long as κ is not too large. Figure 12.12(b)

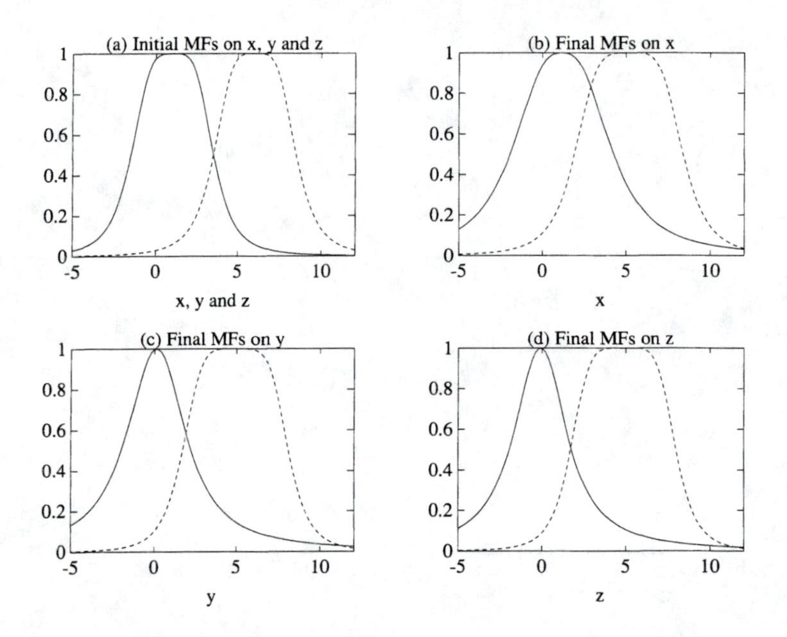

Figure 12.11. *Example 2: (a) MFs before learning; (b), (c), (d) MFs after learning.* (MATLAB file: `trn_3in.m`)

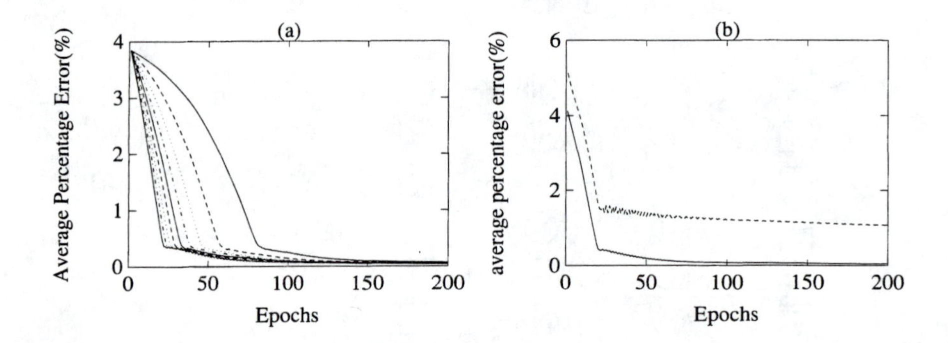

Figure 12.12. *Error curves of Example 2: (a) nine training error curves for nine initial step sizes from 0.01 (rightmost) to 0.09 (leftmost); (b) training (solid line) and checking (dashed line) error curves with initial step size equal to 0.1.* (MATLAB file: `trn_3in.m`)

shows the training and checking error curves with initial step size equal to 0.1. After 199.5 epochs, the final results were $APE_{\text{trn}} = 0.043\%$ and $APE_{\text{chk}} = 1.066\%$, which are listed in Table 12.2 along with the results of other earlier work [14, 34]. Here ANFIS achieves the best performance at the cost of requiring more training data.

Table 12.2. *Example 2: comparisons with earlier work. The last three rows are from ref. [34].*

Model	Training error	Checking error	Param. number	Training data size	Checking data size
ANFIS	0.043%	1.066%	50	216	125
GMDH model [14]	4.7%	5.7%	—	20	20
Fuzzy model 1 [34]	1.5%	2.1%	22	20	20
Fuzzy model 2 [34]	0.59%	3.4%	32	20	20

12.6.4 Example 3: On-line Identification in Control Systems

Here we repeat a simulation example from ref. [28], where a 1-20-10-1 backpropagation MLP is employed to identify a nonlinear component in a control system, except that here we use ANFIS instead to show its superiority. The plant under consideration is governed by the following difference equation:

$$y(k+1) = 0.3y(k) + 0.6y(k-1) + f(u(k)), \tag{12.16}$$

where $y(k)$ and $u(k)$ are the output and input, respectively, at time step k. The unknown function $f(\cdot)$ has the form

$$f(u) = 0.6\sin(\pi u) + 0.3\sin(3\pi u) + 0.1\sin(5\pi u). \tag{12.17}$$

In order to identify the plant, a series-parallel model governed by the difference equation

$$\hat{y}(k+1) = 0.3\hat{y}(k) + 0.6\hat{y}(k-1) + F(u(k)) \tag{12.18}$$

was used, where $F(\cdot)$ is the function implemented by ANFIS and its parameters are updated at each time step. Here the ANFIS architecture has seven MFs on its input (thus seven rules with 35 fitting parameters) and the on-line learning paradigm adopted has a learning rate $\eta = 0.1$ and a forgetting factor $\lambda = 0.99$. The input to the plant and the model was a sinusoid $u(k) = sin(2\pi k/250)$; the adaptation started at $k = 1$ and stopped at $k = 250$. As shown in Figure 12.13, the output of the model follows the output of the plant almost immediately, even after the adaptation was stopped at $k = 250$ and the $u(k)$ was changed to $0.5sin(2\pi k/250) + 0.5sin(2\pi k/25)$ at $k = 500$. In comparison, the MLP in ref. [28] failed to follow the plant when the adaptation was stopped at $k = 500$, and the identification procedure had to continue for $50,000$ time steps using a random input. Table 12.3 summarizes the comparison.

In the preceding simulation, the number of rules is determined by trial and error. If the number of MFs is below seven, then the model output will not follow the plant output satisfactorily after 250 adaptations. By using the more effective off-line learning, we can decrease the number of rules. Figures 12.14, 12.15, and 12.16

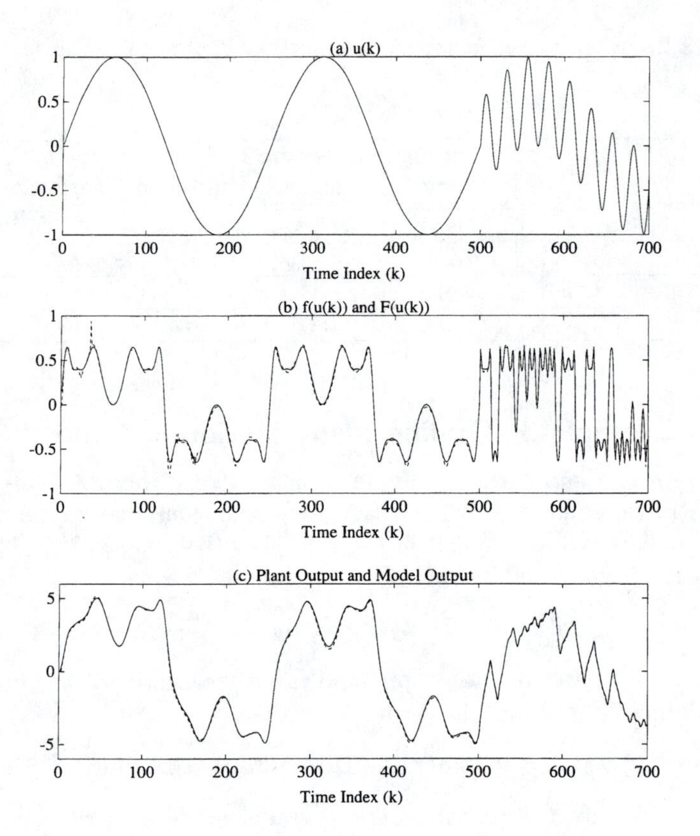

Figure 12.13. *Example 3: (a) $u(k)$; (b) $f(u(k))$ and $F(u(k))$; (c) plant and model outputs.*

Table 12.3. *Example 3: comparison with MLP identifier [28].)*

Method	Parameter number	Time steps of adaptation
MLP	261	50000
ANFIS	35	250

show the results after 49.5 epochs of off-line learning when the number of MFs is 5, 4, and 3, respectively. From these figures, it is obvious that ANFIS is a good model even when as few as three MFs are used. However, as the number of rules becomes smaller, the relationship between $F(u)$ and each rule's output becomes less clear, in the sense that it is harder to sketch $F(u)$ from each rule's output. In

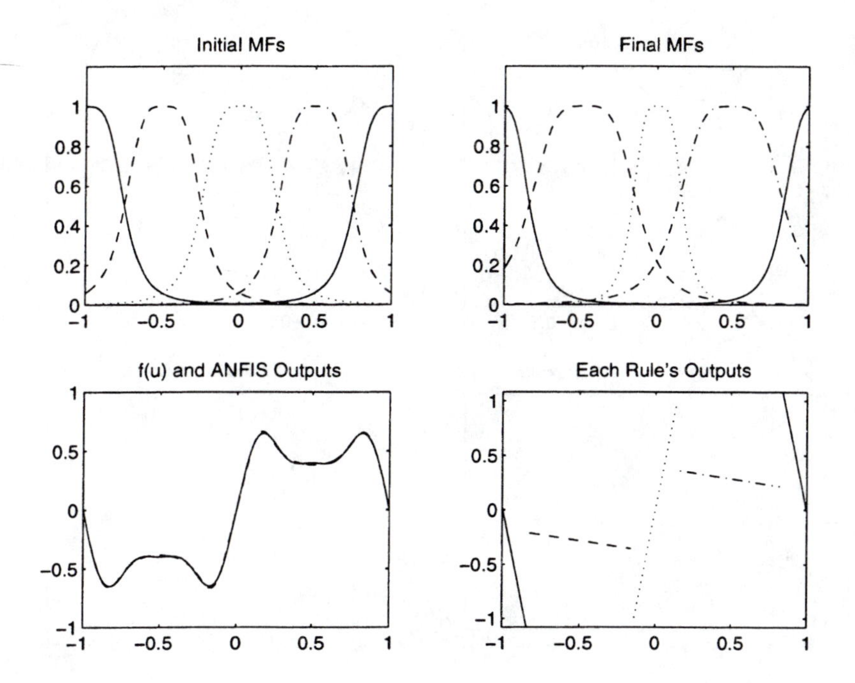

Figure 12.14. *Example 3: off-line learning with five MFs.* (MATLAB command: `trn_1in(5)`)

other words, when the number of parameters is reduced moderately, ANFIS usually still does a satisfactory job, but at the cost of sacrificing its semantics in terms of the local-description nature of fuzzy if-then rules. In this case, ANFIS is less of a structured knowledge representation and more like a black-box model, such as a backpropagation MLP.

12.6.5 Example 4: Predicting Chaotic Time Series

Examples 1, 2, and 3 demonstrate the capability of ANFIS for modeling nonlinear functions. In this example, we demonstrate how ANFIS can be employed to predict future values of a chaotic time series. The performance obtained in this example is compared with the results of a cascade-correlation neural network approach [7] and the conventional auto-regressive (AR) model.

The time series used in our simulation is generated by the chaotic Mackey-Glass differential delay equation [23] defined as

$$\dot{x}(t) = \frac{0.2x(t-\tau)}{1+x^{10}(t-\tau)} - 0.1x(t). \tag{12.19}$$

The prediction of future values of this time series is a benchmark problem that has been used and reported by a number of connectionist researchers, such as Lapedes

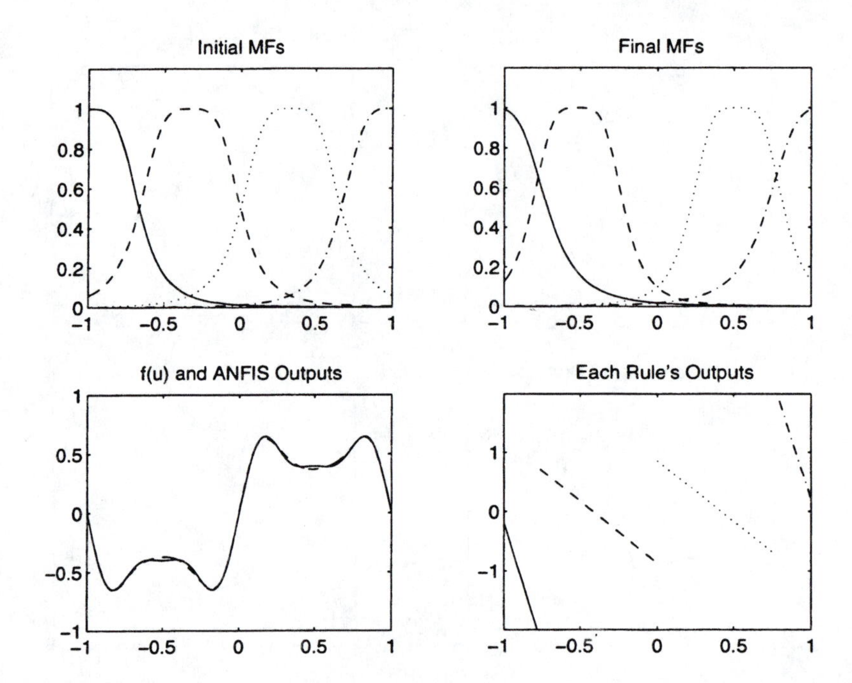

Figure 12.15. *Example 3: off-line learning with four MFs.* (MATLAB command: `trn_lin(4)`)

and Farber [16], Moody [24, 26], Jones et al. [9], Crowder [7], and Sanger [33].

The goal of the task is to use past values of the time series up to time t to predict the value at some point in the future $t + P$. The standard method for this type of prediction is to create a mapping from D points of the time series spaced Δ apart—that is, $[x(t - (D - 1)\Delta), \ldots, x(t - \Delta), x(t)]$, to a predicted future value $x(t + P)$. To allow comparison with earlier work (Lapedes and Farber [16], Moody [24, 26], Crowder [7]), the values $D = 4$ and $\Delta = P = 6$ were used. All other simulation settings were arranged to be as close as possible to those reported in ref. [7].

To obtain the time series value at each integer time point, we applied the fourth-order Runge-Kutta method to find the numerical solution to Equation (12.19). The time step used in the method was 0.1, initial condition $x(0) = 1.2$, and $\tau = 17$. In this way, $x(t)$ was thus obtained via numerical integration for $0 \leq t \leq 2000$. [We assume $x(t) = 0$ for $t < 0$ in the integration.] From the Mackey-Glass time series $x(t)$, we extracted 1000 input-output data pairs of the following format:

$$[x(t - 18), x(t - 12), x(t - 6), x(t); x(t + 6)], \tag{12.20}$$

where $t = 118$ to 1117. The first 500 pairs were used as the training data set for ANFIS, while the remaining 500 pairs were the checking data set for validating

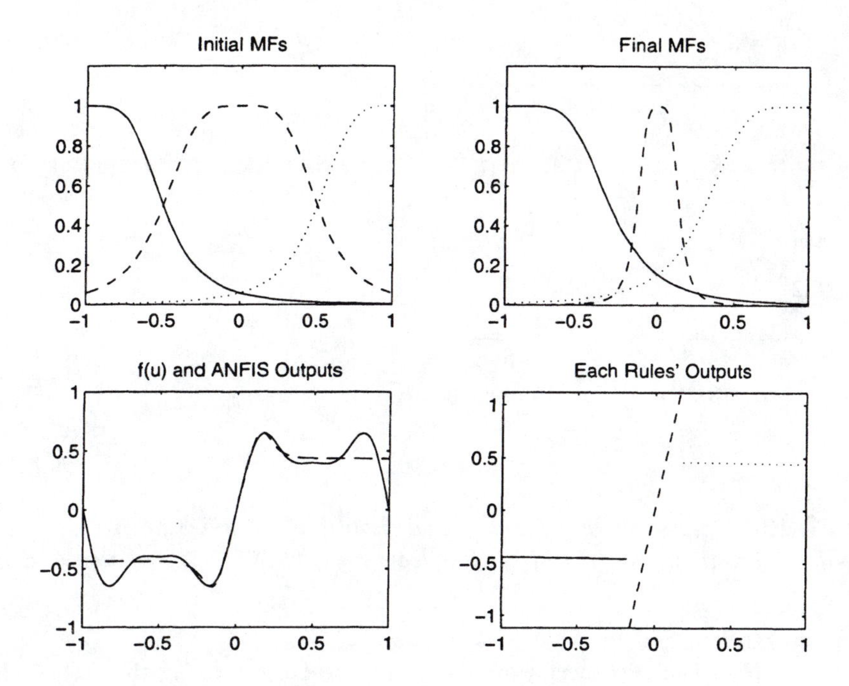

Figure 12.16. *Example 3: off-line learning with three MFs.* (MATLAB command: `trn_1in(3)`)

the identified ANFIS. The number of MFs assigned to each input of the ANFIS was set to two, so the number of rules is 16. Figure 12.17(a) depicts the initial membership functions for each input variable. The ANFIS used here contains a total of 104 fitting parameters, of which 24 are premise (nonlinear) parameters and 80 are consequent (linear) parameters

After 499.5 epochs, we had $RMSE_{\text{trn}} = 0.0016$ and $RMSE_{\text{chk}} = 0.0015$, which are much better than the results of the other approaches, as will be explained later. The desired and predicted values for both training data and checking data are essentially the same in Figure 12.18(a); the differences between them can only be seen on a much finer scale, as shown in Figure 12.18(b). Figure 12.17(b) is the final membership functions; Figure 12.19 shows the RMSE curves, which indicate that most of the learning was done in the first 100 epochs. It is unusual to observe that $RMSE_{\text{trn}} > RMSE_{\text{chk}}$ during the training process, as is the case here. (If we change the role of the training and checking data set, then we have the usual situation in which $RMSE_{\text{trn}} < RMSE_{\text{chk}}$ during the learning process.) Since both RMSEs are both very small, we conjecture that (1) the ANFIS used here has captured the essential components of the underlying dynamics; and (2) the training data contain the effects of the initial conditions [remember that we set $x(t) = 0$ for $t \leq 0$ in the integration], which might not be easily accounted for by the essential

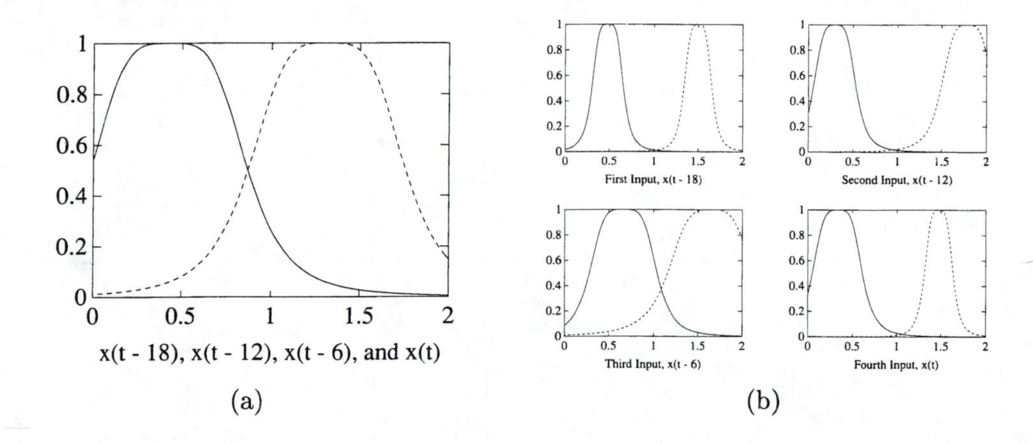

Figure 12.17. *Membership functions in chaotic time series prediction: (a) initial MFs for all four inputs; (b) MFs after learning.* (MATLAB file: `trn_4in.m`)

components identified by ANFIS.

As a comparison, we performed the same prediction using the AR model with the same number of parameters:

$$x(t+6) = a_0 + a_1 x(t) + a_2 x(t-6) + \cdots + a_{103} x(t - 102 * 6), \tag{12.21}$$

where there are 104 fitting parameters a_k, $k = 0$ to 103. From $t = 712$ to 1711, we extracted 1000 data pairs, of which the first 500 were used to identify a_k and the remainder were used for checking. The results obtained through the standard least-squares method were $RMSE_{\text{trn}} = 0.005$ and $RMSE_{\text{chk}} = 0.078$, which are much worse. Figure 12.20 shows the predicted values and the prediction errors. Obviously, the over-parameterization of the AR model causes over-fitting in the training data, which produces large errors in the checking data. To search for an appropriate AR model in terms of the best generalization capability, we tried different AR models with the number of parameters varying from 2 to 104. Figure 12.21 displays the results; the AR model with the best generalization capability is obtained when the number of parameters is 45. Using this AR model, we repeated the generalization test. Figure 12.22 shows the results; in this case, there is no over-fitting, at the price of larger training errors.

The nonlinear ANFIS obviously outperforms the linear AR model. However, identification of the AR model took only a few seconds, while the ANFIS simulation took about 1.5 hours on an HP (Hewlett-Packard) Apollo 700 Series workstation.

Table 12.4 lists the generalization capabilities of other methods, which were measured by using each method to predict 500 points immediately following the training set. Here the **non-dimensional error index (NDEI)** [16, 7] is defined as

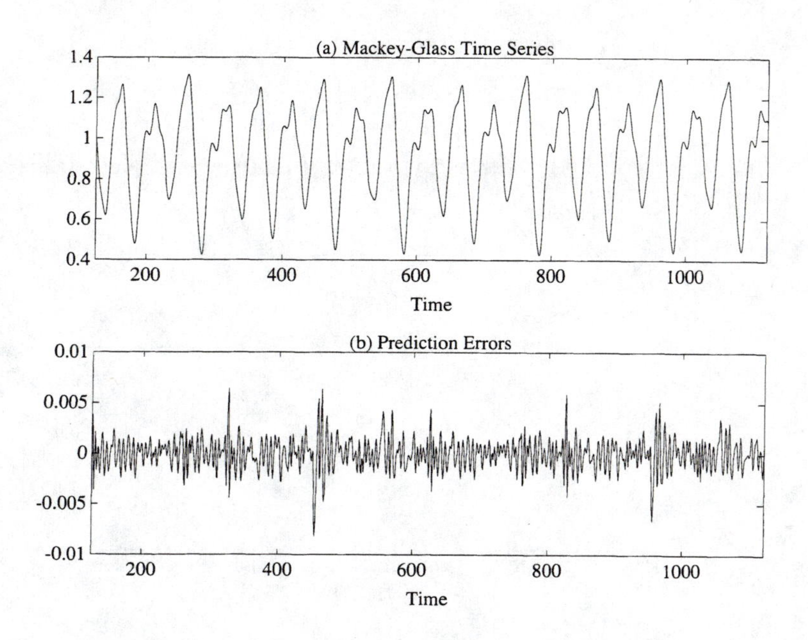

Figure 12.18. *Example 3, (a) Mackey-Glass time series from* $t = 124$ *to* 1123 *and six-step-ahead prediction (which is indistinguishable from the time series here); (b) prediction error.* (MATLAB file: `trn_4in.m`)

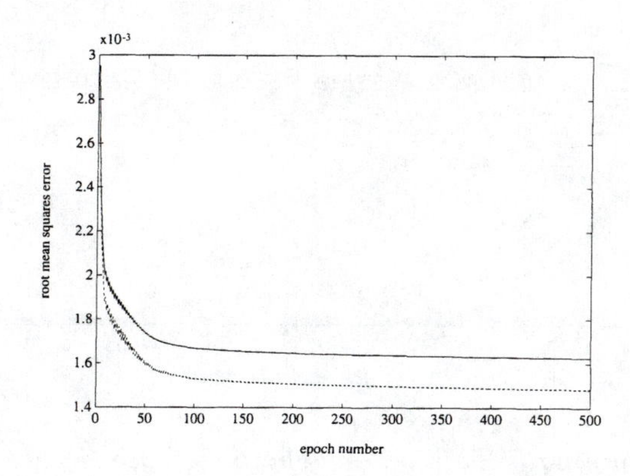

Figure 12.19. *Training (solid line) and checking (dashed line) RMSE curves for ANFIS modeling.* (MATLAB file: `trn_4in.m`)

the root mean square error divided by the standard deviation of the target series. (Note that the **average relative variance** used in refs. [42, 43] is equal to the square of NDEI.) The remarkable generalization capability of ANFIS, we believe, is derived from the following facts:

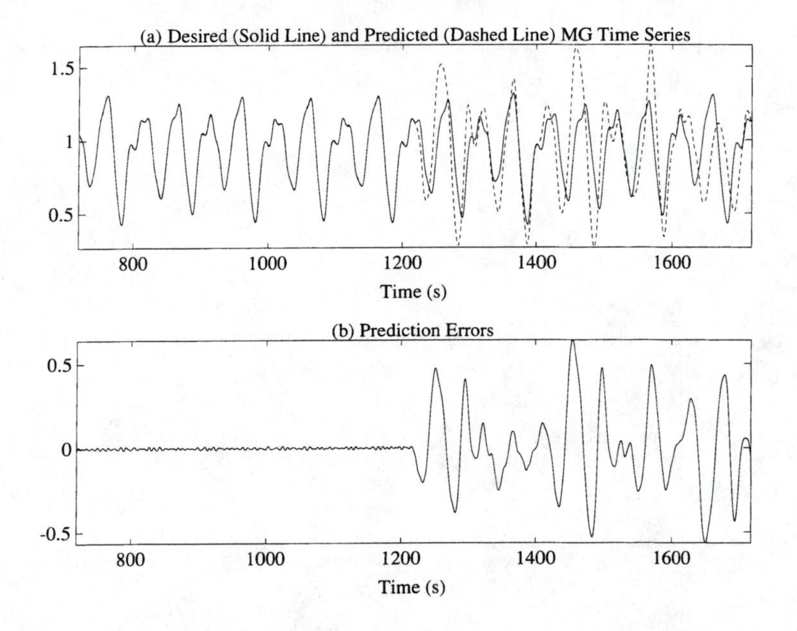

Figure 12.20. *(a) Mackey-Glass time series (solid line) from* $t = 718$ *to* 1717 *and six-step-ahead prediction (dashed line) by AR model with parameter* = 104; *(b) prediction errors. (The first 500 data points are training data, while the remaining are for validation.)*

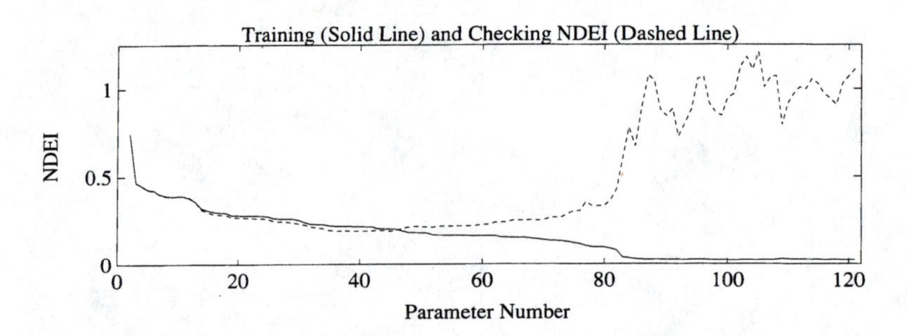

Figure 12.21. *Training (solid line) and checking (dashed line) errors of AR models with numbers of parameters varying from 2 to 104.*

- ANFIS can achieve a highly nonlinear mapping, as shown in Examples 1, 2, and 3. Therefore, it is superior to common linear methods in reproducing nonlinear time series.

- The ANFIS used here has 104 adjustable parameters, far fewer than those used in the cascade-correlation NN (693, the median) and backpropagation

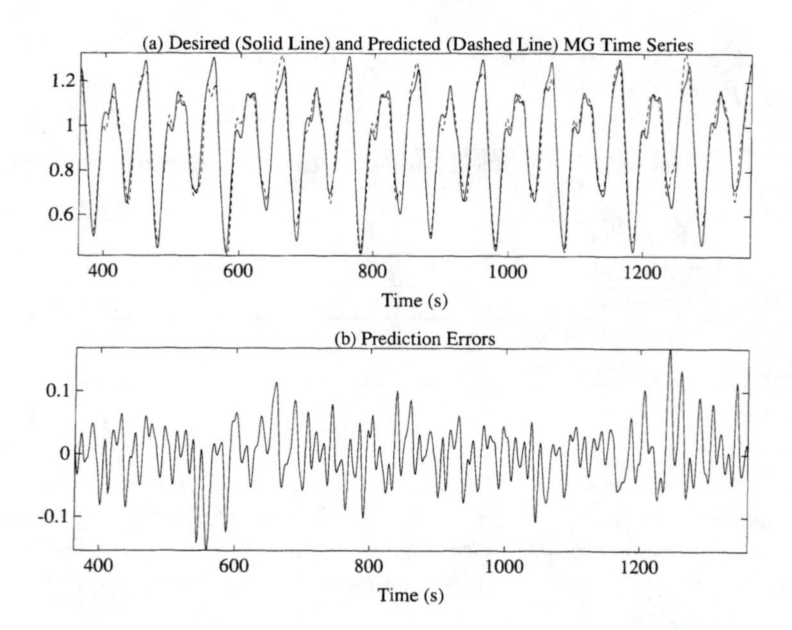

Figure 12.22. *Example 3, (a) Mackey-Glass time series (solid line) from* $t = 364$ *to* 1363, *and six-step-ahead prediction (dashed line) by the best AR model with 45 parameters; (b) prediction errors.*

MLP (about 540) listed in Table 12.4.

- Although not based on *a priori* knowledge, the initial parameters of ANFIS are intuitively reasonable and all the input space is covered properly; this results in fast convergence to good parameter values that captures the underlying dynamics.

- ANFIS consists of fuzzy rules which are local mappings (which are called local experts in ref. [10]) instead of global ones. These local mappings facilitate the **minimal disturbance principle** [45], which states that the adaptation should not only reduce the output error for the current training pattern but also minimize disturbance to response already learned. This is particularly important in on-line learning. We also found that the use of least-squares method to determine the output of each local mapping is of particular importance. Without using LSE, the learning time would have been 5 to 10 times longer.

Table 12.5 lists the results of the more challenging generalization test, in which P is 84 and 85 for rows 1 through 6 and 7 through 10, respectively. The results of the first six rows were obtained by iterating the prediction of $P = 6$ until $P = 84$. ANFIS still outperformed these statistical and connectionist approaches in all cases

Table 12.4. *Comparison of generalization capability for $P = 6$. (The last four rows are from [7].)*

Method	Training cases	Non-dimensional error index
ANFIS	500	0.007
AR model	500	0.19
Cascaded-correlation NN	500	0.06
Backpropagation MLP	500	0.02
6th-order polynomial	500	0.04
Linear predictive method	2000	0.55

Table 12.5. *Comparison of generalization capability for $P = 84$ (the first six rows) and 85 (the last four rows). Results for the first six methods are generated by iterating the solution at $P = 6$. Results for localized receptive fields (LRFs) and multiresolution hierarchies (MRHs) are for networks trained for $P = 85$. The last eight rows are from ref. [7].*

Method	Training cases	Non-dimensional error index
ANFIS	500	0.036
AR model	500	0.39
Cascaded-correlation NN	500	0.32
Backpropagation MLP	500	0.05
6th-order polynomial	500	0.85
Linear predictive method	2000	0.60
LRF	500	0.10 – 0.25
LRF	10000	0.025 – 0.05
MRH	500	0.05
MRH	10000	0.02

except where a substantially larger amount of training data were used (e.g., the last row of Table 12.5). Figure 12.23 depicts the generalization test results for ANFIS when $P = 84$.

12.7 EXTENSIONS AND ADVANCED TOPICS

Because of the extreme flexibility of adaptive networks, ANFIS can be generalized in a number of different ways. For instance, the membership functions can be changed to any of the parameterized MFs described in Section 2.4 of Chapter 2.

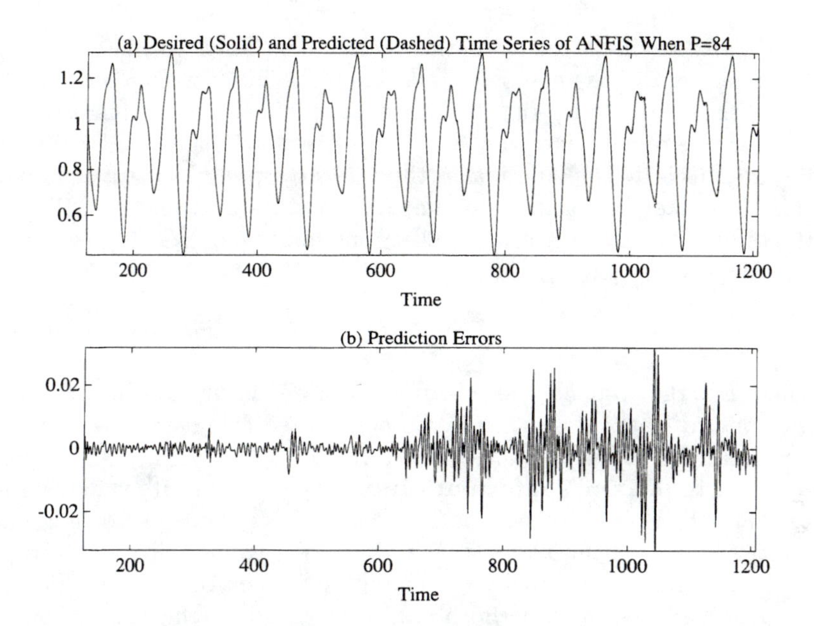

Figure 12.23. *Generalization test of ANFIS for* $P = 84$.

Furthermore, we can replace the Π nodes in layer 2 with the parameterized T-norm (see Section 2.5.3 of Chapter 2) and let the learning rule decide the best T-norm operator for a specific application. Moreover, the realization of rules with OR'ed antecedents, linguistic hedges (Section 3.3.1), and multiple outputs can be put into ANFIS accordingly. An extended ANFIS model, CANFIS, is discussed in Chapter 13.

Another important issue in the training of ANFIS is how to preserve some intuitive features that make the resulting fuzzy rules easy to interpret. These features include ϵ-completeness [17, 18], moderate fuzziness, and reasonably shaped membership functions. Although we did not pursue these directions in our discussion, most of these features can be preserved by maintaining certain constraints or by modifying the error measure, as explained next.

- The requirement of ϵ-completeness ensures that for any given value of an input variable, there is at least an MF with grade greater than or equal to ϵ. This guarantees that the whole input space is covered properly if ϵ is greater than zero. The ϵ-completeness can be maintained by the constrained gradient descent [48]. For instance, suppose that $\epsilon = 0.5$ and the adjacent membership functions are of the generalized bell MF in Equation (12.2) with parameter sets $\{a_i, b_i, c_i\}$ and $\{a_{i+1}, b_{i+1}, c_{i+1}\}$. Then ϵ-completeness is satisfied if $c_i + a_i \geq c_{i+1} - a_{i+1}$, and the satisfaction of this constraint is guaranteed throughout the training if the constrained gradient descent is employed.

- Moderate fuzziness refers to the requirement that within most regions of the input space, there should be a dominant fuzzy if-then rule with a firing strength close to unity that accounts for the final output, instead of multiple rules with similar firing strengths. This prevents neighboring MFs from having too much overlap and makes the rule set more informative. In particular, this eliminates one of the most unpleasant situations that an MF goes under the other one. An simple way to keep moderate fuzziness is to use a modified error measure

$$E' = E + \beta \sum_{i=1}^{P} [-\overline{w}_i \ln(\overline{w}_i)], \tag{12.22}$$

  where E is the original squared error; β is a weighting constant; P is the size of the training data set; and $\overline{w}_i$ is the normalized firing strength of the ith rule [see Equation (12.4)]. The second term $\sum_{i=1}^{P}[-\overline{w}_i \ln(\overline{w}_i)]$ in the preceding equation is Shannon's **information entropy** [5], and its value is minimized whenever there is a $\overline{w}_i$ equal to one. Since this modified error measure is not based on data fitting alone, the ANFIS thus trained also has a potentially better generalization capability. The improvement of generalization by using an error measure based on both data fitting and weight elimination has been reported in the neural network literature [42, 43].

- The easiest way to maintain reasonably shape for each MF is to parameterize the MF correctly to reflect adequate constrains. For one thing, we want all the MFs to remain bell-shaped regardless of their parameter values. This is not true for the generalized bell MF in Equation (12.2) if $b_i < 0$; a quick fix is to replace b_i with $b_i^2 + k$, where k is a positive fixed constant.

Throughout this chapter, we have assumed that the structure of ANFIS is fixed and that the **parameter identification** is solved through the hybrid learning rule. However, to make the whole approach more complete, the ***structure identification*** [34, 35] (which is concerned with the selection of an appropriate input-space partition style, the number of membership functions on each input, and so on) is equally important to the successful application of ANFIS, especially for modeling problems with a large of inputs. Effective partitioning of the input space can decrease the number of rules and thus increase the speed in both the learning and application phases. Advances in neural network structure identification [4, 20] may shed some light on this.

Fuzzy control is by far the most successful application of fuzzy set theory and fuzzy inference systems. The adaptive capability of ANFIS makes it almost directly applicable to adaptive control and learning control. In fact, ANFIS can replace almost any neural network in a control system and perform the same function. We shall give a detailed coverage of neuro-fuzzy control in Chapters 17 and 18.

The active role of neural networks in signal processing [15, 47] also suggests similar applications for ANFIS. The nonlinearity and structured knowledge representation of ANFIS are its primary advantages over classical linear approaches in adaptive filtering [6] and adaptive signal processing [46], such as identification, inverse modeling, predictive coding, adaptive channel equalization, adaptive interference (noise or echo) canceling, and so on.

By employing the adaptive network as a common framework, we can construct other adaptive fuzzy models that are tailored for applications such as data classification and feature extraction. These types of adaptive fuzzy models are covered in Chapter 20.

EXERCISES

1. How many premise and consequent parameters are there in the ANFIS architectures shown Figure 12.1(b) and Figure 12.4(a)? (Assume that the generalized bell function is used for all the membership functions.)

2. Which node functions in Figure 12.1 need to be changed if the ANFIS architecture is to become an equivalent structure for a zero-order Sugeno fuzzy model? How many parameters does the resulting ANFIS have?

3. Construct an ANFIS that is equivalent to a two-input two-rule Mamdani fuzzy model with max-min composition and centroid defuzzification. Explain the function you use to approximate the centroid defuzzification. Specify how to convert this function into node functions in the resulting ANFIS.

4. Construct an ANFIS that is equivalent to a two-input two-rule Mamdani fuzzy model with sum-product composition and centroid defuzzification. In particular, the constructed ANFIS should take advantage of Theorem 4.1 and have a simpler structure than the one based on max-min composition in the previous exercise.

5. Prove that the scaled Gaussian function in Equation (12.11) is invariant under the product operator. (In other words, prove that the product of two scaled Gaussian functions is still a scaled Gaussian function.)

6. Let the MFs in Figure 12.6 be specified by four generalized bell functions, each with three parameters $\{a_i, b_i, c_i\}$, $i = 1$ (leftmost) to 4 (rightmost). Describe the relationships (constraints) between these parameters such that any two neighboring MFs always intersect at a height that is (a) equal to 0.5; (b) equal to 0.3; (c) between 0.3 and 0.5.

7. The training data of Example 1 are generated at the beginning of the MATLAB file `trn_2in.m`. Use your favorite neural network program (or the MATLAB program `tanmlp.m`) to learn this mapping; the neural network should have approximately the same number of parameters as that in Example 1. Plot

your results and compare them with Figure 12.7. [Note that there are more than 20 different public domain NN software programs that can be retrieved by anonymous ftp. A list of ftp sites for these codes, including the quick-propagation neural network, can be found in the monthly posting of FAQ (frequently asked questions) on the Usenet newsgroup `comp.ai.neural-nets`. The FAQ file `neural-net-faq` is also archived in the periodic posting archive on host `rtfm.mit.edu`, under the anonymous ftp directory `pub/usenet/news`. For people without ftp access, a mail server can be used as well to obtain this file. For more information, send an e-mail message to `mail-server@rtfm.mit.edu` with `help` and `index` in the body on separate lines.]

8. Repeat the previous exercise, but increase the size of the neural network until it can yield about the same performance as the ANFIS in Figure 12.7. How many parameters are there in your neural network?

9. The training data of Example 2 are generated at the beginning of the MATLAB file `trn_3in.m`. Use your favorite neural network program to learn this mapping; the neural network should have approximately the same number of parameters as that in Example 2. Plot your results and compare them with Figure 12.12.

10. The off-line training data of Example 3 are generated at the beginning of the MATLAB file `trn_1in.m`. Use your favorite neural network program to learn this mapping. Plot your results and compare them with Figures 12.14, 12.15, and 12.16. Vary the number of hidden nodes to see how many hidden units are necessary to achieve a performance level similar to Figure 12.15.

11. The Mackey-Glass time series, as specified by the time delay differential equation (Equation (12.19)), can be obtained via a numerical method called the fourth-order Runge-Kutta method. The file `ts.c` (available via FTP, see page xxiii) is a simple C program that generates the Mackey-Glass time series using this numerical method. Compile this program and run it. Use MATLAB to plot the resulting time series when $\tau = 17$ and 50.

12. Write a MATLAB script to transform the raw time series data (when $\tau = 17$, see the previous exercise) into the format of the training data (and the checking data) shown in Equation (12.20).

13. Write MATLAB scripts to perform polynomial fitting to the training data obtained in the previous exercise. Specifically, try (a) first-order polynomial fitting; (b) second-order polynomial fitting; and (c) third-order polynomial fitting. Verify your model by using both training and checking data, and compare the results with Figure 12.18.

14. Using the polynomial models obtained in the previous exercise, perform iterative prediction until $P = 84$. Compare the results with Figure 12.23.

15. Redo the simulation in Example 4, but swap the training and checking data. Plot the prediction errors and the MFs before and after training.

16. Redo the simulation in Example 4, but set the initial step to 0.1, 0.2, ..., 1.0, respectively. Plot 10 training error curves corresponding to these initial step sizes. Do these training error curves vary a lot?

REFERENCES

[1] S. M. Botros and C. G. Atkeson. Generalization properties of radial basis functions. In D. S. Touretzky, editor, *Advances in neural information processing systems III*, pages 707–713. Morgan Kaufmann, San Mateo, CA, 1991.

[2] S. Chen, C. F. N. Cowan, and P. M. Grant. Orthogonal least squares learning algorithm for radial basis function networks. *IEEE Transactions on Neural Networks*, 2(2):302–309, March 1991.

[3] S. E. Fahlman. Faster-learning variations on back-propagation: an empirical study. In D. Touretzky, G. Hinton, and T. Sejnowski, editors, *Proceedings of the 1988 Connectionist Models Summer School*, pages 38–51, Carnegic Mellon University, 1988.

[4] S. E. Fahlman and C. Lebiere. The cascade-correlation learning architecture. In D. S. Touretzky, G. Hinton, and T. Sejnowski, editors, *Advances in neural information processing systems II*. Morgan Kaufmann, San Mateo, CA, 1990.

[5] Robert G. Gallager. *Information theory and reliable communication*. John Wiley & Sons, 1968.

[6] S. S. Haykin. *Adaptive filter theory*. Prentice Hall, Upper Saddle River, NJ, 2nd edition, 1991.

[7] R. S. Crowder III. Predicting the Mackey-Glass timeseries with cascade-correlation learning. In D. Touretzky, G. Hinton, and T. Sejnowski, editors, *Proceedings of the 1990 Connectionist Models Summer School*, pages 117–123, Carnegic Mellon University, 1990.

[8] J.-S. Roger Jang. ANFIS: Adaptive-Network-based Fuzzy Inference Systems. *IEEE Transactions on Systems, Man, and Cybernetics*, 23(03):665–685, May 1993.

[9] R. D. Jones, Y. C. Lee, C. W. Barnes, G. W. Flake, K. Lee, and P. S. Lewis. Function approximation and time series prediction with neural networks. In *Proceedings of IEEE International Joint Conference on Neural Networks*, pages 649–665 (Volume I), 1990.

[10] M. I. Jordan and R. A. Jacobs. Hierarchical mextures of experts and the EM algorithm. Technical report, M.I.T., 1993.

[11] V. Kadirkamanathan, M. Niranjan, and F. Fallside. Sequential adaptation of radial basis function neural networks. In D. S. Touretzky, editor, *Advances in neural information processing systems III*, pages 721–727. Morgan Kaufmann, San Mateo, CA, 1991.

[12] L. V. Kantorovich and G. P. Akilov. *Functional analysis.* Pergamon, Oxford, 2nd edition, 1982.

[13] M. S. Klassen and Y.-H. Pao. Characteristics of the functional-link net: a higher order delta rule net. In *IEEE Proceedings of the International Conference on Neural Networks*, San Diego, June 1988.

[14] T. Kondo. Revised GMDH algorithm estimating degree of the complete polynomial. *Tran. of the Society of Instrument and Control Engineers*, 22(9):928–934, 1986. (Japanese).

[15] B. Kosko. *Neural networks for signal processing.* Prentice Hall, Upper Saddle River, NJ, 1991.

[16] A. S. Lapedes and R. Farber. Nonlinear signal processing using neural networks: prediction and system modeling. Technical Report LA-UR-87-2662, Los Alamos National Laboratory, Los Alamos, New Mexico 87545, 1987.

[17] C.-C. Lee. Fuzzy logic in control systems: fuzzy logic controller-part 1. *IEEE Transactions on Systems, Man, and Cybernetics*, 20(2):404–418, 1990.

[18] C.-C. Lee. Fuzzy logic in control systems: fuzzy logic controller-part 2. *IEEE Transactions on Systems, Man, and Cybernetics*, 20(2):419–435, 1990.

[19] S. Lee and R. M. Kil. A Gaussian potential function network with hierarchically self-organizing learning. *Neural Networks*, 4(2):207–224, 1991.

[20] T.-C. Lee. *Structure level adaptation for artificial neural networks.* Kluwer Academic, 1991.

[21] C.-T. Lin and C. S. G. Lee. Neural-network-based fuzzy logic control and decision system. *IEEE Transactions on Computers*, 40(12):1320–1336, December 1991.

[22] D. Lowe. Adaptive radial basis function nonlinearities, and the problem of generalization. In *Proceedings of the First IEEE International Conference on Artificial Neural Networks*, pages 171–175, London, UK, 1989.

[23] M. C. Mackey and L. Glass. Oscillation and chaos in physiological control systems. *Science*, 197:287–289, July 1977.

[24] J. Moody. Fast learning in multi-resolution hierarchies. In D. S. Touretzky, editor, *Advances in neural information processing systems I*, chapter 1, pages 29–39. Morgan Kaufmann, San Mateo, CA, 1989.

[25] J. Moody and C. Darken. Learning with localized receptive fields. In D. Touretzky, G. Hinton, and T. Sejnowski, editors, *Proceedings of the 1988 Connectionist Models Summer School*, San Mateo, CA, 1988. Carnegie Mellon University, Morgan Kaufmann.

[26] J. Moody and C. Darken. Fast learning in networks of locally-tuned processing units. *Neural Computation*, 1:281–294, 1989.

[27] M. T. Musavi, W. Ahmed, K. H. Chan, K. B. Faris, and D. M. Hummels. On the training of radial basis function classifiers. *Neural Networks*, 5(4):595–603, 1992.

[28] K. S. Narendra and K. Parthsarathy. Identification and control of dynamical systems using neural networks. *IEEE Transactions on Neural Networks*, 1(1):4–27, 1990.

[29] S. J. Nowlan. Maximum likelihood competitive learning. In D. J. Touretzky, editor, *Advances in Neural Information Processing Systems 2*, pages 574–582, San Mateo, CA, 1989. Morgan Kaufmann.

[30] Y.-H. Pao. *Adaptive pattern recognition and neural networks*, chapter 8, pages 197–222. Addison-Wesley, Reading, MA, 1989.

[31] D. B. Parker. Optimal algorithms for adaptive networks: Second order back propagation, second order direct propagation, and second order Hebbian learning. In *Proceedings of IEEE International Conference on Neural Networks*, pages 593–600, 1987.

[32] H. L. Royden. *Real analysis.* Macmillan, New York, 2nd edition, 1968.

[33] T. D. Sanger. A tree-structured adaptive network for function approximate in high-dimensional spaces. *IEEE Transactions on Neural Networks*, 2(2):285–293, March 1991.

[34] M. Sugeno and G. T. Kang. Structure identification of fuzzy model. *Fuzzy Sets and Systems*, 28:15–33, 1988.

[35] C.-T. Sun. Rulebase structure identification in an adaptive network based fuzzy inference system. *IEEE Transactions on Fuzzy Systems*, 2(1):64–73, 1994.

[36] H. Takagi and I. Hayashi. NN-driven fuzzy reasoning. *International Journal of Approximate Reasoning*, 5(3):191–212, 1991.

[37] T. Takagi and M. Sugeno. Derivation of fuzzy control rules from human operator's control actions. *Proceedings of the IFAC Symposium on Fuzzy Information, Knowledge Representation and Decision Analysis*, pages 55–60, July 1983.

[38] T. Takagi and M. Sugeno. Fuzzy identification of systems and its applications to modeling and control. *IEEE Transactions on Systems, Man, and Cybernetics*, 15:116–132, 1985.

[39] L.-X. Wang. Fuzzy systems are universal approximators. In *Proceedings of the IEEE International Conference on Fuzzy Systems*, San Diego, March 1992.

[40] L.-X. Wang and J. M. Mendel. Back-propagation fuzzy systems as nonlinear dynamic system identifiers. In *Proceedings of the IEEE International Conference on Fuzzy Systems*, San Diego, March 1992.

[41] L.-X. Wang and J. M. Mendel. Fuzzy basis function, universal approximation, and orthogonal least squares learning. *IEEE Transactions on Neural Networks*, 3(5):807–814, September 1992.

[42] A. A. Weigend, D. E. Rumelhart, and B. A. Huberman. Back-propagation, weight-elimination and time series prediction. In D. Touretzky, G. Hinton, and T. Sejnowski, editors, *Proceedings of the 1990 Connectionist Models Summer School*, pages 105–116, Carnegic Mellon University, 1990.

[43] A. S. Weigend, D. E. Rumelhart, and B. A. Huberman. Generalization by weight-elimination with application to forecasting. In D. S. Touretzky, editor, *Advances in neural information processing systems III*, pages 875–882. Morgan Kaufmann, San Mateo, CA, 1991.

[44] D. Wettschereck and T. Dietterich. Improving the performance of radial basis function networks by learning center locations. In J. E. Moody, editor, *Advances in Neural Information Processing Systems 4*, pages 1133–1140, San Mateo, CA, 1992. Morgan Kaufmann.

[45] B. Widrow and M. A. Lehr. 30 years of adaptive neural networks: Perceptron, madline, and backpropagation. *Proceedings of the IEEE*, 78(9):1415–1442, 1990.

[46] B. Widrow and D. Stearns. *Adaptive signal processing.* Prentice Hall, Upper Saddle River, NJ, 1985.

[47] B. Widrow and R. Winter. Neural nets for adaptive filtering and adaptive pattern recognition. *IEEE Computer*, pages 25–39, March 1988.

[48] D. A. Wismer and R. Chattergy. *Introduction to nonlinear optimization: a problem solving approach*, chapter 6, pages 139–162. North-Holland, Amsterdam, 1978.

Chapter 13

Coactive Neuro-Fuzzy Modeling: Toward Generalized ANFIS

E. Mizutani

13.1 INTRODUCTION

As discussed in the preceding chapter, ANFIS enjoys its own hybrid learning strategy, automatically tuning a Sugeno-type inferencing system and generating a single output of a weighted linear combination of the consequents. This chapter highlights extensions of such ANFIS concepts. Specifically, we discuss *multiple-output* ANFIS with *nonlinear* fuzzy rules. ANFIS originally made its debut as an adaptive system with much emphasis on its offering the advantage of being a linguistically interpretable fuzzy inference system (FIS) that allows prior knowledge to be embedded in its construction and allows the possibility of understanding the results of learning.

The extensions emphasize characteristics of a more fused neuro-fuzzy system that enjoys many of the advantages claimed for neural networks (NNs) and the linguistic interpretability of an FIS. As a result, we call this generalized ANFIS "**CANFIS**," which stands for **coactive neuro-fuzzy inference systems**, wherein both NNs and FIS play active roles in an effort to reach a specific goal [27]. In this sense, CANFIS migrates various degrees of the neuro-fuzzy spectrum between the two extremes: a completely understandable FIS and a black-box NN, which is at the other end of the interpretability spectrum. Neuro-fuzzy models can be characterized by the neuro-fuzzy spectrum, in light of linguistic transparency and input-output mapping precision.

In this chapter, using trivial examples, we shall clarify the nature of CANFIS from an NN perspective and survey the modeling methodologies of a generalized ANFIS. In Chapter 22, we shall describe the application of CANFIS to a more complicated problem.

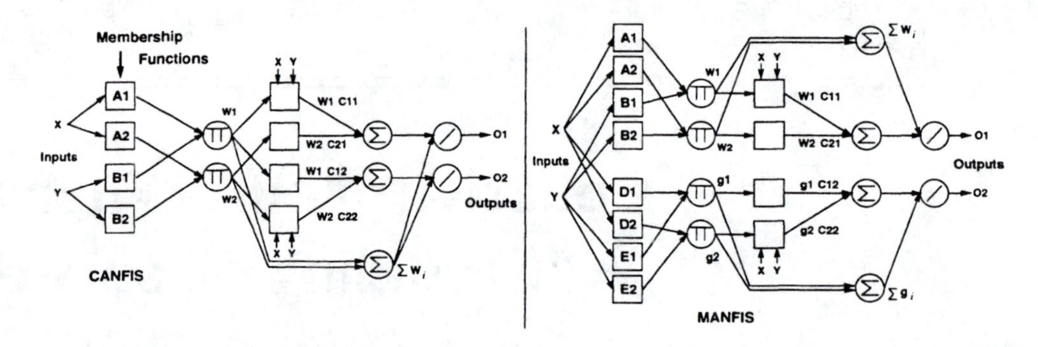

Figure 13.1. Two-output CANFIS architecture with two rules per output (left), and a comparable MANFIS structure (right).

13.2 FRAMEWORK

We delineate the basic outline of CANFIS in connection with a radial basis function network (RBFN) and a well-known backpropagation multilayer perceptron (MLP) from an architectural standpoint. (For functional equivalence between FIS and RBFN, refer to Section 9.5.2.)

13.2.1 Toward Multiple Inputs/Outputs Systems

CANFIS has extended the notion of a single-output system, ANFIS, to produce multiple outputs. One way to get multiple outputs is to place as many ANFIS models side by side as there are required outputs. In this MANFIS (multiple ANFIS) model illustrated in Figure 13.1 (right), no modifiable parameters are shared by the juxtaposed ANFIS models. That is, each ANFIS has an independent set of fuzzy rules, which makes it difficult to realize possible certain correlations between outputs. An additional concern resides in the number of adjustable parameters, which drastically increases as outputs increase. We describe the application of MANFIS to a two-output inverse kinematics problem in Section 19.3.

Another way of generating multiple outputs is to maintain the same antecedents of fuzzy rules among multiple ANFIS models; Figure 13.1 (left) visualizes this CANFIS concept. In short, fuzzy rules are constructed with shared membership values to express correlations between outputs. In the following, we shall elaborate on CANFIS ideas.

13.2.2 Architectural Comparisons

In both CANFIS and RBFN, locality is considered by Euclidean norms between each local center and the input vector, as we will see in Figure 13.3. By comparison, the inner product of each weight vector and input vector is taken in a backpropagation MLP to measure similarity between training patterns. In this section, we compare a simple backpropagation MLP with locally tuning models: CANFIS and RBFN.

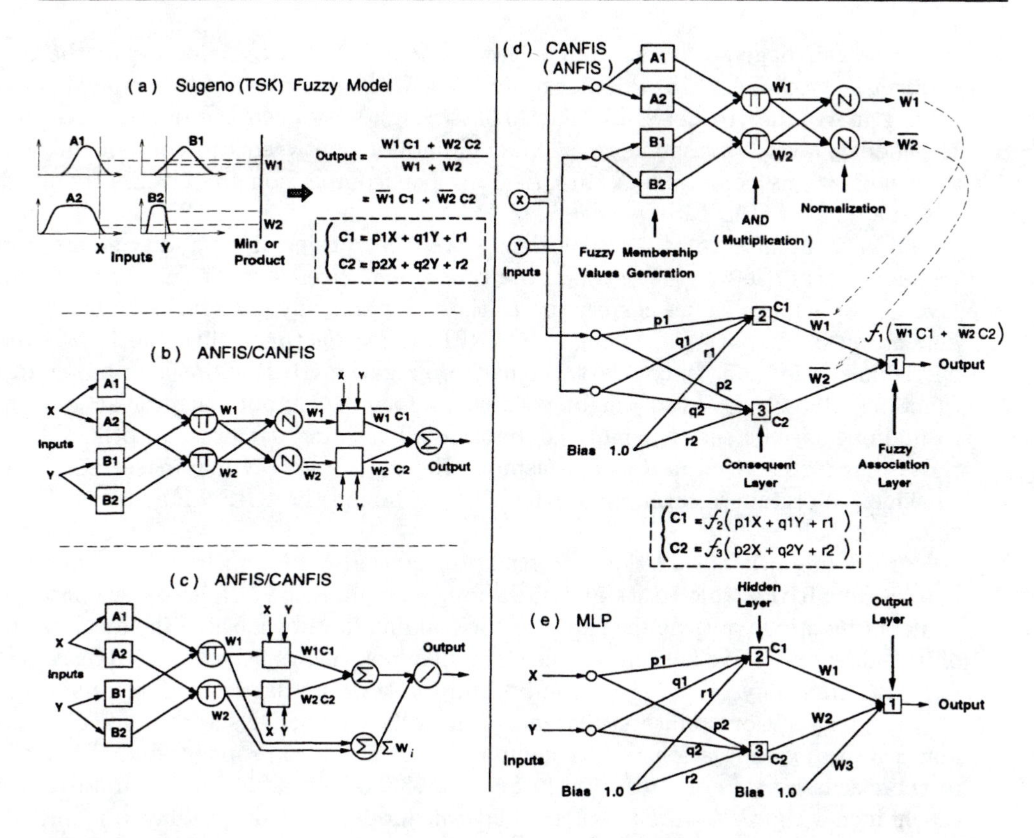

Figure 13.2. *(a) A two-input, one-output Sugeno (TSK) fuzzy model; (b),(c),(d) equivalent ANFIS/CANFIS architectures; (e) a comparable simple backpropagation MLP.*

A single-output CANFIS can be illustrated in the same schematic diagrams of ANFIS in Figures 13.2(b) through 13.2(d). When all three neurons (1, 2, 3) have identity functions in Figure 13.2(d), the presented CANFIS is equivalent to the Sugeno (TSK) fuzzy inference system [43] in Figure 13.2(a), which accomplishes fuzzy if-then rules (linear rules) such as the following:

$$\text{Rule:} \quad \text{If x is } A_1 \text{ and y is } B_1, \quad \text{then } C_1 = p_1\text{x} + q_1\text{y} + r_1.$$

(For more details, see Section 12.2.)

Moreover, in Figure 13.2(d), the consequent parts, C_1 and C_2, are expressed in a network-layered representation to provide a clearer comparison with a simple backpropagation MLP in Figure 13.2(e). This figure also suggests an easy implementation of such a neuro-fuzzy model by modifying an available bareboned backpropagation MLP program. Let us contrast the neuro-fuzzy model to the well-discussed black-box MLP, whose weights are just numeric connection strengths but not good language to us; the hidden layer of the MLP is tantamount to the consequent layer of CANFIS. Putting more hidden nodes in the MLP is equivalent to adding more rules to CANFIS. The MLP's weights between the output layer and the hidden layer correspond to membership values between the consequent layer and the fuzzy association layer in CANFIS. This comparison emphasizes the inside transparency of CANFIS.

From an architectural point of view, CANFIS's powerful capability stems from pattern-dependent weights between the consequent layer and the fuzzy association layer. Membership values correspond to those dynamically changeable weights that depend on input patterns. That is, CANFIS is *locally* tuned, like the RBFN discussed in Section 9.5. In contrast, the backpropagation MLP with sigmoidal neuron functions globally updates weight coefficients for every input pattern, attempting to find one specific set of weights common to all training patterns. In other words, those weights are used in a *global* fashion. Hence, the RBFN may need more data to achieve a certain accuracy than the MLP, although the RBFN may learns faster than the MLP.

Furthermore, it is said that the backpropagation MLP can be a better *extrapolator* than RBFN due to its global nature, and that RBFN fails to estimate the values of functions outside the range of the training data because of the local nature of its hidden receptive fields [6]. This claim may not always be valid; an RBFN with *normalization* may be able to sense beyond the training data set. Thus, questions about extrapolation ability still remain. Locally tuning NNs with *normalization* may lead to extrapolation results comparable to the backpropagation MLP (refer to the discussion of normalization in Section 13.5.5). In addition, when an RBFN has hidden weights C_i (i.e., weights between hidden and output layers) that are expressed in the form of linear functions, $C_i = p_i x + q_i y + r_i$, such an RBFN can be regarded as a sort of compromise between local and global methods because linear functions have global natures just like sigmoidal functions. Both ANFIS and CANFIS are based on this strength as fusion models, deriving from the first-order Sugeno (TSK) FIS.

13.3 NEURON FUNCTIONS FOR ADAPTIVE NETWORKS

We shed light on a diversity of neuron functions for adaptive networks in this section. We also provide fundamental design aspects for rule formation in CANFIS in light of RBFN and modular network features.

In pursuit of a truly adaptive network, there should be no constraints on neuron functions. Various functions are used as basis functions for alternative Gaussian functions in RBFNs. Since neuron functions in the RBFN hidden layer correspond to MFs in CANFIS, we direct our first step toward exploring the NN literature on hidden neuron functions.

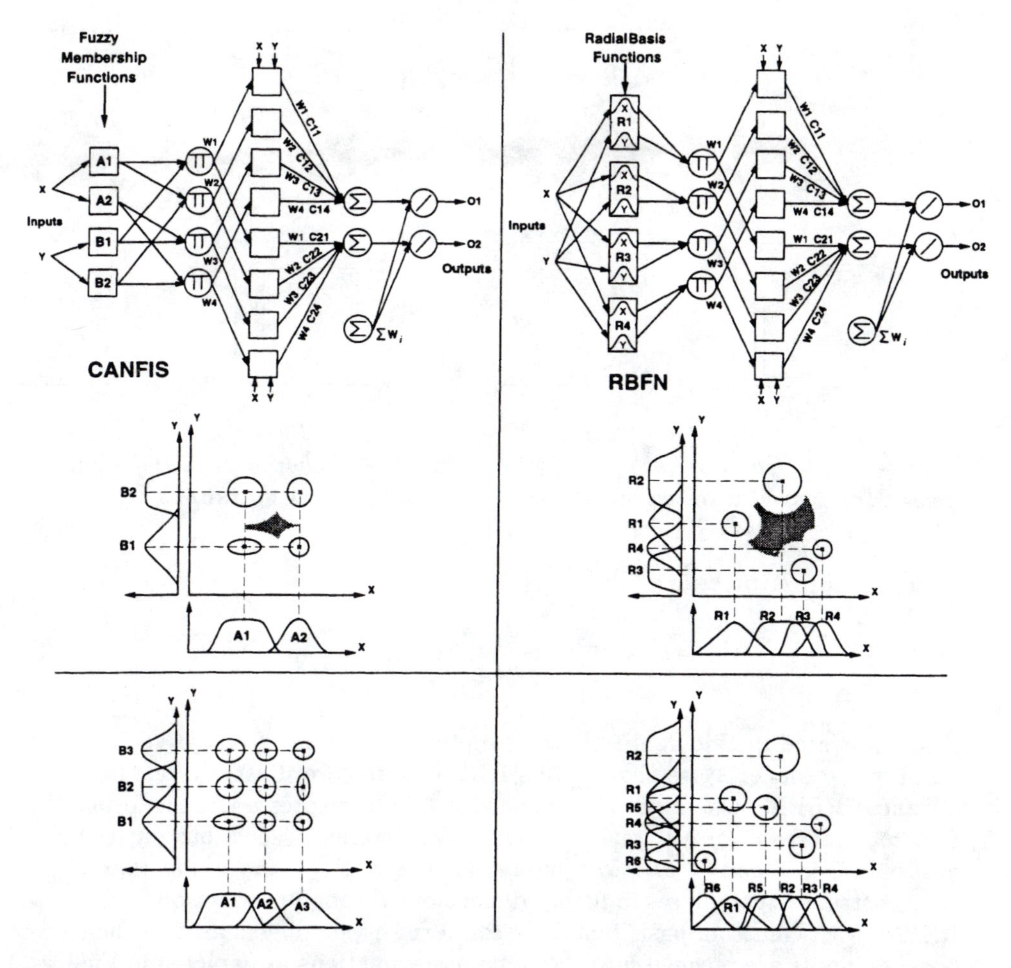

Figure 13.3. *Anatomy of CANFIS and RBFN (top pair), and corresponding contours of their receptive fields (middle pair). Two-output CANFIS with four rules per each output (upper left) and a comparable RBFN with normalization (upper right), which exactly corresponds to Figure 9.10(d). Here both CANFIS and RBFN are in adaptive network representation. For comparison purposes, the two bottom figures illustrate receptive field contours constructed by CANFIS with six MFs (or nine rules) (bottom left), and RBFNs with six basis functions (bottom right).*

13.3.1 Fuzzy Membership Functions versus Receptive Field Units

Figure 13.3 provides an anatomical view of CANFIS in parallel with a corresponding RBFN with normalization, which divides the output of each neuron in the output layer by the sum of all basis functions' outputs. This procedure corresponds to the

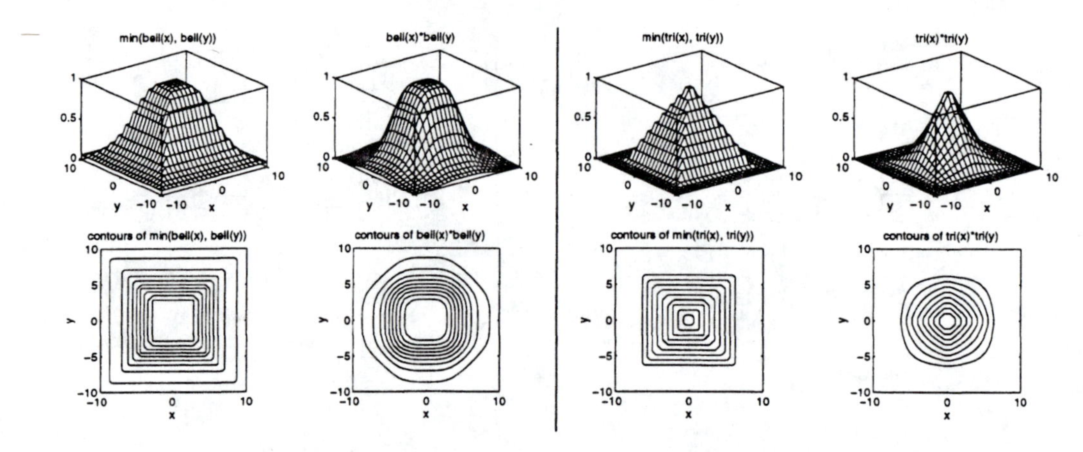

Figure 13.4. *Contours of outputs of generalized bell-shaped MFs (left) and triangular MFs (right) with two inputs using min and product operations.*

following output function:

$$o_j = \sum_{i=1}^{4} w_i C_{ji} / \sum_{i=1}^{4} w_i \quad (j = 1, 2).$$

Also illustrated in this figure are the simplified contours of receptive fields on the input $x - y$ plane. Note that in this RBFN figure, we illustrate one of the most advanced RBFNs, where the hidden weights C_{ji} are expressed in the form of linear functions rather than just real numbers. (For detailed discussions on RBFNs, see Section 9.5 in Chapter 9.) The figures on the $x - y$ plane imply that CANFIS can construct hyperellipses in higher dimensions through product operations, while RBFNs with Gaussian basis functions can form hyperspheres around their centers because inputs are plugged into the same basis functions as depicted in Figure 13.3 (upper right), where inputs, X and Y, both go to the same basis function, R_i. In the RBFN framework, hyperellipses can be built up by using other basis functions [45], such as a generalized Gaussian radial-basis function [23, 33] based on the concept of a weighted norm, and an oriented non-radial-basis function [35]; both functions have modifiable orientation parameters. Notice that in CANFIS, nine elliptical contours are formed on the $x - y$ input plane when three MFs are introduced per input, whereas six round contours are formed when six basis functions are introduced in an RBFN, as illustrated in Figure 13.3 (bottom pair). In other words, CANFIS basically realizes grid partitions while the RBFN realizes scatter partitions (see Section 4.5.1).

To present a clearer illustration of the contours, Figure 13.4 shows the output contours resulting from two bell-shaped MFs and triangular MFs using two common operations, "min" and "product."

An NN with semi-local activation hidden units (or Gaussian-bar hidden units),

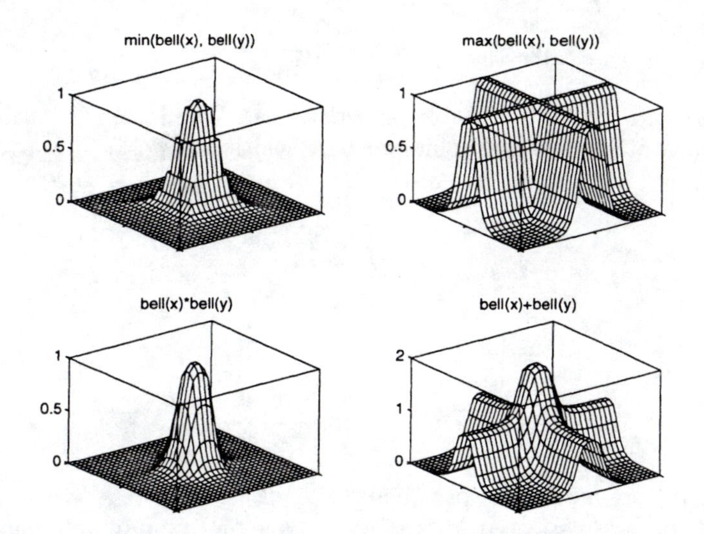

Figure 13.5. *Outputs of generalized bell-shaped MFs obtained using four representative logical operations: min, max, product, and summation.*

which can attain convergence performance comparable to RBFNs, has been reported [5]. The output response of such an ith unit is given by

$$(W_i =)B_i(\|x - u_i\|) = \Sigma_j p_{ij} \exp\left[-\frac{(x_j - u_{ij})^2}{2\sigma_i^2}\right], \tag{13.1}$$

where p_{ij} is a positive parameter. By comparison with this summation unit, the Gaussian hidden unit in an RBFN can be regarded as a product unit:

$$(W_i =)B_i(\|x - u_i\|) = \Pi_j \exp\left[-\frac{(x_j - u_{ij})^2}{2\sigma_i^2}\right]. \tag{13.2}$$

Hence, product operations are indicated conspicuously in the RBFN illustration in Figure 13.3 (upper right). In the fuzzy logic literature, those operations are treated in a systematic way, as detailed in Section 2.5.2. Figure 13.5 shows resulting bell-shaped MF response characteristics, obtained from four representative logical operations: min, max, product, summation. Through the summation operation, we can realize response characteristics similar to those of the semilocal activation unit, defined in Equation (13.1).

Lane [21] discusses an NN with spline receptive field functions; the spline functions are also common as fuzzy MFs [36, 44]. Popular MFs are listed in Section 2.4; here we present a few MFs for the purpose of further discussion:

$$\mu_{\text{bell}}(x) = \frac{1}{1 + |\frac{x-c}{a}|^{2b}}, \tag{13.3}$$

$$\mu_{\text{modbell}}(x) = \max\left\{\frac{2}{1+|\frac{x-c}{a}|^{2b}} - 1, 0\right\}, \tag{13.4}$$

where $\{a,\ b,\ c\}$ is an adjustable parameter set. The latter definition is a modified bell-shaped MF, which has a limited base width (support).

Asymmetric functions are also common in FIS, such as the following two-sided Gaussian MF:

$$\mu_{\text{tsg}}(x) = \begin{cases} g_1(x) & \text{if } x \le c_1, \\ g_2(x) & \text{if } x \ge c_2, \\ 1 & \text{otherwise,} \end{cases} \tag{13.5}$$

where $g_i(x)$ is the Gaussian function:

$$g_i(x) = \exp\left[-\frac{(x-c_i)^2}{2a_i^2}\right],$$

and $\{a_i,\ c_i\}$ are modifiable parameters.

The more sophisticated MFs tend to have more modifiable parameters. Local tuning of parameters within such MFs may be necessary; one of the simplest means of local tuning is to fix center parameters and update shape parameters in the training phase. Another local tuning method uses different learning rates for adjusting those parameters [4].

13.3.2 Nonlinear Rule

In this section, we focus on neuron functions at the consequent layer, such as f_2 and f_3; they form the consequent parts C_1 and C_2 in Figure 13.2(d). (Note that in ANFIS, both f_2 and f_3 are typically identity functions.) When we replace them with nonlinear functions, we have nonlinear consequents. Accordingly, the neuron functions in the consequent layer play an important role in rule formations.

Suppose we have a sigmoidal function as a neuron function in the consequent layer. Then we have a nonlinear consequent, C_{non}:

$$C_{\text{non}} = \frac{1}{1+\exp[-(p_1x + q_1y + r_1)]}. \tag{13.6}$$

In this case, we have a **sigmoidal rule** (see Figure 13.6, upper pair).

Furthermore, when each rule's consequent is realized by an NN, we have **neural rules**, as illustrated in Figure 13.6 (lower left). Note that when the four consequent NNs have sigmoidal output neuron functions with no hidden layers, the four neural rules are reduced to sigmoidal rules. That is, the upper two diagrams are identical in Figure 13.6.

Although the inside of each neural consequent turns out to be a black box, the whole CANFIS model retains the concept of fuzzy reasoning in terms of the behavior of the whole system; it still enjoys transparency with respect to *MF interpretability,* which is discussed in Section 13.4. One possible way of obtaining more precision in a given mapping is to construct the sophisticated neural consequents

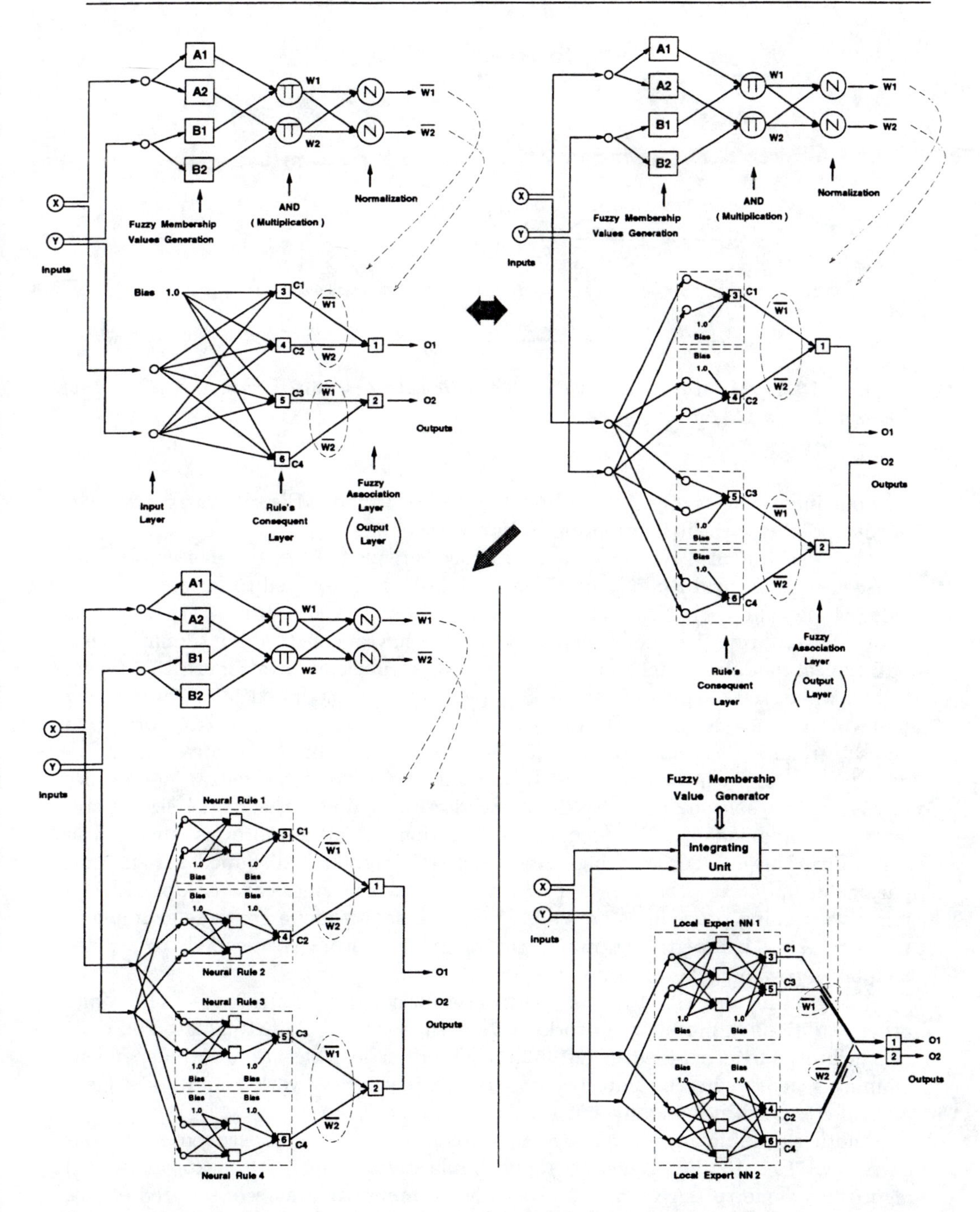

Figure 13.6. *CANFIS with two linear/sigmoidal rules per output (upper pair), CANFIS with two neural rules per output (lower left), and a comparable typical modular network (lower right).*

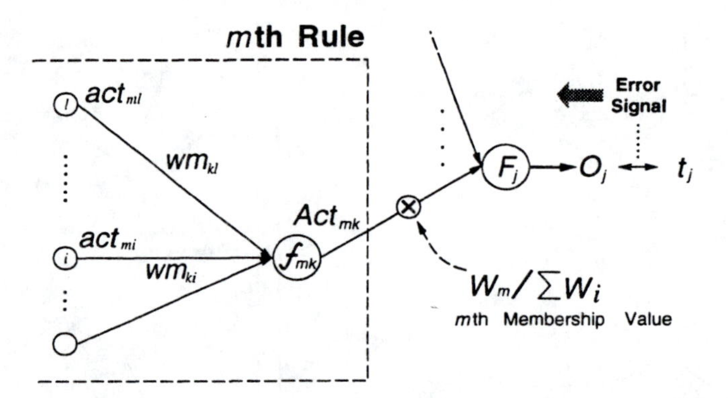

Figure 13.7. *Anatomical graph of the mth rule associated with the jth output neuron, F_j, in CANFIS.*

without increasing fuzzy rules, although commonly more MFs or more fuzzy rules are introduced with little attention to interpretability.

When such neural consequents are further entwined—that is, when two neural consequents, "Neural Rule$_1$" and "Neural Rule$_3$" are combined to form one neural rule (i.e., Local Expert NN$_1$), and "Neural Rule$_2$" and "Neural Rule$_4$" are fused into another neural rule (i.e., Local Expert NN$_2$)—we have a construction similar to that of a typical modular connectionist architecture, as illustrated in Figure 13.6 (lower right), where the outputs of two local expert NNs are mediated by an integrating unit (or typically a gating network) [11, 12, 17, 18, 19, 29, 30]. This formation helps to reduce the number of modifiable parameters. In this context, CANFIS with neural rules can be equivalent to *modular networks*. The central idea resides in *task decomposition*, as discussed in Section 9.6. When the model size is still large, the modified bell MF [defined in Equation (13.4)] is more instrumental in controlling the number of firing rules than the original bell-shaped MF in that parameter-updating procedures for irrelevant or inactive rules can be skipped when iterative training procedures are employed. (We shall show a concrete example in Chapter 22.) An integrating unit in the modular network corresponds to a fuzzy membership value generator in CANFIS.

Takagi et al. described this model from a fuzzy logic point of view [41]. Many other variations of modular networks have been discussed [42, 31, 22, 10, 20]; their design procedures, detailed in ref. [42], are more or less based on an independent training scheme whereby antecedent parts (or gating networks) and consequent parts are trained individually and then put together.

Another training approach is to train both antecedent and consequent parts concurrently [7, 11, 13, 26]. When we apply simple steepest descent (backpropagation) algorithms alone to CANFIS, the procedure to minimize a sum of squared errors, E, is straightforward; Let O_j and F_j be the jth CANFIS output and the jth neuron function at the final output layer, respectively, as depicted in Figure 13.7. We then have

$$O_j = F_j(NET_j),$$

where NET_j denotes net input. The procedure for updating the mth rule's consequent, which has a weight coefficient signified by wm_{ki}, is as follows:

$$\begin{aligned}
\Delta wm_{ki} &= -\eta_{wm}\frac{\partial E}{\partial wm_{ki}} \\
&= -\eta_{wm}\frac{\partial E}{\partial NET_j}\frac{\partial NET_j}{\partial wm_{ki}} \\
&= -\eta_{wm}\frac{\partial E}{\partial O_j}\frac{\partial O_j}{\partial NET_j}\frac{\partial NET_j}{\partial Act_{mk}}\frac{\partial Act_{mk}}{\partial net_{mk}}\frac{\partial net_{mk}}{\partial wm_{ki}} \\
&= -\eta_{wm}\frac{\partial E}{\partial O_j}F_j'(NET_j)\frac{W_m}{\sum_m W_m}f_{mk}'(net_{mk})\frac{\partial}{\partial wm_{ki}}\sum_l wm_{kl}act_{ml} \\
&= -\eta_{wm}\{-(t_j - O_j)\}F_j'(NET_j)\frac{W_m}{\sum_m W_m}f_{mk}'(net_{mk})act_{mi},
\end{aligned}$$

where η_{wm} is a learning rate, t_j is the jth desired output, W_m is the mth membership value assigned to the mth rule, and NET_j is given by the following three equations using W_m and the kth output of the mth rule, Act_{mk}:

$$\begin{aligned}
NET_j &= \frac{\sum_m W_m Act_{mk}}{\sum_m W_m} = \frac{W_1 Act_{1j} + W_2 Act_{2j} + \cdots W_m Act_{mk} + \cdots}{W_1 + W_2 + \cdots + W_m + \cdots}, \\
Act_{mk} &= f_{mk}(net_{mk}), \\
net_{mk} &= \sum_l wm_{kl}act_{ml},
\end{aligned}$$

where f_{mk} is the kth output function in the mth rule, and net_{mk} is the total input. Accordingly, the procedure for updating the antecedent part, which has a parameter denoted by a, can be given by

$$\begin{aligned}
\Delta a &= -\eta_a\frac{\partial E}{\partial a} \\
&= -\eta_a\frac{\partial E}{\partial O_j}\frac{\partial O_j}{\partial a} \\
&= -\eta_a\frac{\partial E}{\partial O_j}\frac{\partial O_j}{\partial NET_j}\frac{\partial NET_j}{\partial W_m}\frac{\partial W_m}{\partial a} \\
&= -\eta_a\frac{\partial E}{\partial O_j}F_j'(NET_j)\left(\frac{\partial}{\partial W_m}\frac{\sum_m W_m Act_{mk}}{\sum_m W_m}\right)\frac{\partial W_m}{\partial a} \\
&= -\eta_a\{-(t_j - O_j)\}F_j'(NET_j)\frac{Act_{mk} - NET_j}{\sum_m W_m}\frac{\partial W_m}{\partial a},
\end{aligned}$$

where η_a is a learning rate.

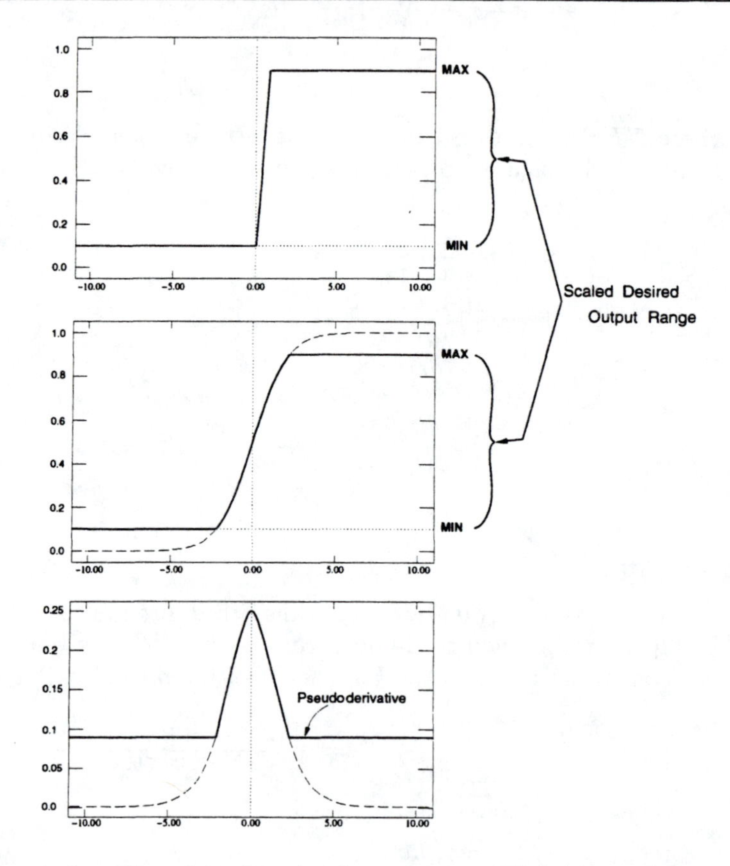

Figure 13.8. *A truncation filter function (top), and a modified sigmoidal function (middle) compared with its derivative (bottom) when MAX is 0.9 and MIN is 0.1. By comparison, dashed curves show a normal sigmoidal logistic function.*

CANFIS with neural rules stands on a sort of *complementary* mixture viewpoint, due to the weighted average of fuzzy membership functions' outputs. In contrast, use of the *softmax* activation function [25, 2] [defined in Equation (9.26)] in modular networks provides a sort of *competitive* mixing perspective. Section 9.6 in Chapter 9 provides more detailed description of this perspective.

13.3.3 Modified Sigmoidal and Truncation Filter Functions

In this section, we discuss the neuron functions placed at the final output layer [i.e., the fuzzy association layer such as f_1 in Figures 13.2(d) and 13.6]. In particular, we introduce a **modified sigmoidal function** and a **truncation filter function**[1].

[1]Here we use the term *truncation filter functions* rather than *piecewise-linear functions*, because the desired output range is purposely changed to match the output range of the truncation

Recall that ANFIS typically possesses the identity function as f_1.

We sometimes expect an NN to specify very small desired outputs. That is, the NN is required to learn extreme values close to the rim of the output range. When normal sigmoidal logistic functions are introduced at the output layer of an NN, it is known that the NN fails to learn such extreme values [34]. One way to improve NN performance is to replace the squared error function with an entropic function [1, 8, 37]; minimizing error can be regarded as maximizing entropy or log likelihood. As an alternative approach, we have introduced a modified sigmoidal function, f_{mod}, and a truncation filter function, f_{trc}, as neuron functions for the output layer [26, 28]. They are easily implemented with small modifications, as follows:

$$f_{trc}(\mathrm{x}) = \begin{cases} \text{MIN} & \text{if } x \le \text{MIN} \\ \text{MAX} & \text{if } x \ge \text{MAX}, \\ \mathrm{x} & \text{otherwise} \end{cases} \tag{13.7}$$

$$f_{mod}(\mathrm{x}) = \begin{cases} \text{MIN} & \text{if } f(x) \le \text{MIN} \\ \text{MAX} & \text{if } f(x) \ge \text{MAX}, \\ f(x) & \text{otherwise} \end{cases} \tag{13.8}$$

where $f(x)$ is the normal sigmoidal logistic function

$$f(x) = \frac{1}{1 + \exp(-x)}. \tag{13.9}$$

This improvement keeps neuron outputs within the desired output range, [MIN,MAX]. We set MIN to 0.1 and MAX to 0.9, and outputs that are above MAX or below MIN are then forced to MAX or MIN.

To take advantage of the modified sigmoidal function and the truncation filter function, we need to change the range of desired outputs (i.e., **output scaling**). Suppose the original range of a desired output (do_i) is between 0.0 and 1.0. It can be linearly changed [24] to [MIN, MAX] by $d_i = do_i$ (MAX$-$MIN) + MIN:

$$0 \le do_i \le 1 \implies \text{MIN} \le d_i \le \text{MAX}.$$

Thus, these functions, $f_{trc}(x)$ and $f_{mod}(x)$, prevent an NN from exceeding the desired output range. We can effectively teach the NN the boundaries of the desired output range by scaling to the interval [MIN, MAX]. Suppose, for instance, that the ith desired output (d_{ip}) of training pattern p is MIN. When the ith output (o_{ip}) is smaller than MIN, the error [either $d_{ip} - f_{trc}(o_{ip})$ or $d_{ip} - f_{mod}(o_{ip})$] is zero, due to the modified functions defined in Equations (13.7) and (13.8). In other words, that pattern p is not used to update the weights associated with the ith output unit, but just skipped.

Another important alteration of these functions resides in the calculation of derivatives. When the neuron output is MIN or MAX, these functions produce

---

filter functions.

a pseudoderivative; for example, a modified sigmoidal function produces a pseudo-derivative whose size is MIN(1 − MIN) or MAX(1 − MAX) as shown in Figure 13.8. We can just stick to the same derivative form as that of the normal sigmoidal logistic function, given by

$$f'(x) = f(x)(1 - f(x)),$$

which has an important computational advantage due to its simple form. The amount of weight change is proportional to the magnitude of the derivative. This modification produces the derivative that is always larger than a certain size, and therefore may have a positive impact on learning extreme values and accelerating learning. Similar modification can be found in a method whereby a small positive *bias derivative* is simply added to the original derivative—that is, $f'(x) + bias$ [6]. Fahlman [3] suggests use of a small magic constant, 0.1, as the *bias derivative*. (Notice that this derivative advantage is valid when the weight-change derivatives are not normalized. For derivative normalization, see Sections 6.7 and 8.3.) The aforementioned modifications can be applied to any sigmoidal function, such as the hyperbolic tangent function.

These functions, $f_{trc}(x)$ and $f_{mod}(x)$, are easily implemented and can surely help an NN to learn the boundaries of the output range and extreme values close to the rim of the output range. Application examples are presented in Section 13.5.5 and in Chapter 22. Again, notice that we use these functions only in the output layer.

13.4 NEURO-FUZZY SPECTRUM

This section explains the concept of **neuro-fuzzy spectrum** in terms of the trade-offs between input-output mapping precision and membership function (MF) interpretability from the fuzzy logic standpoint.

Neuro-fuzzy models allow prior knowledge to be embedded via fuzzy rules with appropriate linguistic labels, and they offer the possibility of understanding the resultant models after learning. On the other hand, black-box neural networks, particularly backpropagation multilayer perceptrons, do not have the same level of ability to do knowledge embedding and extraction.

This observation motivates the concept of neuro-fuzzy spectrum, which is defined on the interpretability-precision plane depicted in Figure 13.9. Ideally, the learning of a neuro-fuzzy model should follow the vertical route to the top in such a way that the mapping precision is being improved while the interpretability maintained. In practice, however, the learning process often follows the diagonal route of improving mapping precision and deteriorating interpretability at the same time. We refer to this situation as the **dilemma between interpretability and precision.**

Adaptive neuro-fuzzy models like ANFIS/CANFIS transit smoothly between the two ends of neuro-fuzzy spectrum: a completely understandable fuzzy inference system and a black-box neural network.

When we apply advanced optimization techniques, the dilemma becomes more

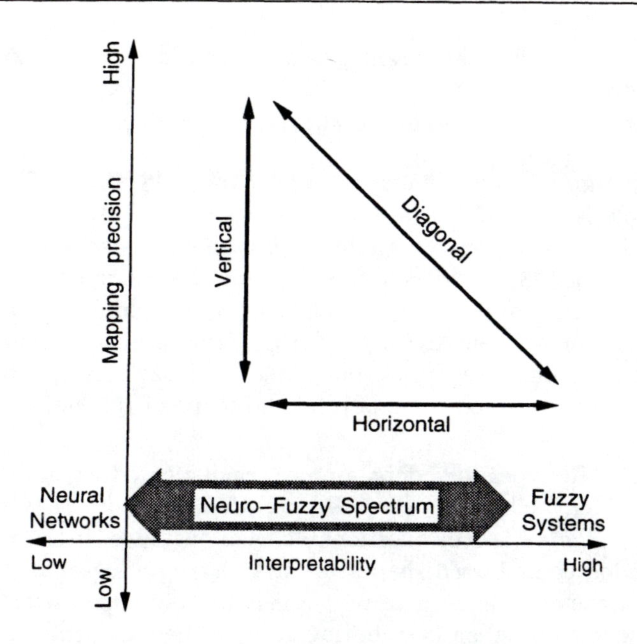

Figure 13.9. *Neuro-Fuzzy spectrum. The plane's axes denote neuro-fuzzy spectrum (horizontal route) and input-output mapping precision (vertical route). Ideally, the learning of a neuro-fuzzy model should follow the vertical route to the top, but it often takes the diagonal route of improving mapping precision at the expense of interpretability.*

conspicuous. For example, given a fixed amount of computation time, the Levenberg-Marquardt (LM) method, discussed in Chapter 6, can achieve higher mapping precision than the steepest descent method. Yet the resultant MFs obtained by the LM method may vary significantly from their initial setups; consequently, we may lose the original interpretability designed for the initial membership functions [16]. In other words, the sophisticated methods may attain a higher input-output mapping precision, but it may lead to meaningless fuzzy rules, accordingly.

Similar observations can be found when the number of rules is increased; the resulting MFs may not lend themselves to good linguistic interpretation. This issue is considered in Sections 13.5.2 and 22.4.2. The MFs should be carefully set up so that fuzzy rules can be held to meaningful limits (see also Section 22.4.1).

If linguistic interpretability is not a concern, we are entitled to choose the most efficient learning algorithm. In this case, such a neuro-fuzzy model stands on the black-box NN endpoint on the neuro-fuzzy spectrum, attempting to achieve as high precision as possible. However, if linguistic interpretability is a concern, then we may need to update MF parameters carefully. Even if we use the simple steepest descent (backpropagation) algorithm, we cannot guarantee the resultant interpretability. This is demonstrated in Section 13.5.2. The hybrid learning rule generally gives interpretable results, as discussed in Section 13.5.3, but this is not always guaranteed.

There are various approaches to alleviating the dilemma:

- Apply another way of interpretation to ANFIS/CANFIS; many other interpretable NN models have structures similar to RBFN or modular networks. Actually, in a variety of studies of Bayesian networks and probabilistic networks [9, 32, 38, 39, 40], some networks had a configuration similar to RBFN. Probabilistic interpretation can also be put on modular networks [29, 18]. This implies that a given ANFIS/CANFIS structure can be interpreted from different viewpoints, regardless of calling it a fuzzy system. But such a different *interpretability spectrum* is beyond the scope of this book.

- Change MF types, or adopt a more sophisticated asymmetric MFs, such as a two-sided bell MFs. (See Section 13.5.3.) Note that the interpretation of *MFs' shapes* is still questionable. For instance, if the resultant MFs end up having complicated shapes, we may have no idea about how to interpret each of them. This is because we usually pay much attention to the global relationships between neighboring MFs, rather than the local peculiarity of a single MF.

- Modify the learning algorithms that maintain MF interpretability. The hybrid learning algorithm has been contrived in this spirit (see Section 13.5.3).

- Alter fuzzy rules' structures by setting up nonlinear rules, as discussed in Section 13.3.2. Particularly, CANFIS with neural rules may have less chance to lose MF interpretability than linear rules, because neural rules have more learning power than linear counterparts. Such rules' learning power may prevent MFs from varying a lot during the learning phase. Although MF interpretability can be maintained, rules' consequent parts may become hard to understand due to their neural network structures (see Figure 13.15 in Section 13.5.4).

- Put proper constraints on neighboring MFs so that resultant MF interpretability can be retained. The most simplest way is to apply some knowledge to fixing the center positions of MFs.

- Formulate a new error measure designed to increase interpretability, such as an error measure with a term similar to Shannon's information entropy, as suggested in ref. [15].

- Transform the input space to another space, in which input values can be treated in a linguistically meaningful way, as discussed in Section 22.4.1.

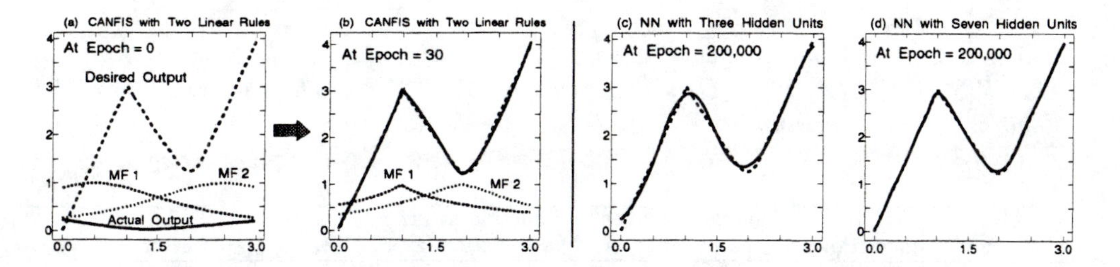

Figure 13.10. *Results of N-shape problem obtained by ANFIS: CANFIS with two linear rules (a, b), and a simple backpropagation MLP with three hidden units (c) and with seven hidden units (d).*

13.5 ANALYSIS OF ADAPTIVE LEARNING CAPABILITY

We shall use the following two examples to clarify further CANFIS's learning capabilities. We also discuss the overall performance of the steepest descent method alone in contrast to the hybrid learning algorithm. These problem examples are trivial, but they provide us an insight into the power of CANFIS. (In Chapter 22, we shall discuss the application of CANFIS to a more difficult problem.)

13.5.1 Convergence Based on the Steepest Descent Method Alone

The first simulation example is simple: fitting an N-shaped letter that has two corners: a pointed top left-hand corner and a rounded right-hand corner. The target letter is shown as a dashed line in Figure 13.10(a).

To see how adaptive capability depends on architecture itself, five CANFIS models (A)–(E) were trained on the basis of the same conventional steepest descent (backpropagation) algorithm with a fixed momentum (0.8) and a small fixed learning rate. (For detailed discussions on the steepest descent methods, refer to Chapter 6.) They are (A) CANFIS with linear rules (i.e., ANFIS), (B) CANFIS with sigmoidal rules, and (C),(D),(E) CANFIS with neural rules; each neural rule has two hidden units (cf. Figure 13.6,lower left). CANFIS (E) has a sigmoidal function at the fuzzy association layer to generate final outputs, whereas CANFIS (C) and (D) have identity functions. The difference between CANFIS (C) and (D) is a neuron function in the output layer put within a neural consequent (e.g., functions 3, 4, 5, and 6 in Figure 13.6,lower left). Specifically, each neural rule or local expert NN has a sigmoidal function to produce the rule's output in CANFIS (C) or has an identity function in CANFIS (D). The results are shown in Table 13.1. By contrast, Table 13.2 shows the results from four simple backpropagation MLPs.

Figure 13.10 shows a comparison of results from CANFIS and simple backpropagation MLPs. CANFIS with two linear rules is able to capture the peculiarity of the N-shape, as shown in Figure 13.10(b), fitting well both the pointed and round corners of the N-shape. A systematic training procedure leads the bell-MFs to the results presented in Figure 13.10(b). (They never cease to amaze us in that their

Table 13.1. *Root-mean-squared errors ($\times 10^{-2}$) of five CANFIS models. Rules of CANFIS models (A, B, C, D) are shown in Figure 13.15. The column "Para. #" denotes "Parameter number."*

CANFIS model	Rule formation	Function at rules' consequent layer	Function at fuzzy association layer	Checking error	Training error	Para. #
(A)	Linear	Identity	Identity	1.38	1.35	10
(B)	Sigmoidal	Sigmoid	Identity	8.22	7.45	10
(C)	Neural	Sigmoid	Identity	2.40	2.08	20
(D)	Neural	Identity	Identity	1.78	1.63	20
(E)	Neural	Identity	Sigmoid	4.57	4.07	20

Table 13.2. *Root-mean-squared errors ($\times 10^{-2}$) of four simple backpropagation MLPs (three single-hidden-layered MLPs and one two-hidden-layered MLP).*

# of hidden units	3	5	7	3 × 3
Checking error	9.91	4.99	2.84	2.05
Training error	8.64	4.75	2.73	1.84
Parameter #	10	16	22	22

neat figures can metamorphose into such unexpected shapes; manually tuned MF shapes may not match them.)

Due to the discontinuity of the left-hand corner of the N-shape, MLPs with a small number of hidden neurons (three or four), possessing sigmoidal neuron functions, do not evolve to piecewise fit the pointed corner, as shown in Figure 13.10(c), in spite of having a comparable number of adjustable parameters (compare the parameter numbers in Tables 13.1 and 13.2). They never reach the fitting level of Figure 13.10(b) within a preset iteration limit (200,000).

These results reinforce the strength of CANFIS with generalized bell MFs defined in Equation (13.3) in the convergence capacity. In the following three sections, we consider CANFIS's learning ability from the standpoint of transparency.

13.5.2 Interpretability Spectrum

When three rules are introduced, it is observed that two MFs (MF2, MF3) transit back and forth at the beginning of their evolution as if struggling to find comfortable niches [see Figure 13.11(d)]. After that, their tracks merge into one. Figure 13.11(c), resulting from Figure 13.11(b), shows that they eventually took almost the same center position. Tuning MFs based on supervised learning algorithms sometimes leads us to meaningless fuzzy rules, as in this example. Because the bell MF used

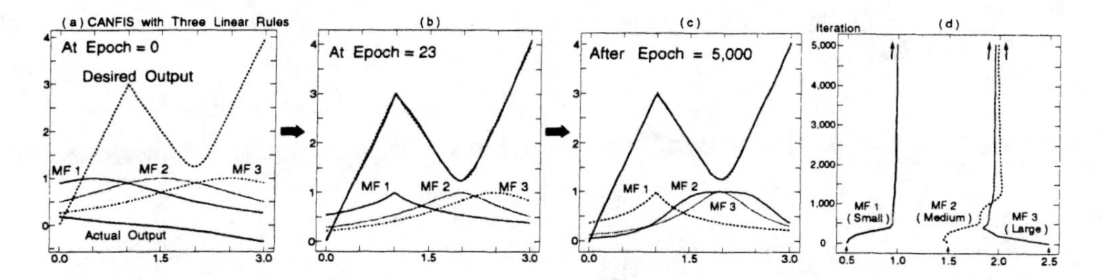

Figure 13.11. *(a),(b),(c) How to adapt to the N-shape when using CANFIS with three linear rules based on the steepest descent method alone. (d) Trajectories of the center positions of three MFs during the training phase. Although actual outputs (solid line) were almost perfectly fitted to the desired N-shape (dashed line) in (c), there is still a question regarding how to extract rules from fuzzy logic standpoint.*

in this simulation has a symmetric neat shape, without having local/independent tuning of its own parameters, the MF may have no choice other than to do unnecessary movement and therefore may end up hiding itself inside another MF, as in Figure 13.11.

Now we encounter the limitation of understandable fuzzy rules. It is actually beyond our expectation that the two antecedents of Rule 2 and Rule 3 ended up being the same. The linguistic labels assigned to them, "medium" and "large," turned out to be meaningless. This exemplifies the **dilemma between precision and interpretability**, discussed in Section 13.4.

This may imply that the initial three fuzzy rules should be two fuzzy rules; either Rule 2 (MF2) or Rule 3 (MF3) may be redundant, which may be a clue about selection of appropriate fuzzy partitions. Indeed, CANFIS with two rules can accomplish the task well, but CANFIS with three rules fits the N-shape better, obtaining higher precision within a fixed amount of computation time. When we split the acquired three rules in Figure 13.11(c) into two combinations—"Rule 1 and Rule 2" and "Rule 1 and Rule 3"—neither output fits the N-shape; obviously, the three rules help each other in improving overall performance. Adding more fuzzy rules may result in lack of interpretability or ill-defined fuzzy rules. This issue will be discussed in another example in Section 22.4.2 of Chapter 22.

13.5.3 Evolution of Antecedents (MFs)

We notice that the resulting MF occupancies depend on a training method as well as an MF type. As in all results in Figures 13.12 and 13.13 obtained by the hybrid learning algorithm, there are no MFs that share the same center position. In this simulation, when original bell MFs were introduced, CANFIS based on the steepest descent method alone unexpectedly converged faster than CANFIS with the hybrid learning procedure. Moreover, CANFIS based on the hybrid learning procedure did not fit the N-shape very well, as Figures 13.12(a) and 13.12(b) show, while CANFIS

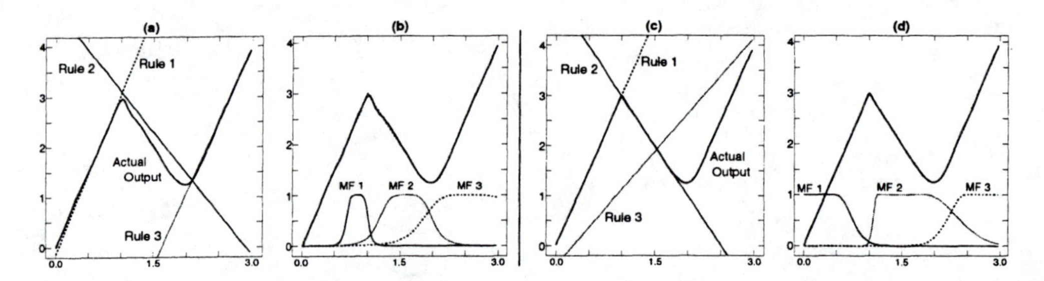

Figure 13.12. *Results of the N-shape problem obtained by CANFIS with three linear rules based on the hybrid learning algorithm using original bell MFs (left) and using asymmetric MFs (right). (a)(b) The rules' outputs. (c)(d) The adapted MFs after the training phase.*

with the steepest descent method alone recognized the features of the N-shape well, as shown in Figure 13.11(c). In many cases, the hybrid learning algorithm works better than the steepest descent method alone, but it may be worth considering possible reasons for this. The hybrid learning procedure strongly predominates where intuitively positioned MFs do not need to evolve very much. Initially, LSE (least-squares estimation) may specialize rules' consequents to a great extent, which may prevent MFs from evolving. LSE can find certain rules' consequent values that have minimal errors with the current MF setup, but after updating coefficients, it may end up losing its way to a better fitting level.

Figures 13.12(c) and 13.12(d) show results obtained using the asymmetric two-sided Gaussian MF defined in Equation (13.5) based on the hybrid learning algorithm; the resulting positions of the three MFs are different from those in Figure 13.11(c). For comparison purposes, Figure 13.13 shows the results obtained by CANFIS with linear rules using four different MF types in accordance with the hybrid learning algorithm. The four types are triangular MFs (trimf), trapezoidal MFs (trapmf), Gaussian MFs (gaussmf), and difference of two sigmoidal MFs (dsigmf). See Section 2.4 for their definitions.

Without *a priori* knowledge, initial MF and rules' consequent arrangements may not be perfect; trusting initial guesses may turn out to be obstacles to obtaining better results. Note that the resultant MFs in Figure 13.10(b) may not match any initial guess.

We now assume a case in which we acquire two rules' consequents from data sets. Is it a good idea to stick to them? Figure 13.14 shows results obtained using the two acquired rules' consequents, which purposely coincide with the two side lines of the N-shape; Figure 13.14(b), resulting from 13.14(a), suggests that clinging to those two consequents may not be useful in obtaining a good fit.

On the other hand, Figure 13.14(d) shows that CANFIS is able to adapt to the N-shape to some extent even when the initial MFs are poorly set up, as in Figure 13.14(c), where there is almost no intersection between the two modified

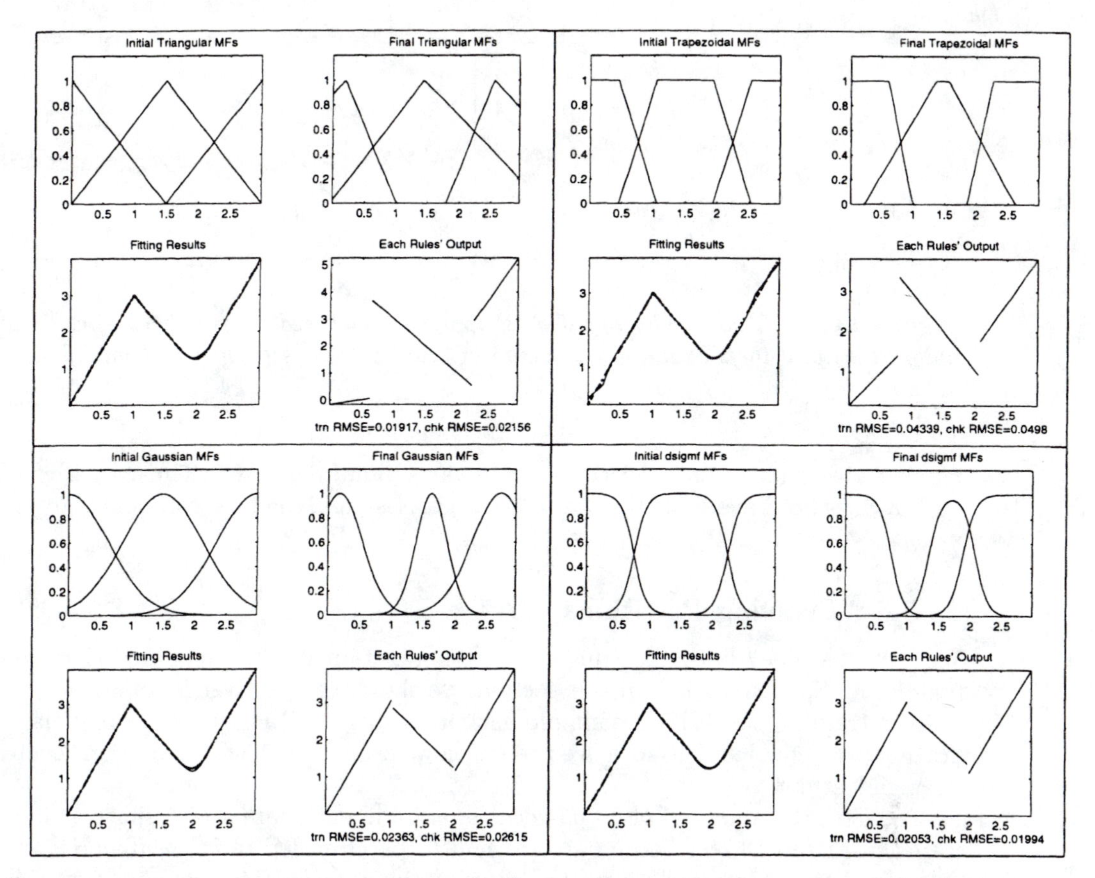

Figure 13.13. *Results of N-shape problem obtained by CANFIS with linear rules using four different MF types based on the hybrid learning algorithm. The four types are triangular MFs (trimf), trapezoidal MFs (trapmf), Gaussian MFs (gaussmf), and difference of two sigmoidal MFs (dsigmf). The training RMSE (root mean squared error) and checking RMSE are also presented together in trn RMSE and chk RMSE.*

bell MFs defined in Equation (13.4). [In Figure 13.14(d), the left-hand base of MF_1 reached the pointed corner of the N-shape, and therefore the evolved MF_1 has a shape different from the MF_1 in Figure 13.10(b).]

13.5.4 Evolution of Consequents (Rules)

There must be some optimal combinations of "shapes" of MFs and "forms" of rules' consequents. Figure 13.15 shows some of them.

Interestingly enough, outputs of the adapted rules' consequents depicted in Figures 13.15(a)–(d) end up being different and far from the desired N-shape. We have

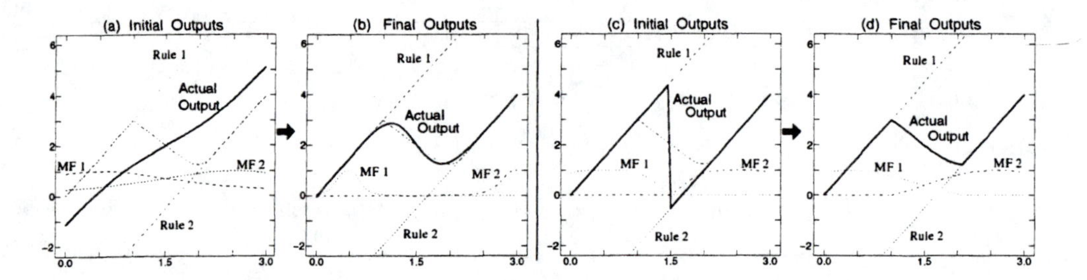

Figure 13.14. *Two results of the N-shape problem based on two fixed (i.e., manually tuned and then frozen) rules using original bell MFs (a),(b) and modified bell MFs (c),(d). Only MFs were trained.*

trained both antecedent and consequent parts simultaneously. Thus, each rule's output does not have to fit the desired output, N-shape; the final combined outputs fit it.

13.5.5 Evolving Partitions

We have discussed both the truncation filter function and the modified sigmoidal function in Section 13.3.3. In this section, we show the effectiveness of those functions in training CANFIS and simple backpropagation MLPs, using a small classification problem. Furthermore, we present how resulting CANFIS partitions evolve through learning.

The data set consisted of 80 patterns in two-dimensional pattern space as illustrated in Figure 13.16. Those patterns had to be classified to four categories; the patterns should be mapped into the following two-dimensional vectors: (*ON, ON*) for class 1, (*ON, OFF*) for class 2, (*OFF, ON*) for class 3, and (*OFF, OFF*) for class 4 where *ON* is set to 0.9 and *OFF* is set to 0.1.

Because we use a common *sum of squared error* measure and a *soft-limiting* neuron function that has sigmoidal characteristics, we specially define the following criterion so that the outputs, O_i ($i = 1, 2$), can be regarded as *ON* (0.9) or *OFF* (0.1):

$$\begin{array}{ll} \text{if } O_i \geq 0.8 & O_i \text{ is classified to } ON \text{ (0.9)}, \\ \text{if } O_i \leq 0.2 & O_i \text{ is classified to } OFF \text{ (0.1)}, \\ \text{otherwise} & O_i \text{ is ``undecided.''} \end{array} \tag{13.10}$$

Note that such classification problems may be treated in various ways, such as by employing an error measure different from sum of squared error measure [14], by using *hard-limiting* activation functions, and so forth. A good discussion of classification problems using the *TLU (threshold logic unit)* can be found in ref. [46].

The classification results are shown in Table 13.3; we tested the four presented models: two CANFIS models with three MFs per input using nine linear rules

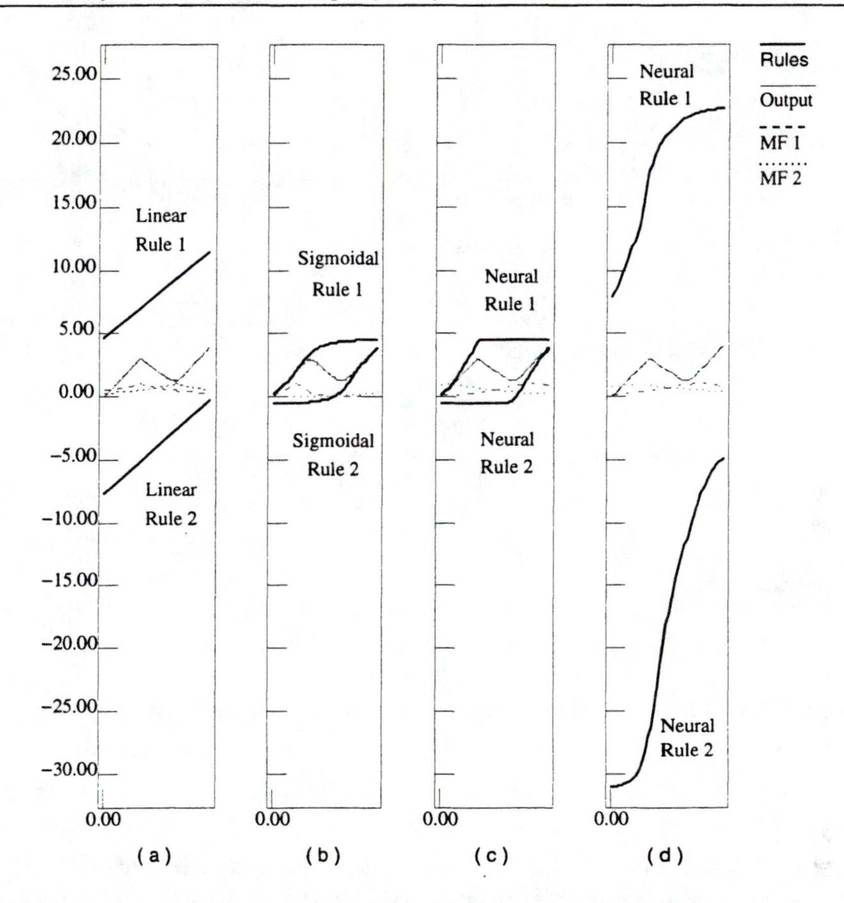

Figure 13.15. *Different outputs of optimized rules' consequents after training phase. They are generated by the four CANFIS models: CANFIS with linear rules (a), sigmoidal rules (b), and neural rules (c) and (d). The corresponding results are tabulated in Table 13.1.*

per output, and two simple backpropagation MLPs with 14 hidden units. All four models have the same number of adjustable parameters, 72. The "stopped epoch" shows when the models classified all 80 patterns correctly according to the preceding criterion defined in Equation (13.10). Training was performed using the steepest descent method alone up to the preset iteration limit, 10,000.

The better results from both $\text{CANFIS}_{\text{trc}}$ and NN_{mod} confirm that the truncation filter function and the modified sigmoidal function help in learning given mappings. (See the exercises at the end of this chapter for more discussion of these results.) Note that the use of both the truncation filter and the modified sigmoidal function are not confined to classification problems; another example can be seen in Chapter 22.

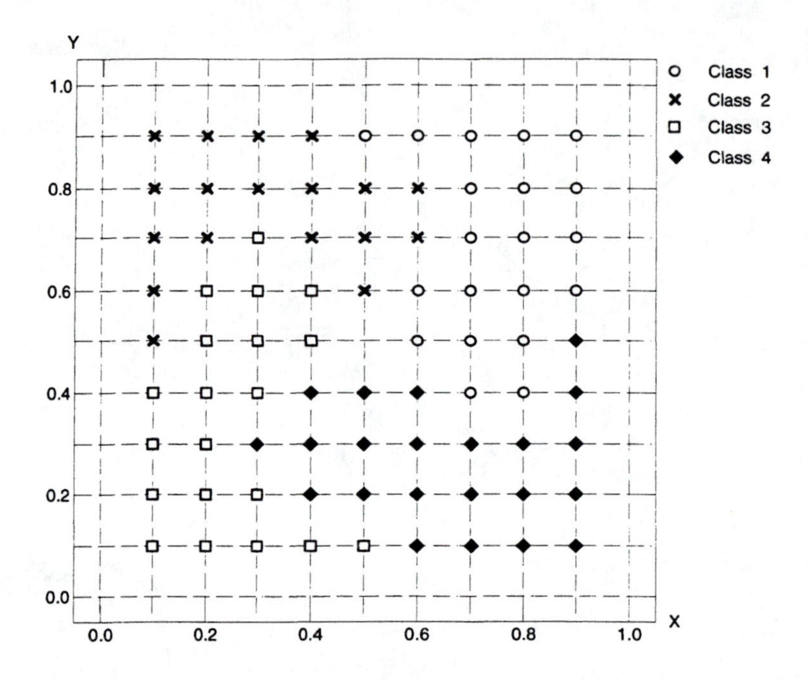

Figure 13.16. *A classification problem example: two-dimensional input space with four classes.*

Table 13.3. *Results of the classification problem (depicted in Figure 13.16); two CANFIS models and two backpropagation MLPs are compared. The values in this table are averaged over seven trials. Our preset limit epoch was 10,000. Note that "incorrect patterns" include "undecided patterns."*

	# of incorrect patterns	Stopped epoch	Neuron function at the output layer
$\text{CANFIS}_{\text{trc}}$	0	855.6	Truncation filter function
$\text{CANFIS}_{\text{id}}$	49.6	10,000.0	Identity function
NN_{mod}	0	1,993.3	Modified sigmoidal function
NN_{norm}	0	5,422.4	Normal sigmoidal function

Figure 13.17 depicts the initial MF setup and the resulting MF setup obtained at a stopped iteration. Figure 13.18 shows the initial partitionings and the resulting partitionings constructed by the obtained six MFs. Also, Figure 13.18 illustrates different constructed surfaces before and after MF normalization.

The initial MFs seem to have excessive overlap to the extent that nine peaks

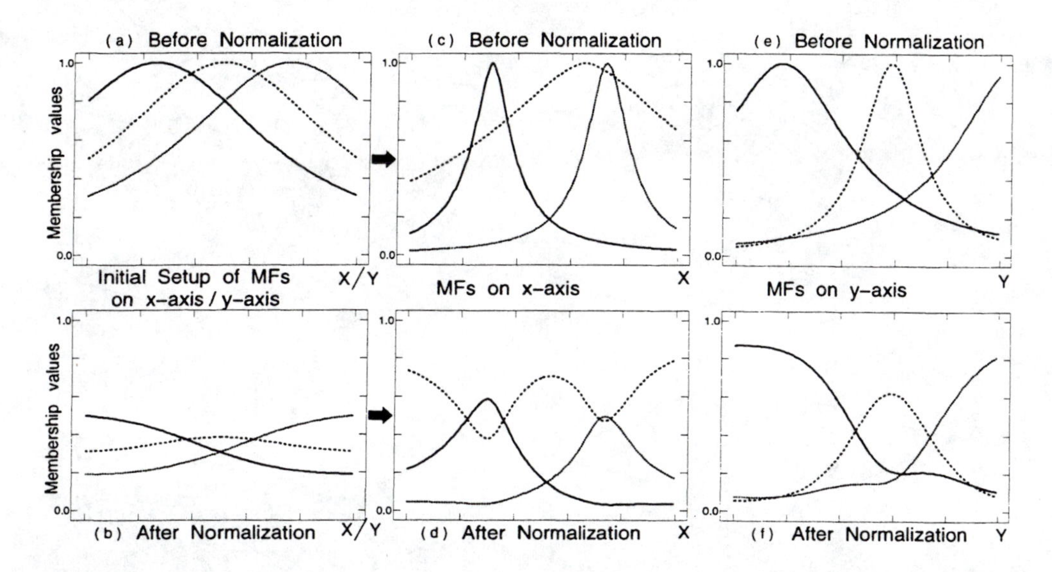

Figure 13.17. *Initial MFs and resulting MFs; initial set up of six MFs on x or y before normalization (a) and after normalization (b); trained MFs on x before normalization (c) and after normalization (d); trained MFs on y before normalization (e) and after normalization (f) .*

are vaguely recognizable in the initial output surfaces in Figure 13.18 (upper left). Remember, however, that we usually have almost no chance of achieving perfect initial MF arrangements; here, we have purposely shown how CANFIS learns from poor MF setups. We can see that the constructed surfaces are very different after normalization. Normalization can help to "see" beyond the range of the training data, as suggested in Section 13.2.2. Note that an RBFN approach requires *nine* basis functions to construct such *nine* peaks as in Figure 13.18 (upper left), as we discussed in Figure 13.3.

13.6 SUMMARY

We have studied in depth extended ANFIS ideas, exploiting various conceptual CANFIS architectures; CANFIS bears a close relationship to the computational paradigms of RBFNs and modular networks. As preparations for use in practical environments, it is worthwhile to investigate CANFIS's strengths and weaknesses by using trivial examples.

Particularly, we have described several problems faced in designing CANFIS in light of the neuro-fuzzy spectrum. *Automatic rule extraction* is a useful aspect of adaptive neuro-fuzzy models. Yet we should pay attention to the limitation behind many successful reports: Acquired rules may sometimes be hard to understand. Also, the initial number of MFs should be carefully determined so that fuzzy rules

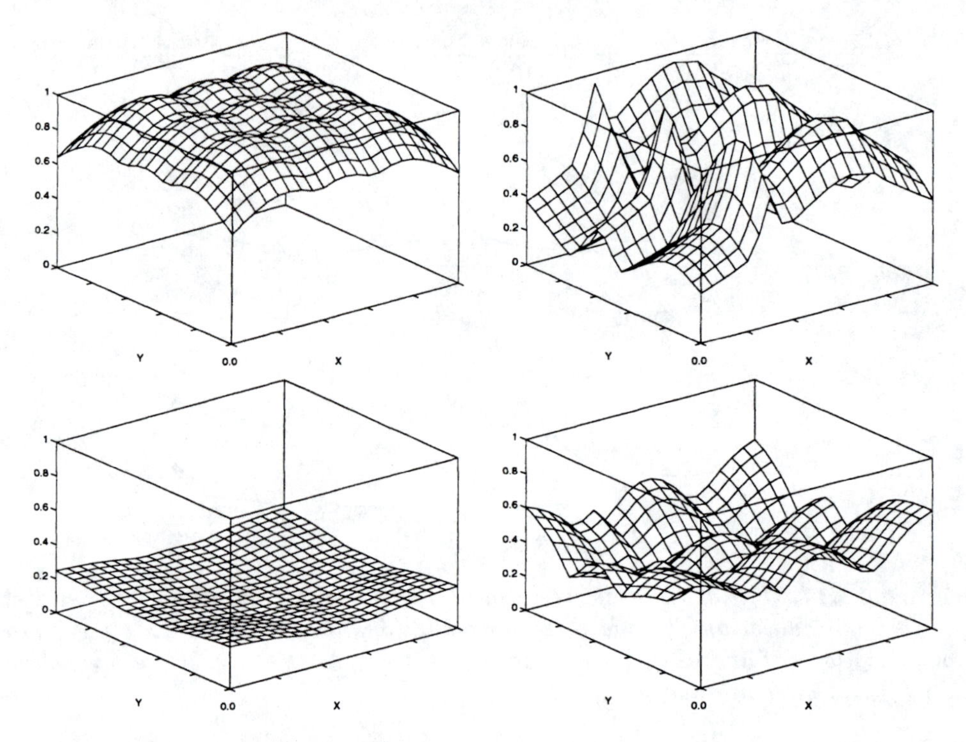

Figure 13.18. *Comparison of output surfaces constructed by the six MFs before normalization and after normalization; the initial output surfaces before normalization (upper left) and after normalization (lower left); the final output surfaces before normalization (upper right) and after normalization (lower right).*

can be held to meaningful limits. In this regard, it may be futile to pursue interpretability from a fuzzy logic standpoint.

Although CANFIS may face such difficulties, most likely it will outperform manually designed systems; the adaptive learning procedure helps. To obtain higher precision within a fixed amount of computation time and to limit the number of MFs so we can also retain interpretability, we may construct *nonlinear rules* such as *neural rules*. Or we may prefer to choose a more sophisticated MF, such as a generalized bell MF and an asymmetric MF. Moreover, manipulating neuron functions may have a beneficial effect on performance enhancement. The question is, what combinations of them are best? Exploring this idea further is a good small step toward finding an ideal model of a truly adaptive network.

To investigate further the empirical observations discussed in this chapter, we shall apply this CANFIS modeling to a more realistic problem in Chapter 22 (see Figures 22.3 and 22.4).

Table 13.4. *Experimental results in finding an optimal learning rate using an MLP with modified sigmoidal functions or normal sigmoidal functions at the output layer. The limit epoch was set to 10,000. Note that "incorrect patterns" include "undecided patterns."*

	Modified sigmoidal function		Normal sigmoidal function	
Learning rate	Stopped epoch	# of incorrect patterns	Stopped epoch	# of incorrect patterns
1.4	6,958	0	10,000	2
1.3	5,657	0	3,925	0
1.2	5,582	0	10,000	15
1.1	3,814	0	9,390	0
1.0	2,628	0	2,515	0
0.9	2,362	0	2,549	0
0.8	1,854	0	2,066	0
0.7	2,035	0	6,578	0
0.6	2,545	0	5,852	0
0.5	2,505	0	2,520	0
0.4	3,006	0	10,000	1
0.3	4,047	0	10,000	1
0.2	3,143	0	5,496	0
0.1	10,000	1	10,000	4
0.05	10,000	10	10,000	16

EXERCISES

Concerning the classification problem discussed in Section 13.5.5, answer the following questions.

1. When we introduce four-dimensional target vectors—*(ON, OFF, OFF, OFF)* for class 1, *(OFF, ON, OFF, OFF)* for class 2, *(OFF, OFF, ON, OFF)* for class 3, and *(OFF, OFF, OFF, ON)* for class 4—what *disadvantage* will CANFIS suffer in contrast to simple backpropagation MLPs?

2. Assuming that we explore various learning rates on the basis of pattern-by-pattern learning, we get the results shown in Table 13.4. (Deciding an optimal learning rate requires a rule of thumb; it is actually state-of-the-art.) What advantage can be gained by using an MLP with modified sigmoidal functions as opposed to an MLP with normal sigmoidal functions?

Table 13.5. *Comparison of actual outputs of the four models ($CANFIS_{trc}$, $CANFIS_{id}$, NN_{mod}, and NN_{norm}) for four patterns. Those sample outputs were picked at their stopped epoch."*

	Pattern	# 1	Pattern	# 2	Pattern	# 3	Pattern	# 4
Ideal outputs	0.9	0.9	0.9	0.1	0.1	0.9	0.1	0.1
$CANFIS_{trc}$	0.9	0.9	0.9	0.1	0.1	0.9	0.1	0.1
$CANFIS_{id}$	0.8788	0.8879	0.9654	−0.0757	−0.1303	−0.0006	−0.0755	−0.1830
NN_{mod}	0.9	0.9	0.9	0.12581	0.1	0.9	0.1	0.1
NN_{norm}	0.9980	0.9999	0.9183	0.1120	0.0744	0.8972	0.0005	0.0016

3. In Table 13.3, $CANFIS_{id}$ (CANFIS with identity functions) shows poor performance. Specify a possible reason, looking at the comparison of actual outputs of the four models appearing in Table 13.5.

4. Suppose we get the RMSE [between the outputs O_i ($i = 1, 2$) and the desired outputs ON or OFF] result shown in Figure 13.19. Can we conclude, in light of Criterion 13.10, that an MLP with modified sigmoidal functions is doing a better job than an MLP with normal sigmoidal functions in classifying given patterns? (If not, why not?)

REFERENCES

[1] E. B. Baum and F. Wilczek. Supervised learning of probability distribution by neural networks. In D. Z. Anderson, editor, *Neural information processing systems*, pages 52–61. American Institute of Physics, New York, 1988.

[2] J. Bridle. Probabilistic interpretation of feedforward classification network outputs, with relationships to statistical pattern recognition. In F. Fogelman and J. Herault, editors, *Neuro-computing: algorithms, architectures, and applications*. Springer-Verlag, London, 1989.

[3] Scott E. Fahlman. Faster-learning variations on back-propagation: An empirical study. In *Proceedings of the 1988 Connectionist Models Summer School*, 1988.

[4] V. Gorrini, T. Salome, and H. Bersini. Self-structuring fuzzy systems for function approximation. In *Proceedings of IEEE International Conference on Fuzzy Systems*, pages 919–926, March 1995.

[5] J. A. Hartman and J. D. Keeler. Predicting the future: Advantages of semilocal units. *Neural Computation*, 3(4):566–578, 1991.

[6] Mohamad H. Hassoun. *Fundamentals of artificial neural networks*. MIT Press, Cambridge, MA., 1995.

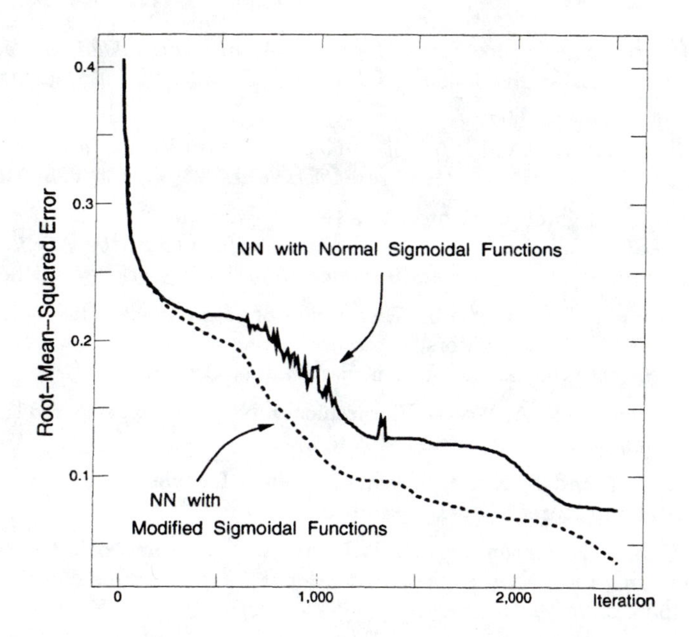

Figure 13.19. *Learning curves of two MLP models (an MLP with modified sigmoidal functions and an MLP with normal sigmoidal functions) for the classification problem.*

[7] Simon Haykin. *Neural networks: a comprehensive foundation.* Macmillan College Publishing, 1994.

[8] Geoffrey. E. Hinton. Connectionist learning procedures. *Artificial Intelligence,* pages 185–234, September 1994.

[9] J. Hollatz and V. Tresp. A rule-based network arichtecture. In I. Aleksander and J. Taylor, editors, *Artificial neural networks 2.* Elsevier Science, 1992.

[10] N. Imasaki, S. Kubo, S. Nakai, T. Yoshitsugu, J. Kiji, and T. Endo. Elevator group control system tuned by a fuzzy neural network applied method. In *Proceedings of IEEE International Conference on Fuzzy Systems,* pages 1735–1740, March 1995.

[11] R. A. Jacobs and M. I. Jordan. A competitive modular connectionist architecture. In R. P. Lippmann, J. E. Moody, and D. J. Touretzky, editors, *Advances in Neural Information Processing Systems 3,* pages 767–773, San Mateo, CA, 1991. Morgan Kaufmann.

[12] R. A. Jacobs, M. I. Jordan, S. J. Nowlan, and G. E. Hinton. Adaptive mixtures of local experts. *Neural Computation,* 3:79–87, 1991.

[13] J.-S. Roger Jang. Fuzzy modeling using generalized neural networks and Kalman filter algorithm. In *Proceedings of the Ninth National Conference on Artificial Intelligence (AAAI-91),* pages 762–767, July 1991.

[14] J-S. Roger Jang. *Neuro-Fuzzy Modeling: Architecture, Analyses, and Applications.* PhD thesis, EECS Department, University of California at Berkeley, Berkeley, CA, 1992.

[15] J.-S. Roger Jang. ANFIS: Adaptive-Network-based Fuzzy Inference Systems. *IEEE Transactions on Systems, Man, and Cybernetics*, 23(03):665–685, May 1993.

[16] J.-S. Roger Jang and E. Mizutani. Levenberg-Marquardt method for ANFIS learning. In *Proceedings of the International Joint Conference of the North American Fuzzy Information Processing Society Biannual Conference*, Berkeley, California, June 1996.

[17] M. I. Jordan and R. A. Jacobs. Hierarchies of adaptive experts. In J. Moody, S. Hanson, and R.Lippmann, editors, *Advances in neural information processing systems IV*, pages 985–993. Morgan Kaufmann, San Mateo, CA., 1992.

[18] M. I. Jordan and R. A. Jacobs. Hierarchical mixtures of experts and the EM algorithm. *Neural Computation*, 6:181–214, 1994.

[19] M. I. Jordan and L. Xu. Convergence results for the EM approach to mixtures of experts architectures. *Neural Networks.* in press.

[20] R. J. Kuo, P. H. Cohen, and R. T. Kumara. Neural network driven fuzzy inference system. In *Proceedings of the International Conference on Neural Networks*, pages 1532–1536, June 1994.

[21] S. H. Lane, M. G. Flax, D. A. Handelmanand, and J. J. Gelfand. Multi-layer perceptrons with b-spline receptive field functions. In R. P. Lippmann, J. E. Moody, and D. J. Touretzky, editors, *Advances in Neural Information Processing Systems 3*, pages 684–692, San Mateo, CA, 1991. Morgan Kaufmann.

[22] K. C. Lee and J. Kim. Hybrid neural network-driven reasoning approach to bankruptcy prediction: Comparison with mda, acls, and neural network. In *Proceedings of the International Conference on Neural Networks*, pages 1787–1792, June 1994.

[23] D. Lowe. Adaptive radial basis function nonlinearities, and the problem of generalization. In *Proceedings of the First IEEE International Conference on Artificial Neural Networks*, pages 171–175, London, UK, 1989.

[24] Timothy Masters. *Practical neural network recipe in C++.* Academic Press, Inc., 1993.

[25] P. McCullagh and J. A. Nelder. *Generalized linear models.* Chapman and Hall, 1983.

[26] E. Mizutani, J.-S. R. Jang, K. Nishio, H Takagi, and D. M. Auslander. Coactive neural networks with adjustable fuzzy membership functions and their applications. In *Proceedings of the International Conference on Fuzzy Logic and Neural Networks*, pages 581–582, Iizuka, Japan, August 1994.

[27] E. Mizutani and J-S. Roger Jang. Coactive neural fuzzy modeling. In *Proceedings of the International Conference on Neural Networks*, pages 760–765, November 1995.

[28] E. Mizutani, H. Takagi, and D. M. Auslander. A cooperative system based on soft computing methods to realize higher precision of computer color recipe prediction. In *Proceedings of Applications and Science of Artificial Neural Networks, part of SPIE's International Symposium on OE/Aerospace Sensing and Dual Use Photonics*, pages 303–314, April 1995.

[29] S. J. Nowlan. Maximum likelihood competitive learning. In D. J. Touretzky, editor, *Advances in Neural Information Processing Systems 2*, pages 574–582, San Mateo, CA, 1989. Morgan Kaufmann.

[30] S. J. Nowlan and G. E. Hinton. Evaluation of adaptive mixtures of competing experts. In R. P. Lippmann, J. E. Moody, and D. J. Touretzky, editors, *Advances in Neural Information Processing Systems 3*, pages 774–780, San Mateo, CA, 1991. Morgan Kaufmann.

[31] L. Pavel and M. Chelaru. Neural fuzzy architecture for adaptive control. In *Proceedings of IEEE International Conference on Fuzzy Systems*, pages 1115–1122, March 1992.

[32] Judea Pearl. *probabilistic reasoning in intelligent systems: networks of plausible inference.* Morgan Kaufmann, San Mateo, CA, 1988.

[33] T. Poggio and F. Girosi. Networks for approximation and learning. *The Proceedings of the IEEE*, 78(9):1485–1497, September 1990.

[34] D. E. Rumlhart, G. E. Hinton, and R. J. Williams. Learning internal representations by error propagation. In D. E. Rumlhart and James L. McClelland, editors, *Parallel distributed processing: volume 1*, chapter 8, pages 318–362. MIT Press, Cambridge, MA., 1986.

[35] A. Saha, J. Christian, D. S. Tang, and C-L. Wang. Oriented non-radial basis functions for image coding and analysis. In R. P. Lippmann, J. E. Moody, and D. J. Touretzky, editors, *Advances in Neural Information Processing Systems 3*, pages 728–734, San Mateo, CA, 1991. Morgan Kaufmann.

[36] K. Shimojima, T. Fukuda, and F. Arai. Self-tuning fuzzy inference based on spline function. In *Proceedings of IEEE International Conference on Fuzzy Systems*, pages 690–695, June 1994.

[37] S. A. Solla, E. Levin, and M. Fleisher. Accelerated learning in layered neural networks. *Complex Systems*, 2:25–639, 1988.

[38] D. F. Specht. Probabilistic neural network. *Neural Networks*, 3:109–118, 1990.

[39] Donald F. Specht. A general regression neural network. *IEEE Transactions on Neural Networks*, 2(6):568–576, November 1991.

[40] H. J. Suermondt and G. F. Cooper. A combination of exact algorithms for inference on bayesian belief networks. *Int. J. Approximaate Reasoning*, 5:521–542, 1991.

[41] H. Takagi and I. Hayashi. Nn-driven fuzzy reasoning. *International Journal of Approximate Reasoning*, 5(3):191–212, 1991.

[42] H. Takagi, N. Suzuki, T. Koda, and Y. Kojima. Neural networks designed on approximate reasoning architecture and their application. *IEEE Transactions on Neural Networks*, 3(5):752–759, September 1992.

[43] T. Takagi and M. Sugeno. Fuzzy identification of systems and its application to modeling and control. *IEEE Transactions on Systems, Man, and Cybernetics*, 15:116–132, 1985.

[44] C-H. Wang, W-Y. Wang, T-T. Lee, and P-S. Tseng. Fuzzy b-spline membership function (bmf) and its applications in fuzzy-neural control. *IEEE Transactions on Systems, Man, and Cybernetics*, 25(05):841–851, May 1995.

[45] D. P. Wasserman. *Advanced methods in neural computing.* Van Nostrand Reinhold, New York, 1993.

[46] Jacek M. Zurada. *Introduction to artificial neural systems.* West Publishing, 1992.

Part V

Advanced Neuro-Fuzzy Modeling

Chapter 14

Classification and Regression Trees

J.-S. R. Jang

14.1 INTRODUCTION

We introduced the ANFIS (adaptive neuro-fuzzy inference system) network architecture in Chapter 12, including its configuration, learning rules, and several application examples. However, the learning rules (or any other parameter-level adaptation methods) only deal with **parameter identification**; we still need methods for **structure identification** to determine an initial ANFIS architecture before any parameter-tuning procedures can take over. By having solid methods for both structure and parameter identification, we are completing the cycle for fuzzy modeling introduced in Section 4.5.2.

Structure identification in fuzzy modeling involves the following primary issues:

- Selecting relevant input variables
- Determining an initial ANFIS architecture, including
    1. Input space partitioning (see Section 4.5.1)
    2. Number of membership functions (MFs) for each input
    3. Number of fuzzy if-then rules
    4. Antecedent (premise) parts of fuzzy rules
    5. Consequent (conclusion) parts of fuzzy rules
- Choosing initial parameters for MFs.

In this chapter, we shall assume that the tree partition (see Section 4.5.1) has been adopted for our fuzzy modeling task. Based on the CART (classification and regression tree) algorithm [1], this chapter introduces a quick method for solving the problem of structure identification. The proposed method generates a tree partitioning of the input space, which relieves the "curse of dimensionality" problem (number of rules increasing exponentially with number of inputs) associated with the grid partitioning described in Section 4.5.1. Moreover, the resulting ANFIS architectures based on CART are more efficient in both training and application because of their implicit weight normalization.

We start with a brief introduction to decision trees and the CART algorithm used to derive them. Following that, we explain how to transform CART-derived decision trees into efficient ANFIS network structures with implicit weight normalization.

14.2 DECISION TREES

A **decision tree** partitions the input space (also known as the feature or attribute space) of a data set into mutually exclusive regions, each of which is assigned a label, a value, or an action to characterize its data points. The decision tree mechanism is transparent and we can follow a tree structure easily to explain how a decision is made. Therefore, the decision tree method has been used extensively in machine learning, expert systems, and multivariate analysis; it is perhaps the most highly developed technique for partitioning sample data into a collection of decision rules.

A decision tree is a tree structure consisting of internal and external nodes connected by branches. An **internal node** is a decision-making unit that evaluates a decision function to determine which child node to visit next. In contrast, an **external node**, also known as a **leaf** or **terminal node**, has no child nodes and is associated with a label or value that characterizes the given data that lead to its being visited. In general, a decision tree is employed as follows. First, we present a datum (usually a vector composed of several attributes or elements) to the starting node (or root node) of the decision tree. Depending on the result of a decision function used by an internal node, the tree will branch to one of the node's children. This is repeated until a terminal node is reached and a label or value is assigned to the given input data.

In the case of a binary decision tree, each internal node has exactly two children, so a decision can always be interpreted as either true or false. Of all decision trees, binary decision trees are the most often used because of their simplicity and our extensive knowledge of their characteristics.

Decision trees used for classification problems are often called **classification trees**, and each terminal node contains a label that indicates the predicted class of a given feature vector. In the same vein, decision trees used for regression problems are often called **regression trees**, and the terminal node labels may be constants or equations that specify the predicted output value of a given input vector. Figure 14.1(a) is a typical binary regression tree with two inputs x and y and one output z. As shown in Figure 14.1(b), the decision tree partitions the input space into four non-overlapping rectangular regions, each of which is assigned a label f_i (which could be a constant or an equation) to represent a predicted output value. Note that each terminal node has a unique path that starts with the root node and ends at the terminal node; the path corresponds to a decision rule that is a

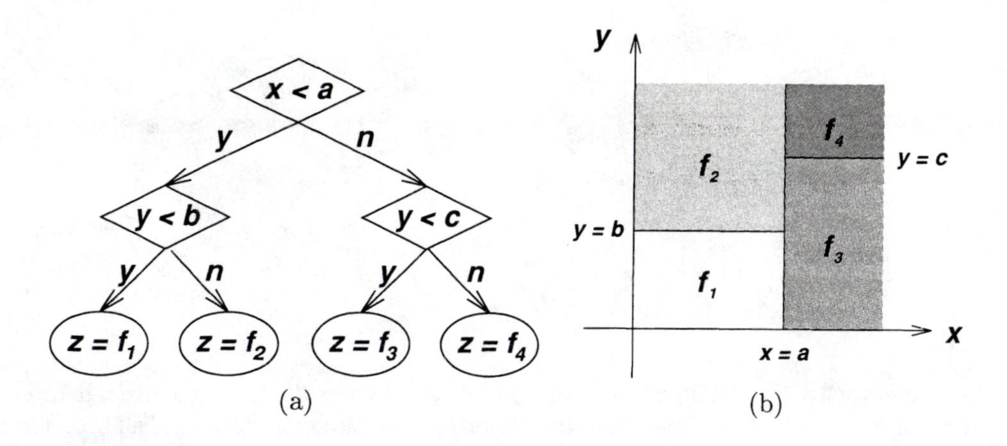

Figure 14.1. *(a) A binary decision tree and (b) its input space partitioning.*

conjunction (AND) of various tests or conditions. For any given input vector, one and only one path in the tree will be satisfied.

Example 14.1 *Input-output surfaces of regression trees*

When the labels of terminal nodes in a regression tree are constants, the resulting input-output mapping looks like several constant-height planes put together with sharp boundaries. Figure 14.2(a) is a typical example that shows the input-output surface of the decision tree in Figure 14.1 with $a = 6$, $b = 3$, $c = 7$, $f_1 = 1$, $f_2 = 3$, $f_3 = 5$, and $f_4 = 9$. On the other hand, if we assign a linear function of the input variables to each terminal node, then the resulting surface is piecewise linear, as shown in Figure 14.2(b), where $f_1 = 2x - y - 20$, $f_2 = -2x + 2y + 10$, $f_3 = 6x - y + 5$, and $f_4 = 3x + 4y + 20$. Apparently, a regression tree is a very easy-to-interpret representation of a nonlinear input-output mapping. However, the discontinuity at the decision boundaries [say, $x = a$ in Figures 14.2(a) and 14.2(b)] is unnatural and brings undesired effects to the overall regression and generalization.

□

Before describing decision-tree induction in the next section, we must first introduce some important nomenclature for binary trees. A typical binary tree, as shown in Figure 14.3(a), is usually denoted as T with the root node t_1. A generic node in T is denoted by t, and the subtree with t as a root node is usually represented by T_t, as shown in Figure 14.3(a), where $t = t_3$. We use $\tilde{T}$ to denote the set of terminal nodes in a tree T; the number of terminal nodes is thus represented by $|\tilde{T}|$ [which is equal to 5 in the case of Figure 14.3(a)]. It is easy to prove that in a complete binary tree (where each node has zero or two children), the number of terminal nodes is always one more than the number of internal nodes.

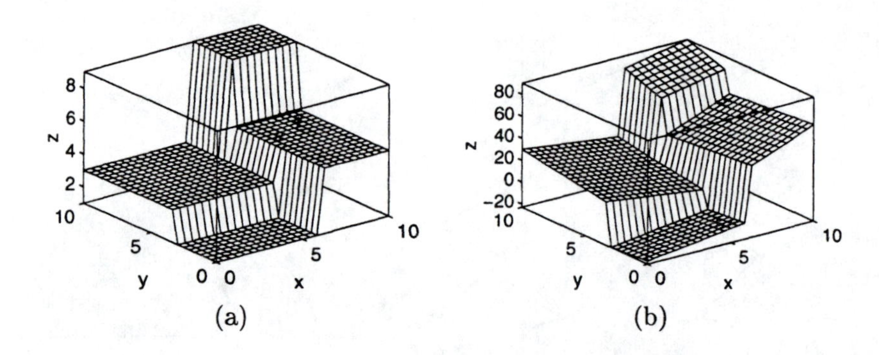

Figure 14.2. *Input-output surfaces of decision trees with terminal nodes characterized by (a) constants and (b) linear equations in Example 14.1 .* (MATLAB files: `tsurf1.m` and `tsurf2.m`)

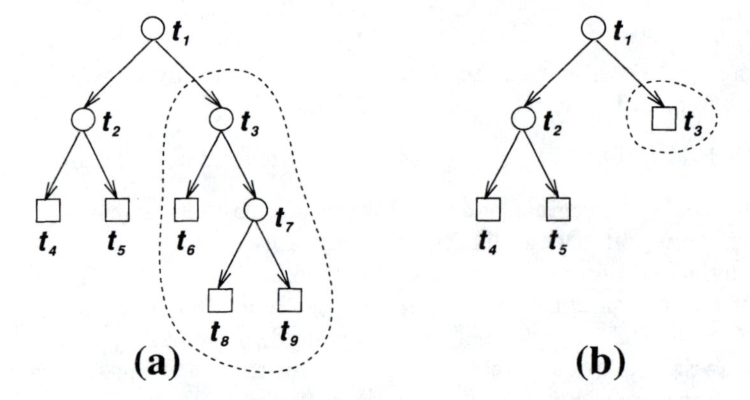

Figure 14.3. *(a) A typical tree T with root node t_1 and a subtree T_{t_3}; (b) $T - T_{t_3}$, the tree after shrinking the subtree T_{t_3} into a terminal node t_3.*

An example of tree pruning is shrinking the subtree T_{t_3} in Figure 14.3 into a terminal node. The tree after pruning, denoted as $T - T_{t_3}$, is a subset of the original tree, and this is usually expressed as

$$T - T_{t_3} \subset T.$$

14.3 CART ALGORITHM FOR TREE INDUCTION

The use of tree-based classification and regression dates back to the AID (Automatic Interaction Detection) program of Morgan and Sonquist [3]. Methods of decision-tree induction from sample data, also known as **recursive partitioning**, have since been an active topic in artificial intelligence (particularly machine learning) and statistics (in particular, multivariate analysis) research communities. In machine learning literature, the most representative method of decision-tree induction are the **ID3** [4] and **C4** [5] procedures proposed by Quinlan; detailed treatments can be found in ref. [6]. Similar problems were approached by statisticians at about the same time, and the most well-known work was published by Breiman et al. [1] in their monograph entitled *Classification and Regression Trees*; thus the methodology is often referred to as the **CART** algorithm. The fundamentals of ID3 and CART are similar; the major distinction is that CART induces strictly binary trees and uses resampling techniques for error estimation and tree pruning, while ID3 partitions according to attribute values.

This section presents a summary of the CART procedure, and we confine our scope to the discussion of binary trees only; the extension to n-ary trees is straightforward. The material presented in this section will serve as a background for understanding the use of CART for structure identification in ANFIS in the next section.

To construct an appropriate decision tree, CART first grows the tree extensively based on a sample (training) data set, and then prunes the tree back based on a minimum cost-complexity principle [1]. The result is a sequence of trees of various sizes; the final tree selected is the tree that performs best when another independent (checking or test) data set is presented. In summary, the CART procedure consists of two parts: tree growing and tree pruning.

14.3.1 Tree Growing

CART grows a decision tree by determining a succession of **splits** (decision boundaries) that partition the training data into disjoint subsets. Starting from the root node that contains all the training data, an exhaustive search is performed to find the split that best reduces an error measure (or cost function). Once the best split is determined, the data set is partitioned into two disjoint subsets accordingly; the subsets are represented by two child nodes originating from the root nodes, and the same splitting method is applied to both child nodes. This recursive procedure terminates either when the error measure associated with a node falls below a certain tolerance level, or when the error reduction resulting from further splitting will not exceed a certain threshold value.

Classification Trees

Classification trees are used to solve classification problems in which attributes of an object are used to determine what class the object belongs to. To grow a classification tree, we need to have an error measure $E(t)$ that quantifies the performance of a node t in separating data (or cases) from different classes. The error measure for classification trees is often referred to as the **impurity function**; for a given node (or equivalent, for a data set), it should attain a minimum at zero when the given data all belong to the same class, and reach a maximum when the data are evenly distributed through all possible classes. A formal definition of the impurity function for J-class problems is given next.

Definition 14.1 *The impurity function for J-class problems*

The impurity function ϕ is a J-place function that maps its input arguments $p_1, p_2, \ldots, p_J$, with $\sum_{j=1}^{J} p_j = 1$, into a non-negative real number, such that

$$\begin{aligned} &\phi(1/J, 1/J, \cdots, 1/J) = \text{maximum}, \\ &\phi(1,0,0,\cdots,0) = \phi(0,1,0,\cdots,0) = \phi(0,0,0,\cdots,1) = 0. \end{aligned} \tag{14.1}$$

The input arguments p_j, $j = 1$ to J, is the probability that a case in a node belongs to class j. Therefore, the impurity function for a given node is largest when all classes are equally mixed in the node, and is smallest when the node contains cases from only one class.

□

By using the impurity function ϕ, the impurity measure of a node t is expressed as

$$E(t) = \phi(p_1, p_2, \cdots, p_J),$$

where p_j is the percentage of cases in node t that belong to class j. Similarly, the impurity measure of a tree T can be expressed as

$$E(T) = \sum_{t \in \tilde{T}} E(t),$$

where $\tilde{T}$ is the set of terminal nodes in tree T.

The best known impurity functions for a J-class classification tree are the **entropy function** and the **Gini diversity index** [1].

$$\begin{array}{lll} \text{Entropy function:} & \phi_e(p_1, \cdots, p_J) = -\sum_{j=1}^{J} p_j \ln p_j; \\ \text{Gini index:} & \phi_g(p_1, \cdots, p_J) = \sum_{i \neq j} p_i p_j = 1 - \sum_{j=1}^{J} p_j^2. \end{array} \tag{14.2}$$

Since $\sum_{j=1}^{J} p_j = 1$ and $0 \leq p_j \leq 1$ for all j, the preceding two functions are always positive unless one of p_j is unity and all the others are zero. Moreover, they reach their maxima when $p_j = 1/J$ for all j. The proof of these properties is left as an exercise.

Example 14.2 *Entropy and Gini function visualization*

Both the entropy function and the Gini index can be expressed as functions of p_j, $j = 1$ to $J - 1$, if we plug $p_J = 1 - \sum_{i=1}^{J-1} p_j$ into Equation (14.2). In other words, both impurity functions can be visualized as curves when J is 2 and as surfaces when J is 3, as shown in Figure 14.4. When J is 3, both functions are not defined for $p_1 + p_2 > 1$, since $p_3 = 1 - p_1 - p_2$ should always be kept non-negative; this is clearly shown in Figures 14.4(b) and 14.4(d).

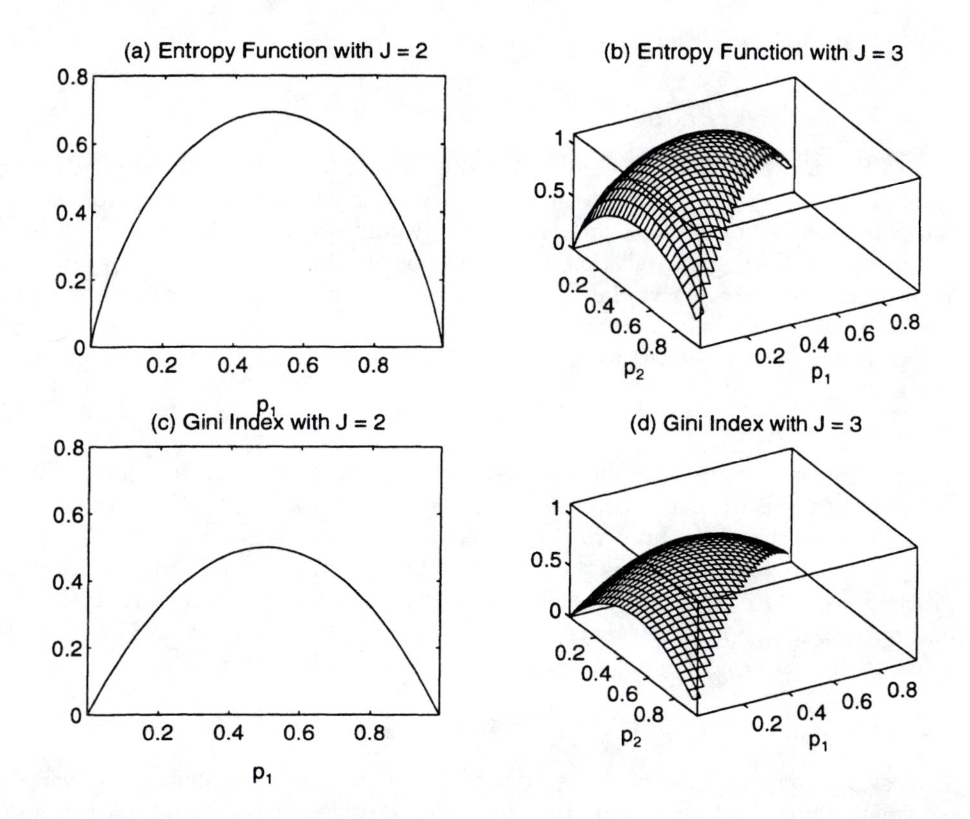

Figure 14.4. *Impurity functions for classification trees: (a) entropy function for two-class problem; (b) entropy function for three-class problem; (c) Gini index for two-class problem; (d) Gini index for three-class problem. Both the curves in (a) and (c) reach their maxima at $p_1 = p_2 = 1/2$; both the surfaces in (b) and (d) reach their maxima at $p_1 = p_2 = p_3 = 1/3$.* (MATLAB file: `impurity.m`)

□

Given an impurity function for computing the *cost* of a node, the tree-growing procedure tries to find an optimal way to split the cases (or objects) in the node such that the cost reduction is the greatest. In a binary tree, the impurity change due to splitting is

$$\Delta E(s,t) = E(t) - p_l E(t_l) - p_r E(t_r), \tag{14.3}$$

where t is the node being split; $E(t)$ is the impurity of the current node t; $E(t_l)$ and $E(t_r)$ are the impurities of the left and right branch nodes; and p_l and p_r are the percentages of cases in node t that branch left and right, respectively. In symbols, the-tree growing procedure tries to find a split s^* for the root node t_1 such that the split gives the largest decrease in impurity:

$$\Delta E(s^*, t_1) = \max_{s \in S} \Delta E(s, t_1),$$

where S is a set of all the possible ways of splitting the cases in node t_1. By using the optimal s^*, t_1 is split into t_2 and t_3, and the same search procedure for the best $s \in S$ is repeated on both t_2 and t_3 separately, and so on.

So far we have supposed that the inputs or attributes under consideration are **numerical** or **ordered variables** that assume numerical values; examples of this kind of variables include temperatures, heights, lengths, and so on. For binary trees, a typical split (or question) for a numerical variable x takes the following form:

$$\text{Is } x \leq s_i?$$

Usually the split value s_i is the average of the x values of two data points that are adjacent in terms of their x coordinates alone. For a data set of size M, the number of candidate splits for a numerical variable is less than or equal to $M - 1$.

For **categorical variables** that assume labels with no natural ordering, the tree-growing procedure given here is still applicable except that splitting a node depends on how to put the possible labels of a variable into two disjoint sets. Therefore, for a binary tree, a typical split (question) takes the following form:

$$\text{Is } x \text{ in } S_1?$$

The set S_1 is a non-empty proper subset of S, the set of all possible labels of variable x. To eliminate duplication due to symmetry, the size of S_1 is usually less than or equal to half the size of S. In general, a categorical variable x with k possible labels has $(2^k - 2)/2 = 2^{k-1} - 1$ candidate splits for this variable. (Why?)

Example 14.3 *Splits for numerical and categorical variables*

If the x attributes of a data set are represented as a set $\{1, 2, 4, 7\}$, then the candidate split values are $\{\frac{1+2}{2}, \frac{2+4}{2}, \frac{4+7}{2}\} = \{1.5, 3, 5.5\}$. In contrast, if a categorical variable *status* assumes four possible labels {single, married, divorced, widowed}, then we have seven candidate splits that divide a data set into two disjoint non-empty subsets based on this variable.

□

The following example demonstrates how to split a node containing five cases, each of which has two numerical attributes.

Example 14.4 *Node-splitting for classification trees*

Suppose that we want to split a node t containing five data points of two attributes x and y, as shown in Figure 14.5, where data from classes 1 and 2 are denoted by crosses and circles, respectively. Apparently, there is a total of eight possible splits, denoted by s_i in Figure 14.5. If we choose the entropy function as the error measure, the impurity of this node is

$$\begin{aligned} E(t) &= -\tfrac{2}{5}\ln\left(\tfrac{2}{5}\right) - \tfrac{3}{5}\ln\left(\tfrac{3}{5}\right) \\ &= 0.6730. \end{aligned}$$

Now we have to evaluate the change of impurity due to each split. For instance, for the split of s_2, we have

$$\begin{aligned} p_l &= \frac{2}{5}, \\ E(t_l) &= -\tfrac{1}{2}\ln\left(\tfrac{1}{2}\right) - \tfrac{1}{2}\ln\left(\tfrac{1}{2}\right) = 0.6983, \\ p_r &= \frac{3}{5}, \\ E(t_r) &= -\tfrac{1}{3}\ln\left(\tfrac{1}{3}\right) - \tfrac{2}{3}\ln\left(\tfrac{2}{3}\right) = 0.6365. \end{aligned}$$

Therefore, the change of impurity due to split s_2 is

$$\begin{aligned} \Delta E(s_2, t) &= E(t) - \tfrac{2}{5}E(t_l) - \tfrac{3}{5}E(t_r) \\ &= 0.0138. \end{aligned}$$

Apparently, this is not a very effective split. Following the same procedure, we can get a list of all split performances:

$$\begin{aligned} \Delta E(s_1, k) &= 0.2231, \\ \Delta E(s_3, k) &= 0.2911, \\ \Delta E(s_4, k) &= 0.1185, \\ \Delta E(s_5, k) &= 0.2231, \\ \Delta E(s_6, k) &= 0.6730, \\ \Delta E(s_7, k) &= 0.2911, \\ \Delta E(s_8, k) &= 0.1185. \end{aligned}$$

Therefore, the best split is s_6, which separates the data most effectively and reduces the impurity to zero.

□

Another criterion, called the **twoing rule**, is to select a split that minimizes

$$\frac{p_l p_r}{4}\left[\sum_j p_j(t_l)p_j(t_r)\right]^2,$$

where $p_j(t_l)$ and $p_j(t_r)$ are the probabilities of a data point in class j, given that the data set comes from the left and right children, respectively. The justification of the twoing rule can be found in the CART monograph [1].

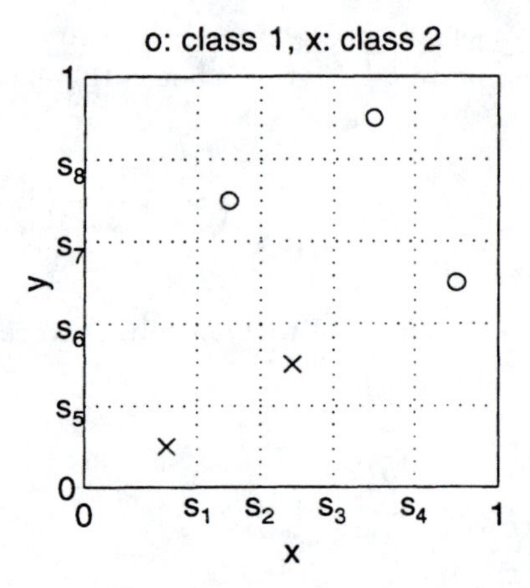

Figure 14.5. *Node-splitting in classification trees; see Example 14.4.* (MATLAB file: `splits.m`)

Regression Trees

We use regression trees to solve regression problems where attributes of an object are used to determine one or more numerical attributes of the object. For a regression tree, the error measure of a node t is usually taken as the squared error, or **residual**, of a local model employed to fit the data set of the node:

$$E(t) = \min_{\boldsymbol{\theta}} \sum_{i=1}^{N(t)} (y_i - d_t(\mathbf{x}_i, \boldsymbol{\theta}))^2, \tag{14.4}$$

where $\{\mathbf{x}_i, y_i\}$ is a typical data point, $d_t(\mathbf{x}, \boldsymbol{\theta})$ is a **local model** (with modifiable parameter $\boldsymbol{\theta}$) for node t and $E(t)$ is the mean-squared error of fitting the local model d_t to the data set in the node. If $d(\mathbf{x}, \boldsymbol{\theta}) = \theta$ is a constant function independent of $\mathbf{x}$, then the minimizing θ of the preceding error measure is the average value of the desired output y_i for the node—that is, $\theta^* = \frac{1}{N(t)} \sum_{i=1}^{N(t)} y_i$. Similarly, if $d(\mathbf{x}, \boldsymbol{\theta})$ is a linear model with linear parameters $\boldsymbol{\theta}$, then we can always use the least-squares methods introduced in Chapter 5 to identify the minimizing $\boldsymbol{\theta}^*$ and $E(t)$ for a given node t.

For any split s of node t into t_l and t_r, the change in error measure is expressed as

$$\Delta E(s, t) = E(t) - E(t_l) - E(t_r). \tag{14.5}$$

The best split s^* is the one that maximizes the decrease in the error measure:

$$\Delta E(s^*, t) = \max_{s \in S} \Delta E(s, t).$$

The strategy for growing a regression tree is to split nodes (or data set) iteratively and thus maximize the decrease in $E(T) = \sum_{t \in \tilde{T}} E(t)$, the overall error measure (or

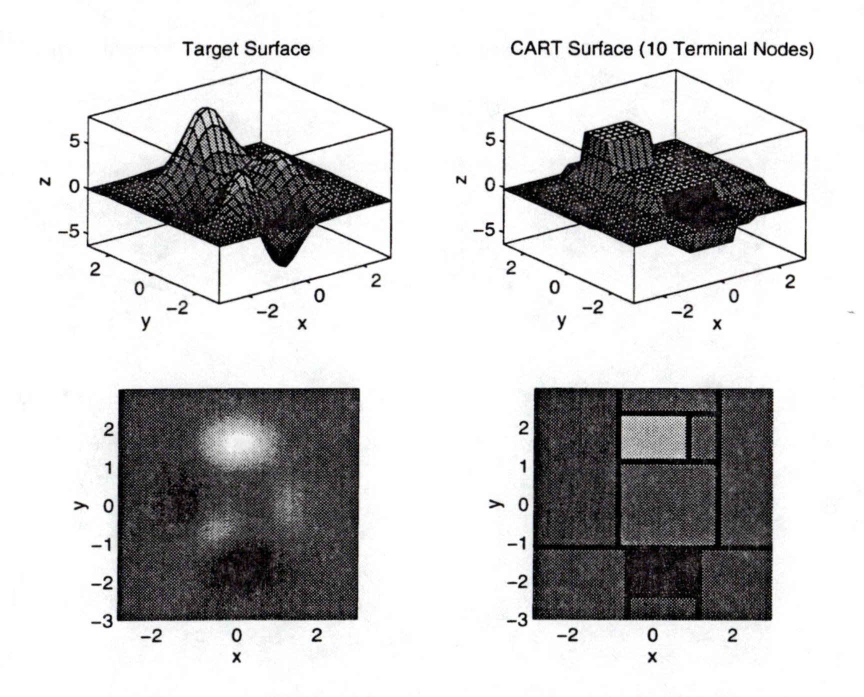

Figure 14.6. *Growing a CART tree. First column: target surface; second column: regression tree surface with 10 rules or terminal nodes.* (MATLAB file: `go_cart.m`)

cost) of the tree. Therefore, the goal of growing either a classification or regression tree is the same: to split nodes (or, equivalently, partition data set or input space) recursively and thus minimize a given reasonable error measure in a greedy, single look-ahead manner.

Example 14.5 *Growing a regression tree*

Figures 14.6 and 14.7 demonstrate the progression of growing a regression tree to match the "peaks" function defined by Equation (7.1) in Chapter 7. The input-output surfaces of the regression tree are snapshots when the number of terminal nodes is equal to 10, 20, and 30, respectively. The boundary plots clearly indicate the square local region governed by each terminal node (or rule).

□

14.3.2 TREE PRUNING

The tree that the preceding growing procedure yields is often too large, and it is biased toward the training data set. Thus, it places an unreliably high degree of accuracy on reproducing desired outputs from the training data. In other words,

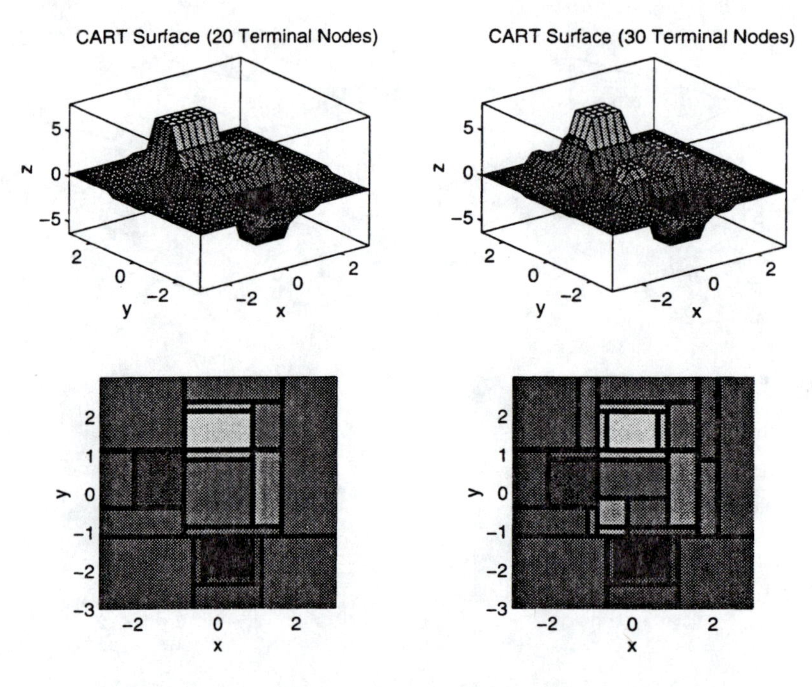

Figure 14.7. *Growing a CART tree (continued). First column: regression tree surface with 20 rules (or terminal nodes); second column: regression tree surface with 30 rules (or terminal nodes).* (MATLAB file: `go_cart.m`)

we may encounter the familiar problem of overfitting and overspecializing toward the training data, and the tree may not generalize well for new cases.

There are several methods to find the tree size that gives a better estimate of the true error measure. One of the most effective methods is based on the principle of **minimum cost-complexity** or **weakest-subtree shrinking**. The first step is to grow a fully expanded tree $T_{\max}$ that has a fairly low apparent error measure based on the training data set. This tree is usually too large, and we want to prune it back consistently by finding the weakest subtree in it. The weakest subtree is found by considering both the training error measure and the number of terminal nodes, which is considered a measure of the tree's complexity.

Definition 14.2 *Cost-complexity measure of decision trees [1]*

For any subtree $T \subset T_{\max}$, define its complexity as $|\tilde{T}|$, the number of terminal nodes in T. Then the **cost-complexity measure** $E_\alpha(T)$ is defined by

$$E_\alpha(T) = E(T) + \alpha|\tilde{T}|, \tag{14.6}$$

where α is a **complexity parameter** that accounts for the cost due to the tree's complexity. Thus $E_\alpha(T)$ is a linear combination of the cost of the tree and its complexity.

□

For each value of α, we can find a minimizing subtree $T(\alpha)$ with respect to the cost-complexity measure for a given α:

$$E_\alpha(T(\alpha)) = \min_{T \subset T\mathrm{max}} E_\alpha(T).$$

If $T(\alpha)$ is a minimizing tree for a given value of α, then it continues to be minimizing as α increases until a jump point α' is reached and a new tree $T(\alpha')$ becomes a new minimizing tree.

Suppose that $T\mathrm{max}$ has L terminal nodes. The idea of progressive upward tree pruning is to find a sequence of smaller and smaller trees $T_L, T_{L-1}, T_{L-2}, \cdots$, and T_1 that satisfies

$$\{t_1\} = T_1 \subset T_2 \subset \cdots \subset T_{L-2} \subset T_{L-1} \subset T_L = T\mathrm{max},$$

where T_i has i terminal nodes. Each T_{i-1} is obtained from T_i as the first minimizing subtree of the cost-complexity measure as α increases from zero.

To find the next minimizing tree for a tree T, we proceed as follows. For each internal node t in T, we first find a value for α that makes $T - T_t$ the next minimizing tree; this value of α, denoted by α_t, is equal to the ratio between the change in error measures and the change in the number of terminal nodes before and after shrinking:

$$\alpha_t = \frac{E(t) - E(T_t)}{|\tilde{T}_t| - 1}.$$

Then we choose the internal node with the smallest α_t as the target node for shrinking. Therefore, a tree-pruning cycle consists of the following tasks:

1. Calculate α_t for each internal node t in T_i
2. Find the minimal α_t and choose $T - T_t$ as the next minimizing tree.

This process is repeated until the tree contains a single root node. Figure 14.3 demonstrates an example of tree pruning by shrinking the subtree T_{t_3} (with the internal node t_3 as its root) into a terminal node. The tree after pruning, denoted as $T - T_{t_3}$, is a subset of the original tree.

By repeating the pruning process, a series of candidate trees can be obtained by shrinking each weakest subtree sequentially, where each shrinkage results in a minimal increase in the value of α in proceeding toward the next minimizing tree. The problem has now been reduced to selecting one of these candidate trees as the optimum-sized tree. There are two general methods for doing this: using an independent test (checking) data set and performing cross-validation. Of the two, use of a test data set is computationally simpler, but cross-validation makes more

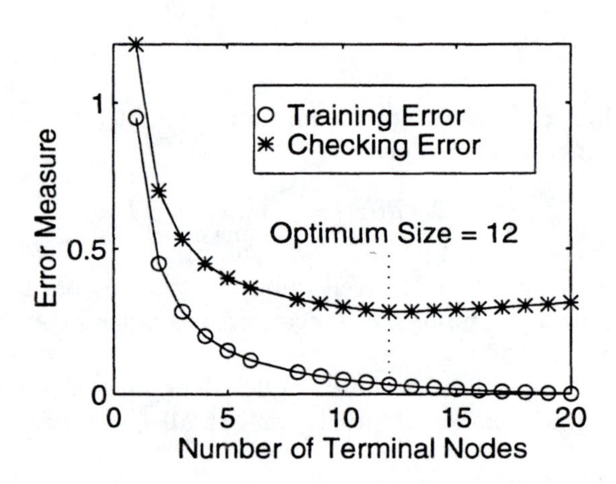

Figure 14.8. *Error measures with respect to tree sizes. The jump from T_8 to T_6 indicates that the shrunken subtree in T_8 has more than two terminal nodes.* (MATLAB file: `carterr.m`)

effective use of all available data. To use an independent test data set, we simply pick the tree that generates the smallest error measure when the test data set is presented. The cross-validation method is more complicated, and the reader is referred to the CART monograph [1] for a complete treatment of it. Figure 14.8 is a typical pattern of tree error measures versus tree sizes for the candidate trees $T_1, T_2, \cdots$, and T_L, $L = 20$, obtained from the preceding tree-pruning procedure. As the tree's complexity (that is, number of terminal nodes) increases, the training error decreases, reaching zero when the tree is fully expanded. In contrast, the checking error decreases initially, reaching a minimum, and then increases gradually due to the tree's overspecialization to the training data. Without any other *a priori* information, we usually take the checking error as a true unbiased estimate of the real error measure and take the corresponding tree as the optimum-sized tree.

14.4 USING CART FOR STRUCTURE IDENTIFICATION IN ANFIS

The CART algorithm is a powerful nonparametric method with the following features:

- Conceptual simplicity
- Computation efficiency
- Applicability to classification and regression problems

- Solid statistic foundation

- Suitability for high-dimensional data

- Ability to identify relevant inputs simultaneously.

In this section, we describe how to use CART for structure identification in ANFIS. That is, we use CART to find the number of ANFIS rules and the initial locations of membership functions (MFs) before training. For simplicity, we confine our scope to regression problems; a similar approach can be also used for classification problems.

To construct a regression tree with constant-output terminal nodes [see Example 14.1 and Figure 14.2(a)], the CART algorithm described earlier can always identify a right-size tree and determine irrelevant inputs not required by the tree. On the other hand, if the terminal nodes are characterized by linear equations [see Example 14.1 and Figure 14.2(b)], more computation is necessary to find relevant inputs. One way to reduce the computation burden is to employ the least-squares estimator (LSE) introduced in Chapter 5; of particular importance is the LSE that is obtained recursively in accommodating new data and new parameters.

It is obvious that the decision tree in Figure 14.1 is equivalent to a set of crisp rules:

$$\begin{cases} \text{If } x < a \text{ and } y < b, \text{ then } z = f_1. \\ \text{If } x < a \text{ and } y \geq b, \text{ then } z = f_2. \\ \text{If } x \geq a \text{ and } y < c, \text{ then } z = f_3. \\ \text{If } x \geq a \text{ and } y \geq c, \text{ then } z = f_4. \end{cases} \tag{14.7}$$

Given an input vector $[x,\ y]$, only a single rule out of the four will be fired at full strength, while the other three will not be activated at all. This crispness reduces the computation required to construct the tree using CART, but it also gives undesirable discontinuous boundaries in the overall input-output mapping. To smooth out the discontinuity at each split, a natural option is to use fuzzy sets to represent the premise parts of the rules in Equation (14.7), thus converting Equation (14.7) into a set of Sugeno-style fuzzy if-then rules, as described in Chapter 4. The resultant Sugeno fuzzy inference model can be of zero order if f_i's are constants, or first order if f_i's are linear equations.

To fuzzify the premise part, the statement $y \geq c$ can be represented as a fuzzy set characterized by, for instance, the sigmoidal MF introduced in Section 2.4:

$$\mu_{y \geq c}(y; \alpha) = \text{sig}(y; \alpha, c) = \frac{1}{1 + \exp[-\alpha(y - c)]}, \tag{14.8}$$

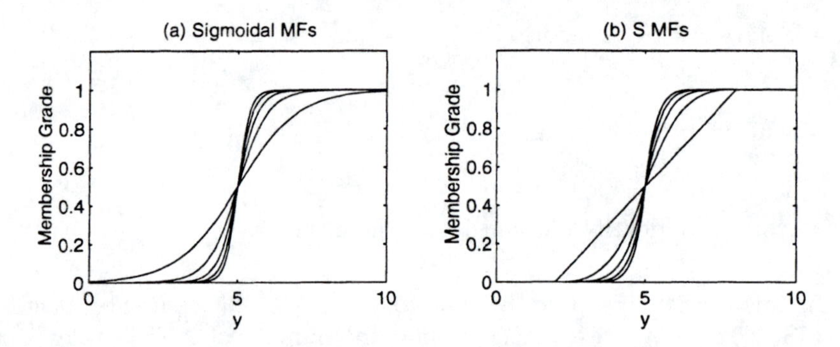

Figure 14.9. *Two types of MFs for $x > c$ (where $c = 5$): (a) sigmoidal MF with different α's; (b) extended S MF with different γ's.* (MATLAB file: `cartmf.m`)

where α and c are modified parameters for the MF. Similarly, we can also employ the S MF (see page 43) to represent the meaning of $y \geq c$:

$$\mu_{y \geq c}(y; w) = S(y; c-w, c+w) = \begin{cases} 0, & \text{if } x \leq c - w, \\ 2\left[\dfrac{x-(c-w)}{2w}\right]^2, & \text{if } c - w < x \leq c, \\ 1 - 2\left[\dfrac{(c+w)-x}{2w}\right]^2, & \text{if } c < x \leq c + w, \\ 1, & \text{if } c + w < x. \end{cases} \tag{14.9}$$

To increase the degree of freedom, we can even use an **extended S MF** with an extra parameter γ:

$$\mu_{y \geq c}(y; w, \gamma) = S_{\text{ext}}(y; c-w, c+w, \gamma) = \begin{cases} 0, & \text{if } y \leq c - w, \\ \dfrac{1}{2}\left|\dfrac{y-(c-w)}{w}\right|^{2\gamma}, & \text{if } c - w < y \leq c, \\ 1 - \dfrac{1}{2}\left|\dfrac{c+w-y}{w}\right|^{2\gamma}, & \text{if } c < y \leq c + w, \\ 1, & \text{if } c + w < y. \end{cases} \tag{14.10}$$

(Note that when $\gamma = 0.5$, the preceding extended S MF becomes a piecewise linear function.) Figure 14.9 shows the sigmoidal and the extended S MFs for the linguistic term $y \geq c$; Figure 14.10 is the fuzzy version of the surface plots in Figure 14.2, where the extended S MF (with $w = 1$ and $\gamma = 1$) is used to define the meaning of $>$. Remember that when $\alpha \to \infty$ in the sigmoidal MF, or when $w = 0$ or $\gamma \to \infty$ in the extended S MF, both MFs reduce to the step function and the fuzzy rules reduce to the original crisp rules.

Based on the fuzzy version of the rules in Equation (14.7), we can derive another class of adaptive network for identifying the premise and consequent parameters of the underlying fuzzy inference system. This ANFIS architecture is depicted in Fig-

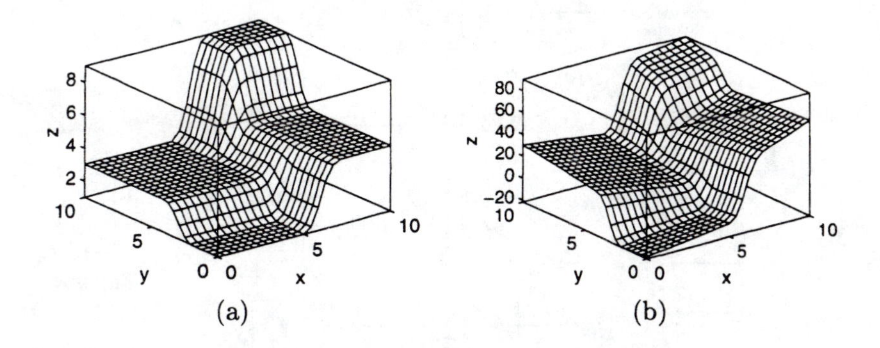

Figure 14.10. *Input-output behaviors of decision trees with terminal nodes characterized by (a) constants and (b) linear equations.* (MATLAB file: **ftsurf1.m** and **ftsurf2.m**)

ure 14.11. Layer 1 calculates the membership grades of given input variables (INV nodes represent negation operators); layer 2 multiplies the given membership grades to find the firing strength of each rule; layer 3 computes the contribution of each rule based on given firing strengths; and layer 4 finds the summation of incoming signals, which is equal to the overall output of this fuzzy inference system. Premise and consequent parameters are contained in layers 1 and 3, respectively; these parameters are fine-tuned according to the fast hybrid learning rules introduced in Section 12.3, or any of the other nonlinear parameter identification methods introduced in Section 6.8 of Chapter 6. Note that the normalization layer (layer 3) in Figure 12.1 is missing from Figure 14.11. This is attributable to the following theorem of **implicit weight normalization** [2].

Theorem 14.1 *Implicit weight normalization in a CART-constructed ANFIS network*

In converting a decision tree to a fuzzy inference system, if (1) $\mu_{x>a}(x)+\mu_{x\leq a}(x) = 1$, where x is any of the input variables and a is any of the splits of x, and (2) multiplication is used as the T-norm operator to calculate each rule's firing strength, then the summation over each rule's firing strength is always equal to unity.

Proof: This theorem can be proved by induction. Let n be the number of rules and w_i, $i = 1, \ldots, n$ be the firing strength of the ith rule. For $n = 2$, we have $w_1 + w_2 = 1$ since w_1 and w_2 are the membership grades for $\mu_{x<a}(x)$ and $\mu_{x\geq a}(x)$ for a certain input x and a certain split value a.

Suppose that $\sum_{i=1}^{n} w_i = 1$ holds when $n = k$. When $n = k+1$, we need to show that $\sum_{i=1}^{k+1} w_i = 1$ still holds. Without loss of generality, we can assume the newly generated rules are k and $k+1$, the result from splitting the previously terminal

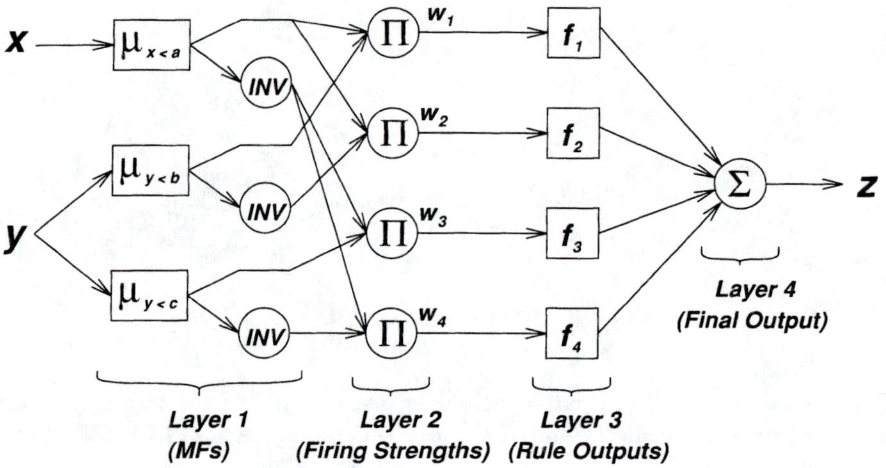

Figure 14.11. *ANFIS architecture corresponding to the fuzzy version of the rule set in Equation (14.7).*

node k (or rule k). Consequently, we have

$$\begin{aligned}\sum_{i=1}^{k+1} w_i &= \sum_{i=1}^{k-1} w_i + w_k + w_{k+1} \\ &= \sum_{i=1}^{k-1} w_i + \hat{w}_k(\mu_{x<a}(x) + \mu_{x\geq a}(x)) \\ &= \sum_{i=1}^{k-1} w_i + \hat{w}_k \\ &= 1,\end{aligned}$$

where $\hat{w}_k$ is the firing strength of rule k before splitting. This concludes the proof.

□

The implicit weight normalization of the ANFIS architecture in Figure 14.11 is maintained throughout training processes; this eliminates the need for another normalization layer and reduces training and application computation time as well as round-off errors.

In summary, fuzzy modeling based on the CART-ANFIS approach consists of two tasks:

Structure identification This is done by CART to find an initial set of crisp rules.

Parameter identification After fuzzifying the premise parts of the initial rules, we can construct an ANFIS architecture to fine-tune the parameters.

The major advantage offered by this approach is that we can quickly determine the roughly correct structure of a fuzzy inference using CART, and then refine the MFs and output functions via an efficient ANFIS architecture that does not need a normalization layer. Note that CART can select relevant inputs and do tree partitioning of the input space, while ANFIS refines the regression and makes it everywhere continuous and smooth. Thus it can be seen that CART and ANFIS are complementary and their combination constitutes a solid approach to fuzzy modeling.

14.5 SUMMARY

This chapter presents the CART (classification and regression tree) algorithm, a fast one-pass approach for multivariate analysis. CART is a popular non-parametric approach for both data classification and regression in statistics, so we devote a whole chapter to it. Because CART can select relevant input variables and partition the input space effectively, it is an ideal tool for structure identification in ANFIS (Chapter 12).

EXERCISES

1. Prove that in a complete binary tree where each node has zero or two children, there is always one more terminal node than there are internal nodes.

2. Prove that the entropy function in Equation (14.2) reaches a maximum when $p_1 = p_2 = \cdots = p_J = 1/J$.

3. Prove that the Gini index in Equation (14.2) reaches a maximum when $p_1 = p_2 = \cdots = p_J = 1/J$.

4. A new impurity function can be defined by

$$\phi(p_1, \cdots, p_J) = 1 - \max\{p_1, \cdots, p_J\}.$$

   Prove that this function satisfies all the requirements in Equation (14.1) for impurity functions.

5. For the preceding impurity function, write a MATLAB script to plot $\phi(p_1, 1-p_1)$ as a curve and $\phi(p_1, p_2, 1-p_1-p_2)$ as a two-dimensional surface.

6. Explain why a categorical variable with k possible labels has $(2^k - 2)/2 = 2^{k-1} - 1$ candidate splits.

7. Why is the $\Delta E(s,t)$ for regression trees [Equation (14.5)] different from that for classification trees [Equation (14.3)]? Suppose that we redefine the regression tree error measure in Equation (14.4) as the mean-squared error:

$$E(t) = \frac{1}{N(t)} \min_{\boldsymbol{\theta}} \sum_{i=1}^{N(t)} (y_i - d_t(\mathbf{x}_i, \boldsymbol{\theta}))^2.$$

How should you modify $\Delta E(s,t)$ in Equation (14.5) accordingly?

8. Express $E(t)$ in Equation (14.4) as an explicit function of $\mathbf{x}_i$ and y_i, assuming that the local model for node t is a modifiable constant $\boldsymbol{\theta} = \theta$—that is, $d_t(\mathbf{x}, \boldsymbol{\theta}) = \theta$.

9. Repeat Exercise 8, but assume that the local model for node t is a linear model—that is, $d_t(\mathbf{x}, \boldsymbol{\theta}) = \theta_0 + x_1\theta_1 + x_2\theta_2 + \cdots + x_n\theta_n$, where $\mathbf{x} = [x_1 \ \cdots \ x_n]$ and $T = [\theta_0 \, \theta_1 \ \cdots \ \theta_n]$.

10. If $d_t(\mathbf{x}, \boldsymbol{\theta}) = \theta$ is a modifiable constant, prove that $\Delta E(s,t)$ in Equation (14.5) is always greater than or equal to zero. When is it equal to zero?

11. Repeat Example 14.5, but change the local model of each terminal node to a linear model. Plot the input-output surfaces and compare them with Figures 14.6 and 14.7.

REFERENCES

[1] L. Breiman, J. H. Friedman, R. A. Olshen, and C. J. Stone. *Classification and regression trees.* Wadsworth, Inc., Belmont, California, 1984.

[2] J.-S. Roger Jang. Structure determination in fuzzy modeling: a fuzzy CART approach. In *Proceedings of IEEE International Conference on Fuzzy Systems*, Orlando, Florida, June 1994.

[3] J. N. Morgan and J. A. Sonquist. Problems in the analysis of survey data, and a proposal. *Journal of American Statistics Association*, 58:415–434, 1963.

[4] J. R. Quinlan. Induction of decision trees. *Machine Learning*, 1:81–106, 1986.

[5] J. R. Quinlan. Simplifying decision trees. *International Journal of Man-Machine Studies*, 27:221–234, 1987.

[6] J. R. Quinlan. Decision trees and decisionmaking. *IEEE Transactions on Systems, Man, and Cybernetics*, 20(2):339–346, 1990.

Chapter 15

Data Clustering Algorithms

J.-S. R. Jang

15.1 INTRODUCTION

Clustering algorithms are used extensively not only to organize and categorize data, but are also useful for data compression and model construction. Chapter 11 describes some of the on-line clustering algorithms that can be realized by unsupervised learning neural networks. This chapter introduces four of the most representative off-line clustering techniques frequently used in conjunction with radial basis function networks and fuzzy modeling: (hard) C-means (or K-means) clustering, fuzzy C-means clustering, the mountain clustering method, and subtractive clustering.

Clustering partitions a data set into several groups such that the similarity within a group is larger than that among groups. Achieving such a partitioning requires a similarity metrics that takes two input vectors and returns a value reflecting their similarity. Since most similarity metrics are sensitive to the ranges of elements in the input vectors, each of the input variables must be normalized to within, say, the unit interval $[0, 1]$. Hence, the rest of this chapter assumes that data set under consideration has already been normalized to be within the unit hypercube.

Clustering techniques are used in conjunction with radial basis function networks or fuzzy modeling primarily to determine initial locations for radial basis functions or fuzzy if-then rules. For this purpose, clustering techniques are validated on the basis of the following assumptions:

1. Similar inputs to the target system to be modeled should produce similar outputs.

2. These similar input-output pairs are bundled into clusters in the training data set.

Assumption 1 states that the target system to be modeled is a smooth input-output mapping; this is generally true for real-world systems. Assumption 2 requires the data set to conform to some specific type of distribution; however, this is not always true. Therefore, clustering techniques used for structure identification in neural or fuzzy modeling are highly heuristic, and finding a data set to which clustering techniques cannot be applied satisfactorily is not uncommon.

15.2 K-MEANS CLUSTERING

The **K-means clustering** [6, 8], also known as **C-means clustering**, has been applied to a variety of areas, including image and speech data compression [3, 7], data preprocessing for system modeling using radial basis function networks [9], and task decomposition in heterogeneous neural network architectures [3].

The K-means algorithm partitions a collection of n vector $\mathbf{x}_j, j = 1, \ldots, n$, into c groups $G_i, i = 1, \ldots, c$, and finds a cluster center in each group such that a cost function (or an objection function) of dissimilarity (or distance) measure is minimized. When the Euclidean distance is chosen as the dissimilarity measure between a vector $\mathbf{x}_k$ in group j and the corresponding cluster center c_i, the cost function can be defined by

$$J = \sum_{i=1}^{c} J_i = \sum_{i=1}^{c} \left(\sum_{k,\, x_k \in G_i} \|\mathbf{x}_k - \mathbf{c}_i\|^2 \right), \tag{15.1}$$

where $J_i = \sum_{k,\, x_k \in G_i} \|\mathbf{x}_k - \mathbf{c}_i\|^2$ is the cost function within group i. Thus, the value of J_i depends on the geometrical properties of G_i and the location of $\mathbf{c}_i$.

In general, a generic distance function $d(\mathbf{x}_k, \mathbf{c}_i)$ can be applied for vector $\mathbf{x}_k$ in group i; the corresponding overall cost function is thus expressed as

$$J = \sum_{i=1}^{c} J_i = \sum_{i=1}^{c} \left(\sum_{k,\, x_k \in G_i} d(\mathbf{x}_k - \mathbf{c}_i) \right). \tag{15.2}$$

For simplicity, the Euclidean distance is used as the dissimilarity measure and the overall cost function is expressed as in Equation (15.1).

The partitioned groups are typically defined by an $c \times n$ binary **membership matrix U**, where the element u_{ij} is 1 if the jth data point $\mathbf{x}_j$ belongs to group i, and 0 otherwise. Once the cluster centers $\mathbf{c}_i$ are fixed, the minimizing u_{ij} for Equation (15.1) can be derived as follows:

$$u_{ij} = \begin{cases} 1 \text{ if } \|\mathbf{x}_j - \mathbf{c}_i\|^2 \leq \|\mathbf{x}_j - \mathbf{c}_k\|^2, \text{ for each } k \neq i, \\ 0 \text{ otherwise.} \end{cases} \tag{15.3}$$

Restated, $\mathbf{x}_j$ belongs to group i if $\mathbf{c}_i$ is the closest center among all centers. Since a given data point can only be in a group, the membership matrix $\mathbf{U}$ has the following properties:

$$\sum_{i=1}^{c} u_{ij} = 1, \forall j = 1, \ldots, n$$

and

$$\sum_{i=1}^{c}\sum_{j=1}^{n} u_{ij} = n.$$

On the other hand, if u_{ij} is fixed, then the optimal center $\mathbf{c}_i$ that minimize Equation (15.1) is the mean of all vectors in group i:

$$\mathbf{c}_i = \frac{1}{|G_i|} \sum_{k, \mathbf{x}_k \in G_i} \mathbf{x}_k, \tag{15.4}$$

where $|G_i|$ is the size of G_i, or $|G_i| = \sum_{j=1}^{n} u_{ij}$.

For a batch-mode operation, the K-means algorithm is presented with a data set $\mathbf{x}_i, i = 1, \ldots, n$; the algorithm determines the cluster centers $\mathbf{c}_i$ and the membership matrix $\mathbf{U}$ iteratively using the following steps:

Step 1: Initialize the cluster center $\mathbf{c}_i$, $i = 1, \ldots, c$. This is typically achieved by randomly selecting c points from among all of the data points.

Step 2: Determine the membership matrix $\mathbf{U}$ by Equation (15.3).

Step 3: Compute the cost function according to Equation (15.1). Stop if either it is below a certain tolerance value or its improvement over previous iteration is below a certain threshold.

Step 4: Update the cluster centers according to Equation (15.4). Go to step 2.

The algorithm is inherently iterative, and no guarantee can be made that it will converge to an optimum solution. The performance of the K-means algorithm depends on the initial positions of the cluster centers, thereby making it advisable either to employ some front-end methods to find good initial cluster centers or to run the algorithm several times, each with a different set of initial cluster centers. Moreover, the preceding algorithm is only a representative one; it is also possible to initialize a random membership matrix first and then follow the iterative procedure.

The K-means algorithm can also be operated in the on-line mode, where the cluster centers and the corresponding groups are derived through time averaging. That is, for a given data point $\mathbf{x}$, the algorithm finds the closest cluster center $\mathbf{c}_i$ and it is updated using the formula

$$\Delta\mathbf{c}_i = \eta(\mathbf{x} - \mathbf{c}_i).$$

This on-line formula is essentially embedded in many learning rules of the unsupervised learning neural networks introduced in Chapter 11.

15.3 FUZZY C-MEANS CLUSTERING

Fuzzy C-means clustering (FCM), also known as **fuzzy ISODATA**, is a data clustering algorithm in which each data point belongs to a cluster to a degree specified by a membership grade. Bezdek proposed this algorithm in 1973 [1] as an improvement over earlier hard C-means (HCM) clustering described in the previous section.

FCM partitions a collection of n vector $\mathbf{x}_i, i = 1, \ldots, n$ into c *fuzzy* groups, and finds a cluster center in each group such that a cost function of dissimilarity measure is minimized. The major difference between FCM and HCM is that FCM employs fuzzy partitioning such that a given data point can belong to several groups with the degree of belongingness specified by membership grades between 0 and 1. To accommodate the introduction of fuzzy partitioning, the membership matrix U is allowed to have elements with values between 0 and 1. However, imposing normalization stipulates that the summation of degrees of belongingness for a data set always be equal to unity:

$$\sum_{i=1}^{c} u_{ij} = 1, \forall j = 1, \ldots, n. \tag{15.5}$$

The cost function (or objective function) for FCM is then a generalization of Equation (15.1):

$$J(U, \mathbf{c}_1, \ldots, \mathbf{c}_c) = \sum_{i=1}^{c} J_i = \sum_{i=1}^{c} \sum_{j=1}^{n} u_{ij}^m d_{ij}^2, \tag{15.6}$$

where u_{ij} is between 0 and 1; $\mathbf{c}_i$ is the cluster center of fuzzy group i; $d_{ij} = \|\mathbf{c}_i - \mathbf{x}_j\|$ is the Euclidean distance between ith cluster center and jth data point; and $m \in [1, \infty)$ is a weighting exponent.

The necessary conditions for Equation (15.6) to reach a minimum can be found by forming a new objective function $\bar{J}$ as follows:

$$\begin{aligned} \bar{J}(U, \mathbf{c}_1, \ldots, \mathbf{c}_c, \lambda_1, \ldots, \lambda_n) &= J(U, \mathbf{c}_1, \ldots, \mathbf{c}_c) + \textstyle\sum_{j=1}^{n} \lambda_j (\sum_{i=1}^{c} u_{ij} - 1) \\ &= \textstyle\sum_{i=1}^{c} \sum_{j=1}^{n} u_{ij}^m d_{ij}^2 + \sum_{j=1}^{n} \lambda_j (\sum_{i=1}^{c} u_{ij} - 1), \end{aligned} \tag{15.7}$$

where λ_j, $j = 1$ to n, are the Lagrange multipliers for the n constraints in Equation (15.5). By differentiating $\bar{J}(U, \mathbf{c}_1, \ldots, \mathbf{c}_c, \lambda_1, \ldots, \lambda_n)$ with respect to all its input arguments, the necessary conditions for Equation (15.6) to reach its minimum are

$$\mathbf{c}_i = \frac{\sum_{j=1}^{n} u_{ij}^m \mathbf{x}_j}{\sum_{j=1}^{n} u_{ij}^m}, \tag{15.8}$$

and

$$u_{ij} = \frac{1}{\sum_{k=1}^{c} \left(\frac{d_{ij}}{d_{kj}} \right)^{2/(m-1)}}. \tag{15.9}$$

Proving these two necessary conditions are as Exercise 1 at the end of this chapter. The fuzzy C-means algorithm is simply an iterated procedure through the preceding two necessary conditions. In a batch-mode operation, FCM determines the cluster centers $\mathbf{c}_i$ and the membership matrix U using the following steps [1]:

Step 1: Initialize the membership matrix U with random values between 0 and 1 such that the constraints in Equation (15.5) are satisfied.

Step 2: Calculate c fuzzy cluster centers $\mathbf{c}_i, i = 1, \ldots, c$, using Equation (15.8).

Step 3: Compute the cost function according to Equation (15.6). Stop if either it is below a certain tolerance value or its improvement over previous iteration is below a certain threshold.

Step 4: Compute a new U using Equation (15.9). Go to step 2.

The cluster centers can also be first initialized and then the iterative procedure carried out. No guarantee ensures that FCM converges to an optimum solution. The performance depends on the initial cluster centers, thereby allowing us either to use another fast algorithm to determine the initial cluster centers or to run FCM several times, each starting with a different set of initial cluster centers.

Figure 15.1 presents a MATLAB demo of the fuzzy C-means clustering method in the Fuzzy Logic Toolbox. The data set, number of clusters, exponent weighting, and several stopping criteria can all be changed via the graphical user interface. Pushing the "Start" button allows for one to observe how the cluster centers move toward the "right" positions. In particular, if the "Label Data" is marked, you will be able to see how each group evolves when the cluster centers move. After the clustering process stops, a cluster center can be selected, which will display the membership grades of all data points toward the selected cluster center. Figure 15.2 illustrates the MF plots with respect to three cluster centers. Recall that the membership grades are only defined on the location of data points; the surface plots in Figure 15.2 are obtained via 2-D interpolation using the MATLAB command `griddata`.

Bezdek's monograph [2] provides a detailed treatment of fuzzy C-means clustering, including its variants and convergence properties. Applications of fuzzy C-means include medical image segmentation [5] and qualitative modeling [10].

15.4 MOUNTAIN CLUSTERING METHOD

The **mountain clustering method**, as proposed by Yager and Filev [11, 13], is a relatively simple and effective approach to approximate estimation of cluster centers on the basis of a density measure called the **mountain function**. This method can be used to obtain initial cluster centers that are required by more sophisticated cluster algorithms, such as fuzzy C-means clustering introduced in the previous section. It can also be used as a quick stand-alone method for approximate clustering. The method is based on what a human does in visually forming clusters of a data set.

The first step involves forming a grid on the data space, where the intersections of the grid lines constitute the candidates for cluster centers, denoted as a set V. A finer gridding increases the number of potential clustering centers, but it also increases the computation required. The gridding is generally evenly spaced, but it

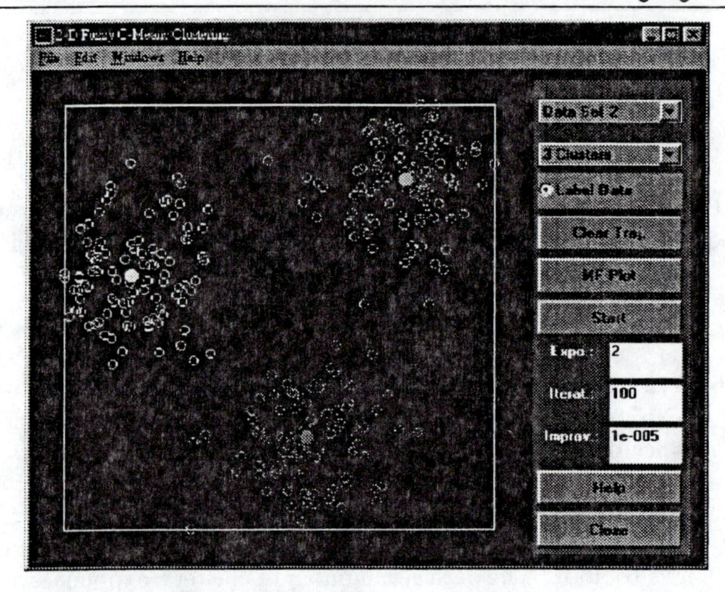

Figure 15.1. *Demo program for fuzzy C-means clustering.* (MATLAB command: `fcmdemo`)

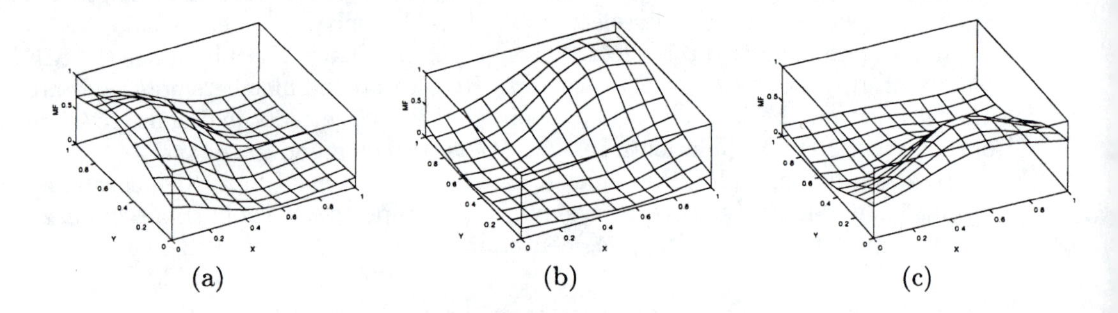

(a) (b) (c)

Figure 15.2. *MF plots for the demo of fuzzy C-means clustering in Figure 15.1.*

is not a requirement. We can have an unevenly spaced gridding to reflect *a priori* knowledge of data distribution. Moreover, if the data set itself (instead of the grid points) is used as the candidates for cluster centers, then we have a variant called subtractive clustering, as discussed in the next section.

The second step entails constructing a mountain function representing a data density measure. The height of the mountain function at an a point $\mathbf{v} \in V$ is equal to

$$m(\mathbf{v}) = \sum_{i=1}^{N} \exp\left(-\frac{\|\mathbf{v} - \mathbf{x}_i\|^2}{2\sigma^2}\right), \tag{15.10}$$

where $\mathbf{x}_i$ is the ith data point and σ is an application-specific constant. The preceding equation implies that each data point $\mathbf{x}_i$ contributes to the height of the mountain function at $\mathbf{v}$ and the contribution is inversely proportional to the distance between $\mathbf{x}_i$ and $\mathbf{v}$. The mountain function can be viewed as a measure of *data density* since it tends to be higher if more data points are located nearby, and lower if fewer data points are around. The constant σ determines the height as well as the smoothness of the resultant mountain function; this is demonstrated in Example 15.1 The clustering results are normally insensitive to the value of σ, as long as the data set is of sufficient size and is well clustered.

The third step involves selecting the cluster centers by sequentially destructing the mountain function. We first find the point in the candidate centers V that has the greatest value for the mountain function; this becomes the first cluster center $\mathbf{c}_1$. (In case of more than one maxima, one of them is randomly selected as the first cluster center.) Obtaining the next cluster center requires eliminating the effect of the just-identified center, which is typically surrounded by a number of grid points that also have high density scores. This is realized by revising the mountain function; a new mountain function is formed by subtracting a scaled Gaussian function centered at $\mathbf{c}_1$:

$$m_{\text{new}}(\mathbf{v}) = m(\mathbf{v}) - m(\mathbf{c}_1)\exp\left(-\frac{\|\mathbf{v}-\mathbf{c}_1\|^2}{2\beta^2}\right) \tag{15.11}$$

The subtracted amount $m(\mathbf{c}_1)\exp\left(-\frac{\|\mathbf{v}-\mathbf{c}_1\|^2}{2\beta^2}\right)$ is inversely proportional to the distance between $\mathbf{v}$ and the just-identified center $\mathbf{c}_1$, as well as being proportional to the height $m(\mathbf{c}_1)$ at the center. Note that after subtraction, the new mountain function $m_{\text{new}}(\mathbf{v})$ reduces to zero at $\mathbf{v} = \mathbf{c}_1$.

After subtraction, the second cluster center is again selected as the point in V that has the largest value for the new mountain function. This process of revising the mountain function and finding the next cluster center continues until a sufficient number of cluster centers is attained. The following example clarifies the concept.

Example 15.1 *Mountain clustering method for 2-D data*

Figure 15.3(a) displays a set of 2-D data, in which three clusters can be observed effortlessly. However, for data sets of higher dimensions (e.g., more than three), no effective visualization techniques are available to determine the clusters visually; therefore clustering techniques described in this chapter must be relied on. In this example, the mountain method is employed to find the cluster centers.

To demonstrate the effects of σ, Figures 15.3(b) through 15.3(d) are the surface plots of the mountain functions with σ equal to 0.02, 0.1, and 0.2, respectively. Obviously σ affects the mountain function's height as well as its smoothness; therefore, the value of σ should be chosen cautiously considering both the data size and input dimension.

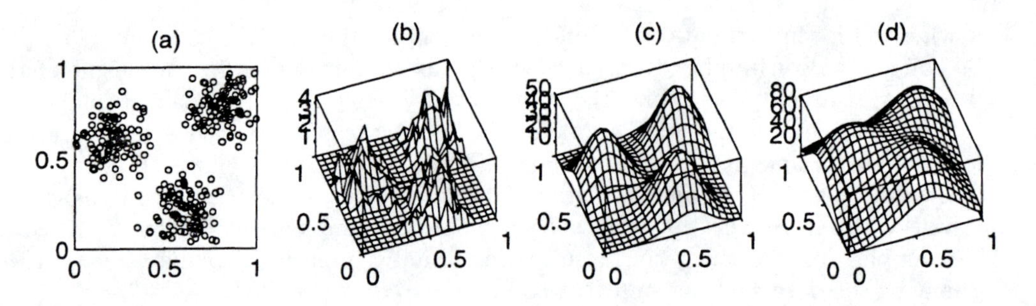

Figure 15.3. *Mountain construction: (a) 2-D data set, with the corresponding mountain function with σ equal to (a) 0.02, (b) 0.1, and (c) 0.2.* (MATLAB command: `mount1`)

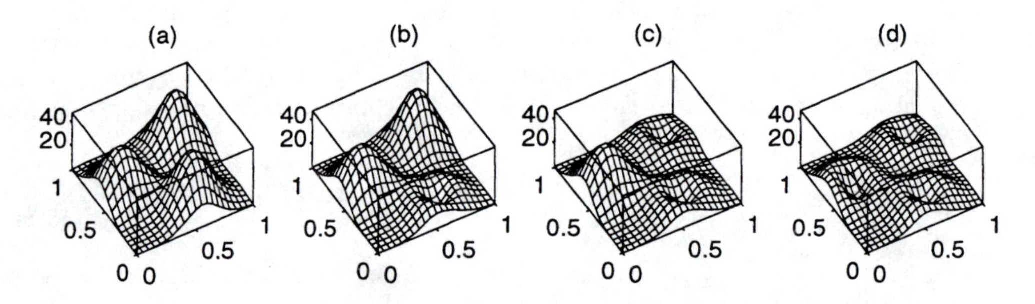

Figure 15.4. *Mountain destruction with $\beta = 0.1$: (a) the original mountain function with $\sigma = 0.1$; (b) mountain function after the first reduction; (c) mountain function after the second reduction; (d) mountain function after the third reduction.* (MATLAB command: `mount2`)

Once the σ is determined (0.1 in this example) and the mountain function is constructed, we begin to select clusters and revise the mountain function sequentially. This is shown in Figures 15.4(a), (b), (c), and (d), with β equal to 0.1 in Equation (15.11).

□

Yager and Filev [12, 13] also applied mountain clustering to the structure identification of fuzzy modeling. They used a training data set (including inputs and desired outputs) to find cluster centers $(\mathbf{x}_i, y_i)$ via mountain clustering first, and then formed a zero-order Sugeno fuzzy modeling in which the ith rule is expressed as

$$\text{If X is close to } \mathbf{x}_i \text{ then Y is close to } y_i.$$

Restated, the ith rule is based on the ith cluster centers identified by the mountain clustering method. After the structure is determined, backpropagation-type gradient descent and other optimization schemes can be applied to proceed with parameter identification. Several examples can be found in ref. [13].

15.5 SUBTRACTIVE CLUSTERING

The mountain clustering method described in the previous section is relatively simple and effective. However, its computation grows exponentially with the dimension of the problem because the method must evaluate the mountain function over all grid points. For instance, a clustering problem with four variables and each dimension having a resolution of 10 grid lines would result in 10^4 grid points that must be evaluated. An alternative approach is **subtractive clustering** proposed by Chiu [4], in which data points (not grid points) are considered as the candidates for cluster centers. By using this method, the computation is simply proportional to the number of data points and independent of the dimension of the problem under consideration.

Consider a collection of n data points $\{\mathbf{x}_1, \ldots, \mathbf{x}_n\}$ in an M-dimensional space. Without loss of generality, the data points are assumed to have been normalized within a hypercube. Since each data point is a candidate for cluster centers, a *density measure* at data point $\mathbf{x}_i$ is defined as

$$D_i = \sum_{j=1}^{n} \exp\left(-\frac{\|\mathbf{x}_i - \mathbf{x}_j\|^2}{(r_a/2)^2}\right),$$

where r_a is a positive constant. Hence, a data point will have a high density value if it has many neighboring data points. The radius r_a defines a neighborhood; data points outside this radius contribute only slightly to the density measure.

After the density measure of each data point has been calculated, the data point with the highest density measure is selected as the first cluster center. Let $\mathbf{x}_{c_1}$ be the point selected and D_{c_1} its density measure. Next, the density measure for each data point $\mathbf{x}_i$ is revised by the formula

$$D_i = D_i - D_{c_1} \exp\left(-\frac{\|\mathbf{x}_i - \mathbf{x}_{c_1}\|^2}{(r_b/2)^2}\right),$$

where r_b is a positive constant. Therefore, the data points near the first cluster center $\mathbf{x}_{c_1}$ will have significantly reduced density measures, thereby making the points unlikely to be selected as the next cluster center. The constant r_b defines a neighborhood that has measurable reductions in density measure. The constant r_b is normally larger than r_a to prevent closely spaced cluster centers; generally r_b is equal to $1.5r_a$, as suggested in ref. [4].

After the density measure for each data point is revised, the next cluster center $\mathbf{x}_{c_2}$ is selected and all of the density measures for data points are revised again. This process is repeated until a sufficient number of cluster centers are generated. A more sophisticated stopping criterion for automatically determining the number of clusters can be found in ref. [4].

When applying subtractive clustering to a set of input-output data, each of the cluster centers represents a prototype that exhibits certain characteristics of the system to be modeled. These cluster centers would be reasonably used as the centers for the fuzzy rules' premise in a zero-order Sugeno fuzzy model, or radial basis functions in an RBFN. For instance, assume that the center for the ith cluster is $\mathbf{c}_i$ in an M dimension. The $\mathbf{c}_i$ can be decomposed into two component vectors $\mathbf{p}_i$ and $\mathbf{q}_i$, where $\mathbf{p}_i$ is the input part and it contains the first N element of $\mathbf{c}_i$; $\mathbf{q}_i$ is the output part and it contains the last $M - N$ elements of $\mathbf{c}_i$. Then, given an input vector $\mathbf{x}$, the degree to which fuzzy rule i is fulfilled is defined by

$$\mu_i = \exp\left(-\frac{\|\mathbf{x} - \mathbf{p}_i\|^2}{(r_a/2)^2}\right).$$

This is also the definition of the ith radial basis function if we adopt the perspective of modeling using RBFNs. Once the premise part (or the radial basis functions) has been determined, the consequent part (or the weights for output unit in an RBFN) can be estimated by the least-squares method. After these procedures are completed, more accuracy can be gained by using gradient descent or other advanced derivative-based optimization schemes (Chapter 6) for further refinement.

15.6 SUMMARY

This chapter presents four of the most representative off-line clustering techniques frequently used in conjunction with radial basis function networks and fuzzy modeling: (hard) C-means clustering, fuzzy C-means clustering, the mountain clustering method, and subtractive clustering. These clustering techniques provide batch-mode approaches to finding prototypes characterizing a data set; these prototypes are then used as the centers for radial basis functions in RBFNs (Chapter 9) or fuzzy rules in ANFIS (Chapter 12). For data compression, these prototypes are used as a codebook in vector quantization.

Some on-line clustering algorithms implemented by unsupervised learning neural networks are explained in Chapter 11.

EXERCISE

1. Differentiate Equation (15.7) to obtain Equations (15.8) and (15.9).

REFERENCES

[1] Jim C. Bezdek. *Fuzzy Mathematics in Pattern Classification.* PhD thesis, Applied Math. Center, Cornell University, Ithaca, 1973.

[2] Jim C. Bezdek. *Pattern recognition with fuzzy objective function algorithms.* Plenum Press, New York, 1981.

[3] Chedsada Chinrungrueng. *Evaluation of Heterogeneous Architectures for Artificial Neural Networks.* PhD thesis, University of California at Berkeley, May 1993.

[4] S. L. Chiu. Fuzzy model identification based on cluster estimation. *Journal of Intelligent and Fuzzy Systems*, 2(3), 1994.

[5] L. O. Hall, A. M. Bensaid, L. P. Clarke, R. P. Velthuizen, M. S. Silbiger, and J. C. Bezdek. A comparison of neural network and fuzzy clustering techniques in segmenting magnetic resonance images of the brain. *IEEE Transactions on Neural Networks*, 3(5):672–682, 1992.

[6] P. R. Krishnaiah and L. N. Kanal, editors. *Classification, pattern recognition, and reduction of dimensionality*, volume 2 of *Handbook of Statistics*. North-Holland, Amsterdam, 1982.

[7] S. P. Lloyd. Least squares quantization in PCM. *IEEE Transactions on Information Theory*, (2):129–137, 1982.

[8] J. Makhoul, S. Roucos, and H. Gish. Vector quantization in speech coding. *Proceedings of the IEEE*, 73(11):1551–1588, 1985.

[9] J. Moody and C. Darken. Fast learning in networks of locally-tuned processing units. *Neural Computation*, 1:281–294, 1989.

[10] M. Sugeno and T. Yasukawa. A fuzzy-logic-based approach to qualitative modeling. *IEEE Transactions on Fuzzy Systems*, 1(1):7–31, February 1993.

[11] R. R. Yager and D. P. Filev. Approximate clustering via the mountain method. *IEEE Transactions on Systems, Man, and Cybernetics*, 24:1279–1284, 1994.

[12] R. R. Yager and D. P. Filev. Generation of fuzzy rulels by mountain clustering. *Journal of Intelligent and Fuzzy Systems*, 2:209–219, 1994.

[13] Ronald R. Yager and Dimitar P. Filev. *Essentials of fuzzy modeling and control.* John Wiley & Sons, Inc, 1994.

Chapter 16

Rulebase Structure Identification

C.-T. Sun

16.1 INTRODUCTION

We can perform fuzzy modeling by extracting knowledge from human experts and by transforming the expertise into rules and membership functions. However, depending on human introspection and experience results in some difficulties. First, human's knowledge is often incomplete and episodic rather than systematic. Moreover, there is no formal and effective way of knowledge acquisition. As a result, researchers have been trying to automatize the neuro-fuzzy modeling process based on numerical training data. In general, neuro-fuzzy modeling is a branch of system identification (Chapter 5) and it also involves two phases: **structure identification** and **parameter identification**. The former is related to finding a suitable number of rules and a proper partition of the feature space. The latter is concerned with the adjustment of system parameters, such as the membership functions, linear coefficients, and so on.

The problems related to parameter identification in fuzzy modeling are covered in Chapters 5, 6, 12, and 13. For structure identification in fuzzy modeling, Chapters 14 and 15 give several heuristic but practical and systematic approaches. However, the problem of structure identification in fuzzy modeling is by no means solved; there are many problems in practice remain to be addressed. This chapter raises these problem and suggests potential solutions from a high-level point of view.

As mentioned in Chapter 4, if we do not use any structure identification techniques in fuzzy modeling, we have to accept the simple grid partitioning of the input space, as shown in Figure 4.13(a) in page 87. However, this leads to "curse of dimensionality" when the number of input variables becomes large. One way to alleviate this problem is to employ input selection schemes to choose relevant inputs for the modeling task. A simple input selection scheme that can be embedded in the parameter identification phase is described in Section 16.2.

Another way to relieve the problem of an exponentially-grown rulebase is to have a sophisticated partitioning of the input space. A hill-climbing method based on *k-d trees* is described in Section 16.3 to implement a tree partitioning of the input space. Two objective functions, a *density* measure and a *typicality* measure, are used in the partition process to find a proper starting point for the parameter identification phase.

However, when we face a complex system, a simple, global, and effective partition is difficult to find. This is the dilemma between modeling *accuracy* and learning/operation *efficiency*. A method of rule organization to solve this problem is discussed in Section 16.4. It employs the concept of divide-and-conquer to achieve modeling accuracy with many *small* rules each of them covering a small local region. Then we build a binary *fuzzy boxtree* out of the rules based on a *similarity* measure between their antecedent patterns. A branch-and-bound algorithm can do the pattern matching job for firing appropriate rules with logarithmic efficiency so that the big number of rules will cause no trouble for the entire system. Moreover, to maintain fuzzy rulebased systems' advantage in parallel processing, a parallel algorithm is proposed to meet the requirement.

Another alternative to cope with sophisticated systems is to use *rule combination* so that we can apply a simplified rulebase. This is applicable whenever there is a *focus area* in the application domain. The algorithm for rule combination with respect to a boxtree is given in Section 16.5.

The ANFIS architecture discussed in Chapter 12 is general, and it can incorporate advanced fuzzy pattern-matching techniques such as weights of importance and fuzzy quantifiers in a fuzzy reasoning process. Further discussion on structure identification is based on the ANFIS model.

16.2 INPUT SELECTION

Up to now, the discussion of neuro-fuzzy modeling has been based on the implication that the input variables are of equal importance. However, in applications of pattern recognition, time series prediction, or multi-criteria decision making, usually this assumption is not true. In other words, the first step in a general modeling scheme should be **input** or **feature selection**, which can identify a subset of all possible inputs as the actual inputs for ANFIS/CANFIS modeling. These identified inputs should possess more discriminating power to produce better regression or classification results.

In this section, we shall apply the concept of **weight of importance** in fuzzy pattern matching to proceed with our input selection scheme. Due to the flexibility of adaptive networks (Chapter 8), the input selection scheme can be embedded into the ANFIS architecture, so we can perform parameter identification and input selection simultaneously.

Assume that an input variable x is associated with an **importance measure** $\sigma \in [0, 1]$. If a premise construct "x is A" is used in a fuzzy rule with an AND-connected premise (IF) part, then the MF grade rescaled by the input importance measure σ is denoted by

$$\begin{aligned} s &= (1-\sigma) * 1 + \sigma * \mu_A(x) \\ &= 1 - \sigma * [1 - \mu_A(x)]. \end{aligned}$$

The preceding equations represents a linear interpolation between 1 and $\mu_A(x)$. If the input variable x is of full importance ($\sigma = 1$), s reduces to $\mu_A(x_i)$. On the other hand, if x is of no importance ($\sigma = 0$), then s is 1 and it plays no part in computing the firing strength of an AND rule—that is, rule with an AND-connected premise (IF) part.

Similarly, to incorporate the input importance measure into an OR rule or fuzzy rule with an OR-connected premise (IF) part, we can define a new MF grade:

$$\begin{aligned} s &= (1-\sigma) * 0 + \sigma * \mu_A(x) \\ &= \sigma * \mu_A(x). \end{aligned}$$

This is a linear interpolation between 0 and $\mu_A(x)$. If σ is close to 1, then s is close to $\mu_A(x)$ to assume its regular role. If σ is close 0, then s is close to 0 and it plays little part in computing the firing strength.

The preceding two equations are based on the concept of weights of importance; a detailed discussion can be found in ref. [2].

Through the concept of parameter sharing, discussed in Section 8.2, it is rather easy to incorporate the input importance measure σ, as an extra parameter associated with the input x, to the ANFIS architecture in Figure 12.1 of page 336. The configuration of the ANFIS architecture with an input importance measure is left as Exercise 1.

The initial value of an input importance measure is defaulted at 1.0 but can be assigned to any value in $(0, 1]$ by users based on their heuristic judgment. During the training process, once an input importance measure is stablized below a certain threshold value, the corresponding variable is considered unimportant in the system to be modeled. Thus, the variable can be neglected and a simplified structure/parameter identification process can be resumed to find an even better solution.

16.3 INPUT SPACE PARTITIONING

As mentioned in Section 4.5.1 of Chapter 4, the premise part of a fuzzy inference system implements a fuzzy partition in the multidimensional input (feature) space. In Figure 4.13, we have seen three partition schemes frequently used in modeling a multidimensional system. A recapitulation plus some additions are shown in Figure 16.1.

Figure 16.1(a) is a common grid partitioning in which there is no adaptation method is used to change the premise part of a fuzzy inference system. An ANFIS model based on the grid partitioning forms the adaptive grid partitioning in Figure 16.1(b). In other words, at the beginning of training, a uniformly partitioned

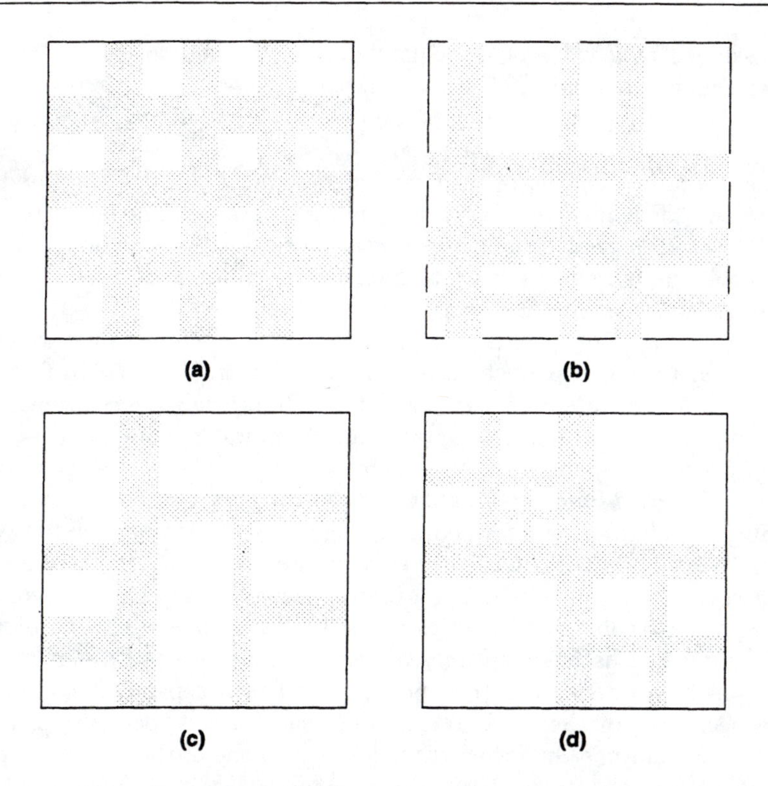

Figure 16.1. *Four fuzzy multidimensional data structures. (a) fuzzy grid, (b) adaptive fuzzy grid, (c) fuzzy k-d tree, (d) multi-level fuzzy grid. Shaded area represents overlapping among fuzzy regions. In two-dimensional space, the structure in (d) is also called a fuzzy quad tree.*

grid is taken as the initial state. As the parameters in the premise membership functions are adjusted, the grid evolves. The steepest descent method (or any other optimization techniques described in Chapters 6 and 7) finds the optimal location and size of the fuzzy regions and the degree of overlapping among them. Two problems exist in this scheme. First, the number of linguistic terms for each input variable is predetermined and is highly heuristic. Second, the learning complexity suffers an exponential explosion as the number of inputs increases.

Grid partitioning gives the most restricted structures; on the other extreme, fuzzy clustering algorithms [1, 8] (see Section 15.3 of Chapter 15) based on training data result in the most flexible scatter partitioning shown in Figure 4.13(c) (page 87). However, this approach has its own problems. First, the resulting structures are not necessarily hyper-rectangles, they need refinement. To map a fuzzy cluster to a set of bell-shaped functions, we can consider c_i in Equation (2.23) (page 26) as the coordinate of the cluster center, $\vec{h}$, in the ith dimension; meanwhile,

a_i is assigned to a value of the cluster radius, defined to be the longest distance from the center to a point, $\vec{p_i}$, with nonzero membership in the ith dimension. The value of b_i is interpreted as a slope and can be determined as a linear function of the membership of the boundary point $\vec{p_i}$.

The second weak point of clustering algorithms is that the cost tends to be high, because the efficiency of convergence is not guaranteed. Third, the total number of rules (clusters) is predetermined in these algorithms. When the resulting partition is not good enough and we want to increase the number of rules, we have to rerun the clustering algorithm from beginning.

The essential point here is this: in the context of adaptive network training, we do not need to find a perfect clustering since our goal is just to find a *satisfiable* initial state for the adaptive network to tune. In other words, since we have verified the validity of the parameter identification mechanism using adaptive networks, it makes no sense to spend a lot of time in optimizing the cluster criteria, no matter what the objective functions might be.

Thus, we adopted an intermediately flexible partitioning, the **fuzzy k-d trees** [Figure 16.1(c), or Figure 4.13(b)] for structure identification. In the following, we explore several ways of providing an input space partition based on fuzzy k-d trees.

A k-d tree results from a series of **Guillotine cuts**. By a Guillotine cut, we mean a cut which is made entirely across the subspace to be partitioned; each of the regions so produced can then be subjected to independent Guillotine cutting. At the beginning of the ith iteration step, the feature space is partitioned into i regions. Now another Guillotine cut is applied to one of the regions to partition the entire space further into $i+1$ regions.

There are various strategies to decide which dimension to cut and where to cut it at each step; some of them are based merely on the distribution of training examples; others take the parameter identification methods into consideration. We list and briefly discuss several strategies in the following before introducing a hill-climbing method based on fuzzy clustering objective functions.

Balanced-sampling criterion The simplest tactic is to cut the dimension in which the training data associated with the region are most spread out and to cut it at the median value of those samples in that dimension [3]. The expected shape of the regions under this procedure is asymptotically cubical because the long dimension is always cut. In general, this method produces homogeneously distributed localized receptive fields.

Information gain The cutting procedure can be viewed as a method of building a decision tree, see Chapter 14. Quinlan [5] proposed a method based on information theory and defined the concept of *information gain* at each branch. Therefore, the rule of thumb is to choose a cut with the most information gain at each step.

Regional linearity This strategy is suitable when the consequence part of a rule is represented as a linear combination of the input features, such as in the first-order ANFIS model. Conceptually, we want to use a hyperplane to approximate the training samples in a region and minimize the mean-square error. Thus, we can apply LSE (Chapter 5) to identify a set of linear coefficients, and the cut resulting in the least error will be selected under this criterion.

Direct evaluation The most direct, but inefficient, way to evaluate a partition is to feed the resulting structure into the parameter identification phase and use the final performance to choose the best cut. Sugeno and Kang [10] used this approach together with lots of heuristics to find the proper structure.

The preceding methods are all crisp partitions, so the result needs to be fuzzified. Furthermore, the evaluation functions used in these hill-climbing algorithms are either too shallow, without considering the need of parameter identification, or too deep, and thus bothered by a massive amount of computation. A compromising method is to use two fuzzy clustering objective functions, a typicality measure and a density measure. The basic assumption of this approach is simply that a good fuzzy rule is usually represented by a cluster which has a prototypical center and a strong support from the samples.

Our method is still an n-step hill-climbing approach, where n is the desired number of rules, because of the consideration of efficiency. At each step a fuzzy set is defined for each cluster i with the following membership function, which will be used in the objective functions:

$$\mu_{ik} = \prod_{j=1}^{V} \frac{1}{1 + \left| \frac{x_{kj} - c_{ij}}{a_{ij}} \right|^{2b_{ij}}}, \tag{16.1}$$

where μ_{ik} denotes the membership of the kth point ($\vec{x}_k$) in the ith cluster, V is the number of variables, x_{kj} is the jth coordinate of $\vec{x}_k$, and $\prod$ stands for a fuzzy conjunctive operator. The calculation of parameters, a_{ij}, b_{ij} and c_{ij}, is dependent on the hyper-rectangle defined by a cluster resulted from Guillotine cuts. Let $\vec{h}_i$ be the physical center of the ith hyper-rectangle, c_{ij} is defined as the jth coordinate of $\vec{h}_i$; a_{ij} is calculated as half of the length along the hyper-rectangle's jth dimension; and b_{ij} is determined by the desired degree of overlapping between fuzzy regions. At the end of the structure identification phase, a's, b's and c's are fed into the adaptive network as the initial parameters.

Now we define two objective functions for the best-first search process. As analyzed by Bezdek [1], various objective functions can suggest radically different substructures in the same data set. To achieve a meaningful structure for a fuzzy rulebase, we have to select appropriate measures. In our approach, we use two objective functions, one is a *density* measure (J_D), the other is a *typicality* measure (J_T).

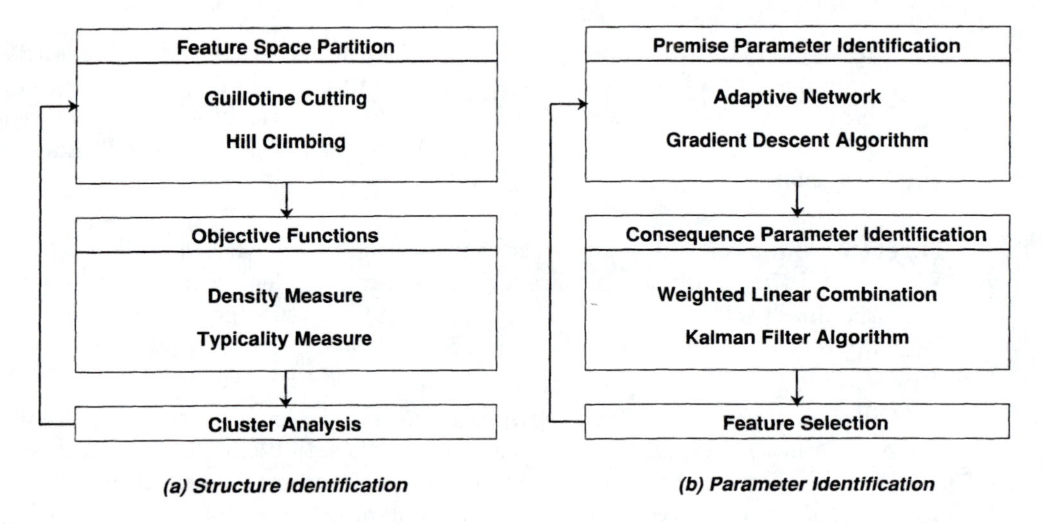

Figure 16.2. *A general fuzzy modeling scheme.*

J_D was proposed by Ruspini [7]:

$$J_D = \sum_{j=1}^{P} \sum_{k=1}^{P} \left\{ \left[\sum_{i=1}^{C} (\mu_{ij} - \mu_{ik})^2 \right] - d_{jk}^2 \right\}^2, \tag{16.2}$$

where P is the number of training data, C is the number of clusters (rules), d_{jk} is the distance (or the measure of dissimilarity) between sampling points j and k, and μ_{ij} is the membership of point j in cluster i. As pointed out by Ruspini, J_D is a measure of cluster quality based on local density, because J_D will be small when the terms in Equation (16.2) are individually small; in turn, this will occur when close pairs of points have nearly equal fuzzy memberships in the C clusters.

J_T is a variation of the *least-square functional* proposed by Bezdek [1]:

$$J_T = \sum_{k=1}^{P} \sum_{i=1}^{C} \mu_{ik}^2 d_{ik}^2 \tag{16.3}$$

where d_{ik} is the distance from point k to the center (or prototype) $\vec{h}_i$ of cluster i. We call J_T a typicality measure because it will be small when points in a cluster adhere tightly (have small d_{ik}'s) to their cluster center $\vec{h}_i$.

Density and typicality are important measures because they are closely related to two important characteristics of linguistic terms: the *support* and the *core*, respectively. The *support* is the range of nonzero membership values ($\mu > 0$), whereas the *core* is the range of full membership ($\mu = 1$). In general, we want a linguistic term to have a *strong support* (high density, or small J_D) and a *representative core* (good prototype, or small J_T). Thus, it is reasonable to choose $J_D + J_T$ to be our

Figure 16.3. *A clustering without salient discriminating features.*

objective function. In other words, for each possible Guillotine cut, we calculate $J_D + J_T$ of the resulting partition. Then we select the partition with the least $J_D + J_T$ value as our next hypothesis to continue the hill-climbing process.

Figure 16.2 summarizes our adaptive network–based fuzzy rulebase modeling scheme. As mentioned previously, the cutting procedure can be viewed as a method of building a decision tree. Efficient decision tree construction depends on the existence of salient discriminating features. If it is not the case, for example, there are no adequate Guillotine cuts in Figure 16.3, and the mechanism of partitioning the feature space is no longer suitable for structure identification. Thus, usually in a complicated system, we are forced to use a large number of *small* rules. By a small rule we mean a rule whose antecedent part covers a relatively small region in the feature space. In the next section we propose a method of rule organization to cope with the resulting computational complexity.

16.4 RULEBASE ORGANIZATION

In a fuzzy inference system, the rules can also be viewed as a set of **fuzzy points** which as a whole approximate a compatibility relation. As more rules are involved, finer approximation as well as better modeling accuracy are likely to be achieved (see Figure 16.4). However, when modeling accuracy is the major concern and a massive amount of rules is used to main the model's accuracy, then we must deal with the problem of computational complexity. A basic assumption here is that massively parallel implementation of an intelligent system with learning ability will still be costly, if not impossible, in the near future; thus, we have to realize the proposed ANFIS model on traditional computer systems.

In this section, we introduce a data structure called a fuzzy boxtree to organize rules so that pattern matching can be performed in logarithmic time. The mechanism includes the following steps:

1. Use a divide-and-conquer data structure, the multi-level fuzzy grid, to partition the feature space and fine-tune a large amount of small rules so that accurate local mappings are achieved.

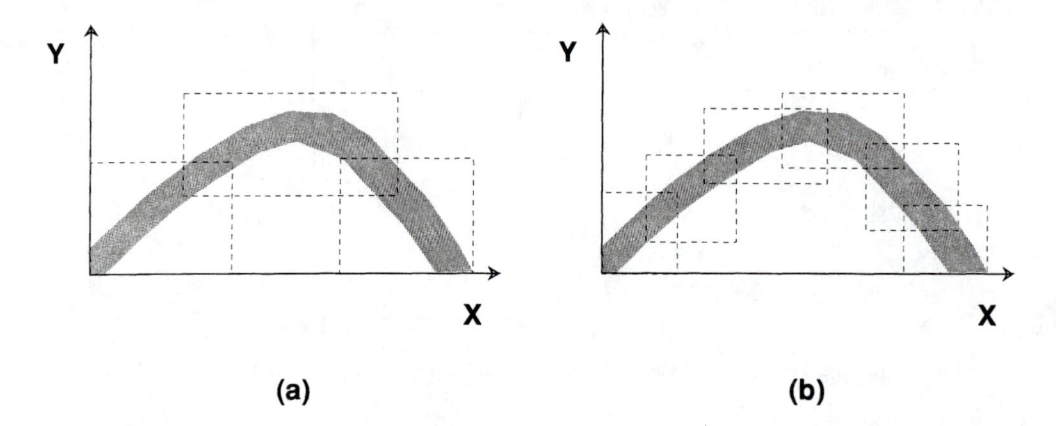

Figure 16.4. *Fuzzy points approximating a compatibility relation. Various numbers of fuzzy points result in different degrees of information granularity: (a) coarse, (b) finer. The interpretation of fuzzy inference as a compatibility relation was discussed in detail by Ruspini [9].*

2. Define a fuzzy boxtree on antecedents of rules and provide a linear-time algorithm to construct it.

3. Introduce a branch-and-bound algorithm for pattern matching in logarithmic time.

4. Provide a parallel algorithm to maintain the advantage of parallel processing presumed in fuzzy inference systems.

Since we are going to use a large number of rules to achieve modeling accuracy and take different degrees of local complexity as well as unbalanced sample distribution into consideration, we adopt a multi-level fuzzy grid [see Figure 16.1(d)] as the structure to partition the feature space. The top level grid coarsely partitions the whole space into equal-sized and evenly spaced fuzzy boxes, which can be further partitioned by finer fuzzy grids. This straightforward partitioning continues until a terminating condition is met. Two criteria can be used as the terminating condition. The first is the balanced sampling criterion (i.e., the resulting boxes should contain similar numbers of training examples). The alternative is to use an application-dependent evaluation. For example, if we assume that each output is a linear combination of the inputs, as in the first-order ANFIS model, we can use LMS methods to evaluate the fitness of each grid. When the mean square error is below a threshold, we stop the partitioning process.

Now we can apply the learning model based on adaptive networks to identify the parameters in each region. Because of the small size of the regions and the resulting local linearity, the learning efficiency is expected to be good. Moreover, since the entire feature space is still covered by the overlapping regions, the smoothness among regions will not be affected although the regions are now separately trained. The only problem to be solved is the computational complexity in operation time due to the resulting large number of rules. To solve the problem, we construct a boxtree to put the rules together.

A binary **fuzzy boxtree**, T, is a rooted tree in which each internal node has two children. Let R denote the set of nodes of T. Each node $r \in R$ is a fuzzy set with a membership function $\mu_r(u)$ such that

If s is a child of r then $\mu_s(u) \le \mu_r(u)$, $\forall u \in U$; in other words, $s \subseteq r$.

In our application, each leaf stands for a fuzzy pattern which represents the antecedent part of a certain fuzzy rule. Moreover, each membership function is bell shaped and is determined by six parameters:

$$\mu_A(u) = \begin{cases} \dfrac{1}{1 + \left|\dfrac{u - c_1}{a_1}\right|^{2b_1}} & \text{if } u < c_1, \\ 1 & \text{if } c_1 \le u \le c_2, \\ \dfrac{1}{1 + \left|\dfrac{u - c_2}{a_2}\right|^{2b_2}} & \text{if } u > c_2, \end{cases} \tag{16.4}$$

where $c_1 \le c_2$. Note that the 3-parameter generalized bell membership function defined in (2.23) (page 26) is a special case of this function with $a_1 = a_2, b_1 = b_2, c_1 = c_2$.

The similarity measure between two fuzzy patterns, A and B, is given by the following formula:

$$S(A, B) = \prod_i S(A_i, B_i), \tag{16.5}$$

that is, a conjunctive aggregation of partial similarity measures, $S(A_i, B_i)$'s, in individual feature dimensions. $S(A_i, B_i)$ is in turn calculated by

$$S(A_i, B_i) = sup_{u \in U}\{\min(\mu_{A_i}(u), \mu_{B_i}(u))\}. \tag{16.6}$$

(See Figure 16.5.) The boxtree construction algorithm repeatedly finds the two boxes (patterns) with the largest similarity degree, makes them siblings, and inserts the parent as an internode.

A membership function in an internode C is defined as a combination of the corresponding functions in its child nodes, A and B. For example, if A is specified by parameter set $\{a_{1_A}, b_{1_A}, c_{1_A}, a_{2_A}, b_{2_A}, c_{2_A}\}$ and B by $\{a_{1_B}, b_{1_B}, c_{1_B}, a_{2_B}, b_{2_B}, c_{2_B}\}$, and if $(c_{1_A}, c_{2_A}) \le (c_{1_B}, c_{2_B})$ in *Pareto ordering*, we use the following parameters to characterize C:

$$a_{1_C} = a_{1_A}, b_{1_C} = b_{1_A}, c_{1_C} = c_{1_A}, a_{2_C} = a_{2_B}, b_{2_C} = b_{2_B}, c_{2_C} = c_{2_B}. \tag{16.7}$$

Figure 16.6 shows the idea and the resulting inclusion relation among fuzzy sets.

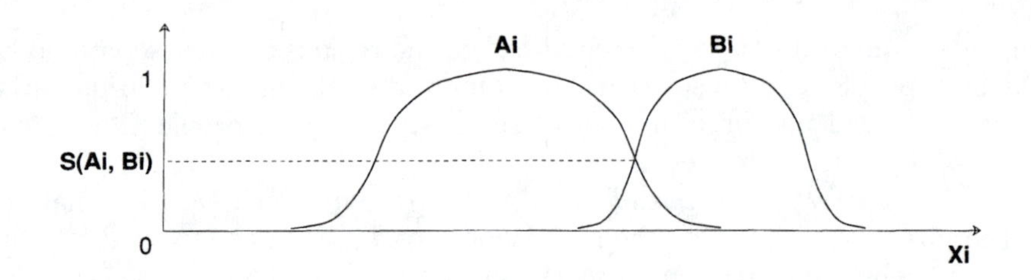

Figure 16.5. *Similarity measure between two fuzzy sets.*

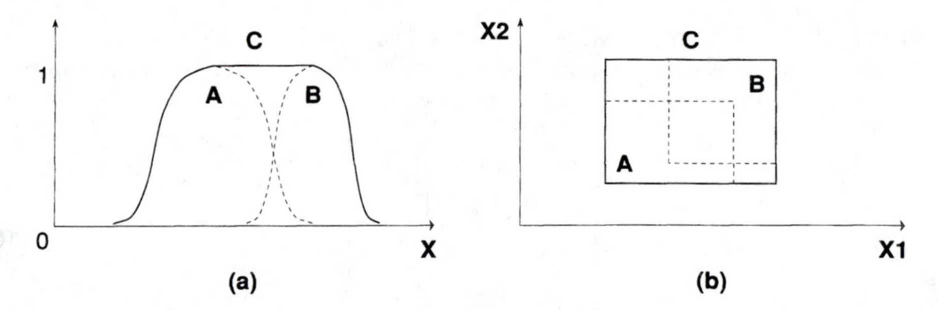

Figure 16.6. *Constructing an internode in a boxtree.* (a) covering of membership functions, (b) corresponding inclusion of boxes.

Since an internode inherits the (fuzzy) boundaries defined by its children, the preceding construction can be realized in linear time by employing the famous greedy algorithm for finding a maximum spanning tree, given the similarity measure between each pair of fuzzy patterns. Figure 16.7 shows a boxtree constructed.

For $r \in R$ that is a leaf, $\mu_r(u)$ is the compatibility measure of an input u against the pattern r; for an internode r, $\mu_r(u)$ is the upper bound of compatibility of the subtree it defines. This property provides a data structure to apply the basic branch-and-bound algorithm in searching optimal solutions. For example, if we want to find all rules which have a firing strength larger than a specified value, the boxtree structure allows a search from the root to prune any subtrees whose root function is smaller than that value.

Thus, we can use the following algorithm to find the best rule against which an input u is matched. In this algorithm, F stands for a *frontier* of expanded nodes, B is the upper bound to a certain point.

Algorithm 16.1 *Branch-and-bound algorithm to find the best-matched rule*

1. $F \leftarrow \{\text{Root}\}$; $B \leftarrow -\infty$;

2. while $F \neq \emptyset$ do
   select a set of nodes $S \subseteq F$;
   expand the internodes in S to get the set of their children, $L(S)$;

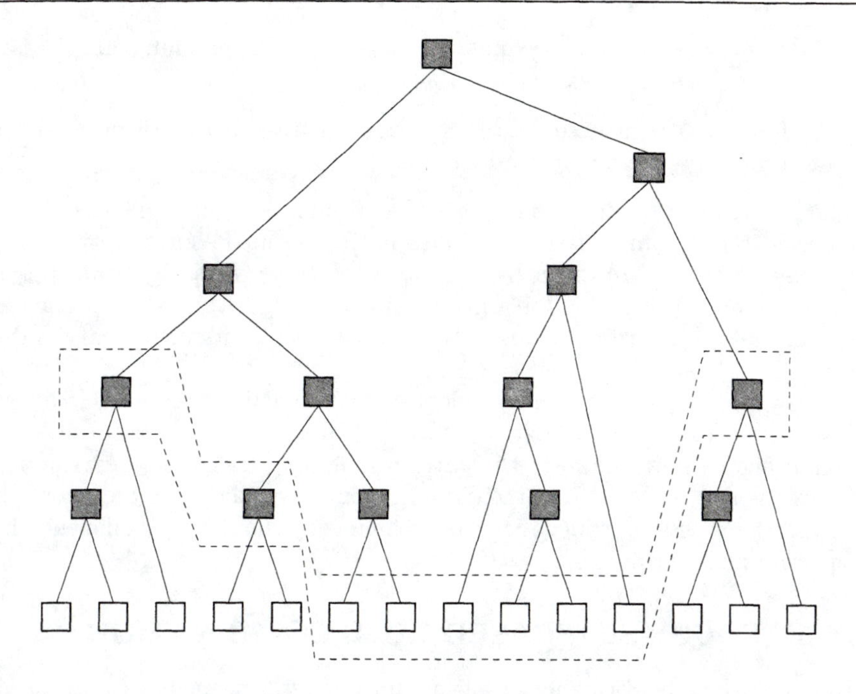

Figure 16.7. *A boxtree. Each leaf corresponds to a fuzzy rule. The dotted lines define a frontier in a boxtree. Each frontier can be considered as a compressed rulebase.*

$F \leftarrow \{F - S\} \cup L(S)$; $B \leftarrow \max(\{B\} \cup \{\mu_v(u) : v \in S \text{ and } v \text{ is a leaf}\})$;
$F \leftarrow \{v \in F : \mu_v(u) \geq B\}$.

□

Usually, in a fuzzy inference system, it is not necessary to find all rules with a firing strength greater than zero. Instead, we are satisfied with the best k rules whose antecedents are compatible to the input. Algorithm 16.1 can be generalized to this case by keeping a priority queue of size k and using the kth best value in pruning. This algorithm is of $O(\log_2 R)$ efficiency in pattern matching for a fuzzy rulebase with R rules.

The advantage of parallel processing is presumed for fuzzy rulebased inference systems because the rules are considered to be independent of each other in the pattern matching process. If we organize the rules into a structure, the boxtree, can we still claim the benefit? In other words, if we have p processors instead of one, can we decrease the processing time to $O(\frac{1}{p})$, or achieve a linear speedup? The answer is positive.

Let each of the p processors maintain a local frontier, F_i, and a local priority queue B_i. At each step every processor i does one of two things:

1. If $F_i \neq \emptyset$ then it expands the node of best matching in F_i and sends its children to processors chosen at random.

2. If $F_i = \emptyset$ then it sends the message "there is a rule of firing strength s" to processors chosen at random.

The processors then update the sets F_i and queues B_i on the basis of the messages received. The computation continues until all sets F_i are empty. At this point, the best matches are given by the merge of B_i's. This algorithm provides a linear speedup. The details of a general parallel algorithm for a branch-and-bound procedure are described in ref. [11]. The analysis of its complexity is discussed in ref. [6].

In summary, given a fuzzy rulebase with R rules that models an application system, we can build a boxtree of $2R - 1$ nodes and use a parallel branch-and-bound algorithm to perform the pattern-matching task with logarithmic efficiency. Consequently, with the boxtree data structure, we can use many more rules in the modeling process to achieve high performance without losing efficiency in the later pattern-matching process.

16.5 FOCUS SET–BASED RULE COMBINATION

To improve the performance of self-organized systems, such as those based on adaptive networks, **dynamic skeletonization** is usually necessary. By skeletonization we mean trimming the redundant or the less important part of a complicated system, as suggested in ref. [4]. However, we claim that skeletonization should be done under a dynamic relevance criterion (i.e., which part to trim or to simplify should be determined by the current situation). In this section we discuss a method of skeletonization, or **rulebase compression**, for adaptive network–based fuzzy inference systems.

Note that in Figure 16.7, every frontier in a boxtree can be viewed as a fuzzy rulebase because it covers the entire feature space. For a frontier containing internodes, we can use either of the following two methods to determine the consequent as well as the antecedent parameters. The first way, the local approach, is to adopt the antecedent parameters as specified in the boxtree and to use LMS methods of finding a hyperplane to approximate the training data covered by individual regions. The alternative, the global approach, is to use the antecedent parameters as the initial values for the ANFIS model and rerun the entire training process. The latter way is much more time consuming and should be used only when the goal is to find a merged rulebase permanently. If we consider the feature space partitioning method introduced Section 16.3 as a *top-down* approach, rule merging can be viewed as a *bottom-up* way of identifying a system structure.

As asserted before, a fuzzy rulebased system can be used to solve various interpolation and classification problems. However, to be of practical use for system representation and communication purposes (e.g., in medical application of image archiving systems), dynamic rulebase compression becomes essential. With rule compression those components of the system are (temporarily) simplified that are supposed to be of less relevance for the current use of the system. The more relevant a component, the more rules are used for that component. The degree of relevance is specified by a focus set.

A **focus set**, or a **focus window**, is a fuzzy set defined on the feature space which indicates the focus of our current interest. Given a focus window W, the similarity gain, $G(r)$, defined on an internode r is calculated as

$$G(r) = S(r_1, W) + S(r_2, W) - S(r, W), \tag{16.8}$$

where r_1, r_2 are the two children of r, and S is the similarity measure defined before.

Now we can use the following algorithm to find the most suitable frontier containing n rules to approximate the original rulebase with respect to W.

Algorithm 16.2 *Algorithm to find the best-matched rulebase with respect to a focus set*

1. $F \leftarrow \{\text{Root}\}$; calculate similarity gain $G(\text{Root})$ for Root;

2. while $|F| < n$ do
   select an internode $r \in F$ with the largest similarity gain $G(r)$;
   expand r to get the set of its children, $L(r)$;
   calculate similarity gain for nodes in $L(r)$;
   $F \leftarrow \{F - r\} \cup L(r)$.

□

This algorithm is of linear efficiency.

A compressed rulebase with respect to a focus window is shown by dotted lines in Figure 16.7. Once the structure is determined, we apply the local approach mentioned previously to identify the consequent parameters. Thus, a simplified but still proper rulebase is constructed. It can be used for applications like image coding and hierarchical pattern matching. When higher resolution is required for a simplified region, the corresponding internode can be expanded to provide a finer sub-rulebase.

16.6 SUMMARY

Structure identification in general is realized with two different approaches. The top-down method partitions the input (feature) space by Guillotine cuts under the guidance of fuzzy clustering objective functions. The two measures, density and typicality, we choose to evaluate clusters have a sound theoretical background in fuzzy sets.

The bottom-up approach emphasizes modeling accuracy and uses many small rules. The rules are structured into a fuzzy binary boxtree to speed up the pattern matching process when the rulebase is in operation. A parallel algorithm described in this chapter can be used to maintain the advantage of fuzzy systems in parallel processing.

The rules can also be merged or combined according to a dynamically defined focus set. Algorithms for identifying a suitable frontier in a boxtree and determining the parameters are described in this chapter. The sub-rulebases thus found can be organized into a hierarchy so that rulebases with various granularities can be employed in accordance with different demands of accuracy and efficiency.

EXERCISE

1. Modify the two-input two-rule ANFIS architecture in Figure 12.1 to incorporate two importance measures σ_1 and σ_2 for both inputs. Explain clearly the node functions for your new ANFIS architecture.

REFERENCES

[1] Jim C. Bezdek. *Pattern recognition with fuzzy objective function algorithms.* Plenum Press, New York, 1981.

[2] Didier Dubois, Henri Prade, and Claudette Testemale. Weighted fuzzy pattern matching. *Fuzzy Sets and Systems*, 28:313–331, 1988.

[3] Jerome H. Friedman, Jon Louis Bentley, and Raphael Ari Finkel. An algorithm for finding best matches in logarithmic expected time. *ACM Transactions on Mathematical Software*, 3(3):209–226, 1977.

[4] Michael C. Mozer and Paul Smolensky. Skeletonization: a technique for trimming the fat from a network via relevance assessment. Technical Report CU-CS-421-89, Department of Computer Science and Institute of Cognitive Science, University of Colorado, 1989.

[5] J. R. Quinlan. Induction of decision trees. *Machine Learning*, 1:81–106, 1986.

[6] Abhiram Ranade. A simpler analysis of the Karp-Zhang parallel branch-and-bound method. Technical Report UCB/CSD 90/586, Computer Science Division, University of California, Berkeley, 1990.

[7] Enrique H. Ruspini. Numerical methods for fuzzy clustering. *Information Sciences*, 2:319–350, 1970.

[8] Enrique H. Ruspini. Recent development in fuzzy clustering. In *Fuzzy set and possibility theory*, pages 133–147. North Holland, 1982.

[9] Enrique H. Ruspini. On the semantics of fuzzy logic. *International Journal of Approximate Reasoning*, 5:45–88, 1991.

[10] M. Sugeno and G. T. Kang. Structure identification of fuzzy model. *Fuzzy Sets and Systems*, 28:15–33, 1988.

[11] Yanjun Zhang. Parallel algorithms for combinatorial search problems. Technical Report UCB/CSD 89/543, Computer Science Division, University of California, Berkeley, 1989.

Part VI

Neuro-Fuzzy Control

Chapter 17

Neuro-Fuzzy Control I

J.-S. R. Jang

17.1 INTRODUCTION

Application of fuzzy inference systems to automatic control was first reported in Mamdani's paper [17] in 1975, where, based on Zadeh's proposition [32], a **fuzzy logic controller** (**FLC**) was used to emulate a human operator's control of a steam engine and boiler combination. Since then, **fuzzy logic control** [12, 14, 15, 25] has gradually been recognized as the most significant and fruitful application for fuzzy logic and fuzzy set theory. In the past few years, advances in microprocessors and hardware technologies have created an even more diversified application domain for fuzzy logic controllers, which ranges from consumer electronics to the automobile industry. Indeed, for complex and/or ill-defined systems that are not easily subjected to conventional automatic control methods, FLCs provide a feasible alternative since they can capture the approximate, qualitative aspects of human reasoning and decision-making processes. However, without adaptive capability, the performance of FLCs relies exclusively on two factors: the availability of human experts, and the knowledge acquisition techniques to convert human expertise into appropriate fuzzy if-then rules and membership functions. These two factors substantially restrict the application domain of FLCs.

On the other hand, investigation into using neural networks in automatic control systems did not receive much attention until the backpropagation learning rule was reformulated by Rumelhart et al. [24] in 1986. Since then, research of neural control has evolved quickly and a number of neural controller design methods have been proposed in the literature [6, 23, 28].

As explained in Chapters 8, 9, and 12, supervised learning neural networks and fuzzy inference systems are special instances of adaptive networks, which in certain ways are the most general form of modeling and computing-structure construction. Consequently, a neural control design approach can usually be carried over directly to the design of fuzzy controllers, unless the design method depends directly on the specific architecture of the neural network used (which is rare). This **portability** endows us with a number of design methods for fuzzy controllers which can easily take advantage of *a priori* human information and expertise in the form of fuzzy if-then rules. The resulting methodologies, often referred to as **neuro-fuzzy control**, are the topics of this and the next chapter.

Generally speaking, these methodologies can be classified into two categories. The first category consists of design methods obtained directly from neural control literature directly, such as expert (mimicking) control, inverse learning, specialized learning, backpropagation through time, and real-time recurrent learning; these approaches are discussed in this chapter. The second category contains design methods that are not directly or necessarily related to neural-like learning; these methods are explained in the next chapter. Some of the design methods in the second category take advantage of conventional control techniques such as gain scheduling, feedback linearization, adaptive control, and sliding mode control; others apply derivative-free optimization techniques (Chapter 7) or reinforcement types of learning (Chapter 10).

As usual, there is no single best way to design a fuzzy controller under all circumstances. We shall summarize the pros and cons of each of these methods and provide simple guidelines for choosing an appropriate method for a specific application in this and the next chapter.

17.2 FEEDBACK CONTROL SYSTEMS AND NEURO-FUZZY CONTROL: AN OVERVIEW

17.2.1 Feedback Control Systems

Figure 17.1 is a block diagram of a typical **feedback control system**, where the **plant** (or **process**) represents the dynamical system to be controlled and the **controller** employs a control strategy to achieve a control goal. Here we shall denote the state variables of the plant as a vector $\mathbf{x}(t)$; these variables are usually governed by a set of **state equations** (usually differential equations) that characterize the dynamic behavior of the plant. Since the state variables are internal to the plant, some of them may not be directly measurable from the external world. The measurable quantities of the plant, also known as its outputs, are denoted as a vector $\mathbf{y}(t)$ and are usually a static function of the state variables. Unless otherwise specified, we shall assume that all states are measurable; thus the output of the plant $\mathbf{y}(t)$ is equal to the state $\mathbf{x}(t)$. The preceding state equation will be used extensively in subsequent discussions.

The state equation for a general nonlinear time-invariant plant can be expressed in the matrix notation

$$\dot{\mathbf{x}}(t) = \mathbf{f}(\mathbf{x}(t), \mathbf{u}(t)) \quad \text{(plant dynamics)}, \tag{17.1}$$

where $u(t)$ is the controller's output at time t, and the size of the vector $\mathbf{x}(t)$ is called the **order** of the plant. A general control goal is to find a controller with a static function $\phi(\cdot)$ that maps an observed plant output $\mathbf{x}(t)$ to a control action $\mathbf{u}$—that is, $\mathbf{u}(t) = \phi(\mathbf{x}(t))$—such that the plant output $\mathbf{x}(t)$ can follow some given desired

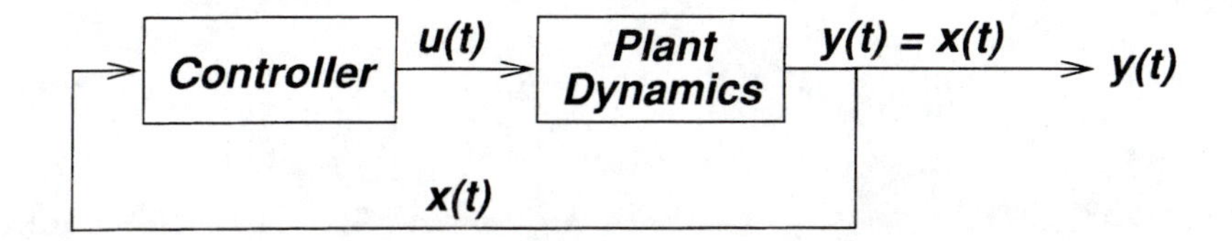

Figure 17.1. *Block diagram for a continuous-time feedback control system.*

output signal $\mathbf{x}_d(t)$ as closely as possible. If $\mathbf{x}_d(t)$ is a constant vector (usually the origin in state space), then the control problem is referred to as a **regulator problem**, where the plant states are directly fed back to the controller. This is actually what Figure 17.1 shows. On the other hand, if the desired trajectory $\mathbf{x}_d(t)$ is a time-varying signal, then we have a **tracking problem** in which an error signal, defined as the difference between desired and actual outputs, is fed back to the controller. If $\mathbf{f}$ is unknown, we need to perform system identification first to find an appropriate model for the plant. Moreover, if $\mathbf{f}$ is time varying, it is desirable to make $\phi(\cdot)$ adaptive to respond to the changing characteristics of the plant.

If we are dealing with a linear feedback control system, the plant and controller can be reformulated as the following equations:

$$\begin{aligned} \dot{\mathbf{x}}(t) &= \mathbf{A}\mathbf{x}(t) + \mathbf{B}\mathbf{u}(t) && \text{(plant dynamics),} \\ \mathbf{u}(t) &= \mathbf{K}\mathbf{x}(t) && \text{(linear controller).} \end{aligned} \tag{17.2}$$

The treatment of linear feedback control systems is relatively complete in the literature (for example, see refs. [2, 33]) and will not be discussed separately here.

In contrast, control systems without feedback loops are called **open-loop** control systems and lack certain advantages that feedback control systems provide. They are relatively more sensitive to unexpected external disturbances and internal changes in system characteristics. Thus, they are often bypassed in favor of feedback control systems in real-world applications, despite the latter's tendency toward instability due to overcorrecting in response to feedback signals. All the following discussions are therefore based on feedback control systems unless otherwise indicated.

A simple example of a nonlinear feedback control system is the inverted pendulum system. We shall use this system repeatedly throughout the rest of this chapter and the next one.

Example 17.1 *The inverted pendulum system*

Figure 17.2 shows an inverted pendulum system (also called a "cart-pole system"), a classic example of a nonlinear feedback control system. A rigid pole is attached to a cart with a hinge, a free joint with only one degree of freedom. The cart can move to the right or left on rails when a force is exerted on it. This dynamic system is characterized by four state variables: θ (angle of the pole with respect to the vertical axis), $\dot{\theta}$ (angular velocity of the pole), z (position of the cart on the track)

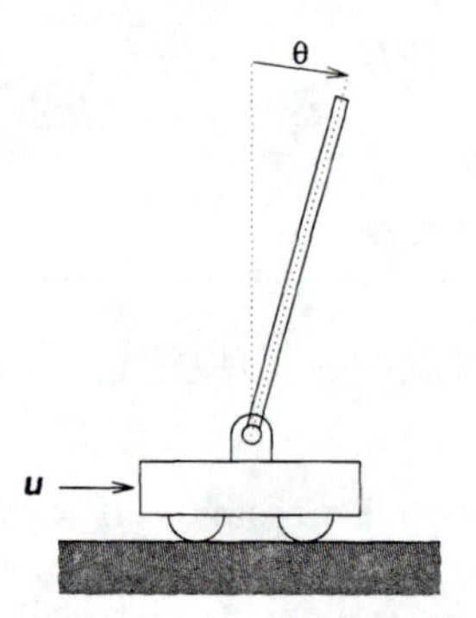

Figure 17.2. *The inverted pendulum system.*

and $\dot{z}$ (velocity of the cart); these state variables are governed by the following second-order differential equations [3, 13]:

$$\ddot{\theta} = \frac{g \sin\theta + \cos\theta \left(\dfrac{-u - ml\dot{\theta}^2 \sin\theta}{m_c + m} \right)}{l \left(\frac{4}{3} - \dfrac{m \cos^2\theta}{m_c + m} \right)}, \tag{17.3}$$

$$\ddot{z} = \frac{u + ml(\dot{\theta}^2 \sin\theta - \ddot{\theta}\cos\theta)}{m_c + m}, \tag{17.4}$$

where g is the acceleration due to gravity (usually 9.8 meter/sec^2), m_c is the mass of the cart, m is the mass of the pole, l is the half-length of the pole, and u is the applied force in Newtons. By defining the state vector $[x_1\ x_2\ x_3\ x_4]^T$ as $[\theta\ \dot{\theta}\ z\ \dot{z}]^T$, we can put the preceding equations into the standard format for state equations:

$$\dot{\mathbf{x}} = \begin{bmatrix} \dot{x}_1 \\ \dot{x}_2 \\ \dot{x}_3 \\ \dot{x}_4 \end{bmatrix} = \mathbf{f}(\mathbf{x},u) = \begin{bmatrix} x_2 \\ \dfrac{g \sin x_1 + \cos x_1 \left(\dfrac{-u - mlx_2^2 \sin x_1}{m_c + m} \right)}{l \left(\frac{4}{3} - \dfrac{m \cos^2 x_1}{m_c + m} \right)} \\ x_4 \\ \dfrac{u + ml(x_2^2 \sin x_1 - \dot{x}_2 \cos x_1)}{m_c + m} \end{bmatrix}. \tag{17.5}$$

(Note that $x_1 = \theta$ and $\dot{x_1} = \dot{\theta}$. Also, $x_2 = \dot{\theta}$ by definition, so we have $\dot{x_1} = x_2$. Similarly, $\dot{x_3} = x_4$.)

As control engineers, our mission is to find a controller $u = \phi(\mathbf{x})$ that maps a state vector $\mathbf{x}$ (or an error signal $\mathbf{x}_d - \mathbf{x}$) into an appropriate force u, such that a control goal can be achieved in a satisfactory manner. Usual control goals for the inverted pendulum system include the following:

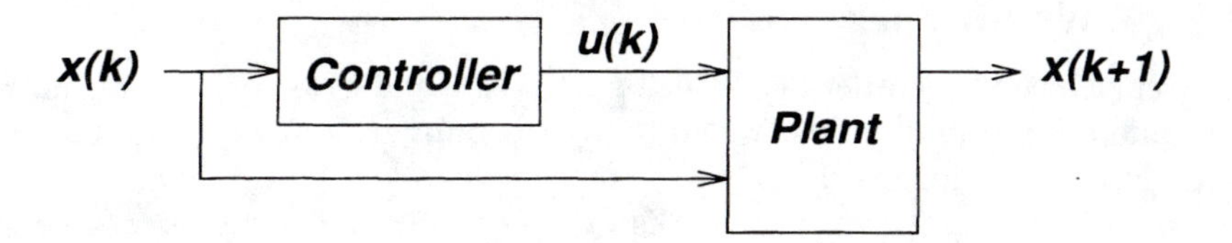

Figure 17.3. *Block diagram for a discrete-time feedback control system.*

- Keep the pole balanced, regardless of the cart position.
- Keep the pole tracking a desired signal, regardless of the cart's position.
- Keep the pole balanced and limit the cart's position to a track of limited length.
- Keep the pole balanced while the cart is tracking a desired signal.

The preceding control goals are listed according to their degrees of difficulty. We shall see how some of these goals may be achieved in subsequent discussions.

□

A general block diagram of a feedback control system in a discrete-time domain is shown in Figure 17.3; $\mathbf{x}(k)$ and $\mathbf{u}(k)$ are the state vector and control action, respectively, at time k. (When we say "time is k" in a discrete-time sense, what we really mean is "time is kT," where T is the sampling period of the underlying computer-controlled system.) Note that the inputs to the plant block include the control action $\mathbf{u}(k)$ and the previous plant output $\mathbf{x}(k)$ (assuming that the plant state vector is equal to the plant output vector), so the plant block now represents a static mapping. In symbols, we have

$$\begin{cases} \mathbf{x}(k+1) &= \mathbf{f}(\mathbf{x}(k), \mathbf{u}(k)) \quad \text{(plant)}, \\ \mathbf{u}(k) &= \mathbf{g}(\mathbf{x}(k)) \quad \text{(controller)}. \end{cases} \tag{17.6}$$

Again, the control problem becomes that of finding the mapping $\phi(\cdot)$ for the controller such that the resulting overall system exhibits certain desired behavior.

We shall use Equations (17.1) and (17.6) extensively in the following discussion. So we reiterate the assumptions behind these formulas:

- The order of the plant (that is, the number of state variables) is known.
- All state variables in the plant are measurable, that is, all states are also output variables.

17.2.2 Neuro-Fuzzy Control

If we replace the controller blocks in Figures 17.1 and 17.3 with neural networks or fuzzy inference systems, then we end up with **neural** or **fuzzy control systems**, respectively. In other words, neural or fuzzy control design methods are systematic ways of constructing neural networks or fuzzy inference systems, respectively, as controllers intended to achieve prescribed control goals. In the same vein, **neuro-fuzzy control** refers to the design methods for fuzzy logic controllers that employ neural network techniques. In particular, we shall concentrate on design methods for ANFIS (adaptive neuro-fuzzy inference systems; see Chapter 12); thus ANFIS and neuro-fuzzy controllers will be used interchangeably in this book.

As demonstrated in previous chapters, fuzzy inference systems (FISs) and multilayer perceptrons (MLPs) are special instances of a more general computing framework called the adaptive networks, therefore, both of these instances inherit the backpropagation learning ability of the adaptive network. However, the fuzzy inference system is superior to the multilayer perceptron in that the former can represent structured knowledge while the latter is more or less like a black box. As a result, we can identify some unique properties of ANFIS controllers:

1. Learning ability
2. Parallel operation
3. Structured knowledge representation
4. Better integration with other control design methods

Note that a multilayer perceptron also has properties 1 and 2, but not 3 and 4. In the rest of this chapter and the next one, we shall introduce several neuro-fuzzy design methods for constructing an ANFIS controller. Note that while some of the methods are unique to ANFIS, most of them are derived directly from methods for neural controller design, and these methods usually apply directly to more general cases of adaptive network controller design.

Most neural or fuzzy controllers are nonlinear; thus rigorous analysis for neuro-fuzzy control systems is difficult and remains a challenging area for further investigation. On the other hand, a neuro-fuzzy controller usually contains a large number of parameters; it is thus more versatile than a linear controller in dealing with nonlinear plant characteristics. Therefore, neuro-fuzzy controllers almost always surpass pure linear controllers if designed properly.

17.3 EXPERT CONTROL: MIMICKING AN EXPERT

The original purpose of a fuzzy logic control, as proposed in Mamdani's seminal paper in 1975, was to mimic the behavior of a human operator able to control a complex plant satisfactorily. The complex plant in question could be a chemical reaction process, a subway train, or a traffic signal control system. After more than

20 years, the ultimate goal of fuzzy controllers stays the same—that is, to automate an entire control process by replacing a human operator with a fuzzy controller made up of computer software/hardware, or a single silicon chip that costs little, responds quickly, behaves consistently, and works around the clock.

To construct a fuzzy controller, we need to perform **knowledge acquisition**, which takes a human operator's knowledge about how to control a system and generates a set of fuzzy if-then rules as the backbone for a fuzzy controller that behaves like the original human operator. Usually we can obtain two types of information from a human operator: **linguistic information** and **numerical information**.

Linguistic information An experienced human operator can usually summarize his or her reasoning process in arriving at final control actions or decisions as a set of fuzzy if-then rules with imprecise but roughly correct membership functions; this corresponds to the linguistic information supplied by human experts, which is obtained via a lengthy interview process plus a certain amount of trial and error.

Numerical information When a human operator is working, it is possible to record the sensor data observed by the human and the human's corresponding actions as a set of desired input-output data pairs. This data set can be used as a training data set in constructing a fuzzy controller.

Prior to the emergence of neuro-fuzzy approaches, most design methods used only the linguistic information to build fuzzy controllers; this approach is not easily formalized and is more of an art than an engineering practice. Following this approach usually involves manual trial-and-error tweaking processes to fine-tune the membership functions. Successful fuzzy control applications based on linguistic information plus trial-and-error tuning include steam engine and boiler control [17], Sendai subway systems [31], container ship crane control [30], elevator control [16], nuclear reaction control [1], automobile transmission control [9], aircraft control [5], and many others [25].

Now, with learning algorithms, we can take further advantage of the numerical information (input-output data pairs) and refine the membership functions in a systematic way. In other words, we can use linguistic information to identify the structure of a fuzzy controller, and then use numerical information to identify the parameter such that the fuzzy controller can reproduce the desired action more accurately. Full exploitation of linguistic/numerical information is expected to expand greatly the application possibilities of fuzzy controllers.

Note that mimicking a human expert is not only good for control applications. If the target system to be emulated is a human physician or a credit analyst, then the resulting fuzzy inference systems become **fuzzy expert systems** for diagnosis and credit analysis, respectively.

The capacity to use linguistic information is specific to fuzzy inference systems; it is hard for multilayer perceptrons to take advantage of this information and directly encode it into the network's structure. Using numerical data to train neural networks to emulate a human expert has also achieved some success; examples include the unmanned vehicle developed at Carnegie-Mellon University [21, 22].

17.4 INVERSE LEARNING

17.4.1 Fundamentals

The development of **inverse learning** [29], also known as **general learning** [23], for designing neuro-fuzzy controllers involves two phases. In the learning phase, an on-line or off-line technique is used to model the inverse dynamics of the plant. The obtained neuro-fuzzy model, which represents the inverse dynamics of the plant, is then used to generate control actions in the application phase. These two phases can proceed simultaneously, hence this design method fits in perfectly with the classical adaptive control scheme.

By assuming that the order of the plant (that is, the number of state variables) is known and all state variables are measurable, we have

$$\mathbf{x}(k+1) = \mathbf{f}(\mathbf{x}(k), u(k)), \tag{17.7}$$

where $\mathbf{x}(k+1)$ is the state at time $k+1$, $\mathbf{x}(k)$ is the state at time k, and $u(k)$ is the control signal at time k. [For pedagogic purposes, we assume here that $u(k)$ is a scalar.] Similarly, the state at time $k+2$ is expressed as

$$\mathbf{x}(k+2) = \mathbf{f}(\mathbf{x}(k+1), u(k+1)) = \mathbf{f}(\mathbf{f}(\mathbf{x}(k), u(k)), u(k+1)). \tag{17.8}$$

In general, we have

$$\mathbf{x}(k+n) = \mathbf{F}(\mathbf{x}(k), \mathbf{U}), \tag{17.9}$$

where n is the order of the plant, F is a multiple composite function of $\mathbf{f}$, and $\mathbf{U}$ is the control actions from k to $k+n-1$, which is equal to $[u(k),\ u(k+1),\ \ldots,\ u(k+n-1)]^T$. The preceding equation points out the fact that given the control input u from time k to $k+n-1$, the state of the plant will move from $\mathbf{x}(k)$ to $\mathbf{x}(k+n)$ in exactly n time steps. Furthermore, we assume that the inverse dynamics of the plant do exist, that is, $\mathbf{U}$ can be expressed as an explicit function of $\mathbf{x}(k)$ and $\mathbf{x}(k+n)$:

$$\mathbf{U} = \mathbf{G}(\mathbf{x}(k), \mathbf{x}(k+n)). \tag{17.10}$$

This equation essentially says that there exists a unique input sequence $\mathbf{U}$, specified by mapping $\mathbf{G}$, that can drive the plant from state $\mathbf{x}(k)$ to $\mathbf{x}(k+n)$ in n time steps. The problem now becomes how to find the inverse mapping $\mathbf{G}$.

Let us look at a special case in which the state equation in Equation (17.7) is linear.

Example 17.2 *Inverse dynamics of a linear system*

In terms of linear systems, Equation (17.7) can be written as

$$\mathbf{x}(k+1) = \mathbf{A}\mathbf{x}(k) + \mathbf{B}u(k), \tag{17.11}$$

where $\mathbf{A}$ and $\mathbf{B}$ are $n \times n$ and $n \times 1$ matrices, respectively. By repeating the preceding equation, we obtain the state at $k+n$:

$$\mathbf{x}(k+n) = \mathbf{A}^n\mathbf{x}(k) + \mathbf{W}\mathbf{U}, \tag{17.12}$$

where $\mathbf{W} = [\mathbf{A}^{n-1}\mathbf{B} \ \cdots \ \mathbf{AB} \ \mathbf{B}]$ is the controllability matrix. If $\mathbf{W}$ is nonsingular, then the system is controllable and $\mathbf{U}$ can be calculated as

$$\mathbf{U} = \mathbf{W}^{-1}[\mathbf{x}(k+n) - \mathbf{A}^n\mathbf{x}(k)]. \tag{17.13}$$

In other words, the controllability in a linear system is equivalent to the inverse condition mentioned earlier.

□

Although the inverse mapping $\mathbf{G}$ in Equation (17.10) exists by assumption, it does not always have an analytically close form. Therefore, instead of looking for methods of solving Equation (17.10) explicitly, we can use an adaptive network or ANFIS with $2n$ inputs and n outputs to approximate the inverse mapping $\mathbf{G}$ according to the generic training data pairs

$$[\mathbf{x}(k)^T, \mathbf{x}(k+n)^T; \mathbf{U}^T]. \tag{17.14}$$

Figure 17.4 illustrates the situation in which n is equal to 1. Figure 17.4(a) shows a plant block in which the plant output $x(k+1)$ is a function of a previous state $x(k)$ and input $u(k)$; we use z^{-1} block to represent the unit-time delay operator. Figure 17.4(b) is the block diagram during the training phase; Figure 17.4(c) is the block diagram during the application phase.

Assume that the adaptive network truly imitates the input-output mapping of the inverse dynamics $\mathbf{G}$. Then, given the current state $\mathbf{x}(k)$ and the desired future state $\mathbf{x}_d(k+n)$, the adaptive network will generate an estimated $\hat{\mathbf{U}}$:

$$\hat{\mathbf{U}} = \hat{\mathbf{G}}(\mathbf{x}(k), \mathbf{x}_d(k+n)). \tag{17.15}$$

After n steps, this control sequence can bring the state $\mathbf{x}(k)$ to the desired state $\mathbf{x}_d(k+n)$, assuming that the adaptive network function $\hat{\mathbf{G}}$ is exactly the same as the inverse mapping $\mathbf{G}$. This application phase is shown in the block diagram of Figure 17.4(b). If the future desired state $\mathbf{x}_d(k+n)$ is not available in advance, we can use the current desired state $\mathbf{x}_d(k)$ instead in Figure 17.4(b). This implies that the current desired state will appear after n time steps and the whole system behaves like a pure n-step time delay system.

When $\hat{\mathbf{G}}$ is not close to $\mathbf{G}$, the control sequence $\hat{\mathbf{U}}$ cannot bring the state to $\mathbf{x}_d(k+n)$ in exactly the next n time step. As more data pairs are used to refine the parameters in the adaptive network, $\hat{\mathbf{G}}$ will become closer to $\mathbf{G}$ and the control will be more and more accurate as the training process goes on.

For off-line applications, we have to collect a set of training data pairs and then train the adaptive network in the batch mode. For on-line applications to deal

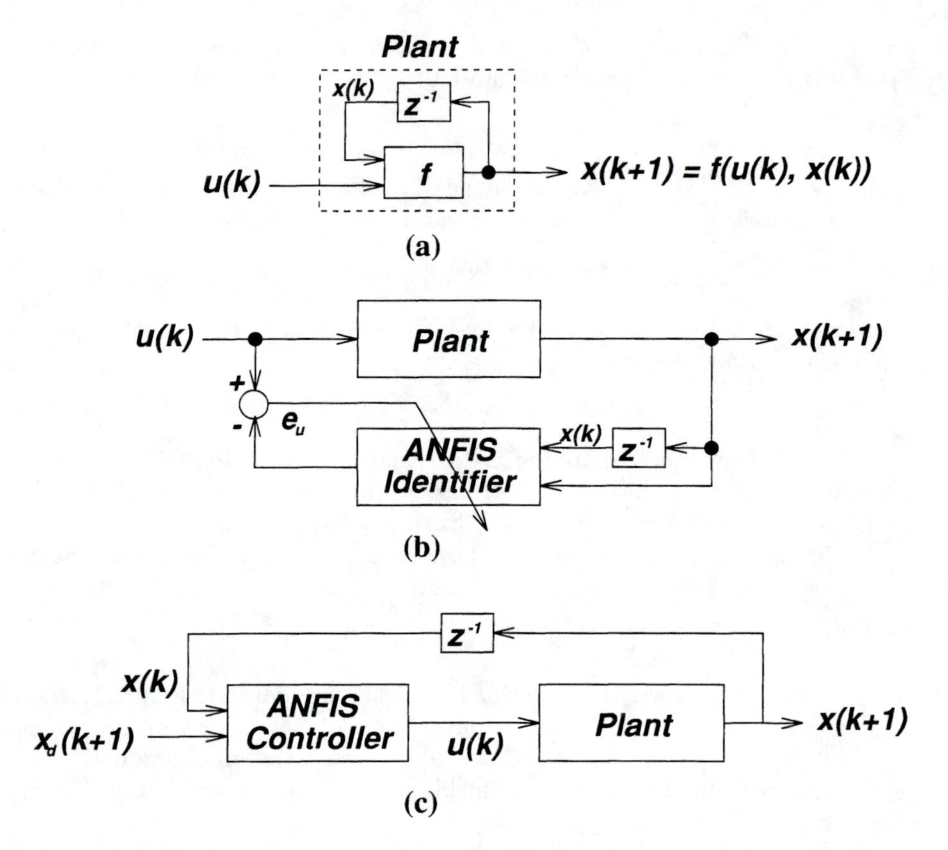

Figure 17.4. *Block diagram for the inverse learning method: (a) plant block; (b) training phase; (c) application phase.*

with time-varying systems, the control actions in Equation (17.15) are generated every n time steps while on-line learning occurs at every time step. Alternatively, we can generate the control sequence at every time step and apply only the first component to the plant. Figure 17.5 is a block diagram for on-line learning when n is equal to 1. The dashed line in the figure indicates that the two ANFIS blocks are exact duplicates of each other. (For simplicity, we have removed the unit-time delay operator z^{-1}.)

The rationale behind inverse learning seems straightforward. However, it assumes the existence of inverse dynamics for a plant, which is not generally valid. Moreover, minimization of the **network error** $||\mathbf{U} - \hat{\mathbf{U}}||^2$ does not guarantee minimization of the overall **system error** $||\mathbf{x}_d(k) - \mathbf{x}(k)||^2$.

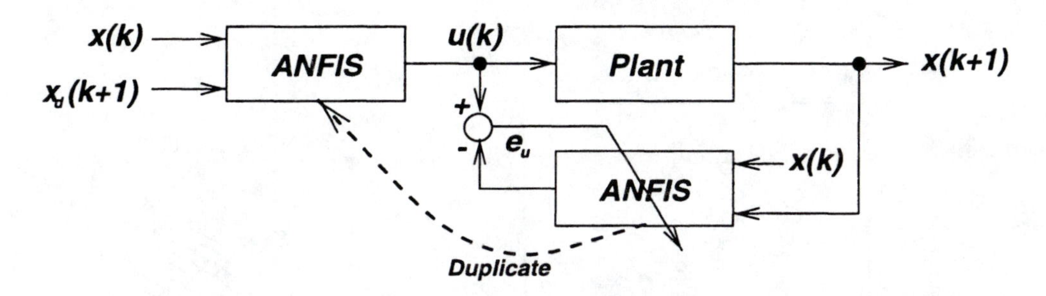

Figure 17.5. *Block diagram for on-line inverse learning.*

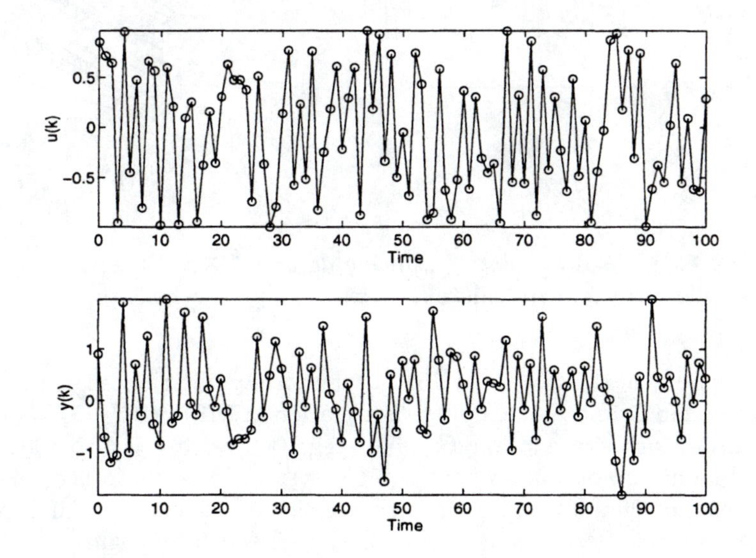

Figure 17.6. *Collecting training data for inverse control, where the upper plot is $u(k)$ and the lower one is $y(k)$.* (MATLAB file: `inv_sig.m`)

17.4.2 Case Studies

Suppose that a plant is described by the following discrete dynamical equation:

$$y(k+1) = \frac{y(k)u(k)}{1+y^2(k)} - \tan(u(k)), \tag{17.16}$$

where $y(k)$ and $u(k)$ are the state and control action, respectively, at time step k. Here we assume the dynamics of the plant to be unknown, and we are going to build an ANFIS that maps given a given input pair $[y(k),\ y(k+1)]$ to a desired control action $u(k)$. This mapping is not easily expressed as an analytical formula, even if the preceding equation governing the plant dynamics is already known.

To train the ANFIS, we need to collect training data pairs. This is done by

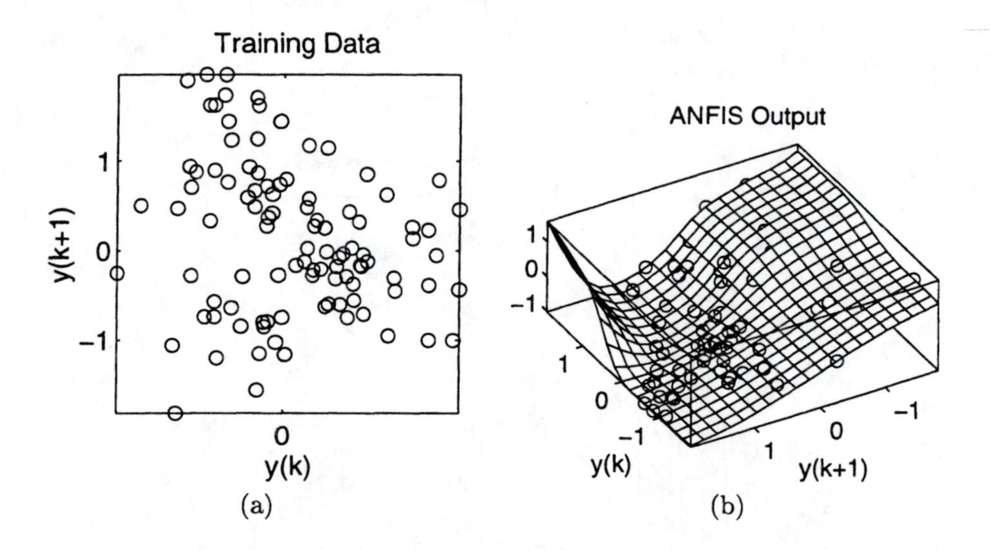

Figure 17.7. *(a) Scatter plot of training data; (b) ANFIS surface after training.* (MATLAB files: `inv_sig.m` and `inv_fc.m`)

choosing inputs $u(k)$, $k = 1$ to 101, as uniformly distributed random numbers between -1 and 1, and then employing Equation (17.16) [with $y(1) = 0$] to find 100 training data pairs of the form $[y(k), y(k+1); u(k)]$, with $k = 1$ to 100. Figure 17.6 shows the input and output sequences of the system to be controlled; Figure 17.7(a) is a scatter plot of the training data thus collected. After 30 training epochs, an ANFIS with nine rules exhibits a control surface shown in Figure 17.7(b). For the desired output specified by the equation

$$y_d(k) = 0.6\sin(2\pi k/250) + 0.2\sin(2\pi k/50),$$

the ANFIS controller achieves good performance, as shown in Figure 17.8, where the left-hand plot indicates desired and actual outputs of the plant and the right-hand the difference between them.

This simple example serves to illustrate the concept of inverse control. Some remarks regarding the simulation are in order to highlight the strengths and shortcomings of this method.

- We do not need to know the plant dynamics in advance; identification of them is embedded in the training of ANFIS to find the inverse model.
- Our simulation was based on off-line learning only. It is possible to turn on on-line learning to cope with time-varying plant dynamics. In general, the best approach is to use off-line learning to find a working controller for a

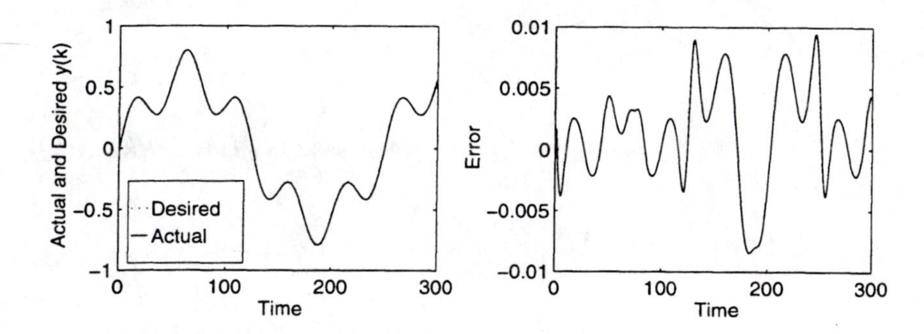

Figure 17.8. *Performance of ANFIS controller for inverse control. Due to the small error, it is hard to see the desired $y(k)$, which is almost overlaid by the actual desired $y(k)$.* (MATLAB file: `inv_fc.m`)

nominal plant, and then use on-line learning to fine-tune the controller if the plant is time varying.

- We assume that at time step k, $y_d(k+1)$ is available and it is used as an input to the ANFIS controller. If $y_d(k+1)$ is not available until time step $k+1$, then we can use $y_d(k)$ as an input to the ANFIS controller at time step k, and the resulting overall system will behave like a unit-delay system.

- Before using inverse learning, one should make sure the system to be controlled by this technique has a unique inverse. This is not so easy as the order of the plant dynamics increases.

- The distribution of the training data could also pose a problem for this method. Ideally, we would like to see the training data distributed across the input space of the controller in a somewhat uniform manner. However, this may not be possible due either to the scarcity of the data (especially when there are many inputs) or to the limits imposed by the underlying plant dynamics. In our simulation, the lack of data at the upper right and lower right corners of the input space [see Figure 17.7(a)] is primarily the result of the underlying plant dynamics. This causes a sharp ascent and descent at each of the corners, as can be seen clearly in Figure 17.7(b).

17.5 SPECIALIZED LEARNING

A major problem with inverse learning is that an inverse model does not always exist for a given plant. Moreover, inverse learning is an indirect approach that tries to minimize the network output error instead of the overall system error (defined as the difference between desired and actual trajectories). **Specialized learning** [23]

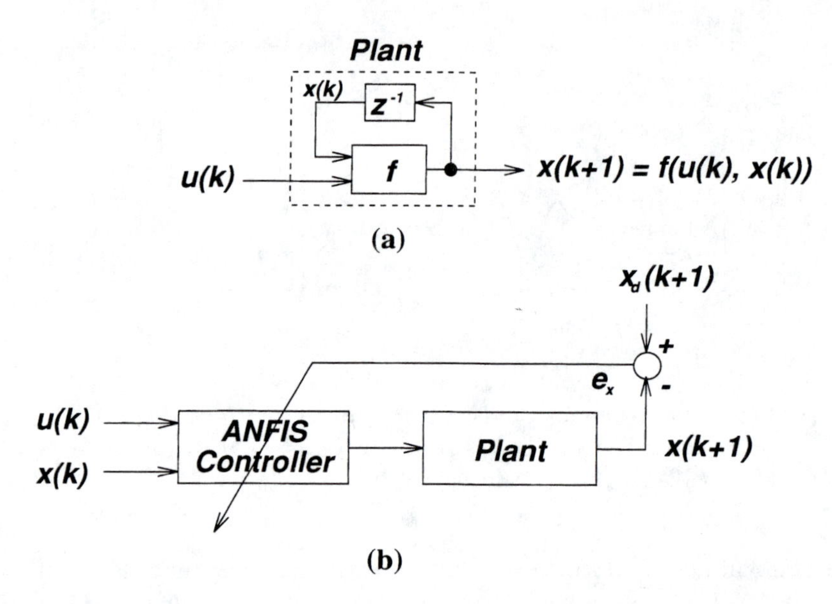

Figure 17.9. *(a) Plant block; (b) specialized learning using desired trajectory.*

is an alternative method that tries to minimize the system error directly by back-propagating error signals through the plant block. The price is that we need to know more about the plant under consideration.

Figure 17.9 illustrates the most basic type of specialized learning, Figure 17.9(a) is the plant block (assuming its order is 1), and Figure 17.9(b) indicates the training of the ANFIS controller. The ANFIS parameters are updated to reduce the system error $\mathbf{e}_{\mathbf{x}}(k)$, which is defined as the difference between the system's output $\mathbf{x}(k)$ and the desired output $\mathbf{x}_d(k)$.

To be more specific, let the plant dynamics be specified by

$$\mathbf{x}(k+1) = \mathbf{f}(\mathbf{x}(k), v(k))$$

and the ANFIS output be denoted as

$$\hat{v}(k) = F(\mathbf{x}(k), u(k), \boldsymbol{\theta}), \tag{17.17}$$

where $\boldsymbol{\theta}$ is a parameter vector to be updated. (Without loss of generality, we assume the plant has a single scalar input $u(k)$.) If we set the ANFIS output as the plant's input, then $v(k) = \hat{v}(k)$ and we have a closed-loop system specified by

$$\mathbf{x}(k+1) = \mathbf{f}(\mathbf{x}(k), F(\mathbf{x}(k), u(k), \boldsymbol{\theta})).$$

The objective of specialized learning is to minimize the difference between the closed-loop system and the desired model. Hence we can defined an error measure:

$$J(\boldsymbol{\theta}) = \sum_k \|\mathbf{f}(\mathbf{x}(k), F(\mathbf{x}(k), u(k), \boldsymbol{\theta})) - \mathbf{x}_d(k+1)\|^2. \tag{17.18}$$

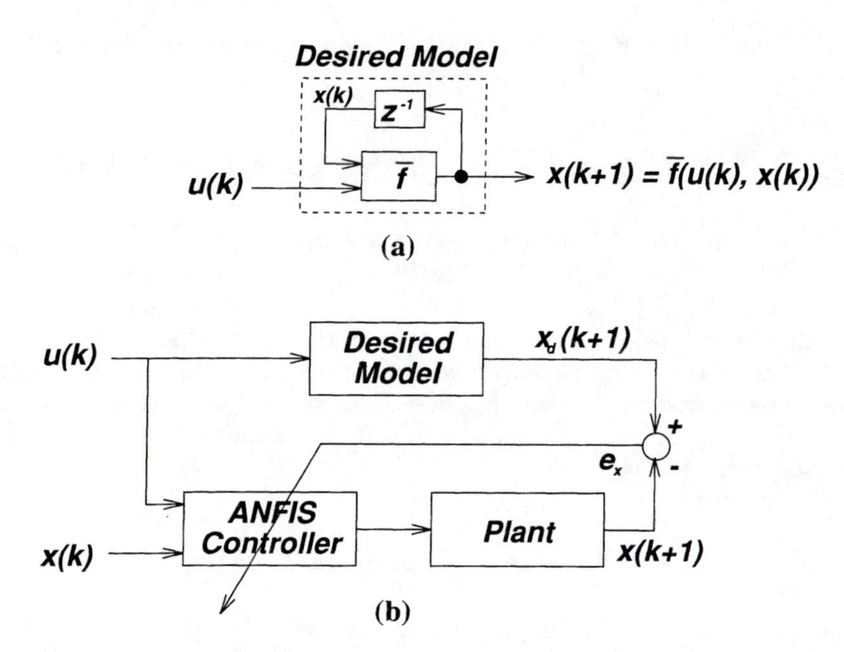

Figure 17.10. *(a) Desired model block; (b) specialized learning with model referencing.*

As usual, we are relying on backpropagation or steepest descent to update $\boldsymbol{\theta}$ to minimize the above error measure. To find the derivative of $J(\boldsymbol{\theta})$ with respect to $\boldsymbol{\theta}$, we need to know the derivative of $\mathbf{f}(\cdot, \cdot)$ with respect to its second argument. In other words, to backpropagate error signals through the plant block in Figure 17.10(b), we need to know the **Jacobian matrix** of the plant, where the element at row i and column j is equal to the derivative of the plant's ith output with respect to its jth input. This usually implies that we need a model for the plant and the Jacobian matrix obtained from the model, which could be a neural network, an ANFIS, or another appropriate mathematical description of the plant.

For a single-input plant, if the Jacobian matrix is not easily found directly, a crude estimate can be obtained by approximating it directly from the changes in the plant's input and output(s) during two consecutive time instants. Other methods that aim at using an approximate Jacobian matrix to achieve the same learning effects can be found in refs. [4, 10, 27].

It is not always convenient to specify the desired plant output $\mathbf{x}_d(k)$ at every time instant k. As a standard approach in model reference adaptive control (), the desired behavior of the overall system can be implicitly specified by a (usually linear) model that is able to achieve the control goal satisfactorily. This is shown in Figure 17.9(b), where the desired output $\mathbf{x}_d(k+1)$ is generated via the desired model depicted in Figure 17.9(a). Let the desired model be specified by

$$\mathbf{x}(k+1) = \bar{\mathbf{f}}(\mathbf{x}(k), u(k)).$$

Then the error measure in Equation (17.18) becomes

$$\begin{aligned} J(\boldsymbol{\theta}) &= \textstyle\sum_k \|\mathbf{f}(\mathbf{x}(k), \hat{v}(k)) - \mathbf{x}_d(k+1)\|^2 \\ &= \textstyle\sum_k \|\mathbf{f}(x(k), F(x(k), u(k), \boldsymbol{\theta})) - \bar{\mathbf{f}}(\mathbf{x}(k), u(k))\|^2. \end{aligned} \tag{17.19}$$

Again, we still need the Jacobian matrix of the plant to do backpropagation.

Under certain circumstances, we do not need the Jacobian matrix of the plant to proceed backpropagation. Suppose that we are only interested in one element $x(k)$ of the output vector $\mathbf{x}(k)$. If $x(k)$ is feedback-linearizable system, its dynamics can be expressed as

$$\begin{aligned} x(k+1) &= f(\mathbf{x}(k), v(k)) \\ &= g(\mathbf{x}(k)) + h(\mathbf{x}(k))v(k). \end{aligned} \tag{17.20}$$

If both $g(\cdot)$ and $h(\cdot)$ are known perfectly, the desired model, denoted by $x_d(k+1) = \bar{f}(\mathbf{x}_d(k), u(k))$, can be achieved by setting the input $v(k)$ in Equation (17.20) as follows:

$$v(k) = \frac{\bar{f}(\mathbf{x}(k), u(k)) - g(\mathbf{x}(k)}{h(\mathbf{x}(k))}.$$

If $g(\cdot)$ is unknown and $h(\cdot)$ is known, we can use an ANFIS to approximate $g(\cdot)$ directly; the desired input-output pair for ANFIS training is

$$[\mathbf{x}(k); x_d(k+1) - h(\mathbf{x}(k))u(k)].$$

Similarly, if $h(\cdot)$ is unknown and $g(\cdot)$ is known, we can use an ANFIS to approximate $h(\cdot)$ directly; the desired input-output pair for ANFIS training is

$$[\mathbf{x}(k); (x_d(k+1) - g(\mathbf{x}(k)))/u(k)].$$

In either of these two situations, the training of ANFIS does not required the Jacobian matrix of the plant. Also to make the training data as rich as possible, the input signal $u(k)$ is preferably a random signal.

Note that the ANFIS controller in Equation (17.17) represents the most general situation. More commonly, the ANFIS controller is a function of $\mathbf{x}(k)$ and $\boldsymbol{\theta}$ only and the input to the plant $v(k)$ is expressed as the difference between the command signal $u(k)$ and ANFIS output, as follows:

$$\hat{v}(k) = u(k) - F(\mathbf{x}(k), \boldsymbol{\theta}).$$

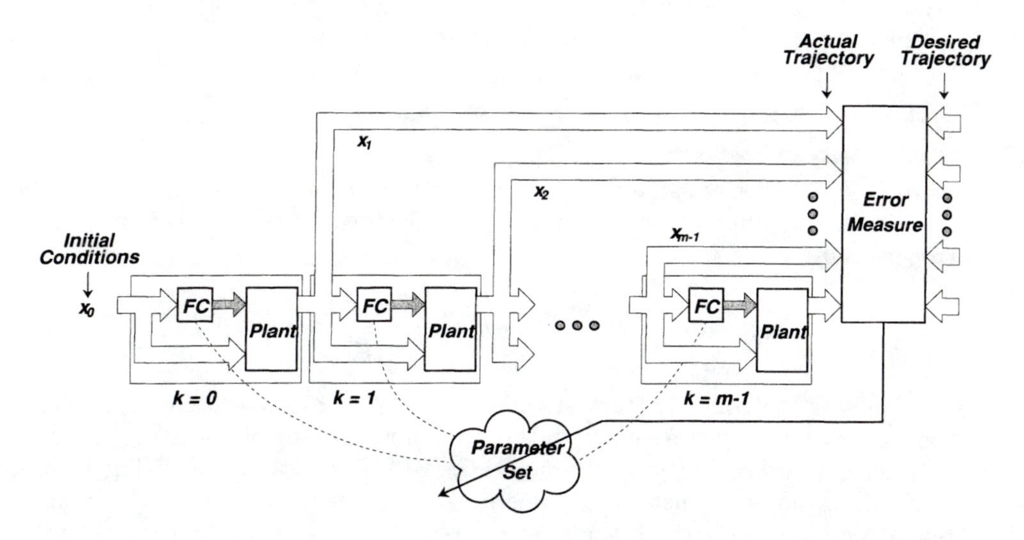

Figure 17.11. *A trajectory adaptive network for control application (FC stands for "fuzzy controller").*

17.6 BACKPROPAGATION THROUGH TIME AND REAL-TIME RECURRENT LEARNING

17.6.1 Fundamentals

If we replace the controller and the plant block in Figure 17.1 with two adaptive networks, the feedback control system becomes the recurrent adaptive network discussed in Section 8.4. Assuming that synchronous operation is adopted here (which virtually converts the system into a discrete-time domain), we can apply the concept of unfolding of time introduced in Chapter 8 to obtain a feedforward network, and then use the same backpropagation learning algorithm to identify a set of parameters to generate satisfactory trajectories.

To obtain the state trajectory, we cascade the block diagram (or adaptive network, if both the controller and the plant are replaced with appropriate adaptive networks) in Figure 17.3 to obtain the **trajectory adaptive network** shown in Figure 17.11. In particular, the inputs to the trajectory adaptive network are the initial conditions of the plant; the outputs are the state trajectories from $k = 1$ to $k = m$. The adjustable parameters all pertain to the FC (fuzzy controller) block implemented as an ANFIS. Although there are m FC blocks, all of them refer to the same parameter set. For clarity, this parameter set is shown explicitly in Figure 17.11 and is updated according to the output of the error measure block.

Each entry in the training data set for the trajectory network is of the following format:

$$\text{(initial condition; desired trajectory)},$$

and the corresponding error measure to be minimized is

$$E = \sum_{k=1}^{m} \|\mathbf{x}(k) - \mathbf{x}_d(k)\|^2,$$

where $\mathbf{x}_d(k)$ is a desired state vector at time step k (which corresponds to real time kT; T is the sampling period). If we take control efforts into consideration, a revised error measure would be

$$E = \sum_{k=1}^{m} \|\mathbf{x}(k) - \mathbf{x}_d(k)\|^2 + \lambda \sum_{k=0}^{m-1} \|\mathbf{u}(k)\|^2,$$

where $\mathbf{u}(k)$ is the control action at time step k. By proper selection of λ, a compromise between trajectory error and control efforts can be obtained.

Note that backpropagation through time (BPTT) is usually an off-line learning algorithm in the sense that the parameters will not be updated until the sequence ($k = 1$ to m) is completed. If the sequence is too long, or if we want to update the parameters in the middle of the sequence, we can always apply real-time recurrent learning (RTRL), as introduced in Chapter 8.

Use of BPTT to train a neural network to back up a tractor-trailer system was reported in ref. [19]. The same technique was used to design an ANFIS controller for balancing an inverted pendulum [8]; this is discussed thoroughly next.

17.6.2 Case Studies: the Inverted Pendulum System

In this section, we present a detailed description of our simulation that used BPTT to find a fuzzy controller for the inverted pendulum system, as reported in ref. [8]. For simplicity, we consider the pole dynamics only. Thus, the plant has only two state variables—x_1 and x_2—representing, respectively, the pole angle and angular velocity. The differential equation for the pole dynamics is provided in Equation (17.3). Note that this is a feedback linearizable system and there exists other advanced nonlinear control design methods (e.g., sliding mode control, see Section 18.5). Here we shall use the pole system as a simple application example of BPTT, which does not exploit the feedback linearizability of the system.

To use BPTT, the first thing we have to do is system identification (see Chapter 5)—that is, find an adaptive network representation of the plant block in Figure 17.3. In fact, we can choose whatever function approximators that can best represent the input-output behavior of the plant, as long as the chosen approximator is piecewise differentiable. Potential candidates for the plant approximator include conventional linear or nonlinear difference equations and unconventional network structures (neural networks [24], radial basis function networks [18], GMDH (group method of data handling) structures [7], functional-link networks [11, 20], ANFIS, and so on). This model-insensitive attribute is mostly due to the extreme flexibility of adaptive networks which allows implementation of all kinds of piecewise differentiable functions.

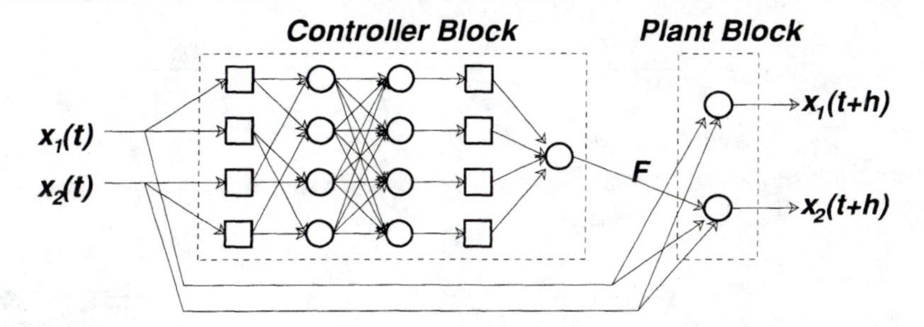

Figure 17.12. *Network implementation of the discrete control block diagram in Figure 17.3.*

If the plant can be modeled as a set of n (= number of state variables) first-order difference equations, then the plant block can be replaced with n nodes, each of which uses one difference equation to obtain the state variable at the next time step. For simplicity, we assume here that the plant is represented by the difference equations

$$\begin{cases} x_1(k+1) &= h\dot{x}_1(k) + x_1(k), \\ x_2(k+1) &= h\dot{x}_2(k) + x_2(k), \end{cases} \tag{17.21}$$

where $\dot{x}_1$ and $\dot{x}_2$ are specified in Equation (17.5). These two equations are the node functions of the plant block in Figure 17.12.

The controller block in Figure 17.11 is implemented as an ANFIS with two inputs, each of which is assigned two membership functions, so it is a fuzzy controller with four fuzzy if-then rules of Sugeno's type [26]. (See the controller block in Figure 17.12.)

As a result of replacing blocks with adaptive networks, the block diagram in Figure 17.3 becomes an adaptive network containing two subnetworks, the FC block (ANFIS) and the plant block. This adaptive network is referred to as a **stage adaptive network** at time stage k. The trajectory adaptive network shown in Figure 17.11 contains m replicas of stage adaptive networks at different time steps.

For the controller block, we assume that no domain knowledge (from a human expert) about the inverted pendulum system is available. Without any domain knowledge, we have to set the initial parameters for the ANFIS controller in a general and unbiased way. The consequent parameters are all set at zero, which means the control action is zero initially, as shown in Figure 17.14(a). The premise parameters are set in such a way that the membership functions can cover the domain intervals (or universe of discourse) completely with sufficient overlapping of each other. Figures 17.13(a) and Figures 17.13(b) illustrate the generalized bell membership functions before training; the domain intervals for θ (degrees) and $\dot{\theta}$ (deg/s) are assumed to be $[-20, 20]$ and $[-50, 50]$, respectively.

We employ 100 stage adaptive networks to construct the trajectory adaptive network, and each stage adaptive network corresponds to a time transition of 10

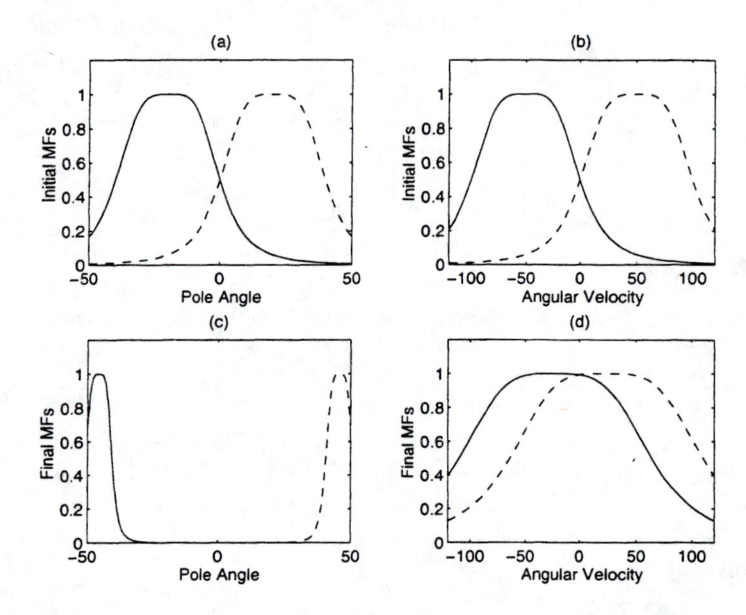

Figure 17.13. *(a)(b) Initial membership functions; (c)(d) final membership functions.*

microseconds. That is, the sampling period T used is 10 ms, and the trajectory adaptive network corresponds to a time interval from $t = 0$ to $t = 1$ sec. If T is too small, a large network has to be built to cover the same time span, which increases the signal propagation time and thus delays the whole learning process. On the other hand, if T is too big, then the linear approximation of the plant's behavior may not be good enough, requiring that a more precise difference equation for the plant be used.

The training data set contains desired input-output pairs of the format

$$\text{initial condition; desired trajectory}), \tag{17.22}$$

where the initial condition is a two-element vector that specifies the initial condition of the pole; the desired trajectory is a 100-element vector that contains the desired pole angle at each time step. In our simulation, only two training data entries are used: the initial conditions are (10, 0) and (−10, 0), respectively, and the desired trajectory is always a zero vector of size 100. In other words, we expect the controller to be able to bring the pole back to the upright position starting from either +10 or −10 degrees. The error measure used here is

$$E = \sum_{k=1}^{100} \theta^2(0.01k) + \lambda \sum_{k=0}^{99} f^2(0.01k), \tag{17.23}$$

where $f(0.01k)$ is the controller's output force and λ (= 10) accounts for the relative

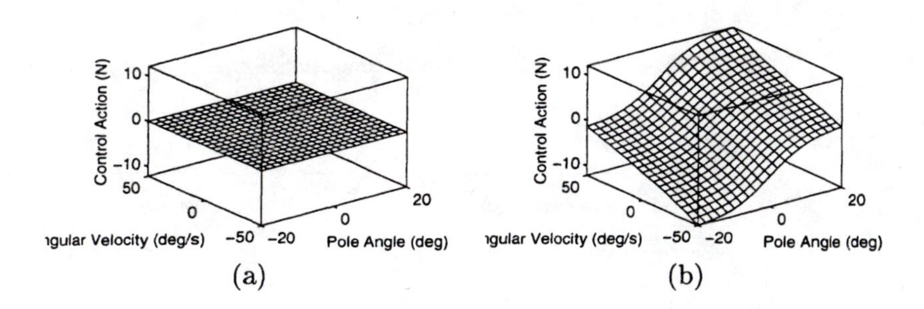

Figure 17.14. *Control action surfaces: (a) before training; (b) after training.*

unit cost of the control effort.

To speed up convergence, we follow a strict steepest descent in the sense that each parameter update leads to a smaller error measure. If the error measure increases after a parameter update, we back up to the original point in the parameter space and decrease the current step size by half. This process is repeated until a parameter update leads to a smaller error measure. However, this step size update procedure tends to use a small step size if the error measure surface encountered in the first few updates is smooth. We therefore multiply the step size by 4 after observing three consecutive updates without any backing-up action. The initial step size in the simulation is 20, and the learning process stops whenever the number of successful parameter updates (which is equal to the number of reductions in error measure) reaches 10. We did play around with the initial step size and found that if the initial step size was too small, the training process converged prematurely to a set of parameters, presumably a local minimum, that do not really minimize the error measure.

All the aforementioned simulation settings are referred to as the *reference setting*; other simulations are based on this setting with minor changes. In the learning task with the reference setting, it is observed that the FC balance the pole right after the first parameter update and keep on refining the controller (minimizing the error measure) until the 10th successful parameter update. Figures 17.13(a) and Figures 17.13(b) show the initial membership functions for pole angle and angular velocity; Figure 17.13(c) and Figure 17.13(d) show the final membership functions. If θ is in degrees and $\dot{\theta}$ is in deg/s, the initial fuzzy if-then rules are

$$\begin{cases} \text{if } \theta \text{ is } A_1 \text{ and } \dot{\theta} \text{ is } B_1, \text{ then } force = 0; \\ \text{if } \theta \text{ is } A_1 \text{ and } \dot{\theta} \text{ is } B_2, \text{ then } force = 0; \\ \text{if } \theta \text{ is } A_2 \text{ and } \dot{\theta} \text{ is } B_1, \text{ then } force = 0; \\ \text{if } \theta \text{ is } A_2 \text{ and } \dot{\theta} \text{ is } B_2, \text{ then } force = 0; \end{cases} \tag{17.24}$$

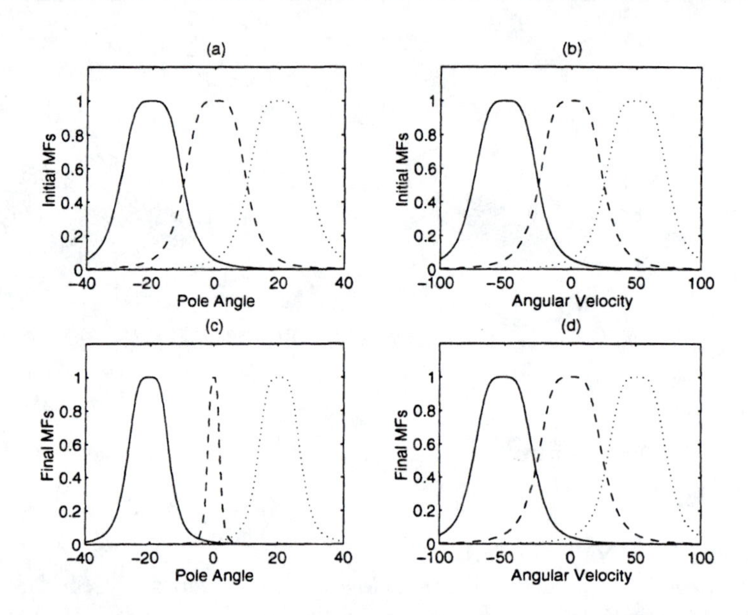

Figure 17.15. *(a)(b) Initial membership functions and (c)(d) final membership functions of a nine-rule fuzzy controller.*

where A_1, A_2, B_1, and B_2 are the linguistic labels characterized by the generalized bell MF parameters (20, 2, −20), (20, 2, 20), (50, 2, −50), and (50, 2, 50), respectively. Figure 17.14(a) is the initial control action surface.

The final fuzzy if-then rules derived from the reference settings are as follows:

$$\begin{cases} \text{if } \theta \text{ is } A_1 \text{ and } \dot{\theta} \text{ is } B_1, \text{ then } force = 0.0502\theta + 0.1646\dot{\theta} - 10.09; \\ \text{if } \theta \text{ is } A_1 \text{ and } \dot{\theta} \text{ is } B_2, \text{ then } force = 0.0083\theta + 0.0119\dot{\theta} - 1.09; \\ \text{if } \theta \text{ is } A_2 \text{ and } \dot{\theta} \text{ is } B_1, \text{ then } force = 0.0083\theta + 0.0119\dot{\theta} + 1.09; \\ \text{if } \theta \text{ is } A_2 \text{ and } \dot{\theta} \text{ is } B_2, \text{ then } force = 0.0502\theta + 0.1646\dot{\theta} + 10.09; \end{cases} \tag{17.25}$$

where A_1, A_2, B_1, and B_2 are the linguistic labels characterized by the generalized bell MF parameters (−1.59, 2.34, −19.49), (−1.59, 2.34, 19.49), (85.51, 1.94, −23.21), and (85.51, 1.94, 23.21), respectively. Figure 17.14(b) is the final control action surface.

Figure 17.13 indicates that the final membership functions for θ are quite different from the initial membership functions. Note that there appears to be no membership functions covering the interval $[-25, 25]$ of θ, making linguistic interpretation of the fuzzy rules difficult. (In fact, MF grades in the interval $[-25, 25]$ are never zero; they are just too small to be noticeable in the plot.) However, since we are utilizing ANFIS as a function approximator that can generate a required nonlinear mapping, linguistically desirable features (such as enough overlap between neighboring membership functions and total coverage of the whole input domain) do not have to be one of the fuzzy controller's attributes in this case. If we

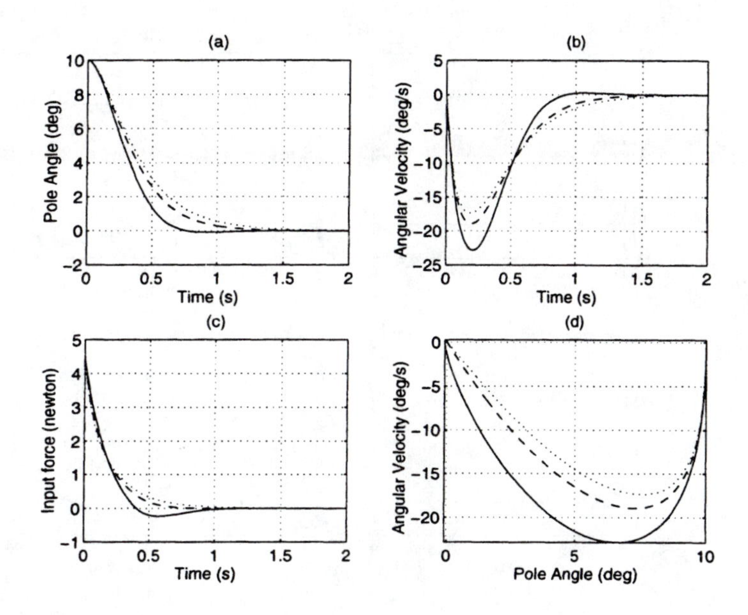

Figure 17.16. *(a) Pole angle; (b) pole angular velocity; (c) state space; (d) input force. [Solid, dashed, and dotted curves correspond to λ = 10, 40, and 100, respectively.]*

want to keep these desirable features, we can either impose some constraints on the premise parameters or simply increase the number of MFs to give the neuro-fuzzy controller more degrees of freedom. Figure 17.15 shows the membership functions of a nine-rule fuzzy controller that has about the same performance as the four-rule fuzzy controller. Due to its greater number of degrees of freedom, the premise parameters of the nine-rule fuzzy controller do not have to change a lot to minimize the error measure; therefore, the final membership functions clearly cover all the domain intervals with desirable overlapping.

Solid curves in Figure 17.16 demonstrate the state variable trajectories at the reference setting: (a), (b), and (d) show the pole angle (degrees), angular velocity (deg/s), and control actions (N) from $t = 0$ to $t = 2$ s; (c) is the state-space plot that shows how the trajectory approaches the origin from the initial point (10, 0). Dashed and dotted curves in Figure 17.16 correspond to λ equal to 40 and 100, respectively. From Figure 17.16(a), it can be seen that a smaller λ (solid curve) achieves the control goal faster since the controller can apply a larger force to balance the pole. For a large λ (dotted curve), the controller's output has to be kept small, thus slowing down the approach to the goal.

To demonstrate how the fuzzy controller can survive substantial changes in plant parameters, we used poles of different lengths to test the controller obtained from the reference setting. The results are shown in Figure 17.17, where solid, dashed, and

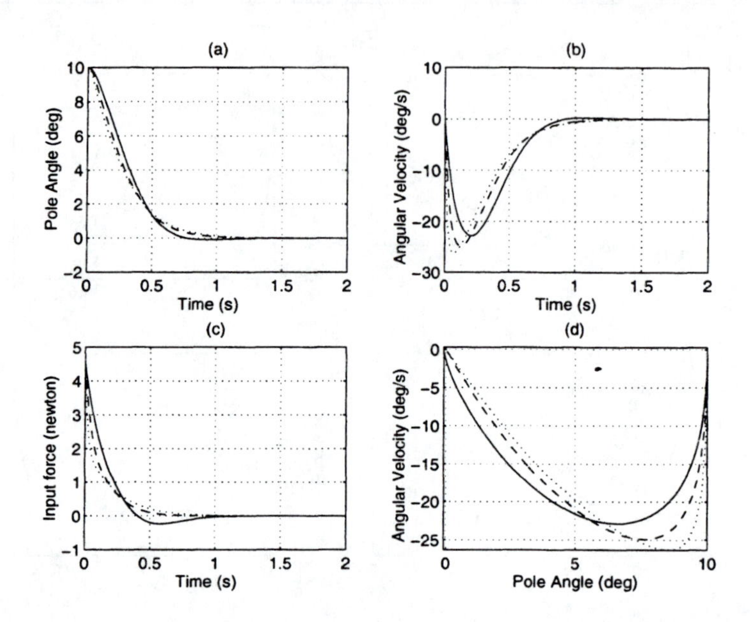

Figure 17.17. *(a) Pole angle; (b) pole angular velocity; (c) state space; (d) input force. [Solid, dashed, and dotted curves correspond to half pole lengths of 0.5, 0.25, and 0.125 m, respectively.]*

dotted curves correspond to half-lengths of the pole equal to 0.5 (reference setting), 0.25, and 0.125 m, respectively. The controller obtained from the reference setting can handle the shorter pole easily and gracefully.

In the learning phase, we supply only two training data corresponding to initial conditions (10, 0) and (−10, 0) of the pole. It would be interesting to know how the FC (obtained from the reference setting) deals with other initial conditions. So we monitor the pole behavior starting from other initial conditions which make the control goal even harder. Figure 17.18 shows the results; the solid solid, dashed, and dotted curves correspond to the initial conditions (10, 20), (15, 30), and (20, 40), respectively. Again, the same fuzzy controller can perform the control task starting from the unseen initial conditions not used for training. Figure 17.18 and Figure 17.17 reveal the robustness and fault tolerance of the fuzzy controller obtained via backpropagation through time.

17.7 SUMMARY

This chapter presents five design techniques for neuro-fuzzy controllers: expert control, inverse learning, specialized learning, backpropagation through time, and real-time recurrent learning. Most of these techniques are derived directly from neural control literature. Other design techniques that are not close coupled with

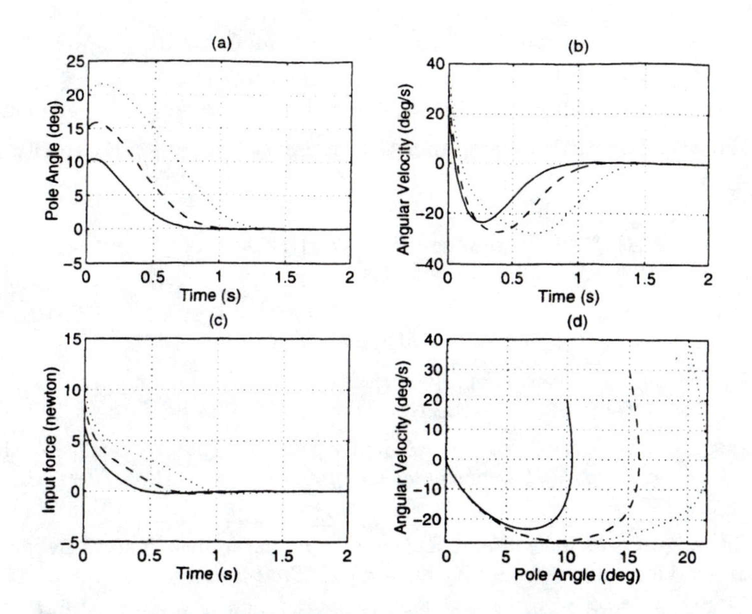

Figure 17.18. *(a) Pole angle; (b) pole angular velocity; (c) state space; (d) input force. [Solid, dashed, and dotted curves correspond to initial conditions* (10, 20), (15, 30), *and* (20, 40), *respectively.]*

neural networks are described in the next chapter.

EXERCISES

1. In Example 17.2, if the control signal in the state equation in Equation (17.11) is a vector of size m and matrix B is $n \times m$, how do you modify the controllability (and thus invertibility) condition?

2. Assume that the inputs to the state equation in Equation (17.7) is a vector of size m. How does the multi-input plant change the learning and application stages discussed in Section 17.4?

REFERENCES

[1] J. A. Bernard. Use of rule-based system for process control. *IEEE Control Systems Magazine*, 8(5):3–13, 1988.

[2] W. Brogan. *Modern control theory.* Prentice Hall, Upper Saddle River, NJ, 3rd edition, 1991.

[3] R. H. Cannon. *Dynamics of physical systems.* McGraw-Hill, New York, 1967.

[4] V. C. Chen and Y. H. Pao. Learning control with neural networks. In *Proceedings of International Conference on Robotics and Automation*, pages 1448–1453, 1989.

[5] S. Chiu, S. Chand, D. Moore, and A. Chaudhary. Fuzzy logic for control of roll and moment for a flexible wing aircraft. *IEEE Control Systems Magazine*, 11(4):42–48, 1991.

[6] W. T. Miller III, R. S. Sutton, and P. J. Werbos, editors. *Neural networks for control.* MIT Press, 1990.

[7] A. G. Ivakhnenko. Polynomial theory of complex systems. *IEEE Transactions on Systems, Man, and Cybernetics*, 1(4):364–378, October 1971.

[8] J.-S. Roger Jang. Self-learning fuzzy controller based on temporal back-propagation. *IEEE Transactions on Neural Networks*, 3(5):714–723, September 1992.

[9] Y. Kasai and Y. Morimoto. Electronically controlled continuously variable transmission. In *Proceedings of International Congress Transportation Electronics*, Dearborn, Michigan, 1988.

[10] M. Kawato, K. Furukawa, and R. Suzuki. A hierarchical neural network model for control and learning of voluntary movement. *Biological Cybernetics*, 57:169–185, 1987.

[11] M. S. Klassen and Y.-H. Pao. Characteristics of the functional-link net: a higher order delta rule net. In *IEEE Proceedings of the International Conference on Neural Networks*, San Diego, June 1988.

[12] B. Kosko. *Neural networks and fuzzy systems: a dynamical systems approach.* Prentice Hall, Upper Saddle River, NJ, 1991.

[13] H. Kwakernaak and R. Sivan. *Linear optimal control systems.* Wiley-Interscience, New York, 1972.

[14] C.-C. Lee. Fuzzy logic in control systems: fuzzy logic controller-part 1. *IEEE Transactions on Systems, Man, and Cybernetics*, 20(2):404–418, 1990.

[15] C.-C. Lee. Fuzzy logic in control systems: fuzzy logic controller-part 2. *IEEE Transactions on Systems, Man, and Cybernetics*, 20(2):419–435, 1990.

[16] Fujitec Company Limited. FLEX-8800 series elevator group control system, 1988. Osaka, Japan.

[17] E. H. Mamdani and S. Assilian. An experiment in linguistic synthesis with a fuzzy logic controller. *International Journal of Man-Machine Studies*, 7(1):1–13, 1975.

[18] J. Moody and C. Darken. Fast learning in networks of locally-tuned processing units. *Neural Computation*, 1:281–294, 1989.

[19] D. H. Nguyen and B. Widrow. Neural networks for self-learning control systems. *IEEE Control Systems Magazine*, pages 18–23, April 1990.

[20] Y.-H. Pao. *Adaptive pattern recognition and neural networks*, chapter 8, pages 197–222. Addison-Wesley, Reading, MA, 1989.

[21] D. A. Pomerleau. Efficient training of artificial neural networks for autonomous navigation. *Neural Computation*, 3:88–97, 1991. 1991.

[22] D. A. Pomerleau. *Neural network perception for mobile robot guidance.* PhD thesis, Department of Computer Science, Carnegie Mellon Univeristy, 1992.

[23] D. Psaltis, A. Sideris, and A. Yamamura. A multilayered neural network controller. *IEEE Control Systems Magazine*, 8(4):17–21, April 1988.

[24] D. E. Rumelhart, G. E. Hinton, and R. J. Williams. Learning internal representations by error propagation. In D. E. Rumelhart and James L. McClelland, editors, *Parallel distributed processing: explorations in the microstructure of cognition, volume 1*, chapter 8, pages 318–362. MIT Press, Cambridge, MA., 1986.

[25] M. Sugeno, editor. *Industrial applications of fuzzy control.* Elsevier Science, 1985.

[26] T. Takagi and M. Sugeno. Derivation of fuzzy control rules from human operator's control actions. *Proceedings of the IFAC Symposium on Fuzzy Information, Knowledge Representation and Decision Analysis*, pages 55–60, July 1983.

[27] K. P. Venugopal, R. Sudhakar, and A. S. Pandya. An improved scheme for direct adaptive control of dynamical systems using backpropagation neural networks. *Journal of Circuits, Systems and Signal Processing*, 1994. Forthcoming.

[28] P. J. Werbos. An overview of neural networks for control. *IEEE Control Systems Magazine*, 11(1):40–41, January 1991.

[29] B. Widrow and D. Stearns. *Adaptive signal processing.* Prentice Hall, Upper Saddle River, NJ, 1985.

[30] S. Yasunobu and G. Hasegawa. Evaluation of an automatic container crane operation system based on predictive fuzzy control. *Control Theory and Advanced Technology*, 2(2):419–432, 1986. 1986.

[31] S. Yasunobu and S. Miyamoto. Automatic train operation by predictive fuzzy control. In M. Sugeno, editor, *Industrial applications of fuzzy control*, pages 1–18. North-Holland, Amsterdam, 1985.

[32] L. A. Zadeh. Outline of a new approach to the analysis of complex systems and decision processes. *IEEE Transactions on Systems, Man, and Cybernetics*, 3(1):28–44, January 1973.

[33] L. A. Zadeh and C. A. Desoer. *Linear system theory: the state space approach.* McGraw-Hill, New York, N.Y., 1963.

Chapter 18

Neuro-Fuzzy Control II

18.1 INTRODUCTION

In the previous chapter, we introduced neuro-fuzzy control and some design approaches that use neuron-like learning directly; these include expert control, inverse learning, specialized learning, backpropagation through time, and real-time recurrent learning.

In this chapter, we shall introduce more design methods that do not rely totally on neural-like learning. Instead, some of them employ derivative-free optimization (see Chapter 7) or reinforcement learning techniques (see Chapter 10); others employ conventional control techniques (such as adaptive control, feedback linearization, sliding mode control, gain scheduling, and so on) to accomplish their tasks.

18.2 REINFORCEMENT LEARNING CONTROL

Reinforcement learning plays an important role in the adaptive control field. It surely helps, especially when no explicit teacher signal is available in the environment (or world) where an interacting agent must learn to perform an optimal control action. The world informs the agent of a *reinforcement signal* associated with the performed control action and of the resulting new state (see Figure 18.1).

This section provides a brief description of reinforcement learning control and neuro-fuzzy reinforcement control systems. (For more a thorough discussion on general aspects of reinforcement learning, refer to Chapter 10.)

18.2.1 Control Environment

A basic problem in feedback control is that of determining an appropriate control action at each time instant to optimize a long-term objective. For a control goal explicitly defined as an objective function, this is achievable by using any of the supervised learning methods. However, such an explicit objective function is not always available; occasionally the only information about the agent's performance is a scalar score (usually called *reinforcement*) indicating how good the current action is; or even just a binary signal indicating whether the action is *right* or *wrong*. The

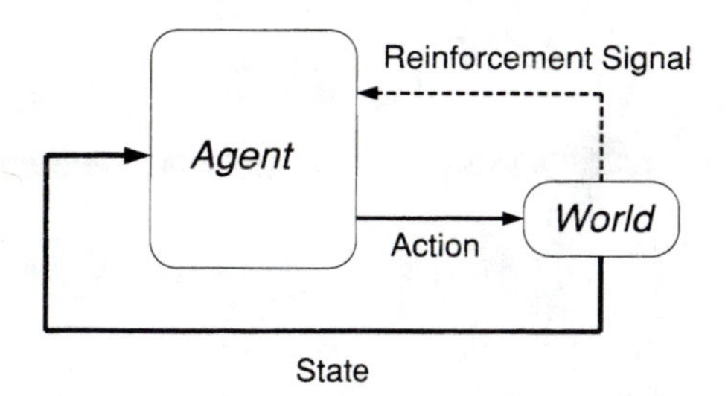

Figure 18.1. *An interactive learning agent.*

reinforcement signal is obviously a low-quality feedback signal. In other words, it is *evaluative* rather than *instructive*. (The learning agent receives either *reward* or *punishment* according to such evaluations.) Furthermore, the signal is often delivered infrequently and delayed—it is not available at each time instant; and when it is available at a certain moment, it represents the results of a series of control actions probably performed over a lengthy period of time. Note that a plant model is not necessarily required for this type of learning; the attempt of achieving a given control objective often results in a longer learning time.

In the reinforcement learning control literature, the pole-balancing control problem has been widely explored by many researchers [1, 2, 4, 16, 20, 23, 29]. (See also Section 10.5.1 in Chapter 10.) In this control problem, the only training signal available is the knowledge that the cart has reached a certain maximum displacement or that the pole has reached a maximum angle of deviation.

18.2.2 Neuro-Fuzzy Reinforcement Controllers

The basic idea behind *fuzzy* reinforcement learning is to apply a fuzzy partitioning scheme to the continuous state space and to introduce linguistic interpretation. Such averaging over neighboring partitioned subspaces can create generalization abilities; previously unknown states can thus be evaluated unlike the Boxes system [20] we discussed in Section 10.5.1 of Chapter 10. The Boxes system divided the entire state space into 162 non-overlapping digitized subspaces (boxes).

There are two representative neuro-fuzzy reinforcement learning models: GARIC (Generalized Approximate Reasoning for Intelligent Control), from Berenji and Khedkar [4, 3], and RNN-FLCS (Reinforcement Neural-Network-based Fuzzy Logic Control System) from Lin and Lee [16]. These models basically realize the AHC (adaptive heuristic critic) idea illustrated in Figure 18.2; the AHC architecture typically consists of an action (or control) module and of an critic (or evaluation) module. (For more details on AHC, refer to Chapter 10.)

GARIC has three components [4]: the action selection network (ASN), the action

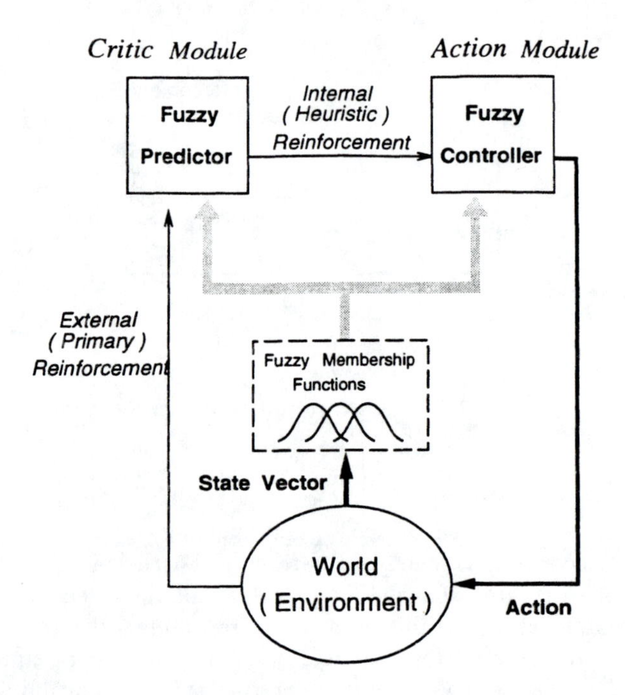

Figure 18.2. *A neuro-fuzzy AHC model.*

Table 18.1. *Comparison of the three AHC models: AHCON, GARIC, and RNN-FLCS. In this table, neuro means a multilayer perceptron.*

AHC models	Critic module	Action module
AHCON	Neuro	Neuro
GARIC	Neuro	Neuro-fuzzy
RNN-FLCS	Neuro-fuzzy	Neuro-fuzzy

evaluation network (AEN), and the stochastic action modifier (SAM). It has basically the same architecture as Lin's AHCON (AHC connectionist) model [17, 18], except that the ASN is expressed in a neuro-fuzzy framework. Lin and Lee's RNN-FLCS consists of fuzzy controller (action NN) and fuzzy predictor (value NN); they viewed the "stochastic action selection" function as part of the action NN. (To treat the stochastic action selection function as an individual component or not is inevitably a matter of individual judgment.) The structural comparison of these three models is presented in Table 18.1. Several variations have already been proposed; for instance, the critic module can be replaced by an RBFN [14] or a CMAC [15, 34, 35]. (Pedrycz discussed a fuzzy controller as a CMAC component [24].) By contrast, the whole RNN-FLCS is expressed in a neuro-fuzzy framework; both critic (fuzzy predictor) and action module (fuzzy controller) share the antecedent parts of the fuzzy rules. That is, both fuzzy functional modules share the same input fuzzy membership functions, just like CANFIS, discussed in Chapter 13. In GARIC, on the other hand, the ASN and the AEN do not share the antecedent parts because the AEN is expressed as a two-layer feedforward NN with sigmoidal functions except in the output layer. The ASN part of GARIC is a five-layer neuro-fuzzy model that has almost the same conceptual framework as our proposed ANFIS/CANFIS.

18.3 GRADIENT-FREE OPTIMIZATION

Most of the learning algorithms for automatic control applications introduced so far are derivative-based optimization techniques. However, the calculation of gradient vectors can be messy if the plant under consideration is complicated or the process is lengthy due to sluggish dynamics (see Figure 17.11). If this is the case, then derivative-free optimization schemes are preferable alternatives for optimization-based control designs. As stated in Chapter 7, four of the most popular optimization methods of this kind are genetic algorithms [7, 8], simulated annealing [11], the random optimization method [19], and the downhill Simplex method [21]. To apply any of these algorithms to control applications involves the following three steps:

1. Define a parameterized controller. It could be a linear controller, a neural network, a fuzzy controller, etc. (Since we are not going to rely on the gradient, the controller could even be non-differentiable with respect to its parameters.)
2. Define an objective function that relates to the control goal. Usually we minimize the objective function to achieve the control goal. In other words, the smaller the objective function, the better the control performance.
3. Find the objective function (usually by simulation) and update the controller's parameters by any of the derivative-free optimization methods; repeat until the objective function is below a given value or the computing time exceeds a specified upper bound.

In step 3, we need a plant model to find the objective function, and the plant model is assumed to be correct throughout the entire design process. So basically this is an off-line design method and it relies directly on the correctness of the plant model.

Note that the preceding steps also apply to modeling tasks, except that the objective function is defined as an error measure that describes the discrepancy between desired outputs and model's output.

Any of the four derivative-free methods can be used in step 3 of the preceding design procedure. However, for the rest of this section, we shall focus our discussion on genetic algorithms, for the following reasons:

- Exploitation of GAs for neural or fuzzy control has been around for a while and thus there is a more extensive body of literature than on the other three methods. Examples of using GAs for neural network controllers can be found in ref. [36]; for fuzzy logic controllers, see refs. [9, 10, 13].

- GAs are parallel in nature; this could be a decisive advantage when used with parallel machines.

- Above all, the application of GAs is not as straightforward as the other three methods. The definitions of the coding scheme and genetic operators (crossover and mutation) could be trickier than they appear. If these are not defined appropriately to match the nature of the problem to be solved, GAs behave somewhat like a parallel random search method.

Moreover, we shall further narrow our scope by discussing GAs for fuzzy control only, although similar approaches can be used for neural control as well. Specifically, we shall explain how to define coding schemes and genetic operators for fuzzy control, and how to embed *a priori* knowledge into GAs.

18.3.1 GAs: Coding and Genetic Operators

Successful use of GAs depends heavily on coding strategies for underlying applications; fuzzy control is no exception. The coding scheme for a fuzzy inference system (FIS) refers to the way of arranging the parameters of the FIS into a bit-string representation (or **chromosome**) such that the representation preserves certain good properties after recombination specified by genetic operators like crossover and mutation.

Figure 18.3 illustrates the hierarchical structure of GAs for fuzzy inference systems. The topmost level indicates that each generation contains a number of FISs as individuals in a population; the lowerest level demonstrates that each parameter is represented by a unit of eight bits, called a **gene**, in a long bit-string representation of a chromosome.

The coding scheme in Figure 18.3 is straightforward, but it is too simple to be of practical value because we did not pay any attention to the rule structure or the relationship between neighboring membership functions. This simple-minded coding scheme may result in problems such as the following:

- The crossover operator produces some random child FISs that do not inherit good properties from their parents.

- The crossover and/or mutation operators produces incomplete or ill-defined FISs. For instance, there could be a "hole" in the input space that is not covered by any membership functions.

- Structure-level adaptation that changes the number of rules and number of MFs are not explicitly accounted for.

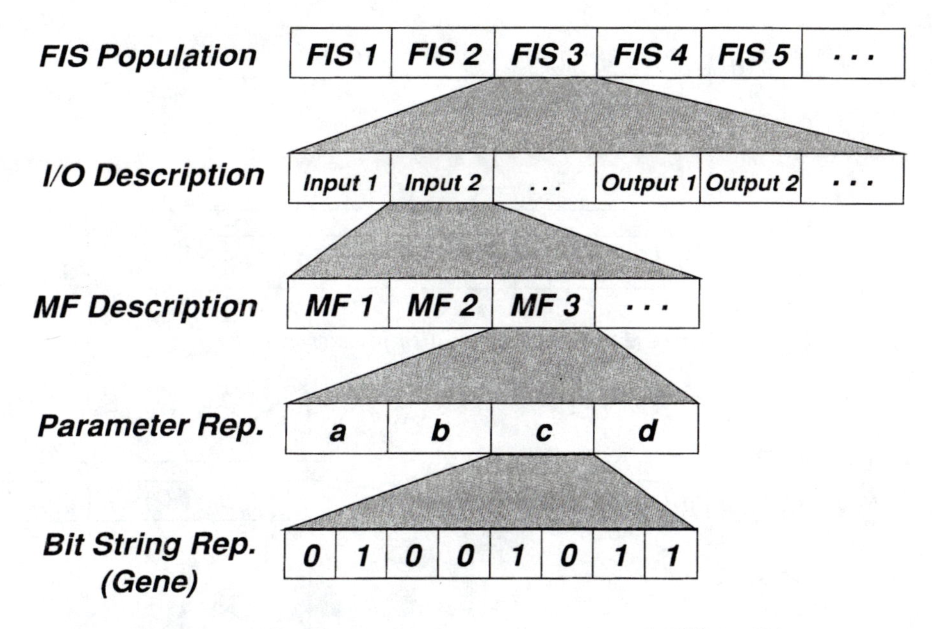

Figure 18.3. *Hierarchical representation of FIS in GA.*

- There could be difficulty accommodating *a priori* knowledge about the target system, such as symmetries, regularities, and homogeneities, that are not easily encoded in general. Genetic operators would then be likely to rupture these good properties in child FISs.

There are many coding schemes that can deal with these problems. Here we give two examples to show the general flavor of a good coding strategy.

Example 18.1 *Coding scheme for orthogonal MFs [22]*

Figure 18.4 demonstrates a coding scheme for membership functions (MFs), where the center position of each MF is represented by "1". Usually the first and the last bits of the string representation are "1" to ensure coverage of the boundaries. Figure 18.5 illustrates the effects of the crossover and mutation operators. The child strings after crossover, as shown in Figure 18.5(a), still preserve the condition of orthogonality (that is, the sum of all MF values for a specific input value is always equal to unity, see the definition in Chapter 2), thus eliminating the risk of accidentally introducing a "hole" in the input domain. The effects of mutation, as shown in Figure 18.5(b), are equivalent to adding an MF (when $0 \rightarrow 1$) or deleting one (when $1 \rightarrow 0$) while keeping orthogonality. Obviously, the genetic operators in this example change both shapes as well as numbers of MFs; this amounts to both parameter- and structure-level adaptation.

□

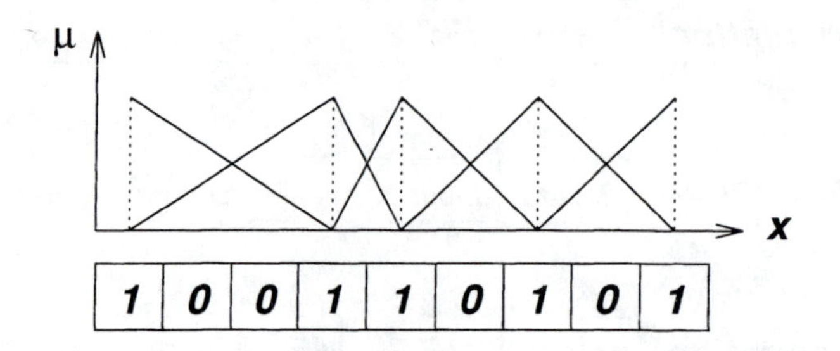

Figure 18.4. *Orthogonality-preserved coding.*

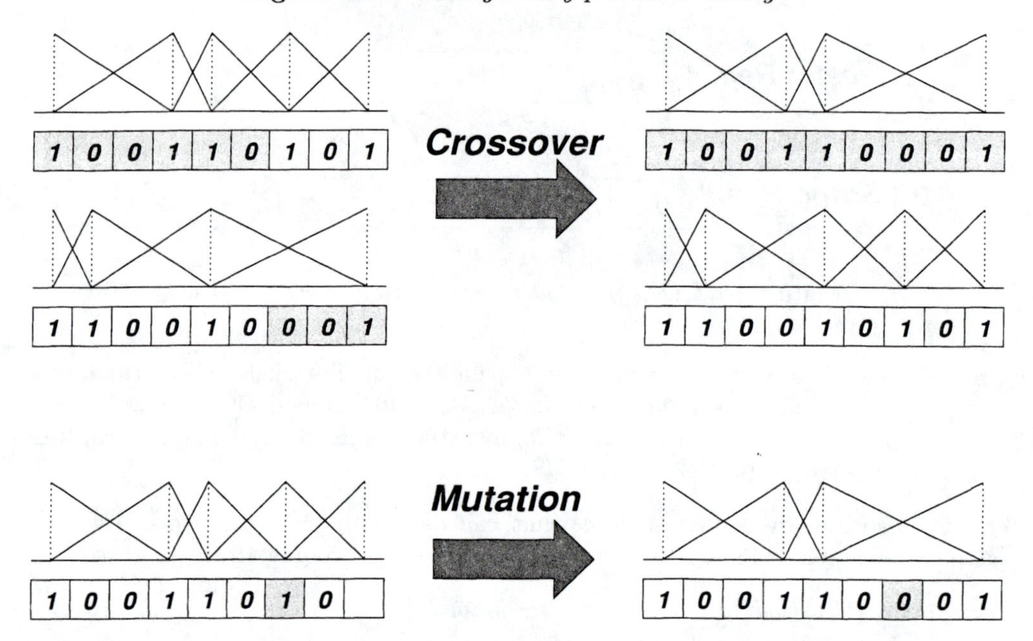

Figure 18.5. *Crossover and mutation for orthogonality-preserved coding.*

Example 18.2 *Genetic operators for input space partitioning*

Input space partitioning determines the premise part of a fuzzy rule set. For instance, the tree-style partitioning of Figure 18.6 divides the input space into six regions, each of which defines the premise part of a rule. This input space partitioning can be represented as a string $C_1C_2\cdots C_{n-1}$, where n is the number of rules and each C_i encodes the information for a straight-line cut, which includes the position of the cut and the dimension in which it is executed. This string becomes a common chromosome when each C_i is replaced by a bit string. Figure 18.7 illustrates one way of defining the crossover and mutation operators. The crossover operator essentially restricts the crossover points to those points separating C_i's.

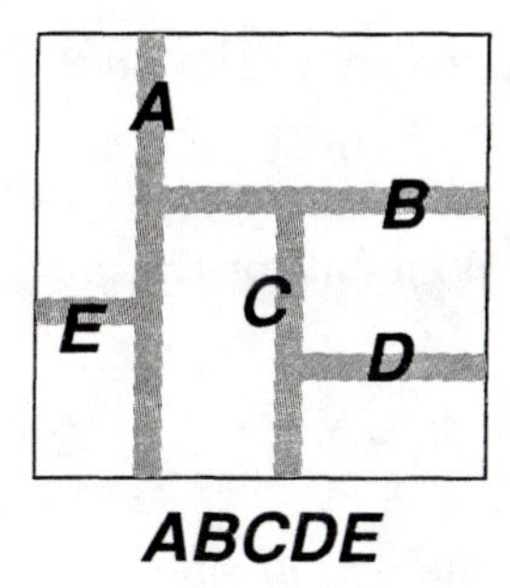

Figure 18.6. *String representation for input space partitioning.*

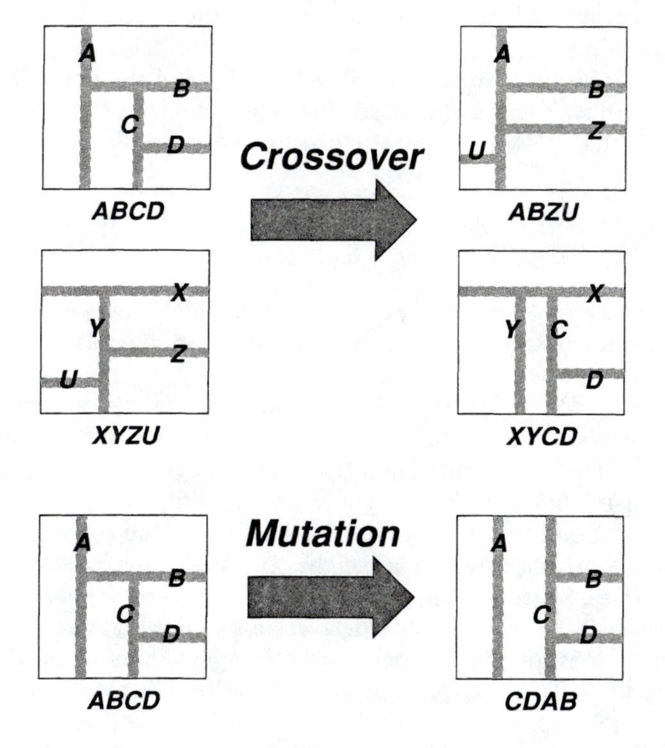

Figure 18.7. *Crossover and mutation for input space partitioning.*

The mutation operator shifts the string in a cyclic manner; the amount is a random number between 1 and $n-1$. Note that when there is more than one way to execute a cut, we always choose the longest way.

□

18.3.2 GAs: Formulating Objective Functions

To apply any optimization method to control design, we need to formulate an objective function directly related to the control goal to be achieved. In optimum control literature, a general format for the objective function is [5]

$$J = S(\mathbf{x}(N)) + \sum_{k=0}^{k=N-1} L(\mathbf{x}(k), \mathbf{u}(k)), \tag{18.1}$$

where $\mathbf{x}(k)$ and $\mathbf{u}(k)$ are the output and control actions, respectively, at time k; and N is the stop time for the process under consideration.

If we set $S = 0$ and $L = \mathbf{u}^T\mathbf{u}$, then $J = \sum \mathbf{u}^T\mathbf{u}$ is a measure of control effort (or energy). The minimization of J is called the least-effort problem. If we set $S = \|\mathbf{x}(N) - \mathbf{x}_d(N)\|^2$ and $L = 0$, then minimizing J is called the minimum terminal-error problem since J specifies the square of the norm of the error between the final state $\mathbf{x}(N)$ and a desired final state $\mathbf{x}_d(N)$. In particular, for a linear system with the following quadratic objective function

$$J = \mathbf{x}(N)\mathbf{M}\mathbf{x}(N) + \sum_{k=0}^{k=N-1} [\mathbf{x}^T(k)\mathbf{Q}\mathbf{x}(k) + \mathbf{u}^T(k)\mathbf{R}\mathbf{u}(k)], \tag{18.2}$$

where $\mathbf{M}$ and $\mathbf{Q}$ are symmetric positive semi-definite matrices and $\mathbf{R}$ is a symmetric positive-definite matrix, there exists a unique linear controller that solves this finite-time regulator problem analytically [5, 12].

When using derivative-free optimization methods, we can employ an even more complex objective function than Equation (18.1). This means that we can incorporate structure-level information into the objective function and let the derivative-free optimization methods to do the whole job: finding the best structure, as well as the optimal parameters; in neural networks, finding the correct number of neurons, as well as the proper connection weights; in fuzzy controllers, finding the correct number of rules, as well as the proper MF parameters. This seems too good to be true. However, keep in mind that derivative-free methods are slow and they could take a tremendous amount of time to obtain a less-than-optimal solution.

Let the original objective function in Equation (18.1) be denoted as J'. Then a new objective function that also takes the number of MFs into consideration could be

$$J = J' + \delta(\text{total number of MFs}), \tag{18.3}$$

where δ is a constant that specifies the importance of minimizing the number of MFs. Usually, more MFs implies more rules, so minimizing J also indirectly reduces the number of rules.

In the same vein, we could define a new objective function as

$$J = \frac{(\text{settling time}) + k_1 J'}{(\text{number of rules}) + k_2} \tag{18.4}$$

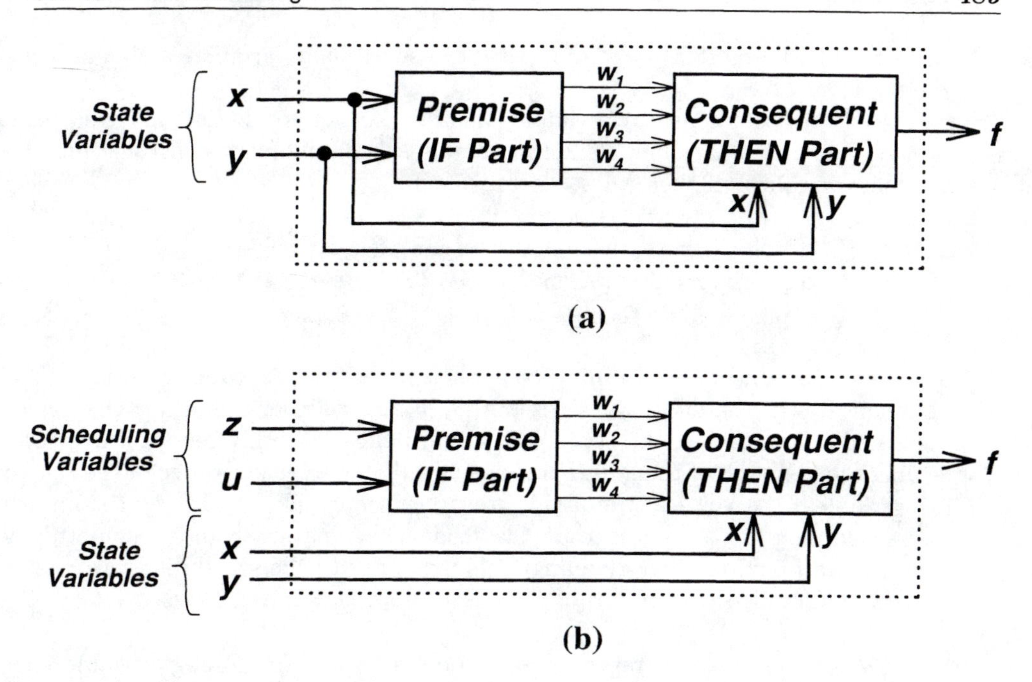

Figure 18.8. *Schematic diagram for (1) Sugeno fuzzy controller; (b) gain-scheduling fuzzy controller.*

in which k_1 and k_2 are constants. This objective function tries to reduce both the settling time and the number of rules. This was used in the fuzzy controller design in ref. [13].

18.4 GAIN SCHEDULING

18.4.1 Fundamentals

A regular first-order Sugeno fuzzy controller uses its inputs both in the premise part to determine firing strengths, and in the consequent part to determine each rule's output; this is shown in Figure 18.8(a), where both the premise and consequent parts receive the same inputs. For certain applications, it suffices to use only some of the inputs for the premise part and the others for the consequent part, as shown in Figure 18.8(b). Obviously, Figure 18.8(b) is a special case of 18.8(a), but is very useful in designing fuzzy controllers based on the concept of gain scheduling.

Specifically, the inputs to a **gain-scheduling fuzzy controller** contain two types of variables: scheduling and state. Scheduling variables are used in the premise part to determine what mode or characteristics the plant has. Once the plant mode or characteristics have been determined, the corresponding rule is fired with an output equal to the state variables multiplied by appropriate state feedback gains.

In terms of Figure 18.8(b), z and u are the scheduling variables and x and y are the state variables.

This is best explained by a simple example. For a hypothetical inverted pendulum system with a varying pole length, a gain-scheduling fuzzy controller may have the following fuzzy if-then rules:

$$\begin{cases} \text{If pole is short, then } f_1 = k_{11}\theta + k_{12}\dot{\theta} + k_{13}z + k_{14}\dot{z}. \\ \text{If pole is medium, then } f_2 = k_{21}\theta + k_{22}\dot{\theta} + k_{23}z + k_{24}\dot{z}. \\ \text{If pole is long, then } f_3 = k_{31}\theta + k_{32}\dot{\theta} + k_{33}z + k_{34}\dot{z}. \end{cases} \tag{18.5}$$

This is actually a gain-scheduling controller, where the scheduling variable is the pole length and the state variables are $[\theta,\ \dot{\theta},\ z,\ \dot{z}]$. Depending on the value of the scheduling variable, the control action switches smoothly among three sets of feedback gains, each of them designed specifically for a certain range of the scheduling variable. The key feature of this controller that makes it different from other gain-scheduling controllers is that the feedback gains are blended smoothly via membership functions that give linguistic meanings to the scheduling variable.

A more detailed description of the design approach is as follows:

1. Determine a set of representative points in the scheduling variable space. These points should be distributed more or less uniformly throughout the scheduling variable domain.

2. Construct MFs for the scheduling variables such that each representative point fires a rule at maximum strength and the number of rules is equal to the number of representative points.

3. Find the feedback gains at each representative point. This can be achieved by any conventional linear control technique, such as pole placement, quadratic optimal design, and gain/phase margin methods. The feedback gains thus found are used in the output equation of each rule corresponding to a representative point.

In the preceding approach, we assumed that the number of representative points was small, so we could construct fuzzy rules directly. This corresponds to the **interpolation problem**, in which the controller should use exactly the same specified feedback gains at each representative point and use the interpolated results between representative points. On the other hand, if we have enough computing power to deal with a large number of representation points, we can always use ANFIS to fit desired control actions to a gain-scheduling fuzzy controller; this corresponds to the **approximation problem**. Figure 18.9 shows the situation when we have two scheduling variables.

Examples of applying this method to both one-pole and two-pole inverted pendulum systems with varying pole lengths are explained in the following section.

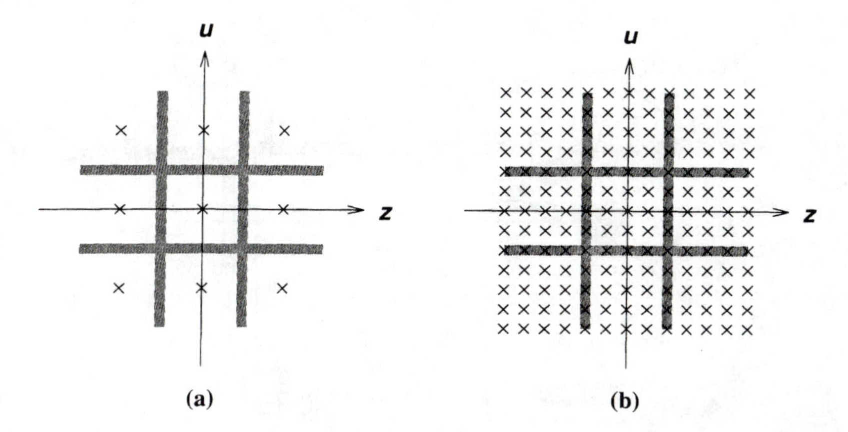

Figure 18.9. *Applying gain scheduling to the design of a fuzzy controller: (a) interpolation when the number of operating points is small; (b) approximation when the number of operating points is large .*

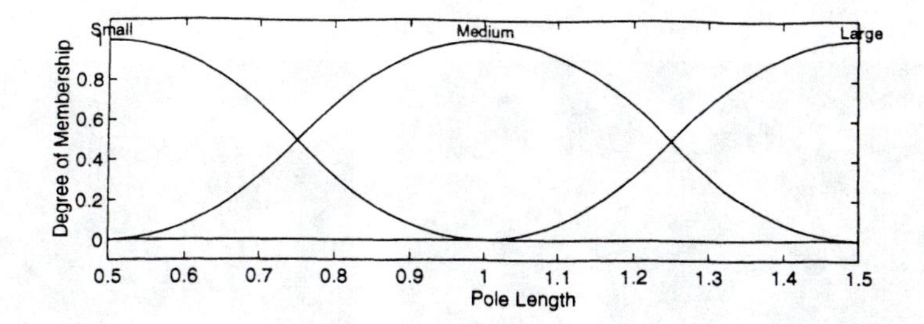

Figure 18.10. *MFs for scheduling variable in the CP system with a varying pole length.*

18.4.2 Case Studies

Cart and Pole System with a Varying Pole Length

Here the pole length follows a sinusoidal wave that changes between 0.5 and 1.5 m. We assume that the pole has a constant density, so the changes in pole length also imply changes in pole mass. We used three representative points (0.5, 1, and 1.5) of the scheduling variable (that is, pole length) to construct three fuzzy rules, as described in Equation (18.5). The membership functions for the pole length are Π functions shown in Figure 18.10. The feedback gains of each rule were obtained via the linear quadratic optimal design method (with the MATLAB command `lqr`), where the linearized model was derived (with the MATLAB command `linmod`) at the origin and the representative point for the state and the scheduling variables, respectively.

Figure 18.11 is the SIMULINK block diagram of this system. Once the simula-

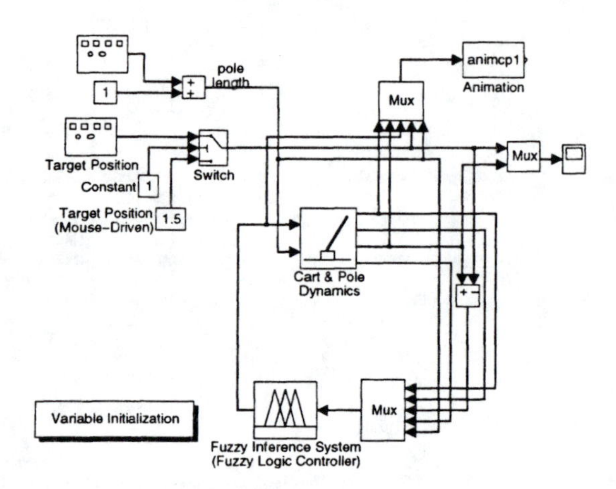

Figure 18.11. *Block diagram for the CP system with a varying pole length.*

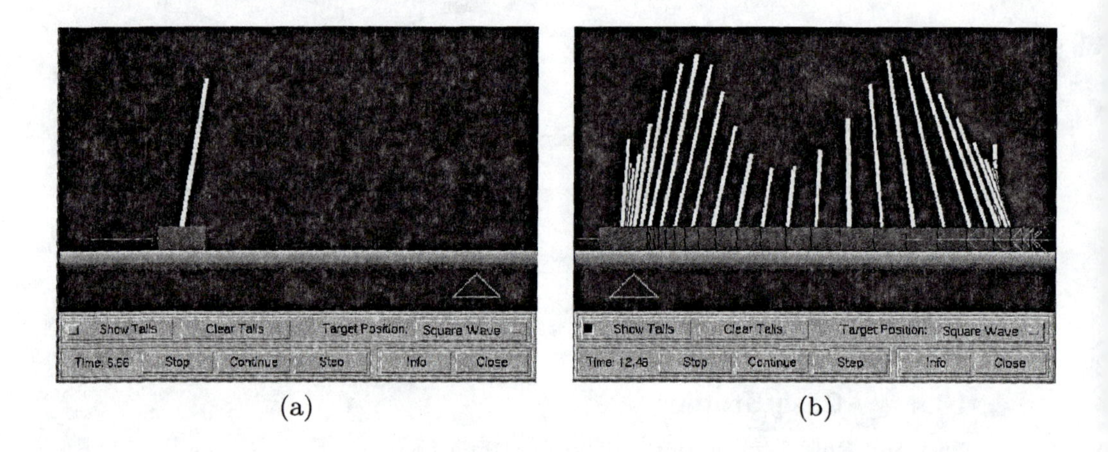

Figure 18.12. *Animation for CP system: (a) animated graphic representation; (b) time-lapse plot.*

tion starts, an animation window [Figure 18.12(a)] displays the motion of the cart and the pole; the triangle is the desired cart position and the arrow indicates the direction and the magnitude of the applied force. If we take a snapshot at each time step, it can clearly be seen that the pole length follows a sinusoidal wave, as shown in Figure 18.12(b).

We found that this system is extremely robust; the controller can balance the pole as well as move the cart to a desired position even when the pole length is

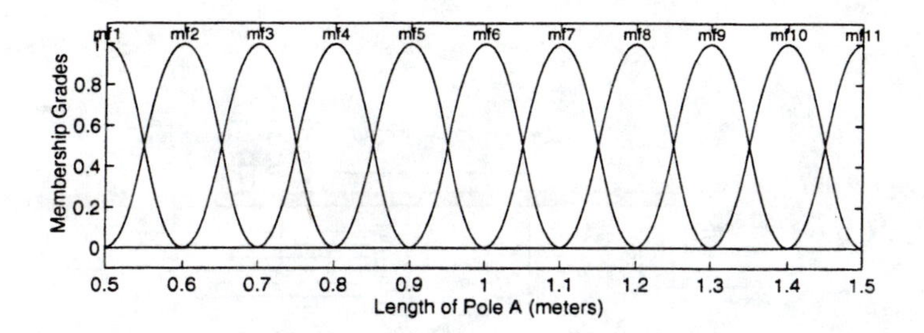

Figure 18.13. *MFs for scheduling variable in the CPP system with a varying pole length.*

changed randomly within the interval [0.5, 1.5] m. However, this is not the case for the next example.

Cart and Parallel Poles System with a Varying Pole Length

The CPP system has two poles (A and B) on the same cart; the control task is to balance both poles and at the same time move the cart to a desired position. Note that the length of pole B is fixed at 1 m, but that of pole A is time varying between 0.5 and 1.5 m. This makes the control task difficult since the system is not controllable when both poles have the same length and the same mass. (Consider the situation when A and B are exactly the same. Then due to symmetry, it is impossible to drive two poles to zero angles if the initial conditions are, say, -10 degrees for A and $+10$ degrees for B.) Moreover, it is obvious that the control strategy when pole A is longer than pole B is very different from the one when pole A is shorter than pole B.

Our design approach was almost the same as in the CP system discussed previously, except that we had to use more rules to deal with the extreme sensitivity of this system. Here we used 11 rules; the Π MFs for the scheduling variable (length of pole A) are shown in Figure 18.13. Again, we used `linmod` to find the linearized model at each representative point and `lqr` to obtain feedback gains for each rule.

Figure 18.14 is the SIMULINK block diagram of this system. Figure 18.15(a) displays the initial conditions of the system; 18.15(b) is a time-lapse plot that shows how the length of pole A changes with time.

Unlike the CP system described earlier, the CPP system is extremely sensitive and it could easily become unstable if the fuzzy rules are too few or the length of pole A is too close to that of pole B for a too long period of time.

18.5 FEEDBACK LINEARIZATION AND SLIDING CONTROL

The equations of motion of a class of dynamic systems in a continuous-time domain can be expressed in the canonical form:

$$x^{(n)}(t) = f(x(t), \dot{x}(t), \cdots x^{(n-1)}(t)) + bu(t), \tag{18.6}$$

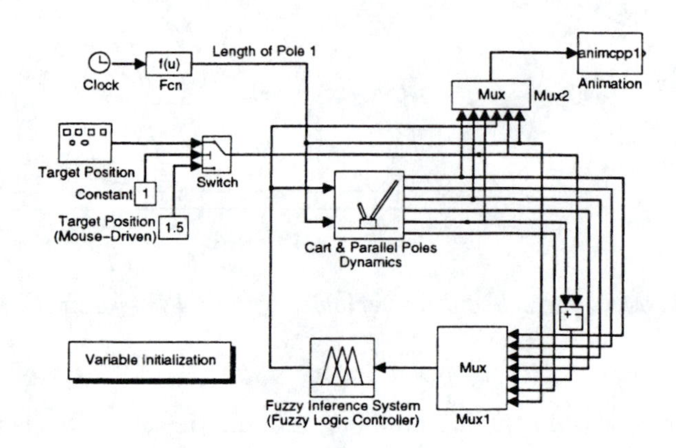

Figure 18.14. *Block diagram for the CPP system with a varying pole length.*

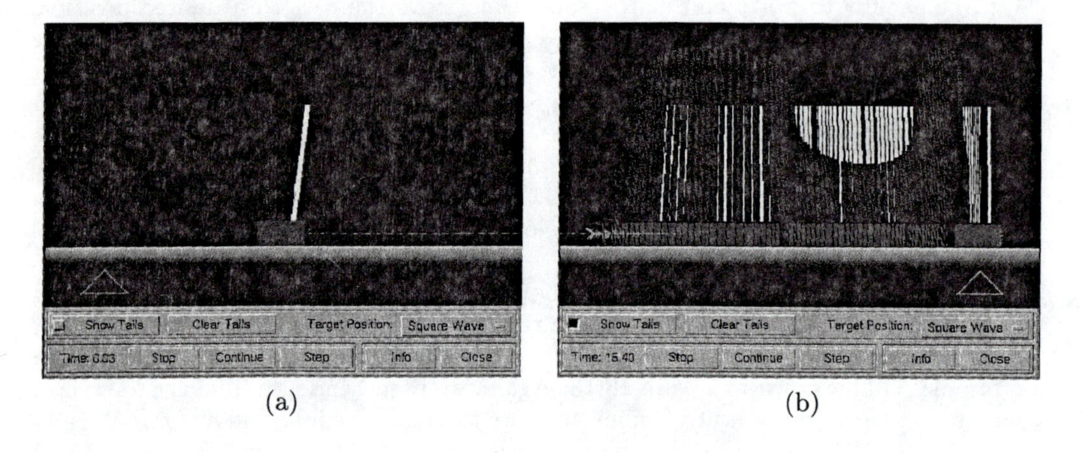

Figure 18.15. *Animation for CPP system: (a) animated graphic representation; (b) time-lapse plot.*

where f is an unknown continuous function, b is the control gain, and $u \in R$ and $y \in R$ are the system's input and output, respectively. The control objective is to force the state vector $\boldsymbol{x} = [x, \dot{x}, \ldots, x^{(n-1)}]^T$ to follow a specified desired trajectory $\boldsymbol{x_d} = [x_d, \dot{x}_d, \ldots, x_d^{(n-1)}]^T$. If we define the tracking error vector as $\mathbf{e} = \mathbf{x} - \mathbf{x}_d$, then the control objective is to design a control law $u(t)$ which ensures $\boldsymbol{e} \rightarrow 0$ as $t \rightarrow \infty$. (For simplicity, we assume $b = 1$ in the following discussion.)

Equation (18.6) is a typical **feedback linearizable** system since it can be reduced to a linear system if f is known exactly; specifically, the control law

$$u(t) = -f(\boldsymbol{x}(t)) + x_d^{(n)} + \boldsymbol{k}^T\boldsymbol{e} \tag{18.7}$$

would transform the original nonlinear dynamical equation into a linear one:

$$e^{(n)}(t) + k_1 e^{(n-1)} + \cdots + k_n e = 0, \tag{18.8}$$

where $\mathbf{k} = [k_n, \ldots, k_1]^T$ is an appropriately chosen vector that ensures satisfactory behavior of the closed-loop linear system in Equation (18.8).

Since f is unknown, an intuitive candidate for u would be

$$u = -F(\boldsymbol{x}, \boldsymbol{p}) + x_d^{(n)} + \boldsymbol{k}^T\boldsymbol{e} + v, \tag{18.9}$$

where v is an additional control input to be determined later, and $F(\cdot)$ is a parameterized function (such as an ANFIS, neural network, or any other type of adaptive network) that has enough degrees of freedom to approximate $f(\cdot)$. Using this control law, the closed-loop system becomes

$$e^{(n)} + k_1 e^{(n-1)} + \cdots + k_n e = (f - F) + v. \tag{18.10}$$

Now the problem is divided into two tasks:

- How to update the parameter vector $\boldsymbol{p}$ incrementally so that $F(\boldsymbol{x}, \boldsymbol{p}) \approx f(\boldsymbol{x})$ for all $\boldsymbol{x}$.
- How to apply v to guarantee global stability while F is approximating f during the whole process.

The first task is not too difficult as long as F is equipped with enough parameters to approximate f. For the second task, we need to apply the concept of a branch of nonlinear control theory called **sliding mode control** [27, 32]. The standard approach is to define an error metrics as

$$s(t) = (\frac{d}{dt} + \lambda)^{n-1} e(t), \text{ with } \lambda > 0. \tag{18.11}$$

The equation $s(t) = 0$ defines a time-varying hyperplane in R^n on which the tracking error vector $\boldsymbol{e}(t) = [e(t), \dot{e}(t), \ldots, e^{n-1}(t)]^T$ decays exponentially to zero, so that perfect tracking can be obtained asymptotically. Moreover, if we can maintain the following condition:

$$\frac{d|s(t)|}{dt} \leq -\eta, \tag{18.12}$$

then $|s(t)|$ will approach the hyperplane $|s(t)| = 0$ in a finite time less than or equal to $|s(0)|/\eta$. In other words, by maintaining the condition in Equation (18.12), $s(t)$ will approach the sliding surface $s(t) = 0$ in a finite time, and then the error vector $\boldsymbol{e}(t)$ will converge to the origin exponentially with the time constant $(n-1)/\lambda$.

From Equation (18.11), s can be rearranged as follows:

$$s = (\lambda + \frac{d}{dt})^{n-1} e = [\lambda^{n-1}, (n-1)\lambda^{n-2}, \ldots, 1]\mathbf{e}. \quad (18.13)$$

Differentiating the preceding equation and plugging in $e^{(n)}$ from Equation (18.10), we obtain

$$\begin{aligned} \frac{ds}{dt} &= e^{(n)} + [0, \lambda^{n-1}, (n-1)\lambda^{n-2}, \cdots, \lambda]\mathbf{e} \\ &= f - F + v - [k_n, k_{n-1}, \cdots, k_1]\mathbf{e} + [0, \lambda^{n-1}, (n-1)\lambda^{n-2}, \cdots, \lambda]\mathbf{e}. \end{aligned} \quad (18.14)$$

By setting $[k_n, k_{n-1}, \cdots, k_1] = [0, \lambda^{n-1}, (n-1)\lambda^{n-2}, \cdots, \lambda]$, we have

$$\frac{ds}{dt} = f - F + v,$$

and

$$\begin{aligned} \frac{d|s|}{dt} &= \frac{ds}{dt}\text{sgn}(s) \\ &= (f - F + v)\text{sgn}(s). \end{aligned}$$

That is, Equation (18.12) is satisfied if and only if

$$(f - F + v)\text{sgn}(s) \leq -\eta.$$

If we assume that the approximation error $|f - F|$ is bounded by a positive number A, then the preceding equation is always satisfied if

$$v = -(A + \eta)\text{sgn}(s).$$

In summary, if we choose the control law as

$$u(t) = -F(\mathbf{x}, \mathbf{p}) + x_d^{(n)} + [0, \lambda^{n-1}, (n-1)\lambda^{n-2} \ldots \lambda]\mathbf{e} - (A + \eta)\text{sgn}(s),$$

where $F(\mathbf{x}, \mathbf{p})$ is an adaptive network that approximates $f(\text{x})$ and A is the error bound, then the closed-loop system can achieve perfect tracking asymptotically with global stability.

This approach uses a number of nonlinear control design techniques and possesses rigorous proofs for global stability. However, its applicability is restricted to feedback linearizable systems. The reader is referred to ref. [27] for a more detailed treatment of this subject. Applications of this technique to neural and fuzzy control can be found in refs. [26] and [33], respectively.

18.6 SUMMARY

This chapter presents four design techniques for neuro-fuzzy controllers: reinforcement learning, derivative-free optimization (GA in particular), gain-scheduling approaches, and feedback linearization in conjunction with sliding control. Note that we are not attempting to present an exhaustive review here. Some other design and analysis approaches do exist, though they are less common than those described in this chapter. Some of the methods not described here include cell-to-cell mapping techniques [6, 28], the model-based design method [30], and self-organizing controllers [25, 31].

EXERCISES

1. The crossover operator in Example 18.1 preserves the total number of MFs. Explain why briefly.

2. Redefine the crossover operator in Example 18.1 such that the child FISs have the same numbers of MFs as their parents.

3. Redefine the mutation operator in Example 18.1 such that the number of rules is preserved.

4. In Example 18.2, we always choose the longest cut when there is more than one way to execute a cut. Redraw Figure 18.7 for a policy that chooses the shortest cut.

5. Devise a coding scheme for a multilayer perceptron and define its crossover and mutation operators.

REFERENCES

[1] C. W. Anderson. Strategy learning with multilayer connectionist representations. In P. Langely, editor, *Proceedings of the Fourth International Workshop on Machine Learning*, pages 103–114, San Mateo, CA, 1987. Morgan Kaufmann.

[2] A. G. Barto, R. S. Sutton, and C. W. Anderson. Neuronlike adaptive elements that can solve difficult learning control problems. *IEEE Transactions on Systems, Man, and Cybernetics*, 13(5):834–846, 1983.

[3] H. R. Berenji. Fuzzy systems that can learn. In J. M.Zurada, R. J. Marks, and C. J. Robinson, editors, *Computational Intelligence: Imitating Life*, pages 23–30. IEEE Press, 1994.

[4] H. R. Berenji and P. Khedkar. Learning and tuning fuzzy logic controllers through reinforcements. *IEEE Transactions on Neural Networks*, 3(5):724–740, 1992.

[5] W. Brogan. *Modern control theory*. Prentice Hall, Upper Saddle River, NJ, 3rd edition, 1991.

[6] Y.-Y. Chen and T.-C. Tsao. A description of the dynamic behavior of fuzzy systems. *IEEE Transactions on Systems, Man, and Cybernetics*, 19(4):745–755, July 1989.

[7] D. E. Goldberg. *Genetic algorithms in search, optimization, and machine learning.* Addison-Wesley, Reading, MA, 1989.

[8] J. H. Holland. *Adaptation in natural and artificial systems.* University of Michigan Press, Michigan, 1975.

[9] C. L. Karr. GAs for fuzzy controllers. *AI Expert*, 6(2):26–33, February 1991.

[10] C. L. Karr and E. J. Gentry. Fuzzy control of pH using genetic algorithms. *IEEE Transactions on Fuzzy Systems*, 1(1):46–53, February 1993.

[11] S. Kirkpatrick, C. D. Gelatt, and M. P. Vecchi. Optimization by simulated annealing. *Research Report 9335 IBM T. J. Watson Center*, 1983.

[12] H. Kwakernaak and R. Sivan. *Linear optimal control systems.* Wiley-Interscience, New York, 1972.

[13] M. A. Lee and H. Takagi. Integrating design stages of fuzzy systems using genetic algorithms. In *Proceedings of the second IEEE International Conference on Fuzzy Systems*, pages 612–617, San Francisco, 1993.

[14] C. S. Lin, Y. H. E. Cheng, and H. Kim. Radial basis function networks for adaptive critic learning. In *Proceedings of IEEE International Conference on Neural Networks*, pages 903–906, July 1994.

[15] C. S. Lin and H. Kim. CMAC-based adaptive critic self-learning control. *IEEE Transactions on Neural Networks*, 2(5):530–533, September 1991.

[16] C.-T. Lin and C.-S. G. Lee. Reinforcement structure/parameter learning for neural-network-based fuzzy logic control systems. *IEEE Transactions on Fuzzy Systems*, 2(1):46–63, 1994.

[17] Long-Ji Lin. Self-improving reactive agents: Case studies of reinforcement learning frameworks. In *Proceedings of the First International Conference on Simulation of Adaptive Behavior: From Animals to Animats*, pages 297–305, 1990.

[18] Long-Ji Lin. Self-improvement based on reinforcement learning, planning and teaching. In L. A. Birnbaum and G. C. Collins, editors, *Machine Learning: Proceedings of the Eighth International Workshop*, pages 323–327, San Mateo, CA., 1991. Morgan Kaufmann.

[19] W. S. Meisel. *Computer-oriented approaches to pattern recognition*, volume 83 of *Mathematics in science and engineering.* Academic Press, New York, 1972.

[20] D. Michie and R. A. Chambers. Boxes: An experiment in adaptive control. In J. T. Tou and R. H. Wilcox, editors, *Machine intelligence*, pages 137–152. Oliver and Boyd, Edinburgh, 1968.

[21] J. A. Nelder and R. Mead. A simplex method for function minimization. *Computer Journal*, 7:308–313, 1964.

[22] H. Nomura, I. Hayashi, and N. Wakami. A self-tuning method of fuzzy reasoning by genetic algorithm. In *Proceedings of International Fuzzy Systems and Intelligent Control Conference*, pages 236–245, Louisville, Kentuky, March 1992.

[23] M. O. Odetayo and D. R. McGregor. Genetic algorithm for inducing control rules for a dynamic system. In *Proceedings of the Third International Conference on Genetic Algorithms*, pages 177–182, 1989.

[24] W. Pedrycz. Developments of fuzzy controllers - fuzzy-neural-network approach. In *Fuzzy control and fuzzy systems*, chapter 9, pages 249–270. Research Studies Press Ltd, 1993. 2nd extended edition.

[25] T. J. Procyk and E. H. Mamdani. A linguistic self-organizing process controller. *Automatica*, 15:15–30, 1978.

[26] R. M. Sanner and J. J. E. Slotine. Gaussian networks for direct adaptive control. *IEEE Transactions on Neural Networks*, 3:837–862, 1992.

[27] J.-J. E. Slotine and W. Li. *Applied nonlinear control.* Prentice Hall, Upper Saddle River, NJ, 1991.

[28] S. M. Smith and D. J. Comer. Automated calibration of a fuzzy logic controller using a cell state space algorithm. *IEEE Control Systems Magazine*, 11(5):18–28, August 1991.

[29] B. Soucek. Neural networks in real-time applications. In *Neural and Concurrent Real-Time Systems: The Sixth Generation*, chapter 4. John Wiley & Sons, New York, 1989.

[30] K. Tanaka and M. Sugeno. Stability analysis and design of fuzzy control systems. *Fuzzy Sets and Systems*, 45:135–156, 1992.

[31] R. Tanscheit and E. M. Scharf. Experiments with the use of a rule-based self-organizing controller for robotics applications. *Fuzzy Sets and Systems*, 26:195–214, 1988.

[32] V. I. Utkin. Variable structure systems with sliding mode: a survey. *IEEE Transactions on Automatic Control*, 22:212, 1977.

[33] L.-X. Wang. Stable adaptive fuzzy control of nonlinear systems. *IEEE Transactions on Fuzzy Systems*, 1(1):146–155, 1993.

[34] D. A. White and D. A. Sofge. Neural network based control for composite manufacturing. In *Intelligent processing of materials.* ASME Publications, New York, November 1990.

[35] D. A. White and D. A. Sofge, editors. *Artificial Neural Networks in Manufacturing and Process Control.* Van Nostrand Reinhold, New York, 1992.

[36] Alexis P. Wieland. Evolving controls for unstable systems. In D. Touretzky, G. Hinton, and T. Sejnowski, editors, *Proceedings of the 1990 Connectionist Models Summer School*, pages 91–102, Carnegie Mellon University, 1990.

Part VII

Advanced Applications

Chapter 19

ANFIS Applications

J.-S. R. Jang

19.1 INTRODUCTION

This chapter describes several applications of ANFIS to a variety of domains. Some of these representative applications employ real-world data, some use synthetic data; all of them shed light on how to tackle different tasks with similar natures.

Applications covered in this chapter fall into several categories:

- Pattern recognition: printed character recognition
- Robotics: inverse kinematics
- Nonlinear regression: automobile miles per gallon (MPG) prediction
- Nonlinear system identification: furnace modeling
- Adaptive signal processing: channel equalization and noise cancellation.

ANFIS applications to automatic control are not described here; this is covered extensively in Chapters 17 and 18.

19.2 PRINTED CHARACTER RECOGNITION

In this section, we describe a straightforward design method for a fuzzy inference system to solve pattern recognition problems; this is based loosely on the concept of **nearest-neighbor classification** or **case-based reasoning**. Iterated training is not mandatory for this design method, but the method does require some representative, noise-free data points from the recognition system to be modeled. We revisit the Exclusive-OR (XOR) problem to demonstrate the concept behind this design method, and then we apply the method to printed character recognition problems.

As explained in Section 9.2.2, to solve a binary XOR problem, we need to classify a binary input vector to class 0 if the vector has an even number of 1s; otherwise, it is assigned to class 1. The desired behavior of the two-input XOR problem is described by the following truth table:

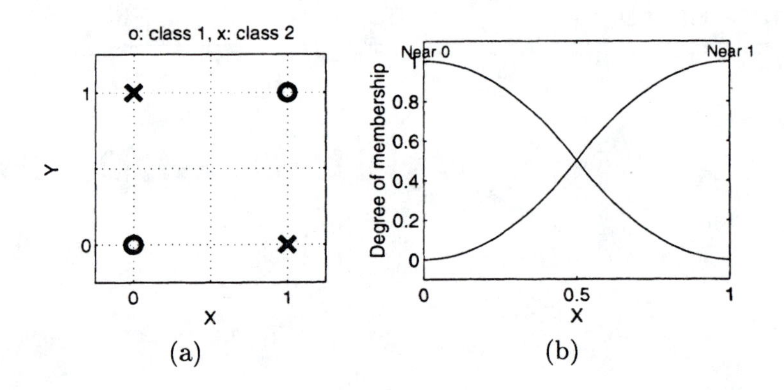

Figure 19.1. *(a) Training data for XOR problem* (MATLAB file: `xordata.m`); *(b) Z and S Membership functions for "near 0" and "near 1," respectively.* (MATLAB file: `xormf.m`)

	X	Y	Class
Desired i/o pair 1	0	0	0
Desired i/o pair 2	0	1	1
Desired i/o pair 3	1	0	1
Desired i/o pair 4	1	1	0

From the training data plot in Figure 19.1(a), it is obvious that the XOR problem is not linearly separable and cannot be solved by a single-layer perceptron. To use an MLP (multilayer perceptron) with a hidden layer to solve it, we need to train the network.

By noting that these training data are representative and noise free, we can use them as **prototypes** for the fuzzy logic design approach based on nearest-neighbor classification or case-based reasoning (see also the *interpolation RBFN* in Section 9.5.3). For a given set of prototypes, the underlying rationale for classifying a new data point is simple: Find the prototype nearest to the new data point and assign the point to that prototype class. To do this, we need a **similarity measure** that quantifies the meaning of *near*. This is done in terms of membership functions (MFs). In Figure 19.1(b), for instance, the meaning of "near 0" and "near 1" can be expressed as the Z and S MFs, respectively.

We still need to know the meaning of closeness between the input data $[x,\ y]$ and one of the prototypes, say, [0, 1]. If we take "$[x,\ y]$ is near [0, 1]" to mean that "x is near 0 AND y is near 1," then all we need to do is assign an appropriate operator to AND. The most popular fuzzy AND operators are "product" and "min". Figure 19.2 demonstrates the use of "product" and "min" in generating two-dimensional MFs for "$[x,\ y]$ is near [0, 1]". The use of "product" generates concentric contours, while the use of "min" generates square contours. See Figure 19.2

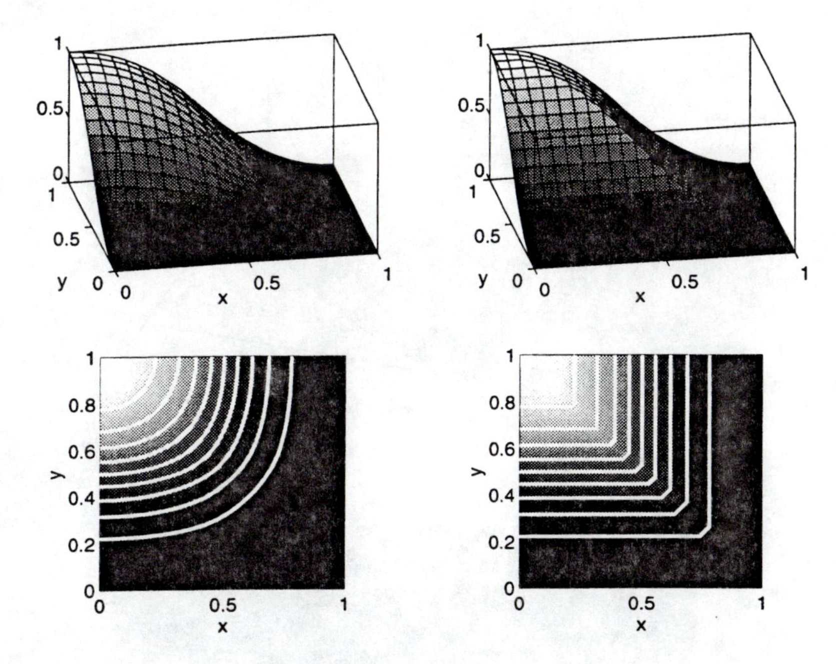

Figure 19.2. *Two-dimensional MFs for "[x, y] is near [0, 1]," with the use of "product" (left column) and "min" (right column) for the fuzzy AND operator.* (MATLAB file: `xor2dmf.m`)

for the composite two-dimensional MFs and their contours.

Creating a fuzzy rule set for solving the XOR problem is now obvious:

Rule 1: IF x is near 0 AND y is near 0 THEN output = 0.

Rule 2: IF x is near 0 AND y is near 1 THEN output = 1.

Rule 3: IF x is near 1 AND y is near 0 THEN output = 1.

Rule 4: IF x is near 1 AND y is near 1 THEN output = 0.

In other words, if input data $[x, y]$ is close to one of the prototypes, it is then assigned to that prototype class. We can display the input-output behavior of the constructed fuzzy inference system as a two-dimensional surface, as shown in Figure 19.3.

Now we can move on to a more challenging problem: printed character recognition (PCR), in which each of 26 letters is defined as a 7×5 pixel matrix, as shown in Figure 19.4. The challenge is to build a fuzzy inference system that can classify a given set of 35 ($= 7 \times 5$) pixels to one of the 26 alphabet characters. Again, these 26 prototypes are noise free, and we can employ the concept referred to previously in designing a fuzzy inference system.

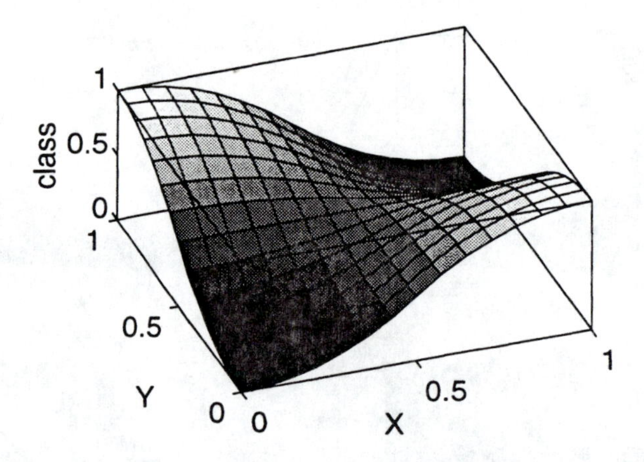

Figure 19.3. *XOR surface.* (MATLAB file: `xorsurf.m`)

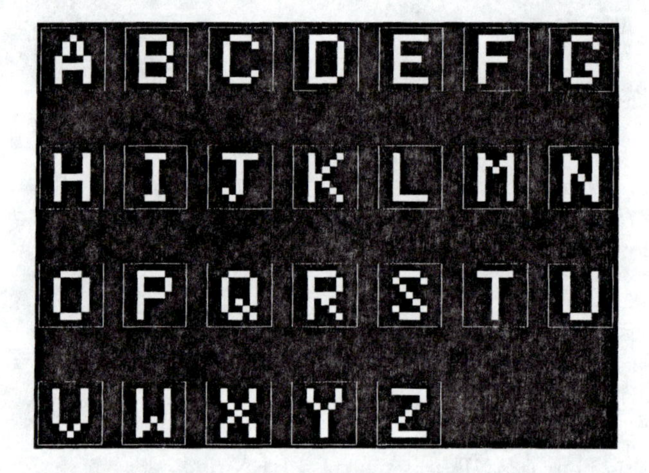

Figure 19.4. *Twenty six printed alphabet characters.* (MATLAB file: `pchar.m`)

- Construct MFs for each of the 35 inputs. Note that in the prototypes, each pixel is either 0 or 1, so we can set up MFs for "near 0" and "near 1" in the same way as we did in Figure 19.1(b).
- Set up rules. Each prototype represents a rule, so we have 26 rules, each of them an AND rule with 35 preconditions. Each rule's output is not critical, and we can set it to be an arbitrary constant (in a Sugeno fuzzy model) or MF (in an Mamdani fuzzy model).
- Use the fuzzy inference system. Note that the output is a categorical variable for which no numerical order is assumed, so it would be wrong to interpret the final numerical output of the fuzzy inference system directly. (The final

output is wrong anyway since we set the outputs of all rules to be an arbitrary constants or MFs.) Instead, distance measure information is embedded in each rule's firing strength—the larger the firing strength of a rule is, the closer a given input is to the prototype of that rule. Therefore, we obtain 26 firing strengths; the alphabet corresponding to the maximal firing strength is then selected as the predicted class.

To test the fuzzy inference system, we can assign various noise levels to the input pattern, as shown in Figure 19.5. The fuzzy PCR system thus obtained performs comparably to a similar system using an MLP (multilayer perceptron). Other factors that make fuzzy PCR a better choice are as follows:

- It does not required any training.
- It is a knowledge representation, and each rule in the system represents our insight into the problem we want to solve.

In this approach, we did not use any optimization schemes. We may invoke derivative-based (Chapter 6 or derivative-free (Chapter 7) optimization techniques if the described approach fails to classify noisy characters recognizable by humans correctly. Since the described method already gives us a roughly correct fuzzy inference system, the training time required to fine-tune membership functions is likely to be much shorter than that for an MLP starting with random weights.

Note that training a fuzzy inference system for pattern recognition is not exactly the same as the ANFIS training described in Chapter 12 . Taking the fuzzy inference system used here as an example, we are only interested in the firing strengths, not the final outputs after weighted average of defuzzification. Therefore, the error measure should be a function of the discrepancy between desired and actual firing strengths; there is no need to calculate the final output of the fuzzy inference system. This corresponds to tuning premise parameters only, and it is faster than the full-scale ANFIS training introduced in Chapter 12.

19.3 INVERSE KINEMATICS PROBLEMS

In this section, we use ANFIS to model the inverse kinematics of the two-joint planar robot arm shown in Figure 19.6. This problem involves learning to map from an endpoint Cartesian position (x, y) to joint angles (θ_1, θ_2), and it requires that the end effector ("hand") be able follow the reference signal without being given the joint angles. The forward kinematics equations from (θ_1, θ_2) to (x, y) are straightforward:

$$\begin{cases} x = l_1 \cos(\theta_1) + l_2 \cos(\theta_1 + \theta_2), \\ y = l_1 \sin(\theta_1) + l_2 \sin(\theta_1 + \theta_2), \end{cases}$$

where l_1 and l_2 are arm lengths; and θ_1 and θ_2 are their respective angles (see Figure 19.6). However, the inverse mappings from (x, y) to (θ_1, θ_2) are not as clear.

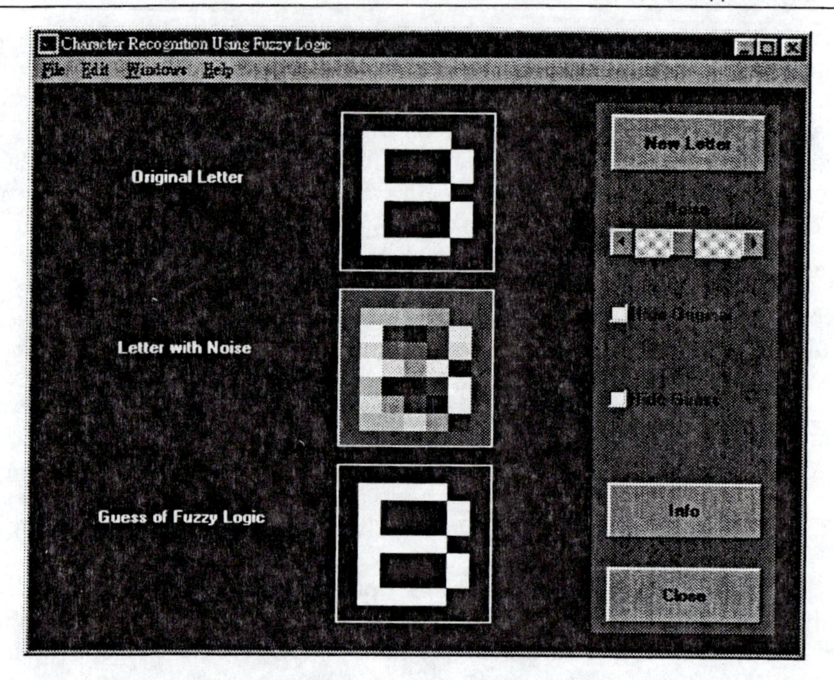

Figure 19.5. *Fuzzy PCR demo.* (MATLAB file: `fuzpcr.m`)

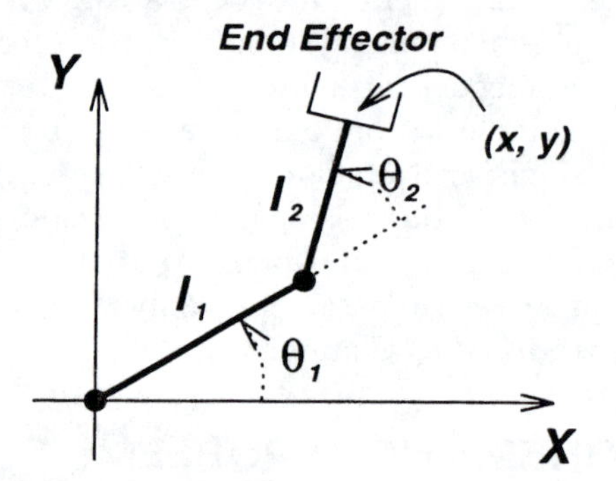

Figure 19.6. *Two-joint planar robot arm.*

In this case, it is possible to find the inverse mappings algebraically, but the solutions are not generally available for a multiple-joint robot arm in 3-D space. Instead of solving the equations directly, we use two ANFIS systems to learning these inverse mappings. Figure 19.7 demonstrates the forward mappings from (θ_1, θ_2) to (x, y) (the first row) and the inverse mappings from (x, y) to (θ_1, θ_2) (the second row). Here we assume that $l_1 = 10$, $l_2 = 7$, and the values of θ_2 are restricted to $[0, \pi]$. Note that when $\sqrt{x^2 + y^2}$ is greater than $l_1 + l_2$ or less than $|l_1 - l_2|$, there is no corresponding (θ_1, θ_2). This is called the **unreachable workspace**, and it can be seen clearly in the plots in the second row of Figure 19.7.

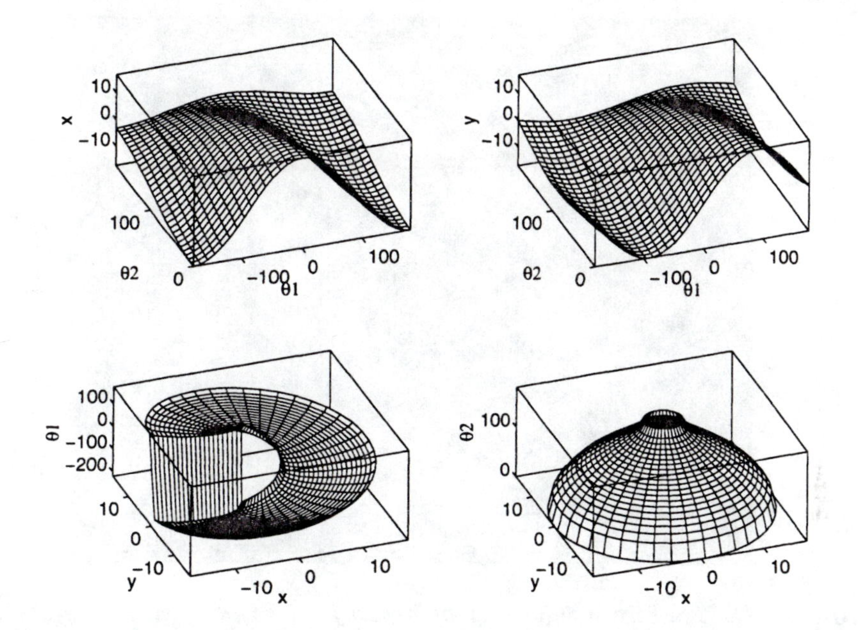

Figure 19.7. *Direct (the first row) and inverse (the second row) kinematics of a two-joint planar robot arm. (MATLAB file:* `invsurf.m`*)*

From the first quadrature, we collected 229 training data pairs of the form (x, y, θ_1) and (x, y, θ_2), respectively, to train two ANFIS systems. (This corresponds to the MANFIS architecture described in Section 13.2.) We used three MFs for each input; thus there were nine rules and 45 parameters for each ANFIS. We trained both ANFIS systems for 50 epochs. Figure 19.8 shows the test results for when an ellipse was chosen as the reference path. The dashed line shows how the end effector follows the path based on the inverse mappings learned by the two ANFIS systems; the crosses indicate the locations of the training data. Note that as long as the ellipse was inside the region covered by the training data, both ANFIS systems performed almost perfectly in following the desired trajectory. However, when some part of the ellipse was outside the region covered by the training data, the robot arm behaved unpredictably when the desired trajectory reached the "untrained" parts. This shows that ANFIS is very good at *interpolation* when data are abundant, but not so good at *extrapolation* when data are scarce. This phenomenon is common to all regression models, and it is more important to understand your data rather than select a fancy model.

The approach described here is similar to the inverse control introduced in Chapter 17; we can apply other on-line approaches to force the end effector to follow the trajectory more closely over time.

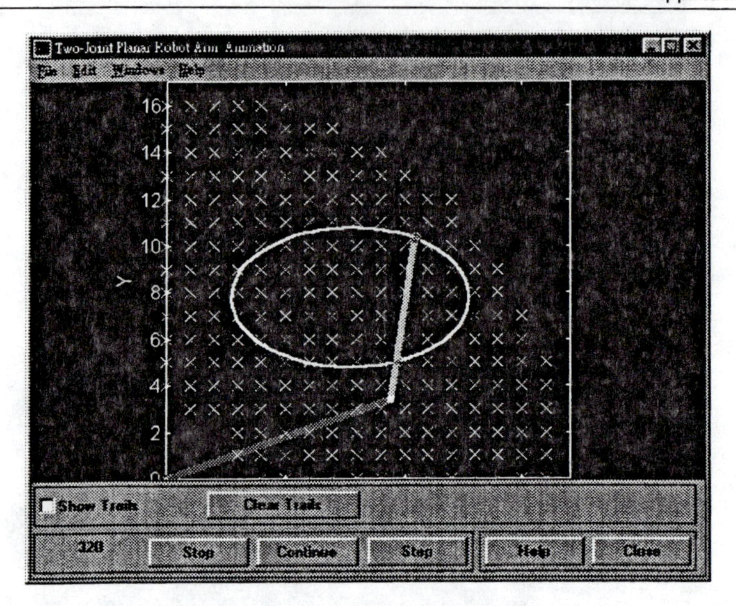

Figure 19.8. *Trajectory following of a two-joint robot arm, where the crosses indicate the training data locations, the solid line is the desired trajectory, and the dashed line is the actual trajectory. You can move the ellipse by clicking inside it and dragging it around. (*MATLAB*file:* `invkine.m`*)*

19.4 AUTOMOBILE MPG PREDICTION

This section describes the use of ANFIS for nonlinear regression. In particular, we address the issue of **input selection** for finding important input variables and reducing training data dimensions. We shall use automobile MPG (miles per gallon) prediction as a case study, in which an automobile's fuel consumption in terms of MPG is predicted by ANFIS based on several given characteristics, such as number of cylinders, weight, model years, and so on.

The automobile MPG prediction problem is a typical nonlinear regression problem, in which several attributes (input variables) are used to predict another continuous attribute (output variable). In this case, the six input attributes includes profile information about the automobiles:

No. of cylinders:	multi-valued discrete
Displacement:	continuous
Horsepower:	continuous
Weight:	continuous
Acceleration:	continuous
Model year:	multi-valued discrete

The attribute to be predicted in terms of the preceding six input attributes is the fuel consumption in MPG. Table 19.1 provides a list of seven instances selected at random from the data set.

Table 19.1. *Samples of the MPG training data set. (The last column is used for reference only and not for prediction.)*

Cyl.	Disp.	HP	Weight	Accel.	Year	MPG	Car name
8	307	130	3504	12	70	18	Chevrolet Chevelle Malibu
6	198	95	2833	15.5	70	22	Plymouth Duster
4	90	75	2108	15.5	74	24	Fiat 128
8	260	110	4060	19	77	17	Oldsmobile Cutlass Supreme
4	89	62	2050	17.3	81	37.7	Toyota Tercel
4	107	75	2205	14.5	82	36	Honda Accord
4	120	79	2625	18.6	82	28	Ford Ranger

The data set is available from the UCI (University of California at Irvine) Repository of Machine Learning Databases and Domain Theories[1]. More historical information about the data set can be found there. After removing instances with missing values, the data set was reduced to 392 entries. Our task was then to use this data set and ANFIS to construct a fuzzy inference system that could best predict the MPG of an automobile given its six profile attributes.

To apply ANFIS to MPG prediction, we needed to take care of two problems first: the scarcity of data points and the style of input space partition.

- **Data scarcity**: For a single-input data-fitting problem of medium complexity, we usually need 10 data points to come up with a good model. Similarly, for a two-input data-fitting problem, we need $10^2 = 100$ data points to get approximately the same performance. Therefore, for a six-input problem, such as the MPG prediction, ideally we should have $10^6 = 1{,}000{,}000$ data points. However, this is prohibitively large for any common modeling problem. Considering that we have only 392 data instances (which corresponds to $\sqrt[6]{392} = 2.5$ data points for single-input data fitting), the use of these data becomes an important issue. This data scarcity dilemma is ubiquitous in multivariate regression. A commonly used solution, already explained in Chapter 12, is to divide the data set into training and test data sets; the training set is used for model building, while the test set is used for model validation. Thus, the resultant model is not biased toward the training data set and it is likely to have a better generalization capacity to new data.

- **Input space partitioning**: Grid partitioning is the most frequently used input partitioning method. However, for a problem with six inputs, grid partitioning leads to at least $2^6 = 64$ rules, which results in $(6+1) \times 64 = 448$ linear parameters if we want to stick to the first-order Sugeno fuzzy model. This implies that we have too many fitting parameters, and the resultant

[1]FTP address: `ftp://ics.uci.edu/pub/machine-learning-databases/auto-mpg`

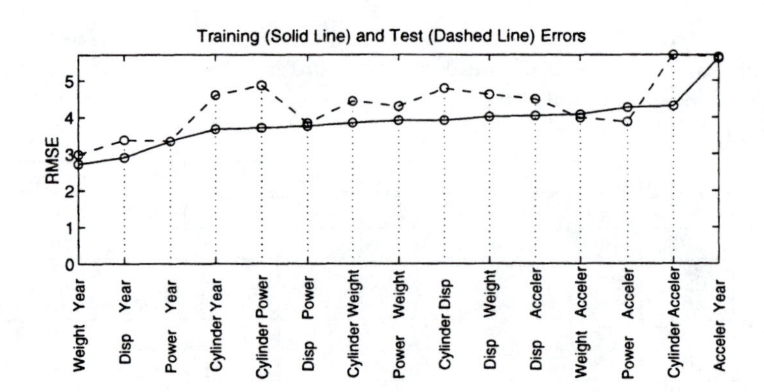

Figure 19.9. *Fifteen two-input fuzzy models for automobile MPG prediction.* (MATLAB file: `mpgpick2`)

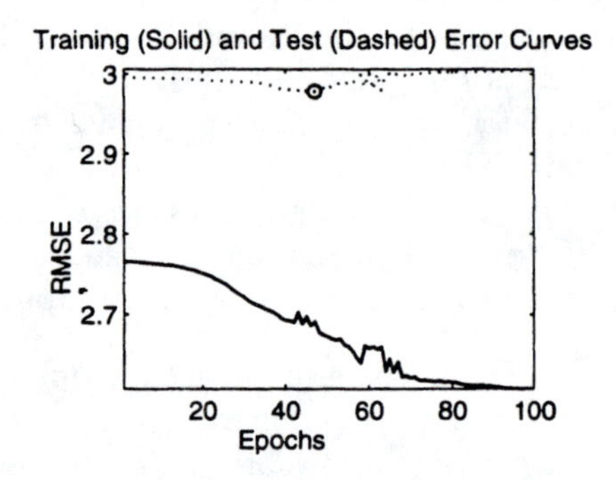

Figure 19.10. *Error curves obtained by training a fuzzy inference system to predict MPG.* (MATLAB file: `mpgtrain.m`)

model is not reliable for unforeseen inputs. To deal with this, we can either select certain inputs that have more prediction power instead of using all the inputs, or choose tree or scatter partitioning using the structure identification techniques described in Chapters 14 and 15, respectively. Here we consider only input dimension reduction.

Before training a fuzzy inference system, we divide the data set into training and test sets. The training set is used to train (or tune) a fuzzy model, while the test set is used to determine when training should be terminated to prevent overfitting. The 392 instances are randomly divided into training and test sets of equal size (196).

If we only want to select the two most relevant inputs as predictors, we can cycle through all the inputs and build $C_2^6 = 15$ fuzzy models trained by the `anfis` command in the Fuzzy Logic Toolbox. The `anfis` command utilizes iterative op-

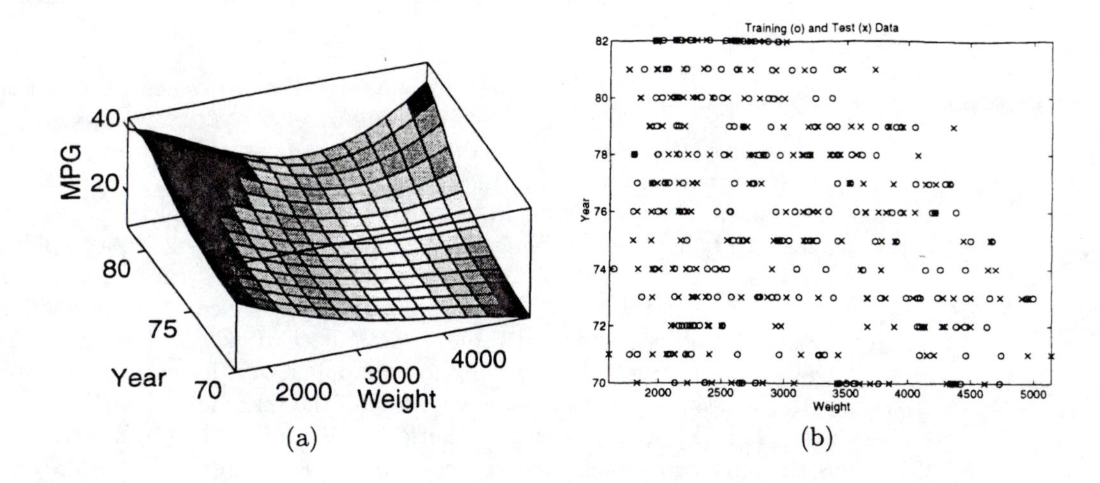

Figure 19.11. *Membership functions in chaotic time series prediction: (a) ANFIS surface for MPG prediction* (MATLAB file: `mpgtrain.m`); *(b) training and checking data distribution.* (MATLAB file: `mpgdata.m`)

timization techniques to fine-tune parameters and the training process could be lengthy. Fortunately, an efficient least-squares method is employed in the inner loop of `anfis`, and the performance after the first epoch is usually a good index of how well the fuzzy model will perform after further training. Based on this heuristic observation, we built 15 fuzzy models each with a single epoch of ANFIS training; the results are shown in Figure 19.9, with two curves representing training and test RMSE (root-mean-squared errors). We reordered these 15 models according to their training errors. Obviously, the best model takes "weight" and "model year" as the input variables, which is reasonable. In this case, both error curves are more or less consistent; this implies that the training and test data were evenly distributed across the original data set. In particular, we will end up with the same model if we pick the one with the smallest test error. Note that Figure 19.9 is based on only one epoch of training; more reliable results can be obtained if more training epochs are allotted to each of the 15 models.

Once we have selected the model with "weight" and "model year" as inputs, we can refine its performance via extended training by the `anfis` command. Figure 19.10 shows the error curves for 100 epochs of training. The training error decreases all the way, but the test error, after decreasing initially, reaches a plateau, oscillates a bit, and then increases. Usually we use the test error as a true measure of the model's performance; therefore, the best model we can achieve occurs when the test error is minimal. This corresponds to the circle in Figure 19.10; although further training beyond this point decreases the training error, it will degrade the performance of the fuzzy inference system on unforeseen inputs.

As a comparison, we now look at the result of linear regression, where the model is expressed as

$$\text{MPG} = a_0 + a_1 * \text{cyl} + a_2 * \text{disp} + a_3 * \text{hp} + a_4 * \text{weight} + a_5 * \text{accel} + a_6 * \text{year},$$

with a_0, a_1, $\cdots$, a_6 being seven modifiable linear parameters. The optimum values of these linear parameters were obtained directly by the least-squares method described in Chapter 5; the training and test errors are 3.45 and 3.44, respectively. In contract, after 100 epochs of training, the minimal test error is 2.98, at which the training error is 2.61. It is worth noting that the linear model takes all six inputs into consideration, but the error measures are still high since MPG prediction is nonlinear. On the other hand, our input selection technique of choosing the two most relevant inputs can result in a nonlinear mapping with lower error measures.

Figure 19.11(a) is a three-dimensional surface of the fuzzy model with the smallest test error. This is a smooth nonlinear surface, but it raises a legitimate question: Why does the surface increase toward the right upper corner? This is an apparently spurious result that states that heavy old cars have higher MPG ratings. The anomaly can be explained by the scatter plot of the data distribution in Figure 19.11(b), in which it is obvious that the lack of data (due to the tendency of automobile manufacturers to begin building small compact cars instead of big heavy ones during mid-1970s) is responsible. In other words, our trained fuzzy inference system is good at interpolation, but not at extrapolation, as explained in the previous section. Therefore, it is advisable for us to understand the data and qualify the scope of their validity before interpreting ANFIS output.

19.5 NONLINEAR SYSTEM IDENTIFICATION

This section applies ANFIS to nonlinear system identification, using the well-known Box and Jenkins gas furnace data [1] as the training data set. This is a time-series data set for a gas furnace process with gas flow rate $u(t)$ as the furnace input and CO_2 concentration $y(t)$ as the furnace output. We want to extract a dynamic process model to predict $y(t)$ using 10 candidate inputs to ANFIS: $y(t-1)$, $y(t-2)$, $y(t-3)$, $y(t-4)$, $u(t-1)$, $u(t-2)$, $u(t-3)$, $u(t-4)$, $u(t-5)$, and $u(t-6)$. The original data set contains 296 $[u(t), y(t)]$ data pairs; converting the data so that each training data point consists of $[y(t-1), \cdots, y(t-4), u(t-1), \cdots, u(t-6); y(t)]$ (the last one is the desired output) reduces the number of effective data points to 290. We use the first 145 data points as the training set, and the remaining 145 as the test set.

Since we have 10 candidate input variables for ANFIS, it is reasonable to do input selection first to rate variable priorities and reduce the input dimension. For dynamic system modeling, the inputs selected for ANFIS must contain elements from both the set of historical furnace outputs $\{y(t-1), y(t-2), y(t-3), y(t-4)\}$ and the set of historical furnace inputs $\{u(t-1), u(t-2), u(t-3), u(t-4), u(t-5), u(t-6)\}$. For simplicity, we assume that there are two inputs for ANFIS: one is

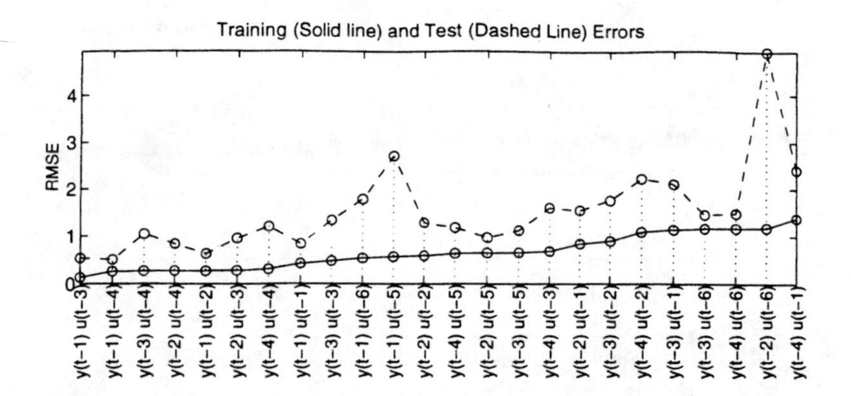

Figure 19.12. *Input selection for Box-Jenkins data.* . (MATLAB file: `bjpick2.m`)

from the historical furnace outputs, the other from the historical furnace inputs. In other words, we have to build 24 (= 4 × 6) ANFIS models with various input combinations, and then choose the one with the smallest training error for further parameter-level fine-tuning. We could have chosen the ANFIS with the smallest test error, but this would have led to *indirect training on test data*. Figure 19.12 shows the performance of these 24 ANFIS models; they are listed according to their training errors. Note that each ANFIS has four rules, and the training took only one epoch each to identify linear parameters. If computing power is not a problem, we could then assign more training epochs to each ANFIS.

In Figure 19.12, we can see that the ANFIS with $y(t-1)$ and $u(t-3)$ as inputs has the smallest training error, so it is reasonable to choose this ANFIS for further parameter tuning. Figure 19.13 shows the result of training this ANFIS for 100 epochs. In particular, Figure 19.13(a) display the training and test error curves; the optimal ANFIS parameters were obtained at the time when the test error reached the minimum indicated by a small circle. Figure 19.13(b) shows the data distribution; it demonstrates that the training and test data do not cover the same region. Better performance can be expected if they cover roughly the same region; this can be achieved by using other schemes to divide the original data set. (For instance, the training and test sets can be interleaved in the original data set.) Figure 19.13(c) displays the desired curve and ANFIS prediction; the performance for time index from 1 to 145 is better since this is the domain from which the training data were extracted. Figure 19.13(d) is the ANFIS surface; it is cut off at the maximum and minimum of the desired output.

Our input selection criterion for nonlinear system identification is straightforward and provides satisfactory results. Other more advanced criteria and model validation techniques can generate more accurate results; see ref. [4] for details.

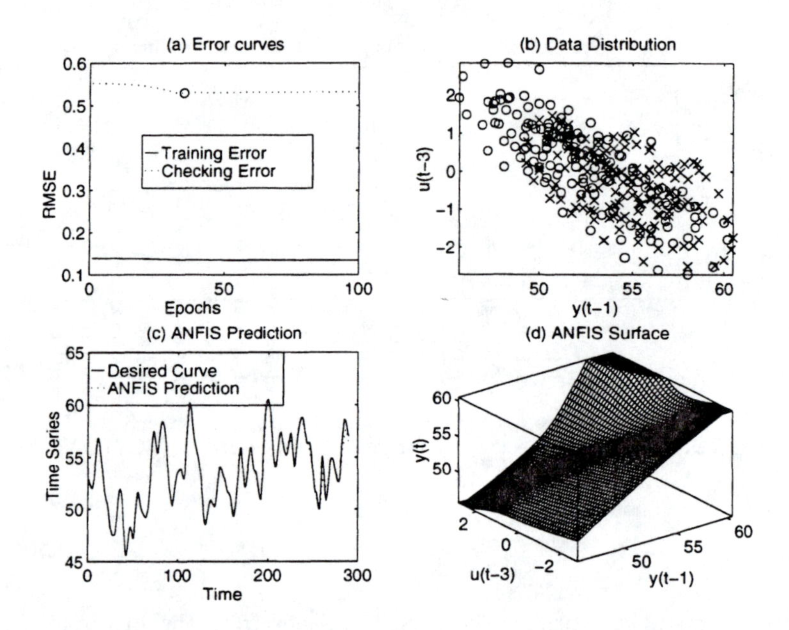

Figure 19.13. *ANFIS for Box-Jenkins data: (a) training and checking error curves; (b) training and checking data distribution; (c) desired system response and ANFIS prediction; (d) ANFIS surface.* . (MATLAB file: `bjtrain.m`)

19.6 CHANNEL EQUALIZATION

This section introduces the channel equalization problem, which arises frequently in dispersive digital communication channels, and proposes a way of using ANFIS to tackle this specific classification problem in signal processing. Since ANFIS employs an efficient least-squares method, little training time is required required to do the task.

The concept of a digital communication system is simple: A sequence of binary signals $s(t)$ is transmitted from one place to another via a communication channel. Ideally, the signal $x(t)$ at the receiving end should be exactly the same as $s(t)$, with a slight time delay. However, complications emerge in the real world because

- communication channels are never perfect; cross coupling, interference, and attenuation tend to disperse and weaken signals during transmission;
- noise is everywhere and is easily added to the transmitted signals.

Figure 19.14 is a schematic diagram of a digital communications system in which a random binary sequence $s(t)$ is transmitted through a linear, dispersive channel denoted by $H(z)$, and then corrupted by additive noise $e(t)$. The term $s(t)$ is usually assumed to be an independent sequence with an equal probability of being -1 or

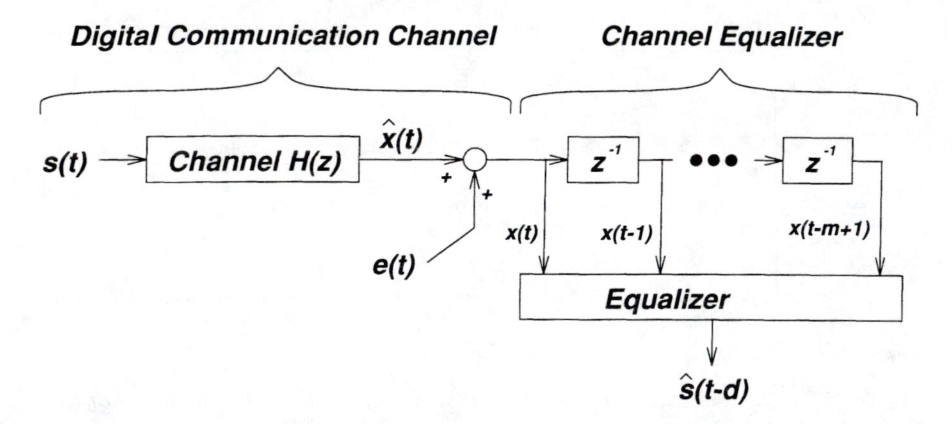

Figure 19.14. *Data transmission system and channel equalizer.*

1. The task of the channel equalizer in Figure 19.14 is to estimate the input signals using the information contained in the observations $\mathbf{x}(t) = [x(t), \ldots, x(t-m+1)]$, where m is known as the **order** of the equalizer. Often a **delay** d is introduced into the equalizer, so that at time t, the equalizer estimates the input signal at $t-d$. When $m = 2$ and $d = 1$, the task of the channel equalizer becomes one of estimating the input signal $s(t)$ by using the observed signals $\mathbf{x}(t) = [x(t), x(t-1)]$. Without loss of generality, we assume that $m = 2$ and $d = 0$ throughout the following discussion.

When there is no noise, let us define

$$\begin{aligned} P_+ &= \{\mathbf{x} \in R^2 | s(t) = 1\}, \\ P_- &= \{\mathbf{x} \in R^2 | s(t) = -1\}. \end{aligned}$$

P_+ and P_- represent two sets of possible channel output vectors that can be produced from sequences of channel inputs containing $s(t) = 1$ and -1, respectively. It is easy to show that both P_+ and P_- are finite in size, since we assume that there is no noise. The task of the equalizer is to decide whether an observation $\mathbf{x}(t)$ represents a noise-corrupted version of an element in either P_+ or P_-, and thus to determine the input signal $s(t)$.

One structure often used for this purpose is the **linear transversal equalizer**, which estimates the input signal $s(t)$ by

$$\hat{s}(t) = \mathrm{sgn}(b + k_0 x(t) + k_1 x(t-1)),$$

where $\mathrm{sgn}(\cdot)$ is the signum function defined by

$$\mathrm{sgn}(y) = \begin{cases} 1, & \text{if } y \geq 0, \\ -1, & \text{if } y < 0. \end{cases}$$

A linear transversal equalizer can estimate the input signal $s(t)$ correctly, if and only if sets P_+ and P_- are linearly separable—that is, we can use a single straight

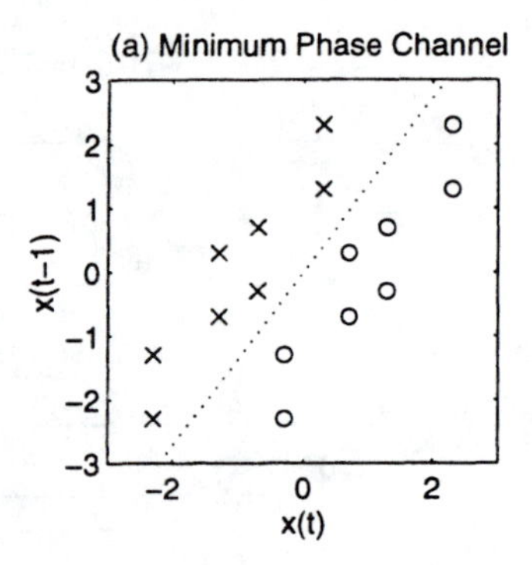

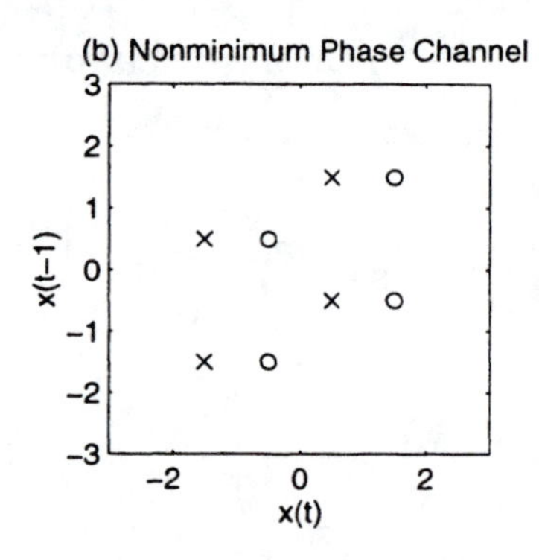

Figure 19.15. *Channel outputs without noise: (a) minimum phase channel* $H(z) = 1.0+0.8z^{-1}+0.5z^{-2}$*; (b) nonminimum phase channel* $H(z) = 0.5+z^{-1}$. (MATLAB file: `nonoise.m`)

line to separate P_+ from P_-. Suppose that the communication channel is modeled as a finite impulse response filter with the following transfer function:

$$H(z) = a_0 + a_1 z^{-1} + \cdots + a_k z^{-k} = \sum_{i=0}^{k} a_i z^{-k}.$$

Then for P_+ and P_- to be linearly separable, the roots of the polynomial

$$a_0 z^k + a_1 z^{k-1} + \cdots + a_k$$

must lie strictly within the unit circle on the complex plane. Linear channels whose transfer functions satisfy this condition are referred to as **minimum phase**; otherwise, they are said to be **nonminimum phase**. For instance, $H(z) = 1.0+0.8z^{-1}+0.5z^{-2}$ is minimum phase since the corresponding roots, $-0.4 \pm 0.58i$, lie strictly within the unit circle on the complex plane. We can plot P_+ and P_-, denoted as "o" and "x", respectively, on a two-dimensional plane, as shown in Figure 19.15(a). The plot indicates that P_+ and P_- are linearly separable, and one possible decision boundary is shown. The nonminimum-phase channel $H(z) = 0.5 + z^{-1}$ shown in a similar plot in Figure 19.15(b) illustrates that P_+ and P_- are not linearly separable, and we cannot expect a linear transversal equalizer to solve the problem satisfactorily.

If the channel is indeed represented by a finite impulse response

$$H(z) = \sum_{i=1}^{n} a_i z^{-i},$$

and the additive noise $e(t)$ is a Gaussian sequence, then an optimal equalizer with a minimum bit error rate can be derived via the concept of the two-state Bayes decision rule [5]:

$$\begin{aligned}\hat{s}(t) &= \operatorname{sgn}(f_{de}(\mathbf{x}(t))) \\ &= \operatorname{sgn}(f_+(\mathbf{x}(t)) - f_-(\mathbf{x}(t))),\end{aligned}$$

where $f_+(\mathbf{x})$ and $f_-(\mathbf{x})$ are the conditional density functions of $\mathbf{x}(t)$ given $s(t) = 1$ and -1, respectively. Symbolically,

$$\begin{aligned}f_+(\mathbf{x}) &= \textstyle\sum_{\mathbf{x}_+ \in P_+} \exp(-\frac{1}{2}(\mathbf{x}-\mathbf{x}_+)^T \mathbf{S}^{-1}(\mathbf{x}-\mathbf{x}_+)), \\ f_-(\mathbf{x}) &= \textstyle\sum_{\mathbf{x}_- \in P_-} \exp(-\frac{1}{2}(\mathbf{x}-\mathbf{x}_-)^T \mathbf{S}^{-1}(\mathbf{x}-\mathbf{x}_-)),\end{aligned}$$

where $\mathbf{S}$ is the covariance matrix of the Gaussian noise e(t):

$$\mathbf{S} = \begin{bmatrix} E[e^2(t)] & E[e(t)e(t-1)] \\ E[e(t-1)e(t)] & E[e^2(t-1)] \end{bmatrix},$$

and $\mathbf{x}_+$ and $\mathbf{x}_-$ are elements in P_+ and P_-, respectively. Therefore, the optimal decision boundary consists of the set of points

$$\{\mathbf{x} \in R^2 | f_{de}(\mathbf{x}) = 0\}.$$

For the nonminimum phase channel characterized by $H(z) = 0.5 + z^{-1}$, the optimum decision surface and boundary are shown in Figure 19.16, where the covariance matrix is assumed to be

$$\begin{bmatrix} 0.2 & 0 \\ 0 & 0.2 \end{bmatrix}.$$

It is obvious that if the noise source is of an unknown nature instead of being uncorrelated Gaussian noise, the decision boundary will be affected in a way that cannot be directly analyzed. Moreover, if the channel characteristics are not linear, then the preceding formulation of an optimal decision boundary is not valid.

We shall use ANFIS to approximate the optimal decision boundary in a channel equalization problem in which the channel characteristics are nonminimum and are described by $H(z) = 0.5 + z^{-1}$. Since ANFIS estimates the decision boundary by sample data directly, we do not need any assumptions about the nature of the channel characteristics or the noise signal.

Before training a fuzzy inference system, we need to collect sample data first. This is done by feeding a sequence of binary random signals $s(t)$ into the impulse response function defined by

$$H(z) = 0.5 + z^{-1}.$$

Figure 19.17 plots the signals involved in this data collection step; there are 500 total training data points. Since the order of the equalizer is 2, we can display the data as a 2-D scatter plot, as shown in Figure 19.18(a). [Figures 19.18(b) through 19.18(d) are explained later in this section.]

For simplicity, we trained ANFIS for only a single epoch. This corresponds to identifying output coefficients of fuzzy rules in a Sugeno-style fuzzy inference system using the least-squares method; the membership functions (MFs) are not modified.

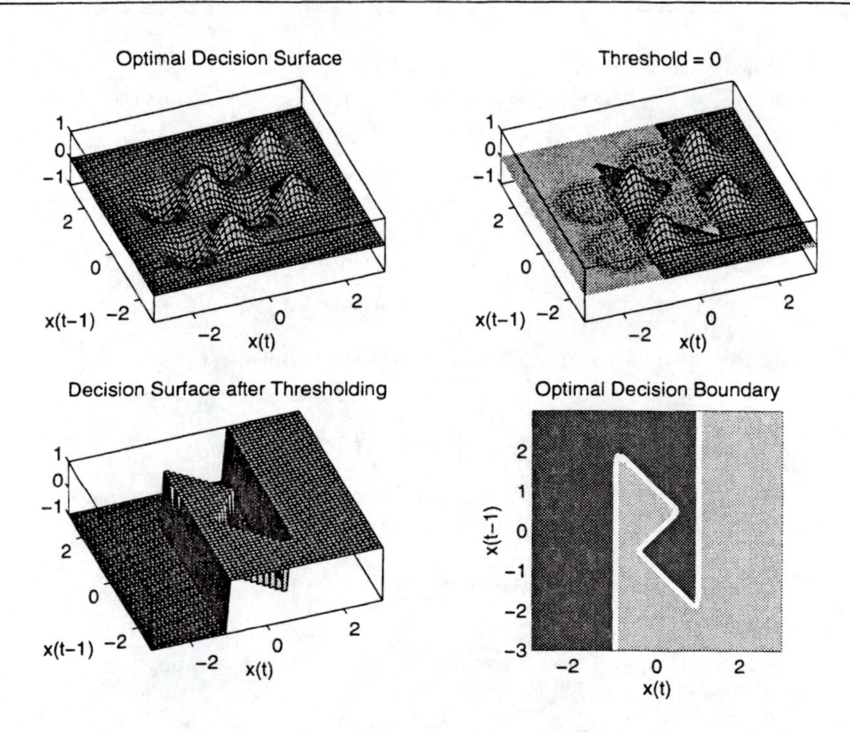

Figure 19.16. *Optimum decision surface and boundary for a nonminimum phase channel characterized by* $H(z) = 0.5 + z^{-1}$. (MATLAB file: `optideci.m`)

A four-rule ANFIS (with two MFs on each input) after one training epoch exhibits the surface shown in Figure 19.19(a). Figure 19.19(b) illustrates how to do thresholding at zero; 19.19(c) is the surface after thresholding; and 19.19(d) is the decision boundary. It is obvious that with only a single epoch of training, ANFIS can construct a nonlinear decision boundary to separate P_+ and P_- correctly.

If we increase the number of MFs on each input, the number of rules also increases and the performance should improve since we are allowing the ANFIS equalizer more degrees of freedom to match the given training data. Figure 19.20 demonstrates similar plots for a nine-rule ANFIS. The performance is slightly better than that of the four-rule ANFIS, but we can also see some spurious decisions on the left-hand side of Figure 19.20(d). If we look back at Figure 19.18(a), it becomes clear that the wrong decisions made by ANFIS occurred in an area where training data is scarce. The lack of data leads to wrong predictions by the nine-rule ANFIS, which the four-rule ANFIS did not make because it had fewer parameters. On the other hand, we do need to have enough tuning parameters to match the optimal decision boundary. This is an inherent trade-off between modeling and generalization capabilities.

The paucity of data can be defined as the **data density** according to the fol-

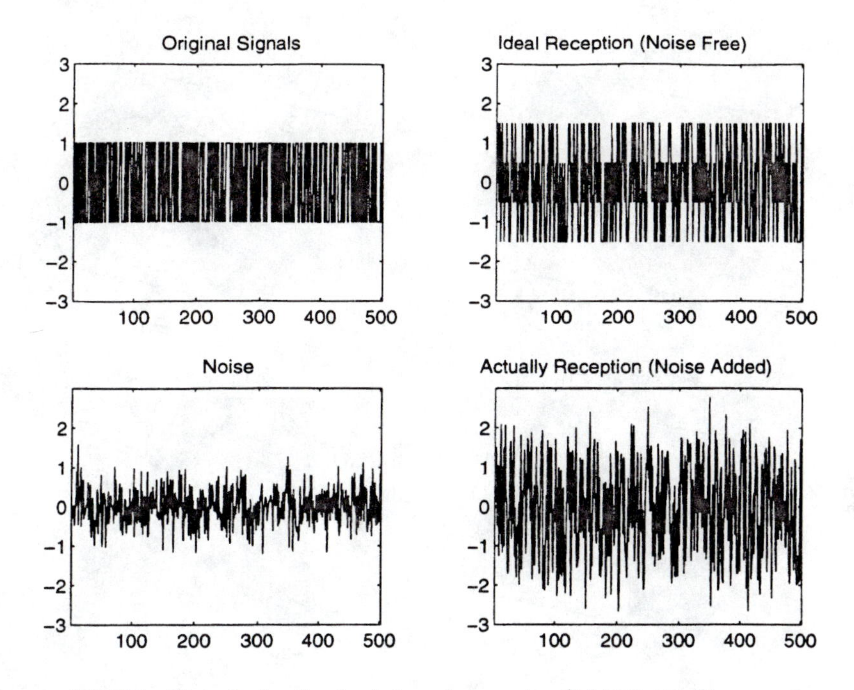

Figure 19.17. *Signals in the training data set.* (MATLAB file: `equdata.m`)

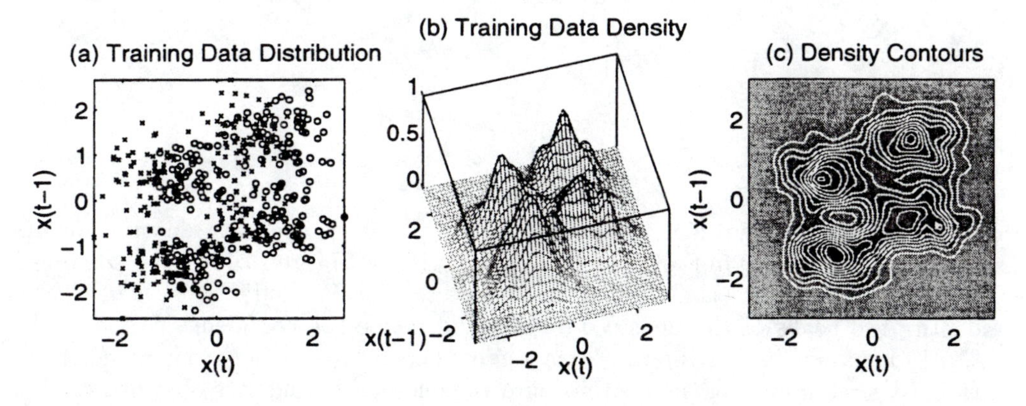

Figure 19.18. *Training data distribution and density.* (MATLAB file: `equdensi.m`)

lowing formula:

$$\text{density}(\mathbf{x}) = \sum_{\mathbf{d}_i \in D} \exp(-\frac{\|\mathbf{x} - \mathbf{d}_i\|^2}{2\sigma}),$$

where $\mathbf{d}_i$'s are the input portions of the training data, and σ is an effective radius.

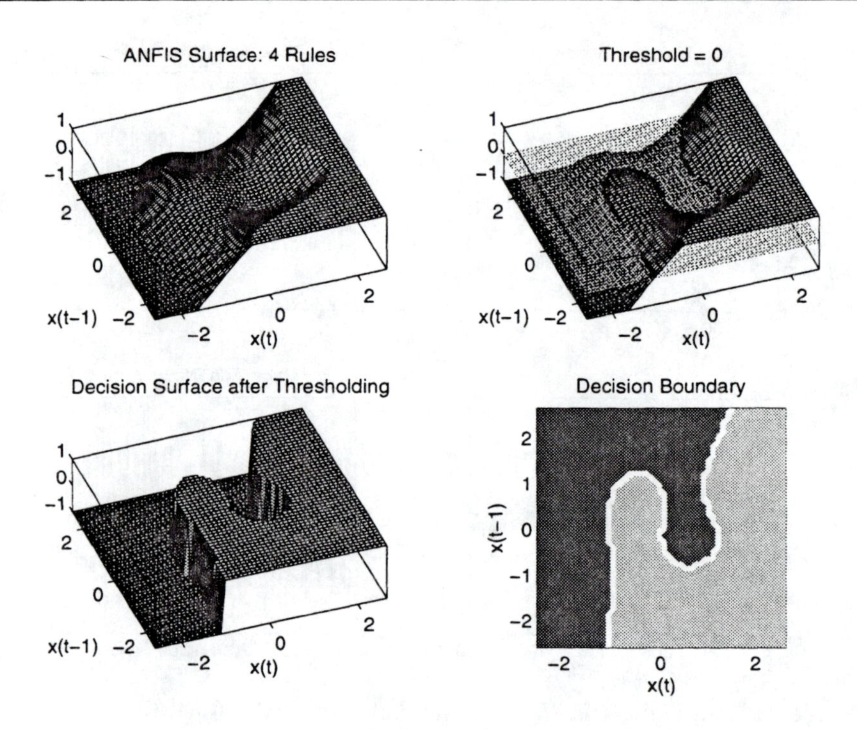

Figure 19.19. *four-rule ANFIS equalizer.* (MATLAB command: `eqtrain(2)`)

After normalizing the density function to within [0, 1], the data density and its contours are as shown in Figure 19.18(b) and 19.18(c); FIgure 19.18(d) is a combined plot. Since the noise is Gaussian around vectors in P_+ and P_-, we can clearly identify the peaks of the density function and define the confidence region of the ANFIS equalizer as the internal area surrounded by the the contour at height = 0.05. By superimposing the contour onto the decision boundaries obtained earlier, we obtain Figure 19.21. It is now clear that the wrong predictions made by the nine-rule ANFIS are definitely outside the confidence region. In general, when input vectors fall outside the ANFIS confidence region, we should re-examine its results using information from other sources.

In summary, ANFIS is effective in solving channel equalization problems, however, its results should not be taken too literally—understanding the data is always crucial.

Other fuzzy modeling and neural network approaches to channel equalization problems can be found in refs. [2, 3, 6].

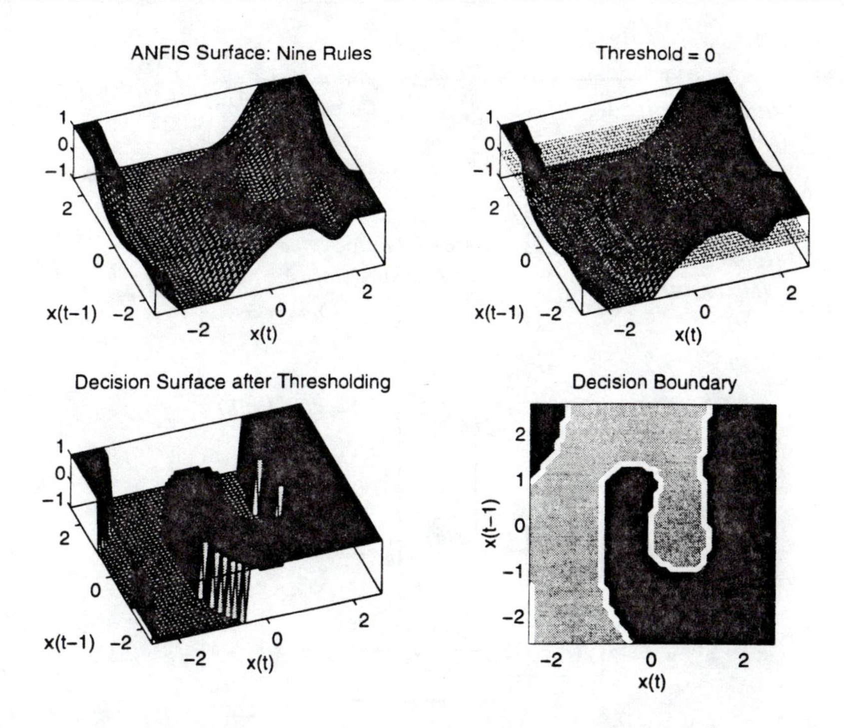

Figure 19.20. *nine-rule ANFIS equalizer.* (MATLAB command: `eqtrain(3)`)

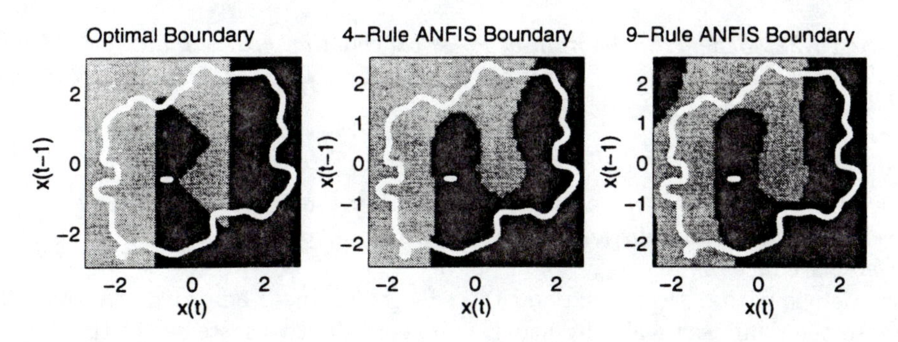

Figure 19.21. *Decision boundaries plus confidence region.* (MATLAB file: `equdec.m`)

19.7 ADAPTIVE NOISE CANCELLATION

Adaptive noise cancellation was first proposed by Widrow and Glover in 1975 [7]; the objective is to filter out an interference component by identifying a linear model between a measurable noise source and the corresponding unmeasurable interference. Adaptive noise cancellation using linear filters has been used successfully in

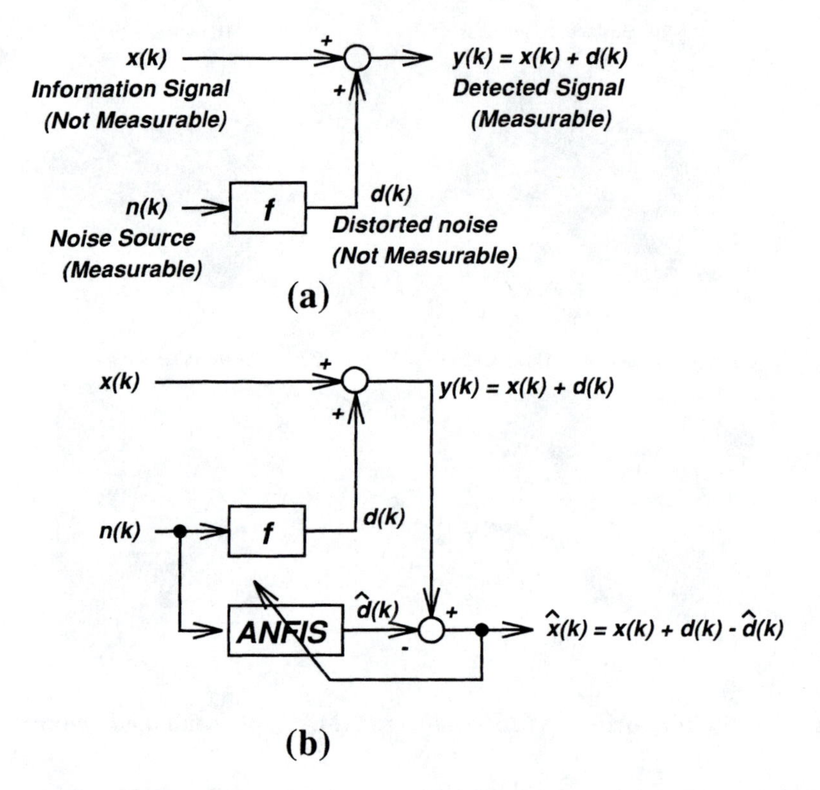

Figure 19.22. *Schematic diagram of noise cancellation: (a) without ANFIS filtering; (b) with ANFIS filtering. [Note that the inputs to blocks f and ANFIS could contain past values of $n(k)$ not shown here.]*

real-world applications such as interference canceling in electrocardiograms (ECGs), echo elimination on long-distance telephone transmission lines, and antenna sidelobe interference canceling [8].

It is obvious that we can expand the concept of linear adaptive noise cancellation into the nonlinear realm by using nonlinear adaptive systems. In this section, we shall show how ANFIS can be used to identify an unknown nonlinear passage dynamics that transforms a noise source into an interference component in a detected signal. Under certain conditions, the proposed approach is sometimes more suitable than noise elimination techniques based on frequency-selective filtering.

Figure 19.22(a) shows the schematic diagram of an ideal situation to which adaptive noise cancellation can be applied. Here we have an unmeasurable information signal $x(k)$ and a measurable noise source signal $n(k)$; the noise source goes through unknown nonlinear dynamics to generate a distorted noise $d(k)$, which is then added to $x(k)$ to form the measurable output signal $y(k)$. Our task is to retrieve the information signal $x(k)$ from the overall output signal $y(k)$, which consists of the information signal $x(k)$ plus $d(k)$, a distorted and delayed version of $n(k)$.

An example of noise cancellation is the suppression of maternal ECG component in fetal ECG [8]. Suppose that we want to measure the fetal ECG $x(k)$ during labor. If we record signals from a sensor placed in the abdominal region, the obtained signal $y(k)$ is inevitably noisy due to the mother's heartbeat signal $n(k)$, which can be measured clearly via a sensor at the thoracic region. However, the heartbeat signal $n(k)$ does not appear directly in $y(k)$. Instead, $n(k)$ travels through the mother's body and arrives delayed and distorted to appear in the overall measurement $y(k)$. In symbols, the detected output signal is expressed as

$$y(k) = x(k) + d(k) = x(k) + f(n(k), n(k-1), n(k-2), \cdots). \tag{19.1}$$

The function $f(\cdot)$ represents the passage dynamics that the noise signal $n(k)$ goes through. If $f(\cdot)$ were known exactly, it would be easy to recover the original information signal by subtracting $d(k)$ from $y(k)$ directly. However, $f(\cdot)$ is usually unknown in advance and could be time varying due to changes in the environment. Moreover, the spectrum of $d(k)$ may overlap that of $x(k)$ substantially, invalidating the use of common frequency-domain filtering techniques.

To estimate the distorted noise signal $d(k)$, we need to pick up a a clean version of the noise signal $n(k)$ that is independent of the information signal. However, we cannot access the distorted noise signal $d(k)$ directly since it is an additive component of the overall measurable signal $y(k)$. Fortunately, as long as the information signal $x(k)$ is zero mean and not correlated with the noise signal $n(k)$, we can use the detected signal $y(k)$ as the desired output for ANFIS training, as shown in Figure 19.22(b).

More specifically, let the output of ANFIS be denoted by $\hat{d}(k)$. The learning rule of ANFIS tries to minimize the error

$$\begin{aligned} \|e(k)\|^2 &= \|y(k) - \hat{d}(k)\|^2 \\ &= \|x(k) + d(k) - \hat{d}(k)\|^2 \\ &= \|x(k) + d(k) - \hat{f}(n(k), n(k-1), n(k-2), \cdots)\|^2, \end{aligned} \tag{19.2}$$

where $\hat{f}$ is the function implemented by ANFIS. Since $x(k)$ is not correlated with $n(k)$ or its history, ANFIS has no clue how to minimize the error component attributable to x. In other words, the information signal x serves as an uncorrelated "noise" component in the data fitting processing, so ANFIS can do nothing about it except pick up its steady-state trend. Instead, the best that ANFIS can do is to minimize the error component attributable to $d(k)$—that is, $\|d(k) - \hat{f}(n(k), n(k-1), n(k-2), \cdots)\|^2$—and this happens to be the desired error measure; it is as if we could measure $d(k)$ directly. To make this clear, we can expand Equation (19.2) to

$$\|e(k)\|^2 = \|x(k)\|^2 + \|d(k) - \hat{d}(k)\|^2 + 2x(k)d(k) - 2x(k)\hat{d}(k). \tag{19.3}$$

Taking expectations from both sides of Equation (19.3) and realizing that $x(k)$ is

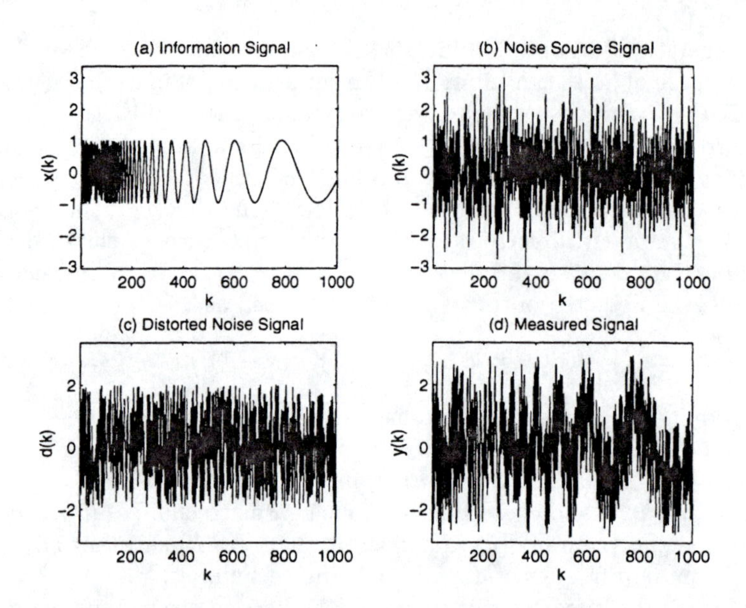

Figure 19.23. *Various signals for noise cancellation: (a) information signal $x(k)$; (b) noise signal $n(k)$; (c) distorted noise signal $d(k)$; (d) measurable output signal $y(k)$.* (MATLAB file: `noise1.m`)

not correlated with $\hat{d}(k)$ yields

$$E[e^2] = E[x^2] + E[(d - \hat{d})^2] - 2E[x\hat{d}]. \tag{19.4}$$

If $x(k)$ is a random signal with a zero mean, then ANFIS has no way to model it and $\frac{1}{n}\sum_{k=1}^{n} x(k)\hat{d}(k)$ approaches zero as n goes to infinity. This implies $E[x\hat{d}] = 0$, and we have

$$E[e^2] = E[x^2] + E[(d - \hat{d})^2], \tag{19.5}$$

where $E[x^2]$ is not affected when ANFIS is adjusted to minimize $E[e^2]$. Therefore, training ANFIS to minimize the total error $E[e^2]$ is equivalent to minimizing $E[(d - \hat{d})^2]$, such that the ANFIS function $\hat{f}(\cdot)$ can be as close as possible to the passage dynamics $f(\cdot)$ in a least-squares sense.

Note that $x(k)$ is the information we want to recover, but it also serves as additive "noise" in ANFIS training. To simplify the following discussion, let us assume that

1. $x(k)$ is a zero signal for all k, and

2. we have fixed premise parameters and updated consequent parameters of ANFIS using the least-squares method.

Assumption 1 implies that we can obtain perfect training data that are subject only to measurement noise; assumption 2 states that we are using ANFIS with

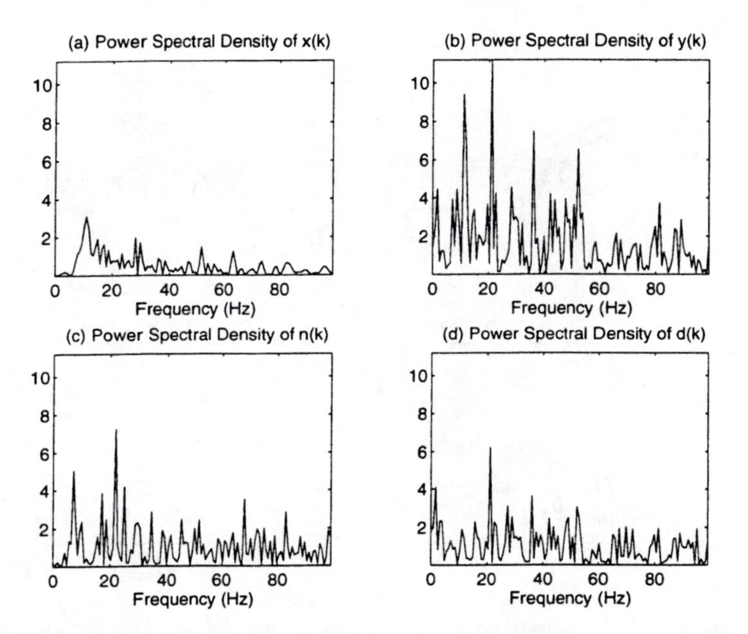

Figure 19.24. *Spectral density distributions: (a) information signal $x(k)$; (b) noise signal $n(k)$; (c) distorted noise signal $d(k)$; (d) measurable output signal $y(k)$, for $k = 0$ to 255.* (MATLAB file: `noise1.m`)

linear parameters only. Even with perfect training data, ANFIS (with modifiable linear parameters only) would produce a fitting error $e(k)$ equal to the difference between a desired output and the ANFIS output; this error term is attributable to measurement noise and/or modeling errors. Statistically, if the error term $e(k)$ is zero mean, then the consequent parameters ANFIS obtains via the least-squares method are unbiased. This is a well known property of the linear least-squares estimator (LSE) and is stated in the Gauss-Markov theorem in Section 5.7.

We now want to relax our previous assumptions and see how well the Gauss-Markov theorem fits. Assumption 1 states that $x(k)$ is a zero signal, which is unrealistic. Fortunately, however, $x(k)$ is an additive component, and the new error term becomes $e(k) + x(k)$. Therefore, as long as $x(k)$ is zero mean, we can still identify unbiased consequent parameters using LSE.

Assumption 2 requires ANFIS to update its consequent parameters only. In our simulations, we applied the proposed hybrid learning rule to update both premise (nonlinear) and consequent (linear) parameters. This made ANFIS a nonlinear model and the Gauss-Markov theorem no longer held. However, we do possess the capacity to reduce modeling errors further.

If we replace the ANFIS block in Figure 19.22(b) with a linear filter, we then have the original adaptive linear noise cancellation settings proposed by Widrow and Glover in 1975 [7]. By enhancing the linear filter with a nonlinear ANFIS filter,

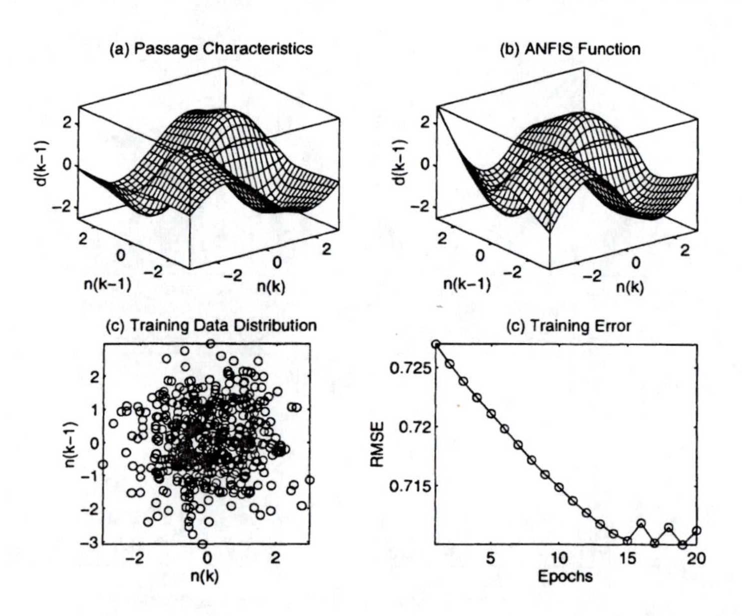

Figure 19.25. *Using ANFIS for noise cancellation: (a) actual nonlinear passage dynamics $f(\cdot)$; (b) ANFIS function $\hat{f}$; (c) training data distribution; (d) RMSE curve.* (MATLAB file: `noise1.m`)

we are able to deal with a wide range of nonlinear passage dynamics.

Before presenting simulation results, we shall reiterate the conditions under which adaptive noise cancellation is valid:

- The noise signal $n(k)$ should be available and independent of the information signal $x(k)$.
- The information signal $x(k)$ must be zero mean.
- The order of the passage dynamics is known. (This determines the number of inputs to the ANFIS filter.)

In our experiments, we applied ANFIS to two nonlinear passage dynamics of orders 2 and 3, respectively. In the first experiment, the unknown nonlinear passage dynamics were assumed to be defined as

$$d(k) = f(n(k), n(k-1)) = \frac{\sin(n(k))\, n(k-1)}{1 + [n(k-1)]^2}, \tag{19.6}$$

where $n(k)$ is a noise source and $d(k)$ denotes the resultant from the nonlinear passage dynamics $f(\cdot)$ attributable to $n(k)$ and $n(k-1)$. Figure 19.25(a) displays function $f(\cdot)$ as a three-dimensional surface. Since $f(\cdot)$ is unknown, we use ANFIS

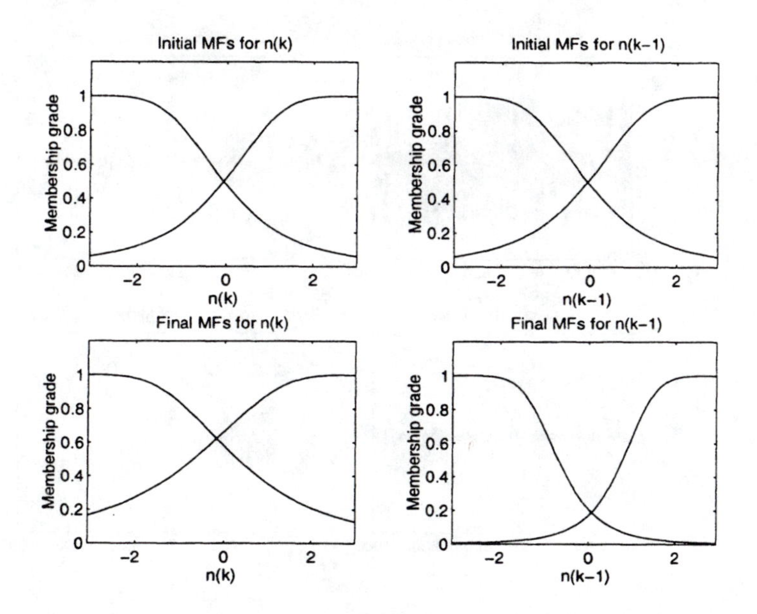

Figure 19.26. *MFs before and after training.* (MATLAB file: `noise1.m`)

to approximate this function under the assumption that we do know that $f(\cdot)$ is of order 2.

We assume that the information signal $x(k)$ is expressed as

$$x(k) = \sin\left(\frac{8000}{k+20}\right), \tag{19.7}$$

where k is a step count, and the sampling period is equal to 5 μs. Figure 19.23(a) demonstrates $x(k)$ when k runs from 0 to 1,000 (or when time runs from 0 to 5 s). We assume that the measurable noise source is Gaussian with zero mean and unity variance, as shown in Figure 19.23(b). The resulting distorted noise $d(k)$ produced by the nonlinear dynamics in Equation (19.6) is shown in Figure 19.23(c). The measurable signal at the receiving end, denoted as $y(k)$, is equal to the sum of $x(k)$ and $d(k)$ and is demonstrated in Figure 19.23(d). Due to the nonlinear passage dynamics of $f(\cdot)$ and the large amplitude of $d(k)$, it is hard to correlate $y(k)$ and $x(k)$ in the time domain.

Before we move on, we should first examine how these signals behave in the frequency domain. Figures 19.24(a) through 19.24(d) demonstrate the spectral density distributions of $x(k)$, $n(k)$, $d(k)$, and $y(k)$, respectively, for the first 256 points. Obviously, the spectra of $x(k)$ and $d(k)$ overlap each other considerably, making it impossible to employ frequency-domain filtering techniques to remove $d(k)$ from $y(k)$.

To use ANFIS in this situation, we collected 500 training data pairs of the

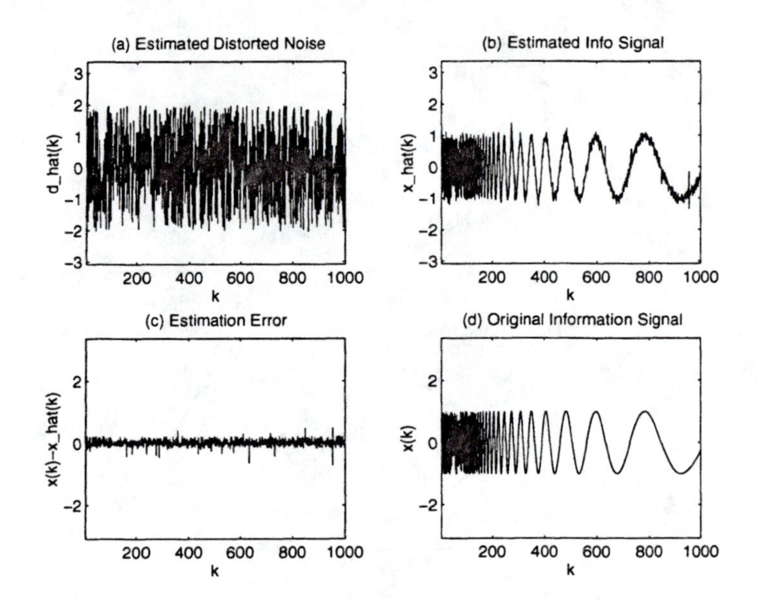

Figure 19.27. *Results of using ANFIS for noise cancellation: (a) ANFIS output $\hat{d}(k)$; (b) estimated information signal $\hat{x}(k)$; (c) estimation error $\hat{x}(k) - x(k)$; (c) original information signal $x(k)$.* (MATLAB file: `noise1.m`)

following form:

$$[n(k), n(k-1); y(k)], \tag{19.8}$$

with k runs from 1 to 500. We used a four-rule ANFIS to fit the training data, in which each of the two inputs was assigned two generalized bell membership functions. Figure 19.25(b) is the ANFIS surface $\hat{f}(\cdot)$ after 20 epochs of batch learning; 19.25(c) is the scatter plot of the training data; 19.25(d) is the RMSE (root-mean-squared error) curve through 20 epochs. The starting point of the RMSE curve shows the error when only the linear parameters have been identified by LSE. By also updating the nonlinear parameters, we were able to decrease the error further; Figure 19.26 shows the MFs before and after training, reflecting changes in premise (nonlinear) parameters. Note that the error cannot be minimized to zero; the minimum error is regulated by the information signal $x(k)$, which appears as fitting noise.

By using ANFIS, the estimated resultant $\hat{d}$ from the nonlinear passage is expressed as $\hat{d}(k) = \hat{f}(n(k), n(k-1))$, as shown in Figure 19.27(a). Thus, the estimated information signal $\hat{x}(k)$, derived as $y(k) - \hat{d}(k)$, is shown in Figure 19.27(b). The difference between $x(k)$ and $\hat{x}(k)$ is shown in Figure 19.27(c). Note that $\hat{x}(k)$ is already fairly close to $x(k)$; the estimation error in Figure 19.27(c) is expected to decrease if more training data are used over more training epochs.

In our second experiment, we used real-world audio signals for simulation.

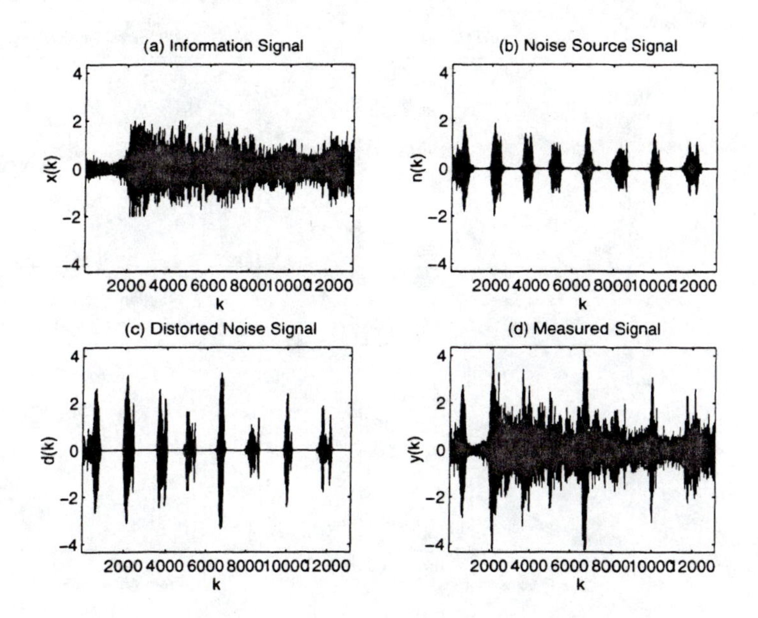

Figure 19.28. *Various signals for noise cancellation: (a) information signal $x(k)$; (b) noise signal $n(k)$; (c) distorted noise signal $d(k)$; (d) measurable output signal $y(k)$.* (MATLAB file: `noise2.m`)

The audio signals were obtained from the MATLAB sound files `handel.mat` and `chirp.mat`. When these two files are loaded into MATLAB and played by the command `sound`, `handel.mat` is a piece of music of composer George Frideric Handel's "The Hallelujah Chorus" and `chirp.mat` the sound of a bird's chirping.

We used `handel.mat` as the information signal $x(k)$ and `chirp.mat` as the noise source $n(k)$. These audio signals were sampled at 8190 Hz. The nonlinear passage dynamics $f(\cdot)$ is represented by the following equation:

$$d(k) = f(n(k), n(k-1), n(k-2)) = \frac{8\ \sin(n(k)\, n(k-1)\, n(k-2))}{1 + [n(k-1)]^2 + [n(k-2)]^2}, \tag{19.9}$$

where $d(k)$ is the distorted noise signal. Figure 19.28(a) demonstrates $x(k)$ when k runs from 0 to 13,128 (or when time runs from 0 to 1.6 s). The measurable noise signal $n(k)$ is assumed to be the chirping sound, as shown in Figure 19.28(b). The corresponding distorted noise $d(k)$ due to the nonlinear dynamics in Equation (19.9) is shown in Figure 19.28(c). The measurable signal at the receiving end, denoted as $y(k)$, is equal to the sum of $x(k)$ and $d(k)$ and is shown in Figure 19.28(d).

Figures 19.29(a) through 19.29(d) show the spectral density distributions for $x(k)$, $n(k)$, $d(k)$, and $y(k)$, respectively, for the first 1,000 points. Again, the spectra of $x(k)$ and $d(k)$ overlap each other considerably, and we cannot use frequency-selective filters to remove $d(k)$ from $y(k)$.

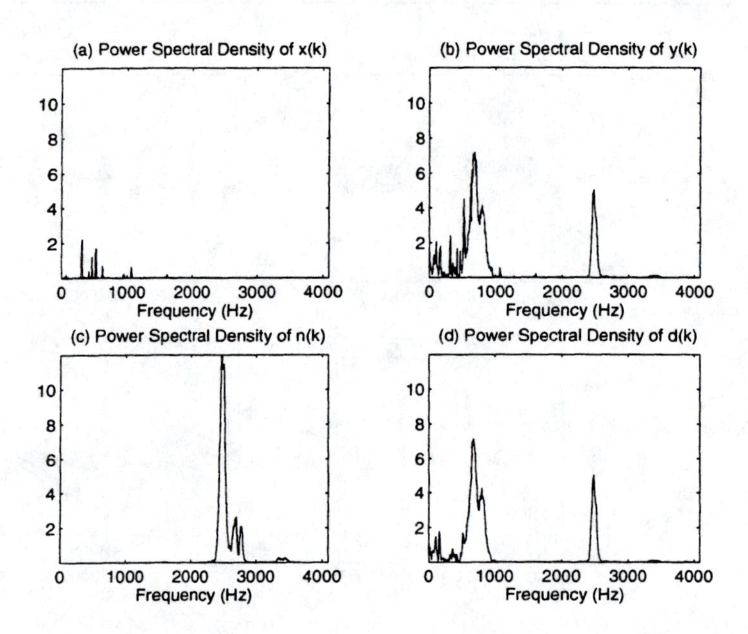

Figure 19.29. *Spectral density distributions: (a) information signal $x(k)$; (b) noise signal $n(k)$; (c) distorted noise signal $d(k)$; (d) measurable output signal $y(k)$, for $k = 0$ to 999.* (MATLAB file: `noise2.m`)

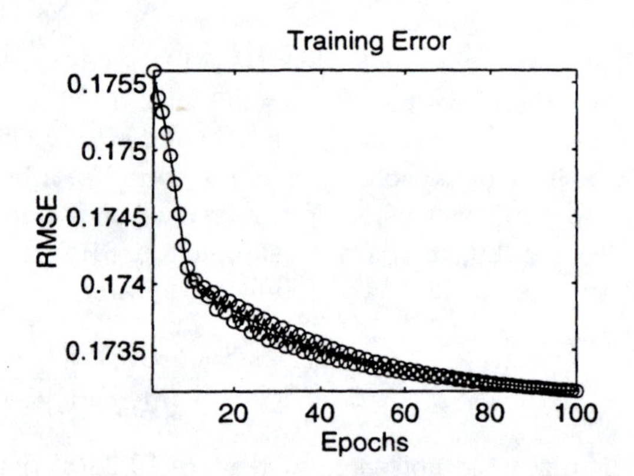

Figure 19.30. *RMSE curve for ANFIS learning.* (MATLAB file: `noise2.m`)

To model $f(\cdot)$ using ANFIS, we collected 1,000 training data pairs of the following form:

$$[n(k), n(k-1), n(k-2); y(k)], \tag{19.10}$$

with k runs from 2 to 1001. We used an eight-rule ANFIS to fit the training data, in which each of the three inputs was assigned two generalized bell membership

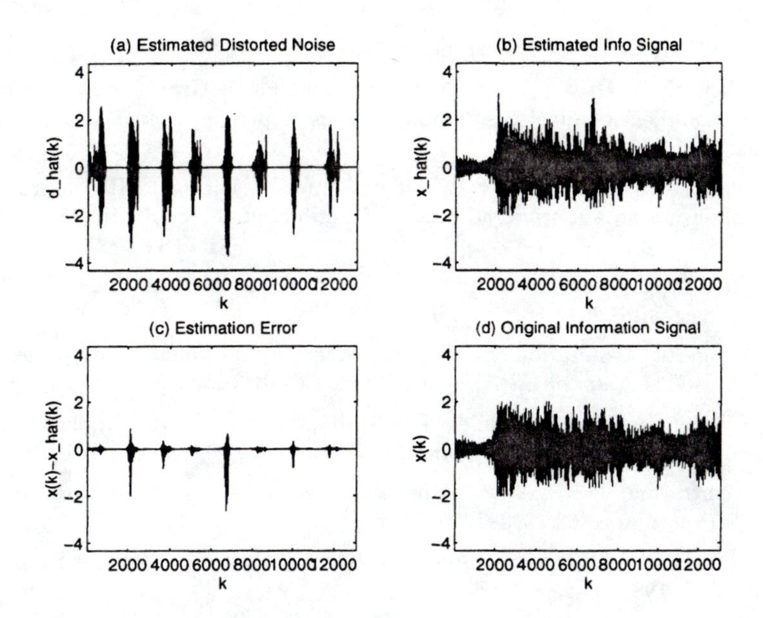

Figure 19.31. *Results of using ANFIS for noise cancellation: (a) ANFIS output $\hat{d}(k)$; (b) estimated information signal $\hat{x}(k)$; (c) estimation error $\hat{x}(k) - x(k)$; (c) original information signal $x(k)$.* (MATLAB file: `noise2.m`)

functions. Figure 19.30 is the RMSE curve through 100 training epochs.

By using ANFIS, the estimated output $\hat{d}$ of the nonlinear passage is expressed as $\hat{d}(k) = \hat{f}(n(k), n(k-1), n(k-2))$, as shown in Figure 19.31(a). Thus the estimated information signal $\hat{x}(k)$, derived as $y(k) - \hat{d}(k)$, is shown in Figure 19.31(b). The difference between $x(k)$ and $\hat{x}(k)$, as shown in Figure 19.31(c), was small when k ranged from 0 to 999 since the training data were obtained from this interval. The difference is most pronounced when k is equal to 6700 or so; improvement is expected when we collect training data from a longer interval. Note that collecting data from a longer interval usually generates an excessive number of training data pairs. (To keep training time reasonably low in this case, we usually need to perform some kind of data reduction to extract representative data pairs and weed out redundant ones.)

By actually playing the audio signals $x(k)$, $y(k)$ and $\hat{x}(k)$, we found that ANFIS did a good job of removing the unknown distorted noise signal $d(k)$ from the measured signal $y(k)$.

REFERENCES

[1] G. E. P. Box and G. M. Jenkins. *Time series analysis, forcasting and control*, pages

532–533. Holden Day, San Francisco, 1970.

[2] S. Chen, G. J. Gibson, C. F. N. Cowan, and P. M. Grant. Adaptive equalization of finite nonlinear channels using multilayer perceptrons. *Signal Processing*, 20:107–119, 1990.

[3] S. Chen, G. J. Gibson, C. F. N. Cowan, and P. M. Grant. Reconstruction of binary signals using an adaptive radial-basis-function equalizer. *Signal Processing*, 22:77–93, 1991.

[4] Stephen L. Chiu. Selecting input variables for fuzzy models. *Journal of Intelligent and Fuzzy Systems*, 1996. Forthcoming.

[5] D. F. Specht. Generation of polynomial discriminant functions for pattern recognition. *IEEE Transactions on Electron. Comput.*, EC-16(3):308–319, 1967.

[6] L.-X. Wang and J. M. Mendel. Fuzzy adaptive filters, with application to nonlinear channel equalization. *IEEE Transactions on Fuzzy Systems*, 1(3):161–170, 1993.

[7] B. Widrow and J. R. Glover. Adaptive noise cancelling: Principles and applications. *IEEE Proceedings*, 63:1692–1716, 1975.

[8] B. Widrow and D. Stearns. *Adaptive signal processing.* Prentice Hall, Upper Saddle River, NJ, 1985.

Chapter 20

Fuzzy-Filtered Neural Networks

C.-T. Sun

20.1 INTRODUCTION

This chapter introduces a neuro-fuzzy model, called a **fuzzy-filtered neural network**, for adaptive learning and feature detection. Its applications in the domains of plasma spectrum analysis, and pattern recognition are described. The issues involved in using a neural network to process massive amount of possibly redundant input information are analyzed and solutions are suggested. In particular, we provide three architectures for fuzzy-filtered neural networks, which employ one-dimensional fuzzy filters, two-dimensional fuzzy filters, and genetic algorithm–based fuzzy filters, respectively. The validity, efficiency, and generality of the proposed models are verified by experimental results.

A considerable share of recent research on machine recognition has focused on the use of neural networks. An important step in most existing NN-based methods is extraction of the features of the patterns to be recognized. The goal of automatic feature extraction performance is twofold: reducing the complexity of the NN architecture and increasing system adaptability. A better feature extraction mechanism not only leads to a higher recognition rate but also implies a simpler NN structure in most cases. Simpler architectures have a better chance of avoiding the problem of **overfitting** or **overlearning**, which is frequently encountered in complicated adaptive systems.

The complexity and limitations of traditional neural networks are largely due to the lack of an effective way to extract meaningful information from the learned configuration. This problem becomes more intractable when the number of physical sensors used for measurement increases. For example, an image to be processed contains thousands of pixels, which is far more than the number usually needed for pattern recognition. In addition, the drifting of sensory equipment and variations in samples would cause any fixed-position feature detector to fail to recognize the resulting altered signal. To cope with these two factors and the general problem of background noise, a form of signal filtering is needed. In the following we shall introduce a mechanism for fuzzy filtering to cope with the complexity of feature extraction.

Moreover, even when a set of tightly structured fuzzy filters is used, redundant information is still read in. This issue cannot be handled well by simple adaptation because of the frequent occurrence of local optima. Consequently, we need another technique for global search. Genetic algorithms (GAs), as described in Section 7.2 of Chapter 7, are a general method for optimization. GA-based fuzzy filters are highly flexible because the size, shape, and location of the filters can be identified separately. Of course, the cost of employing a GA must also be taken into account if we are to achieve better performance. We shall describe an effective mechanism to compensate for the computation cost introduced by using GAs.

20.2 FUZZY-FILTERED NEURAL NETWORKS

Fuzzy filtering is the task of partitioning a large number of physical data channels into much fewer **fuzzy channels**. These channels, adaptable during a training process, are employed for both noise filtering and feature detection. The boundary between two neighboring meaningful channels is assumed to be a continuous, overlapping area in which a physical channel has partial membership in both fuzzy channels. A fuzzy channel defines a range of input signal intensity characterized by an appropriate membership function. The position and shape of this membership function are adjusted during a learning process so that the system error is minimized. At the end of training, the fuzzy channels are expected to hook onto salient physical channels to provide a meaningful interpretation of the qualitative aspects of patterns.

Using fuzzy filters as a mechanism for feature extraction has several advantages. First, the features detected are insensitive to variation in samples, and the effect of noise is negligible. Second, duplicated features are automatically combined. Third, this approach considerably reduces the complexity of the architecture, so that not only is the training efficiency improved but the possibility of overfitting is decreased. Finally, this model provides a meaningful interpretation of the detected features.

Biological evidence suggests that a positional feature detector should preferably behave as a *localized receptive field* [4]. Mathematically, localized receptive fields can be represented as *radial basis functions*, as described in Section 9.5 of Chapter 9. We have shown that there exists a type of functional equivalence between radial basis function networks and fuzzy inference systems [3]. Furthermore, localized receptive field–based architectures are more efficient than standard neural networks in terms of learning [5]. The preceding background justifies the choice of a fuzzy neural network as a feature extraction tool for problems with a massive amount of sensory input.

We use a multilayer feedforward adaptive network (Chapter 8) to implement the concept of fuzzy filtering. Figure 20.1 depicts a fuzzy-filtered adaptive network in which the x_i's are inputs and the y_j's are outputs. The nodes in the same layer have the same type of node function.

Layer 1 is the input layer. Each node in layer 2 is associated with a generalized

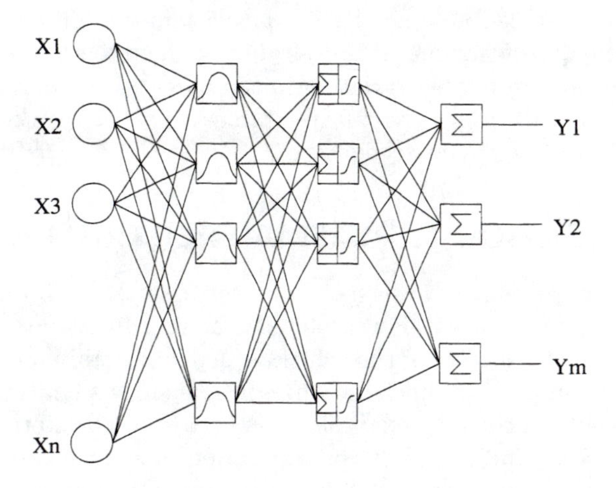

Figure 20.1. *Fuzzy-filtered neural network.*

bell membership function:

$$\mu_A(x_i) = \frac{1}{1 + \left|\frac{x_i - c_i}{a_i}\right|^{2b_i}}, \tag{20.1}$$

where x_i stands for the position (or, equivalently, frequency) of a physical channel, A is the linguistic term associated with this node function, and $\{a_i, b_i, c_i\}$ is the parameter set. The node output is a normalized weighted sum:

$$\frac{\sum_i \mu_A(x_i) f(x_i)}{\sum_i \mu_A(x_i)}, \tag{20.2}$$

where $f(x_i)$ is the intensity of input channel x_i. Thus, a fuzzy filter behaves as a bandpass filter with the added capability of learning.

The initial values of the MF parameters are set in a way similar to that in Figure 12.6 on page 345. As mentioned in Section 12.6, although these initial MFs are set heuristically and subjectively, they provide an easy interpretation that parallels the human thinking process. The parameters are then tuned using steepest descent in a learning process based on the training data set.

The nodes in layer 3 perform the same function as the hidden layer in a standard multilayer perceptron (Section 9.4); each node takes the weighted sum of inputs and produces a transferred output through a sigmoidal function. Layer 4 is similar except that the nonlinear transfer function is not used.

In summary, fuzzy filters can be considered a type of locally receptive field which emphasizes local positional features. As a result, it is natural to use fuzzy filters to divide a massive number of physical input channels into a much smaller number of fuzzy channels. Since we do not have *a priori* information about the shapes and positions of these fuzzy channels, it is desirable for them to be adaptable by training. Consequently, they can handle variation in samples and redundant information. In the following, we will apply the fuzzy-filtered neural networks in two domains: plasma analysis and character recognition.

20.3 APPLICATION 1: PLASMA SPECTRUM ANALYSIS

The importance of plasma analysis has long been asserted by both the scientific and the engineering communities. For problems ranging from outer space physics to medical diagnosis, the success of the solutions depends highly on our understanding of primitive but complicated information: spectral signals emitted from plasma.

In VLSI manufacturing, for example, it is necessary to determine the endpoints of an etching process and to detect contamination in a chemical chamber. Existing recipes or rules of thumb are mostly heuristic and therefore not guaranteed to be reliable for all possible scenarios. Moreover, these approaches depend highly on the knowledge of which species (that is, spectral intensities at some frequencies) are indicative to the process; they cannot be used to identify the important species. The most direct way of monitoring a chemical process should be to use the full range of optical signals (the spectra) generated by an optical emission spectrometer to determine the actual status of the chemical reaction.

Because of drifting of equipment, clouding of the chamber's window, and variations in the plasma, any single-wavelength detector is likely to miss an important signal at a specific frequency; therefore, some type of signal filtering is needed. Furthermore, we know that the number of essential species in the chamber is much less than the number of channels. Taking all these factors into consideration, we decided to use the fuzzy-filtered neural network as a suitable model for coping with the complexity and uncertainty of plasma analysis.

20.3.1 Multilayer Perceptron Approach

For comparison, we first used a standard multilayer perceptron (MLP) to achieve the desired mapping. Our MLP for plasma analysis is a three-layer (731-8-4) network with linear outputs. It examines 731 optical channels and uses backpropagation to adjust the weights so that the inputs are associated with four control variables of an oxide etching process: power, chamber pressure, and two gas (H_2 and CF_3) flows. The training data set was generated according to the *center cubic experimental design*. We recorded the plasma's spectral data when the etching is working desirably. The differences across spectra are so subtle that a process engineer cannot detect them.

Due to the large size of inputs (731), we had to be careful in selecting the initial values of the weights connecting the input and hidden layers. If these weights are median, some of the hidden unit might be driven into saturation and the training is sluggish. On the other hand, if these weights are too small, then the training is also slow due to a small gradient vector. Meanwhile, compensatory learning rates are set to match the small weights resulting from the large number of input channels so that fast convergence can still be achieved. With the carefully chosen initial weights and learning rates, the MLP can effectively identify small signal changes in a noisy environment. After 200 training epochs on 30 spectra, the root-mean-square error was driven to below 0.3%. The receding analysis can be applied to an on-line training situation.

20.3.2 Fuzzy-Filtered Neural Network Approach

Although a careful selection of initial weights can overcome the saturation problem introduced by a massive number of input channels, it is not the best solution. A simple fact is that we simply do not need to use all the 731 inputs in the first place. We should be able to single out the important inputs and combine the redundant ones. Furthermore, the standard MLP architecture is not able to tell us which input channels (species) are important in analyzing this chemical process because the weight distribution in the trained network does not show any clear patterns. Thus, we want to employ the fuzzy-filtered neural network to achieve three goals at once. First, we want to improve the learning performance. Second, we need to avoid possible overfitting due to a limited number of training data because etching experiments are expensive. Third, we aim to provide a meaningful interpretation of the trained network.

The fuzzy filtering mechanism described in the last subsection simplifies the network architecture because far fewer system parameters need to be adjusted. The 731-8-4 network mentioned in the previous subsection has about $731 \times 8 + 8 \times 4 = 5880$ weights to fine-tune; in comparison, a 731-15-15-4 fuzzy-filtered network has only about $3 \times 15 + 15 \times 15 + 15 \times 4 = 330$ parameters. Obviously, this benefits learning efficiency and reduces the chance of overfitting. A more important point is that the fuzzy approach provides a meaningful interpretation for the training results so that we can obtain a better understanding of the complex nature of plasma emission.

A spectrum is shown in Figure 20.2, with the x-axis representing the optical channel number, which corresponds to the wavelength of a certain species, and the y-axis representing the (scaled) intensity of the plasma emission signal. As mentioned previously, we employed a 731-15-15-4 network to handle the data. Three channels, among the initial 15 channels, were lost in the learning process (driven out of the wavelength range); the remaining ones are shown in the figure.

For an etching process, the important channels are typically those that have the highest intensities or change the most dramatically as the etching proceeds. We can match the wavelengths to the species that emit them. The fuzzy channels automatically identified two hydrogen peaks and a CO emission line (see Figure 20.2). The nitrogen lines indicate a possible leak in the vacuum system.

In brief, the network learns on its own from the training data and selects the most pertinent channels without any guidance from human experts on plasma diagnostics. The ability to detect features is one of the most important advantages of

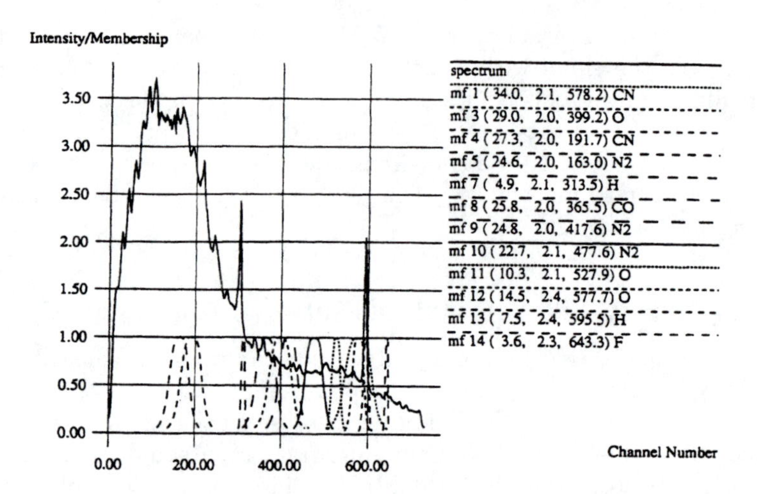

Figure 20.2. *A typical spectrum and fuzzy channels after training. The three values associated with each membership function are a, b, c in Equation (20.1), respectively.*

the fuzzy filtering mechanism. This capability is significant because human experts can actually learn from the network and gain a better understanding of the plasma discharge and thus develop better manufacturing recipes.

20.4 APPLICATION 2: HAND-WRITTEN NUMERAL RECOGNITION

To verify the generality of the fuzzy-filtered NN model, we applied it to another problem: hand-written numeral recognition. We used a numeral data set in the public domain as a benchmark. This set contained 3471 numerals from 49 different writers. The numerals have been spatially normalized to a 32×32 frame of pixels and shifted to the left. We used 3000 numerals for training and the remaining 471 for testing. Most of the testing numerals were written by persons different from those who wrote the numerals in the training set. This testing arrangement accurately simulates real applications, such as zip-code recognition. Figure 20.3 shows some examples of the numerals in the testing set.

For comparison, we initially tested the data set on a standard three-layer (1024-30-10) feedforward neural network. It examined 3000 training numerals and used backpropagation to adjust the weights. After 250 training epochs, we tested it with the remaining 471 numerals. The recognition rate was 85%. Further, because of the network's complicated architecture, its learning efficiency was poor.

Figure 20.3. *Some examples of hand-written numerals.*

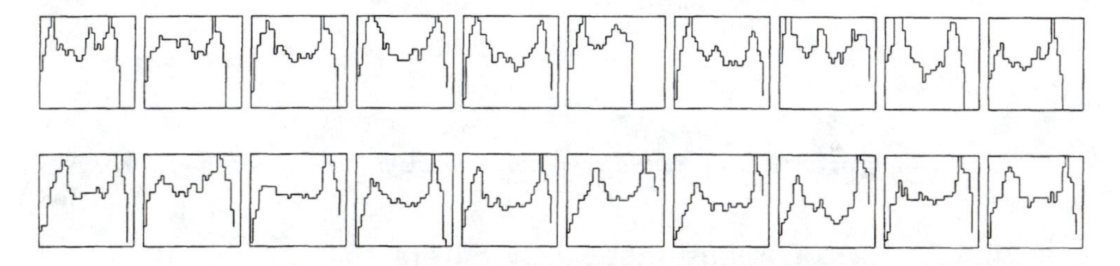

Figure 20.4. *The x-axis and y-axis spectra of some training images. Upper row, left to right: the x-axis spectra of digit 0 to digit 9; lower row: the y-axis spectra of the same digits.*

20.4.1 One Dimensional Fuzzy Filters

To use the fuzzy filtering mechanism described in the previous section, we have to transform the numerals to spectrum-like signals by taking projections on both the x-axis and y-axis. In other words, each 32×32 image must be transformed to a 64-channel spectrum with each channel representing the number of black pixels in a column or in a row of the image. Figure 20.4 shows the x-axis and y-axis spectra of some training images. We used a 64-20-10-10 fuzzy-filtered network to learn the training patterns. The recognition rate was 90%, a better result than that yielded by the pure neural network.

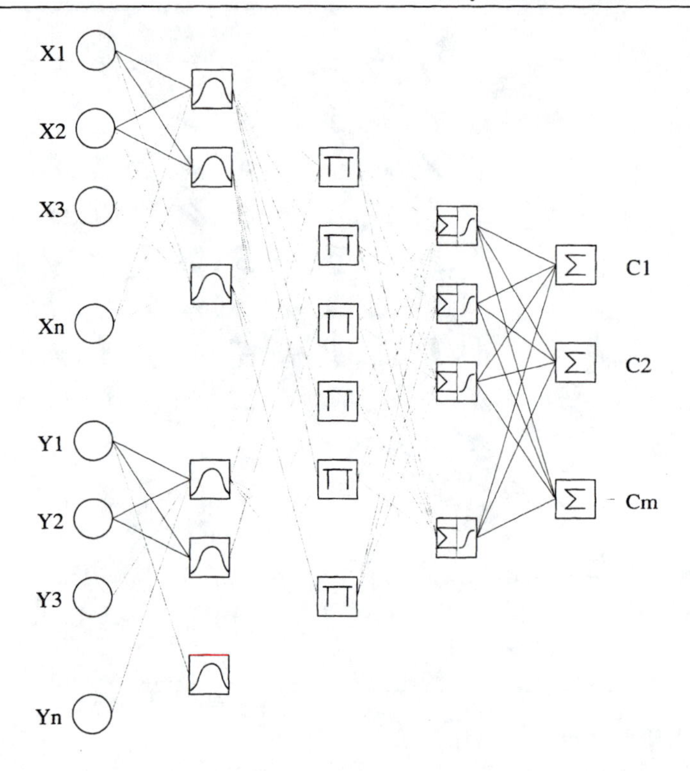

Figure 20.5. *A two-dimensional fuzzy-filtered neural network.*

20.4.2 Two Dimensional Fuzzy Filters

Because the nature of image patterns is different from one-dimensional spectra, such as those used in plasma analysis, it is desirable to have fuzzy filters which handle two-dimensional data appropriately. It is easy to generalize the fuzzy-filtered NN model to a 2-D version, as shown in Figure 20.5. In this architecture, x-membership and y-membership are integrated by a T-norm operator (e.g., multiplication). The combined membership is then used as the filtering weight.

The initial distribution of membership functions is represented by the grids in Figure 20.6(a). After training, the grids may be adjusted to a distribution like that shown in Figure 20.6(b) to capture the important pieces of information. We used 36 two-dimensional fuzzy filters in the experiment. The recognition rate was increased to 92%, compared with the rate of 90% obtained using 40 one-dimensional filters.

The learning efficiency was improved. However, because the fuzzy filters are arranged as an array, the degree of freedom in terms of adaptability of one filter is constrained by its neighbors. In other words, the positions of these fuzzy filters are fixed to a certain extent because it is difficult for the gradient descent learning algorithm to overcome local optimal points. Apparently, we need a more flexible

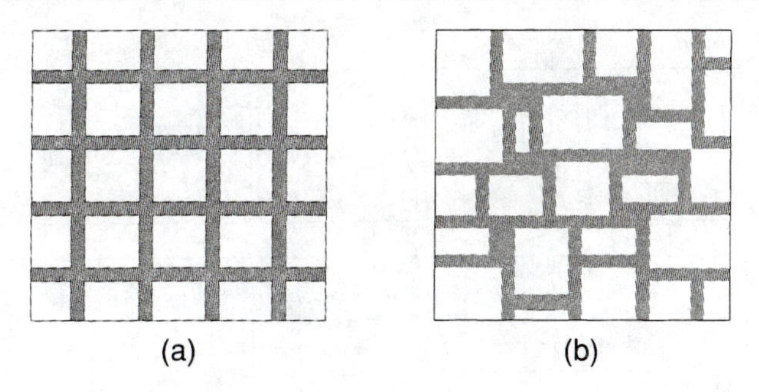

Figure 20.6. *2-D fuzzy filters. (a) Before training, (b) after training.*

mechanism to identify shape and position in complex applications. In the next section, we describe an advanced fuzzy filter model based on genetic algorithms that achieves better performance without sacrificing learning efficiency.

20.5 GENETIC ALGORITHM–BASED FUZZY FILTERS

Genetic algorithms have been used in classification problems because of their ability to identify the weights of importance among features. GAs have been employed for feature selection in hybrid models with K-nearest neighbor algorithms [6] or with feature partitioning [1]. In this section we explore the possibility of integrating GAs with the fuzzy filtering model.

For pattern recognition problems, the important positional features are typically those that have the highest or the lowest intensity and change the most dramatically across different numerals. As in the previous sections, we want to use only this kind of low-level, positional feature because for the machine to learn in a realistic sense, high-level human involvement should be avoided.

In this study, various methods were developed for this purpose. Most of them achieved high recognition rates. We found that many factors affected performance in terms of recognition rate. In the following, we will describe all the methods we have tried and make some comparisons.

20.5.1 A General Model

In pursuing higher flexibility, we employed GAs to determine a set of rectangular regions; each will later be transformed into a fuzzy filter for each digit. We want these regions to represent a particular digit and not to duplicate the regions for other digits. In other words, we expect these regions to catch the positional features of a digit. The selection procedure of this GA was guided by the following evaluation

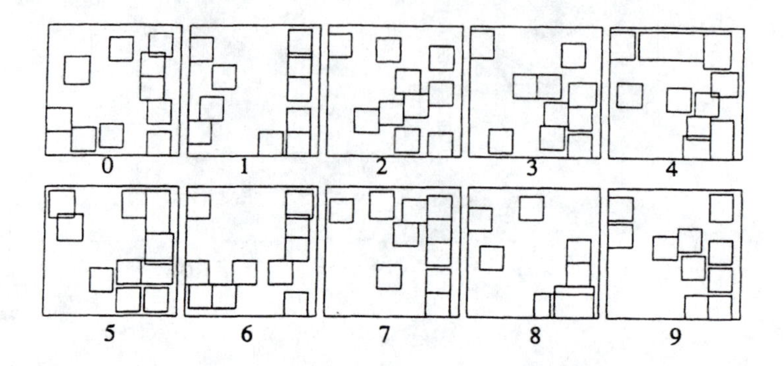

Figure 20.7. *Rectangular feature detectors.*

function:

$$\sum_n \sum_r \frac{b(r)}{w(r)^2 \times z} \times 10 + \sum_{other\ digits} \sum_n \sum_r \frac{w(r)}{b(r)^2 \times z}, \qquad (20.3)$$

where n stands for a numeral, r stands for a region, $b(r)$ is the number of black pixels in a region, and $w(r)$ is the number of white pixels in that region. This formula says that to be a discriminant feature a region should have a high density of black pixels for one digit and low density for most other digits. Because we do not want the rectangles to overlap each other too much, we introduce the denominator term z, which is defined as one plus the total number of overlapped pixels of the rectangles represented by a chromosome string. In other words, the higher the degree of overlapping, the lower the evaluation score.

We grouped together all x coordinates of the lower-left corners of the rectangles as the first section of the chromosome. The sections of y coordinates, heights, and widths follow in that order, as shown in Figure 20.9(a). After 400 generations, with a crossover rate of 0.005 per bit and a mutation rate of 0.005, the GA converged. It produced 10 rectangular positional feature detectors for each digit, as shown in Figure 20.7.

We then fuzzified the rectangles by using two Gaussian functions to approximate the wall and the base of each rectangle. In other words, we created a membership function with its mean at the middle point of one side and with its deviation equal to half of the height or the width of a certain rectangle. Finally, we put these GA-generated fuzzy feature detectors in front of a standard neural network for classification training. The mechanism of this integrated GA-fuzzy-neural approach is shown in Figure 20.8.

We found that the preceding architecture gave us the best performance up to this point (i.e., a 95.8% recognition rate). In other words, of the 471 numerals in the testing set, only 20 were misclassified.

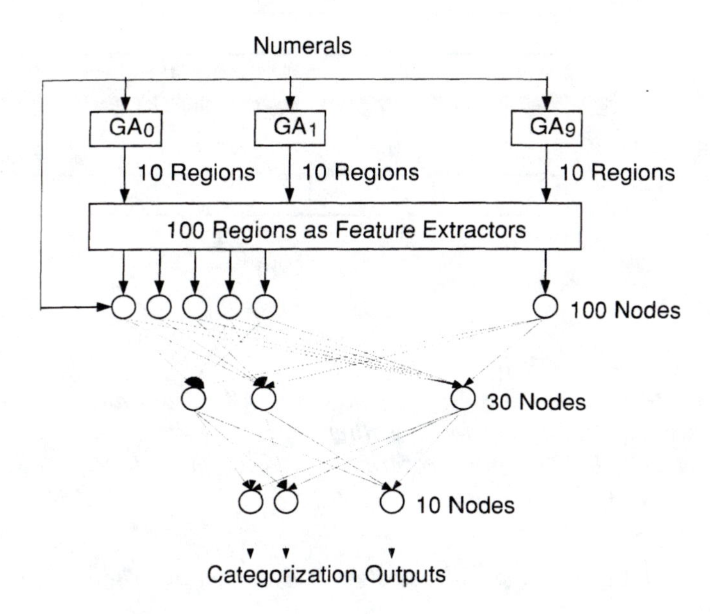

Figure 20.8. *GA-determined fuzzy-filtered neural network. This figure shows a two-phase training procedure. The numerals are first used to identify 100 regions by using 10 GAs, one for each digit. After that, we train the GA-determined fuzzy-filtered NN by using the same numerals to fine-tune the parameters.*

20.5.2 Variations and Discussion

Since there are many design parameters for GAs, we developed several GA versions to optimize the fuzzy-filtered NN model and compared their performance. In the following, we will describe the design motivation behind these variations of the basic model as well as the mechanisms they employed, including gene encoding schemes and choice of evaluation functions.

We tried two gene encoding schemes for this model, as shown in Figure 20.9. The first scheme is the one discussed in the previous subsection. In the second scheme we have the four attributes (i.e., x coordinate, y coordinate, height, and width of a rectangle) grouped together as a unit. Table 20.1 shows the number of misclassified numerals of each digit. Some of the misclassified patterns of one scheme do not appear in the other schemes.

Although the second scheme seems more natural for encoding individual rectangles, we found that the first one yielded to better performance. A possible reason is that diversity of feature locations is at least as important as individual features.

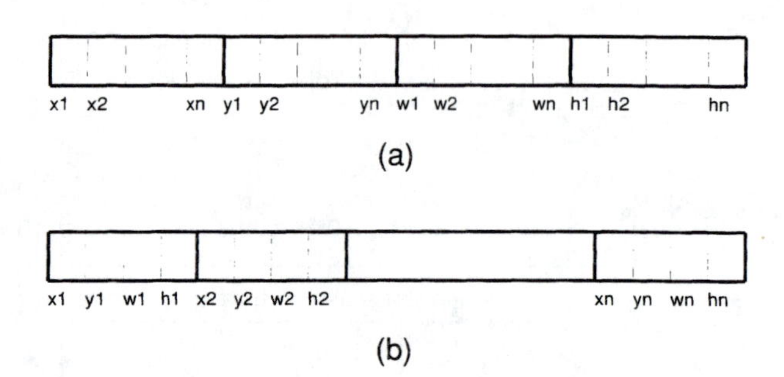

Figure 20.9. *Gene encoding schemes. (a) All x coordinates of the lower-left corners of the rectangles are grouped together, and then the y coordinates, the heights, and the widths. (b) The four features of a rectangle—its x coordinate, y coordinate, height—and width, are grouped together.*

Table 20.1. *Number of misclassified numerals when different gene encoding schemes were used.*

Digit	*0*	*1*	*2*	*3*	*4*	*5*	*6*	*7*	*8*	*9*
First scheme	1	1	1	0	2	5	3	2	6	2
Second scheme	1	0	2	0	3	8	5	2	7	5

The first encoding scheme takes this into account better than the other two because of the one-point crossover mechanism that we used.

In this study, we tried several evaluation functions to guide the selection procedure for GAs before we found the one described in Equation (20.3). Different evaluation functions, as expected, resulted in different filter patterns, which in turn affected the overall performance. For instance, when we employed only the first term in Equation (20.3) as our evaluation function, which is

$$\sum_n \sum_r \frac{b(r)}{w(r)^2 \times z}, \tag{20.4}$$

we obtained the set of fuzzy filters illustrated in Figure 20.10. This set of fuzzy filters looks similar to the written numerals. However, its recognition rate is lower than that produced by the set shown in Figure 20.7. The reason for this is straightforward: When we consider positional features, the existence of pixels does not necessary imply discrimination between different digits.

Since our model is a GA-NN combination, we also tried a hybrid training pro-

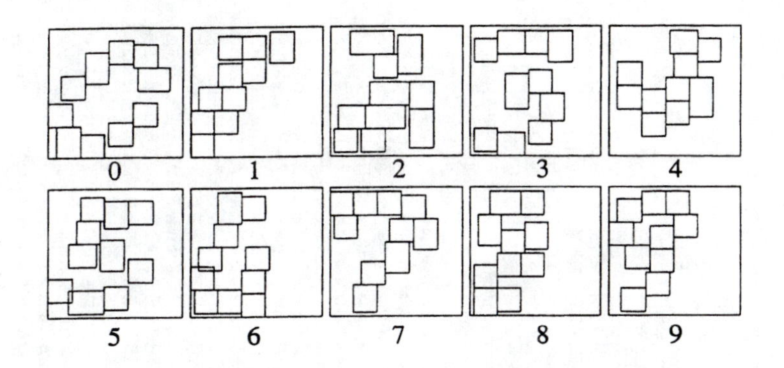

Figure 20.10. *Another set of rectangular feature detectors.*

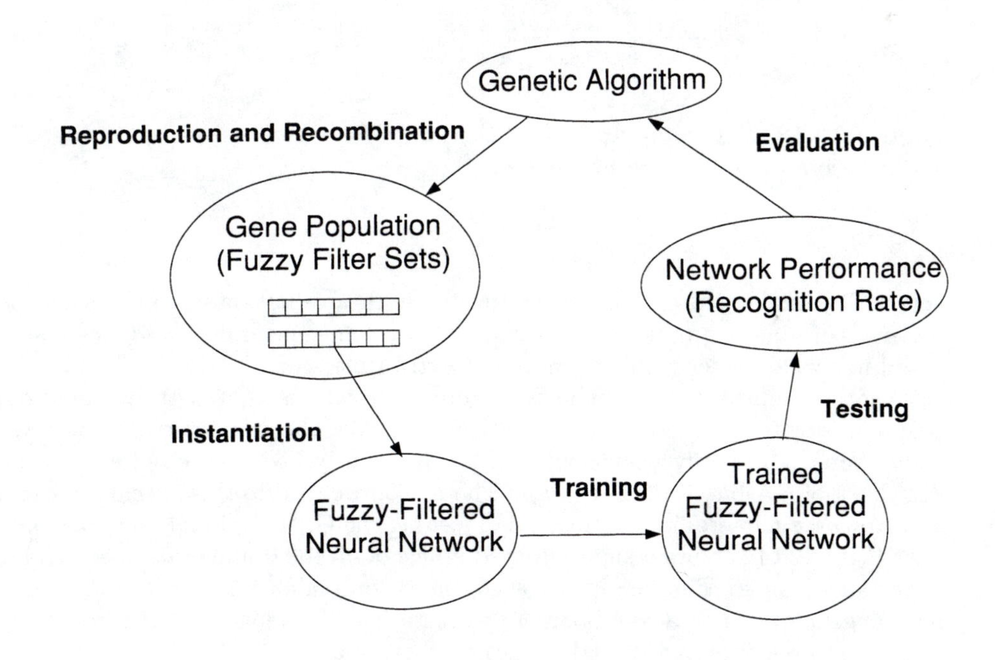

Figure 20.11. *Hybrid GA-NN learning cycle. The original idea comes from ref. [2].*

cedure. The learning cycle is illustrated in Figure 20.11. In other words, after the GA has selected regions, we construct a fuzzy-filtered neural network to finish the training process. Consequently, the NN training error can be used as an evaluation

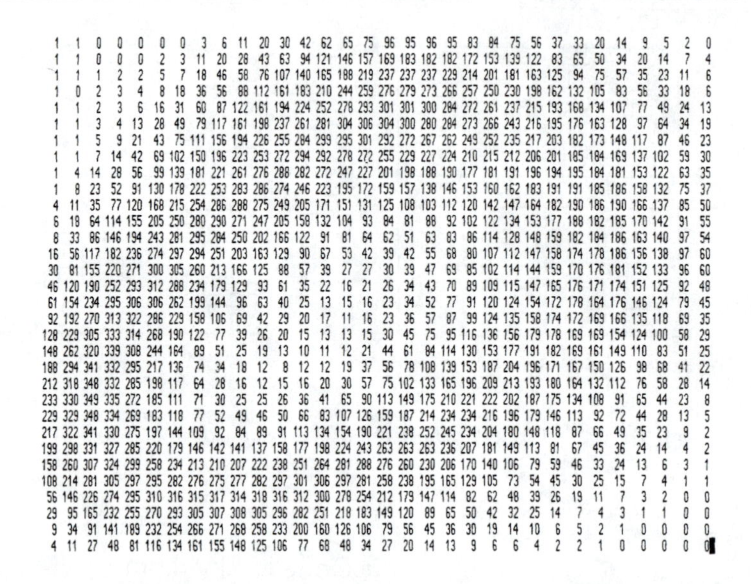

1	1	0	0	0	0	0	3	6	11	20	30	42	62	65	75	96	95	96	95	83	84	75	56	37	33	20	14	9	5	2	0
1	1	0	0	0	2	3	11	20	28	43	63	94	121	146	157	169	183	182	182	172	153	139	122	83	65	50	34	20	14	7	4
1	1	1	2	2	5	7	18	46	58	76	107	140	165	188	219	237	237	237	229	214	201	181	163	125	94	75	57	35	23	11	6
1	0	2	3	4	8	18	36	56	88	112	161	183	210	244	259	276	279	273	266	257	250	230	198	162	132	105	83	56	33	18	6
1	1	2	3	6	16	31	60	87	122	161	194	224	252	278	293	301	301	300	284	272	261	237	215	193	168	134	107	77	49	24	13
1	1	3	4	13	28	49	79	117	161	198	237	261	281	304	306	304	300	280	284	273	266	243	216	195	176	163	128	97	64	34	19
1	1	5	9	21	43	75	111	156	194	226	255	284	299	295	301	292	272	267	262	249	252	235	217	203	182	173	148	117	87	46	23
1	1	7	14	42	69	102	150	196	223	253	272	294	292	278	272	255	229	227	224	210	215	212	206	201	185	184	169	137	102	59	30
1	4	14	28	56	99	139	181	221	261	276	288	282	272	247	227	201	198	188	190	177	181	191	196	194	195	184	181	153	122	63	35
1	8	23	52	91	130	178	222	253	283	286	274	246	223	195	172	159	157	138	146	153	160	162	183	191	191	185	186	158	132	75	37
4	11	35	77	120	168	215	254	286	288	275	249	205	171	151	131	125	108	103	112	120	142	147	164	182	190	186	190	166	137	85	50
6	18	64	114	155	205	250	280	290	271	247	205	158	132	104	93	84	81	88	92	102	122	134	153	177	188	182	185	170	142	91	55
8	33	86	146	194	243	281	295	284	250	202	166	122	91	81	64	62	51	63	83	86	114	128	148	159	182	184	186	163	140	97	54
16	56	117	182	236	274	297	294	251	203	163	129	90	67	53	42	39	42	55	68	80	107	112	147	158	174	178	186	156	138	97	60
30	81	155	220	271	300	305	260	213	166	125	88	57	39	27	27	30	39	47	69	85	102	114	144	159	170	176	181	152	133	96	60
46	120	190	252	293	312	288	234	179	129	93	61	35	22	16	21	26	34	43	70	89	109	115	147	165	176	171	174	151	125	92	48
61	154	234	295	306	306	262	199	144	96	63	40	25	13	15	16	23	34	52	77	91	120	124	154	172	178	164	176	146	124	79	45
92	192	270	313	322	286	229	158	106	69	42	29	20	17	11	16	23	36	57	87	99	124	135	158	174	172	169	166	135	118	69	35
128	229	305	333	314	268	190	122	77	39	26	20	15	13	13	15	30	45	75	95	116	136	156	179	178	169	169	154	124	100	58	29
148	262	320	339	308	244	164	89	51	25	19	13	10	11	12	21	44	61	84	114	130	153	177	191	182	169	161	149	110	83	51	25
188	294	341	332	295	217	136	74	34	18	12	8	12	12	19	37	56	78	108	139	153	187	204	196	171	167	150	126	98	68	41	22
212	318	348	332	285	198	117	64	28	16	12	15	16	20	30	57	75	102	133	165	196	209	213	193	180	164	132	112	76	58	28	14
233	330	349	335	272	185	111	71	30	25	25	26	36	41	65	90	113	149	175	210	221	222	202	187	175	134	108	91	65	44	23	8
229	329	348	334	269	183	118	77	52	49	46	50	66	83	107	126	159	187	214	234	234	216	196	179	146	113	92	72	44	28	13	5
217	322	341	330	275	197	144	109	92	84	89	91	113	134	154	190	221	238	252	245	234	204	180	148	118	87	66	49	35	23	9	2
199	298	331	327	285	220	179	146	142	141	137	158	177	198	224	243	263	263	263	236	207	181	149	113	81	67	45	36	24	14	4	2
158	260	307	324	299	258	234	213	210	207	222	238	251	264	281	288	276	260	230	206	170	140	106	79	59	46	33	24	13	6	3	1
108	214	281	305	297	295	282	276	275	277	282	297	301	306	297	281	258	238	195	165	129	105	73	54	45	30	25	15	7	4	1	1
56	146	226	274	295	310	316	315	317	314	318	316	312	300	278	254	212	179	147	114	82	62	48	39	26	19	11	7	3	2	0	0
29	95	165	232	255	270	293	305	307	308	305	296	282	251	218	183	149	120	89	65	50	42	32	25	14	7	4	3	1	1	0	0
9	34	91	141	189	232	254	266	271	268	258	233	200	160	126	106	79	56	45	36	30	19	14	10	6	5	2	1	0	0	0	0
4	11	27	48	81	116	134	161	155	148	125	106	77	68	48	34	27	20	14	13	9	6	6	4	2	2	1	0	0	0	0	0

Figure 20.12. *Superimposed image: Digit 0 (358 written numerals) represented by the intensity value on each position.*

score. The smaller the training error, the higher the evaluation score. This approach looks natural and complete; however, it is very time-consuming, and, surprisingly, it did not yield better results despite the extra time cost.

Another important benefit of the proposed model is that with a simple data-preprocessing technique we can dramatically improve the training efficiency of GAs, which has traditionally been poor and has frequently been criticized as a primary drawback of GA-based optimization. The reason we can do this is that we can use *superimposed* numeral images instead of feeding the GA individual numerals one by one. Figure 20.12 shows a superimposed image of all the 0 numerals in our training set as an example. The GA in our study needs only about 30 seconds to converge for 400 generations on a Sun Sparc-10 workstation. This justifies the incorporation of GAs into our integrated model for their flexibility.

The last point concerns the number and size of regions. We used 10 regions for each digit, which is good enough for this application. All the aforementioned methods use a variable rectangle size with a minimal value of 5×5. Our evaluation functions prefer small-sized rectangles because they tend to have higher pixel density. As a result, we put a lower bound on the size of the rectangles because it needs to be large enough to extract meaningful features. Generally speaking, when the difference between numerals of a digit is larger, we need larger regions for that digit. However, larger regions reduce the difference across digits. Thus, determining a proper lower bound for region size involves a design trade-off.

20.6 SUMMARY

The banner of fuzzy logic, as pointed out by its inventor, Professor Lotfi Zadeh, is to exploit the tolerance for imprecision [7]. Because high precision entails high cost and low tractability, reasonable solutions to problems encountered in daily life usually employ knowledge at a compromised degree of precision granularity. In this sense, feature detection can be defined as an effort to extract essential attributes from a massive amount of information so that a pattern recognition problem can be solved efficiently.

This chapter introduces a general methodology for adaptive feature extraction. The neuro-fuzzy model, called a *fuzzy-filtered neural network*, described here is not only capable of learning and adapting to variations in training samples, it also identifies pertinent features effectively.

We first investigate the use of the proposed model to monitor a plasma environment through optical signals. Simulations on experimental spectra substantiate the effectiveness of fuzzy channels. The location and shape of membership functions provide new insight into the complicated chemical reaction. This helps give us an idea of what wavelengths are of the greatest interest. Once the critical wavelengths are identified, further automation goals become achievable, such as endpointing, contamination monitoring, and process control.

We also validate the generality of the fuzzy-filtered neural network by using it to identify important positional features for recognition of hand-written numerals. We explore three different architectures: one-dimensional fuzzy filters, two-dimensional fuzzy filters, and GA-based fuzzy filters. Again, experiments on a large-scale data set validate the effectiveness of the model. In this case the location and shape of membership functions are used to identify positional features for pattern recognition problems with a very low level of human involvement. In both applications our model gives us an idea of what kinds of features the machine could use to simulate human decision making. Once the features are identified, neural network learning become much easier and more efficient.

This technique can also be used for pattern analysis in other fields, such as medical diagnosis or global change, in which the explanation of the detected features plays a role as important as the applicability of the working model. In brief, complicated patterns are not self-explanatory; they require a good deal of interpretation. Fuzzy-filtered neural networks successfully serve this purpose.

In summary, the network described here actually learns on its own from the training data and selects the most pertinent positional features without any high-level guidance from human experts. This ability to detect features is the primary advantage of the fuzzy filtering mechanism. This capability is significant, because human experts can actually learn from the network and gain a better understanding of the problem it treats and thus produce better solutions.

The primary limitation of this model is that it handles positional features exclusively. This model treats inputs as a multichannel spectrum, regardless of whether the spectrum is of single dimension or multiple dimensions. This method is very effective for simple pattern recognition. However, for the task of analyzing complicated patterns, such as in Chinese character recognition, this method is less effective. In this type of task, it is preferable to use our model as a perception-level, subsymbolic module in conjunction with other human expertise–intensive models.

REFERENCES

[1] H. Altay Güvenir and İzzet Şirin. A genetic algorithm for classification by feature partitioning. In *Proceedings of the Fifth International Conference on Genetic Algorithms*, pages 543–548, 1993.

[2] S. A. Harp, T. Samad, and A. Guha. Towards the genetic synthesis of neural networks. In *Proceedings of the Third Conference on Genetic Algorithms*, 1989.

[3] Jyh-Shing Jang and Chuen-Tsai Sun. Functional equivalence between radial basis function networks and fuzzy inference systems. *IEEE Transactions on Neural Networks*, 4(1):156–159, 1993.

[4] John Moody and Christian Darken. Learning with localized receptive fields. Technical Report YALEU/DCS/RR-649, Department of Computer Science, Yale University, 1988.

[5] Stephen M. Omohundro. Geometric learning algorithms. Technical Report TR-89-041, International Computer Science Institute, 1989.

[6] W. F. Punch, E. D. Goodman, Min Pei, Lai Chia-Shun, P. Hovland, and R. Enbody. Further research on feature selection and classification using genetic algorithms. In *Proceedings of the Fifth International Conference on Genetic Algorithms*, pages 557–564, 1993.

[7] Lotfi A. Zadeh. Unpublished lectures. Department of Electrical Engineering and Computer Sciences, University of California at Berkeley.

Chapter 21

Fuzzy Sets and Genetic Algorithms in Game Playing

C.-T. Sun

21.1 INTRODUCTION

This chapter describes two approaches to integrating *fuzzy set theory* and *genetic algorithms* and explores their application to game playing. The proposed models indicate a promising direction for adaptation in a changing environment. We first discuss the realization of *diversified selection* by employing multiple coaches in a game-playing program with a genetic algorithm–based learning module. We show that when several coaches are used, the collective learning result is better than when only a single coach is involved, regardless of the ability of the coach.

Next, we introduce two ways of incorporating fuzzy set theory into GAs. At different stages of a game, players usually focus on different features, which are used in static evaluation of board configurations. Thus, characterizing the features by means of *membership functions* is a natural extension of the basic model. This extension leads to better performance, as expected. Moreover, membership functions based on human expertise are not sufficient to cope with situations in an ever-changing environment. Consequently, we introduce the concept of *fuzzy stages* as an alternative solution. We integrate fuzzily divided game-playing stages with a technique called *genetic structural expansion* by employing a multistage chromosome coding scheme in a genetic algorithm. In particular, this chapter compares three chromosome coding schemes—*haploidy, triploidy* (a special case of *polyploidy*), and *structural expansion*—and discusses their impact on multiple fuzzy-stage game-playing strategies.

21.2 VARIANTS OF GENETIC ALGORITHMS

As mentioned in Section 7.2 of Chapter 7, genetic algorithms (GAs) were invented to simulate evolutionary processes observed in nature so that the goal of survival or optimization in a changing environment could be achieved. A GA manipulates *chromosomes* which encode a set of parameters of a target system to be optimized. A GA uses three operators—*selection* (or *reproduction*), *mutation*, and *crossover*—to achieve the goal of evolution. The single piece of information a GA receives from the environment is a scalar indicator that evaluates the performance of each chromosome. The GA then uses that evaluation to bias the selection of chromosomes so that those with better scores tend to reproduce more often than those with worse scores. In addition, GAs use mutation and crossover to create children that differ from their parents.

It is well known that *diversity* helps a population survive under changing environmental conditions, both in nature and in evolutionary computation. In other words, a successful evolutionary strategy should force chromosomes to exhibit diversity so that the evolutionary process will not suffer from early saturation brought on by uniformity. In brief, it is important to *keep the gene pot boiling* to achieve the goal of continual evolution. Although this issue has long been addressed by researchers, very few satisfactory strategies have been suggested because of the difficulty of achieving a balance between fitness and diversity.

The simplest way to achieve diversity is by increasing mutation rates or by introducing more radical mutation operators, such as the *partial complement operator* discussed in ref. [3]. It has been found, however, that this approach usually results in poor performance. Increasing the population size to compensate for this side effect does not work well either, because it results in poor learning efficiency. Other mutation mechanisms include *random immigrants* and *triggered hypermutation* [1].

Another method of enhancing diversity is to consider the distribution of individuals as an evaluation factor in the selection process; see ref. [5] for an example of such a method. This concept was also employed in the *rank-space method* suggested by Patrick Winston in his AI textbook [15]. However, it is generally difficult to define distance adequately in a multidimensional search space. Recently, a systematic discussion of *disruptive selection* has been presented that provides us with a better understanding of the behavior of a nonmonotonic fitness function [8]. A GA employing disruptive selection can be used successfully to solve problems such as optimizing a *needle-in-a-haystack* function, which is traditionally GA-hard. Nevertheless, in general, we cannot establish a balance between fitness and diversity merely by selecting some of the most unfit members.

A third approach was pioneered by Goldberg and Smith's seminal work [4], in which chromosome structures are expanded to meet new challenges from the environment. Goldberg and Smith investigated the genetic mechanisms of *diploidy* and *dominance* and their effects on tracking the optimum chromosome of a changing environment. In this chapter, we follow in their steps by employing *fuzzified polyploid chromosomes* in response to different stages of a dynamic problem, such as game playing.

A topic related to polyploidy is *chromosome structure expansion* (i.e., expansion aimed at developing more complicated gene structures to meet varying conditions in the environment, exploiting a certain *niche*, or achieving *co-evolution* [3]). Although structural expansion has been considered as a means of increasing diversity to benefit the GA optimization process, it has seldom been studied in the context of a multistage reinforced environment.

It is natural to apply different strategies at different stages of a problem. In most cases, stages in a dynamic system overlap each other (i.e., the boundaries between them are fuzzy rather than crisp). Consequently, we use *membership functions* [17] to characterize different stages in a temporally reinforced environment. Examples in this chapter are *open game*, *midgame*, and *end-play* in game playing. Thus, different strategies, encoded in polyploid chromosomes, should be integrated through a fuzzy combination scheme to provide a smooth transition between stages. Ever since Karr's work on employing genetic algorithms to determine fuzzy rules [7], researchers have been trying to integrate these two paradigms, in different ways, to obtain optimal solutions and maintain flexibility at the same time [2, 11, 12]. Here we propose another way of applying fuzzy logic in evolutionary computation.

21.3 USING GENETIC ALGORITHMS IN GAME PLAYING

We test the combined concept of fuzzy set theory and GAs in the domain of *Othello*, which is a challenging game for human players because of the difficulty of envisioning the drastic board changes that result from moves. On the other hand, this game is of reasonable complexity for the task of learning because of its moderate branching factor. World-class Othello programs have been developed by Rosenbloom [13] and Lee and Mahajan [10], who applied techniques such as iterative deepening, move ordering, pattern classification, and Bayesian learning in their work. Their approach requires a great deal of human expertise for both game-playing strategies and learning mechanisms. The goal of this chapter, on the other hand, is to use fuzzy set theory and GAs to improve the learning capability of game-playing programs.

Before explaining how GAs can be applied to the adaptation of game-playing strategies, we shall briefly describe the domain of Othello. The game is played by two players, Black and White, on an 8 by 8 board, which is initially set up as shown in Figure 21.1(a). Black starts the game by placing a black piece on any empty square on the board adjacent to one or more of White's pieces. By this move, Black captures the adjoining white pieces, which are then flipped over to show their black side. Figures 21.1(b) and 21.1(c) show an example. There are two restrictions, however: (1) One of the bracketing pieces must be the piece just placed on the board, and (2) a move must flip at least one of the opponent's pieces. When a player does not have any legal moves, he or she loses a turn. The players take turns placing pieces on the board until neither player can make another move. The player with the most pieces on the board is then declared the winner. (Refer to ref. [9] for further details.)

Most successful game-playing programs employ a heuristic search that uses a *static evaluation function* to guide the direction of the search. A typical linear

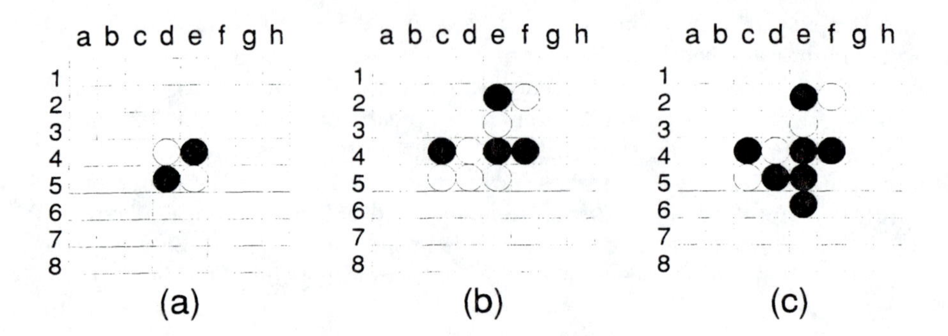

Figure 21.1. *Game of Othello. (a) The initial setup. After Black plays to e6 on (b), the board configuration changes to (c).*

evaluation function has the following form:

$$h = \sum_{i=1}^{n} w_{f_i} f_i, \tag{21.1}$$

where h denotes the *static evaluation function* of a game board configuration, the f_i's are *features*, n of them in total, that play important roles in game-playing strategies, and w_{f_i}'s are the corresponding *weights* that indicate the relative importance of the features. With such an evaluation function, we can apply the well-known *minimax search* algorithm as well as *alpha-beta pruning* techniques. The power of a game-playing program is thus determined by two factors: how the discriminating features are selected and how the weights are assigned. These two factors have been the focus of a great deal of research since Arthur Samuel published his seminal work on machine learning [14]. In this chapter we concentrate on the second factor by using genetic algorithms.

Genetic algorithms avoid local minima or suboptimal results. Consequently, GAs are ideal for searching the weight space of the heuristic function h used in game-tree algorithms. In the design of GA learning algorithms, we take into consideration many important factors (e.g., those summarized in ref. [6]). To play Othello games, our coach programs and GA learning programs employ features such as *position, piece advantage, mobility,* and *stability.* We treat a chromosome string as a vector of real-valued parameters which represent the coefficients of the game-playing heuristic function.

Before beginning this research, we organized a local computer Othello tournament with about 60 competitors. The top five players in the tournament were selected as the five coach programs. A full-width minimax search with alpha-beta pruning was commonly used in the coach programs. To deal with time constraints,

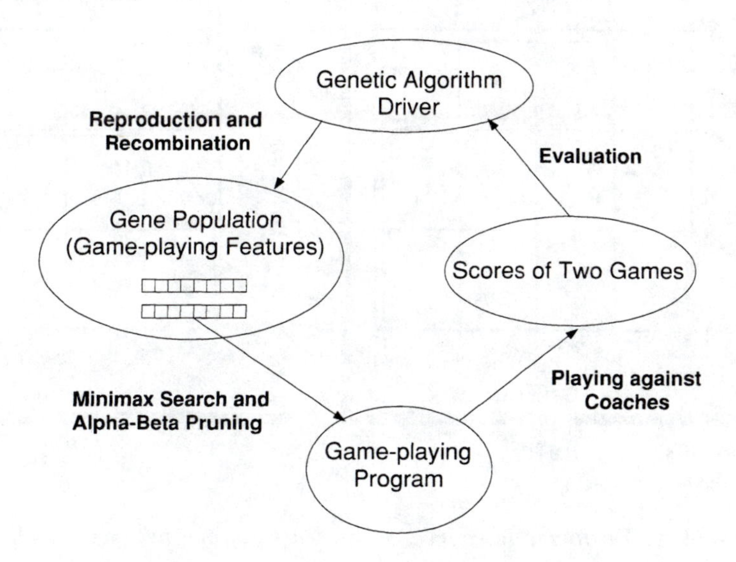

Figure 21.2. *GA training of Othello players.*

some of the programs employed the *iterative deepening* strategy (i.e., they performed a full N-ply search before attempting an $N+1$-ply search). We will call the coaches Coach_1, Coach_2, Coach_3, Coach_4, and Coach_5, respectively, in order of increasing power. In other words, Coach_5 was the program that won the tournament. The rationale behind employing multiple coaches is twofold. First, it is difficult to choose a single evaluation standard for success in a real-world, multicriteria environment such as game playing. A necessary condition for a player to be "good" is that he or she can beat more than one "good" rival. Second, if we use only one coach for training, it is easy for individuals to overcommit themselves to that coach's weak points and thus fall into local optima.

All the programs were coded in C++. The population size was set to 200. Each member of the population took about 15 s on a 486/DX2-66 PC to finish a game. Evolutionary behavior was visible even with the relatively small population size. We employed fitness-proportional selection in this project. Each member of a certain generation played two games with one coach. Each pair of opponents took turns starting the two games. The fitness score was defined as the sum of the piece advantages of the two games. In terms of reproduction, we found that recombination, especially crossover, was very productive in yielding good offspring. Figure 21.2 summarizes the GA training process.

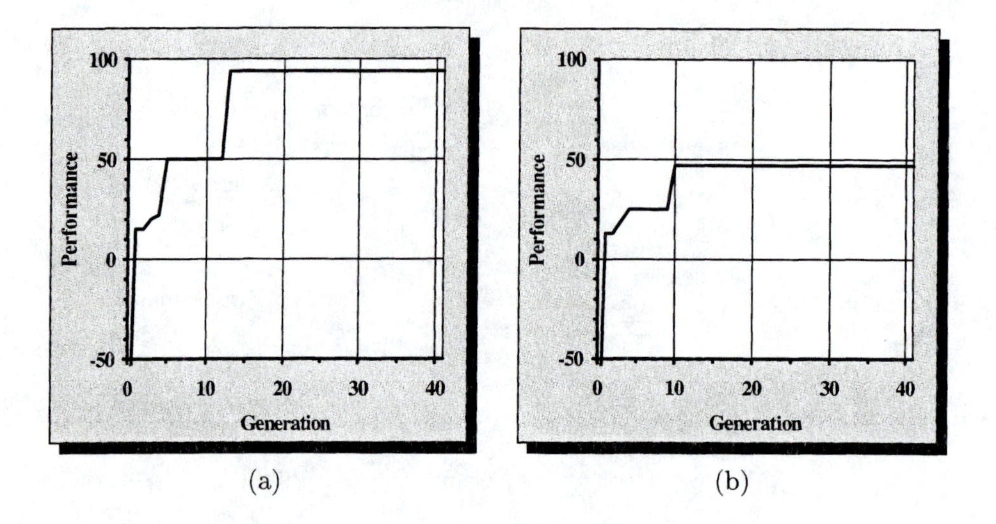

Figure 21.3. *Performance curves when learning against (a) $Coach_1$; (b) $Coach_5$.*

21.4 SIMULATION RESULTS OF THE BASIC MODEL

To test the validity of the proposed model, we applied it in several learning situations. In our first try, we used a game-playing heuristic function with six features: board position measure, piece advantage, current mobility (defined as the difference in the number of possible moves between a GA player and a coach), stability (number of unflippable pieces), potential mobility 1 (number of opponent's pieces adjacent to empty squares), and potential mobility 2 (number of empty squares adjacent to opponent's pieces).

For comparison, we initially let the GA player learn from a single coach. Learning curves demonstrating the evolutionary processes against $Coach_1$ and $Coach_5$ are shown in Figure 21.3(a) and 21.3(b), respectively. As expected, the best individual in each case learned to beat the corresponding coach because it successfully found a good weight combination for its heuristic function. The GA player won by a larger margin against $Coach_1$ than against $Coach_5$, since $Coach_5$ was the tougher of the two coaches.

Now we come to an interesting question: If the GA player has an opportunity to learn from all five coaches, will the result be better than that in cases where only a single coach is involved, regardless of whether the coach is the champion or an ordinary player? This question is not trivial, because we know that learning with multiple goals at the same time may result in useless interpolation. The only way to answer this question is by doing experiments.

In the initial population for training with multiple coaches, each individual was scheduled to play two games with $Coach_1$. In any later generation, only those

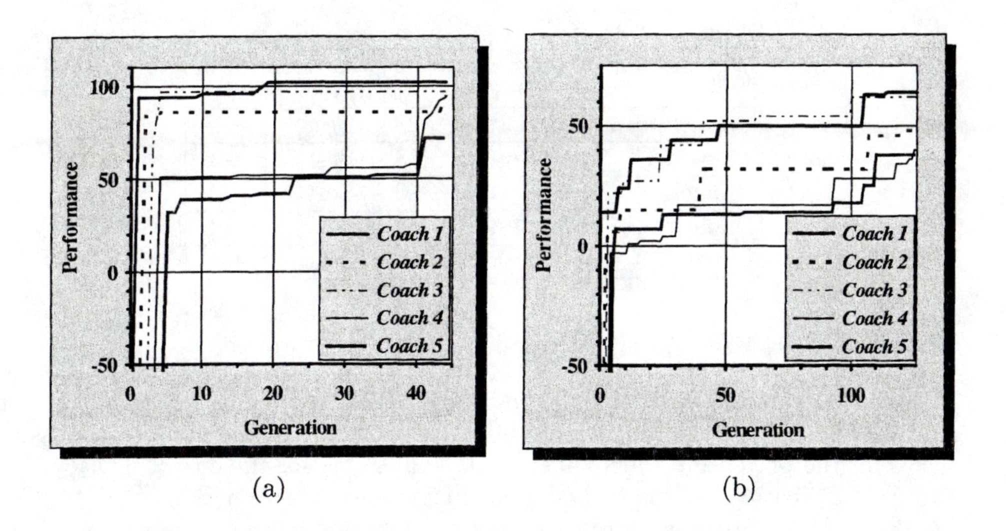

Figure 21.4. *Performance curves when learning against all five coaches: (a) basic models; (b) with expanded position features.*

individuals who had beaten Coach_i and were not changed by the reproduction operators were scheduled to play against Coach_{i+1}. The evolutionary learning results are shown in Figure 21.4(a).

Obviously, the learning effect was better than that with single-coached training. The best player in the final population against Coach_1 achieved a total score of more than 100, which was better than the result shown in Figure 21.3(a). The most remarkable result was that the best player against Coach_5 was able to beat the coach by a margin of greater than 70. Note that the final score in Figure 21.3(b) was saturated below 50. Thus, keeping chromosomes with greater variety helped produce a performance breakthrough. From the learning curves we can observe that every time an individual developed a better weight combination, the result was likely to ripple out to affect its performance against other coaches. Moreover, we observe that the best players against different coaches were in general different individuals, thus the ripple effect was caused by crossover.

After we found that the GA produced satisfactory results, we tried to enhance its potential ability to identify lower-level features. The current board position measurement plays a pivotal role in Othello. In the previous design the importance of each square was determined by human experience. To avoid this kind of high-level human involvement in learning, we broke the configuration feature of the whole board down into individual position features. In the expanded chromosome encoding, we represented the board position measure as 10 values, taking advantage of the symmetry of the Othello board. Furthermore, we found that a single feature of relative mobility was not enough to reflect the importance of the total number

Table 21.1. *Play against Master. Two games were played. The table shows the sum of the final scores of our program players against Reversi Master.*

Coach_1	+38	Best_1^6	+28	Best_1^{17}	−38
Coach_2	+6	Best_2^6	−38	Best_2^{17}	−6
Coach_3	+8	Best_3^6	+6	Best_3^{17}	−60
Coach_4	−10	Best_4^6	−12	Best_4^{17}	+2
Coach_5	+20	Best_5^6	+11	Best_5^{17}	−18

of legal moves in different game-playing stages, such as opening moves and closing moves. Hence, we broke that feature into two subfeatures: the number of possible moves for the player and the number of possible moves for the coach. Thus we had a total of 17 features in the revised game-playing heuristic function.

The learning curves with this expanded chromosome format are shown in Figure 21.4(b). As expected, the learning speed was reduced because it was time-consuming to find the correct setting of importance for each position. However, interaction between coach curves occurred much more frequently than before.

Another interesting point is that the order of the coaches in terms of power, from the GA player's point of view, was not quite the same as the ranking found in the tournament. For example, the GA player beat Coach_3 by a larger margin than Coach_2, whereas the former beat the latter in the tournament. Moreover, as pointed out in ref. [16], an interesting relationship could develop between average human players (Hs), expertise-based game-playing programs (Es), and GA-based learning programs (Ls). This relationship would be cyclic, just like that between the scissors, paper, and stone in the well-known children's game: Although Es outperform Hs, Ls learn to beat Es, and Hs are, in most cases, able to beat Ls easily. The apparent reason for this relationship is that Ls overcommit themselves to the weak points of Es and develop a narrow strategy that is not good for generalization.

Since it is interesting to see whether training with multiple coaches enables Ls to avoid this drawback, we had the programs play against the commercial Othello program Reversi for $Windows^{TM}$. There are four levels in Reversi: *Beginner*, *Novice*, *Expert*, and *Master*. We chose the Master level for our program players, including all the coaches and the best GA players against each coach. Since we had two versions of GA encoding, we use the superscripts 6 and 17 to denote the number of genes used in the chromosome format. For example, Best_2^{17} denotes the best player against Coach_2 using 17 features. We gave our programs 0.3 seconds for each move; this was in general less than the time consumed by *Reversi*, which typically used 2 to 3 seconds for its midgame search. The results are summarized in Table 21.1.

If we consider the Best_i^{17}'s, the cyclic relationship is clear. Best_i^{17} beat Coach_i, Coach_i beat Master, and Master beat Best_i^{17}, with Best_4^{17} the only exception. However, if we consider the Best_i^6's, which employed more human knowledge in terms of position evaluation than the Best_i^{17}'s, the outcome was different. In this case, we had three GA players, Best_1^6, Best_3^6, and Best_5^6, who beat both their coaches and the Reversi Master program. Although these results enhanced our belief in the learning ability of GAs, they also left us many questions to answer in the future.

21.5 USING FUZZILY CHARACTERIZED FEATURES

In an Othello game there are at most 60 moves. We know that different strategies should be employed in different game stages. However, in most cases, the stages can only be described linguistically. For example, "keeping high mobility is important *at the beginning*" and *"in the endgame stage* one should capture as many pieces as possible" are two obvious strategic rules in playing a game. Linguistically characterized game features are suitable for representing strategies. In this section, we will realize this concept directly using *fuzzily characterized features*. In the next section, we will propose a general model based on *fuzzy polyploid chromosomes*.

The general form of the static evaluation function $h(t)$ is now as follows:

$$h(t) = \sum_{i=1}^{n} \mu_{f_i}(t) w_{f_i} f_i, \tag{21.2}$$

where $\mu_{f_i}(t)$ is the membership function characterizing feature f_i, with t denoting the ply number in a game.

We employed the expanded seventeen features in this set of experiments. Each feature has its own function to indicate its importance as a game proceeds. The membership functions characterizing these features are shown in Figure 21.5.

The 10 positional features are largely time invariant. In other words, the importance of a position in Othello does not change dramatically as a game proceeds. However, as the end of a game approaches, players' freedom of choice concerning which position to take decreases steadily. Thus, the corresponding curve drops down near the end. The same observation applies to the two potential mobility measures. Because at the final stage of a game there are fewer and fewer empty squares left on the board, the relative importance of potential mobility should decrease.

In contrast, experience tells us that in an Othello game we should not be too greedy (i.e., we should not emphasize the piece advantage in the early stages of a game). As the game proceeds, we will gradually increase our focus on this feature to capture as many pieces as possible.

On the other hand, because we also emphasize the number of possible moves that our opponent can make, with the aim of forcing him or her to pass in the final stage of the game, we treat mobility as an important feature no matter what stage of the game we are currently in. Similarly, because we do not want our pieces to be flipped at any time, we use a function with the unit value everywhere to indicate the importance of stability.

The learning curves against the five coaches of the GAs employing fuzzified fea-

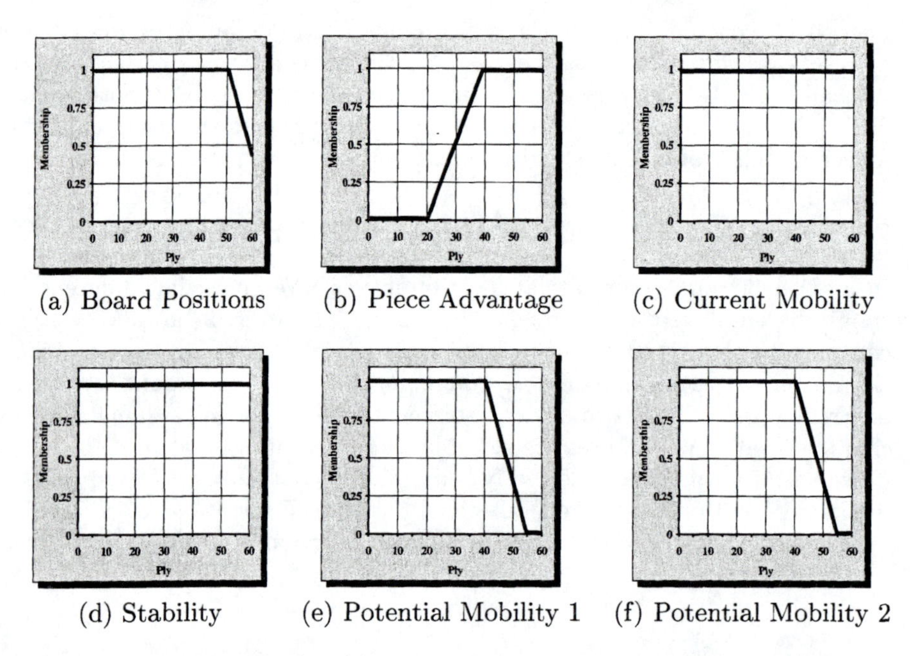

(a) Board Positions (b) Piece Advantage (c) Current Mobility

(d) Stability (e) Potential Mobility 1 (f) Potential Mobility 2

Figure 21.5. *Membership functions characterizing game-playing features.*

tures are shown in Figure 21.6. Compared to Figure 21.4(b), the overall performance and the learning speed of the GAs with fuzzified features are clearly better than those with pure GA training alone. This result should not be surprising because we actually put more knowledge into the system than before.

In summary, employing fuzzified game-playing features allows the static evaluation function to emphasize different features at different stages, so that the evaluation score of the board configuration is more appropriate than before. However, since the membership functions characterizing the features are determined by human experts, there is still room for improvement in terms of automatic learning.

21.6 USING POLYPLOID GA IN GAME PLAYING

After discussing *fuzzily characterized game features*, we now introduce an even more flexible approach by using *fuzzily characterized game stages.* By dividing a game into several fuzzy stages, we actually consider the features in a dynamic way, which is intuitive and likely to result in better performance.

Take a three-stage game as an example. In this case, we employ three membership functions to characterize the three stages of the game: *open game, midgame*, and *end game*, as shown in Figure 21.7. Since it takes 60 plies to finish an Othello game, we use the thirtieth ply as the core of the midgame membership function.

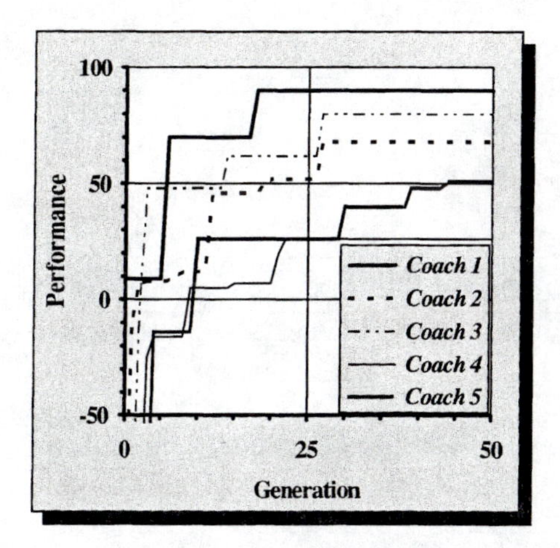

Figure 21.6. *Performance curves: Learning against five coaches, with fuzzified features.*

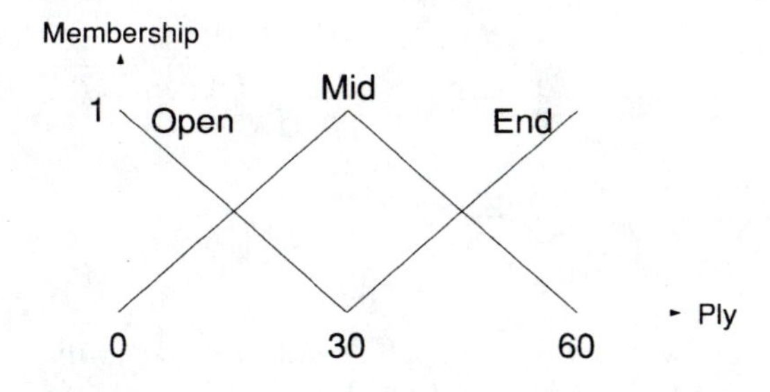

Figure 21.7. *Three Stages of an Othello Game Characterized by Fuzzy Membership Functions.*

Note that both open game and end game overlap with midgame, so a natural transition is supported in this fuzzy scheme.

Each stage needs a corresponding static evaluation function, which is represented by a chromosome. Consequently, we have three chromosomes for each member in a population, as shown in Figure 21.8(b). This chromosome coding scheme is called *triploidy*, which is a special case of *polyploidy*. The overall value of the static evaluation now is the weighted average, governed by the membership grades of the current ply, of the individual evaluation functions represented by the individual

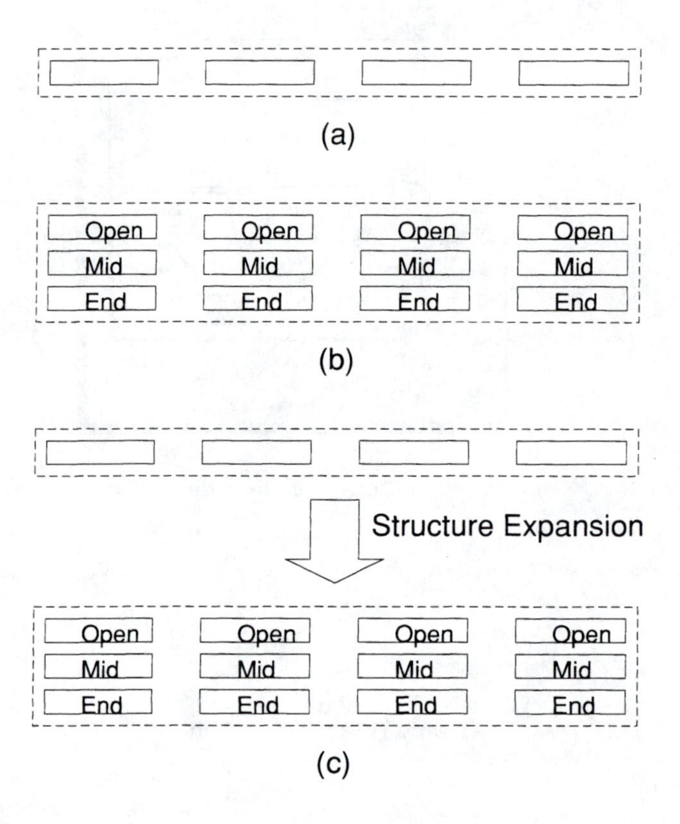

Figure 21.8. *Gene structures. (a) Haploid chromosomes, (b) triploid chromosomes, (c) structural expansion. Each dashed-line box represents a population. Each solid-line box stands for a chromosome, which is a collection of game-playing features.*

chromosomes. Formally, the static evaluation function $h(t)$ can be written as

$$h_m(t) = \sum_{k=1}^{m} \left[\mu_k(t) \sum_{i=1}^{n} w_{f_i}(k) f_i \right], \tag{21.3}$$

where k stands for a fuzzy stage (m stages in total) and $\mu_k(t)$ is the membership value of ply t in stage k. Since we have a set of genes (features) for each stage in this case, we use $f_i(k)$ to denote the ith feature in stage k. Compare the preceding formula with Equations (21.1) and (21.2) to see the different focus of each method.

In nature, polyploidy is usually accompanied by a type of *dominance* mechanism so that at any time only one chromosome acts as the surrogate of the group. The chromosomes other than the dominant one are called *recessive* chromosomes. Our approach treats polyploid chromosomes in a different way. Their relationship is one of cooperation rather than dominance. In other words, each chromosome's role is not one of all-or-nothing participation but rather one of being part of a weighted combination guided by membership functions.

An interesting question to consider is whether polyploidy is an original form or is itself a result of evolution. In other words, in the context of our application, should we employ triploid chromosomes at the very beginning or somewhere further on in the process of evolution? To answer this question, in addition to the traditional haploidy in Figure 21.8(a) and the proposed triploidy in Figure 21.8(b), we also used a third scheme: *structural expansion* at a certain point during the evolutionary process. Figure 21.8(c) illustrates this concept. In practice, we chose to combine the haploid members in the twentieth generation to form triploid chromosomes. Since long chromosomes usually slow down learning considerably, we used *delayed structure expansion* to alleviate the effect of expansion on learning speed. Further, at the twentieth generation the members are largely elite according to the judgment of our five coach programs. Thus combinations of these individuals have a better chance of producing improved triploid members than the combinations produced by the random initialization procedures employed in simple triploid experiments.

The learning curves against the five coaches of the GAs employing triploidy and structural expansion are shown in Figures 21.9(a) and 21.9(b), respectively. On average, structural expansion produced better performance than simple triploidy, as expected. After the twentieth generation, the point of structural expansion, there are obvious breakthroughs on the curves in Figure 21.9(b). Moreover, the interaction between curves is also relatively frequent in the case of structural expansion, which can be considered an indicator of diversity.

To compare the performance achieved with haploidy, triploidy, and structural expansion in more detail, we summarize the scores of the fifty-sixth generation and that of the one-hundred-fortieth generation corresponding to the three chromosome coding schemes in Tables 21.2 and Table 21.3, respectively. Although the character of the coaches affected the variance to a certain degree, from the data in these tables we can still draw two important conclusions. First, simple triploidy did not produce satisfactory improvement because randomly initiated longer chromosomes had more difficult evolving. Second, structural expansion yielded the best results in terms of final performance and learning efficiency.

In our final experiment, we pushed the concept of structural expansion to its limit. This time we used a GA that started with haploid chromosomes, and then, after every 10 generations, increased the length of each chromosome by one unit. The method we used was to duplicate one unit in each of the polyploid chromosomes. This concept is demonstrated in Figure 21.10. The increased stages were characterized by corresponding multistage membership functions as before. In an Othello game we cannot, of course, expand the chromosome structure endlessly because it is not reasonable to divide a game into too many stages. Thus, we stopped the expansion mechanism at the one-hundredth generation (i.e., 10 fuzzy game-playing

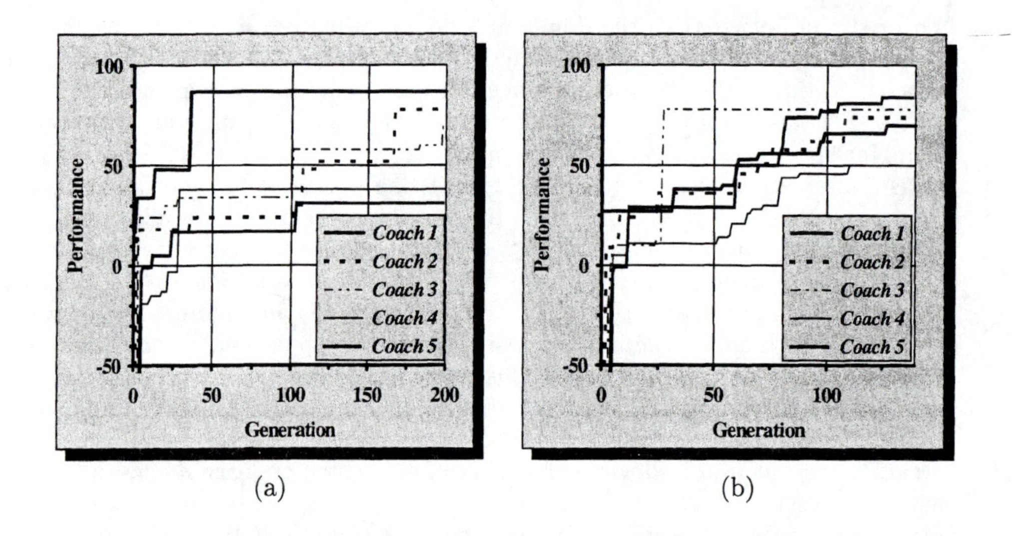

Figure 21.9. *Performance curves when learning against five coaches: (a) with triploid chromosomes; (b) with structurally expanded chromosomes.*

Table 21.2. *Comparison: fifty-sixth generation. Two games were played between each player and each coach. The table shows the sum of the final scores of program players against coaches. Three chromosome structures are compared in the table: haploid, triploid, and structural expansion. The scores were taken from the fifty-sixth generation of evolution in each case.*

	Haploid	Triploid	Expansion
Coach_1	+50	+40	+40
Coach_2	+32	+46	+36
Coach_3	+52	+27	+78
Coach_4	+17	+9	+14
Coach_5	+13	+70	+29

stages). The learning curves against the five coaches of the GAs employing this type of structural expansion are shown in Figure 21.11. Obviously, this mechanism produced better results than any of the triploid cases.

21.7 SUMMARY

Maintaining diversity in GAs is an important issue worthy of further investigation. In addition to the traditional approach of trying to construct a balanced criterion

Table 21.3. *Comparison: one-hundred-fortieth generation. The specifics are the same as in Table 21.2. The scores were taken from the one-hundred-fortieth generation of evolution in each case.*

	Haploid	Triploid	Expansion
Coach_1	+68	+113	+70
Coach_2	+51	+46	+74
Coach_3	+69	+44	+78
Coach_4	+49	+15	+50
Coach_5	+38	+84	+84

between chromosome diversity and evolutionary convergence, we can also explore the promising alternative of using multiple and dynamic standards of survival or success. Game-playing provides a fertile ground for experiments in this area. In this chapter we have focused primarily on two aspects of game playing: learning behavior with multiple coaches and the effects on learning of using fuzzified features.

This chapter describes two methods of integrating fuzzy set theory and genetic algorithms. The first employs heuristically determined membership functions to indicate changes in the importance of features as a game proceeds. The second, even more flexible, approach explores the use of polyploidy, which encodes game stages in chromosome structures. This allows us to apply a different set of feature weights in different game stages. Furthermore, the stages are fuzzily characterized so that no abrupt jump occurs across stage boundaries. Both of the aforementioned mechanisms produced satisfactory results.

The attributes and mechanisms of polyploidy remain important issues worthy of further study. Possible future work includes adaptation of membership functions so that a higher degree of flexibility in search of optima can be achieved. It is also possible to vary the degree of polyploidy to explore the relationship between complexity and performance. Another direction is to introduce Lamarckian properties into the paradigm used here. After a fast learning algorithm is applied to individuals in a generation, the resulting weight vector could be encoded back into the chromosomes so that the adapted behavior of the parents could be passed on to their children.

REFERENCES

[1] Helen G. Cobb and John J. Grefenstette. Genetic algorithms for tracking changing environments. In *Proceedings of the Fifth International Conference on Genetic Algorithms*, pages 523–530, 1993.

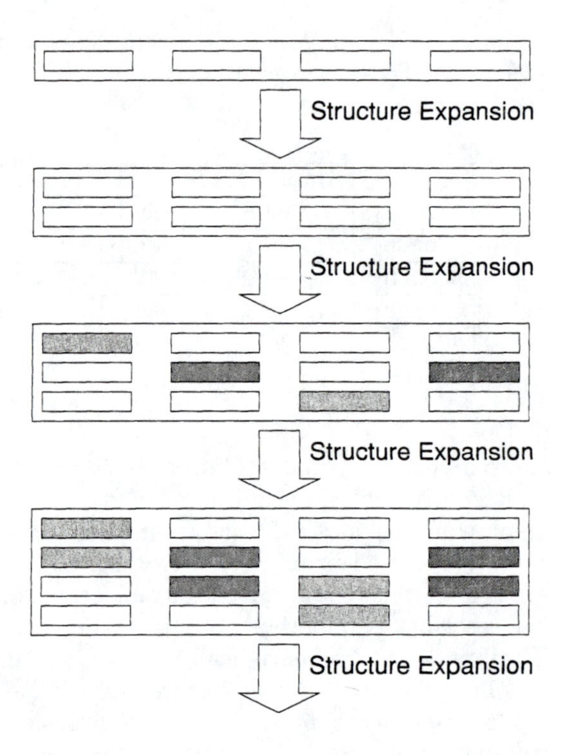

Figure 21.10. *Evolutionary gene structures. Each chromosome expands itself by one unit after every 10 generations. One unit in the polyploidy chromosome duplicates itself in the structure. The transition from the 3rd stage to the 4th stage demonstrates duplication.*

[2] David S. Feldman. Fuzzy network synthesis with genetic algorithms. In *Proceedings of the Fifth International Conference on Genetic Algorithms*, pages 312–317, 1993.

[3] D. E. Goldberg. *Genetic algorithms in search, optimization, and machine learning.* Addison-Wesley, Reading, MA, 1989.

[4] David E. Goldberg and Robert E. Smith. Nonstationary function optimization using genetic dominance and diploidy. In *Proceedings of the Second International Conference on Genetic Algorithms*, pages 59–68, 1987.

[5] Yoshiaki Ichikawa and Yoshikazu Ishii. Retaining diversity of genetic algorithms for multivariable optimization and neural network learning. In *Proceedings of the IEEE International Conference on Neural Networks*, pages 1110–1114, 1993.

[6] Kenneth De Jong and William Spears. On the state of evolutionary computation. In *Proceedings of the Fifth International Conference on Genetic Algorithms*, pages 618–623, 1993.

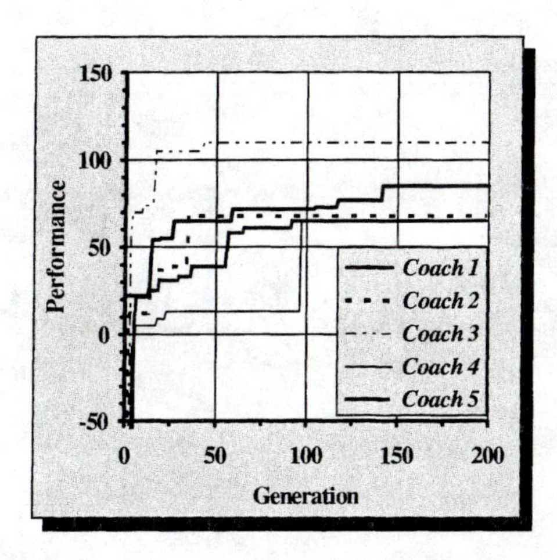

Figure 21.11. *Performance curves: Learning against five coaches, with 10-stage chromosome expansion.*

[7] Chuck Karr. Applying genetics to fuzzy logic. *AI Expert*, 6(3):38–43, 1991.

[8] Ting Kuo and Shu-Yuen Hwang. A genetic algorithm with disruptive selection. In *Proceedings of the Fifth International Conference on Genetic Algorithms*, pages 65–69, 1993.

[9] Kai-Fu Lee and Sanjoy Mahajan. A pattern classification approach to evaluation function learning. *Artificial Intelligence*, (36):1–25, 1988.

[10] Kai-Fu Lee and Sanjoy Mahajan. The development of a world class Othello program. *Artificial Intelligence*, (43):21–36, 1990.

[11] Michael A. Lee and Hideyuki Takagi. Dynamic control of genetic algorithms using fuzzy logic techniques. In *Proceedings of the Fifth International Conference on Genetic Algorithms*, pages 76–83, 1993.

[12] Alexandre Parodi and Pierre Bonelli. A new approach to fuzzy classifier systems. In *Proceedings of the Fifth International Conference on Genetic Algorithms*, pages 223–230, 1993.

[13] Paul S. Rosenbloom. A world-championship-level Othello program. *Artificial Intelligence*, (19):279–320, 1982.

[14] Arthur L. Samuel. Some studies in machine learning using the game of checkers. *IBM Journal of Research and Development*, 3(3):210–229, 1959.

[15] Patrick Henry Winston. *Artificial intelligence*. Addison-Wesley, Reading, MA, 1992.

[16] Edge C. Yeh, G. K. Ruan, and Y. Y. Hsu. Development of game strategy by genetic algorithm for "five pieces in a line". In *Proceedings of the First National Symposium on Fuzzy Set Theory and Applications*, pages 543–550, 1993. In Chinese.

[17] L. A. Zadeh. Fuzzy sets. *Information and Control*, 8:338–353, 1965.

Chapter 22

Soft Computing for Color Recipe Prediction

E. Mizutani

22.1 INTRODUCTION

Colors give meaning and value to our daily lives; for example, painting a room the proper color can enliven it and make it more comfortable. We often need to specify to painters the favorite color coming from a pigment of our imagination. Using **color recipe prediction** as a practical application of **soft computing** in the paint industry, this chapter introduces readers to a neuro-fuzzy methodology we discussed in Chapter 13 and another **computational intelligence** approach that combines a knowledge base (KB) and three principal *soft computing* components: fuzzy systems (FSs), neural networks (NNs), and genetic algorithms (GAs). They function complementarily when put together; their synergism consequently presents unexpected performance enhancements.

We shall demonstrate how the fusion of techniques surpasses the individual capacity of any one technique: *Color matching* is an excellent test of these methods because it is difficult even for skilled human operators to do well, yet human color perception is sensitive, and therefore the matching must be done well to meet acceptable standards. In the next section, we briefly introduce the color recipe prediction task. We then present a simple backpropagation multilayer perceptron (MLP) approach in Section 22.3. After that, we discuss a neuro-fuzzy approach in Section 22.4 and subsequently describe a genetic neuro-fuzzy approach in Section 22.5. Finally, this chapter concludes with a futuristic picture of *soft computing*; emerging *soft computing intelligence* may revolutionize approaches to developing industrial applications.

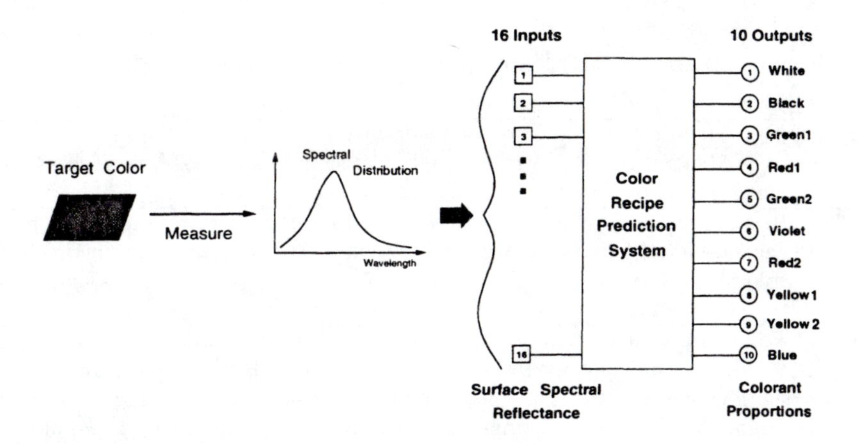

Figure 22.1. *Input-output relation in a typical color recipe prediction system.*

22.2 COLOR RECIPE PREDICTION

Color recipe prediction relates the surface spectral reflectance of a target color to a list of several required colorant proportions that are needed to produce the same color as the reference color (see Figure 22.1). In a practical situation, it is necessary to examine the color match in daylight as well as in artificial light. This is an arduous task even for professional colorists. A succinct description of the main concerns in the recipe prediction is presented in Table 22.1.

In our prediction task, we had 1446 training samples of Munsell color chips and 302 checking samples of standard paint color chips from the Japan Paint Manufacturers Association. Both data sets were sampled by surface spectral reflectance of target colors at 16 points in the visible range of the color spectrum between 400 nm and 700 nm in wavelength (20-nm intervals). With regard to (P2) in Table 22.1, the desired average number of colorants required to produce any color was about 4 out of 10 colorants, as presented in Table 22.2. Those 10 types of colorants included three pairs of the same types of colorants (i.e., green, yellow, and red ones) and also complementary colorants such as "green and red" and "blue and yellow" (see the 10 output units in Figure 22.1). (That is, we carefully determine which colarants to use, avoiding use of the same colorant types and complementary colorants at the same time.) All subsequent experiments were conducted using the same data sets.

22.3 SINGLE MLP APPROACHES

Since the conventionally used Kubelka-Munk theory requires certain assumptions that limit the situations in which the theory may be applied [14], a simple backpropagation MLP approach has been introduced as an alternative method to overcome practical obstacles in color recipe prediction [2, 9, 11].

Two types of simple MLPs, NN_{norm} and NN_{mod}, have been applied as a

Table 22.1. *Main concerns in color recipe prediction.*

(P1)	It is difficult to predict precise colorant concentrations. We sometimes need to predict proportions with enough precision to specify levels such as 0.01%, which is the desired minimal colorant proportion level.
(P2)	It is necessary to specify use of a limited number of colorants to use for acceptable cost performance requirements. At the same time, in the choice of colorants, we need to avoid the use of complementary colorants and of the same types of colorants.
(P3)	The magnitude of mean-squared error of colorant vectors may not correspond exactly to that of color difference. The question is which colorant has the most significant impact on the entire color. For instance, if the target color is very bright, we have to determine carefully the concentrations of dark-colored pigments.
(P4)	It is important to consider human visual sensitivity to *color difference*, which is closely related to perceptual attributes of color (i.e., *lightness*, *hue*, and *chroma* [4, 14]).
(P5)	Some different combinations of colorants may have the same *perceptual* attributes of color as seen by humans.

Table 22.2. *Number of data classified by the desired number of colorants required to produce color in data sets.*

	Two colorants desired	Three colorants desired	Four colorants desired	Desired average # of colorants to use
1446 training data	4	60	1,382	3.95
302 test data	0	13	289	3.96

touchstone to the recipe prediction to fathom the intrinsic difficulty of the task [9]; NN_{norm} has normal sigmoidal functions and NN_{mod} has *modified sigmoidal functions* in the output layer. Both NN_{norm} and NN_{mod} have the same model size ($16 \times 18 \times 21 \times 10$ neurons), mapping surface spectral reflectance of a target color (16 sampled inputs) to a list of required colorant concentrations (10 outputs) (see Figure 22.1).

As indicated in (P2) of Table 22.1, we need to specify which colorants to use. Table 22.2 shows the desired number of colorants in our data sets. The average number of colorants required to produce any color is fewer than five; this means that 6 of the 10 final outputs should be zero. In addition, we sometimes need

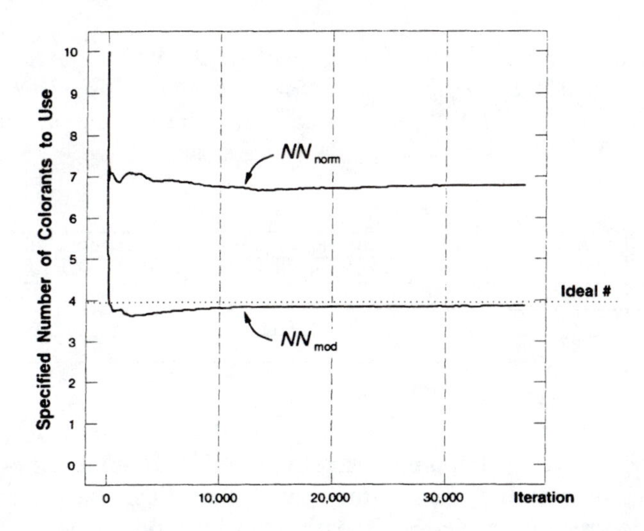

Figure 22.2. *Number of necessary colorants predicted by an MLP with the normal sigmoidal functions (NN_{norm}), and an MLP with the modified sigmoidal functions (NN_{mod}).*

to predict proportions with enough precision to specify levels such as 0.01% [see Table 22.1 (P1)]. It is an important concern in color recipe prediction to specify such output range extremities [2].

To handle these concerns, we have introduced *modified sigmoidal functions* and *truncation filter functions* in the output layer [8, 9] (refer to Section 13.3.3 for details). These functions prevent an NN from exceeding the desired output range. Thus the outputs are further processed to eliminate redundant colorants at the minimum of the desired output range.

The effects of the modified sigmoidal functions can be seen more clearly in Figure 22.2. The NN_{norm} tends to specify use of more colorants than necessary; it averages almost seven specified colorants, which is far from the ideal of about four. On the other hand, in Figure 22.2, the NN_{mod} shows that the predicted number of colorants asymptotically approached the ideal number of colorants as iterations progressed. The comparison of prediction accuracy between NN_{norm} and NN_{mod} is detailed in Table 22.3; NN_{mod} was more effective in avoiding use of the same types of colorants and of complementary colorants than NN_{norm}.

Although NN_{mod} did a better job than NN_{norm}, greater precision in concentration specification is desired. This has inspired us to construct hybrid systems.

22.4 CANFIS MODELING FOR COLOR RECIPE PREDICTION

In practice, we sometimes encounter severe standards which may be hard to meet by employing a backpropagation MLP alone. We contend that another approach, such

Table 22.3. *Performance comparison of single MLP approaches:* NN_{norm} *and* NN_{mod}, *using 302 checking data.* NN_{norm} *is a simple backpropagation MLP, and* NN_{mod} *is an improved* NN_{norm}. *A column, "Error," denotes the average colorant error.*

	Ave. # of colorants	Error $\times 10^{-2}$	No. of test data which outputs same or complementary color				
			2Green	2Yellow	2Red	Red & Green	Yellow & Blue
NN_{norm}	6.66	2.616	97	154	125	198	129
NN_{mod}	3.90	2.031	14	7	5	0	1
Ideal	3.96	0	0	0	0	0	0

as fuzzy modeling, must complement simple MLPs to enhance overall performance. In this section, we show how neuro-fuzzy models can be generalized for application to color recipe prediction; the neuro-fuzzy approaches are expressed within the framework of CANFIS (Coactive Neuro-Fuzzy Inference Systems), detailed in Chapter 13.

To find an ideal adaptive model for this task, we have investigated a variety of structures. They feature knowledge-embedded architectures and an adaptive FS, which serves to determine color selection. They have enormous potential for augmenting prediction capacity.

22.4.1 Fuzzy Partitionings

In fuzzy modeling, it is important to determine a reasonable number of membership functions (MFs) to maintain appropriate linguistic meanings. In the ANFIS simulation examples in Chapter 12, MFs were set up for all inputs using grid partitions, but this is questionable; the color recipe prediction problem has 16 surface spectral reflectance inputs and 10 colorant proportion outputs as depicted in Figure 22.1. When we pick 16 values ($X_1, \dots X_{16}$) from the surface spectral reflectance curve of a given target color, can we specify any rule for each value? Or is it necessary to establish MFs for each input value? We must have the following 16 fuzzy rules:

Rule 1: If X_1 (at 400 nm) is A_1, then use a rule, C_1.
Rule 2: If X_2 (at 420 nm) is A_2, then use a rule, C_2.
... ...
Rule 16: If X_{16} (at 700 nm) is A_{16}, then use a rule, C_{16}.

In these rules, A_i is a fuzzy linguistic label. (Note that the visible color spectrum is 400 nm to 700 nm.) These rules may not make sense since we do not have such explicit knowledge per wavelength. Without explicit domain knowledge, adaptive learning mechanisms enable ANFIS/CANFIS to build up fuzzy rules automatically [5]. But if the initial MF setup has no meanings, it is futile to extract fuzzy rules from a fuzzy logic point of view. Blindly applying fuzzy MFs to all scalar inputs may turn out to be meaningless. Also, if there are a great many inputs, *the curse of dimensionality* problem arises. The number of MFs should be carefully determined so that fuzzy rules can be held to meaningful limits. Fortunately, there is a formula for transforming the surface spectral reflectance of color to perceptual attributes, "lightness," "hue," and "chroma" [4, 14] (see also Section 22.5.5). These three values must be more suitable for treating color in a linguistically meaningful way than the 16 spectral values mentioned previously, and so we use them as our MF inputs. When we invert the 3-D partitions in the color attribute space to the 16 dimensions of the spectral input space, certain complicated partitions must be constructed in the 16-D input space. In this way, we realize a complicated fuzzy partitioning.

22.4.2 CANFIS Architectures

First, we consider just one perceptual attribute of color hue as a linguistic variable. Using hue alone, we build up fuzzy MFs on the polar coordinates that define five color regions: red, yellow, green, blue, and violet (see the *membership value generator* in Figure 22.3). Specifically, fuzzy rules in the if-then format serve to determine color selection. For instance,

Yellow rule: If the target color is "yellow," then use a "yellow" rule, C_y.

Each color MF specifies the degree of membership of a color region and assigns the degree value to each color rule (rule's consequent) as the firing strength. In the preceding yellow rule, the firing strength (W_y) is determined by the yellow MF.

To introduce more MFs may lead us to better results. Hence we also consider a case in which each color region has three MFs to express its three degrees of color. For example, concerning the yellow region between green and red area, the following rules apply:

Yellow rule 1: If the target color is "greenish yellow,"
then use a "greenish yellow" rule, C_{gy},
Yellow rule 2: If the target color is "very yellow,"
then use a "very yellow" rule, C_{vy},
Yellow rule 3: If the target color is "reddish yellow,"
then use a "reddish yellow" rule, C_{ry}.

In this case, we have 15 linguistic values (e.g., "greenish yellow") on one linguistic variable (hue) alone [see CANFIS (b) in Tables 22.4 and 22.6]. To introduce more than 15 MFs onto only hue may result in less interpretability from a fuzzy logic standpoint, and resulting fuzzy rules may be ill defined or hard to understand simply because of the difficulty of specifying the difference between greenish yellow and very yellow that humans perceive by saying "slightly greenish yellow" or using

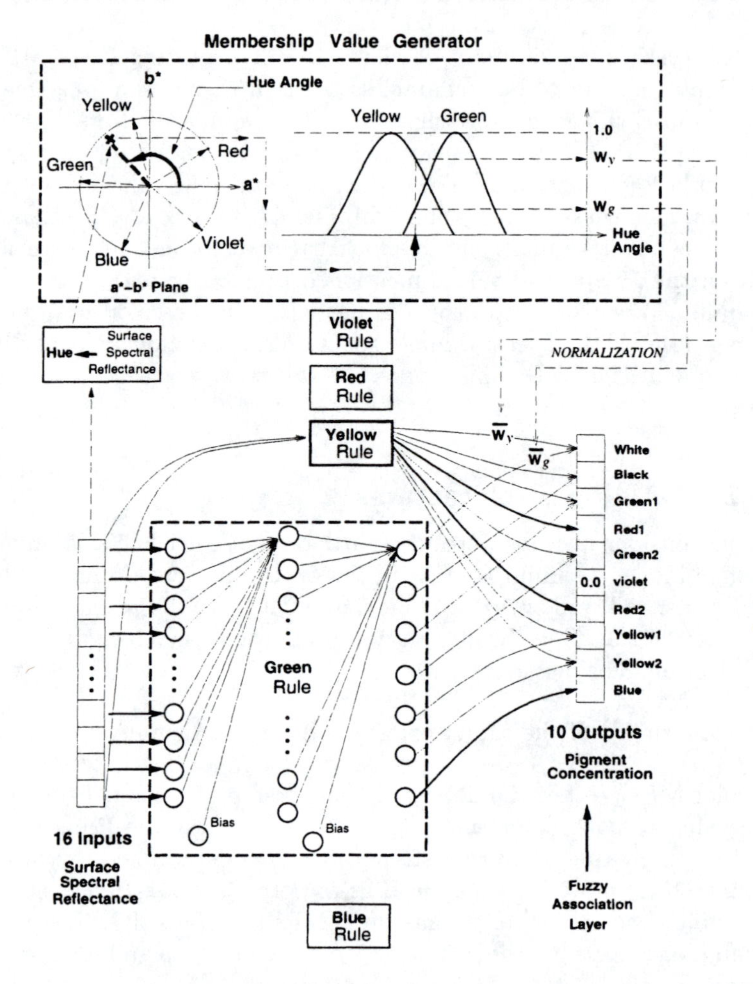

Figure 22.3. *CANFIS with five color rules for color recipe prediction.*

some other vague description.

Instead of increasing MFs, we can construct more sophisticated rules' consequents, such as neural rules (or local color expert NNs), as we have discussed in Chapter 13. Figure 22.3 illustrates such a CANFIS with five color rules [which correspond to CANFIS (d) and (e) in Tables 22.4 and 22.6]; one color MF is positioned for one color region. This CANFIS model can be viewed as a variation of the modular network we previously discussed in Sections 13.3.2 and 9.6. The given prediction task is decomposed into five color rules or five local color experts, which form rules' consequents. In Figure 22.3, the "green rule" is expressed in a neural rule with 16 spectral reflectance inputs. Each rule can be a linear rule, a sigmoidal rule, or a neural rule.

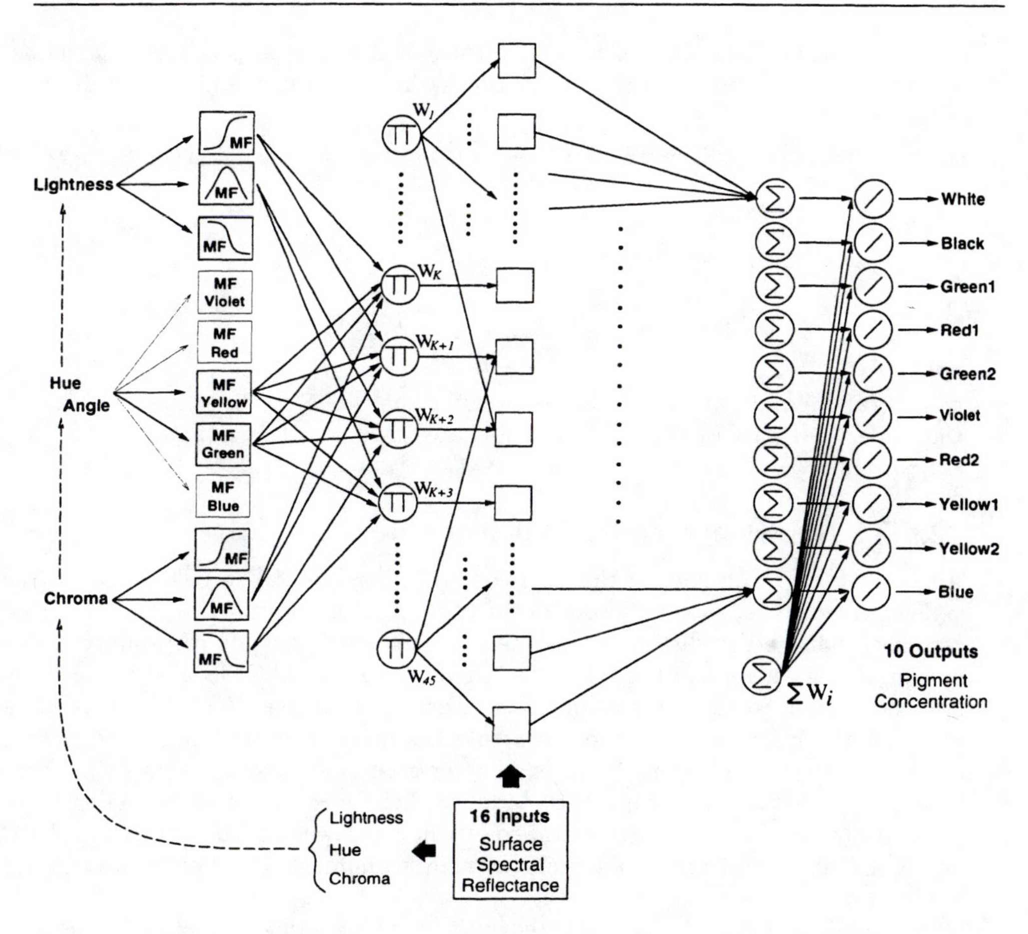

Figure 22.4. *CANFIS with 11 MFs (45 fuzzy rules) for color recipe prediction.*

So far, we have discussed how to build up a CANFIS with several MFs for the hue aspect alone. Next, we take into account all three perceptual attributes of color—lightness, hue, and chroma—to alleviate the problem (P4) in Table 22.1. Specifically, in our experiments, we set up three MFs for lightness, and chroma, respectively, and five color MFs for hue. Hence, we have CANFIS with 45 fuzzy rules, as illustrated in Figure 22.4 [see CANFIS (c) in Tables 22.4 and 22.6].

The discussed CANFIS architectures may have too many adjustable parameters. To accelerate learning, we can employ the modified bell MFs defined in Equation (22.1) to control the number of firing rules (i.e., local experts). This may be useful because it may prove unnecessary to use more than two color rules at the same time: For instance, when the target color is in a region between green and yellow (that is, when the yellow rule and the green rule are fired), the neighboring red rule and blue rule are not necessarily fired because of the *yellow-blue* and *green-red* complementary color relationships. Several weight-updating procedures for unnecessary or inactive rules can then be skipped when iterative training procedures are employed.

The definitions of the modified bell MF and the original bell-shaped MF are presented next for the purpose of subsequent discussion:

$$\mu_{\text{mod}}(x) = \max\left\{\frac{2}{1+|\frac{x-c}{a}|^{2b}} - 1, 0\right\}, \tag{22.1}$$

$$\mu_{\text{original}}(x) = \frac{1}{1+|\frac{x-c}{a}|^{2b}}, \tag{22.2}$$

where {a, b, c} is an adjustable parameter set. The modified bell MF is just the upper half part of the original bell-shaped MF and has a limited base width (support).

22.4.3 Knowledge-embedded structures

Adaptive fuzzy MFs specify the degree of membership of five color regions (red, yellow, green, blue, violet) according to perceptual attributes of color. They determine what weight should be assigned to each rule's output to produce a final output. We have applied the colorist's judgment to the CANFIS architecture; several connections to the fuzzy association layer can be pruned. This idea is pictured in Figure 22.3; for instance, the *green* rule has no connection line to *red* units at the fuzzy association layer. That is, a green rule (weighted by a green MF) has no effect on red colorant proportions because of the *green-red* complementary color relationship. In this way, each neural color rule has fewer output units than the 10 final CANFIS output units. The unit (or neuron) numbers are clearly presented in Table 22.5.

As previously stated, the predicted number of colorants should be about four; this means that 6 of the 10 final outputs should be zero. Reducing the number of zero outputs through the pruning procedure can have a positive impact on the construction of the desired input-output mappings inside CANFIS. This modification is intended mainly to eliminate the problems of (P1) and (P2) in Table 22.1.

22.4.4 CANFIS Simulation

We explored CANFIS with linear or nonlinear rules with different MF setups. Table 22.4 shows five representative CANFIS descriptions in the simulation. In any CANFIS, we used the truncation filter function in the output layer.

Although we implemented CANFIS (b) with 15 linear rules extensively, we did not obtain better results than CANFIS (a) with five sigmoidal rules. When the rule number was increased, difficulty in determining initial parameter setups was encountered.

When we use CANFIS with five neural rules, as depicted in Figure 22.3, there may be many possible optimal rule formations; Table 22.5 shows one of them, which

Table 22.4. *Five representative CANFIS models for color recipe prediction.*

(a)	CANFIS with 5 sigmoidal rules, as shown in Figure 22.3, with pruned connections 5 bell-shaped MFs are set up for hue angle alone (i.e., 5 rules are for five color regions)
(b)	CANFIS with 15 linear rules with no pruned connections 15 bell-shaped MFs are set up for hue angle alone
(c)	CANFIS with 45 rules, as shown in Figure 22.4, with no pruned connections 3 bell-shaped MFs are set up for lightness 3 bell-shaped MFs are set up for chroma 5 bell-shaped MFs are set up for hue angle
(d)	CANFIS with 5 neural rules, as shown in Figure 22.3, with no pruned connections 5 modified bell MFs are set up for hue angle alone 5 neural color rules have the same model size (i.e., each neural rule has 22 hidden units)
(e)	CANFIS with five neural rules, as shown in Figure 22.3, with pruned connections 5 modified bell MFs are set up for hue angle alone 5 neural color rules are heuristically optimized independently Those rules' model sizes are specified in Table 22.5

was found by a process of trial and error. [This CANFIS corresponds to CANFIS (e) in Tables 22.4 and 22.6.] Because a different amount of training data goes into each local neural color rule, each of the neural color rules can be optimized for its own territory; that is, each can have a different model size. The initial data classified into five color regions are shown in Table 22.5. Note that when the modified bell MFs are used, color MFs metamorphose as learning progresses, and therefore the amount of data in the five color categories changes. On the other hand, there is another idea that each color expert should have the same model size, so we tested CANFIS with neural color rules that have the same model size [see CANFIS (d) in Tables 22.4 and 22.6].

We show the results from five representative CANFIS models in Table 22.4. Table 22.6 shows a performance comparison among those CANFIS models as well as MLP models discussed in Section 22.3.

22.5 COLOR PAINT MANUFACTURING INTELLIGENCE

The discussed CANFIS served to build the color recipe prediction system shown in Figure 22.1. When we see the paint production environment around the prediction

Table 22.5. *Optimal structures of five color neural rules in CANFIS (e) in Tables 22.4 and 22.6, and their initial number of training/test data. The structures were heuristically optimized. Each neural rule's output units are fewer than the final 10 CANFIS output units.*

Five color neural rules	Model size	Training data	Checking data
Rule_{Red}	$16 \times 16 \times 16 \times 8$	650	138
$\text{Rule}_{\text{Yellow}}$	$16 \times 16 \times 17 \times 8$	707	200
$\text{Rule}_{\text{Green}}$	$16 \times 21 \times 7$	521	105
$\text{Rule}_{\text{Blue}}$	$16 \times 15 \times 8$	363	65
$\text{Rule}_{\text{Violet}}$	$16 \times 17 \times 6$	409	48

Table 22.6. *Performance comparison between single MLP models: NN_{norm} and NN_{mod}, and five representative CANFIS models. Table 22.4 details the five CANFIS models. NN_{norm} was a simple MLP approach, and NN_{mod} was an improved NN_{norm}. A column, "Error," denotes the average checking error. A column, "Para. #," denotes the total modifiable parameter number.*

CANFIS	# of membership functions			Rule		Error	Specified #	Para.
	hue	lightness	chroma	no.	formation	$\times 10^{-2}$	of pigments	#
(a)	5	0	0	5	sigmoidal	7.99	3.46	852
(b)	15	0	0	15	linear	12.90	2.78	2,595
(c)	5	3	3	45	linear	2.59	3.76	7,683
(d)	5	0	0	5	neural	1.90	3.85	3,035
(e)	5	0	0	5	neural	1.41	4.00	2,691
NN_{norm}	—	—	—	1	neural	2.62	6.66	925
NN_{mod}	—	—	—	1	neural	2.03	3.90	925

system, we notice the color paint manufacturing cycle illustrated in Figure 22.5. Basically, the main focus in recipe prediction should be color difference rather than colorant errors. Practically, the color difference defined in Equation (22.3) between pairs of presented colors should be smaller than about 1.0; human eyes cannot distinguish between smaller color differences. This bird's-eye view of the manufacturing cycle gives us a hint about how to feed back information about color difference to improve prediction accuracy. As summarized in Table 22.1, there are five major concerns in color recipe prediction. It is important to consider perceived color difference during the prediction process; we use an MLP, NN_{Lab}, to cope with the third critical concern (P4) in Table 22.1.

This section presents a cooperative hybrid system to simulate such an entire

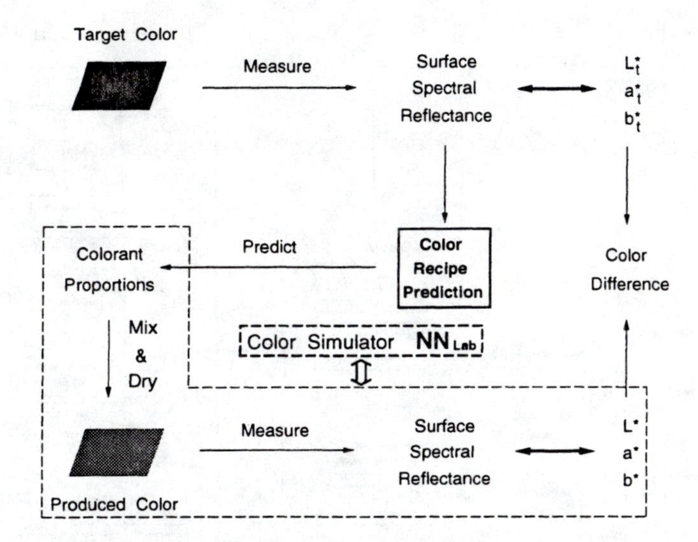

Figure 22.5. *Color paint manufacturing process.*

manufacturing process in an attempt to construct *manufacturing intelligence*[1] for the color paint industry; we integrate the three major elements of soft computing and problem-specific knowledge. That is, NNs, an FS, and a GA with a KB complement each other in obtaining more precise outputs for color recipe prediction through manufacturing simulation based on the entire decision-making process of a professional colorist. Here, the GA plays a leading role in this fusion system by evolving colorant proportion vectors. Because (P2) in Table 22.1 is a kind of combinational problem and an evolutionary framework is necessary, the GA may be a good choice for a leading light at this stage. We shall clarify the evolutionary framework in subsequent sections.

22.5.1 Manufacturing Intelligence Architecture

In the initial stage, the first-generation population or starting points for a GA search are set by a fuzzy population generator and a multi-elite generator using results from the CANFIS and NN approaches. Those results must already be somewhat close to the range of ideal colorant concentrations. In the evolutionary phase, the fusion system tries to improve those encoded proportion members in conjunction with NNs and a KB; that is, two different NNs and a KB are used to make up the fitness function. Genes' colorant concentrations are passed to three functions, which calculate fitness values individually. The three values are combined into the

[1]Here, the term manufacturing intelligence is intended as *soft computing (or computational) intelligence for simulating a paint manufacturing process* in this chapter. In contrast, Wright and Bourne discussed manufacturing intelligence in a broader sense for *the science of creating intelligent systems for manufacturing applications* involving hardware systems [13].

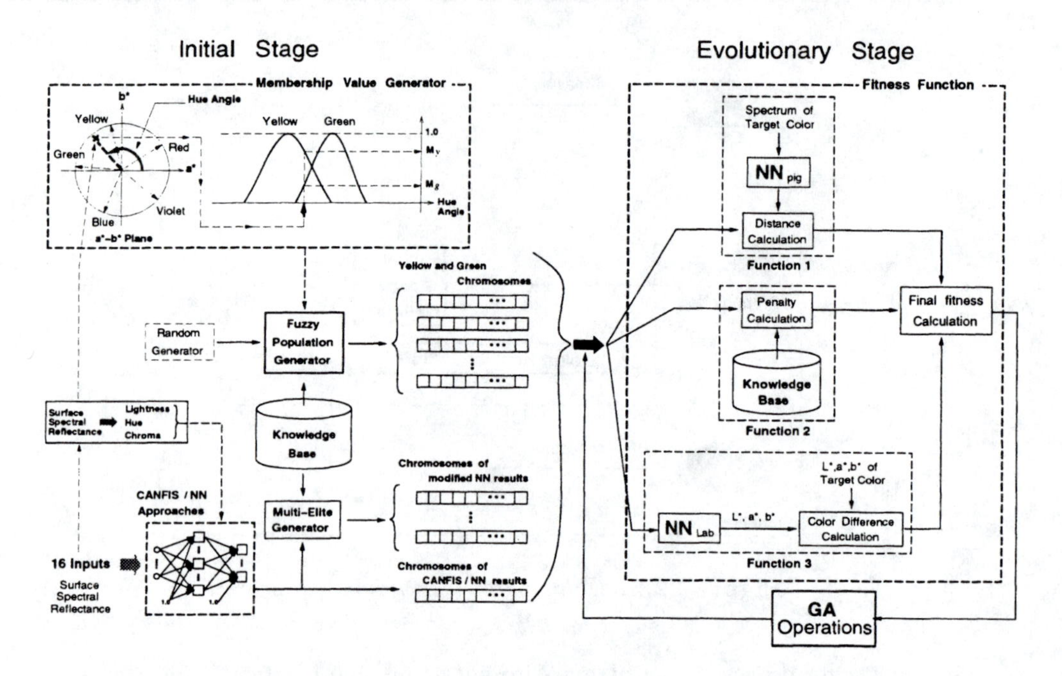

Figure 22.6. *Architecture of color paint manufacturing intelligence.*

final fitness value. In the following subsections, we shed light on more details of this evolutionary mechanism, illustrated in Figure 22.6.

22.5.2 Knowledge Base

Knowledge may be useful in reinforcing some favorable aspects of genetic searches [7]. Performing the color recipe prediction task requires special knowledge. We believe that a KB plays an important role in helping the system evolve to recognize specific features of a target color. The KB has the following main rules:

Rule 1: Keep total proportions of colorants around 100%,
Rule 2: Keep the number of necessary colorants around the ideal number,
Rule 3: Avoid use of complementary colorants: e.g., Red and Green,
Rule 4: Avoid use of the same type of colorants at the same time: (e.g., Red_1 and Red_2).

Note that we have 10 colorants (10 outputs) that include three pairs of the same kind of colorants: green, yellow, and red ones (see Figure 22.1); each pair, such as Red_1 and Red_2, has different characteristics. In this prediction task, the 100% rule (rule 1) was emphasized, as in ref. [11].

22.5.3 Multi-elites Generator

CANFIS and NN approach results are encoded into the initial population as elite members. Then a multi-elite generator produces more elites by modifying those results according to rule 4 in the KB. That is, the concentrations of the same type of colorants are summed into one or another of them (e.g., $\text{Red}_1 + \text{Red}_2 \Rightarrow \text{Red}_1$, or $\text{Red}_1 + \text{Red}_2 \Rightarrow \text{Red}_2$). This is derived from the fact that the simple backpropagation MLP, NN_{norm}, tends to specify use of more than six colorants (see Table 22.3), although the desired number of colorants to produce any color in our data sets is fewer than five. Again, it is important to keep the number of colorants used at a practical level. Multiple elite colorant vectors offer several different starting points for GA searches. The number of encoded elites depends on the quality of the CANFIS/NN results; we take the results of three approaches (NN_{norm}, NN_{mod}, CANFIS), and so we have at least three elite members at the initial stage. The combination of several solutions may be effective in finding the optimal solution [6]. The other members are initialized by a fuzzy population generator. This seeding procedure is shown in Figure 22.6 (left).

22.5.4 Fuzzy Population Generator

The idea is to generate the initial population according to the fuzzy classification of a target color, which serves to determine color selection. First, we classify the target color into one of five color categories (red, yellow, green, blue, and violet) on the a*-b* plane, which shows hue and chroma [4, 14] (see also Section 22.5.5), and decide to what extent the desired color belongs to each color category using fuzzy MFs, as discussed in Section 22.4.1. We then generate initial color chromosomes by modifying chromosomes generated by a random number generator according to rules in the KB. For example, when a target color looks greenish yellow, green chromosomes and yellow ones are generated; green chromosomes have zero values in either Green_1 or Green_2 colorant concentrations and in red colorant concentrations because of the red-green complementary color relationship (see rule 3 and rule 4 in Section 22.5.2). It is effective to inactivate some genes which have information on the same type of colorants and complementary colorants to eliminate redundant colorants at the initial stage.

The number of green chromosomes (Num_{Green}) and that of yellow ones (Num_{Yellow}) are decided according to the following calculations:

$$Pop_{\text{rest}} = Pop_{\text{total}} - Pop_{\text{NN}},$$

$$Num_{\text{Green}} = \frac{M_g Pop_{\text{rest}}}{M_y + M_g},$$

$$Num_{\text{Yellow}} = \frac{M_y Pop_{\text{rest}}}{M_y + M_g},$$

where two membership values, M_y and M_g, signify to what extent the target color belongs to the yellow category and the green one, respectively. Pop_{total} denotes

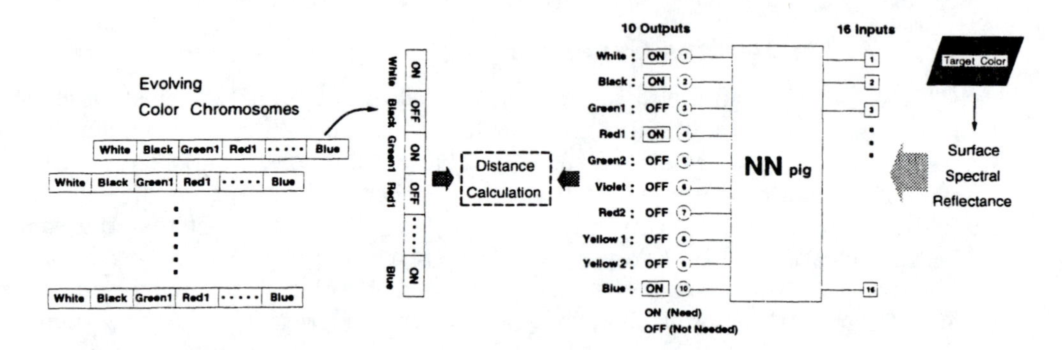

Figure 22.7. *A component of the fitness function based on* NN_{pig}.

the total population number, and Pop_{NN} signifies the number of elite chromosomes from the CANFIS/NN results, including the chromosomes generated by the multi-elite generator.

22.5.5 Fitness Function

The fitness function consists of three functions: two neural fitness functions (function 1 and function 3) and the KB-based fitness function (function 2).

Function 1

Using NN_{pig}, the first function evaluates genes' colorant concentration vectors according to the specified use of colorants. The NN_{pig} ($16 \times 18 \times 21 \times 10$ neurons) maps surface spectral reflectance to a list of required colorants (see Figure 22.7). It gives just ON/OFF values to each output unit to predict which colorants should be used to produce the same color as the target color, where ON means "colorant needed" and OFF means "not needed." Function 1 evaluates each chromosome by calculating a distance in binary space (ON/OFF) after each chromosome's representation has been transformed into the ON/OFF format. Figure 22.7 describes this procedure. Table 22.7 shows the capability of this trained NN_{pig}.

Function 2

The second function calculates a fitness value based on the KB described in Section 22.5.2. The fitness value depends on the extent to which genes' colorant concentration vector obeys the rules in the KB. To keep the GA search moving in a consistent direction, the KB is used in both the initial stage and in the calculation of fitness values, as illustrated in Figure 22.6.

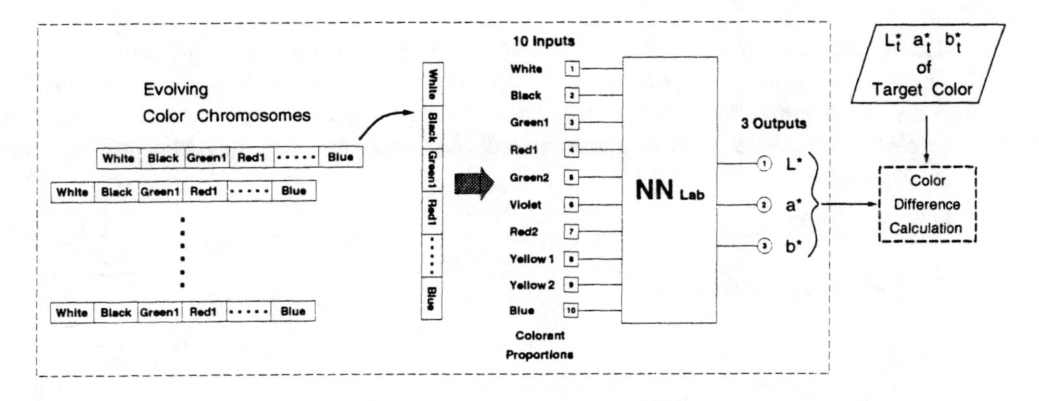

Figure 22.8. *A component of the fitness function based on* NN_{Lab}.

Function 3

The third function, based on NN_{Lab}, generates a fitness value with respect to color difference between a target color and each member's color, whose colorant concentrations are predicted by the system. Because it is time-consuming to manufacture an actual color by mixing genes' specified values (see Figure 22.5), the NN_{Lab} plays a crucial role as a color simulator to predict what color will be produced. The NN_{Lab} ($10 \times 11 \times 14 \times 3$ neurons) maps colorant concentrations to L^*, a^*, and b^*; that is, by plugging each member's colorant proportions into NN_{Lab}, we can obtain L^*, a^*, and b^* to calculate the color difference between a target color and an individual color (see Figure 22.8).

We adopted CIE 1976 (L^*, a^*, b^*)-space [4, 14]. This defines the color difference and perceptual attributes of color—lightness, hue, and chroma—as detailed here:

$$\begin{array}{ll} \text{Color difference} & \sqrt{(L_t^* - L^*)^2 + (a_t^* - a^*)^2 + (b_t^* - b^*)^2} \\ \text{Lightness} & L^* \\ \text{Hue} & \arctan(b^*/a^*) \\ \text{Chroma} & \sqrt{(a^*)^2 + (b^*)^2}, \end{array} \tag{22.3}$$

where L^*, a^*, and b^* are calculated according to surface spectral reflectance and (L_t^*, a_t^*, b_t^*) are the values of a target color. Note that any color can be uniquely identified by its surface spectral reflectance curve (i.e., its physical color attribute).

The calculated color difference shows how satisfactorily the predicted color matches the reference color. The use of NN_{Lab} provides a way to take into account human visual sensitivity to color difference. Table 22.8 shows the potential of the color simulator NN_{Lab}.

Function 3 determines the fitness value (fitness_3) of each chromosome, according to the calculated color difference, E:

$$\text{fitness}_3 = \exp(-E). \tag{22.4}$$

Table 22.7. *Capabilities of different NN approaches in specifying necessary colorants. NN_{norm} is a simple backpropagation MLP, and NN_{mod} is an improved NN_{norm} as discussed in Section 22.3; CANFIS is a neuro-fuzzy model described in Section 22.4. NN_{pig} is a special NN that predicts necessary colorants as shown in Figure 22.7.*

	NN_{norm}	NN_{mod}	CANFIS	NN_{pig}
# of unmatched patterns in 302 test patterns	299	74	73	27
# of unmatched units in 3020 output units	911	106	98	48
Predicted ave. # of required colorants	6.66	3.90	3.89	3.96

Table 22.8. *Average color difference predicted by NN_{Lab} using ideal colorant concentrations, and the results of three NN approaches: CANFIS, NN_{mod}, and NN_{norm}. CANFIS is a neuro-fuzzy model described in Section 22.4; NN_{norm} is a simple backpropagation MLP; and NN_{mod} is an improved NN_{norm} as discussed in Section 22.3. This table shows the potential capability of NN_{Lab} for 302 checking data.*

Colorant vectors	Ideal	CANFIS	NN_{mod}	NN_{norm}
Color difference	0.567	1.976	2.847	5.921

This function was designed to produce higher fitness inversely proportional to the magnitude of color difference.

22.5.6 Genetic Strategies

Genetic operations have a significant impact on the quality of solutions. We have embodied some ideas in both crossover operations and mutation operations.

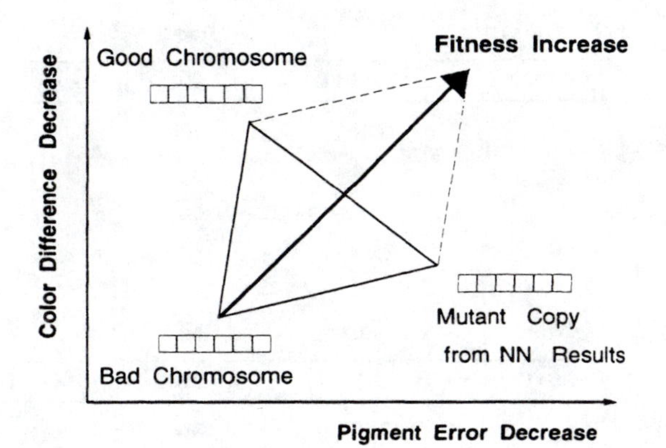

Figure 22.9. *GA search control by the modified simplex crossover.*

Modified Simplex Crossover

We have modified the selection scheme in performing *simplex crossover* operations[2] Our modified *selection* uses the following three procedures:

1. Select one good chromosome with respect to fitness value.
2. Pick, with high probability, an elite member (i.e., one of the mutant copies from the initial CANFIS/NN results) as a good chromosome.
3. Choose one bad chromosome with respect to fitness value.

The procedures share an idea of the *downhill simplex method* [10], based on a reflection away from a bad chromosome. This method may provide a better GA search direction, as illustrated in Figure 22.9. When we have a neural fitness function, we may have a problem; a trained NN may not be a perfect fitness function. Indeed, both NN_{pig} and NN_{Lab} in the fitness function are not perfect, as shown in Tables 22.7 and 22.8, but they may be able to direct a blind GA search to a better region of the search space. Procedure 1 lights a direction toward minimizing color difference. In accordance with NN_{Lab}, a chromosome with higher fitness may

[2]The *simplex crossover* proposed by Bersini and Seront [1] consists of the following *selection* and *bit-assignment* to produce a new child chromosome C_{new}:

- Selection

  Choose randomly three chromosomes, and arrange them in the decreasing order with respect to their fitness values. (Name them C_1, C_2, and C_3 in that order.)

- Bit-assignment

  If the ith bit of C_1 is equal to the ith bit of C_2, it will be assigned to the ith bit of C_{new}. Otherwise, the inverse of the ith bit of C_3 will be assigned to the ith bit of C_{new}.

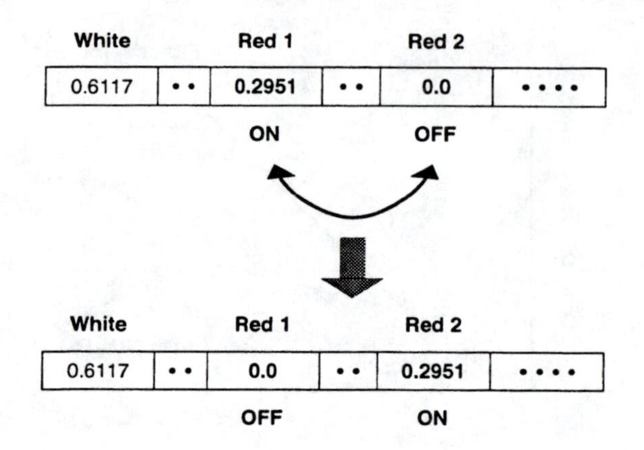

Figure 22.10. *Exchanging mutation.*

have smaller color difference. In this GA search, it is desirable to find a direction that minimizes both color difference and colorant errors. The problem is that we cannot calculate colorant errors directly. Yet the CANFIS/NN results provide a clue about better colorant concentrations since they must already be within some range of the ideal colorant concentrations. That is why mutant copies from the CANFIS/NN results, including ones originally generated by the multi-elite generator, should be involved in guiding the search toward better colorant proportion vectors, as in procedure 2.

Mutation Strategy

Usual mutation operation, as in a simple GA [3, 12], is applied to all members with a changeable mutation rate scheme such that a fixed mutation rate (0.01) is adopted with a probability of 0.4, and otherwise a mutation rate ranging from 0.09 to 0.69 is decided using a random number. Moreover, the following modified operations are also considered:

- *Chromosome template.* To avoid specifying the use of more colorants than necessary, we set out to inactivate some genes using the fuzzy population generator, as described in Section 22.5.4. This has made it possible to use a chromosome itself as a template to do the mutation operation. Namely, before the mutation operation, it is decided whether to mutate an inactivated gene or not; the mutation is applied with low probability (0.1) to inactivated genes, which have zero values of concentrations after decoding the genes' binary representations into colorant concentrations. If the mutation is applied to an inactivated gene, this leads to an increase in the number of necessary colorants.
- *Local search and preservation of multi-elites.* Multi-elites (i.e., chromosomes from the results of CANFIS/NN approaches) are mutated only at the lower bits of each gene to keep traits similar to the NN results; those mutant copies

Table 22.9. *Results of computational prediction in colorant error* ($\times 10^{-2}$) *and the corresponding color difference predicted by* NN_{Lab} *using 111 checking data.* NN_{norm} *is a simple backpropagation MLP, and* NN_{mod} *is an improved* NN_{norm}*, as discussed in Section 22.3; CANFIS is a neuro-fuzzy model described in Section 22.4; GNF is a genetic neuro-fuzzy model.*

	Ideal	NN_{norm}	NN_{mod}	CANFIS	GNF_{ALL}
Colorant error ($\times 10^{-2}$)	0	2.312	1.543	1.139	0.643
Color difference predicted by NN_{Lab}	0.588	6.661	3.165	2.019	0.267

of the multi-elites may stay in the vicinity of the original multi-elites. In this way, local search of the NN results is realized. In addition, the offspring of multi-elites always advances to the next generation; the mutant copies of multi-elites are preserved throughout the entire evolution. Note that this manipulation of low-order bits is applied only to multi-elites.

- *Exchanging mutation.* After the usual mutation, with low probability, members are subjected to another mutation: exchanging genes that have the same type of colorant information. This mutation is illustrated in Figure 22.10. Among 10 output colorant proportions, we have three pairs of the same types of colorants, such as Red_1 and Red_2, which have different natures; we must decide which one to use. This exchanging mutation helps us to explore such colorant choices. This may lead to an escape from local optima in the initial CANFIS/NN and NN_{pig} results; their choices may not match the final choice determined by the system. The agreement with NN_{pig} in Table 22.10 shows how much the predicted choice of colorants optimized by the system matched the colorant choices specified by NN_{pig}.

22.6 EXPERIMENTAL EVALUATION

To evaluate the capability of the hybrid system, color paint manufacturing intelligence, we used 111 randomly selected checking data. The configuration of the GA was as follows:

Population size	80 members
Mutation rate	flexible (see Section 22.5.6)
Crossover method	simplex crossover [1]
Simplex crossover rate	0.85
Maximum generations	10,000.

Table 22.10. *Performance evaluation in computational prediction of 111 checking data. GNF_{ALL} shows the maximal ability of manufacturing intelligence in the prediction task. Parenthesized values denote potential capabilities with respect to colorant errors; they were obtained when colorant errors were minimized. The columns, from left to right, indicate "components of fitness functions," "average number of generations when the solution chromosome appeared," "colorant error ($\times 10^{-2}$)," "accord with NN_{pig}," and "color difference predicted by the color simulator, NN_{Lab}," respectively. Note that the column, "Accord with NN_{pig}," shows how much the predicted choice of colorants optimized by the GNF models match the colorant choice specified by NN_{pig}.*

	Fitness functions			Ave. # of gen.	Error $\times 10^{-2}$	Ave. # of col.	Accord w/ NNpig	Col. dif. by NNLab
	NNLab	NNpig	KB					
GNFALL	○	○	○	5058.7 (4759.5)	0.643 (0.213)	3.90 (3.88)	79.28% (79.28%)	0.267 (0.809)
GNFvoid	○	○	○	4134.4 (3082.8)	72.209 (36.165)	3.94 (4.89)	50.45% (21.62%)	48.800 (34.637)
GNFCP	○	○	×	4915.4 (4358.8)	1.190 (0.206)	4.02 (4.02)	78.38% (74.77%)	0.121 (0.659)
GNFCK	○	×	○	4559.3 (4655.5)	1.695 (0.215)	3.88 (3.89)	74.77% (77.48%)	0.202 (0.656)
GNFC	○	×	×	4604.3 (4742.4)	2.802 (0.191)	5.35 (4.36)	28.83% (55.86%)	0.060 (0.567)

Table 22.9 shows the comparison of our proposed model, GNF_{ALL}, and some other approaches; NN_{norm} was a simple MLP approach, and NN_{mod} was an improved MLP model that had modified sigmoidal functions in the output layer [9]. Both had the same model size (16 × 18 × 21 × 10 neurons) (see Section 22.3). GNF_{ALL}, with all three components of the fitness function, employed the results of three approaches—NN_{norm}, NN_{mod}, and CANFIS—in producing the initial population. According to the corresponding color difference predicted by NN_{Lab}, only the result of GNF_{ALL} was good enough to reach a satisfactory level of color difference where human eyes could not tell the difference between presented colors. (Again note that the desired color difference had to be smaller than about 1.0.)

Furthermore, to demonstrate the validity of each of the three components in the fitness function, we tested GNF_C, GNF_{CP}, and GNF_{CK}. GNF_C had NN_{Lab} as the only component of the fitness function. GNF_{CP} had both NN_{Lab} and NN_{pig} as two components of the fitness function. GNF_{CK} had the KB as well as NN_{Lab} as two components. (GNF_{ALL} had all three components.) Note that NN_{Lab} played an important role as a color simulator, and so it always had to stay in the fitness function. Table 22.10 shows how each component contributed to the

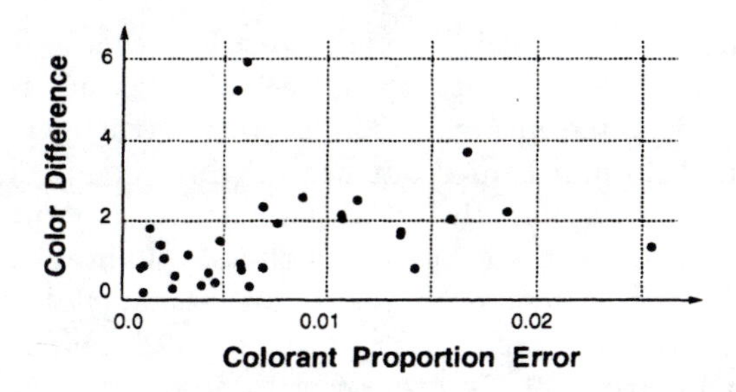

Figure 22.11. *Complicated relationship of 32 color samples between actual color difference and colorant errors; these 32 sample color paints were actually manufactured.*

prediction, and how they complemented each other.

Additionally, to exhibit how indispensable CANFIS/NN results are at the initial seeding stage, we examined GNF_{void}, which had no multi-elites from CANFIS/NN results, but had the same fitness function as GNF_{ALL}. It started the GA search from the randomly initialized points of colorant proportion vectors. In Table 22.10, we see the potential capability of the five models; the values in parentheses signify the best performance with respect to colorant errors, regardless of fitness; they were obtained when colorant errors were minimized.

22.7 DISCUSSION

Usually, it is difficult to construct an ideal "fine-tuner" by using the GA; we discuss a fusion technique and specific modifications applied to the fitness function and GA operations. The GA alone may not be a good fine-tuner, yet the complete resulting system can be viewed as a fine-tuner, overcoming individual limitations; the key idea is that components complement each other. In this context, the GA may not be the only choice to play the central role in conjunction with other components of soft computing.

When NN_{Lab} acted alone as the fitness function employed in GNF_C, the system tended to go too far toward minimizing color difference, and therefore the average number of required colorants was larger. In addition, the specified colorants did not match well those designated by NN_{pig}, as indicated in the low percentage of accord with NN_{pig} in Table 22.10. NN_{pig} was supposed to learn the characteristics of colorant compositions, such as complementary color relationships, yet it did not perform perfectly (see again Table 22.7). Thus, the KB surely helps the system evolve to recognize more colorant features. It must be emphasized that they functioned synergistically.

Both NN_{pig} and NN_{Lab} as components of the fitness function made some prediction errors, as shown in Tables 22.7 and 22.8. Hence, enhancing the overall performance may result from improving the accuracy of such neural fitness functions. Even though NN_{Lab} lacks high precision, the performance of GNF_{ALL} is still better than those of other approaches; this verifies that the strategy of the search direction depicted in Figure 22.9 is trustworthy.

GNF_{void} had no multi-elites but had all three components of the fitness function, starting from the randomly initialized colorant concentration vectors. Its poor performance emphasizes the existence of the multi-elites (i.e., mutant copies from the CANFIS/NN approach results); without them, we cannot draw any advantage from the search direction based on the idea of the downhill Simplex method, summarized in Figure 22.9. In other words, seedings from CANFIS or other NN approaches are indispensable in enabling manufacturing intelligence to function efficiently within a reasonable amount of computation time. Also, such extraordinarily poor performance in predicted color difference implies that Equation (22.4) in the fitness calculation of color difference may be altered for such no-seeding cases.

As shown by the values in parentheses in Table 22.10, the system did not put the highest fitness on the best chromosome in terms of colorant errors. In this simulation, we did not use the elitist selection method since the fitness function could not calculate colorant errors, which may suggest that even if a better child chromosome in terms of colorant error appears, the elitist strategy may jeopardize its chance of advancement to the next generation [9]. Practically, in our strategy, the system selects a chromosome with the highest fitness as the final solution over a preset number of generations. The column "Ave. # of gen.," or average number of generations, in Table 22.10 indicates when the solution chromosome appeared during the evolutionary process.

The colorant errors in parentheses in Table 22.10 show almost the same error level (0.2×10^{-2}), except for that of GNF_{void}, although the colorant errors of the final solutions of the system are different. Actually, only 71 patterns among the 111 checking patterns were improved in terms of colorant error. This may be partly because the CANFIS models did a good job in prediction, so their results may be hard to improve on, but partly also because the system may happen to find another colorant composition solution. The presented manufacturing intelligence may find another solution if a color simulator, NN_{Lab}, learns much of the mapping from colorant compositions to perceptual attributes of color (L^*, a^*, and b^*).

Figure 22.11 shows an interesting fact that the real perceived color difference did not exactly correspond to the magnitude of colorant errors; those data were collected by manufacturing 32 color paint samples. Such complicated relationships between colorant errors and actual color differences may imply that the mapping from surface spectral reflectance to a list of colorants may not be a one-to-one correspondence. [As stated in Table 22.1, we may need to take care of the (P3) and (P5) problems; different colorant compositions may produce the same or almost the same color to human perception.]

One possible explanation of this finding can be seen in the following chromosome examples. Suppose we need to produce a target color whose ideal colorant compositions include White, Red_1, and Yellow_2, and two colorant proportion strings (Candidate_1 and Candidate_2) exist in the population. Further suppose Candidate_2

has a smaller color difference predicted by NN_{Lab} than Candidate$_1$, but Candidate$_2$ has a bigger colorant error due to its different colorant compositions: White, Red$_2$, and Yellow$_1$, as shown in the following table:

	White	Red$_1$	Red$_2$	Yellow$_1$	Yellow$_2$	Colorant error	Color difference
Ideal vector	0.8101	0.0552	0	0	0.1347	0	0
Candidate$_1$	0.7821	0.1631	0	0	0.0548	Smaller	Larger
Candidate$_2$	0.7954	0	0.0719	0.1327	0	Larger	Smaller

In this case, Candidate$_2$ most likely will have a higher fitness value than Candidate$_1$ despite its colorant choice variation from the ideal. Therefore, the system may pick Candidate$_2$ as a solution. This is the trick. If the resulting manufactured color of Candidate$_2$ has a small enough color difference that human eyes cannot discern, Candidate$_2$ will be acceptable. While MLP and CANFIS approaches cannot deal with this "another solution problem" without suitable modifications, the manufacturing intelligence can handle it. In other words, the resulting system can function beyond the fine-tunner's reach. To draw more convincing conclusions, we must explore further resultant colorant vectors by checking if different combinations of colorants really have the same perceptual attributes of color as seen by humans [i.e., (P5) in Table 22.1].

This section concludes with one notice of accuracy of the color difference formula defined in Equation (22.3): The adopted CIE 1976 (L^*, a^*, b^*)-space is not perfect. In color science, it is still important to characterize the nature of human color vision.

22.8 CONCLUDING REMARKS AND FUTURE DIRECTIONS

In Section 22.4, we demonstrated the strength of a knowledge-embedded CANFIS. By constructing MFs in color attribute space, this neuro-fuzzy approach allows us to express and realize meaningful and concise representations of colorists' knowledge. Of course, we must consider whether there is a more effective way to represent human visual sensitivity to color in the space of perceptual attributes than the use of bell-shaped MFs; we may need to contrive a more sophisticated MF. In structural terms of CANFIS, when we used CANFIS with 11 MFs in Figure 22.4, we had 45 rules, and therefore it seems difficult to introduce neural rules or local color expert NNs. Imagine a huge CANFIS construction with 45 neural rules. Here it must be important to determine what rule formation is appropriate in the sense of practical feasibility. To confront these concerns must be our next step, which may endow a breakthrough in understanding the neuro-fuzzy modeling. We believe such efforts will pave the way for a new generation of CANFIS.

In Section 22.5, we presented manufacturing intelligence based on a unique blend of principal components of soft computing, where a GA with a KB plays a leading role in pursuit of predictions, linking an FS and NNs; they function complementarily as a system rather than competitively. In light of both potential versatility and

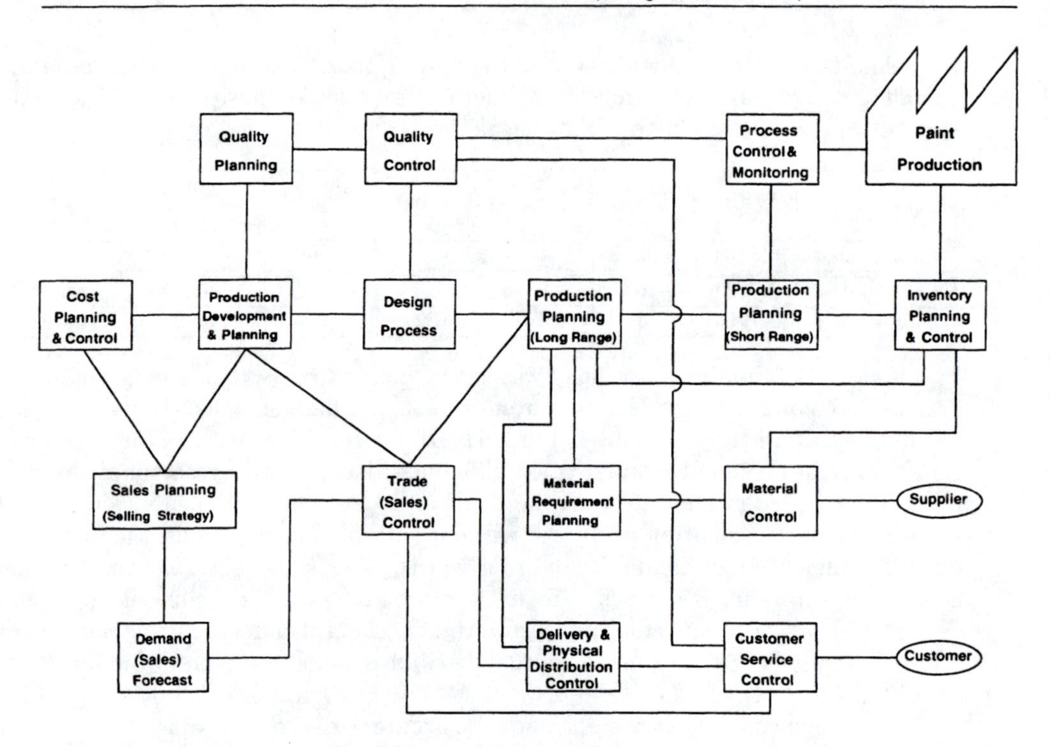

Figure 22.12. *A rough sketch of the paint industry.*

practical validity, such hybrid systems can yield advantages over other individual approaches in that they can evolve their results.

To focus on color difference, we endeavored to simulate the manufacturing cycle of color paint, which involved the color recipe prediction as a process. The manufacturing intelligence has a mechanism for checking predicted perceptual color difference with an embedded color simulator, NN_{Lab}. It realized a higher degree of prediction precision by evolving the results of other approaches. This finding confirms the concept of manufacturing intelligence based on soft computing and provides a small but potentially significant impact on our future research.

We are not claiming a tremendous success in the small application example presented in this chapter. Look at the rough sketch of the paint industry in Figure 22.12, where lots of new challenges can be sensed. The color recipe prediction application is actually a tiny cogwheel in the machine of the paint industry. In a practical industry, there must be huge numbers of applications for computational intelligence.

Now let us step back, and see our whole painted picture of *soft computing intelligence* in this chapter; at present, it may look premature. Yet it is felt to be growing steadly toward making a lasting impact on future technology. We believe that such computational intelligence and technological wizardry must be a match for any gauntlet the industrial world may throw down. Moreover, we hope that new ideas emerging from these studies will eventually stimulate engineers and scientists in ways we cannot now imagine, and that such computationally intelligent systems will pass stringent tests with flying colors.

REFERENCES

[1] H. Bersini and G. Seront. In search of a good evolution-optimization crossover. In R. Manner and B. Mandrick, editors, *Parallel problem solving from nature 2*, volume 2, pages 479–488. Elsevier Science, 1992.

[2] J. M. Bishop, M. J. Bushnell, and S. Westland. Application of neural networks to computer recipe prediction. *Color Research and Application*, 16(1):3–9, 1991.

[3] David E. Goldberg. *Genetic algorithms in search, optimization, and machine learning.* Addison-Wesley, Reading, MA, 1989.

[4] R. W. G. Hunt. *Measuring color.* Ellis Horwood Limited, West Sussex, England, 2nd edition, 1991.

[5] J.-S. Roger Jang. Rule extraction using generalized neural networks. In *Proceedings of the 4th IFSA World Congress*, pages 82–86 (in the Volume for Artificial Intelligence), July 1991.

[6] T. Kido, H. Kitano, and M. Nakanishi. A hybrid search for genetic algorithms: Combining genetic algorithms, tabu search, and simulated annealing. In *Proceedings of fifth International Conference on Genetic Algorithms*, page 640, July 1993.

[7] N. Mansour and G. C. Fox. A hybrid genetic algorithm for task allocation in multicomputers. In *Proceedings of Fourth International Conference on Genetic Algorithms*, pages 466–473, July 1991.

[8] E. Mizutani, J.-S. R. Jang, K. Nishio, H Takagi, and D. M. Auslander. Coactive neural networks with adjustable fuzzy membership functions and their applications. In *Proceedings of the International Conference on Fuzzy Logic and Neural Networks*, pages 581–582, Iizuka, Japan, August 1994.

[9] E. Mizutani, H. Takagi, and D. M. Auslander. A cooperative system based on soft computing methods to realize higher precision of computer color recipe prediction. In *Proceedings of Applications and Science of Artificial Neural Networks, part of SPIE's International Symposium on OE/Aerospace Sensing and Dual Use Photonics*, pages 303–314, April 1995.

[10] J. A. Nelder and R. Mead. A simplex method for function minimization. *The Computer Journal*, 7:308–313, 1965.

[11] J. Spehl, M. Wolker, and J. Pelzl. Application of backpropagation nets for color recipe prediction as a nonlinear approximation problem. In *Proceedings of the International Conference on Neural Networks*, pages 3336–3341, June 1994.

[12] M. Srinivas and L. M. Patnaik. Genetic algorithms: A survey. *IEEE Computer*, pages 17–26, June 1994.

[13] P. K. Wright and D. A. Bourne. *Manufacturing intelligence.* Addison-Wesley, Reading, MA, 1988.

[14] G. Wyszecki and W. S. Stiles. *Color science: concepts and methods, quantitative data and formulae.* John Wiley & Sons, New York, 2nd edition, 1982.

Appendix A

Hints to Selected Exercises

Chapter 2

15. and **16.**

$$\text{width}(A_\alpha) = 2a\left(\frac{1-\alpha}{\alpha}\right)^{\frac{1}{2b}}.$$

20. (a). Let

$$y = (a^{-p} + b^{-p} - 1)^{-1/p}.$$

Then find $\lim_{p\to 0} \ln y$.

Chapter 3

5. Modify the MATLAB file `complv.m`.

6. Modify the MATLAB file `complv.m`.

11. Modify the MATLAB file `implicat.m`.

Chapter 4

9. Modify the MATLAB file `sug2.m`.

10. Modify the MATLAB file `sug2.m`.

Chapter 5

3. Remember that $\boldsymbol{\theta}^T\mathbf{A}\boldsymbol{\theta} = \boldsymbol{\theta}^T\mathbf{B}\boldsymbol{\theta}$ if $\mathbf{B} = (\mathbf{A} + \mathbf{A}^T)/2$.

8. Show that

$$\mathbf{P}_{k+1}\mathbf{a}_{k+1} = \frac{\mathbf{P}_k\mathbf{a}_{k+1}}{1 + \mathbf{a}_{k+1}^T\mathbf{P}_k\mathbf{a}_{k+1}}.$$

Chapter 6

1. Show $E(\boldsymbol{\theta}_{\text{now}} + \eta \mathbf{d}) - E(\boldsymbol{\theta}_{\text{now}}) < 0$.

4. See Algorithm 6.2.

Chapter 7

3.
- Vectorize the calculation of the distance matrix.
- Calculate dE based only on the links that have been changed.

Chapter 9

1. Consider how to represent the three boundary lines of area T: $X = 0$, $Y = 0$, and $X + Y - 1 = 0$.

12. Try the MATLAB file `rbfn.m` available via FTP or WWW (see Preface).

Chapter 12

15. Modify the MATLAB file `trn_4in.m`.

16. Modify the MATLAB file `trn_4in.m`.

Chapter 13

1. It may suffer from slow convergence due to an increase of adjustable parameters.

Chapter 14

10. Use the Schwartz inequality

$$(A^2 + B^2)(C^2 + D^2) \geq (AC + BD)^2.$$

11. Modify `go_cart.m` and other related files. (These related files can be found by pattern-matching commands; for instance, "`grep CART *.m`" under UNIX.) Note that to change to local linear models, you need to reorganize the main book-keeping table `CART_table`. See `cartmain.m` for more details.

Chapter 17

1. For the input signal to be able to drive the plant from any initial states $\mathbf{x}(k)$ to any final states $\mathbf{x}(k+r)$ in r steps, the controllability matrix

$$\mathbf{W} = [\mathbf{A}^{r-1}\mathbf{B}\ \mathbf{A}^{r-2}\mathbf{B}\ \cdots\ \mathbf{AB}\ \mathbf{B}]$$

must be of rank n.

Chapter 18

2. The crossover must be executed at a point where either the left substrings or the right substrings have the same numbers of 1's.

Appendix B

List of Internet Resources

Due to the rapid advances of network technology, the Internet has become a major source of all kind of information needs. This appendix lists information resources of neuro-fuzzy and soft computing that are accessible via the Internet. The information listed here is by no means complete; an up-to-date version is available from the book's homepage at

`http://www.cs.nthu.edu.tw/~jang/soft.htm`

ENTRY POINTS

The following URL addresses provide convenient entry points for the search of information on neuro-fuzzy and soft computing:

This Book's Homepage: `http://www.cs.nthu.edu.tw/~jang/soft.htm`

Fuzzy Logic and Neurofuzzy Resources:
`http://www-isis.ecs.soton.ac.uk/research/nfinfo/fuzzy.html`

PEOPLE

This is a list of "fuzzy" people in random order.

Who's Who in Fuzzy Community: `ftp://ftp.abo.fi/pub/iamsr/who.txt`

Lotfi Zadeh: `http://http.cs.berkeley.edu/csbrochure/faculty/zadeh.html`

Ronald R. Yager: `http://www.Iona.edu/rry.htm`

George Klir: `http://ssie.binghamton.edu/people/klir.html`

Jerry Mendel: `http://sipi.usc.edu/faculty/mendel.html`

Bart Kosko: `http://sipi.usc.edu/faculty/kosko.html`

James M. Keller: `http://www.missouri.edu/~ecewww/keller.html`

Kevin M. Passino: `http://eewww.eng.ohio-state.edu/~passino/`

Martin Brown: `http://www-isis.ecs.soton.ac.uk/people/m_brown.html`

John Yen: `http://www.cs.tamu.edu/faculty/yen/`

Reza Langari: `http://ACS.TAMU.EDU/~meen/2375.html`

Hao Ying: `http://www.cs.tamu.edu/research/CFL/people/ying.html`

Yung-Yaw Chen: `http://ipmc.ee.ntu.edu.tw/~yychen/`

Li-Xin Wang: `http://www.ee.ust.hk/~eewang/`

Michael Lee: `http://HTTP.CS.Berkeley.EDU/~leem/`

J.-S. Roger Jang: `http://www.cs.nthu.edu.tw/~jang/`

NEWSGROUPS

comp.ai.fuzzy: Newsgroup about fuzzy logic and fuzzy set theory.

comp.ai.neural-nets: Newsgroup about neural networks.

comp.ai.genetic: Newsgroup about genetic algorithms.

comp.soft-sys.matlab: Newsgroup about MATLAB and SIMULINK.

SOFTWARE

This is a partial list of commercial and public-domain software for fuzzy logic and neuro-fuzzy applications. A comprehensive list can be found at

`http://www-isis.ecs.soton.ac.uk/research/nfinfo/fzsware.html`

Fuzzy Logic Toolbox: General information at `http://www.mathworks.com/fuzzytbx.html`, User's page at `http://www.mathworks.com/fuzzyupdate.html`

NEFCLASS: A neuro-fuzzy classifier at `http://www.cs.tu-bs.de/ibr/projects/nefcon/nefclass.htm`

Machine Learning Library in C++: `http://www.sgi.com/Technology/mlc`

FIR/TDNN Toolbox Toolbox for finite impulse response (FIR) and TDNN (time-delay neural networks); `http://www.eeap.ogi.edu/~ericwan/fir.html`

NN-based System Identification Toolbox: System ID toolbox using neural networks; `http://kalman.iau.dtu.dk/Projects/proj/nnsysid.html`

GENESIS Neural Simulator: A neural networks simulation package at
`http://www.bbb.caltech.edu/GENESIS`

PDP++ Software: NN Simulation System in C++, at
`http://www.cs.cmu.edu/Web/Groups/CNBC/PDP++/PDP++.html`

DATA SETS

UCI Machine Learning Repository:
`http://www.ics.uci.edu/~mlearn/MLSummary.html`

Time Series Repository:
`http://www.cs.colorado.edu/~andreas/Time-Series/TSWelcome.html`

Face Detection Data Set:
`http://www.ius.cs.cmu.edu/IUS/dylan_usr0/har/faces/test/index.html`

DELVE Data Set:
`http://www.cs.utoronto.ca/neuron/delve/delve.html`

JOURNALS

IEEE Transactions on Fuzzy Systems:
`http://www.ieee.org/pub_preview/fuzz_toc.html`

IEEE Transactions on Neural Networks:
`http://www.eeb.electronic.tue.nl/neural/contents/ieee_trans_on_nn.html`

Fuzzy Sets and Systems:
`http://www.elsevier.nl/catalogue/SAE/515/08410/08417/505545/505545.html`

International Journal of Approximate Reasoning:
`http://seraphim.csee.usf.edu/Nafips/ijar.html`

International Journal of Neural Systems:
`http://www.wspc.co.uk/wspc/Journals/ijns/ijns.html`

International Journal of Uncertainty, Fuzziness and Knowledge-Based Systems:
`http://www.wspc.co.uk/wspc/Journals/ijufks/ijufks.html/`

RESEARCH GROUPS

There are more than 100 research groups all over the world working on neuro-fuzzy and soft computing. A detailed list can be found at

`http://www-isis.ecs.soton.ac.uk/research/nfinfo/fzrgroup.html`

Appendix C

List of MATLAB Programs

This is a list of major MATLAB programs used in this book. These MATLAB programs may invoke other auxiliary MATLAB programs not listed here but available via the same FTP or WWW mentioned on page xxiii. Files labeled with stars indicate that they rely on functions in the Fuzzy Logic Toolbox, which are not freely available.

Appendix D

List of Acronyms

This is a list of acronyms used in this book:

ACE: Adaptive Critic Element
AEN: Action Selection Network
AHC: Adaptive Heuristic Critic
AHCON: AHC Connectionist
AI: Artificial Intelligence
ANFIS: Adaptive Neuro-Fuzzy Inference System
APE: Average Percentage Error
AR: Auto-Regressive
ART: Adaptive Resonance Theory
ASE: Associative Search Element
ASN: Action Selection Network
BOA: Bisector Of Area
CANFIS: CoActive Neuro-Fuzzy Inference System
CART: Classification And Regression Tree
CFR: Calculus of Fuzzy Rules
CMAC: Cerebellar Model Arithmetic Computer
COA: Centroid Of Area
CPP: Cart and Parallel Poles
CQ learning: Compositive Q learning
CRBP: Complementary Reinforcement BackPropagation
DP: Dynamic Programming
ECG: ElectroCardioGram
EECS: Electrical Engineering and Computer Science
EM learning: Expectation-Maximization learning
ERL: Evolutionary Reinforcement Learning
ES: Expert System
FAM: Fuzzy Associative Memory
FAQ: Frequently Asked Question
FC: Fuzzy Control or Fuzzy Controller
FIS: Fuzzy Inference System

FL: Fuzzy Logic
FLC: Fuzzy Logic Controller
FS: Fuzzy System
GA: Genetic Algorithms
GARIC: Generalized Approximate Reasoning for Intelligent Control
GDR: Generalized Delta Rule
GMDH: Group Method of Data Handling
GMP: Generalized Modus Ponent
GRNN: General Regression Neural Network
IEOR: Industrial Engineering and Operations Research
KB: Knowledge Base
LMS: Least Mean Squares
LRF: Localized Receptive Fields
LS: Least-Squares
LSE: Least-Squares Estimator
LVQ: Learning Vector Quantization
MANFIS: Multiple ANFIS
MATLAB: MATrix LABoatory
MENACE: Matchbox Educable Naughts And Crosses Engine
MF: Membership Function
MIQ: Machine Intelligence Quotient
MLP: MultiLayer Perceptron
MOM: Mean Of Maxima
MPG: Miles Per Gallon
MRAC: Model Reference Adaptive Control
MRH: Multi-Resolution Hierarchies
NC: NeuroComputing
NDEI: Non-Dimensional Error Index
NP: Non-Polynomial
OCR: Optical Character Recognition
PC: Personal Computer
PCA: Principal Component Analysis
PCR: Printed Character Recognition
PR: Pattern Recognition or Probabilistic Reasoning
RBFN: Radial Basis Function Network
RL: Reinforcement Learning
RMSE: Root-Mean-Squared Error
RNN-FLCS: Reinforcement Neural Network–based Fuzzy Logic Control System
RTRL: Real-Time Recurrent Learning
SA: Simulated Annealing
SAM: Stochastic Action Modifier
TLS: Total Least Squares
TLU: Threshold Logic Unit
TSP: Traveling Salesperson Problem

VFSR: Very Fast Simulated Reannealing
VLSI: Very Large Scale Intergrated circuits
XOR: Exclusive OR

Index